PHYSICAL SCIENCE WITH MODERN APPLICATIONS

Mt. Hood, Oregon, with snow-covered trees and a view of Mt. Jefferson. (Craig Tuttle/The Stock Market)

PHYSICAL SCIENCE

WITH MODERN APPLICATIONS

Fifth Edition

MELVIN MERKEN
Worcester State College

Saunders Golden Sunburst Series

SAUNDERS COLLEGE PUBLISHING

Harcourt Brace Jovanovich College Publishers

Fort Worth   Philadelphia   San Diego   New York   Orlando   Austin
San Antonio   Toronto   Montreal   London   Sydney   Tokyo

Text Typeface: Caledonia
Compositor: Progressive Typographers
Publisher: John Vondeling
Developmental Editor: Jennifer Bortel
Managing Editor: Carol Field
Project Editor: Maureen Iannuzzi
Copy Editor: Becca Gruliow
Manager of Art and Design: Carol Bleistine
Art Director: Christine Schueler
Associate Art Director: Doris Bruey
Art Assistant: Caroline McGowan
Text Designer: Rebecca Lemna
Cover Designer: Lawrence R. Didona
Text Artwork: Tasa Graphic Arts, Inc.
Layout Artist: Dorothy Chattin
Director of EDP: Tim Frelick
Production Manager: Jay Lichty
Marketing Manager: Marjorie Waldron

Cover Credit: Mountains: Wiley/Wales\Profiles West.
Clouds (cover and interior design): copyright 1992 W. Cody/West-light.

Printed in the United States of America

PHYSICAL SCIENCE WITH MODERN APPLICATIONS, Fifth Edition

ISBN 0-03-096010-X

Library of Congress Catalog Card Number: 92-050706

2345 032 987654321

(Andrew J. Martinez/Photo Researchers, Inc.)

To Shirley
for whom another volume would be required to list the contributions to this work.

(Andrew J. Martinez/Photo Researchers, Inc.)

To Shirley
for whom another volume would be required to list the contributions to this work.

PREFACE

Welcome to the fifth edition of *Physical Science with Modern Applications.*

Like its predecessors, this edition is addressed to nonscience majors in universities, four-year colleges, and two-year colleges, and to students in programs of continuing education, who look forward to learning more about the natural world around them.

OBJECTIVES

My objective in writing this book has been to present physical science in an interesting and intelligible way to readers for whom this may be the last formal contact with the subject: the business managers, health workers, teachers, artists, lawyers, government employees, members of the armed services, journalists, and other citizens of the 21st century. Every person who is living in a democratic culture that is grounded in science and technology should acquire a degree of scientific literacy in order to live a fulfilling life in that culture.

Toward that end, I have endeavored to write the book at a level and with an approach that any college student, regardless of background, should be able to understand and perhaps even to enjoy. My fondest hope is that the student will come to realize that science is a challenging and rewarding human enterprise and become excited about experiencing science in his or her daily life.

FEATURES

Coverage. This book presents some of the major concepts, theories, and principles of physics, chemistry, astronomy, geology, oceanography, and meteorology, indicating in many cases how they were developed and applying them to the world around us and to topics of current interest. Some examples of the latter are the status of the stratospheric ozone hole; global warming, the Hubble Space Telescope's study of a "white spot" on Saturn; COBE's discovery of evidence that supports the big bang theory; spacecraft Galileo's encounter with asteroid Gaspra; weather prediction; volcanic eruptions in the Philippines and Japan; earthquakes in Armenia and California; the effects of Hurricane Hugo; cold fusion; and the emerging technology of active noise control. The number and variety of topics covered allow the instructor considerable latitude in the design of courses of varying lengths and emphases.

Style. I have endeavored to make this book as reader friendly as possible. The emphasis throughout is on the nature of science as a creative human enterprise; the key role that it plays in modern society; its relationship to technology; its open-ended character as reflected in the dynamic nature of scientific concepts and theories; and on the human qualities of scientists. Thus, when a significant natural law or phenomenon is discussed, the reader will also learn about the scientist who made the discovery and what led to it as well as some of its applications and limitations. Such understanding and appreciation, I believe, are desirable outcomes of a course in physical science.

Organization. A broad overview of the scientific enterprise is presented in Chapter 1, which develops such topics as what is science?; the humanist and the scientist; the limitations of science; science and society; science and technology; and scientists and their work. Chapter 2 lays the groundwork for the quantitative aspects of the course, taking up measurement; standardization; the development of the metric system; SI units; multiples and submultiples; exponential notation; and dimensional analysis. The remainder of the book is divided into four parts. Part I: Physics comprises Chapters 3–12; Part II: Chemistry, Chapters 13–19; Part III: Astronomy, Chapters 20–21; and Part IV: Earth Science, Chapters 22–24.

Perspective Essays.   Each of the four parts into which the book is divided is preceded by a Perspective on . . . essay which presents some highlights of the subject matter to be covered in that section and some of the research at the frontiers, and relates them to such basic themes as matter, energy, force, and motion. These Perspective essays also emphasize the connections among the parts, so that the student will view physical science as scientists themselves do, as a seamless whole rather than a field of study characterized by disparate divisions.

Guest Essays.   Ten essays by guest authors are interspersed throughout the book. These essays amplify subjects discussed in the text, such as in "Ultrasound and its Applications," or consider interesting topics from a fresh angle, such as in "The Energy-Efficient House of the Future" and "C_{60} and the Fullerenes."

Level.   The limited background in science and mathematics that the nonscience major often brings to this course should pose no problem. I assume no prior work in science, and the mathematical level is generally no more advanced than arithmetic and simple algebra. I have made a particular effort to keep the language and the mathematics as clear and intelligible as possible. Scientific laws are presented first in verbal and then in mathematical form. Each chapter is divided into 10–15 numbered sections, on the average, to present the material in manageable bits. The relaxed, conversational writing style, with references to and anecdotes about scientists and their work, is intended to humanize the subject and to foster readability and comprehension.

Problem-Solving Strategies.   A large number of worked examples set off from the body of the text are included as models to aid the student in understanding how the principles are applied. Mathematical operations are discussed fully as needed. Each worked example is followed by an extension that illus-

trates the same principle from another perspective. This makes the text more interactive with the student by encouraging the application of the skills that have been presented. The answer to each extension is given for immediate reinforcement. Should the instructor so desire, these features may be de-emphasized in courses of a less quantitative character.

Illustrations and Photos.   Hundreds of superb four-color diagrams and striking color photographs are used throughout this text to enhance the verbal presentation by capturing reader interest and contributing to the assimilation of scientific concepts, theories, and principles. I have used color to emphasize those aspects upon which the reader should focus special attention. Color adds clarity to the artwork and the photos besides being visually pleasing in itself. I have used color coding throughout the text to identify specific physical quantities. The various color schemes are noted on the inside front cover of the book.

Key Terms.   New and important terms are printed in boldface and defined immediately upon introduction. A comprehensive section of Key Terms with accompanying definitions is provided at the end of each chapter as reinforcement to promote student comprehension. This feature motivates students to review important terms as they read through the text.

Things to Do.   Each chapter has a Things to Do section consisting on the average of five to ten home or dormitory laboratory projects that usually require only very simple materials. Students can use the scientific concepts they are learning, even if only on a limited basis. Students of any age know that science is more fun when there are things to *do* rather than to watch or read about (witness the resounding success of hands-on exhibits in museums of science and technology), and these activities add a practical dimension to the study of physical science in courses that do not have a laboratory component.

Questions and Exercises.   At the end of each chapter there is a set of questions and exercises grouped according to topics within the chapter that are designed to test the student's grasp of the material in the chapter. The number of conceptual questions has been increased in this edition, as has the number of problems. As an additional aid, the answers to all numerical exercises are given at the end of each chapter to encourage student learning and to reduce the possible level of frustration by providing immediate access to correct answers.

Chapter Summaries.   In some chapters, there is an end-of-chapter summary—both in words and mathematical symbols—of laws and principles, which should facilitate the student's mastery of these aspects of physical science.

Suggestions for Further Reading.   An annotated, selected, updated, and comprehensive section of Suggestions for Further Reading is another end-of-chapter feature. References are given to articles from several accessible and appropriate journals, including *Journal of Chemical Education, Scientific American, American Journal of Physics, Smithsonian, National Geographic,* and *Physics Today.* The references may serve as a basis for student reports or to broaden the student's knowledge of a particular area. Even a cursory reading of the annotated suggestions can convey a sense of the dynamic nature of science and its applications—a basic theme of this book.

Appendix.   The Appendix contains useful tables for easy reference: SI units and derived units, conversion factors, periodic table of the elements, physical constants, planetary physical data, and atomic masses. Many other tables are interspersed throughout the text.

SIGNIFICANT CHANGES IN THE FIFTH EDITION:

- A section entitled *To the Student—How to Study Science* has been added.
- Measurement has been moved from the Appendix to Chapter 2 and has been updated.
- Questions and exercises at the end of the chapters have been added and are grouped by topic rather than randomly listed.
- Four Perspective essays have been added to introduce and provide interconnections for each of the four parts of the book.
- Ten guest essays covering interesting topics have been added throughout the four parts.

- All of the line art has been redrawn in color and much new art has been added.
- Most of the numerous photographs are new and in full color.
- Statements of laws, theories, concepts, and examples are displayed more effectively through screening, italicizing, and the greater use of boldface.
- New sections on tides, winds, earthquakes, volcanoes, weather forecasting, the ozone hole, and hurricanes have been added.
- Suggestions for further reading have been updated and most of those before the 1970s have been dropped.
- Some of the latest findings in space science have been added.
- The Appendix has been revised to make the contents even more useful.

ANCILLARY MATERIALS

The following ancillary materials are available with this text:

Instructor's Manual and Test Bank.   This manual contains complete solutions and answers to the end-of-chapter exercises. The test bank contains approximately 50 multiple-choice questions per chapter. The format allows instructors to duplicate pages for distribution to students. Also included is a chart correlating the *Investigations in Physical Science* laboratory manual to the text.

ExaMaster™ Computerized Test Bank.   Available for the IBM PC and Macintosh computers, the test bank contains approximately 1200 multiple-choice problems, representing every chapter of the text. The test bank allows instructors to customize tests by rearranging, editing, or adding questions. The software program solves all problems and prints the answers on a separate grading key.

Overhead Transparency Acetates.   This collection of 100 transparencies consists of numerous full-color figures and photographs from the text to enhance lectures; they feature large print for easy viewing in the classroom.

Investigations in Physical Science. This laboratory manual offers 30 experiments designed to supplement the learning of basic principles in the course. The manual features pre-lab assignments to increase student preparation. Every effort has been made to keep laboratory costs to a minimum. A chart correlating the text to the laboratory manual appears in the instructor's manual.

Instructor's Manual to Accompany Investigations in Physical Science. This manual contains discussions of the experiments and teaching hints for the instructor.

ACKNOWLEDGMENTS

This book owes its final shape and form to the creativity and hard work of many people, whose dedication to the project is gratefully acknowledged. Foremost among them is the dean of American scientific publishers, John J. Vondeling, who has expertly shepherded this work through five editions. I am grateful for his wisdom, encouragement, and support.

I owe sincere thanks for the outstanding contributions of the entire talented staff at Saunders in producing the most attractive book in its field. I especially wish to thank Jennifer L. Bortel, Developmental Editor, with whom it has been a distinct pleasure to work on this most thoroughgoing revision, for her ideas, enthusiam, infinite patience, and skills in keeping the project well organized; Maureen R. Iannuzzi, Project Editor, who competently guided the book through the intricacies of production and deadlines; Christine Schueler, Art Director, and Doris Bruey, Associate Art Director, whose talents have created a book that is visually engaging; Jay Lichty, Production Manager, and Tim Frelick, Director of Editing, Design, and Production, both of whom did an outstanding job in ensuring a product of the highest quality. I would like to thank Robert Irion at the University of California at Santa Cruz for writing the excellent Perspective on Essays. I also wish to thank scientific illustrator Dennis Tasa of Tasa Graphic Arts, Inc., for the excellent color line-art program which he provided.

Special thanks go to the many instructors who have used earlier editions of this text. I have been significantly aided by their thoughtful criticisms and suggestions, many of which are responsible for improvements in the present edition. I am particularly grateful to the following colleagues who have contributed to the evolution of this work over five editions and shared their insights with me at various stages in its development.

Steven R. Addison, *University of Central Arkansas*
Alvin Aurand, *Broward Community College, Central Campus*
Maurice V. Barnhill III, *University of Delaware*
B. J. Bateman, *Troy State University*
S. Leslie Blatt, *Ohio State University*
W. H. Breazeale, Jr., *Francis Marion College*
John Breshears, *Southeastern Community College*
Richard Brill, *Honolulu Community College*
Iva D. Brown, *The University of Southern Mississippi*
William J. Brown, *Montgomery College*
Robert Childers, *Tulsa Junior College, Southeast Campus*
Lowell O. Christensen, *American River College*
G. Rolland Cole, *University of Delaware*
Gene F. Craven, *Oregon State University*
Basil Curnette, *Kansas State University*
Perry Doyle, *Maple Woods Community College*
Richard B. Ewing, *University of Delaware*
P. Joseph Garcia, *Bloomsburg State College*
Leonard H. Greenberg, *University of Regina (Canada)*
Vallie Guthrie, *North Carolina A&T State University*
James H. Hill, *Los Angeles Valley College*
Keith R. Honey, *West Virginia Institute of Technology*
Malcolm Hults, *Ball State University*
Kenneth C. Jacobs, *University of Virginia*
John Kalko, *Rancho Santiago College*
Stanley G. Kullman, *University of New York College at Cortland*
William L. Masterton, *University of Connecticut*
Kay C. McQueen, *Tulsa Junior College*
Kenneth S. Mendelson, *Marquette University*
T. Morishige, *Central State University*
Vedula S. Murty, *Texas Southern University*
Carl Naegele, *Michigan State University*
James W. Pendleton, *University of Maine at Gorham*
Walter Placek, *Wilkes University*

Hans. S. Plendl, *The Florida State University*

David Robertson, *Alcorn State University*

Paul Sanderfer, *Winthrop College*

Duane Sea, *Bemidji State University*

Sarah A. Smith, *University of North Alabama*

Leon D. Stancliff, *Middle Tennessee State University*

Christine Staver, *County College of Morris*

Francis Tam, *Frostburg State University*

Aaron W. Todd, *Middle Tennessee State University*

Robert T. Walden, *Mississippi Gulf Coast Junior College*

Joel S. Watkins, *University of Texas*

Irving Williams, *Nassau Community College*

Jerry D. Wilson, *Lander College*

I wish to express my appreciation to President Kalyan K. Ghosh of Worcester State College, to my colleagues in the Department of Natural and Earth Sciences, to my students, and to the members of the faculty and administrative staff for their interest and encouragement during the revision of the text. Last, but most important of all, I thank my wife, Shirley, for her love and support, and it is to her that I dedicate this book.

Melvin Merken
October 1992

TO THE STUDENT

HOW TO STUDY SCIENCE

Students taking a science course often ask their instructor, "How can I do better in science?" Although there is no Rosetta stone that will guarantee success, students who do well in a science course usually have a workable system.

You can devise your own recipe for success by considering the techniques described here and discovering what works for you. Based on what you know about how you learn, give the suggestions that are offered a fair chance by trying them more than once. They should help you to learn science effectively, bolster your confidence, and very probably enable you to earn high grades in your courses. When you have perfected your own system, you will be on your way to learning science with pleasure and success.

A Strategy for Success. As a general rule, you should schedule two hours of study time for every hour in class to study some science nearly every day, in a one- or two-hour block of time. Regular study produces best results. Work in an environment that is comfortable, well-illuminated, conducive to concentrated study, and free from distractions such as a stereo and TV. Be sure to assemble your textbook, lecture notes, syllabus, handouts, and writing materials so that you can work without interruption.

Read the syllabus for the course and follow the schedule set by the instructor. Prepare yourself for each lecture by reading the relevant sections of the text within 24 hours of the lecture. Take complete and organized notes on the lectures. For maximum effectiveness, review the lecture notes within 24 hours of the lecture. Use your text to clarify and complete your notes. Integrate homework, lab work, and additional reading into this cycle of prereading, notetaking, and review. A weekly or biweekly review will help you gain a clearer picture of the course as a whole and enable you to prepare for quizzes and exams.

Be specific in setting up your weekly study schedule. Indicate which sections of your text you plan to study, which problems to work out, which topics to review for an approaching exam. To summarize:

1. Attend all lectures.
2. Preread, take notes, review.
3. Construct a study schedule a week at a time.
4. Keep up with the course syllabus.
5. Do all assigned readings and exercises.
6. Get help quickly, if you must, from your instructor, tutor, or classmates.

Getting the Most Out of Your Textbook. Reading a science text is far different from reading a historical novel. Realize that you cannot possibly absorb the full import of scientific material following just one reading. If an assignment is general, such as, "Read chapter 6 this week, and do exercises 5 through 10," decide how to organize your studying for best results.

Many students make the mistake of reading an entire chapter once and then attempting to do the exercises. They often find that the exercises are too difficult. Rather than reading all parts of the text with equal care, develop the skills of (1) reading selectively and (2) reading actively. Your lecture notes and assigned exercises can help determine your reading priorities.

Begin by surveying the chapter to get an overview of the material. Read the introduction; see how this chapter relates to previous chapters. Note the section headings and subheadings; examples; diagrams, illustrations, and their captions; marginal notes; tables; summary; and key terms. Pay particular attention to the concepts that are involved in the examples and exercises, and look for their definitions and explanations as you read the text.

When you study a section of the chapter in greater detail, read actively. Think constantly about what you are reading, relate one paragraph to another, a new concept to previous concepts. Turn headings into questions, and search for answers to your questions. Asking and answering questions will increase your retention of the material and enable you to perform well on examinations. Make a note of the parts of the text that are not clear to you and seek extra help.

When you have read a section or paragraph actively, either go back and mark it selectively, or take notes in outline form on the text material. Underline or use your hi-liter sparingly. Focus on the main ideas, formulas, and summarizing sentences. Your goal should be a visual summary that will be meaningful to you even days or weeks later and when you review for examinations. Try some of the Things to Do at the end of the chapter. Remember that science describes the world around us and that much of what you learn can apply to familiar situations. Scan the Suggestions for Further Reading for interesting follow-ups.

Your text provides you with many worked-out example problems followed by extensions. Even though you may think that you understand the concepts, working through an example and doing the extensions yourself will enable you to apply your conceptual understanding to solving problems. You will acquire a deeper grasp of the concept in the process and increase your chances of remembering how to solve similar problems on an examination.

After completing a section, try to do the exercises assigned to you for that section, even if they are not to be turned in. Make a note of the exercises that you cannot do or that you are unsure of, and seek help from your instructor or others to clear up what you don't understand.

1. Survey the chapter.
2. Read selectively.
3. Read actively.
4. Underline sparingly.
5. Work through the examples.
6. Do the assigned exercises.
7. Try some of the activities in the Things to Do section.
8. Note any parts of the text that are not clear to you.
9. Seek clarification promptly.

CONTENTS

A laser system monitoring the movement of the San Andreas fault at Parkfield, in central California, the most heavily instrumented earthquake region in the world. The laser bounces light off a network of 18 reflectors located several kilometers away. The system will pick up movements of less than 1 millimeter over a distance of 6 km. (David Parker/Science Photo Library)

1

THE SCIENTIFIC ENTERPRISE

In the three centuries that the scientific revolution has been in progress, but especially in our own time, science and technology have transformed the lives of most people in the Western world.

New agricultural methods make it possible for most nations in the West to raise enough food for their own needs. Today, less than 5% of the working population is occupied in food production, compared to 90% in the 1700s. Antibiotics and vaccines have eliminated many diseases, reduced infant mortality drastically, and doubled the average life span.

Television by satellite provides information and entertainment "live" as it occurs anywhere on earth. Nylon, orlon, and many other synthetic materials lend interest, comfort, and color to our clothing and surroundings. Computers, transistors and integrated circuits, microwave ovens, "instant" cameras, videocassette recorders, video games, word processors, compact disc players, facsimile machines, cellular telephones, and photocopiers affect the lifestyles of many people. Research in alternative energy sources has opened up the possibility of providing the world with its energy needs long after the supplies of fossil fuels—coal, oil, and natural gas—have been exhausted.

Through science and technology, improvements in sanitation, construction, power distribution, communication, and other services have occurred. It is now possible to walk on the moon, live for months in skylabs, investigate the depths of the universe, and launch probes into space in a quest for intelligent life (Fig. 1–1).

In this chapter we examine the nature of the scientific enterprise, that powerful force that plays so dominant a role in modern life. In later chapters we explore specific areas from a number of fields of physical science and consider some of their modern applications.

Figure 1–1 The earth as seen from space, showing most of Africa and surrounding oceans. (Source: Robert E. Gabler, *et al., Essentials of Physical Geography,* 4th edition. Philadelphia: Saunders College Publishing, 1991, p. 24, Fig. 2.1. [Courtesy of NASA])

1–1 WHAT IS SCIENCE?

The scientific approach today is essentially that of the founders of the Royal Society of London, the oldest organization for the advancement of science in the world, chartered in 1662. Christopher Wren, Robert Boyle, and the other founders were by and large professional men who were interested in the "new philosophy," or natural science, that was then emerging.

They admired in particular the experimentation and observations made by Nicolaus Copernicus, Galileo Galilei, Sir William Gilbert, and Johannes Kepler. The Fellows, as they were known, conducted their own experiments and observations, which covered many fields, such as the circulation of the blood, the Copernican hypothesis, and the force exerted by ignited gunpowder. Believing that each problem was unique, they searched for an appropriate method to attack it rather than insisting upon following a particular pattern of experimentation.

One of the greatest scientists who ever lived, Albert Einstein, described science as nothing more than a refinement of everyday thinking. There is no distinct scientific method as such. A scientist probing the unknown is like a composer facing a blank page, an artist facing an empty canvas, or an explorer. He or she cannot work through a code of rules to the discovery of new knowledge or to the invention of new theories. No such code has ever been found. According to this view, **basic** or **fundamental science** is a process of investigation of natural phenomena, a quest for understanding. The objective of this aspect of science, science as search, is to describe and to understand nature through observation, experimentation, and theoretical modeling.

Science differs from the fine arts, literature, and philosophy in these important respects: Science is *self-testing, self-correcting,* and *objective.* Scientists discover things and publish their work in scientific journals, where they expose their data to the scrutiny of other scientists, who may repeat the experiments with essentially the same results or quite different results. Errors in fact or interpretation are discussed openly. The final arbiter is evidence obtained through experimentation or observation, not authority. Self-testing and self-correction in science, however, do not necessarily result in universal agreement. It is impossible to achieve complete certainty and universal agreement in science, since the discovery of a single fact might call the theory into question. The most that can be said for a scientific truth is that it has a high degree of probability.

Science has another aspect, which is a product of its first: It is a body of knowledge. Yet it is more than a mere compendium of facts; it is a search for relationships among them. Understanding is often followed by the ability to predict natural phenomena and ultimately to control them. For example, the discovery of the electrical nature of matter led eventually to practical methods of harnessing electrical energy, such as the battery and the generator.

Understanding in science is achieved through its laws, theories, and models, which create order out of the chaos of raw facts. **Scientific laws** are statements of relationships based on human perceptions of natural phenomena, and they generally can be stated in mathematical relationships. Newton's second law of motion, for example, describes the behavior of a particle when it is subjected to a force. Although laws are based on past experiences, their utility lies in their ability to predict the unknown. Nature is reliable. Given the same set of circumstances, nature can be relied on to operate in the same manner.

Laws, in turn, are explained by means of **theories.** A theory that is introduced to explain one law and is found to explain others as well is particularly valuable. Scientific theories, though, are regarded as tentative. They can be replaced by simpler, more comprehensive, or more rational theories that provide better explanations of the laws of nature. For example, the highly successful caloric theory of heat eventually gave way to the theory of heat as a mode of motion.

The laws of one branch of science have never been found to conflict with the laws of another. There seems to be an underlying unity in nature that applies to both living things and inanimate objects. The laws and theories in one field of science often turn out to be applicable to other fields.

Theories are often made easier to grasp by means of **models.** A model is the product of a scientist's creative imagination, to be retained as long as it remains useful. When it has served its purpose, it may be replaced by one that is more adequate. In this manner, structures or processes that cannot be observed directly may be visualized. Occasionally, as in the case of theories of nuclear structure, a variety of models may exist side by side—the liquid-drop model, the shell model, and others—since each is uniquely suited to describe one aspect of nuclear structure, but no one model alone can account for all of the observed properties.

1–2 THE HUMANIST AND THE SCIENTIST

During the nineteenth century, according to C. P. Snow, a gap developed between the traditional culture and the explosive new sciences. It was marked by two contrasting approaches to understanding the world. By the beginning of the twentieth century, this gap had become almost unbridgeable. The underlying causes for the polarization of society into two cultures and ways of closing the gap have been the subject of long and often heated debates.

The **humanities** are a group of studies consisting of the languages and literature, the fine arts (art, music, drama, and dance), philosophy, religion, and logic. They are presented in terms of qualitative arguments and concepts such as the good, the true, and the beautiful. The **sciences** include mathematics, the natural sciences (physics, chemistry, biology, astronomy, and geology), and the social and behavioral sciences (psychology, sociology, anthropology, economics, government, and history) (Table 1–1). Scientific laws, theories, and models must be supported by empirical evidence to achieve acceptance (see Fig. 1–2).

Although the humanist and the scientist are concerned with the same world, they look at different aspects of it and see it with different eyes. Compare how a poet and a physicist—the extreme representatives of humanist and scientist—might describe a symphony or a sunset. The artist's approach is to express a personal vision through poetry, music, painting, or some other medium, whereas the scientist attempts to achieve **objectivity** by eliminating personal bias and subjective views (although the models scientists use are often chosen subjectively).

The humanist and scientist use the same conceptual tools, analysis and synthesis, but they differ in their concepts of truth and how it may be attained. The humanist communicates the **qualitative** aspects of experience. In expressing an individual view of things, the humanist appeals to the common humanity of the audience. The "truths" do not involve tests or corrections and may be accepted as certain in themselves. The scientist, on the other hand, is concerned with those aspects of phenomena that are ver-

Table 1–1 THE REALMS OF KNOWLEDGE AND EXPERIENCE

Natural Sciences

Mathematics
Astronomy
Physics
Chemistry
Biology
Geology
Meteorology
Oceanography

Social Sciences

Psychology
Sociology
Anthropology
Economics
Government
History

Humanities

Literature
Fine arts
Music
Drama
Philosophy
Religion

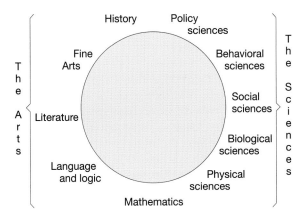

Figure 1–2 The arts (humanities) and sciences defined with reference to university disciplines. (Source: Cassidy, H. G., "Problem of the Sciences and the Humanities." *Journal of Chemical Education,* Vol. 38, No. 9, September, 1961, p. 444.)

ifiable and, at least in theory, subject to universal agreement. Scientists aim for **quantitative** precision; their data are replicable and open to test and correction. Scientific "truths" are tentative rather than eternal and absolute.

Science, more than art, is **cumulative,** the product of many individuals. The physics of Galileo and Newton, for example, is more comprehensive than that of Aristotle and Archimedes but less so than that of Planck and Einstein. The edifice of science grows from the contributions of scientists through the ages and, hence, is "many-brained." The humanities, on the other hand, are "single-brained." Their products are unique. A painting by Rembrandt is not less beautiful today than one by Picasso. Einstein said that even if he himself had never lived, we would still have had some form of the theory of relativity, but if Beethoven had never lived, we would not have had an "Eroica Symphony."

Although the humanities and sciences are so different, the great ages of science have been the great ages of the arts. Since both depend on the exercise of imagination, art and science flourish, or languish, together. A striking example is the contributions of Galileo and Shakespeare, who were born in the same year (1564) and grew into greatness in the same age.

1–3 THE LIMITATIONS OF SCIENCE

Anything beyond the limits of our senses and the instruments we build to extend them is outside the limits of science. Within these broad limits, the scope of science includes everything known to exist or to happen in the universe.

In principle, the arts and other humanities can be studied as psychological, anthropological, and biological phenomena. A Mozart sonata or a rock-and-roll hit can be described by periodic oscillations and nerve impulses. The scientific description can enhance one's appreciation of the music just as an understanding of musical structure does, although they cannot do full justice to the direct experience of the music.

Science has transformed the world by its selection and abstraction from nature of the measurable and quantifiable, but this does not mean that science deals with all of reality. Other human experiences, such as love, loyalty, beauty, and courage, are as real as any measurable experiences of our five senses. Most sci-

entists do not subscribe to the doctrine of **scientism,** the belief that only scientific knowledge is worthwhile and that all other knowledge is nonsense.

Decisions as to how scientific knowledge shall be used are based on judgments that lie beyond the province of science. Scientific knowledge is ethically neutral, but it can be used for good or evil purposes. Polio vaccine and napalm were both created in the laboratory, but society determined how each would be used. Scientists can help society reach decisions, but in their capacity as responsible citizens rather than as scientists.

1–4 SCIENCE AND SOCIETY

Modern society faces difficult problems: the threat of nuclear war, an energy crisis, air and water pollution, depletion of natural resources, urban blight, poverty, famine, and drug dependency (Fig. 1–3). Many sci-

Figure 1–3   The city epitomizes the problems of modern life: congestion, pollution, transportation, urban blight. (Photograph by Michael S. Yamashita, Woodfin Camp & Associates)

entists hold that some of society's problems can be solved only through more research—pure and applied—and more technology. Take pollution as an example. This is a problem that has obvious scientific and technological components. It is difficult to think of a solution to photochemical smog or oil spills without an understanding of the fundamental aspects of the problem. A knowledge of more science, not less, is required.

Frequently, the political, economic, and social aspects of a problem outweigh other considerations. Scientific solutions may be available that society is not ready to accept. An example is our current transport systems, in which science and technology are not employed to their full potential. Such measures as staggered work schedules, exclusion of automobiles from a city's central business district during certain hours, a greater reliance on mass transportation, and the requirement of antipollution devices on vehicles all encounter various degrees of resistance from the public.

Assuming that it is possible to apply the methods of natural science to social or political problems, more collaboration will be needed than has been evident in the past among basic and applied scientists, engineers, social scientists, lawyers, and politicians. One proposal is the creation of national institutes that would be broadly interdisciplinary and team-oriented and would apply the methods of science to our difficult social problems.

1–5 SCIENCE AND TECHNOLOGY

Through **technology,** the findings of science are translated into new or improved products or services: high-speed computers, fertilizers that can double the size of a crop, transistors, antibiotics, organ transplants, and jet travel (Fig. 1–4). The advances of technology, many believe, have yielded benefits that on the whole vastly outweigh the injuries they have caused.

The phenomenal rate of change that a technology may introduce is evident in the field of transportation. The horse was the most rapid means of locomotion until the invention of the railroad, but we have moved from the propeller to the jet-propelled supersonic aircraft in one generation. The products and processes of science become the world's most powerful agents of change through technology.

Until recently, technology was considered innocent until proved guilty. For example, DDT, a chemical compound first synthesized in 1874, was used during World War II to protect American soldiers against disease-carrying insects. Later it was used to protect civilian populations as well. Those who worked with DDT did not think of applying it to control insects infesting crops, livestock, or forests. But farmers and foresters, believing that what killed insects on people would also kill insects on plants, soon turned DDT into a massive assault on the envi-

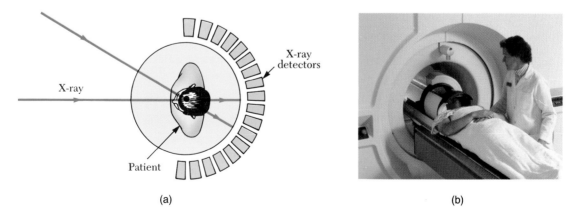

(a)   (b)

Figure 1–4   A patient undergoing a CAT scan. (Source: Raymond A. Serway and Jerry S. Faughn, *College Physics*, 3rd edition. Philadelphia: Saunders College Publishing, 1992, p. 1059, Fig. 33.11. [Jay Fries/The IMAGE Bank])

ronment. If the use of DDT had been restricted to the protection of people, it would not have become an environmental hazard.

Now the burden of proof is shifting to technology. The innovator is expected not only to show the impact that the innovation will have, but also to prove that it will not produce harmful side effects. To test new products and methods for potentially dangerous effects, a new field has developed. **technology assessment.**

Technology assessment means trying to understand in advance some of the human, social, and environmental consequences of introducing new technologies or diffusing existing ones into new areas. It functions as an early-warning device, based on the principle that technology exists to serve people. Even the products of technology now available— automobiles, pesticides, detergents—will have to prove their desirability (Fig. 1–5). Those that seem on balance to be beneficial will be encouraged, whereas those that seem to have risks that outweigh their benefits will be discouraged.

Figure 1–5   Traffic congestion in a major city. Are there better ways to move people? (George Hall/Woodfin Camp & Associates)

Although technology assessment may help avoid many unpleasant surprises, it will be difficult even to guess at some of the future adverse effects of a new technology. On the other hand, failure to make assessments is almost sure to produce unpleasant surprises. To suggest that we turn our backs on further progress is somewhat like suggesting decapitation as a cure for headache. Once a society begins to move forward in technology, it must accept the accompanying responsibilities. The United States was the first to realize this fact when it established the Office of Technology Assessment (OTA) in 1973. Other countries are now following suit.

1–6 SCIENTISTS AND THEIR WORK

Unlike medicine or law, science is not a licensed profession. A person does not have to be a PhD to be a scientist. The National Register of Scientific and Technical Personnel defines a scientist as anyone who is eligible for membership in a recognized professional society, such as the American Chemical Society. The scientific community includes a wide range of individuals: basic scientists in universities, research institutes, and government and industrial laboratories; engineers; workers in medicine and public health; graduate students and technicians working on scientific problems; and teachers of science in colleges and schools.

Scientists are ordinary human beings with essentially the same virtues and faults that most people have (Figs. 1–6 and 1–7). There seem to be no special qualities of personality, intelligence, background, or upbringing that enter into the making of a scientist. Science offers unusual opportunities for the exercise of the creative, imaginative, and manipulative talents, while at the same time enabling one to contribute to the common good.

As recently as the 1930s, **"little science"** was done by individuals pursuing the "art of the soluble," as the molecular biologist Peter Medawar described it, working on problems that they felt they could solve successfully with the very limited material means at their disposal. Samuel Goudsmit, a physicist, referred nostalgically to those days of "string and sealing wax," when basic science in the universities received little public or private support. Impelled by their own curiosity, scientists carried on their researches in basic science, very often with far-reaching results, working on problems that they themselves selected within a

Figure 1–6 A great scientist of the twentieth century, Albert Einstein, relaxes on a visit to California. (Source: Jay M. Pasachoff, *Astronomy: From the Earth to the Universe*, 2nd edition. Philadelphia: Saunders College Publishing, 1983, p. 316, Fig. 19–35. [Courtesy of The Archives, California Institute of Technology]

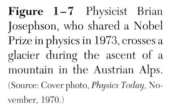

Figure 1–7 Physicist Brian Josephson, who shared a Nobel Prize in physics in 1973, crosses a glacier during the ascent of a mountain in the Austrian Alps. (Source: Cover photo, *Physics Today*, November, 1970.)

narrow field of knowledge. This is how penicillin, nuclear fission, and the genetic code were discovered.

Much of today's science, on the other hand, is conducted by teams of scientists. This **"big science"** is as characteristic of *basic* or "pure" science, which aims at extending knowledge, as it is of **applied science,** which is concerned with practical problems such as the development of radar or a moon rocket. It is centered in large laboratories that were first created around the turn of this century in industry and government, such as the General Electric Research Laboratory and the AT&T Bell Telephone Laboratories (Fig. 1–8).

In dealing with a problem as complex as energy, a scientist soon finds that certain aspects of the solution lie beyond his or her immediate discipline. Such complex problems require teams of scientists representing various disciplines—physics, biology, chemistry, and so on—or different branches of the same discipline, such as physical chemistry, analytical chemistry, and biochemistry. The team, with its competencies in several fields, commands a greater amount of the knowledge necessary for progress than any individual could. Table 1–2 illustrates the specialties that exist in just one major field of science, chemistry.

Important new scientific discoveries are made in many ways (Fig. 1–9). Serendipity, the making of an unexpected discovery by accident, played a part in Wilhelm Roentgen's discovery of X-rays. Precise measurements of the density of air led William Ramsay to discover the presence of the gases argon and krypton in air. On some occasions, systematic analysis is the key; at other times, a sudden thought, a flash of insight, even a vision may lead to discovery (Fig. 1–10). The structure of the benzene molecule occurred

Figure 1–8 Researchers conferring at Bell Laboratories. (Sepp Seitz/Woodfin Camp & Associates)

Table 1–2   DIVISIONS OF THE AMERICAN
             CHEMICAL SOCIETY

Agricultural and Food Chemistry
Analytical Chemistry
Biological Chemistry
Carbohydrate Chemistry
Cellulose, Paper, and Textile Chemistry
Chemical Education
Chemical Health and Safety
Chemical Information
Chemical Marketing and Economics
Colloid and Surface Chemistry
Computers in Chemistry
Environmental Chemistry
Fertilizer and Soil Chemistry
Fluorine Chemistry
Fuel Chemistry
Geochemistry
History of Chemistry
Industrial and Engineering Chemistry
Inorganic Chemistry
Medicinal Chemistry
Microbial and Biochemical Technology
Nuclear Chemistry and Technology
Organic Chemistry
Organic Coatings and Plastics Chemistry
Pesticide Chemistry
Petroleum Chemistry
Physical Chemistry
Polymer Chemistry
Professional Relations
Rubber, Inc.
Small Chemical Businesses

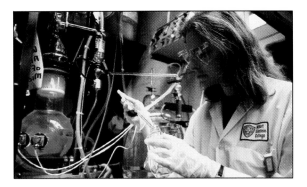

Figure 1–10   A researcher in a molecular biology laboratory. (Guy Gillete/Science Source/Photo Researchers, Inc.)

to August Kekulé in a dream. To benefit from these
flashes of genius, however, the scientist must be familiar with the knowledge of the past. As Louis Pasteur said, "Chance favors the prepared mind" (Fig.
1–11).

1–7 COMMUNICATION IN SCIENCE

More than the arts or literature, science is a coopera-
tive activity. No scientist works alone, not even scientific geniuses. One of them, Sir Isaac Newton, freely
admitted his debt to others when he wrote to Robert
Hooke, "If I have seen further (than you and De-
scartes) it is by standing upon the shoulders of giants."

There were more connections between Newton
and his predecessors and contemporaries than there
were between Beethoven and Mozart in the field of
music or Picasso and Renoir in art. In our own day,

Figure 1–9   Dr. Linus Pauling, who won the Nobel Prize
in chemistry in 1954, holding a model of water molecules.
(Thomas Hollyman/Photo Researchers, Inc.)

Figure 1–11   As a graduate research student at Cambridge University Observatory, England, astronomer Joce-
lyn Bell Burnell discovered the first known pulsar in 1967.
(Source: George Abell, *Exploration of the Universe,* 4th edition. Philadelphia: Saunders College Publishing, 1982, p. 546. [Courtesy of *Sky and Telescope*])

two or three persons, frequently from different countries, may share a Nobel Prize for work for which they are jointly responsible.

Cooperation in science includes the publication of papers in scientific journals. Many appear under joint authorship, frequently bearing the names of three or more contributors. Books, letters, and oral presentations at scientific meetings are other means of communication among scientists. Through these means a scientist describes the results of experimental or theoretical work.

Scholarly journals frequently have expert referees to screen manuscripts, a system that helps sift out the good papers from the bad. Some of the more prestigious journals reject 80% to 90% of the articles submitted. The self-testing and self-correcting features of science are put into operation in this way. Scientists expect colleagues to be scrupulously honest. If the work is significant, others will soon attempt to replicate it. If other laboratories are unable to reproduce the original results, the scientist's reputation may suffer. The basic difference between science and other endeavors is that inherent policing weeds out even the thought of fraud in science, although a few frauds have been perpetrated.

A new phenomenon is the use of modern inventions as instruments of scientific communication. Computer conferences and electronic mail systems are examples. Laboratories in Cambridge, Massachusetts, are in touch with those in New York, Chicago, and Pasadena all on the same day. By the time a paper is published in a journal, people working in that field in laboratories around the world seem to know about the work. Laboratories set at great distances from each other often collaborate as closely as if they were located on the same corridor.

Information travels almost with the speed of light. A bit of information picked up over lunch or in a corridor conversation in New York may be reported almost instantaneously in Houston. The system seems to function with amazing accuracy and with candor. The possibility of being scooped seems less of a problem today. Scientists realize there is so much to be learned that there can never be enough researchers. We have not reached the end of knowledge but have only begun.

Since knowledge is the important outcome of science, great importance has been attached from Newton's day to the present to the issue of **priority** —to being the first to make a particular discovery and to be recognized as such. We now know that Isaac Newton developed the calculus from 1664 to 1666, nine years before Gottfried von Leibniz independently discovered it in 1675. Yet Leibniz has priority of publication in 1684, since Newton did not publish his mathematical works until 1704. The dispute over priority of discovery went on for decades. A frank account, based on his own experience, of the extreme measures a scientist may take to ensure priority has been given by James Watson, who shared a Nobel Prize with Francis Crick and Maurice Wilkins for their work on the DNA molecule. In his book, *The Double Helix*, Watson conveys the flavor of modern science as few others have done. The book is one of the most refreshing documents that any scientist has produced and is required reading for anyone wishing to understand the human dimension of the scientific enterprise.

KEY TERMS

Applied Science   The application of science to the solution of practical problems.

Basic Science   The pursuit of pure science that aims at extending human knowledge.

"Big Science"   The conduct of science by interdisciplinary teams of scientists.

Cumulative   The corpus of science increases through successive contributions by scientists in all ages.

Experiment   A process undertaken to test a hypothesis under controlled conditions.

Humanities   A group of studies that includes the languages and literature, the fine arts, philosophy, and religion.

Interdisciplinary   Involving more than one field of knowledge, such as biochemistry.

"Little Science"   The conduct of science by individuals working on problems that they themselves select within a narrow field of knowledge.

Model   A representation of a structure or process that cannot be observed directly.

Objectivity   Based on observable phenomena; uninfluenced by emotion or personal prejudice.

Priority   In science, priority refers to the first to make a particular discovery and to be recognized as such.

Qualitative   A term referring to any characteristic by which something may be identified.

Quantitative   Expressible in terms of measurement of quantity or amount.

Replicable   The repetition of an experiment or observation with similar results.

Science   An approach to solving problems that is based on logic, observation, and experimentation.

Scientific Law   A statement of an unchanging relationship.

Scientism   The belief that only scientific knowledge is worthwhile.

Technology   The application of science to the development of new products or services.

Technology Assessment   The effort to predict some of the consequences of introducing new technologies.

Theory   A statement that is intended to explain a scientific law.

THINGS TO DO

1. While you are taking this course, visit a library on a regular basis, pick up copies of *Science, Scientific American, Smithsonian,* and *National Geographic,* and savor some of the articles that you will find there.

2. Keep a scrapbook of articles of scientific interest from a daily newspaper or weekly news magazine.

3. Spend an hour or more browsing through the natural and earth sciences, astronomy, and technology areas of your library. Look into books that interest you.

4. Look up the biographies of some of the scientists whose work is discussed in this book.

5. Interview someone who is doing scientific research and explore further some of the issues raised in this chapter.

6. Watch some television programs that have scientific themes.

EXERCISES

1-1. Compare the concept of truth in science with truth in another realm of experience.

1-2. If our age is an "age of science," should scientists claim any particular privilege or be charged with special responsibilities not granted to or expected of other citizens?

1-3. Discuss an instance of (a) an abuse of science; (b) an abuse of an art form.

1-4. Is experimentation characteristic of all sciences? Explain.

1-5. Why has the rate of scientific progress in the twentieth century been so explosive?

1-6. In your opinion, has the impact of science on the quality of life been, on the whole, positive or negative? Explain.

1-7. Discuss how the humanities and the sciences have contributed to the problems of our age.

1-8. Illustrate with a specific example the potential of a scientific discovery for good or evil.

1-9. Discuss the merits of the proposal to call a moratorium on scientific research until some of society's problems have been solved.

1-10. Discuss some connections between science and the arts.

1-11. Discuss the role of (a) cooperation and (b) competition in science.

1-12. Cite evidence for and against the existence of "two cultures" within society.

1-13. It has been proposed that there be professional critics of science and technology, just as there are art and drama critics, to help the public achieve a higher level of cultural understanding. What is your position?

1-14. Look up the status of the priority dispute between researchers at the Pasteur Institute in Paris and the National Cancer Institute in Bethesda, Maryland, over whether Luc Montagnier or Robert C. Gallo should be given credit for unraveling the cause of AIDS.

1-15. *Multiple Choice*
   A. Scientific knowledge
      (a) starts anew with each generation.
      (b) becomes dated quickly.
      (c) is cumulative.
      (d) is readily accessible.

B. The methods that scientists use to solve problems in the laboratory
   (a) can contribute little to the solution of practical problems.
   (b) are just as valid in dealing with all other problems.
   (c) are too exacting for general use.
   (d) probably cannot solve all human problems.
C. The use of scientific discoveries
   (a) is controlled by the scientists who made them.
   (b) is of no concern to nonscientists.

(c) is a matter for the people to decide.
(d) is authorized by the government.
D. The future direction of science
   (a) can be predicted with great accuracy.
   (b) cannot be predicted with great accuracy.
   (c) is known by leading scientists.
   (d) depends on government funding.
E. The fact that scientists and artists are often unable to communicate with each other
   (a) is the fault of the scientists.
   (b) is the fault of the artists.
   (c) if of no great concern to society.
   (d) is a loss to both art and science.

SUGGESTIONS FOR FURTHER READING

Brush, Stephen G., "Should the History of Science Be Rated X?" *Science*, 22 March 1974.
Asserts that the way scientists behave (according to historians) might not be a good model for students.

Crick, Francis, *What Mad Pursuit: A Personal View of Scientific Discovery.* New York: Basic Books, 1988.
The co-discoverer of the double helix of DNA writes candidly about his life in science and about how science is done today.

Green, Martin, "The Two Cultures Gap Revisited." *American Journal of Physics*, December, 1979.
Asserts that the problem of communication between scientists and humanists can be resolved.

Hillman, Howard, *Kitchen Science.* Boston: Houghton Mifflin, 1989.
Presents the scientific principles of creative cooking in a question-and-answer format that takes the mystery out of cooking and does so in plain language.

Klaw, Spencer, *The New Brahmins: Scientific Life in America.* New York: William Morrow & Co., 1968.
Conveys a sense of what it is like to be a natural scientist in America.

Kuhn, Thomas S., *The Structure of Scientific Revolutions.* Chicago: University of Chicago Press, 1962.
Analyzes the nature, causes, and effects of revolutions in scientific concepts.

Lederman, Leon M., "The Value of Fundamental Science." *Scientific American.* November, 1984.
Asserts that the support of fundamental (basic) science yields profoundly significant benefits to society.

Neufeld, Peter J., and Neville Colman, "When Science Takes the Witness Stand." *Scientific American*, May, 1990.
Argues that since scientific testimony is often the deciding factor for the judicial resolution of civil and criminal cases, national standards and the regulation of forensic laboratories are needed. Then the powerful new forensic techniques can serve a beneficial role in criminal justice.

Phillips, Melba, "Laboratories and the Rise of the Physics Profession in the Nineteenth Century." *American Journal of Physics*, June, 1983.
Galileo had a small shop, Boyle worked at home, Newton experimented in his college rooms. Not until the nineteenth century were research laboratories established by universities, industry, and governments.

Ritchie-Calder, Lord, "The Lunar Society of Birmingham." *Scientific American*, June, 1982.
An account of a group of manufacturers, inventors, and natural philosophers who transformed science and technology in Britain.

Roberts, Royston M., *Serendipity: Accidental Discoveries in Science.* New York: John Wiley & Sons, 1989.
Recounts many of the fascinating stories in the annals of scientific serendipity.

Rosenberg, Nathan, and L. E. Birdzell, Jr., "Science, Technology and the Western Miracle." *Scientific American*, November, 1990.
Links the unprecedented prosperity of the Western nations to their capacity to translate scientific knowledge into economic productivity.

Snow, C. P., *The Two Cultures: A Second Look.* Cambridge, England: Cambridge University Press, 1964.
Discusses the division of intellectuals into two groups that no longer communicate.

Watson, James, *The Double Helix.* New York: Atheneum, 1968.
Candidly portrays the race for scientific priority, by one who was there.

One can easily measure the period of a swinging pendulum. Today, a second is defined as 9,192,631,770 times the period of one oscillation of a cesium atom. (Robert Mathena/Fundamental Photographs)

2

PHYSICAL SCIENCE AND MEASUREMENT

Measurement is the process of determining "how many" (Fig. 2–1). To measure the length of a table, for example, we select a suitable reference standard —say, a meter stick—and lay it end to end until the comparison is completed. Then we express the measurement as the number of units it takes to equal the length of the table, such as 1.38 meters.

Precise measurement lies at the very heart of science (Fig. 2–2). Words alone often mean differ-

ent things to different people. How big is "big?" . . . or "small?" How hot is "hot?" What does "heavy" mean? Even in everyday life, concepts mean more when they are expressed in numbers. We speak of a 3″ × 5″ file card, a car that averages 26 miles to the gallon, a temperature of 72°F. In this chapter, we look at standards of measurement and some ways of expressing measurements conveniently.

2–1 WHY STANDARDIZATION?

For most of the billions of measurements made each day, we use the foot-pound-quart system for length, mass, and capacity. This system is known as the **En-**

Figure 2–1 A "mile marker" along the side of a highway enables officials to find certain locations accurately. (Gerald F. Wheeler)

Figure 2–2 Measuring with a micrometer that can typically measure lengths to a precision of about 0.01 mm. (Jeffrey Coolidge/ Stockphotos, Inc.)

glish Customary system of weights and measures. Many of the standards were originally derived from a convenient though crude source—the human anatomy. Thus, the inch was the length of the thumb from tip to knuckle; the palm, the width of four fingers; the foot, four palms; a yard, the distance from the tip of the nose to the tip of the middle finger of the outstretched arm. The anatomically based units are not easy to work with. There are 12 inches in a foot; 3 feet in a yard; 5½ yards in a rod; 320 rods in a mile; and 5280 feet in a mile.

A second drawback of customary units is that the same name quite often stands for different quantities from one place to another or from one context to another. The gallon used in the United States today was Queen Anne's (1665–1714) wine gallon. It was smaller than the ale gallon used in her day and different from the imperial gallon eventually adopted. A Canadian imperial gallon, for example, is about 25% larger than an American gallon. An ounce for measuring fluids is not the same as an ounce for weighing. Moreover, the avoirdupois ounce for ordinary weighing is lighter than the troy ounce for weighing precious stones and metals and the apothecaries' ounce for prescriptions and drugs.

As commerce and industry expanded and the world became increasingly interdependent for raw materials and finished products, it became evident that a common and rational system of weights and measures was needed. The emergence of experimental science made it clear that physical measurement was a basic element of studying natural phenomena. Physical science is concerned with measuring the values of quantities. The value of a quantity is expressed as the product of a number and a unit, such as 25 seconds. Well-defined units of measurement, therefore, were essential to the exchange and comparison of experimental results. By the second half of the eighteenth century, there was a movement in several nations to establish a system of measurement based on easily reproducible standards. This drive culminated at the time of the French Revolution with the development of the metric system.

2–2 THE METRIC SYSTEM

When the **metric system** came into existence in the 1790s, standards were important for only a few kinds of measuring units, primarily length and mass. Today,

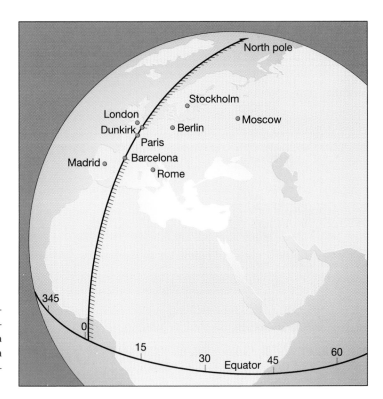

Figure 2–3 The meter was originally intended to be one ten-millionth of the equator-to-pole quadrant of the earth's meridian passing close to Paris. The section between Dunkirk and Barcelona was accurately measured.

we must also have standards for the accurate measurement of temperature, color, electric current, sound and light intensity, and many other physical quantities. Practically all, however, can be derived from four fundamental standards; the **meter** for length, the **kilogram** for mass, the **second** for time, and the **kelvin** for temperature.

The standard for length, the **meter**, was originally defined as one ten-millionth of the distance between the equator and the North Pole—a quadrant of the earth's meridian. The section of the quadrant lying between Dunkirk, France, and a point near Barcelona, Spain, was carefully measured by a surveying team and extrapolated by astronomical observations to the entire quadrant (Fig. 2–3). The unit of mass, the **kilogram**, was originally defined as the mass of a cubic decimeter (a liter) of water at the temperature of its maximum density, 4 degrees Celsius, and then redefined as the mass of a particular platinum-iridium cylinder, Prototype Kilogram No. 1, kept at Sèvres, a suburb of Paris, France. It is equivalent to 2.20 pounds.

The originators of the metric system saw no reason to question the unit of time, the **second**, which had been in use for centuries. The second had long been defined as 1/86 400 of the mean solar day, based on the rotating earth. The day is divided into 24 hours, each hour into 60 minutes, and each minute into 60 seconds.

When problems arose with the metric system as originally conceived, the French government arranged a conference in 1870 to work out standards for a unified measurement system, and in 1875 the **Treaty of the Meter** was signed in Paris by representatives from 17 nations.

The treaty set up an **International Bureau of Weights and Measures** to be the custodian of the standards for an international system of measurement with headquarters at Sèvres. It also established a **General Conference on Weights and Measures**, which meets at least once every six years. Commissions were appointed to design prototype standards of length and mass—the meter and the kilogram—and copies were given to the nations adhering to the treaty.

Soon after the prototype copies were received in the United States, the Treasury Department redefined the yard and the pound (which had been based on copies of English standards) as appropriate fractions of the meter and the kilogram. Along with other

Figure 2–4 Prototype Kilogram No. 20, the national standard of mass for the United States. It is a platinum-iridium cylinder 39 millimeters in diameter and 39 millimeters high. (Courtesy of National Institute of Standards and Technology, U.S. Department of Commerce)

customary units such as the gallon, they are based by law on metric standards.

The American copy of the international standard of mass, known as Prototype Kilogram No. 20, is kept in a vault at the Gaithersburg Laboratory of the National Bureau of Standards (Fig. 2–4). It is removed not more than once a year for checking the values of other standards with a precision balance. Twice since 1889 it has been taken to France for comparison with the master kilogram. In practice, we use secondary working standards of length and mass, which have a high degree of accuracy.

2–3 METRIFY OR PETRIFY?

The United States has been committed to the metric system ever since the **Metric Conversion Act** was signed into law in 1975. The law states that it is U.S. policy to coordinate and plan the increasing use of the metric system. A **National Metric Board** coordinates the voluntary conversion both in and out of government. The key word is "voluntary." No one is going to be put in jail for using feet and pounds.

The *Omnibus Trade and Competitiveness Act*, signed into law in 1988, put teeth into the 1975 law. The new law states that by 1992 Federal agencies

Figure 2–5   This road sign gives distances in English and metric units, miles and kilometers, respectively. (Source: Raymond A. Serway and Jerry S. Faughn, *College Physics*, 3rd edition. Philadelphia: Saunders College Publishing, 1992, p. 11.)

must come up with a plan to use the metric system in their purchases, grants, and other business and report on their progress in each budget submission.

The switch to metric has a strong economic dimension. For example, multinational corporations and exporters were influenced by a directive by the European Economic Community requiring exporters to the Common Market countries to label their products in metric units. Dozens of major corporations organized metric conversion programs. Their products have a better chance of success in foreign coun-

tries if designed to the metric standards observed in most countries. Burma, Liberia, and the United States are the last three countries in the world not using the metric system. Much of the nation's heavy industry has already converted, particularly the auto industry. Thousands of suppliers must either go metric, too, or miss out on the contracts if their products are incompatible or illegal in the rest of the world.

Department store chains are manufacturing their products in metric measures. Football programs list

Figure 2–6   Labels of some grocery store items in English and metric units. (Richard Megna/Fundamental Photographs)

Table 2–1   ENGLISH–METRIC
                CONVERSION FACTORS

Length

1 inch (in.)	= 2.54 centimeters (cm)
1 foot (ft)	= 0.3048 meter (m)
1 yard (yd)	= 0.914 meter (m)
1 mile (mi)	= 1.609 kilometers (km)

Mass

1 grain (gr)	= 64.799 milligrams (mg)
1 ounce (oz)	= 28.35 grams (g)
1 pound (lb)	= 453.6 grams (g)
1 ton (tn)	= 907.2 kilograms (kg)

Volume

1 fluid ounce (fl oz)	= 29.57 milliliters (mL)
1 quart (qt)	= 0.946 liter (L)
1 gallon (gal)	= 3.785 liters (L)

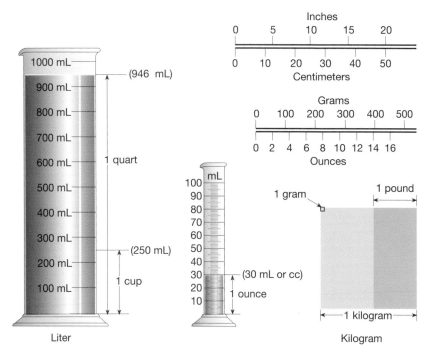

Figure 2–7 Comparison of some English and metric units.

the players' heights and weights in metric as well as customary units. Soft-drink companies have introduced half-liter, liter, and two-liter bottles. In the National Parks, new signs give road distances in metric equivalents (Fig. 2–5). Foot races are commonly expressed in kilometers, such as a "10-k" run.

Grocery store items often include metric-equivalent units on the labels (Fig. 2–6). The vitamin and mineral content of cereals are given in milligrams. Doctors prescribe metrically, and some record patients' weights in kilograms, heights in centimeters or meters, and temperatures in degrees Celsius. Camera hobbyists work with film cut to metric dimension, such as 35 millimeters.

The Olympics and other international athletic events have been using the metric system for years. In some baseball stadiums the dimensions of the playing field are expressed in meters as well as feet. The chemical, optical, pharmaceutical, film, and jewelry industries all have converted to the metric system.

Some common English–metric and metric–English conversion factors are given in Tables 2–1 and 2–2 (Figs. 2–7 and 2–8).

Table 2–2   METRIC–ENGLISH
            CONVERSION FACTORS

Length

1 millimeter (mm)	= 0.03937 inch (in.)
1 meter (m)	= 1.09 yards (yd)
1 kilometer (km)	= 0.621 mile (mi)

Mass

1 gram (g)	= 0.03527 ounce (oz)
1 kilogram (kg)	= 2.20 pounds (lb)

Volume

1 liter (L)	= 1.06 quart (qt)
1 cubic meter (m³)	= 1.308 cubic yards (yd³)

Figure 2–8   Speedometer readings in miles per hour and kilometers per hour. (Paul Silverman/Fundamental Photographs)

2–4 SI UNITS

To be useful, a system of units must change with the times. In 1960, the Eleventh General Conference on Weights and Measures extended and refined the metric system, so that it now goes far beyond the original standards of length and mass. The Conference also officially adopted the **International System of Units** (Système Internationale d'Unités), which is abbreviated **SI** in all languages. This system is built on seven fundamental physical quantities for length, mass, time, temperature, electric current, light intensity, and amount of substance. Table 2–3 gives the basic SI units and their symbols.

The meter was redefined as the length of 1,650,763.73 wavelengths of a certain orange-red line in the emission spectrum of the krypton-86 atom. Today, the meter is defined in terms of another base measure, time, instead of being viewed simply as an arbitrary length. Specifically, the meter is defined as the distance light travels through space in a certain fraction of a second (1/299,792,458 of a second, to be exact).

Refinements based on clocks controlled by certain properties of quartz crystals made it clear that the second as previously defined was not constant but varied, because the earth's rotation was more irregular than had been suspected. Astronomers were using a more reliable time scale, ephemeris time, based on the motion of the earth around the sun, and in 1960, the Eleventh General Conference adopted a new definition of the second.

Exquisite timing accuracy is necessary in such fields as satellite tracking and astronomical observations. In the belief that a second defined in terms of the frequency of vibration of an atom or a molecule would be even more accurate, the Conference urged that work proceed on an atomic definition. The most suitable candidate for timekeeping was found to be a vibration of the cesium-133 atom, which can be determined with great accuracy. In 1967 the Thirteenth General Conference on Weights and Measures redefined the second as 9,192,631,770 cycles of that particular vibration. The accuracy of a cesium clock is such that in 6000 years it would not gain or lose more than a second. The standards of length and time have shifted, then, from the earth to the atom.

The basic SI units and the units derived from them now form a legal basis of measure. The customary system of units is defined in terms of SI units. The yard, for example, is legally established as exactly .91440183 meters. Like most of the world, we are going directly to the metric system, rather than a system derived from it, for commerce, industry, the arts, medicine, agriculture, and engineering, as well as for science.

In practice, scientists use two *subsystems* of SI units: the **centimeter-gram-second** (**cgs**) and the **meter-kilogram-second system** (**mks**). Chemists and others often prefer the former as a matter of convenience, and physicists, the latter. Both subsystems are used in this text.

2–5 MULTIPLES AND SUBMULTIPLES

Since many of the quantities measured in science—the speed of light, the distance of a star, the size of an atom, the quantity of electric charge—are either much larger or much smaller than 1, they are expressed in units appropriate to their dimensions. A big advantage of SI units is their decimal relationship. The prefixes that are used with the units show this. A centimeter is 1/100 of a meter, or 0.01 m; a kilometer is 1000 meters. A megameter is 1 million meters; a kilogram, 1000 grams; a nanosecond, one billionth of a second. The Eleventh General Conference has approved the use of the 16 prefixes listed in Table 2–4, which range from the very large to the very small and are used with any of the fundamental SI units.

2–6 EXPONENTIAL NOTATION

Science and technology have so developed that it is possible to think of a clock having an accuracy of 1 part in 1,000,000,000,000. That is equivalent to 1 sec-

Table 2–3 SI BASE UNITS

Physical Quantity	Name of Unit	Symbol
Length	Meter	m
Mass	Kilogram	kg
Time	Second	s
Thermodynamic temperature	Kelvin	K
Electric current	Ampere	A
Luminous intensity	Candela	cd
Amount of substance	Mole	mol

Table 2–4 SI PREFIXES

Prefix	Symbol	Factor	Meaning	
exa-	E	10^{18}	1,000,000,000,000,000,000	one quintillion
peta-	P	10^{15}	1,000,000,000,000,000	one quadrillion
tera-	T	10^{12}	1,000,000,000,000	one trillion
giga-	G	10^{9}	1,000,000,000	one billion
mega-	M°	10^{6}	1,000,000	one million
kilo-	k°	10^{3}	1,000	one thousand
hecto-	h	10^{2}	100	one hundred
deka-	da	10^{1}	10	ten
Unit-one				
deci-	d	10^{-1}	0.1	one tenth
centi-	c°	10^{-2}	0.01	one hundredth
milli-	m°	10^{-3}	0.001	one thousandth
micro-	μ°	10^{-6}	0.000001	one millionth
nano-	n	10^{-9}	0.000000001	one billionth
pico-	p	10^{-12}	0.000000000001	one trillionth
femto-	f	10^{-15}	0.000000000000001	one quadrillionth
atto-	a	10^{-18}	0.000000000000000001	one quintillionth

° Most commonly used.

ond in 30,000 years. We can deal with a few parts of mass in 1,000,000,000, which is equivalent to the mass of ink in one punctuation period compared with the mass of a whole book. We know that a concentration of phosphorus as small as 0.00000001 g per ml of lake water can cause the lake to begin to die from the overgrowth of algae. It is estimated that there are 602,300,000,000,000,000,000,000 carbon atoms in 12 grams of carbon.

Such numbers are awkward to write, since there are so many zeros to keep track of, making them difficult to work with. (Try calculating the number of carbon atoms in 1 gram of carbon without worrying about the zeros.) It is far better to express such numbers in **exponential notation**—also known as standard notation, **scientific notation**, or **powers of ten**. This compact way of writing numbers makes calculations easier and saves us a good deal of paper and ink in the process (Tables 2–5, 2–6, and 2–7).

Note that the exponential form of a number involves the product of two numbers: a number between 1 and 10, called the **coefficient** (for example, 6.02), times a power of 10, the **exponential** (for example, 10^3). Consider the exponential

$$\underset{\text{base}}{\underline{10}}{}^{3}\} \text{exponent}$$

We call 10 the **base**, and 3, written as a superscript following the 10, the **exponent** or power of the base 10. We read the number 10^3 as "10 cubed" or "10 to

Table 2–5 DISTANCES IN THE UNIVERSE

Meters

10^{28} ——	Visual limit (greatest distance from which light can reach us in the expected age of the universe)
10^{25} ——	
10^{22} ——	Earth to nearest galaxy
10^{19} ——	
	Earth to nearest star
10^{16} ——	
	Distances between planets
10^{13} ——	
	Earth to sun
10^{10} ——	
10^{7} ——	
10^{4} ——	Height of Mt. Everest
10^{1} ——	
	Human height
10^{-2} ——	
10^{-5} ——	Width of paper thickness
10^{-8} ——	Diameter of a virus
	Diameter of a molecule of DNA
10^{-10} ——	Diameter of an atom
10^{-14} ——	Diameter of a nucleus

Table 2–6 RANGE OF TIME INTERVALS

Seconds

10^{18}	—— Age of universe
10^{15}	—— Age of earth
10^{12}	—— Time of appearance of earliest humans to present
10^{9}	—— Human life span
10^{6}	—— One year
	—— One day
10^{3}	—— Time required for light to travel from sun to earth
	——
1	—— Time between heartbeats
10^{-3}	—— Period of a sound wave
10^{-6}	—— Period of a radio wave
	—— Time for electrons to go from source to screen in TV tube
10^{-9}	——
10^{-12}	—— Time for light to penetrate a window
10^{-15}	——
10^{-18}	—— Period of visible light wave
10^{-21}	—— Period of an X-ray
10^{-24}	—— Time for light to cross a nucleus

the third power." The exponent, 3, tells how many times the base, 10, is used as a factor, that is, the number of 10s multiplied together: $10 \times 10 \times 10$. It is also equal to the number of zeros between the 1 and the decimal point in the expanded form of the number. Thus, we could write 10^3 as 1 followed by three zeros: 1000. Table 2–8 gives some positive and negative exponentials. A negative exponent indicates a reciprocal:

$$10^{-3} = \frac{1}{10^3}$$

Table 2–7 RANGE OF MASSES

Grams

10^{50}	——
10^{40}	—— The Milky Way
10^{30}	—— Sun
	—— Earth
10^{20}	—— Moon
10^{10}	—— A ship
1	—— A person
	—— Postage stamp
10^{-10}	—— Red blood cell
10^{-20}	—— Oxygen molecule
10^{-30}	—— Electron

Suppose we wish to express the number 634 in exponential form. We know at the outset that there is a decimal point following the 4. We change 634 to a form consisting of a number between 1 and 10, multiplied by a power of 10, that is, 6.34×10^n. We get the number 6.34 by moving the decimal point two places to the left. The exponent, n, is equal to the number of places the decimal point has been moved, in this case, 2. If the decimal point must be moved to the left, the exponent is positive; if it must be moved to the right, the exponent is negative. Therefore, we express the number 634 as 6.34×10^2. Since $10^2 = 100$, 6.34×10^2 really does equal 634.

For a number less than one, such as 0.000284, we move the decimal point to the right to obtain the coefficient. Moving the decimal point four places to the right, we get the coefficient 2.84. The exponential, therefore, is 10^{-4}. So

$$0.000284 = 2.84 \times 10^{-4}.$$

Since

$$10^{-4} = 0.0001,$$
$$2.84 \times 10^{-4} = 2.84 \times 0.0001 = 0.000284.$$

The numbers below are expressed in exponential form. Check them.

$$62.4 = 6.24 \times 10^1$$
$$19,357,000 = 1.9357 \times 10^7$$
$$2740 = 2.740 \times 10^3$$
$$58,900 = 5.8900 \times 10^4$$
$$0.00361 = 3.61 \times 10^{-3}$$
$$0.000096 = 9.6 \times 10^{-5}$$
$$0.459 = 4.59 \times 10^{-1}$$
$$0.0000058 = 5.8 \times 10^{-6}$$

2–7 MULTIPLICATION AND DIVISION OF EXPONENTIAL NUMBERS

One of the bonuses we get from exponential notation is that we can readily multiply and divide numbers expressed in this way. All we have to do is to apply a few rules.

(A) Multiplication. To multiply two exponential numbers, *add the exponents*. If one or both exponents

Table 2–8   EXPONENTIALS

$$10^6 = 10 \times 10 \times 10 \times 10 \times 10 \times 10 \qquad = 1,000,000$$
$$10^5 = 10 \times 10 \times 10 \times 10 \times 10 \qquad = 100,000$$
$$10^4 = 10 \times 10 \times 10 \times 10 \qquad = 10,000$$
$$10^3 = 10 \times 10 \times 10 \qquad = 1,000$$
$$10^2 = 10 \times 10 \qquad = 100$$
$$10^1 = 10 \qquad = 10$$
$$10^0 \qquad = 1$$

$$10^{-1} = \frac{1}{10^1} = \frac{1}{10} \qquad = 0.1$$

$$10^{-2} = \frac{1}{10^2} = \frac{1}{10 \times 10} = \frac{1}{100} \qquad = 0.01$$

$$10^{-3} = \frac{1}{10^3} = \frac{1}{10 \times 10 \times 10} = \frac{1}{1,000} \qquad = 0.001$$

$$10^{-4} = \frac{1}{10^4} = \frac{1}{10 \times 10 \times 10 \times 10} = \frac{1}{10,000} \qquad = 0.0001$$

$$10^{-5} = \frac{1}{10^5} = \frac{1}{10 \times 10 \times 10 \times 10 \times 10} = \frac{1}{100,000} \qquad = 0.00001$$

$$10^{-6} = \frac{1}{10^6} = \frac{1}{10 \times 10 \times 10 \times 10 \times 10 \times 10} = \frac{1}{1,000,000} = 0.000001$$

happen to be negative, follow the same procedure and add them algebraically. The general rule is

$$10^a \times 10^b = 10^{(a+b)}$$

Therefore,

$$10^2 \times 10^3 = 10^{(2+3)} = 10^5$$

This is because

$$10^2 = 10 \times 10 \qquad \text{and} \qquad 10^3 = 10 \times 10 \times 10$$

Thus,

$$10^2 \times 10^3 = (10 \times 10)(10 \times 10 \times 10) = 10^5$$
$$= 10^{(2+3)}$$

Similarly,

$$10^5 \times 10^{-2} = 10^{[5+(-2)]} = 10^{(5-2)} = 10^3$$

and

$$10^{-3} \times 10^{-5} = 10^{[(-3)+(-5)]} = 10^{(-3-5)} = 10^{-8}$$

To multiply one exponential number by another, first multiply the coefficients together, then add the exponents algebraically. If the product of the coefficients is a number greater than 10 (for example, $8 \times 7 = 56$), it may be expressed in standard exponential notation ($56 = 5.6 \times 10^1$), and the exponentials combined.

EXAMPLE 2–1

Multiply $30,000 \times 200,000$.

SOLUTION

1. Change to exponential form: $(3 \times 10^4) \times (2 \times 10^5)$
2. Rearrange, separating the coefficients from the exponentials: $(3 \times 2) \times (10^4 \times 10^5)$
3. Multiply the coefficients together and add the exponents: $= 6 \times 10^{(4+5)}$
$$= 6 \times 10^9$$

■

EXAMPLE 2–2

Multiply 400×0.002.

SOLUTION

Change to exponential form: $(4 \times 10^2) \times (2 \times 10^{-3})$
Rearrange: $(4 \times 2) \times (10^2 \times 10^{-3})$

Solve: $= 8 \times 10^{2+(-3)}$
$= 8 \times 10^{(2-3)}$
$= 8 \times 10^{-1}$

∎

EXAMPLE 2-3

Multiply $600,000 \times 7,000,000$.

SOLUTION
Change to exponential form: $(6 \times 10^5) \times (7 \times 10^6)$
Rearrange: $(6 \times 7) \times (10^5 \times 10^6)$
Solve: $= 42 \times 10^{(5+6)}$
$= 42 \times 10^{11}$
$= 4.2 \times 10^1 \times 10^{11}$
$= 4.2 \times 10^{12}$

∎

(B) Division. To divide, *subtract the exponents*. The general rule is

$$\frac{10^a}{10^b} = 10^{(a-b)}$$

Therefore,

$$\frac{10^5}{10^2} = 10^{(5-2)} = 10^3$$

This is because

$10^5 = 10 \times 10 \times 10 \times 10 \times 10$ and
$10^2 = 10 \times 10$

Thus,

$$\frac{10^5}{10^2} = \frac{10 \times 10 \times 10 \times 10 \times 10}{10 \times 10} = 10^3$$
$$= 10^{(5-2)}$$

Similarly,

$$\frac{10^2}{10^6} = 10^{(2-6)} = 10^{-4}$$

and

$$\frac{10^3}{10^{-2}} = 10^{3-(-2)} = 10^{3+2} = 10^5$$

To divide one exponential number by another, separate the coefficients from the exponential terms, divide the coefficients, and subtract the exponents.

EXAMPLE 2-4

Divide $60,000$ by 0.003

SOLUTION
1. Change to exponential form: $\dfrac{6 \times 10^4}{3 \times 10^{-3}}$

2. Rearrange: $\dfrac{6}{3} \times 10^{[4-(-3)]}$

3. Divide the coefficients and subtract the exponents: $= 2 \times 10^{(4+3)}$
$= 2 \times 10^7$

∎

EXAMPLE 2-5

Evaluate $\dfrac{9 \times 10^6}{4.5 \times 10^4}$

SOLUTION
Rearrange: $\dfrac{9}{4.5} \times 10^{(6-4)}$

Solve: $= 2 \times 10^2$

∎

EXERCISES
Carry out the following operations.
1. $(4.00 \times 10^3) \times (2.00 \times 10^4)$
2. $(6.00 \times 10^5) \times (1.50 \times 10^{-3})$
3. $(5.00 \times 10^{-4}) \times (1.60 \times 10^{-5})$
4. $(9.50 \times 10^6) \times (3.00 \times 10^{-2})$
5. $35,000 \times 200,000$
6. $6400/80,000$
7. $20,000/0.004$
8. $(7.20 \times 10^5)/(3.60 \times 10^{-3})$
9. $(4.80 \times 10^{-3})/(6.00 \times 10^6)$
10. $\dfrac{(4.20 \times 10^4) \times (2.00 \times 10^{-5})}{(7.00 \times 10^6)}$

2-8 DIMENSIONAL ANALYSIS

It is meaningless to say that the length of a swimming pool is 50. Fifty feet? Yards? Meters? Until we include the dimensional unit, the number has no meaning. Most physical quantities can be expressed in terms of

Figure 2–9 Analytical balance with digital readout. The sample and container together weigh 46.289 g. (Marna G. Clarke)

length, mass, and time, or combinations of these (Fig. 2–9). But lengths, as we have seen, can be expressed in a variety of units. We often have to know how to convert from one unit to another. This is very easy to do in SI. Since all subunits are related by powers of 10, we merely move a decimal point or change an exponent. Thus, 1 meter = 10^2 centimeters = 10^3 millimeters. As long as customary units are used as well, we have to be "bilingual." We must be at home with both customary and SI units and know how to convert from one unit to another.

Such conversions are easily made with **dimensional analysis**. In this process, the dimensional unit in which a measurement is expressed is treated as an algebraic term that may be multiplied or divided.

A **conversion factor** is the numerical relationship between any two units. In the customary system, 1 mile = 5280 feet; 1 foot = 12 inches; and so on. For SI units, 1 kilometer = 1000 meters; 1 meter = 100 centimeters; and so on. The fraction 5280 feet/mile actually equals 1, so that when we multiply a number by this conversion factor, we are multiplying by 1. Doing this does not change the value of a measurement, only the units. To get the number of feet in 5.00 miles, we use the conversion factor 5280 feet/mile. We begin with the given 5.00 miles and multiply by the conversion factor. Note how the miles cancel out, leaving the answer in the desired unit, feet.

$$5.00 \text{ miles} \times \frac{5280 \text{ feet}}{1.00 \text{ mile}} = 26{,}400 \text{ feet}$$

EXAMPLE 2–6

Determine the equivalent in feet of a 50.0-meter Olympic swimming pool.

SOLUTION
The question may be rephrased as, 50.0 meters are equal to how many feet? We start with the measurement of 50.0 meters. Then we use the appropriate conversion factors, one after another, canceling units as we progress. In this calculation, we might go from meters to yards, and then to feet. Finally, we do the mathematical operations.

$$50.0 \text{ m} \times \frac{1.09 \text{ yd}}{1.00 \text{ m}} \times \frac{3.00 \text{ ft}}{1.00 \text{ yd}} = 164 \text{ ft}$$

■

EXAMPLE 2–7

Find the number of centimeters in 1.00 kilometer.

SOLUTION
We know that 1 kilometer is equal to 1000 meters and that 1 meter contains 100 centimeters. Using these equivalents as conversion factors, we multiply the given unit by them. In the process, the kilometers and meters cancel out, and the desired unit, centimeters, is the answer.

$$1.00 \text{ km} \times \frac{10^3 \text{ m}}{1.00 \text{ km}} \times \frac{10^2 \text{ cm}}{1.00 \text{ m}} = 10^5 \text{ cm}$$

■

EXERCISES
Carry out the following conversions:

1. 5.00 cm to in.
2. 60 km to mi
3. 150 lb to kg
4. 760 mm to in.
5. 1 min to s
6. 3000 mi to km
7. 1 year to s
8. 100 yd to m
9. 55 mi to km
10. 1 oz to mg

2–9 ACCURACY AND SIGNIFICANT FIGURES

Although accurate measurements are an important aspect of physical science, no measurement is absolutely precise (Fig. 2–10). There is **uncertainty** associated with every measurement. Uncertainty arises from various sources: the limited accuracy of the measuring instrument, human error, the inability to read an instrument more accurately than some fraction of the smallest division shown (Fig. 2–11).

Suppose that we were to use a centimeter ruler to measure the area of a sheet of paper. Assume that the uncertainty of any measurement with this ruler is ± 0.1 cm ("plus or minus 0.1 cm"), the smallest division on the ruler, since it is difficult to interpolate between the smallest divisions and the ruler has probably not been manufactured to an accuracy much better than this.

We can write the length as 26.5 ± 0.1 cm, meaning that the actual value lies between 26.4 cm and 26.6 cm. Likewise, its width can be expressed as 8.2 ± 0.1 cm, meaning that the actual value lies between 8.1 cm and 8.3 cm. Or we can express the measurements merely as 26.5 cm and 8.2 cm, re-

spectively, without specifying the uncertainty explicitly. In so doing, it is generally accepted that the uncertainty is approximately one or two units in the last digit. The number of reliably known digits in a measurement, including the one uncertain digit, is called the number of **significant figures**. Thus, there are three significant figures in 26.5 cm and two significant figures in 8.2 cm.

To calculate the area of the sheet of paper, we multiply the length by the width, 26.5 cm by 8.2 cm. You will note that the result of the multiplication, 217.3 cm², contains four significant figures. As a general rule, the final answer of a multiplication or division should have only as many digits as the least accurate of the quantities used in the calculation. In this case, the answer for the area can have only two significant figures since the 8.2-cm dimension has only two significant figures. Thus the answer for the area rounded off to two significant figures (the zero locates the decimal point) is 220 cm². We realize that the value of the area, using the outer limits of the assumed uncertainty, could be between (26.4 cm)(8.1 cm) = 213.84 cm² and (26.6 cm)(8.3 cm) = 220.78 cm².

If the presence of zeros in a measurement or calculation is ambiguous because it is not known

Figure 2–10   Some common laboratory equipment. The orange gives you a sense of size. (Source: John C. Kotz and Keith F. Purcell, *Chemistry and Chemical Reactivity*, 2nd edition. Philadelphia: Saunders College Publishing, 1991, p. 16, Fig. 1.9 [Charles D. Winters])

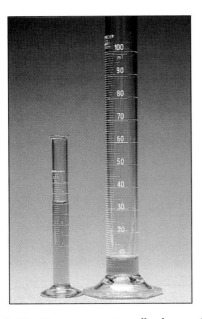

Figure 2–11 To measure out small volumes of liquids precisely, a 10-mL graduated cylinder is more effective than one with a volume of 100 mL. The uncertainty associated with a measurement depends upon the nature of the measuring device. (Marna G. Clarke)

whether the zeros represent significant figures in the measurement or whether they are being used to locate the decimal point, scientific notation is commonly used to indicate the number of significant figures. A mass of 1200 g, for example, would be expressed as 1.2×10^3 g if there are two significant figures in the measured value and 1.20×10^3 g if there are three significant figures.

When numbers are added or subtracted, the number of decimal places in the result should equal the smallest number of decimal places of any factor in the calculation. For example, the answer to $65.32 + 3.6$ would be 68.9 and not 68.92.

In this book, we shall assume in general that the data given are precise enough to yield an answer having *three significant figures*. Thus, if we state that a car travels a distance of 20 km, it is understood that the distance covered is 20.0 km. In some instances, more significant figures may be assumed and answers may be reported to more significant figures.

KEY TERMS

Base In exponential notation, exponents are used to show repeated multiplication of a number referred to as the base. Thus, the 10 in 10^3 is the base, the 10 is used as a factor 3 times: $10 \times 10 \times 10$.

cgs System A subsystem of SI units, the centimeter-gram-second system.

Coefficient In exponential notation, a number between 1 and 10 that is multiplied by 10 raised to a power; for example, 6.022 in 6.022×10^{23}.

Conversion Factor The numerical relationship between any two units.

Dimensional Analysis A process of converting from one unit to another by treating the dimensional unit in which a measurement is ex-

pressed as an algebraic term that may be multiplied or divided.

English Customary System The foot-pound-quart system of weights and measures.

Exponent In the number 10^3, the 3 is the exponent and shows how many times the base 10 is used as a factor: $10 \times 10 \times 10$.

Exponential A number expressed in scientific notation, or powers of ten.

Exponential Notation Also called standard notation, scientific notation, and powers of ten. Expressing a number as the product of two numbers; one, the coefficient, is between 1 and 10 and is multiplied by 10 raised to a power, the exponential. For example, 5.63×10^6.

International Bureau of Weights and Measures The custodian of the standards for the international system of measurement, with headquarters at Sèvres, France.

Kelvin The SI standard of temperature.

Kilogram The SI unit of mass equal to the mass of a platinum-iridium cylinder kept at the International Bureau of Weights and Measures.

Meter The standard for length in the SI system. It is defined as the distance that light travels through space in 1/299,792,458 of a second.

Metric Conversion Act Signed into law in 1975, the Act commits the United States to coordinating and increasing the use of the metric system.

Metric System   A decimal system of weights and measures, based on the meter as the unit of length and the kilogram as the unit of mass.

mks System   A subsystem of SI units, the meter-kilogram-second system.

National Metric Board   A board established by the Metric Conversion Act to coordinate the voluntary conversion both in and out of government to the metric system.

Omnibus Trade and Competitiveness Act   Signed into law in 1988, this Act requires Federal agencies to report on their progress in using the metric system in their purchases, grants, and other business in each budget submission.

Powers of Ten   See Exponential Notation.

Scientific Notation   A number expressed as the product of two numbers: a number between 1 and 10 times a power of 10.

Second   An international unit of time.

SI units   The International System of Units (Système Internationale d'Unités), an extension and refinement of the metric system.

Significant Figures   The number of reliably known digits plus one that is uncertain in a measured quantity.

Standard Kilogram   The mass of a platinum-iridium cylinder kept at the International Bureau of Weights and Measures near Paris.

Standard Meter   The distance between two scratches on a bar of platinum-iridium alloy, kept at the International Bureau of Weights and Measures, calibrated in terms of the wavelength of light.

Standard Notation   See Exponential Notation.

Treaty of the Meter   An international treaty signed in Paris in 1875 that set up the International Bureau of Weights and Measures.

Uncertainty   Every measurement has an element of uncertainty associated with it, making it somewhat less than absolutely precise.

U.S. Customary System   The traditional system of units of measurement, based on feet, pounds, and quarts.

EXERCISES

Scientific Notation

2–1   Express the following numbers in scientific notation.
(a) 440.74
(b) 1,200,000
(c) 0.000263
(d) 0.00145
(e) 27,600
(f) 0.182
(g) 0.0000319
(h) 83.7
(i) 60,322,000
(j) 7840

2–2   Write out the following numbers in ordinary notation.
(a) 8.92×10^2
(b) 5.63×10^{-4}
(c) 3.59×10^{-6}
(d) 7.482×10^3
(e) 6.21×10^5
(f) 8.02×10^{-7}
(g) 4.274×10^6
(h) 2.003×10^{-4}
(i) 9.70×10^8
(j) 3.065×10^6

2–3   Carry out the indicated operations and express the answers in scientific notation.
(a) $(10^3)(10^2)(10^6)$
(b) $\dfrac{(1.8 \times 10^3)(2 \times 10^5)}{4 \times 10^9}$
(c) $\dfrac{(500)(6200)}{20\ 000}$
(d) $(1.67 \times 10^{-27})(6.02 \times 10^{23})$
(e) $\dfrac{1.44 \times 10^{10}}{(3 \times 10^4)(4 \times 10^7)(6 \times 10^{-3})}$
(f) $(24) \times (3,000,000) \times (1/600,000)$
(g) $\dfrac{(3 \times 10^8)(2 \times 10^{-3})}{6 \times 10^{11}}$
(h) $(8 \times 10^5)/(4 \times 10^3)$
(i) $\dfrac{(2 \times 10^{-43})(6 \times 10^8)}{(8 \times 10^{-29})(4 \times 10^{-17})}$
(j) $\dfrac{(10^2)(10^{-3})(10^5)}{(10^6)(10^{-9})}$

Significant Figures

2–4   Calculate the following and report the answers to the correct number of significant figures.
(a) $x = (7.874 \text{ cm})(0.0263 \text{ cm})(0.00054 \text{ cm})$

(b) $x = 12.2$ g $+ 0.38$ g

(c) $x = 36.247$ g $- 33.227$ g

(d) $x = \dfrac{12.2 \text{ g} + 7.258 \text{ g}}{20.30 \text{ cm}}$

(e) $x = \dfrac{34.0300 \text{ g}}{21.09 \text{ cm}^3}$

SI Units

2–5 Rewrite the following in scientific notation, using the SI base units of Table 2–3.
 (a) 2.2 centimeters
 (b) 3.0 milligrams
 (c) 64 kilometers
 (d) 34.2 nanoseconds
 (e) 29 megakelvins

2–6 Express the customary measurements in this saying in terms of their SI equivalents: "An ounce of prevention is worth a pound of cure."

2–7 (a) Give your height and weight in customary units.
 (b) Express each in SI units.

Conversion of Units Using Dimensional Analysis

2–8 Carry out the following conversions
 (a) 7.50 g to mg
 (b) 8.00 in to cm
 (c) 1050 g to oz
 (d) 106 lb to kg
 (e) 75.0 watts to kilowatts
 (f) 100 mm to in.
 (g) 186,000 mi to m
 (h) 2000 lb to g
 (i) 1.00 ft to cm
 (j) 1.00 gal to mL

2–9 The 10,000-meter run is one of the events in the Olympic games. What is this distance in miles?

2–10 The ordinary chair seat is about 450 mm high. How many inches is this?

2–11 A person with 6.00 liters of blood has 1.20×10^3 mg of cholesterol per liter of blood. What is the mass of cholesterol in the blood in (a) grams; (b) ounces?

2–12 How many liters of gasoline does a 21.0-gallon tank hold?

2–13 A speed limit of 55.0 miles per hour is the same as how many kilometers per hour?

2–14 According to a highway sign, you are 150 kilometers from your destination. How many miles is this?

2–15 Select the smaller member of each pair.
 (a) 64.25 g or 64.25 kg

(b) 352 nm or 3.52×10^{-3} km

(c) 405 m or 0.400 km

(d) 300 kg or 0.0300 g

2–16 The distance between carbon atoms in a diamond is 0.154 nanometers (1 mm $= 1 \times 10^{-9}$ m). What is this distance in m? in cm?

2–17 The speed limit in a city is 70 km/h. How many miles per hour is this?

2–18 On certain flights, the maximum weight of luggage allowed a passenger is 20.0 kg. If you travel on one of these flights, what is the weight in pounds you could take with you?

2–19 To produce a ton of steel, 5.00×10^4 gal of water are required. Write this number in its expanded form and give it its appropriate name (for example, 10^2 is one hundred; 10^3 is one thousand; and so on).

2–20 *Multiple Choice*
 A. The prefix "kilo-" means
  (a) one thousandth.
  (b) one tenth.
  (c) ten.
  (d) one thousand.
 B. Of the following, the smallest measurement is
  (a) 500 mm.
  (b) 5.00 m.
  (c) 0.005 km.
  (d) 0.002 mi.
 C. The meter was originally based on the dimensions of the
  (a) arm.
  (b) foot.
  (c) ocean.
  (d) earth.
 D. The prefix that does not match the number is
  (a) centi-, 10^{-1}.
  (b) kilo-, 10^3.
  (c) milli-, 10^{-3}.
  (d) micro-, 10^{-6}.
 E. A yard equals
  (a) 1.1 m.
  (b) 30.48 m.
  (c) 39.37 m.
  (d) 0.92 m.
 F. The number of grams in 2000 pounds is
  (a) 9.068×10^5.
  (b) 1.60×10^4.
  (c) 436.
  (d) 6.24×10^{-2}.
 G. The product of $(1.30 \times 10^{-4}) \times (6.00 \times 10^5)$ is
  (a) 7.80×10^{-9}.
  (b) 7.80×10^{-1}.

(c) 7.80×10^9.

(d) 7.80×10^1.

H. A proper unit for designating volume is

(a) mg.

(b) mm.

(c) mL.

(d) m.

I. The result of dividing (8×10^3) by (2×10^{-5}) is

(a) 4×10^{15}.

(b) 4×10^{-2}.

(c) 4×10^2.

(d) 4×10^8.

J. The number 4.36×10^{-4} is

(a) 0.0436.

(b) 43,600.

(c) 0.000436.

(d) 0.00000436.

SUGGESTIONS FOR FURTHER READING

Adamson, Arthur W., "SI Units? A Camel Is a Camel." *Journal of Chemical Education*, October, 1978.

Argues that the SI system is not convenient to physical chemistry or superior in consistency.

Astin, Allen V., "Standards of Measurement." *Scientific American*, June, 1968.

Discusses the standards of length, mass, time, and temperature.

Heilbron, J. L., "The Politics of the Meter Stick." *American Journal of Physics*, November, 1989.

Illustrates that it is relatively easy to change a country's government but all but impossible to change its weights and measures.

National Bureau of Standards, "Policy for NBS Usage of SI Units." *Journal of Chemical Education*, September, 1971.

Encourages the use of SI units to facilitate communication.

Ritchie-Calder, Lord, "Conversion to the Metric System." *Scientific American*, July, 1970.

Shows what a country may expect when it makes the transition to the metric system.

Socrates, G., "SI Units." *Journal of Chemical Education*, November, 1969.

Discusses the development and use of SI units in virtually every country.

ANSWERS TO NUMERICAL EXERCISES

2–1 (a) 4.4074×10^2

(b) 1.2×10^6

(c) 2.63×10^{-4}

(d) 1.45×10^{-3}

(e) 2.76×10^4

(f) 1.82×10^{-1}

(g) 3.19×10^{-5}

(h) 8.37×10^1

(i) 6.0322×10^7

(j) 7.840×10^3

2–2 (a) 892

(b) 0.000563

(c) 0.00000359

(d) 7482

(e) 621,000

(f) 0.000000802

(g) 4,274,000

(h) 0.0002003

(i) 970,000,000

(j) 3,065,000

2–3 (a) 10^{11}

(b) 9×10^{-2}

(c) 1.55×10^2

(d) 1.01×10^{-3}

(e) 2

(f) 1.2×10^2

(g) 1×10^{-6}

(h) 2×10^2

(i) 3.75×10^{10}

(j) 10^7

2–4 (a) .00011

(b) 12.6 g

(c) 0.020 g

(d) 0.959 g/cm

(e) 01.614 g/cm³

2–5 (a) 2.2×10^{-2} m

(b) 3.0×10^{-6} kg

(c) 6.4×10^4 m

(d) 3.42×10^{-8} s

(e) 2.9×10^{-5} K

2–6 0.02835 kg; 0.4536 kg

2–8 (a) 7.50×10^3 mg

(b) 20.3 cm

(c) 37.03 oz

(d) 48.2 kg

(e) 0.075 kw

(f) 3.94 in

(g) 2.99×10^8 m

(h) 9.072×10^5 g

(i) 30.5 cm

(j) 3.785×10^3 mL

2–9 6.21 mi

2–10 17.7 in

2–11 (a) 6.84 g   (b) 0.24 oz

2–12 79 L

2–13 88 km

2–14 94 mi

2–15 (a) 64.25 g
       (b) 352 nm
       (c) 0.400 km
       (d) 0.0300 g

2–16 1.54×10^{-10} m;
       1.54×10^{-8} cm

2–17 43.47 mi

2–18 44 lb

2–19 50 thousand

The Space Shuttle Atlantis lifting off under the power of main engines and solid rocket boosters. A five-member crew conducted a mission for the Department of Defense. (Courtesy of NASA)

I

PHYSICS

PART-OPENING ESSAY

PERSPECTIVE ON . . . PHYSICS

Our everyday world resides near the center of the universe, not in terms of location but in terms of *size*. Babies and books, grains of sand and jumbo jets—these are neither small nor large, as far as the cosmos is concerned. They are just average. And most of us live woefully provincial lives, robbed by our limited vision of the chance to appreciate the minuscule and the immense, the realms of nature's most astonishing wonders.

A fortunate few have equipped themselves with instruments to overcome this myopia. They are the astronomers, gathering ancient light to peer at galaxies and beyond, and the particle physicists, unraveling the Chinese puzzle of the atom to ever deeper layers. In recent years, these explorers have realized an incredible truth: Their quests to understand the smallest and the largest of all are one and the same. The dances of particles, the laws obeyed by the tiniest imaginable flecks, are the very rules that determined the future course of the universe in the cauldron of its birth billions of years ago.

This is the most dramatic example of the continuity between physics and astronomy in modern science. Many of today's physicists spend their careers smashing together pieces of atoms in gigantic machines in the search for the ultimate units of matter. Others concoct new theories about how particles and forces behave at high temperatures and speeds. By so doing, these physicists are trying to construct a complete picture of our universe— precisely what astronomers also hope to do when they aim their telescopes into space to look back further and further toward the Big Bang.

As you read this book, you will encounter many other ways in which physics, chemistry, astronomy, and earth sciences overlap. Indeed, the lines between these disciplines are often blurred. Although scientists usually specialize in one area of research, most are familiar with the basic principles of other areas. Earth scientists need to know the chemistry of how minerals react inside the planet; chemists must understand the physics of collisions between molecules; and so on.

Physics is a logical starting point because the concepts of matter, energy, motion, and force underlie all of the physical sciences. Learning these concepts takes time, but the effort pays a handsome reward: a richer "world view," an ability to look around and say, "Ah, *that's* how that works!"

For instance, strike your fist against your desk. Your hand and the desk top seem solid enough. But in fact, they are almost entirely empty space. Each atom in ordinary matter consists of a nucleus, which contains virtually all of the atom's mass, surrounded by a swarm of ghostly electrons. Compared to the total sizes of the atoms, the nuclei are incredibly small and very far apart, like ball bearings at opposite ends of a football stadium. Fortunately, the forces of repulsion between the electrons are strong enough to prevent your hand from passing right through the desk—and to keep you on top of the ground instead of plunging to the center of the earth.

Now take your math textbook and throw it across the room. Instead of flying in a straight line, the book follows a curve governed by your strength, the angle at which you threw it, and the force of gravity. That same force holds you in your chair, makes apples drop from trees, and reaches its invisible tentacles upward to keep birds, satellites, and the moon from escaping into deep space. Gravity goes even further, slowing the very expansion of the universe as galaxies feel the pull of their neighbors. And yet, amazingly, gravity is by far the weakest of nature's four basic forces—a million billion billion billion times weaker than the force that glues an atomic nucleus together.

If even these simple actions can inspire such thoughts once you understand matter, energy, motion, and force, imagine what will happen when you *really* start to think about your world. Glance at the sun and see not just a hot yellow ball but Einstein's famous mass-energy conversion equation, $E = mc^2$, in action. Watch gymnasts and skaters use the conservation of angular momentum to control their fantastic spins. Walk down the street and consider the particles and waves that constantly bombard and wash through your body: light from the sun, natural radiation from the earth, sounds from all directions, a thicket of TV and radio signals, even cosmic rays from stars that exploded long ago. To think in this way all the time surely would drive one mad. But to be *able* to ponder such things, to no longer take for granted that which has always seemed mundane, is powerful indeed.

The study of physics is also fascinating for its history. The scientific curiosity of the ancient Greeks, epitomized by Aristotle, led to an impressive world view marred by an understandable flaw: They believed the earth, and hence humankind, was the center of the universe. Nearly 2000 years passed until Copernicus and then Galileo contradicted this view—for which Galileo gave up his freedom.

Isaac Newton, born in the year Galileo died, was the father of what is now called "classical physics." His *Principia* stands alongside the works of Charles Darwin and Albert Einstein as one of the most influential in the history of science. Three hundred years later, Newton's laws of motion and universal gravitation are still valid in almost all circumstances, a remarkable achievement when one considers the technological revolution that has occurred since then.

Cracks began to appear in the armor of Newton's laws when scientists scrutinized the behaviors of atoms. At such tiny scales, things get weird. The framework of quantum mechanics, developed over many years, maintains that only certain motions and energies are "allowed"; all others are "forbidden." Imagine being forced to hop down a staircase two or three steps at a time instead of just one; essentially, that is a quantum mechanical "rule" for electrons in energized atoms. Quantum mechanics is a complex but versatile tool. For example, scientists rely on its rules to identify chemical elements in the laboratory or in distant stars. The key is the distinct signature of light emitted by an element's electrons as they jump from one allowed energy level to the next.

Einstein shook up physics even more with his extraordinary theories of relativity. In part, these theories claim that an object accelerating to a very high speed—close to the speed of light—experiences some other-worldly effects. For instance, the object gets shorter, its mass increases, and its own "clock" literally slows down. Although such exotic effects are beyond the scope of a basic course, they are critical to research in physics and astronomy today.

Clearly, we live in a complicated universe. It is also clear that even with the new technology of recent decades, we still have much to learn. At times in the history of physics, scientists believed they had discovered nearly everything there was to know. It is not likely we will ever think that way again. Some physicists believe, for instance, that they will never find the smallest units of matter and that nature consists of wheels within wheels within wheels. Perhaps such a universe is forever beyond our comprehension. But if we lived in a world we understood completely, society itself might wither away—for what better to drive human ambition than the quest to fathom one's own origins and fate?

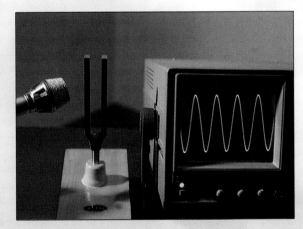

Sound waves produced by a tuning fork shown on an oscilloscope. (Leonard Lessin/Peter Arnold, Inc.)

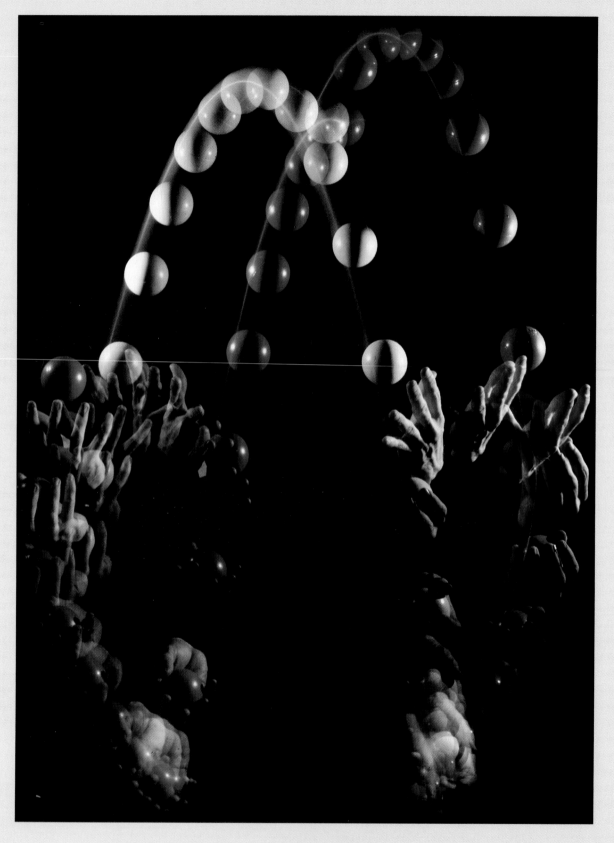

These juggling hands demonstrate parabolic trajectory and constant acceleration. (Richard Megna/
Fundamental Photographs)

3

DESCRIPTION OF MOTION

When an object changes its position relative to some other object or objects, we say it has moved. Motion fascinated even our prehistoric ancestors, who captured the beauty of wild beasts in motion in works of art (Fig. 3–1). Down through the ages, some of the best minds have been drawn to this subject. The development of the laws of motion in the sixteenth and seventeenth centuries signaled the birth of the modern scientific revolution. As we explore the realm of motion in this chapter, we will also be tracing the origins of modern science.

3–1 THE RESTLESS UNIVERSE

The universe is filled with things in motion, from majestic galaxies, stars, and planets to atoms. The motions may be fast or slow, smooth or erratic, simple or complex (Figs. 3–2 and 3–3). The words that de-

Figure 3–1 Prehistoric painting of beasts in motion in the Hall of Bulls, Lascaux Caves, Dordogne, France. (Courtesy of the French Government Tourist Office)

Figure 3–2 Marcel Duchamp's painting, *Nude Descending a Staircase, No. 2*, conveys a sense of motion. (Philadelphia Museum of Art: the Louise and Walter Arensberg Collection)

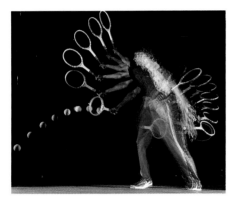

Figure 3–3   Stroboscopic photograph of a tennis player executing a backhand stroke shows successive positions of the tennis racket. (J. Zimmerman/FPG International)

scribe motion suggest an awareness of its variety: soaring, flying, leaping, running, galloping (Figs. 3–4, 3–5, 3–6, and 3–7). Challenged by the limitations imposed on us by nature, we invent things to move

Figure 3–4   Physicist J. Robert Oppenheimer jumping. Might jumping in a laboratory identify a potentially great scientist? (Photograph by Philippe Halsman)

Figure 3–5   Ballet dancers leaping through the air. (John Terence Turner/FPG International)

faster, farther, higher, and deeper. The approximate speeds of some objects are listed in Table 3–1.

Some motions are simple. An athlete races down a track, going from start to finish in the shortest possible time. He or she is undergoing motion in a straight line, the simplest kind of motion. A record-player turntable undergoes circular motion, repeating the same path in a cycle in the same time. This motion, too, has an element of simplicity. The motion of a

Table 3–1   SOME APPROXIMATE SPEEDS

	Speed	
Object	**Miles/Hour**	**Meters/Second**
Earth in orbit	67,000	30,000
Sun in galaxy	43,000	19,400
Earth satellite	18,000	8,300
Moon in orbit	2,200	1,000
Supersonic airplane	2,000	900
Racing car	200	180
Bird flying	60 to 180	27 to 82
Running elephant	25	11
Person running	6 to 15	2.7 to 6.8
Person walking briskly	3 to 4	1.4 to 1.8
Snail's pace	0.0025	0.0011

Figure 3–6   Two teams of skydivers forming two rings. (Vandystadt/Photo Researchers, Inc.)

pendulum is repetitive, like that of the turntable. The pendulum bob swings to and fro but never completely retraces its path. The swings become shorter and shorter until the bob is finally at rest again. Yet if we measure the time required for each swing, we find that it is almost exactly the same. How can we describe the motion of a ball thrown horizontally through the air, or straight up, or dropped straight down from a high spot? Before we discuss such questions, we take up a very useful tool for analyzing motion and certain other physical quantities, *vector analysis.*

3–2 VECTOR ANALYSIS

Certain physical quantities are described in terms of magnitude or size and dimension. They are known as **scalar quantities.** Examples are mass (the amount of matter in an object), time, volume (the amount of space taken up by an object), length, temperature, energy, and density. Scalars are governed by the ordinary mathematical processes of addition, subtraction, multiplication, and division. For example, two objects with masses of 10 kg and 5 kg have a total mass equivalent to an object with a mass of 15 kg.

Other physical quantities, however, are described completely only when direction is specified as well as magnitude. These are known as **vector quantities.** *Displacement*, the shortest (straight-line) distance between two points, is an example, as are velocity, acceleration, force, momentum, the electric field, and the magnetic field.

Figure 3–7   A bicycle race. (Eunice Harris/Photo Researchers, Inc.)

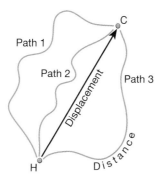

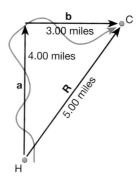

Figure 3–8   Illustration of the distinction between distance traveled over three alternate routes from H to C and the net displacement (represented by the arrow).

Figure 3–9   The net displacement, **R**, is equal to the sum of the displacements **a** and **b**.

Let us suppose that you commute by automobile from your home, H, to class, C (Fig. 3–8). The distance you actually travel on a given trip can be calculated from the difference in odometer readings of your automobile between the beginning and the end of the trip. It is expressed in the number of miles traveled on the particular road chosen for the trip.

However, there may be several alternate routes that you could take, with varying distances. If we ignore the actual route you traveled and consider only the net change in position, or **displacement,** from your home to class, represented by a straight arrow extending from H to C, then your displacement will be the same no matter which route you take. The simplest way to represent your displacement is to draw an arrow from your home to class on a map. This arrow, a vector representing 5.00 miles, is characterized by its length (magnitude), direction, and dimension.

Imagine that on your way to class you stop to pick up a classmate. The displacement from your home to your classmate's, represented by **a** in Figure 3–9, is 4.00 miles, and the displacement from your classmate's home to class, represented by **b**, is 3.00 miles. The sum of these two displacements is equal to your net displacement of 5.00 miles, although the distance you actually traveled is greater than this. Thus, a new method of dealing with quantities having both direction and magnitude is needed. This method is called **vector algebra.**

We denote a vector by an arrow representing the magnitude at the same time that it specifies direction in space. If we adopt a suitable scale, such as 1 cm = 1

mile, then a displacement of 3.00 miles is represented by an arrow 3.00 cm long. A vector may be symbolized conveniently in handwriting by a letter with an arrow above it, such as $\vec{a}$, and its magnitude by the letter "a" without the arrow. In printing, vectors are symbolized by a boldface symbol, such as **a**. If we want to add vector **a** to vector **b**, the rule is to place the tail of one of the vectors at the head of the other (it does not matter whether the tail of **b** is placed at the head of **a**, or vice versa) (Fig. 3–10). The resultant vector, **c**, is drawn from the tail of the first to the head of the second and stands for the vector sum of the two vectors **a** and **b**. To add more than two vectors, the same process is continued, the tail of each successive vector being placed at the head of the last. The resultant or sum vector starts at the tail of the first and ends at the head of the last vector. This particular method of vector addition is called the **polygon** or **"head-to-tail" method.**

$$\mathbf{R} = \mathbf{a} + \mathbf{b} + \mathbf{c}$$

EXAMPLE 3–1

A helicopter travels 30.0 miles due east, then changes direction and travels 40.0 miles due north. How far and in what direction is it from its point of origin?

SOLUTION

The problem is to find the displacement—the straight-line distance and direction between the origin and destination. To do this, we can use vector addition. We begin by setting up a coordinate system that enables us to specify direction; a useful one consists of horizontal (x) and vertical (y) axes. Along the

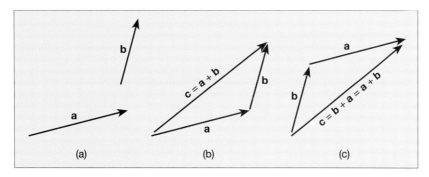

Figure 3–10 Polygon rule of vector addition. The tail of vector **b** is placed at the head of vector **a** in (a) and (b). The resultant vector **c** is drawn from the tail of **a** to the head of **b** and is equal to **a** plus **b**. Placing the tail of vector **a** at the head of vector **b**, as in (c), gives the same resultant.

x-axis, the direction to the right of the origin (the intersection of the x- and y-axes) corresponds to east, or 0°; the direction along the y-axis above the origin is north, or 90°; to the left, west, or 180°; and below the origin, south, or 270°.

Using a scale that is neither too small nor too large, such as 1 cm = 10.0 miles, construct a vector 3 cm long to the right of the origin along the x-axis, corresponding to a displacement of 30.0 miles due east (Fig. 3–11). From this arrowhead, construct a vector 4 cm long in the vertical direction, corresponding to the displacement of 40.0 miles due north, and place the arrowhead at the end of the line. The resultant displacement is the vector from the origin to the last arrowhead. Measure its length, 5 cm, corresponding to 50.0 miles. Determine the direction by measuring the angle with a protractor; it is 36.9°. The resultant displacement of the helicopter, then, is 50.0 miles in a direction 36.9° east of north from its point of origin, or 53.1° north of east.

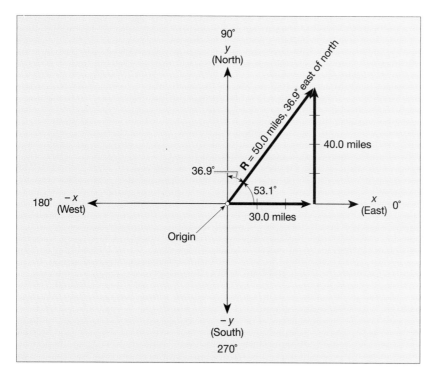

Figure 3–11 Determining a resultant displacement, **R**, by vector addition.

EXTENSION

A football player runs 30.0 yards down the field and then 20.0 yards toward the sideline. How far is he from the starting point? (Answer: 36.0 yards)

■

3–3 RESOLUTION OF VECTORS

At times it is useful to replace a given vector, **R**, by two other vectors, **a** and **b**, whose sum is **R**. The vectors **a** and **b** are called the **components** of **R**, and this process is called the **resolution of R into its components.** Although a vector may be resolved into its components in any two directions, most often the components in the x- and y-directions are desired.

EXAMPLE 3–2

A ship sails a straight course of 100 miles northeast (45°) from its origin. How far is it (a) to the east of its origin; (b) to the north of its origin?

SOLUTION

Construct an arrow 10 cm long (scale: 1 cm = 10.0 miles) in a direction of 45° (or 45° east of north) (Fig. 3–12). From the arrowhead of **R**, drop a vertical line

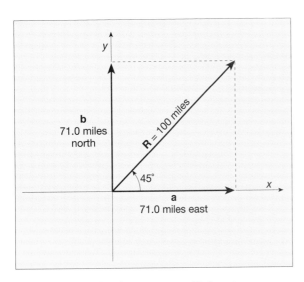

Figure 3–12.   Resolving a vector, **R**, into two components: horizontal and vertical.

to the x-axis, construct an arrowhead there, and measure the length **a**. Next, construct a line parallel to the x-axis, and measure the length **b**. The length **a** represents the distance east the ship is from its origin, 71.0 miles, and the length **b** the distance north from its origin, also 71.0 miles.

EXTENSION

A plane travels 150 miles in a direction 30° north of east. Determine its (a) easterly and (b) northerly displacements. (Answer: (a) 130 mi; (b) 70.0 mi)

■

3–4 SPEED AND VELOCITY

Since motion involves a change in position, we are interested in such questions as the following in dealing with a moving body, whether it is a bowling ball, an automobile, or a spacecraft: Where is it? How fast is it traveling? Where is it going? Is it slowing down or speeding up? Is it moving steadily? Is it changing direction? For the sake of convenience, we treat the moving body as if it were a point particle moving in space. The point particle is a model of whatever object is in motion.

The key concepts of **kinematics**—the scientific description of motion, without regard to its cause—are *position, speed, velocity,* and *acceleration*. They, in turn, are derived from two of the fundamental quantities, length and time. The concepts that are presented here and in the following chapters evolved slowly over centuries and eluded even the greatest minds before their full significance was grasped.

We know that an automobile is moving along a highway from observing changes in its position (displacement) over time. **Speed** is a measure of how far a body travels in a given time, that is, the rate of change of position. If we know the distance traveled and the period of time required, we can calculate the average speed from the formula

$$\text{Average speed} = \frac{\text{Distance traveled}}{\text{Elapsed time}}$$

or, in symbols,

$$\bar{v} = \frac{s}{t} \tag{3–1}$$

(In physics, a bar placed over a symbol, such as $\bar{v}$, usually indicates an average value of that quantity.)

The greater the distance traveled in a given time, the greater the speed. An automobile traveling 50.0 miles in 1 hour has a lower speed than a jet airplane covering 500 miles in the same time. In either case, however, it is unlikely that a uniform speed was maintained for 1 hour. The automobile may have had to slow down on curves or hills or to stop for red lights or cattle crossings; conversely, it may have picked up speed on open stretches and while passing. In other words, we should say that the average speed is 50.0 miles/hour. In the course of the trip, the speedometer may have fluctuated between 0 and 60.0. A particular reading, such as 50.0, means that the speed at a given instant is 50.0 miles/hour. If this speed were maintained for 1 hour, the distance traveled would be 50.0 miles.

EXAMPLE 3–3

A racer in the famous Indianapolis 500 travels 500 miles in 3 hours and 40 minutes. What is the average speed in miles per hour?

SOLUTION

With a problem of this type, it is good practice to tabulate the information given and the question asked in a column on the left, using appropriate symbols and units. Then search for some relation connecting the concepts involved, in this case, speed, distance, and time, and express that relation as an equation in symbols on the right. If necessary, rearrange the equation so that the unknown is to the left and other quantities are to the right of the equal sign. Next, substitute the specific values given, keeping the units, and solve. Express the result by both a number and a unit. In this example, 40 minutes is converted to its equivalent in hours through dimensional analysis.

$$s = 500 \text{ mi} \qquad \bar{v} = \frac{s}{t}$$

$$40 \text{ min} = (40 \text{ min})\left(\frac{1 \text{ h}}{60 \text{ min}}\right) \qquad = \frac{500 \text{ mi}}{3.67 \text{ h}}$$

$$= 0.67 \text{ h}$$

$$t = 3 \text{ h } 40 \text{ min} \qquad = \boxed{136 \, \frac{\text{mi}}{\text{h}}}$$

$$= 3.67 \text{ h}$$

$$\bar{v} = ?$$

EXTENSION
Sound travels at about 1100 ft/s. Five seconds after seeing the lightning in a cloud, you hear thunder. How far is the cloud? (Answer: 5500 ft)

■

In the simplest type of motion, a body moves with uniform speed in a straight line; that is, it travels equal distances in equal intervals of time (Fig. 3–13). The average speed of such a body is given by Equation 3–1. When a direction is associated with a speed, however, we have a new quantity—velocity. The **average velocity** of a body in uniform motion in a straight line is defined as the displacement divided by the time during which the displacement occurred:

$$\text{Average velocity} = \frac{\text{Displacement}}{\text{Elapsed time}}$$

$$\bar{\mathbf{v}} = \frac{\mathbf{s}}{t} \qquad (3\text{–}2)$$

Since displacement, $\mathbf{s}$, is a vector quantity, velocity is a vector quantity, symbolized by $\mathbf{v}$.

For uniform motion in a straight line, the magnitude of the displacement is the same as the distance traveled in a given time interval, and the magnitude of the velocity—for example, 30.0 ft/s—is the same as the speed. The difference between velocity and speed is that velocity has associated with it the idea of direction; that is, velocity is a vector quantity, and speed is a scalar quantity. If the line of motion is horizontal, we can treat displacements to the right of the origin as positive and displacements to the left as negative; therefore, velocities to the right are positive and velocities to the left are negative. In Figure 3–10, the velocity is expressed as 30.0 ft/s to the right.

Instantaneous velocity refers to how fast a body is moving at a given instant and in what direction. It is

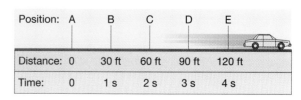

Position:	A	B	C	D	E
Distance:	0	30 ft	60 ft	90 ft	120 ft
Time:	0	1 s	2 s	3 s	4 s

Figure 3–13 An example of uniform motion in a straight line. The car traverses equal distances in equal intervals of time.

defined as the average velocity over a very short time interval, one so short that for all practical purposes the velocity can be considered constant during that time interval. Measurements might indicate, for example, that a car traveled 44.0 ft in 1.00 second. Therefore,

$$\bar{\mathbf{v}} = \frac{\mathbf{s}}{t}$$

$$= \frac{44.0 \text{ ft}}{1.00 \text{ s}}$$

$$= 44.0 \; \frac{\text{ft}}{\text{s}}$$

If the car traveled 0.440 ft in 0.010 s, we would calculate the average velocity to be

$$\bar{\mathbf{v}} = \frac{0.440 \text{ ft}}{0.010 \text{ s}}$$

$$= 44.0 \; \frac{\text{ft}}{\text{s}}$$

This process, continued for shorter and shorter intervals of time, gives the instantaneous velocity.

Few highways are constructed with 50-mile stretches in one direction. On a real highway, an automobile undergoes more or less frequent changes in

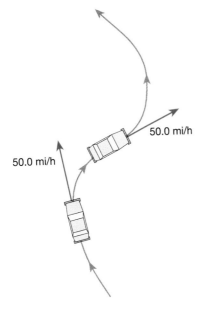

Figure 3–14   Although the speed of the car is constant —50.0 mi/h—the velocity is varying because its direction is changing.

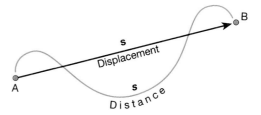

Figure 3–15   The displacement from A to B.

direction. Although the speedometer reading may remain constant for a period of time, the changes in direction mean that the velocity of the automobile is changing (Fig. 3–14). The velocity may be 50.0 miles/hour north at one time and 50.0 miles/hour northeast at another. The numerical values—the magnitude—of speed and velocity are identical: the speed is 50.0 miles/hour regardless of the direction.

Most bodies, in fact, move along paths that are not straight and at rates that vary during the motion (Fig. 3–15). At one instant the body may be at position A and at a later time at position B. During the time interval, the body has traveled along the path a distance s. Its average speed is found by applying Equation 3–1, since it is the distance traveled divided by the time. In this motion, the displacement is given by the vector $\mathbf{s}$, which is different in magnitude from the distance traveled along the path. The average velocity is given by Equation 3–2, the displacement divided by the time, in which t is the time required to travel from A to B. The direction of the average velocity is the same as the direction of the displacement.

3–5 ACCELERATED MOTION

For the motion of a body in which the velocity changes in either magnitude or direction, or both, we define a new quantity: **Acceleration is the rate of change of velocity.**

Acceleration is an aspect of motion that we can often perceive. When riding in a car, even with our eyes shut, we can feel the velocity change. When the car is initially moving forward, we may sink back into our seats; when it stops, we have the sensation of being pushed forward; when it rounds a curve, we experience a sideward push. We can feel the accelerations of a fast elevator at the start and stop of a trip. Each situation in which our velocity is changing rap-

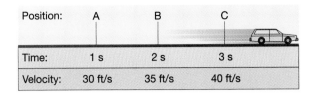

Position:	A	B	C
Time:	1 s	2 s	3 s
Velocity:	30 ft/s	35 ft/s	40 ft/s

Figure 3–16 The velocity of the car is changing. The car is said to be accelerating.

idly involves a physiological awareness of the acceleration.

An automobile is accelerating when it "picks up" speed, slows down (although we usually refer to this as deceleration), or changes direction (Figs. 3–16 and 3–17). If it moves with a velocity, $\mathbf{v}_0$, at point A, and at some time (t) later it is at point B with a velocity $\mathbf{v}$, the average acceleration is the change in velocity divided by the time interval:

$$\text{Acceleration} = \frac{\text{Change in velocity}}{\text{Time taken}}$$

$$= \frac{\text{Final velocity} - \text{Initial velocity}}{\text{Time}}$$

$$\bar{a} = \frac{\mathbf{v} - \mathbf{v}_0}{t} \tag{3-3}$$

where $\mathbf{v}_0$ is the initial velocity, which may or may not be zero, and $\mathbf{v}$ is the final velocity. Acceleration is a vector having the direction of the velocity difference.

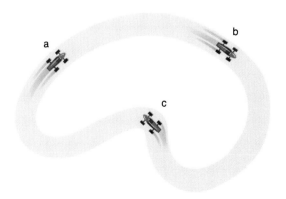

Figure 3–17 Three identical cars moving clockwise on a race track. Although they are maintaining constant speeds, these cars are also accelerating because their direction is changing. (Larry D. Kirkpatrick and Gerald F. Wheeler, *Physics: A World View*, Philadelphia: Saunders College Publishing, 1992, Fig. 3–29, p. 66.)

Rearranging Equation 3–3, we obtain the useful expression

$$\mathbf{v} = \bar{\mathbf{a}}t + \mathbf{v}_0 \tag{3-4}$$

If the initial velocity is zero, that is, if the body starts from rest, $\mathbf{v}_0 = 0$. Equation 3–4 becomes

$$\mathbf{v} = \bar{\mathbf{a}}t \tag{3-5}$$

By considering only motion in which the acceleration is constant in magnitude and direction and the body is moving in a straight line, we shall not have to distinguish between average and instantaneous acceleration. Therefore, the symbol $\mathbf{a}$ without the bar, a vector quantity, will represent the average acceleration from now on.

Except in problems such as those involving automobiles, where speedometer readings are available, it is often difficult to measure velocities. If the acceleration is constant, the distance traveled by an accelerated body has two contributions: (1) the distance that it would have traveled if the velocity $\mathbf{v}_0$ had remained constant, $\mathbf{v}_0 t$; and (2) the additional distance traveled as a result of the change in velocity:

$$\bar{\mathbf{v}} = \frac{\mathbf{v}_0 + \mathbf{v}}{2}$$

$$\mathbf{s} = \bar{\mathbf{v}}t = \left(\frac{\mathbf{v}}{2}\right)t = \left(\frac{\mathbf{a}t}{2}\right)t = \frac{1}{2}\mathbf{a}t^2$$

or

$$\mathbf{s} = \mathbf{v}_0 t + \frac{1}{2}\mathbf{a}t^2 \tag{3-6}$$

Some SI units for acceleration are meters per second per second (m/s²); kilometers per second per second (km/s²); and centimeters per second per second (cm/s²). The kinematics equations are summarized in Table 3–2. The following examples will help familiarize you with those equations and the units of acceleration.

EXAMPLE 3–4

Starting from rest at the end of a runway, an airliner achieves its takeoff velocity of 100 m/s in 50.0 seconds. What is its acceleration?

SOLUTION

Tabulate the data given and the question asked in a column on the left. Do you see that these quantities

Table 3–2   SUMMARY OF EQUATIONS
OF KINEMATICS

Starting from Rest	Intial Velocity Not Zero
1. $\mathbf{s} = \bar{\mathbf{v}}t$	$\mathbf{s} = \bar{\mathbf{v}}t$
2. $\mathbf{s} = \dfrac{\mathbf{v}}{2}t$	$\mathbf{s} = \left(\dfrac{\mathbf{v}_0 + \mathbf{v}}{2}\right)t$
3. $\mathbf{a} = \dfrac{\mathbf{v}}{t}$	$\mathbf{a} = \dfrac{\mathbf{v} - \mathbf{v}_0}{t}$
4. $\mathbf{v} = \mathbf{a}t$	$\mathbf{v} = \mathbf{a}t + \mathbf{v}_0$
5. $\mathbf{s} = \dfrac{1}{2}\mathbf{a}t^2$	$\mathbf{s} = \mathbf{v}_0 t + \dfrac{1}{2}\mathbf{a}t^2$

are related by Equation 3–3? Write down this equation, then substitute and solve.

$$\mathbf{v} = 100 \text{ m/s} \quad \mathbf{a} = \frac{\mathbf{v} - \mathbf{v}_0}{t}$$

$$\mathbf{v}_0 = 0$$

$$t = 50.0 \text{ s} \qquad = \frac{100\,\dfrac{\text{m}}{\text{s}} - 0}{50.0 \text{ s}}$$

$$\mathbf{a} = ?$$

$$= \boxed{\frac{2.00 \text{ m}}{\text{s}^2}}$$

(read "2 meters per second per second," or "2 meters per second squared") in the same direction as the velocity change

EXAMPLE 3–5

A rocket for a space shuttle is given a constant acceleration of 2.00 m/s². What is its velocity 90.0 seconds after liftoff?

SOLUTION
From the tabulated data, we find that we can apply Equation 3–5 to get the velocity.

$$\mathbf{a} = 2.00 \text{ m/s}^2 \qquad \mathbf{v} = \mathbf{a}t$$

$$\mathbf{v}_0 = 0$$

$$t = 90.0 \text{ s} \qquad = \left(2.00\,\frac{\text{m}}{\text{s}^2}\right)(90.0 \text{ s})$$

$$\mathbf{v} = ? \qquad = 180\,\frac{\text{m}}{\text{s}}$$

EXAMPLE 3–6

The driver of a car traveling 75.0 ft/s easterly slams on the brakes. If the car stops in a distance of 200 ft, (a) how long does it take to stop? (b) what is the acceleration?

SOLUTION
As a result of braking, the car is decelerated and comes to a stop. We can calculate the time from Equation 3–2 and the acceleration from Equation 3–3.

$$\mathbf{v}_0 = 75.0 \text{ ft/s}$$

$$\mathbf{v} = 0 \text{ ft/s}$$

$$\mathbf{s} = 200 \text{ ft}$$

$$\mathbf{a} = ?$$

$$t = ?$$

(a) $\mathbf{s} = \bar{\mathbf{v}}t$

$$\mathbf{s} = \left(\frac{\mathbf{v}_0 + \mathbf{v}}{2}\right)t$$

$$\therefore t = \frac{\mathbf{s}}{\left(\dfrac{\mathbf{v}_0 + \mathbf{v}}{2}\right)}$$

$$= \frac{200 \text{ ft}}{\left(\dfrac{75.0\,\dfrac{\text{ft}}{\text{s}} + 0}{2}\right)}$$

$$= \frac{200 \text{ ft}}{37.5\,\dfrac{\text{ft}}{\text{s}}}$$

$$= \boxed{5.33 \text{ s}}$$

(b) $\mathbf{a} = \dfrac{\mathbf{v} - \mathbf{v}_0}{t}$

$$= \frac{0\,\dfrac{\text{ft}}{\text{s}} - 75\,\dfrac{\text{ft}}{\text{s}}}{5.33 \text{ s}}$$

$$= \frac{-14\,\dfrac{\text{ft}}{\text{s}}}{\text{s}}$$

$$= \boxed{-14\,\frac{\text{ft}}{\text{s}^2}}$$

(The minus sign here means deceleration, that is, slowing down, since the direction of the acceleration is opposite to that of the velocities.)

EXTENSION

A downhill skier is accelerating uniformly. If the speed at one instant is 15.0 ft/s and 2.00 s later is 45.0 ft/s, what is the skier's acceleration? (Answer: 15.0 ft/s²)

■

3–6 A THEORY OF MOTION

For nearly 2000 years the ideas of Aristotle (384–322 BC) dominated the attempt to understand motion. Although the theory of motion was not his alone, Aristotle's exposition of it was the most convincing. Modern science began in large part when Aristotle's physics was successfully challenged.

Aristotle considered rest to be the natural state of the universe. Motion, therefore, involved a change in the natural state of order. He taught that all objects on or around the earth are made up of combinations of the four elements—earth, air, fire, and water (Fig. 3–18). Some objects appear to be light and others heavy, depending on the proportion of the different elements in each body.

According to Aristotle, earth was "naturally" heavy and fire "naturally" light; water and air fell between these two extremes. Each of the elements had a tendency to reach its "natural" place, where it remained at rest unless disturbed. If a body was heavy, its natural motion was downward. An apple, being composed mostly of earth, would fall downward when dropped. Smoke, being mostly fire, would rise upward since fire was above earth.

In Aristotle's system, heavy bodies fell because *they* possessed the property of gravity and the earth was their "natural" place. Light bodies were thought to possess the property of levity, which caused them to rise upward to their "natural" place, the sky. Motion was caused by the properties of gravity or levity, which the bodies themselves possessed, rather than by an outside force such as the earth.

The medium through which a body moved played an important role in Aristotle's theory. The denser the medium through which a body moves, the slower the motion (Fig. 3–19). (A rock falls more slowly through water than through air.) In a vacuum, therefore, where no medium exists, all bodies should fall with infinite speed. Since this could not be, Aristotle considered it a strong argument against the possibility of a vacuum. (The first vacuum pumps were invented 2000 years later.) Aristotle also believed that the heavier the body, the greater its ability to overcome the resistance of the medium. If two bodies, one twice as heavy as the other, were dropped from the same height simultaneously, the time for the heavier body to fall would be only half as great as for the lighter body. (That such is *not* the case was actually demonstrated many centuries later.)

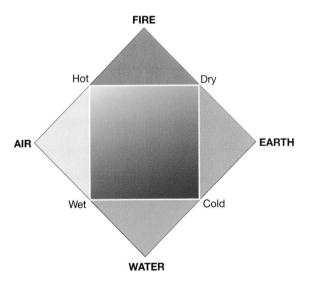

Figure 3–18 The four elements and four qualities of Aristotle.

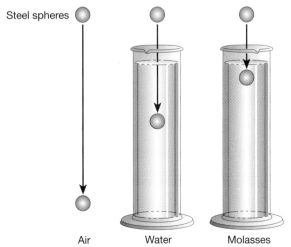

Figure 3–19 The more viscous the medium, the slower the motion of the falling body.

In Aristotle's scheme of the universe, the earth was assumed to be at rest, and the stars, planets, and the sun itself moved around the earth in circles. They described circles because it was their "nature" to do so. The celestial bodies were not made of the same four elements as the earthly bodies, but consisted of a fifth element ("quintessence," or **aether**)—an unchangeable, perfect material unlike the four elements found on earth. It was thought that the natural motion of a body composed of aether had to be circular, the most perfect of all natural motions.

The earth had no "local motion," or movement from one place to another. It was not even supposed to rotate on its axis. Since the earth was composed of "imperfect," corruptible material, it was not "natural" for it to have a circular motion, such as moving about the sun or rotating on its own axis.

Of course, objects very often move in "unnatural" ways—for example, an arrow shot from a bow or a rock thrown straight upward or straight across. Such motion, Aristotle believed, was "violent," or contrary to the nature of the body. It occurred only when some force acted to start and keep the body moving contrary to its nature. When the force was removed from the arrow or rock, the object would fall straight to the earth.

After the initial push supplied by the bow or hand, Aristotle imagined that the medium—the air, in this case—pushed the arrow or rock along by rushing in

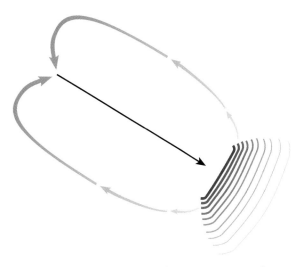

Figure 3–20 Forced or "violent" motion, according to Aristotle. The force is supplied by the medium—the air—rushing in behind the arrow to prevent a vacuum.

behind to prevent a vacuum (Fig. 3–20). Gradually, the push of the medium grew less and less and eventually wore out. At this point, gravity took over and the body dropped under natural motion. Aristotle assumed that forced motion and natural motion did not mix. In a vacuum there could be no medium to serve as a "mover," and for this reason, also. Aristotle did not accept the reality of a vacuum.

On the other hand, Aristotle himself had said that a medium, such as air, resists motion. How, then, could it be the cause of motion? If it is really the air that carries the rock along, why must the hand touch the rock at all? Again, a heavy rock can be thrown much farther than a very light one. This fact is difficult to explain if air is the cause of motion, since air would move light objects more easily than heavy ones. Questions such as these cast doubt on Aristotle's theory of motion.

3–7 GALILEO AND THE EXPERIMENTAL METHOD

In the centuries between Aristotle and Galileo Galilei (1564–1642), the study of motion progressed slowly. Galileo is a transitional figure, standing at the end of the Middle Ages and at the beginning of modern science.

Galileo was not content with qualitative observations. He looked for *quantitative* observations that would describe a phenomenon with mathematical precision. In attacking scientific problems, he emphasized *how* things worked, not why. This approach was radically different from previous ones, and it turned out to be a more fruitful one for answering scientific questions. In this respect, Galileo was the first modern scientist.

While a professor of mathematics at the University of Pisa, Galileo wrote the treatise *On Motion*, in which he discussed the problems of falling bodies and projectiles. Galileo broke with Aristotle's concepts of "heaviness" and "lightness." According to Galileo, all bodies are more or less heavy. Light bodies move upward because heavy bodies fall below them. Whether the medium is air or water, its resistance is a kind of buoyancy that supports lighter bodies more effectively than denser ones. Galileo at first supposed that in a vacuum bodies would fall with speeds proportional to their densities. In a later book, *Dialogues Concerning Two New Sciences*, however, his ideas on motion had changed profoundly. This book laid the foundation for the science of mechanics and modern physical science.

Galileo argued against Aristotle's view that heavy objects falling freely fall faster than light ones. He asked what would happen if a heavy body and a light body were tied together and then dropped from a height. According to Aristotle's view, it could be maintained that the time taken would be the average of the times the two bodies would have taken to fall separately. It could also be argued that the time taken would be the same as that required by a body with their combined weights to fall from the same height. These results are contradictory; therefore, Galileo wrote that they showed Aristotle to be wrong.

There is an uncorroborated story that, to find out what actually does happen when bodies fall, Galileo made a public demonstration from the Leaning Tower of Pisa before the students and faculty of the University (Fig. 3–21). Today we have elaborate instruments and refined clocks to determine how the speed of fall varies with the time of fall or the distance. Since mechanical clocks were unknown in Galileo's day, he found the measurements for a body falling straight down beyond his means. The motion was too rapid to observe directly. If gravity could be "diluted," however, the motion would be slowed down, perhaps enough for measurements to be taken.

Galileo was convinced that smooth, metallic spheres rolling down a smooth, graduated inclined plane with measurable speeds were an example of diluted free fall and would follow the same laws as a body falling vertically. His argument was based on the fact that a ball resting on a horizontal surface does not move, whereas one falling parallel to a vertical surface moves as fast as if the surface were not there (Fig. 3–22). A ball on an inclined surface, Galileo reasoned, should therefore roll with an intermediate speed, depending on the angle of inclination.

Galileo let spheres roll down planes tilted at various angles and determined the distances covered in different time intervals, which he measured with a water clock. He found that, regardless of their weights, the spheres all fell through the same distance in the same time. Extrapolating these results to bodies falling vertically, Galileo could say that, irrespective of their weights, all bodies fall with the same speed

Figure 3–21 The Leaning Tower of Pisa. (Alon Reininger/ Woodfin Camp & Associates)

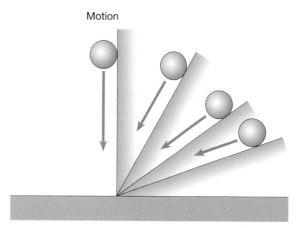

Figure 3–22 Spheres rolling down increasingly steep planes. The vertical plane, at 90°, matches free fall.

Figure 3–23   A feather and apple fall with the same acceleration when air resistance is negligible. (James Sugar/Black Star)

when dropped from the same height (Fig. 3–23). This result contradicted Aristotle's opinion that heavier bodies fall faster than light ones.

In the account of his work, Galileo tells what apparatus he used, how it was set up, and how the experiment was performed. He made many trials to ensure that the results would be as accurate as possible. His experimental conditions were as perfect as he could make them; the plane was polished, and the metal spheres were smooth to minimize the effect of friction. To simplify his study and to get to the fundamentals of his problem, he ignored less important phenomena. He then analyzed his data mathematically and tested his hypothesis with further experimentation. The information he obtained in this way would apply not only to the conditions of the particular experiment, but also more generally to other bodies falling under gravity. The mathematical-experimental method, which characterizes science as we know it today, came to maturity with Galileo.

Galileo concluded that all bodies released from the same height simultaneously fall at the same rate, if air resistance is neglected. He also found that the rate changes with time. We now know that the velocity of a freely falling body increases by an increment of 9.80 meters per second (32 ft/s) each second. The value 9.80 meters per second per second, or 9.80 m/s^2 (32 ft/s^2), is known as the **acceleration due to gravity** and is designated by the symbol **g**. The value of **g** varies slightly from place to place on the earth's surface. Extraordinarily precise measurements of **g** made by Robert H. Dicke and co-workers at Prince-

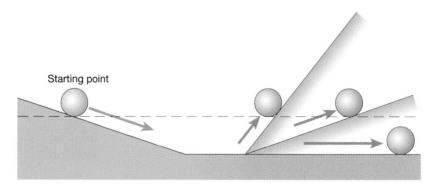

Starting point

Figure 3–24   In the absence of friction, a sphere rolling down one inclined plane would rise to the same height on another inclined plane. Galileo concluded that, if the second plane were horizontal, the sphere would travel indefinitely.

ton University confirm Galileo's conclusion that all bodies fall at precisely the same rate.

3–8 PROJECTILE MOTION

Galileo applied his method to the problem of determining the path traced out by a projectile. He had learned through his experiments with rolling spheres that **projectile motion** could be considered to be made up of *two components*. A sphere that rolled down one incline starts up a second incline with the same speed. As the angle of the second plane decreases, the sphere encounters less resistance to its motion until, when the plane is horizontal, it continues to move with uniform speed, slowed down only by friction (see Fig. 3–24). Galileo concluded that in the absence of friction the sphere would move horizontally indefinitely at uniform speed, and this would be its natural motion. It is as if gravity were switched off when the sphere reached the bottom of the incline, and the *horizontal component* of the ball's motion continued unaffected by gravity.

The conclusion that Galileo reached contradicted Aristotle's doctrine that a constant force is required to keep a body moving with uniform speed; no force that anyone could see kept the sphere moving horizontally. As far as Galileo was concerned, although he did not make the concept explicit, it is natural for a body in a state of uniform motion to remain in that state indefinitely, provided that it is not acted on by an external force such as friction.

Galileo could approximate the motion of a projectile by considering a sphere rolling across a table with uniform speed until it comes to the edge. When it reaches the edge of the table, the sphere, owing to its own weight, acquires a *vertical, accelerated motion* while at the same time maintaining its uniform horizontal motion and traces a curved path to the floor. In the horizontal direction, the constant uniform speed causes the body to cover equal distances in equal times. Vertically, the speed increases with time because of gravity, and the distances covered are proportional to the square of the time. Galileo determined that the form of the path followed by the sphere under these conditions would not be merely a curve of some sort, but a specific curve—a semiparabola (Fig. 3–25). The path of a projectile shot upward from a cannon would be a full parabola (Fig. 3–26).

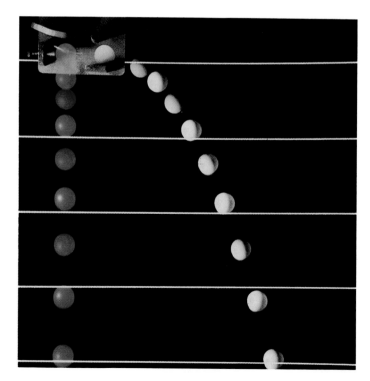

Figure 3–25 A ball dropped from a certain height and one projected horizontally from the same height will hit the ground at the same time. The horizontal motion of a projectile is independent of its vertical motion. (Richard Megna, Fundamental Photographs)

Figure 3-26 The lava particles from a volcanic eruption in Mt. Etna, Sicily, follow parabolic paths. They are projectiles falling in the presence of gravity. (Otto Hahn, Peter Arnold, Inc.)

Galileo's theory of **projectile motion**—that it consists of *horizontal and vertical components* that do not interfere with each other—offered strong support to the hypothesis that the earth moves around the sun. Critics of this hypothesis argued that if the earth moved, a stone dropped from the top of a tower should land behind the tower—not at a point at the base vertically below. The moving earth, it was supposed, would carry the base forward while the stone was falling.

To answer this criticism, Galileo drew the analogy of dropping a stone from the top of a mast of a ship moving with uniform speed (Fig. 3–27). When released, the stone has a horizontal motion equal to that of the ship and unaffected by its vertical motion. As the body falls while the ship moves, the horizontal motion of the stone carries it through the same horizontal distance that the ship travels. The stone and the mast, therefore, arrive at the same point simultaneously. To an observer on the ship who also shares its horizontal motion, the stone appears to fall straight down. In the same way, to an observer standing on the moving earth, a falling stone appears to drop vertically downward.

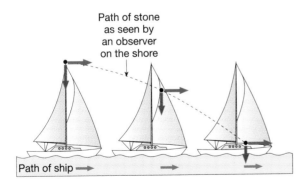

Path of stone
as seen by
an observer
on the shore

Path of ship ➡

Figure 3-27 A stone shares in the uniform horizontal motion of the ship. As the stone falls, it arrives at the same point simultaneously. The trajectory, as seen from the shore, is a semiparabola.

3–9 MOTION IS RELATIVE

To a passenger on a ship moving with uniform speed who shares the ship's horizontal motion, an object falling from the mast appears to fall vertically downward. The trajectory of the object appears curved, however, to an observer on land who does not share the ship's motion; the object appears to move forward with the ship as it drops. Which observer is correct?

Both are, if the **frame of reference** of the observer is considered. When the ship is standing still in the harbor, the trajectory appears to be a straight vertical line to both the passenger aboard ship and the observer on land. To the passenger, the trajectory is the same whether the ship is motionless or moving uniformly. The passenger cannot detect whether the ship is in motion from the trajectory alone. The observer on land sees both components, horizontal and vertical, which combine to give a semiparabola.

Such observations form the basis of the **Galilean relativity principle,** which states that **the laws that describe physical experiments are the same in a reference frame at rest or in a reference frame moving with a constant velocity.** Consequently, physical experiments will give the same results for different observers in reference frames moving at constant but different velocities. An object released from a ship's mast will land at the foot of the mast whether the ship is at rest or moving uniformly, and the laws of falling bodies are the same whether derived from experiments on a ship at rest or moving uniformly.

If the reference frames are moving uniformly with respect to each other (that is, at constant speed in a straight line), there is no preferred reference frame, according to **Einstein's special theory of relativity.** All reference frames are equally correct, although one may be more convenient to use than another. Without a frame of reference, the concept of motion has no meaning; there would be no way of determining whether motion is occurring.

We now know there is no absolute frame of reference by which you can say with confidence that whatever you see moving moves relative to you as you stay at rest. It was thought for centuries that the earth itself was an absolute frame of reference. Experiments have shown, however, that the earth rotates on its axis and undergoes other motions, although we are not aware of any of these motions.

SUMMARY OF EQUATIONS

$$\text{Average speed} = \frac{\text{Distance traveled}}{\text{Elapsed time}}$$

$$\bar{\mathbf{v}} = \frac{s}{t}$$

$$\text{Average velocity} = \frac{\text{Displacement}}{\text{Elapsed time}}$$

$$\bar{\mathbf{v}} = \frac{\mathbf{s}}{t}$$

Average Velocity

$$= \frac{\text{Initial Velocity} + \text{Final Velocity}}{2}$$

$$\bar{\mathbf{v}} = \frac{\mathbf{v}_0 + \mathbf{v}}{2}, \text{ when acceleration is constant}$$

$$\text{Acceleration} = \frac{\text{Change in velocity}}{\text{Time taken}}$$

$$= \frac{\text{Final velocity} - \text{Initial velocity}}{\text{Time}}$$

$$\bar{\mathbf{a}} = \frac{\mathbf{v} - \mathbf{v}_0}{t}$$

KEY TERMS

Acceleration   The rate at which the velocity of an object changes.

Acceleration due to gravity, g
Any freely falling object experiences an acceleration downward, equal to the acceleration due to gravity.

Aether   In Aristotle's physics, the sun, planets, and stars were said to be composed of this perfect material, unlike the four elements found on earth.

Average Velocity   The average velocity of a moving object is equal to the displacement of the object divided by the time interval during which the displacement occurs.

Components The parts into which a vector can be separated.

Displacement Distance traveled in a specified direction; a vector quantity.

Frame of Reference A coordinate system with respect to which the position and motion of an object may be described.

Galilean Relativity Principle The laws that describe physical experiments are the same in a reference frame at rest or in a reference frame moving with a constant velocity.

Instantaneous Velocity Refers to how fast an object is moving at a given instant and in what direction.

Kinematics The scientific description of motion, without regard to the cause.

Polygon Method of Vector Addition A method in which vectors are drawn to scale such that the tail

of each successive vector starts at the head of the last, and the resultant vector is drawn from the tail of the first to the tip of the last.

Projectile Motion Two-dimensional motion that occurs near the surface of the earth when the only force acting on the object is that of gravity; for example, a baseball in motion.

Theory of Relativity In Einstein's special theory of relativity, the idea that there is no preferred reference frame.

Resolution A process by which a vector may be replaced by two other vectors, called components, whose sum is equal to the original vector. Usually, the components in the x and y directions are desired.

Resultant The geometrical sum of two or more vectors.

Scalar Quantity A quantity such as mass, volume, or temperature

that is described by magnitude or size only without regard to direction.

Speed A measure of how fast an object travels over a period of time without regard to direction; distance divided by time.

Trajectory The path that an object such as a projectile describes in space.

Vector Algebra A method of dealing with quantities that have both magnitude and direction, such as velocity and displacement.

Vector Quantity A quantity that has both magnitude and direction, such as velocity, acceleration, or force.

Velocity The speed of an object in a specified direction; a vector quantity.

Violent Motion In Aristotle's physics, motion contrary to the nature of an object.

THINGS TO DO

1. (a) Compare the rates of fall of a sheet of paper and a quarter. (b) Repeat, with the paper slightly crumpled. (c) Repeat again, but with the paper crumpled tightly.

2. Calculate how far away the lightning is during a thunderstorm. Assume that sound travels at 1100 ft/s, and that the time for light to travel from the lightning is negligible.

3. Drop a quarter and a paper clip simultaneously from the same height. Listen for the sound of their impact on the floor. Is the rate of fall dependent on mass?

4. If you have access to a long glass tube that can be evacuated, observe the effect of air resistance on the fall of a

quarter and a feather by quickly inverting the tube (a) before and (b) after exhausting it.

5. With a stopwatch, time a few pitches in a baseball game and determine the speeds in feet per second and miles per hour. Assume that the distance between the pitcher and catcher is 60.0 ft.

6. Flick a quarter off the edge of a table just as you drop a second quarter from the same height. Do they strike the floor simultaneously?

7. Determine your average speed of (a) walking; (b) jogging.

EXERCISES

Vector Analysis

3–1. A man walks 3.00 mi due east, then turns and walks 2.00 mi due north. How far is he from the starting point?

3–2. A plane is climbing with a velocity of 150 m/s at an angle of 30° to the horizontal. Determine the horizontal and vertical components of its velocity.

3–3. A ship sails 75.0 mi due east, then sails in a direction 30° east of south for 60.0 mi, and finally heads due south for 40.0 mi. (a) How far has the ship traveled? (b) How far is it from its starting point?

3–4. A hurricane is drifting at a velocity of 16.0 mi/h in a direction 40° east of north. How fast is it moving in the (a) northerly and (b) easterly directions?

3–5. An airplane travels with a velocity of 200 mi/h in a direction 30° west of north for 1.50 h. How far north and how far west is it from its starting point?

3–6. A submarine is diving at an angle of 30.0° with the surface of the ocean. If its velocity is 5.00 m/s, find (a) the component of velocity toward the bottom; (b) the component of velocity along the surface.

Speed and Velocity

3–7. A driver travels 250 mi from New York to Washington in 5.25 h. What is his average speed in miles per hour?

3–8. A supersonic transport plane is traveling at a speed of 1200 mi/h. How far does it travel in 1.00 min?

3–9. An earth satellite travels 2000 km in 5.00 min. How far does the satellite travel in 24 h?

3–10. A car traveling at 40.0 mi/h skids to rest in 5.00 s. (a) What is its average speed during the stopping process? (b) How far does it go during the stopping process?

3–11. (a) How long does it take sunlight to reach the earth? (The sun is 9.30×10^7 mi away; the velocity of light is 186,000 mi/s.) (b) Moonlight? (The moon is 2.40×10^5 mi from the earth.)

Accelerated Motion

3–12. From a physiological point of view, differentiate between constant velocity and acceleration as experienced in an elevator ride.

3–13. An automobile has a constant speed of 15.0 mi/h. Is it possible for it to have an acceleration at the same time? Explain.

3–14. An automobile traveling at 22.0 ft/s accelerates at a constant acceleration of 4.0 ft/s² for 15.0 s. Determine its velocity at the end of this time.

3–15. Starting from rest, how long must a car accelerate at 5.00 ft/s² to reach a speed of 30.0 mi/h? (Note: A useful conversion factor is 60.0 mi/h = 88.0 ft/s.)

3–16. A truck traveling with an initial speed of 33.0 ft/s accelerates uniformly until it reaches a speed of 77.0 ft/s. If the acceleration is 4.00 ft/s², how much time is required for the truck to reach its final speed?

3–17. Starting from rest, an electron is accelerated in a machine to a velocity of 10^4 m/s in 0.010 s. (a) What is the acceleration? (b) How far does the electron travel in this time?

3–18. A motorcycle produces an acceleration of 7.50 m/s². How fast will it be going after accelerating from rest for 3.50 s?

3–19. The world's land speed record was held until recently by Colonel John P. Stapp, USAF. He rode a rocket-propelled sled that moved down the track at 632 mi/h. He and the sled were brought to rest in 1.4 s. Determine (a) the deceleration he experienced and (b) the distance he traveled during this deceleration.

3–20. Starting from rest, a high-jumper moves with a constant acceleration of 2.00 m/s² for a 5-s run. What is her speed at the end of the run?

Free Fall

3–21. Will a golf ball and a sheet of paper dropped from a bridge simultaneously strike the water at the same time? Explain.

3–22. What is the speed of a freely falling ball 3.00 s after it is released from rest?

3–23. A rock is thrown horizontally with a velocity of 20.0 m/s from a bridge 30.0 m high. (a) How long will the rock be in the air? (b) How far from the bridge will it land?

3–24. A ball is dropped out of the window of a moving car. Does it take longer to reach the ground than one dropped from a car at rest? Explain.

Theory of Motion

3–25. Compare Aristotle's and Galileo's ideas of motion.

3–26. Contrast the vertical and horizontal components of the motion of a thrown football.

3–27. Discuss the features of Galileo's method of dealing with the problem of motion that are relevant to science today.

3–28. Using examples from kinematics, justify the scientist's insistence on precision in terminology.

3–29. Discuss Aristotle's arguments against the existence of a vacuum.

3–30. Explain how Galileo's approach to physics represented an improvement over previous approaches.

3–31. *Multiple Choice*
A. Which one of the following is not a vector?
  (a) displacement   (c) velocity
  (b) speed   (d) acceleration

B. Acceleration is the rate of change of
   (a) displacement.
   (b) position.
   (c) velocity.
   (d) time.
C. Acceleration can be expressed by
   (a) m/s.
   (b) m²/s.
   (c) m/s².
   (d) m²/s².
D. If you cover 3000 mi in 5.00 days on a cross-country trip, 30.0 mi/h could represent the
   (a) instantaneous speed.
   (b) average speed.
   (c) velocity.
   (d) acceleration.
E. According to Galileo,
   (a) rest is the natural state of an object.
   (b) accelerated motion is the natural state of an object.
   (c) if not acted on by a force, a body continues in motion in a straight line.
   (d) heavy objects fall faster than light objects.
F. Which one of the following is a scalar quantity?
   (a) displacement
   (b) time
   (c) velocity
   (d) acceleration

G. When a moving object is brought to rest,
   (a) the average speed is zero.
   (b) the final speed is zero.
   (c) the speed is constant.
   (d) the acceleration is positive.
H. Adding a displacement vector to another displacement vector results in which one of the following?
   (a) a velocity
   (b) an acceleration
   (c) another displacement
   (d) a scalar
I. Two objects of different masses are released simultaneously from the top of a 25.0-m tower and fall to the ground. If air resistance is negligible, which one of the following statements applies?
   (a) The greater mass hits the ground first.
   (b) Both objects hit the ground together.
   (c) The smaller mass hits the ground first.
   (d) No conclusion can be made with the information given.
J. In addition to magnitude, a vector quantity has
   (a) length.
   (b) time.
   (c) position.
   (d) direction.

SUGGESTIONS FOR FURTHER READING

Adler, Carl G., and Byron L, Coulter, "Galileo and the Tower of Pisa Experiment." *American Journal of Physics*, March, 1978.
The story of the Tower of Pisa experiment is probably more legend than fact. If the experiment had been done, would it have worked as well as the legend claims?

Casper, Barry M., "Galileo and the Fall of Aristotle: A Case of Historical Injustice?" *American Journal of Physics*, April, 1977.
Presents Galileo's case against Aristotle's description of freely falling objects.

Drake, Stillman, "Galileo's Discovery of the Law of Free Fall." *Scientific American*, May, 1973.
A manuscript of Galileo's shows how he discovered the correct law of free fall.

Halloun, Ibrahim Abou, and David Hestenes, "Common Sense Concepts about Motion." *American Journal of Physics*, November, 1985.
A survey and analysis of common-sense beliefs of college students about motion and its causes.

Kyle, Chester R., "Athletic Clothing." *Scientific American*, March, 1986.
Attention given to the design of athletic equipment has contributed to new records in the speed sports and to enhanced protection and performance in other contests.

Peterson, Carl W., "High-Performance Parachutes." *Scientific American*, May, 1990.
There are parachutes that can decelerate a capsule from supersonic speeds to a snail's pace in a matter of seconds.

Schroeer, Dietrich, "Brecht's *Galileo:* A Revisionist View." *American Journal of Physics*, February, 1980.
Scientists often claim that Galileo is the first modern physicist. In his play, *Galileo,* Brecht presents Galileo as a symbol for all scientists who reject social responsibility for their work. Are both views distorted?

ANSWERS TO NUMERICAL EXERCISES

3–1. 3.60 mi

3–2. 130 m/s horizontal; 75.0 m/s vertical

3–3. (a) 175 mi   (b) 137 mi

3–4. (a) 12.4 mi/h northerly
(b) 10.0 mi/h easterly

3–5. 250 mi north; 135 mi west

3–6. (a) 2.50 m/s   (b) 4.33 m/s

3–7. 47.6 mi/h

3–8. 20.0 mi

3–9. 5.78×10^5 km

3–10. (a) 20.0 mi/h or 29.0 ft/s
(b) .145 ft

3–11. (a) 8.33 min   (b) 1.29 s

3–14. 82.0 ft/s

3–15. 8.80 s

3–16. 11.0 s

3–17. (a) 10^6 m/s²   (b) 50.0 m

3–18. 26.0 m/s

3–19. (a) -662 ft/s²
(b) 649 ft

3–20. 10.0 m/s

3–22. 29.4 m/s or 96.0 ft/s

3–23. (a) 2.47 s   (b) 49.4 m

Photograph of the earth from the Geostationary Operational Environmental Satellite (GOES). (Courtesy of National Geographic Society)

4

PLANETARY MOTION

Early peoples took an active interest in the skies. Those who were bound to the land learned from the rising and setting of constellations the times for sowing and harvesting (Fig. 4–1). Seafarers used the stars for planning their voyages and for navigation.

As astronomy developed, people mapped the heavens and charted the courses of the planets. The Egyptians and Babylonians were excellent observers of the periodic changes of sunrise and sunset during the year, the phases of the moon, the seasonal appearance and disappearance of constellations, and the more complex motions of the planets. They were aware of months and years and made calendars based on the moon cycle, the sun cycle, or combinations of the two. Until the Greeks, however, people were satisfied to describe the heavens and showed no interest in discovering an underlying regulatory mechanism.

Figure 4–1 Stonehenge, in southern England, was probably an ancient observatory. It appears that it was used to keep track of certain days of the year and for predicting eclipses. (Source: Jay M. Pasachoff, *Astronomy: From the Earth to the Universe*, 4th edition. Philadelphia: Saunders College Publishing, 1991, Fig. 2–20, p. 23.)

4–1 PTOLEMY'S SYSTEM

From the very first, the Greeks tried to *explain* celestial phenomena. To do so, they relied heavily on the observations recorded by the Egyptians and Babylonians.

Long before Columbus, Pythagoras (c. 582–500 BC), a philosopher and mathematician, taught that the earth is a sphere. He may have been guided in this belief by several observations. The surface of the sea is curved, not flat. Because of this, we first see the tallest point of a distant ship as it approaches, then gradually the rest. In an eclipse of the moon, the circular edge of the earth's shadow also suggests that its shape is spherical. The fact that the sun and moon appear to us as circular disks is evidence of their spherical nature. Just as important to Pythagoras, however, was the belief that the sphere was incomparable in symmetry, beauty, and perfection and was therefore the only proper shape for celestial bodies and the earth.

The belief that heavenly bodies move along circular paths was linked to the belief in spherical perfection. The circle, with neither beginning nor end, seemed the right path for heavenly bodies to follow. If some of the planetary motions seemed complicated, it was assumed that analysis would reveal that they could be reduced to uniform circular motions. The earth, however, was imagined to be at rest and its center the center of the universe.

Pythagoras taught that the heavenly bodies were eternal, perfect, and unchangeable and their motions uniform and circular. The earth, on the other hand, was subject to change and decay, and motions occurring on it were unpredictable and irregular. This dualism between the earth and the heavenly bodies had an enduring effect on astronomy.

The first attempt to explain astronomical phenomena in mathematical terms was made by Eudoxus (409–356 BC), a pupil of Plato (427–347 BC). Plato was the founder of the Academy, an institution of higher learning. Eudoxus' **theory of homocentric spheres** was the solution to a problem proposed by Plato to his pupils to "save the phenomena," that is, to account for the motions and positions of the heavenly bodies.

The peculiar motions of the planets baffled observers. At various times the planets seemed to stop, to retrograde, or to trace a curve such as a figure eight (Fig. 4–2). In his system, Eudoxus introduced 27

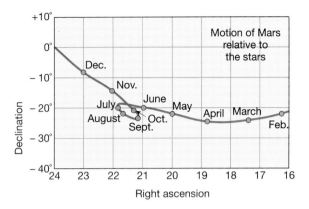

Figure 4–2 Retrograde motion of Mars for 1971. Mars proceeded eastward until July 13, retrograded westward, and resumed its eastward journey on September 11. (Source: Smith, E. V. P., and K. C. Jacobs, *Introductory Astronomy and Astrophysics*. Philadelphia: W. B. Saunders Company, 1973, p. 26.)

spheres to discover a combination of movements that a planet would seem to describe. The sun, moon, and each of the planets in this system were considered to be points on the surface of various interconnected spheres, all concentric ("homocentric"), with the earth at the center. The spheres were thought to turn at different rates, around different axes, and in different directions (Fig. 4–3). Eudoxus did not speculate on the real existence of these spheres or on the causes of their motions; his was a purely mathematical solution.

Admirable though it was, Eudoxus' system of homocentric spheres fell short of "saving the phenomena." It failed to account for variations in the brightness of planets, especially Venus and Mars, and for the difference in size between the sun and moon. Considering the inaccuracies that were present in the data available to him, however, Eudoxus' system was remarkable. Others improved on it by adding more spheres and brought the theory into closer harmony with the observed motions. Aristotle converted the system from a purely geometrical to a mechanical structure that could be presumed to have a physical existence, although it is not certain that Aristotle believed in its physical reality. Even with the refinements and modifications, however, the theory of homocentric spheres still did not "save the phenomena."

The high point of Greek astronomy is found in the work of Hipparchus of Nicaea (c. 190–120 BC) and

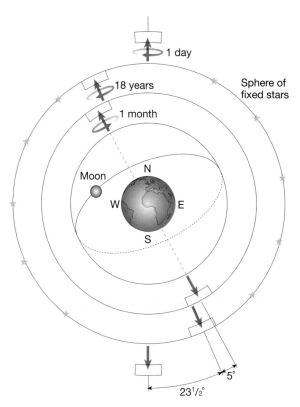

Figure 4–3   With the earth at the center, Eudoxus employed three concentric ("homocentric") spheres to represent the moon's motion. Twenty-seven spheres were necessary for the entire planetary system. (Redrawn from Crawford, F. H., *Introduction to the Science of Physics.* Copyright 1968 by Harcourt Brace Jovanovich, Inc., and reproduced with their permission.)

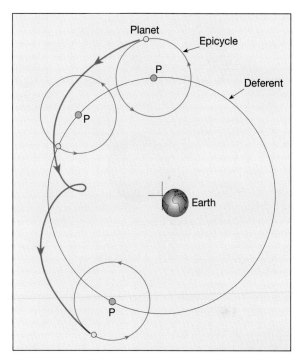

Figure 4–4   Ptolemy's basic devices: the *epicycle* and the *deferent*. A planet moved in a perfect small circle—the epicycle—around a point P, while P revolved in a perfect circle—the deferent. The earth is just off center (*eccentric*). Epicyclic motion saved Plato's perfect circles.

Claudius Ptolemy of Alexandria (c. 90–168 AD). Although Hipparchus' works are lost, he is known through the writings of Ptolemy and others. Ptolemy refers to Hipparchus so frequently in his writings that it takes some effort to recall that the two were separated by three centuries. Nor is it always clear from the context what was original with Hipparchus and what Ptolemy contributed. Both saw their tasks not so much as the taking and recording of observations as the mathematical explanation of the facts that the observations revealed. Their work culminated in Ptolemy's encyclopedic masterpiece, the *Almagest.* The *Almagest* displaced earlier works on astronomy and became the standard treatise on astronomy.

The *Almagest* developed the **Ptolemaic system,** a solar system that was completely earth-centered. The

Ptolemaic system kept the Pythagorean concept of circular motions for the heavenly spheres. It met the deficiencies of the theory of homocentric spheres with a mechanism consisting of eccentrics, epicycles, deferents, and equants (Figs. 4–4 and 4–5). A planet moved in a perfect small circle—the **epicycle**—around a point P, while P revolved in a perfect circle—the **deferent.** The relative simplicity of Eudoxus' system gave way to complexity. The basic devices—epicycles and deferents—were used in various combinations to describe curves in which the planet was close to the center at times, distant at other times, stationary, or even **retrograde.** It was necessary only to choose the proper relative size of epicycle and deferent and speeds of rotation. Like Eudoxus, Ptolemy, in all probability, considered his system to be a mathematical model of the universe having no "real" existence.

For 1400 years, Ptolemy's astronomical system encountered no serious challenge (Fig. 4–6). Its basis in the **geocentric hypothesis**—an immobile earth at

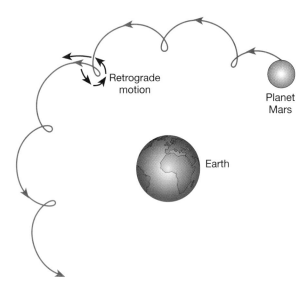

Figure 4–5   In Ptolemy's system, the path of a planet consisted of a series of loops as a consequence of epicyclic motion. A planet's retrograde motion occurs at the bottom of a loop.

the center of a finite universe—was satisfying to most people. The Ptolemaic system was in complete harmony with the Aristotelian physics that it assumed—in particular, with the principle of uniform, circular motion for celestial bodies. Together, the two systems offered a coherent scientific world view that was,

moreover, reconciled with the major religious systems of the West—Judaism, Christianity, and Islam.

4–2 THE COPERNICAN REVOLUTION

As astronomical observations improved, discrepancies showed up between observation and theory. When adjustments were made in certain parts of the Ptolemaic theory to account for these, other discrepancies appeared elsewhere. By the fifteenth century, astronomy was in such a sorry state from the point of view of precision that Nicolaus Copernicus (1473–1543) spoke of a "monster" passing for a system of the world. He felt that no system as cumbersome and inaccurate as the Ptolemaic system had become could possibly be true of nature. The solution to the problem did not seem to involve further tinkering with the Ptolemaic system but a fundamental restructuring of astronomy.

The **Copernican system** was set forth in his work, *On the Revolutions of the Celestial Spheres,* published in 1543. Copernicus received a copy of it on his deathbed. This book reintroduced the idea of a sun-centered, or **heliocentric**, system, proposed 1700 years previously by the notable Alexandrian astronomer, Aristarchus (c. 310–230 BC), whose works describing this theory have not been preserved, and

Figure 4–6   Ptolemy's earth-centered model of the universe reigned supreme for 1400 years. It seemed to account for observations of planetary motions. The epicycles of Mercury and Venus always lie on the line joining the earth and the sun.

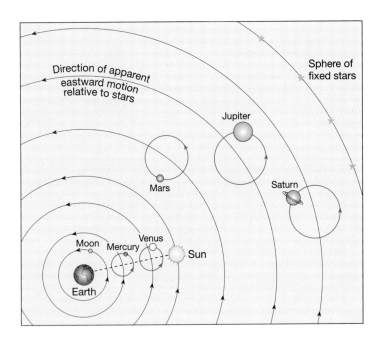

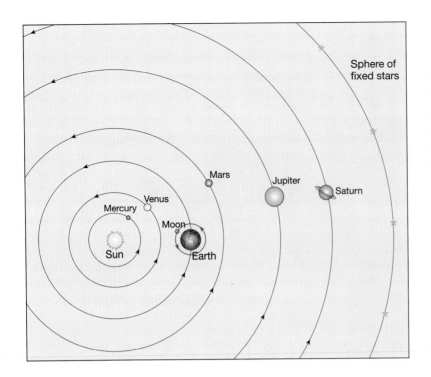

Figure 4–7 The sun-centered universe of Copernicus. The moon travels around the earth, but the earth and the other planets move about the sun. A less complex mechanism is required than in Ptolemy's system.

replaced the notion of an earth at rest with that of one in motion (Fig. 4–7). The earth was described as having three motions: a daily rotation on its axis, an annual revolution in its orbit around the sun, and a wobbly motion like that of a spinning top, called precession (Fig. 4–8).

In effect, Copernicus used the thought patterns of the Greeks and produced the simplest solution to the problem that Plato posed of "saving the phenomena." He kept the mechanism of concentric spheres, epicycles, and deferents and the Pythagorean assumption that only circular, uniform motion is right for heav-

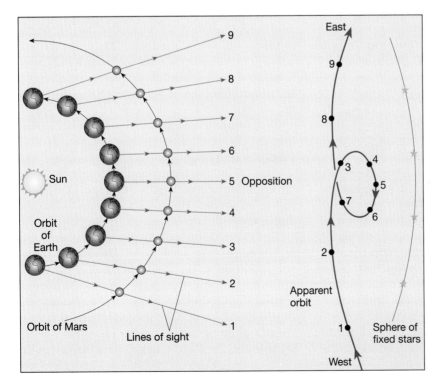

Figure 4–8 Retrograde motion of Mars explained in the Copernican system. The earth moves faster in its orbit than does Mars. Lines of sight drawn from a moving earth to a more slowly moving Mars illustrate how a retrograde loop in Mars' orbit is formed.

enly bodies. The number of spheres required, however, was reduced from Ptolemy's 80 or so to 34.

Copernicus believed firmly in the reality of the earth in motion and made this clear in his preface. In his final illness, the printing of the book was supervised first by George Rheticus (1514–1576), a mathematician and student of Copernicus, and then by Andreas Osiander (1498–1552), a friend. Sensing the explosive contents of the book, Osiander wrote a second preface in which he warned the reader not to take the hypothesis of a moving earth seriously, but to view it only as a device for simplifying mathematical computation. The intent of this preface was to pacify those who would otherwise regard as blasphemous the dethronement of the earth as the center of the universe in favor of the sun. Martin Luther (1483–1546) is quoted as saying, "This fool wishes to reverse the entire science of astronomy." To others, however, Copernicus was regarded as the "restorer of astronomy" and the "Ptolemy of our age."

Astronomers freely used the Copernican system for calculation, and a few even supported the innovative ideas against which Osiander had warned. Popular writers, on the other hand, were hostile. It was difficult to give up the uniqueness of the earth in favor of a system in which the earth was merely another planet, and an insignificant one at that. The spread of Copernican ideas threatened to crack the intellectual structure of the medieval world, changing man's thinking about himself and the universe. It is in this sense that we speak of the Copernican revolution.

4–3 "GATEWAY TO THE SKIES": TYCHO BRAHE

Over the years, tables were developed giving the changing positions of the planets. The Alphonsine planetary tables were based on Ptolemy's geocentric (earth-centered) ideas. Following the appearance of Copernicus' heliocentric (sun-centered) system, the Prutenic tables were compiled and became the standard reference in the sixteenth century. Both tables gave the impression of considerable accuracy, stating planetary positions in degrees, minutes, and seconds of arc. A conjunction of the planets Jupiter and Saturn, however, when they appeared to be closest to each other, in 1563 indicated that both tables were in

error in their predictions, the Alphonsine tables by a full month and the newer Prutenic tables by a number of days. To Tycho Brahe (1546–1601), making his first real observations in 1563 at the age of 16, these discrepancies meant that astronomy was in need of more accurate data. The sudden appearance in 1572 of a new star—what would now be called a **nova** or **supernova**—in the constellation Cassiopeia determined the direction of his career.

The appearance of a nova was not entirely without precedent. Hipparchus had witnessed one, but the fact that more were not recorded in the course of history is testimony to the low esteem in which observational astronomy was held. It was Tycho's opinion that the universe could never be understood until the positions of the stars and planets were known exactly. He was determined to compile the most accurate star catalogue possible, as well as data on the motions of the planets.

The new star of 1572 was as bright as the planet Jupiter at first but gradually faded away in 16 months. Tycho initially thought that it was a meteorological event, since Aristotle had taught that the heavens were perfect and unchangeable. But attempts to measure the **parallax** of the new star failed (Fig. 4–9). If it were below the moon, its relative nearness should have been revealed by its apparent shifts in position in relation to the backdrop of the stars. Tycho and others concluded, therefore, that the new star lay in the sphere of the fixed stars. Here was the first addition to the universe since the records of Hipparchus in the second century BC. It had to be admitted, then, that the heavens did change and therefore were not perfect.

Tycho wrote a book about this event, *On The New Star* (1573), which attracted the attention of King Frederick II of Denmark. The king persuaded Tycho to establish an observatory and conferred on Tycho the feudal lordship of the island of Hveen on which to pursue his studies. He generously supported Tycho during the next 21 years.

On Hveen, Tycho built a castle-observatory exactly in the center of the small island. He called it Uraniborg, "Gateway to the Skies," in honor of the Greek muse of astronomy, Urania. It was a combination of residence, observatory, and school of astronomy and an architectural as well as a scientific treasure. A short distance away he built a smaller observatory called Stjerneborg, the "Star Castle," sunk below ground

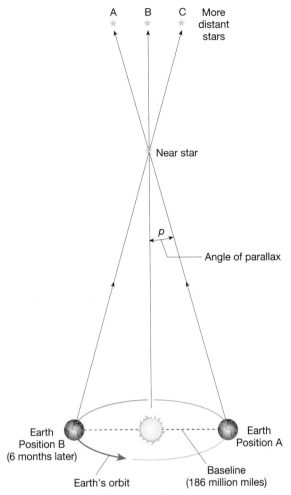

A   B   C   More
distant
stars

Near star

p

Angle of parallax

Earth
Position B
(6 months later)

Earth
Position A

Earth's orbit

Baseline
(186 million miles)

Figure 4–9 Parallax of a star (*p*) is the apparent shift in the position of a star against the backdrop of much more distant stars. As the earth revolves in its orbit, different distant stars are visible behind the near star. The angle of parallax is so small that the first was not measured until 1839 by F. W. Bessel. The parallax of the nearest star is less than one second of arc (0.75″).

level so that the wind would not affect the instruments.

Night after night Tycho peered at the sky, compiling the first massive body of data that astronomy had ever known. His observations were made without the aid of a telescope, which was then unknown. The instruments he used were traditional devices — quadrants, armillary spheres, parallactic rulers — which he improved and made larger than ever before.

He also developed new instruments and techniques and with them attained an accuracy reaching the limits of vision with the human eye. Tycho's celestial catalogue eventually listed positions for more than 700 stars, more than three fourths of those that can be seen with the naked eye from the latitude of Denmark.

One of Tycho's major discoveries concerned the nature of comets, which Aristotle had maintained were meteorological phenomena. Based on his observations of a comet in 1577, Tycho showed that it was not sublunar but more distant than the moon. Like the new star, the comet had to lie in the sphere of the heavens with other true celestial bodies. These regions were thus capable of yet another change. The appearance of other comets confirmed this conclusion; no comet was ever found to be sublunar.

Tycho proved that the **orbits** of the comets penetrated the spheres of the planets. Many astronomers from Aristotle on had assumed that the spheres were "crystalline"–solid and transparent—and that the planets were embedded in them, for otherwise, what kept the planets in their positions? Tycho's conclusion that the heavens could not consist of real celestial spheres did much to sweep away ideas that had been held for centuries but that were not supported by scientific observations.

Tycho himself was not concerned with what kept the planets in their orbits. His use of data in his own theoretical work was negligible, but the data were not wasted. When Frederick's successor, Christian IV, withdrew his support, Tycho left Uraniborg and moved to Prague in Bohemia to serve as imperial mathematician to the Holy Roman Emperor, Rudolph II. There, Johannes Kepler became an assistant to him in one of the most fortunate collaborations in the history of science. Tycho's years of patient observations and Kepler's genius at mathematics combined to complete the Copernican revolution.

4–4 HOW PLANETS MOVE: JOHANNES KEPLER

Johannes Kepler (1571–1630), a mathematician in the Copernican tradition, won recognition with his book, *The Cosmographical Mystery* (1596). He sent copies of it to royalty and to distinguished scientists. Tycho received a copy and immediately recognized

Kepler's mathematical ability. He thought that Kepler would make an ideal assistant at Uraniborg and offered him access to his vast collection of data if Kepler would join him. But Kepler did not accept.

Within three years, however, Kepler's life intersected again with Tycho's. Having lost his patron, Tycho left Denmark in 1599 and settled in Prague in the employ of the emperor, Rudolph II. In 1600, Kepler was forced out of his position as district mathematician and teacher of mathematics and went to Prague to see whether Tycho's offer was still open. The two reached agreement, and Kepler soon had at his disposal the data that Tycho had painstakingly gathered over many years. Their period of collaboration was brief, however; Tycho died in the following year, 1601. Kepler followed Tycho as imperial mathematician and retained access to Tycho's papers.

When Kepler arrived in Prague to work with Tycho, the first task he was assigned was to develop a theory of the orbit of Mars based on Tycho's observations. Kepler brought to this task the belief of an ardent Copernican. He had no doubt that Mars, the earth, and the other planets revolved around the sun. Copernicus, however, had set the sun like a lamp in the middle of the solar system, with no other purpose than to provide heat and light. To Kepler, the sun was much more than just a lamp.

Kepler noted a correlation between a planet's speed and its distance from the sun. He observed that Venus moved more slowly in completing one **revolution** around the sun than Mercury, the closest planet; the earth, at a greater distance than Venus, moved more slowly than Venus; and so on. These facts suggested to Kepler that the sun exerted a force that caused the planets to move and that this force fell off in strength as the distance to the planet increased. Whereas earlier theories had regarded the circular movements of the planets as natural, arising from some inherent property within the body, Kepler was guided by the hypothesis that the circular orbits were the results of forces outside the body.

Initially, Kepler worked out the elements of the orbit of Mars under the influence of the Pythagorean tradition, which required uniform, circular motion for planets. He did develop a circular orbit for Mars that agreed rather closely with Tycho's observations; the discrepancy was only 8 minutes of 1 degree. Kepler knew, however, that Tycho's data were accurate to about 2 minutes, and the discrepancy troubled him.

His respect for the accuracy of Tycho's observations was so great that he abandoned the assumption of a circular orbit and searched for a new hypothesis.

The discrepancy between theory and observation meant to Kepler that the assumption of uniform motion for Mars was in error. The data indicated that Mars moved through its orbit not uniformly but erratically, its speed varying according to no obvious law. The planet moved faster when approaching the sun and more slowly when receding from it. Guided by the inverse relation between speed and distance through hundreds of pages of mathematical calculations, centuries before the introduction of the electronic computer, Kepler arrived at the **law of planetary motion**, known as his **second law**, or the **law of equal areas** (Fig. 4–10):

A line drawn between the sun and a planet (the radius vector) sweeps out equal areas in equal intervals of time.

After finding the law of planetary motion, Kepler continued to work on the shape of the orbit of Mars. He became convinced by the data and by his calculations that the planet could not be traveling in a perfect circle eccentric to the sun. No curve except a perfect ellipse fit the data and the law of equal areas (Fig. 4–11). He was forced to conclude that the orbit was not a circle but an ellipse, with the sun at one focus (Fig. 4–12). He found that the orbits of all the planets, including the earth, are elliptical (Fig. 4–13).

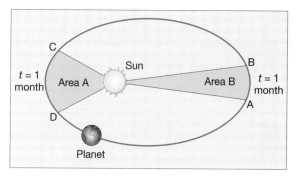

Figure 4–10 *Kepler's second law (law of equal areas):* A line drawn between the sun and a planet sweeps out equal areas in equal intervals of time. Area A = area B. If the planet travels from point A to point B in the same time as from point C to point D, then it travels fastest in its orbit when nearest the sun.

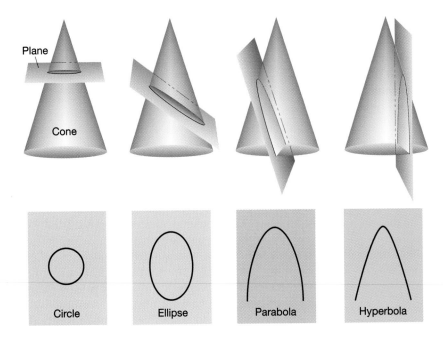

Figure 4–11 An *ellipse* is a conic section—a curve produced by cutting a cone with a plane. If the cut is parallel to the base of the cone, the section is a *circle*. If the cut is made parallel to one side of the cone, the section is a *parabola*. An ellipse is obtained by an intermediate cut. A hyperbola is made by a cut perpendicular to the base.

This discovery is known as **Kepler's first law**, or the **law of ellipses**:

> **Each planet travels in an elliptical orbit with the sun at one focus.**

Kepler announced his first and second laws in one of the great masterpieces of natural science, *The New Astronomy* (1609). Galileo praised it, but few astronomers took his ideas seriously. **Kepler's third law** of

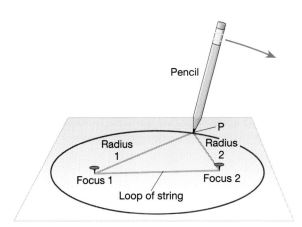

Figure 4–12 Constructing an ellipse. Stick two thumbtacks (the foci) in a piece of paper; place a loop of string loosely around them. Move a pencil within the loop of string, keeping the string taut. The sum of the distances of any point on the ellipse from the two foci (radius 1 and radius 2) is a constant.

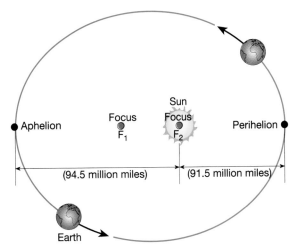

Figure 4–13 *Kepler's first law*: The planets move in elliptical paths, with the sun at one focus. At *perihelion*, the planet is closest to the sun; at *aphelion*, it is most distant from the sun.

planetary motion appeared 10 years later, in *Harmony of the World* (1619):

The square of the period of revolution of a planet (in years) is proportional to the cube of the average distance of the planet from the sun (in astronomical units, AU, the average sun-earth distance).

The third law is written symbolically as follows:

$$T^2 = kR^3 \qquad (3-1)$$

where T is the length of the planet's year—the time required for it to complete one orbit around the sun; R is the planet's average distance from the sun; and k is a constant that has the same value for every planet, 1.00 when these units are used. Table 4–1 indicates how Kepler's third law applies to the planets.

EXAMPLE 4–1

Check Kepler's third law for the planet Mars.

SOLUTION
The period of revolution of Mars, T, in earth-years is 1.88 (see Table 4–1), and its distance from the sun is 1.52 AU. Using Equation 4–1, and setting $k = 1$,

$$T^2 = R^3$$

$T = 1.88$ earth-years $\qquad (1.88)^2 = (1.52)^3$

$R = 1.52$ astronomical units, AU $\qquad \begin{aligned}(1.88)(1.88) = \\ (1.52)(1.52)(1.52)\end{aligned}$

$$3.54 = 3.54$$

Hence, Kepler's third law checks for Mars.

EXTENSION
Check Kepler's third law for Saturn.

■

Although Kepler had no more physical reasoning to support his three laws then Ptolemy had to support his theory, the advantage of Kepler's laws was that they gave a better description of planetary motions. With the use of Kepler's laws, the positions of the planets can be predicted more accurately than with Ptolemy's theory. Tycho, in fact, had retained Kepler primarily to complete the planetary tables and on his deathbed had obtained Kepler's promise that he would do this. Kepler kept his promise. His tables appeared in 1627 as the *Rudolphine Tables,* and they surpassed in accuracy the *Alphonsine Tables* and the *Prutenic Tables.* Kepler's three laws proved indispensable to Isaac Newton when he formulated his system of the world. After 1685, through Newton's work, Kepler's laws became well known and generally accepted.

4–5 GALILEO'S DISCOVERIES WITH THE TELESCOPE

The announcement of Kepler's first and second laws removed the mystery from planetary motion but exerted only a slight impact on the world. The Copernican theory and the discoveries of Kepler had not yet become part of a coherent view of the world comparable to Aristotle's system on which Ptolemy's astronomy was based. To people who thought at all

---

Table 4–1 SOME PLANETARY DATA AND KEPLER'S THIRD LAW

Planet	Distance from Sun (10^6 mi)	Distance from Sun, R (AU)	Revolution Period, T (Earth-years)	T^2	R^3	T^2/R^3
Mercury	36	0.39	0.24	0.058	0.058	1.00
Venus	67	0.72	0.62	0.38	0.38	1.00
Earth	93	1.00	1.00	1.00	1.00	1.00
Mars	142	1.52	1.88	3.54	3.54	1.00
Jupiter	483	5.20	11.86	140.00	140.00	1.00
Saturn	886	9.54	29.46	868.00	868.00	1.00

Figure 4–14   Two of Galileo's telescopes, on display in the Museum of the History of Science, Florence, Italy. (Yerkes Observatory)

about such matters, the quarrel between the supporters of Ptolemy and those of Copernicus concerned purely technical matters.

In 1609, word reached Galileo of the invention in Holland of an optical instrument that made distant objects look closer. Lacking further details, Galileo experimented with combinations of lenses and invented a **telescope** of the type still known as Galilean (Fig. 4–14). With improvements in the design, he constructed an instrument that made objects appear "a thousand times larger and thirty times closer." In the autumn and winter of 1609–1610, Galileo pointed his telescope toward the heavens. What he saw when he viewed the moon, the planets,

the sun, and the stars he discussed in a book, *Sidereal Messenger* (1610), which became an immediate success. A gifted writer, as well as a brilliant mathematician and scientist, Galileo introduced astronomy to a wider public than ever before.

Wherever Galileo looked, he found evidence that either supported the Copernican system or weakened the Ptolemaic system. Thanks to the telescope, he found that the moon resembled the earth more closely than it did a perfect celestial body. Its surface was not smooth and polished, but rough and uneven, full of cavities, mountains, and deep valleys (Fig. 4–15). The moon's close resemblance to the earth refuted Aristotle's division of the universe into celestial

Figure 4–15   Apollo 17 Astronaut Eugene Cernan riding on the Lunar Rover during a mission to the moon. The mountain in the background is the east end of the South Massif. (NASA)

and terrestial regions. Since the moon had a terrestial nature and revolved about the earth, why could not the earth revolve about the sun? Galileo attributed the moon's "secondary light" to the reflection of sunlight from the earth to the moon and back. From this he reasoned that if the earth shined by light reflected from the sun, perhaps the other planets did also. (Prior to this time it had not been known whether the

planets shined by their own light or by light reflected from the sun.)

Galileo observed a difference in the appearance of the planets and stars when viewed through the telescope. The planets appeared round and globular, like little moons with definite boundaries. The stars did not appear very different than they did when seen by the naked eye; they were not much enlarged, only

(a)

(b)

Figure 4–16   The planet Venus exhibits phases similar to those of the moon. (a) Five phases of Venus are shown. Venus is a crescent only when it is in a part of its orbit that is relatively close to the earth, so it appears relatively large. (b) The phases of Venus at eight-day intervals in 1987. The size of Venus changes with its phases. (a, Akira Fujii; b, Jay M. Pasachoff, *Astronomy: From Earth to the Universe*, 4th edition. Philadelphia: Saunders College Publishing, 1991, Fig. 13–15B, p. 38, drawn from a plot made with *Visible Universe.*)

brighter. Even in Galileo's day, the term "stars" applied both to the fixed stars and to the planets; now the telescope revealed differences between them. But he could see countless other stars, "so numerous as to be almost beyond belief." Galileo discovered that the Milky Way, about which there had been many disputes over the ages, consisted of a mass of innumerable stars that had never before been observed.

To Galileo, the most gratifying and important discovery was the four satellites of Jupiter, "never seen from the very beginning of the world up to our own time." This discovery was offered as strong support of the Copernican system. Some objectors to the Copernican system would not accept the idea that the moon revolved around the earth while the earth revolved around the sun. If this were so, to their thinking the earth would lose the moon. But here were four bodies moving around a planet while the planet moved around the sun. It was a model of the Copernican system. If Jupiter could move in an orbit and not lose its four moons, why would it be impossible for the earth to move and not lose its single moon? The fact that there were now two planets with satellites meant that the earth lost yet another claim to being unique; Galileo had proved that there was more than one center of **rotation** in the universe.

Within the year, Galileo discovered yet another similarity between the planets and the earth. Through the telescope, he observed that Venus exhibited phases like those of the moon (Fig. 4–16). The Copernican theory had predicted this, but no phases had ever been observed with the naked eye. Now there was proof that Venus shined by reflected light and was similar in this respect to the earth and the moon. Once again, a distinction between a celestial body and the earth was shown to be imaginary rather than real. As a result of Galileo's discoveries, more and more adherents were won over to Copernican ideas. By the end of the seventeenth century, the stage was set for a new physics, one suitable for a moving earth. That is the subject we examine in the next chapter.

KEY TERMS

Copernican System   The sun-centered, or heliocentric, model of the solar system developed by Nicolaus Copernicus in his book, *On the Revolutions of the Celestial Spheres*, published in 1543.

Deferent   In the Ptolemaic system, the larger circle, centered on the earth, on which the centers of the epicycles revolve.

Ellipse   A conic section on which the sum of the distances of any point from two fixed points, the foci, is constant.

Epicycle   In the Ptolemaic system, a circular orbit of a body, the center of which revolves about a larger circle, the deferent. The planets move on the epicycles.

Geocentric Hypothesis   The idea that the earth is at the center of the universe.

Heliocentric Hypothesis   The idea that the sun is at the center of the universe.

Kepler's Laws   The three laws of planetary motion discovered by Johannes Kepler.

Nova   A star that suddenly increases in brightness.

Orbit   The path in space of a celestial object.

Parallax   The apparent displacement of an object against the background when viewed from two different positions.

Ptolemaic System   An earth-centered model of the universe developed by Claudius Ptolemy in the *Almagest,* the standard treatise on astronomy for 1400 years.

Retrograde Motion   The apparent backward motion of a planet caused by the relative motion of the planet and the earth in their orbits.

Revolution   The orbiting of one celestial body around another.

Rotation   The turning of a body around its own axis.

Supernova   The explosion of a star with the release of tremendous amounts of radiation.

Telescope   An instrument for detecting and observing distant objects.

Theory of Homocentric Spheres A mathematical model of the universe proposed by Eudoxus, a pupil of Plato, in which the sun, moon, and planets were considered to be points on interconnected, concentric spheres with the earth at the center.

THINGS TO DO

1. Observe the moon for several nights. Note its phases and its position in the sky.

2. With the aid of binoculars or a telescope, observe the surface of the moon.

3. Build a model of the solar system according to Copernicus.

4. Draw an ellipse as in Figure 4–12. By moving the tacks closer together or farther away, different-shaped ellipses will be formed.

5. Visit a planetarium or a museum of science and attend a star show, if possible.

6. Observe a twinkling star through binoculars or a telescope.

EXERCISES

The Ptolemaic System

4–1. What were Ptolemy's main objections to a moving earth?

4–2. Explain the epicycle and deferent used by Ptolemy.

4–3. Discuss some connections between the "old physics" and the "old astronomy."

4–4. Account for the fact that the Ptolemaic system reigned unchallenged for 1400 years.

The Copernican System

4–5. What effect did Copernicus' heliocentric (sun-centered) theory of the solar system have on humanity's perception of itself and its place in the universe?

4–6. Did the "publish or perish" dictum apply to Copernicus? Explain.

4–7. What was Copernicus' major contribution?

Tycho Brahe

4–8. Explain the importance of Tycho's conclusions about the comet of 1577 with respect to the Ptolemaic system.

4–9. Compare Tycho's observatory, Uraniborg, with a modern research institute with which you are familiar. (Name the institute.)

4–10. Why is the method of parallax impractical for all but the closest stars?

4–11. How are "big science" and "little science" reflected in the careers of Tycho and Galileo? (Refer to Chapter 1.)

Johannes Kepler

4–12. Give an example of the interplay of theory and observation from the works of Tycho and Kepler.

4–13. Why would Kepler's discovery of the ellipse as the preferred curve for planetary motion be objectionable to supporters of Ptolemy?

4–14. A comet moves much faster when it is near the sun than when it is far away from it. Use Kepler's second law to explain this fact.

4–15. The average distance of Jupiter from the sun is 5.20 AU. Determine the length of Jupiter's year, using Kepler's third law.

4–16. If the earth is farther from the sun in summer than in winter, is it moving around the sun faster in summer or in winter? Explain.

Galileo Galilei

4–17. Which discoveries of Tycho and Galileo posed serious difficulties for the Platonic view of heavenly perfection?

4–18. In what respects do Tycho's observatory and Galileo's application of the telescope to astronomy illustrate an aspect of scientific progress?

4–19. What support did Galileo's discovery of four of Jupiter's moons lend to the Copernican theory?

4–20. List several discoveries that Galileo made with the telescope that met the objections to a moving earth.

4–21. *Multiple Choice*
   A. Ptolemy is known for his
      (a) invention of the telescope.
      (b) study of free-fall.
      (c) earth-centered model of the universe.
      (d) sun-centered model of the universe.
   B. Copernicus restated the
      (a) geocentric theory.
      (b) heliocentric theory.
      (c) theory of horizontal motion.
      (d) theory of relativity.

C. Kepler discovered that planets travel in orbits that are
  (a) circular.
  (b) parabolic.
  (c) elliptical.
  (d) hyperbolic.
D. Tycho Brahe
  (a) built an observatory.
  (b) discovered Mars.

  (c) invented the telescope.
  (d) did all of these.
E. Galileo discovered
  (a) the moons of Jupiter.
  (b) the phases of Venus.
  (c) the rough nature of the moon's surface.
  (d) all of these.

SUGGESTIONS FOR FURTHER READING

Evans, James, "The Division of the Martian Eccentricity from Hipparchos to Kepler: A History of the Approximations to Kepler Motion." *American Journal of Physics,* November, 1988.
Discusses six historically important planetary models. The discussion centers on attempts to provide a model for the motion of Mars.

Gee, Brian, "400 Years: Johannes Kepler." *The Physics Teacher,* December, 1971.
Emphasizes Kepler's collaboration with Tycho Brahe and the development of Kepler's laws.

Gingerich, Owen, "Copernicus and Tycho." *Scientific American,* December, 1973.
Relates the author's discovery of Tycho's personal copy of Copernicus' book, *De revolutionibus,* and its significance for modern science.

Gingerich, Owen, "The Galileo Affair." *Scientific American,* August, 1982.
In arguing that the earth circles the sun, Galileo created new rules of science that have been accepted ever since.

Gingerich, Owen, "Islamic Astronomy." *Scientific American,* April, 1986.
While Europe languished in the Dark Ages, Islamic scholars from the eighth through the fourteenth centuries preserved and transformed Greek science, and from them it passed to Renaissance Europe.

Postl, Anton, "Kepler's Anniversary." *American Journal of Physics,* May, 1972.
Reviews Kepler's life and work and his appreciation for the beauty and harmony of the universe.

Ravetz, Jerome R., "The Origins of the Copernican Revolution." *Scientific American,* October, 1966.
Suggests that Copernicus believed that as an astronomer he had to do more than merely "save the phenomena"; he felt that he should also exhibit the rationality and harmony of God's creation.

Rawlins, Dennis, "Ancient Heliocentrists." *American Journal of Physics,* March, 1987.
Presents evidence suggesting a heliocentrist (sun-centered) origin for geocentrist (earth-centered) Ptolemy's planetary orbit elements and the equant.

Wilson, Curtis, "How Did Kepler Discover His First Two Laws?" *Scientific American,* March, 1972.
Did Kepler begin with the hunch that the ellipse is the path described by planetary motions and then perceive that the calculated distances fitted into an ellipse?

This ellipsoid rollercoaster demonstrates the effects of momentum, inertia, gravity, and centripetal force. (Robert Mathena/Fundamental Photographs)

5

LAWS OF MOTION AND GRAVITATION

Galileo's studies of falling bodies, motion on an inclined plane, and projectile motion did much to clarify the nature of motion. But later scientists were puzzled by the fact that although Galileo apparently understood clearly one of the laws of motion, the law of inertia (that a body in motion or at rest tends to continue in that state), no general statement of this law is to be found in his writings. Moreover, although Galileo was the first to recognize the significance of acceleration, another law of motion that involves acceleration—the force law—also eluded him.

Kepler unraveled the mysteries surrounding the motions of the planets with the discovery of his three laws. Yet each seemed to be distinct from the other two. Although the ellipse replaced the circle

as the curve favored for planetary orbits, the dualism between earthly motions and heavenly motions continued.

With the discoveries of Isaac Newton (1642–1727), we cross the threshold into the scientific age. As he himself freely acknowledged, however, without the work of Galileo, Kepler, and many others who had preceded him, he could not have achieved his synthesis of mechanics and astronomy.

5-1 ISAAC NEWTON'S "MARVELOUS YEAR"

On Christmas day of 1642, the year in which Galileo died, Isaac Newton was born at Woolsthorpe, in Lincolnshire, England (Fig. 5–1). When he reached the age of 18, he entered Trinity College, Cambridge,

Figure 5–1 Birthplace of Sir Isaac Newton at Woolsthorpe, Lincolnshire, England. Newton was born here on Christmas day, 1642. The apple tree in the foreground grew from the stump of the one standing in Newton's time. (Photograph by George O. Abell)

where, except in mathematics, his record was not outstanding. In that field, however, he came under the tutorship of Isaac Barrow (1630–1677), who recognized his extraordinary talents. It was at Cambridge that Newton learned of the work of Copernicus, Kepler, and Galileo.

Newton took the Bachelor of Arts degree in January, 1665, just before the Great Plague, a highly infectious, epidemic disease, forced the temporary closing of Cambridge. He returned to his home in Woolsthorpe and during the next 18 "golden months," between the ages of 23 and 25, made his greatest discoveries in science. He made important contributions to algebra, invented the differential calculus, developed and applied the law of gravitation, carried out experiments with a prism that showed that white light contains all the colors of the spectrum, and invented a reflecting telescope. It may have been the most fruitful 18-month period in the history of creative thought. The rest of his scientific career consisted of an elaboration of the ideas he conceived at Woolsthorpe in that *annus mirabilis*, that "marvelous year."

5–2 THE *PRINCIPIA*

Nearly 20 years after the Woolsthorpe interlude, Newton found himself engaged in another 18-month period of intense concentration and productivity. This developed out of a visit by the astronomer Edmond Halley (1656–1742), of Halley's Comet fame, in August, 1684. Halley had a question about the attraction between the sun and the planets. Halley and Robert Hooke (1635–1703), an experimental philosopher and mathematician, had concluded from Kepler's study of planetary motions that the force of attraction between a planet and the sun must vary inversely with the square of the distance. But they had been unable to prove their hypothesis for elliptical orbits; it worked only for circular ones.

Halley asked Newton what curve would be described by the planets, assuming that gravity diminished as the square of the distance. Newton answered without hesitation, "an ellipse." Halley asked how he knew that. Newton replied that he had calculated it some five years previously. Halley wanted to see the calculations at once, but Newton could not find his notes and promised that he would send the solution to Halley. Unable to find his earlier solution, Newton

reworked the theorems and proofs and sent them to Halley the following November. Urged by Halley to publish his results, Newton devoted the next 18 months to preparing a manuscript for the Royal Society. The publication of the *Principia* in 1687 brought Newton almost immediate fame and made him the leader of science in England.

The *Principia* has been called the greatest scientific work every published. It is divided into an introduction and three "books." In the introduction and the first book, Newton lays down his three laws of motion and considers the motion of bodies under various laws of force. In the second, he treats the motion of bodies in various fluids. In the third book, Newton discusses universal gravitation. Of himself and his achievements, Newton remarked:

I do not know what I may appear to the world, but to myself I seem to have been only like a boy playing on the seashore, and diverting myself in now and then finding a smoother pebble or a prettier shell than ordinary, whilst the great ocean of truth lay all undiscovered before me.

5–3 NEWTON'S FIRST LAW OF MOTION: INERTIA

Newton's first law is an explicit statement of the **law of inertia**:

> **A body at rest or in uniform motion in a given direction remains at rest or in uniform motion in that direction unless acted upon by a net external force.**

Inertia is a property that becomes evident when an external force attempts to change the state of the body (Fig. 5–2). Thus, if a net force acts on the body, the body moves with changing speed, or in a changing direction, or both, but the body's inertia resists such changes. For example, when a moving automobile is suddenly braked to a stop, any loose object will continue in motion. The inertia of motion of the passengers can be overcome by forces exerted by seat belts and shoulder straps. Otherwise, regrettably, the passengers' inertia of motion may be overcome by the forces exerted by the dashboard or windshield with which they may come into contact.

The concept of inertia was borrowed from Galileo, who was aware of it but never stated it explicitly and generally. Newton applied it to both earthly objects,

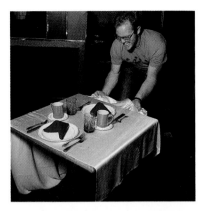

Figure 5–2   Because of the inertia of the objects on this table, the table setting remains intact as the tablecloth is quickly pulled from under it. (David Rogers)

such as projectiles and tops, and to celestial ones—planets and comets—thus giving it universal significance.

Newton stated that motion is no less natural a state for a body than rest, provided that the body when placed in motion encounters no resistance (Fig. 5–3). Gone is Aristotle's physics of "natural" places for earth, air, fire, and water and perfect circular motion for the "quintessence" making up celestial bodies.

Gone, also, is the idea that the steady application of a force is required in "violent" motion, as in shooting an arrow or throwing a rock.

5–4 NEWTON'S SECOND LAW OF MOTION: FORCE

Newton's second law, the **law of force**, is stated as follows:

> **The acceleration of a body is directly proportional to the net force acting on the body and inversely proportional to the mass of the body, and is in the direction of the net force.**

If appropriate units are selected for force, mass, and acceleration, the proportionality can be expressed as an exact equation:

$$\text{Acceleration} = \frac{\text{Force}}{\text{Mass}} \tag{5–1}$$

$$\mathbf{a} = \frac{\mathbf{F}}{m}$$

This vector equation specifies that $\mathbf{a}$ is in the direction of $\mathbf{F}$ (Fig. 5–4). Rewriting this expression, we derive one of the most famous equations in science:

$$\mathbf{F} = m\mathbf{a} \tag{5–2}$$

The role of force in motion had long been puzzling. Aristotle's belief that a constant force was necessary to maintain a body in uniform "violent" motion seemed to be borne out by experience. If a spoon is placed on a table and given a push, it starts to move and its velocity is changed. An automobile traveling at a certain speed on a straight, level road gradually comes to a stop when its engine is disconnected by disengaging the clutch. It does seem that in the absence of a force, all bodies come to rest.

Recall, however, Galileo's investigations of motion on an inclined plane. He found that the smoother the

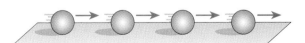

Figure 5–3   An illustration of *Newton's first law*. With no force acting on it in the direction of its motion, a ball on a horizontal plane moves along a straight line with constant speed.

Figure 5–4   *Newton's second law:* A force, $\mathbf{F}$, acting on a block of mass, m, produces an acceleration, $\mathbf{a}$, in the direction of the force, where $\mathbf{F} = m\mathbf{a}$.

Figure 5–5 A "frictionless" system: a dry-ice disk. (Source: *PSSC Physics*. Lexington, Massachusetts: D. C. Heath & Company, 1965.)

Figure 5–7 Forces are applied to the disk through springs mounted on the top and attached to strings. (Source: *PSSC Physics*. Lexington, Massachusetts: D. C. Heath & Company, 1965.)

surface, the greater the distance traveled by a polished metal sphere before slowing down and finally stopping. **Friction**, that is, the resistance to relative motion between two bodies in contact, such as the plane and the sphere, exerts a backward force on the sphere, retarding its motion and finally bringing it to a halt. If a perfectly smooth surface and sphere were available (a *frictionless* system), we would need only to set the sphere in motion and it would continue forever without the application of additional force. Motion is as natural a state as rest.

Figure 5–6 Carbon dioxide escapes through the hole in the bottom of the disk, forming a film of gas on which the disk glides. (Source: *PSSC Physics*. Lexington, Massachusetts: D. C. Heath & Company, 1965.)

It is possible to create approximate frictionless systems with which to study the effects of forces acting on bodies. In one such arrangement, a heavy brass disk, highly polished on the bottom surface, supports a light metal tank carrying a supply of dry ice (solid carbon dioxide) (Fig. 5–5). The dry ice changes into a gas that escapes in all directions through a small hole in the bottom of the disk (Fig. 5–6). The disk and tank apparatus float on a layer of escaping gas, which greatly reduces the friction between the disk and a glass surface. The film of gas supports the system and balances its weight. The only net force acting on the system is the one that we apply.

We can do this through a spring mounted on the top of the disk (Fig. 5–7). When the spring exerts no force on the moving disk, the disk moves with an approximately constant velocity, covering equal distances in equal times. If a constant force is applied, with the spring extended by a constant amount, the system experiences an acceleration (Fig. 5–8). If the same force is applied to disks of different sizes, each disk will have a different acceleration inversely proportional to the mass. Doubling of the force on a given disk gives rise to a doubling of the acceleration (Fig. 5–9).

It is a tribute to Galileo, Newton, and other early physical scientists to whom frictionless systems were not available that they were so successful. They demonstrated that solving idealized problems, such as those involving a nonexistent, ideal, frictionless surface, can shed light on real problems in the physical world.

Figure 5–8   A net force of one unit acting on a disk accelerates it to the right, as shown in this stroboscopic photograph. (Source: *PSSC Physics*. Lexington, Massachusetts: D. C. Heath & Company, 1965.)

Figure 5–9   A net force of two units accelerates the disk to the right. Calculations show that the acceleration is doubled. (Source: *PSSC Physics*. Lexington, Massachusetts: D. C. Heath & Company, 1965.)

5–5 APPLICATIONS OF NEWTON'S SECOND LAW

The quantity m in $\mathbf{F} = m\mathbf{a}$ is called the **inertial mass**, or simply **mass**. It is a quantitative measure of the object's inertia, that is, its resistance to changes in its state of motion. A more massive object such as a loaded oil delivery truck needs more force to move it, stop it, or change its direction than a sports car (Fig. 5–10). Mass is a scalar quantity and has a constant value for an object.

According to Newton's second law, **force** is anything that can change the state of motion of a body, that is, accelerate it (Fig. 5–11). In its simplest sense, a force is a push or a pull. It may start a body moving, speed it up or slow it down, stop it, or change its

direction. In each case, a force produces a change in the state of motion—an acceleration—and the body is accelerated in the same direction as the force is applied. When we throw a ball, we accelerate it, and the acceleration is in the direction of the applied force. Force and acceleration are both vector quantities and are in the same direction.

In SI units, a force of one **newton** (N) is defined as that force that accelerates a mass of 1 kilogram by 1 meter/s² (Fig. 5–12).

$$1 \text{ newton} = 1 \text{ kilogram} \times 1 \frac{\text{meter}}{\text{s}^2}$$

In cgs units, a **dyne** is that force that accelerates a mass of 1 gram by 1 centimeter/s².

Figure 5–10   A truck has greater mass than a bicycle and requires more force to move it. (Tim Davis/Photo Researchers, Inc.)

Figure 5–11   Can you identify some of the forces that influence the motion of this sailboat? (Focus on Sports)

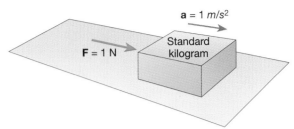

Figure 5–12   Definition of the SI unit of force. A newton is the force that gives a mass of one kilogram an acceleration of one meter per second squared.

$$1 \text{ dyne} = 1 \text{ gram} \times 1 \frac{\text{centimeter}}{\text{s}^2}$$

In the British customary or engineering system, the unit of force is the **pound**, the force that accelerates a mass of 1 slug by 1 foot/s². The units of mass, acceleration, and force are summarized in Table 5–1.

The units newtons and dynes are related in the following way: Since 1 kilogram = 1000 or 10^3 grams, and 1 meter = 100 or 10^2 centimeters, then

$$1 \text{ newton} = 1 \text{ kilogram} \times 1 \frac{\text{meter}}{\text{s}^2}$$

$$= 10^3 \text{ grams} \times 10^2 \frac{\text{centimeters}}{\text{s}^2}$$

$$= 10^5 \frac{\text{gram-centimeters}}{\text{s}^2}$$

$$= 10^5 \text{ dynes}$$

It can also be shown that 1 newton = 0.225 pounds.

EXAMPLE 5–1

What constant net force will give a 1.00-kg disk an acceleration of 5.00 m/s² westward?

SOLUTION
Analyzing the example, we see that the mass and acceleration of a disk are given and that the force is the unknown. We set up the equation that relates the three quantities ($\mathbf{F} = m\mathbf{a}$), substitute, and solve.

$$m = 1.00 \text{ kg} \qquad \mathbf{F} = m\mathbf{a}$$

$$\mathbf{a} = 5.00 \frac{\text{m}}{\text{s}^2} \qquad = \left(1.00 \text{ kg}\right)\left(5.00 \frac{\text{m}}{\text{s}^2}\right)$$

Table 5–1 SUMMARY OF MASS, ACCELERATION, AND FORCE UNITS

	Mass (m)	Acceleration $(\mathbf{a})$	Force $(\mathbf{F} = m\mathbf{a})$	Name of Force Unit
SI Units	1 kilogram	$1 \dfrac{\text{meter}}{\text{s}^2}$	$1 \dfrac{\text{kilogram-meter}}{\text{s}^2}$	newton (N)
cgs Units	1 gram	$1 \dfrac{\text{centimeter}}{\text{s}^2}$	$1 \dfrac{\text{gram-centimeter}}{\text{s}^2}$	dyne
British Engineering	1 slug	$\dfrac{1\ \text{ft}}{\text{s}^2}$	$1 \dfrac{\text{slug-ft}}{\text{s}^2}$	pound (lb)

$$\mathbf{F} = ?$$

$$= 5.00\ \frac{\text{kg-m}}{\text{s}^2}$$

$$= \boxed{5.00\ \text{newtons,}}$$
$$\text{or 5.00 N westward}$$

EXTENSION

A net force of 90.0 newtons eastward acts on a 60.0-kg ice skater. What is the acceleration? (Answer: 1.50 m/s² eastward)

∎

The concept of weight is closely linked to Newton's second law. Not only is a body with a large mass resistant to changes in its motion, but it is also hard to lift. This property is the gravitational force acting on a body, also called the **weight** of the body:

$$\mathbf{w} = m\mathbf{g} \qquad (5\text{–}3)$$

in which $\mathbf{w}$ is the weight or gravitational force, m is the mass, and $\mathbf{g}$ is the acceleration due to gravity. The weight of a body, therefore, is directly proportional to its mass, and the factor of proportionality is numerically equal to $\mathbf{g}$, which is nearly constant near the surface of the earth. The greater the mass of a body, the greater its weight and the harder it is to lift.

EXAMPLE 5–2

Calculate the weight of 1.00-kg mass when it is on the surface of the earth.

SOLUTION

The mass is given and $\mathbf{g}$, the acceleration due to gravity, may be considered a constant near the earth's surface. Knowing how weight is related to these

quantities ($\mathbf{w} = m\mathbf{g}$), we set up the equation, substitute the appropriate quantities, and solve for $\mathbf{w}$.

$$m = 1.00\ \text{kg} \qquad \mathbf{w} = m\mathbf{g}$$

$$\mathbf{g} = 9.80\ \frac{\text{m}}{\text{s}^2} \qquad = \left(1.00\ \text{kg}\right)\left(9.80\ \frac{\text{m}}{\text{s}^2}\right)$$

$$\mathbf{w} = ? \qquad = 9.80\ \frac{\text{kg-m}}{\text{s}^2}$$

$$= \boxed{9.80\ \text{newtons, or 9.80 N}}$$

EXTENSION

A jar of marmalade weighs 4.90 newtons. What is its mass? (Answer: 0.500 kg)

∎

Although mass is a constant, weight is a local property that depends on the value of $\mathbf{g}$ for that locality. The value of $\mathbf{g}$ varies with location on the surface of the earth, increasing with latitude from 9.78 m/s² at the equator to 9.83 m/s² at the North and South poles. It also varies with elevation, decreasing with elevation at a given latitude, as Table 5–2 shows. One's weight, therefore, would increase in traveling from the equator to the North Pole and decrease in going from the surface of the earth at any location to some point above the surface. For example, you would weigh slightly more in Denver, Colorado, than at the top of Pike's Peak because, although these locations are at about the same latitude, Denver is 1600 meters above sea level as compared with 4290 meters for Pike's Peak.

A brick would have just as much mass as it has on the surface of the earth, and it would hurt your toes just as much to kick it in outer space as it would on

Table 5–2 ACCELERATION OF GRAVITY, **g**, AT VARIOUS ALTITUDES ABOVE THE SURFACE OF THE EARTH

Altitude (km)	g, (m/s²)
0	9.80
1000	7.33
2000	5.68
3000	4.53
4000	3.70
5000	3.08
6000	2.60
7000	2.23
8000	1.93
9000	1.69
10,000	1.49

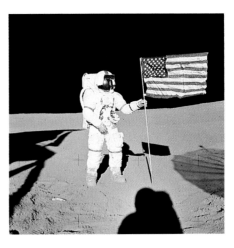

Figure 5–13 The first person to set foot on the moon, Neil Armstrong, took this photograph of his fellow astronaut Edwin E. "Buzz" Aldrin after the *Apollo 11* lunar landing on July 20, 1969. The weight of this astronaut on the moon is less than it is on earth, but his mass remains the same. (Courtesy of NASA)

earth. Yet the brick would not have much weight at all, since the value of **g** in outer space is low (see Table 5–2). If the brick were taken to the moon, its mass would not change, but since the gravitational field on the moon is only one sixth as much as that on earth, it would weigh one sixth as much. When you take your trip to the moon, your mass, too, will be unchanged; but if you weigh 150 pounds on earth, you will weigh only about 25 pounds on the moon (Fig. 5–13).

All the **forces in nature** are of only four basic kinds. In addition to the **gravitational forces** just discussed, there are **electromagnetic forces** and two kinds of **nuclear forces—weak nuclear** and **strong nuclear** (Table 5–3). Recent experiments indicate that there may be two more forces, the super-weak and the super-strong.) Electromagnetism makes it possible for elevators to rise and light bulbs to glow. The weak force causes subatomic particles to shoot out of the nuclei of atoms during the radioactive decay of such unstable elements as uranium. The strong force binds the protons and neutrons in an atomic nucleus. Regardless of its origin, each force plays the same role: It accelerates something that is sensitive to it. Since the number of forces is so small, scientists have wondered whether they can all be unified under one comprehensive theory. The idea of the unity of nature strongly attracts the scientific mind, and the search for a unifying theory is continuing.

The vector properties of forces are further illustrated in the following examples.

EXAMPLE 5–3

A force ($\mathbf{F}_1$) of 10.0 N, acting on a box toward the right, combines with a similar force ($\mathbf{F}_2$) of 10.0 N in the same direction. Find the resultant force.

SOLUTION
The resultant force produces the same acceleration on the box as would a single force of 20.0 N acting toward the right. (See Fig. 5–14).

■

Table 5–3 THE FOUR FORCES

Force	Relative Strength	Particles Affected
Nuclear	1	Nuclear particles
Electric	10^{-2}	Electrically charged particles
Weak	10^{-15}	Nuclear particles
Gravity	10^{-38}	All particles

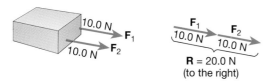

Figure 5–14   Resultant of two forces $\mathbf{F}_1$ and $\mathbf{F}_2$ acting in the same direction.

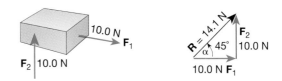

Figure 5–16   Resultant of two forces $\mathbf{F}_1$ and $\mathbf{F}_2$ acting at right angles.

EXAMPLE 5–4

What is the net effect of a 10.0-N force ($\mathbf{F}_1$) pulling to the right, and a 3.00-N force ($\mathbf{F}_2$) pulling to the left?

SOLUTION

The net effect of these two forces is to produce the same acceleration as a single force of 7.00 N to the right. (See Fig. 5–15.)

∎

EXAMPLE 5–5

A 10.0-N force ($\mathbf{F}_1$) acts to the right, and another 10.0-N force ($\mathbf{F}_2$) acts vertically upward on a body. Determine the resultant force. (See Fig. 5–16.)

SOLUTION

We apply the polygon method of vector addition discussed in Chapter 3. From the diagram we determine the magnitude of the resultant $\mathbf{R}$ and measure the angle α. We mean by this that a single force, $\mathbf{R}$, acting on the body at an angle of 45° east of north has the same effect as the two forces, $\mathbf{F}_1$ and $\mathbf{F}_2$, acting in their respective directions. (See Fig. 5–16.)

EXTENSION

A force of 500 N is applied to a car at an angle of 45° to the horizontal. How much force goes into moving the car along the road? (Answer: 350 N)

∎

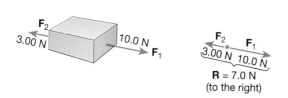

Figure 5–15   Resultant to two forces $\mathbf{F}_1$ and $\mathbf{F}_2$ acting in opposite directions.

5–6 NEWTON'S THIRD LAW OF MOTION: ACTION-REACTION

Newton's third law is stated as follows:

> **Whenever one body exerts a force on another, the second body exerts a force equal in magnitude and opposite in direction on the first body; or, to every action force there is an equal and opposite reaction force.**

Newton's third law is equivalent to stating that forces always occur in pairs. For any force exerted *on* a body, there is a force exerted *by* the body; the two forces are an **action-reaction pair** and act on different objects. Newton came to this conclusion from collision experiments between a moving pendulum and a stationary one. The motion of each pendulum after collision, he decided, resulted from equal forces exerted by each pendulum on the other. Experience has shown that the third law applies to every known kind of force—gravitational, electromagnetic, and nuclear. In each instance, one of the forces is arbitrarily designated the **action force**, and the other equal and opposite force, the **reaction force**.

We experience countless examples of action-reaction force pairs each day. Whenever a force is applied, a reaction force is created (Figs. 5–17 and 5–18). Walking is possible because as we push against the floor, the floor pushes against us (Fig. 5–19). This reaction force accounts for our motion, and we depend on friction to obtain it. Walking on ice is difficult because we cannot easily exert the necessary force on the ice, and therefore the ice cannot exert a reaction force to accelerate us. An ice skater pushes on a wall, and the wall exerts an equal and opposite force on the skater, causing the skater to accelerate away from the wall. The tires of a car push against the road surface, and the road reacts by pushing against the tires (Fig. 5-20). Some of the bruises we suffer arise from action-

Figure 5–17 A multiflash photograph of a hammer striking a nail. The hammer head exerts a force **F** on the nail, and the nail exerts a reaction force − **F** on the hammer head. (© Tom Branch, Science Source)

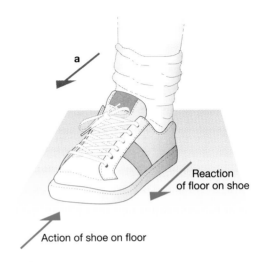

Figure 5–19 An action-reaction pair.

reaction pairs: When we accidentally kick a curbstone, the reaction force exerted by the curbstone on our toes is unpleasant. The same action-reaction principle underlies rocket propulsion: A rocket

pushes gases backward, and the equal and opposite recoil force exerted by the gases on the rocket pushes the rocket forward, even in the vacuum of space (Fig. 5–21). A spaceship maneuvers simply by firing rockets in the proper direction (Fig. 5–22).

5–7 THE "CENTER-SEEKING" FORCE

If you whirl a key at the end of a string, the key tends to fly off in a straight-line path, in accordance with the first law of motion. The fact that the key continues to move in a circle must mean that a force is acting on it that deflects it from its straight-line path (Fig. 5–23). This force, which is supplied by the hand and transmitted by the string and keeps pulling the key into the center, is called **centripetal force**, meaning "center-

Figure 5–18 A person jumping toward shore from a boat. The boat moves opposite the direction the person jumps. (Stan Pantovic/Photo Researchers, Inc.)

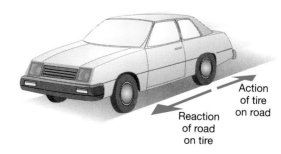

Figure 5–20 As a tire pushes against the road surface, the road reacts by pushing against the tire.

Figure 5–21 The space shuttle *Columbia* lifting off. Thrust is generated by the engines and boosters. Identify the action-reaction forces of the rockets and exhaust gases. (Courtesy of NASA)

Figure 5–23 Centripetal force is needed to produce the inward acceleration that keeps the key moving in a circular path.

seeking." It is possible to show that centripetal force imparts an acceleration directed toward the center of the circle given by the expression

$$\mathbf{a} = \frac{v^2}{r} \qquad (5\text{–}4)$$

in which v is the constant speed, r is the radius of the circle, and $\mathbf{a}$ is the centripetal acceleration. The centripetal force $\mathbf{F}$ may thus be expressed as follows:

$$\mathbf{F} = m\mathbf{a} = \frac{mv^2}{r} \qquad (5\text{–}5)$$

Figure 5–22 An astronaut on a space walk uses a self-maneuvering unit, or hand rocket, that operates on Newton's third law. (Courtesy of NASA)

EXAMPLE 5–6

A racer takes a curve of radius 200 ft at a constant speed of 100 ft/s. What is the centripetal acceleration?

SOLUTION
We may calculate the centripetal acceleration from Equation 5–4:

$$r = 200 \text{ ft} \qquad \mathbf{a} = \frac{v^2}{r}$$

$$v = 100 \, \frac{\text{ft}}{\text{s}} \qquad = \frac{\left(100 \, \dfrac{\text{ft}}{\text{s}}\right)^2}{200 \text{ ft}}$$

$$\mathbf{a} = ? \qquad = \frac{10^4 \, \dfrac{\text{ft}^2}{\text{s}^2}}{2 \cdot 10^2 \text{ ft}}$$

$$= 0.5 \cdot 10^2 \, \frac{\text{ft}}{\text{s}^2}$$

$$= \boxed{50 \, \frac{\text{ft}}{\text{s}^2}}$$

EXTENSION

A car turns a corner of radius 60.0 meters at a speed of 24.0 meters/s. If its mass is 800 kg, what is the centripetal force on the car? (Answer: 7680 N)

◼

At every instant, the centripetal force is exerted at right angles to the direction of motion of the key and imparts an acceleration to it (Fig. 5–24). The centripetal force does not change the speed of the key, which remains constant; it changes only the *direction* of motion. If the key were not subject to the centripetal force, it would follow a straight path at constant speed along a tangent to the circle. In a sense, therefore, the key is always falling toward the center, veering away from the tangent direction that it would otherwise take (Fig. 5–25).

Centripetal force is not a new kind of force, merely a force that keeps a body moving in a curved path. In SI units, it is expressed in newtons, like any other force. It may be exerted by means of cords or springs, electricity, or gravitational attraction (Figs. 5–26 and 5–27). If the string in the preceding example breaks, the centripetal force ceases to act and the key is free to travel in a straight line tangent to the path at that instant, following Newton's first law.

Like other kinds of force, a centripetal force is one member of an action-reaction pair. The reaction force is often called **centrifugal force** (for "center-fleeing"). The two forces, centripetal and centrifugal,

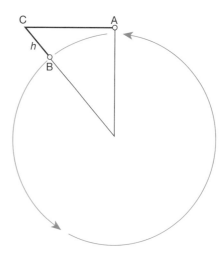

Figure 5–25   The key is always "falling" toward the center. When it moves from A to B, it falls through the distance *h*. If it had moved in a straight line, it would have reached point C.

are equal in magnitude and opposite in direction, in accordance with Newton's third law. They depend on the mass and the speed of the body moving in the circular path and on the radius of curvature.

Only the centripetal force acts on the body. In the example of the string and key, the key exerts a reaction force—centrifugal force—on the string. The two forces, therefore, act on different things: One acts on the key; the other on the string. The centrifugal force, exerted *by* the key, and not *on* the key, has no effect on the key's motion. The steady outward-pulling force felt by the hand—the centrifugal force as transmitted by the string—does not tend to pull the key away from the center of the circle. In the strict sense, there is no such applied force. Because of its inertia, the key is merely exhibiting its tendency to move in a straight line.

When an automobile goes around a curve, centripetal force arising from the friction between the tires and the surface acts on the car and keeps it in the curved path. The reaction force, or centrifugal force, is exerted by the car on the road. If you are a passenger not wearing a seat belt, you may feel pushed against the door—not because centrifugal force pushes you there, but because centripetal force, which would have been transmitted through the car's frame by a seat belt, is not keeping you in circular

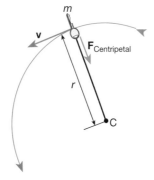

Figure 5–24   An acceleration **a** is imparted to the key, equal to v^2/r, and changes its direction. The centripetal force is exerted at right angles to the motion of the key.

Figure 5–26 People enjoying circular motion about the vertical axis in an amusement park ride. (Henley/Savage, The Stock Market)

Figure 5–27 When rounding a turn, motorcycles depend upon centripetal force to keep moving in a circular path. (Gerry Cranham/Photo Researchers, Inc.)

motion. You therefore tend to go in a straight line while the car tavels in a curved path; consequently, you and the car door meet.

5–8 THE APPLE AND THE MOON

There is a famous story that while sitting in his garden and seeing an apple fall to the ground Newton came to wonder whether the force that the earth exerted on the apple might extend out to the moon. If the story is true, it may have been during the eventful period when Newton waited out the plague at Woolsthorpe and made so many remarkable discoveries. He wrote that "in the same year I began to think of gravity extending to the orb of the moon."

When Newton returned to the problem years later, he found that the force of attraction between two spheres was the same as it would be if the mass of each sphere had been concentrated at the center. He therefore felt justified in considering a sphere as though its whole mass were concentrated at a point at the center. This result enabled him to treat the sun, moon, and planets as point masses and to apply mathematical analysis with great precision to astronomical problems.

The moon in motion tends to follow a straight-line path, according to Newton's first law, the law of inertia. The fact that it follows a curved path meant to Newton that a force must be pulling it out of its straight-line path, just as the key described in Section 5–7 is deflected from its straight-line path.

If the earth, by its gravitational attraction, could cause the moon to "circle" the earth, perhaps it was the same force with which the earth attracted projectiles and caused an apple to fall. A similar force arising from the gravitational attraction of the sun might cause the planets to orbit around it. Newton made use of Galileo's data on falling bodies, Kepler's laws, and other findings, as well as his own experiments, intuition, and analysis, and arrived at the **law of universal gravitation**:

> **Every body in the universe attracts every other body with a force that is directly proportional to the product of their masses and inversely proportional to the square of the distance between their centers of mass.**

Expressing this law symbolically, we obtain the following expression ($\propto$ is a proportion sign):

$$\mathbf{F} \propto \frac{m_1 m_2}{r^2}$$

or

$$\mathbf{F} = \frac{G m_1 m_2}{r^2} \tag{5–6}$$

in which $\mathbf{F}$ = force of attraction between two bodies,

G = gravitational constant,

m_1 = mass of one body,

m_2 = mass of a second body, and

r = distance between the center of mass of the two bodies

Any object near the surface of the earth is approximately 4000 miles from the center (the earth's radius = 4000 miles). Since freely falling bodies near the surface have the same acceleration, 32 ft/s², we may calculate the distance any body falls each second. In the first second of fall, the velocity is 0 at the beginning and 32 ft/s at the end. The average velocity during the first second is therefore expressed as follows:

$$\bar{\mathbf{v}} = \frac{\text{Initial velocity} + \text{Final velocity}}{2}$$

$$= \frac{0 + 32 \dfrac{\text{ft}}{\text{s}}}{2}$$

$$= 16 \frac{\text{ft}}{\text{s}}$$

and the distance traveled in the first second is given by:

$$\text{Distance} = \text{Average velocity} \times \text{Time}$$

$$s = \bar{\mathbf{v}} t$$

$$s = 16 \frac{\text{ft}}{\text{s}} \times 1 \text{ s}$$

$$= 16 \text{ ft}$$

Thus, a freely falling body, such as an apple near the earth's surface, falls through a distance of 16 feet during the first second (Fig. 5–28).

Knowing that the moon's orbit is about 240,000 miles, or 60 earth-radii from the earth, and that the moon revolves around the earth once in 27.3 days, we can compute the moon's velocity in its orbit and its centripetal acceleration, $\mathbf{v}^2/r$. The ratio of the moon's acceleration to the acceleration of an object near the

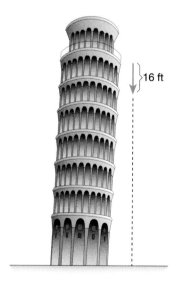

16 ft

Figure 5–28   An apple falls 16 feet during the first second. How far does it fall in two seconds?

earth's surface is found to be about 1 : 3600. The ratio of the moon's distance from the earth to the radius of the earth is 60 : 1, and $60^2 = 3600$. The acceleration, then, falls off with the square of the distance, and the force of gravitation is inversely proportional to the square of the distance.

The moon, like the apple, "falls" toward the earth, but in one second it falls 1/3600 of 16 feet, or about 1/20 of an inch. Yet, like the whirling key at the end of the string, it does not come any closer (Fig. 5–29). It falls away from the straight line it would travel if gravitational force were not acting. Newton felt justified in assigning the same cause—gravity—to the force that causes an apple to fall and that keeps the moon in its orbit.

Newton applied the law of universal gravitation to other heavenly bodies and to other phenomena. The motion of satellites about Jupiter and the motion of the planets around the sun appeared to be similar to the motion of the moon about the earth, and Newton assigned the same cause to them—gravitation. He showed that the orbits of comets were elliptical. He explained ocean tides as resulting from gravitational forces exerted by the sun and moon on the oceans. He showed how small irregularities in the motions of the planets could be explained by forces exerted on them by other planets. The law of universal gravitation and the laws of motion seemed to apply to matter wherever it was found.

5–9 "WEIGHING THE EARTH"

Newton established the law of universal gravitation without being able to measure the small force of attraction between two bodies in the laboratory, but in the *Principia* he outlined a method that was followed successfully a century later. Once the value of G is known, the gravitational attraction between any two spherical bodies, whether bowling balls or planets, can be calculated, given their masses and the distance between their centers. Using this method, it is also possible to "weigh" the earth, the sun, and the other planets.

Henry Cavendish (1731–1810), after whom the Cavendish Laboratory at Cambridge University is named, during the period 1797–1798, carried out the experiment to determine G so meticulously that his value differs by only about 1% from the value accepted today. His experiment is as follows: With a torsion balance, a wooden arm is suspended horizontally by a slender wire of known stiffness (Fig. 5–30).

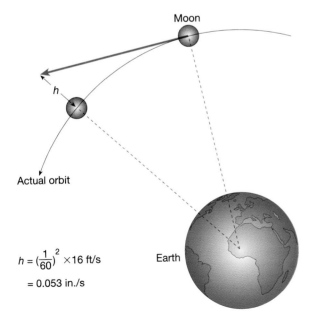

$$h = \left(\tfrac{1}{60}\right)^2 \times 16 \text{ ft/s}$$
$$= 0.053 \text{ in./s}$$

Figure 5–29 The moon is continually "falling" toward the earth. For this reason it remains at the same distance from the earth. If the moon traveled in a straight line, it would be carried further from the earth.

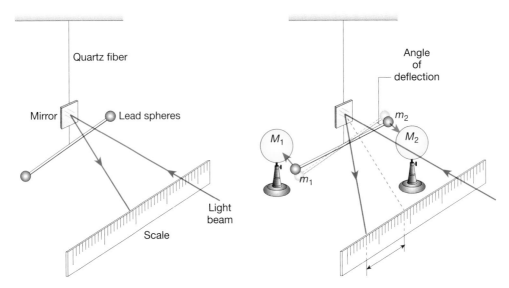

Figure 5–30   A diagram of the torsion balance used by Henry Cavendish. Knowing the masses of the spheres, the distance between their centers, and the force, one may determine the value of G.

A small lead sphere of known mass and radius is mounted on each end of the arm. When large lead spheres, also of known mass and radius, are positioned on either side of the small spheres, the suspension wire is twisted through a measurable angle by the gravitational forces between the spheres. In a separate experiment, the force necessary to twist the wire through a similar angle is determined. Knowing the masses of the spheres, and the force, the value of G—the only unknown—may be determined. The modern value, in SI units, is shown as follows:

$$G = 6.67 \times 10^{-11} \frac{\text{newton-meter}^2}{\text{kilogram}^2}$$

EXAMPLE 5–7

Apply the law of universal gravitation to "weigh the earth." Consider a mass m_1 of 1.00 kilogram on the surface of the earth, where it experiences a force of 9.80 newtons owing to its weight. Assume that its distance from the center of the earth is 6.38×10^6 meters, the mean radius of the earth. Let the mass of the earth be m_2, and determine its value.

SOLUTION
Tabulate the data, then write the force equation that applies. Since the unknown is m_2, rewrite the equa-

tion with m_2 on the left and everything else on the right. Then substitute and solve.

$$\mathbf{F} = 9.80 \text{ N}$$

$$G = 6.67 \times 10^{-11} \frac{\text{N} \cdot \text{m}^2}{\text{kg}^2}$$

$$m_1 = 1.00 \text{ kg}$$

$$r = 6.38 \times 10^6 \text{ m}$$

$$m_2(\text{mass of earth}) = ?$$

$$\mathbf{F} = \frac{Gm_1m_2}{r^2}$$

$$m_2 = \frac{\mathbf{F}r^2}{Gm_1}$$

$$= \frac{9.80 \text{ N} (6.38 \times 10^6 \text{ m})^2}{\left(6.67 \times 10^{-11} \dfrac{\text{N} \cdot \text{m}^2}{\text{kg}^2}\right)(1.00 \text{ kg})}$$

$$= 5.98 \times 10^{24} \text{ kg}$$

EXTENSION
A piece of aluminum of mass 0.0270 kg is 1.00 meter away from a piece of iron that has a mass of 0.0560 kg. What is the gravitational attraction between the two? (Answer: 0.010 N)

■

5–10 NEW NEIGHBORS IN SPACE

Newton's law of universal gravitation, Newton's laws of motion, and Kepler's laws form the basis of the branch of astronomy known as **celestial mechanics**. The motions of the planets can be calculated with such great precision that small discrepancies in their orbits have led astronomers to search for and to discover new planets.

Sir William Herschel (1738–1822), who was to be appointed court astronomer to King George III, discovered a new planet, later named Uranus, in 1781. The details of its elliptical orbit were determined over the years. Small but persistent discrepancies between the predicted orbit and actual observations appeared, however, for which no explanation could be found. Some even believed that the law of universal gravitation did not apply to the far reaches of the solar system.

In the 1840s, it occurred independently to two young astronomers, John Couch Adams (1819–1892) at Cambridge University and Urbain J. J. Leverrier (1811–1877) in Paris, that an unknown planet might be causing the perturbations of Uranus. In the solar system, every body disturbs every other body with "perturbing" forces, which affect the orbit of a planet as predicted from Kepler's laws; the departures from the predicted orbit are known as **perturbations**. Adams and Leverrier spent years at the difficult calculations and predicted the location of a new planet.

In October, 1845, Adams wrote to the Astronomer-Royal at the Greenwich Observatory, advising him that the perturbations of Uranus could be explained by assuming the existence of an outer planet situated then in a specific latitude and longitude and suggesting that a search for the planet be conducted. Nothing came of it, however. Later, Leverrier wrote to J. G. Galle (1812–1910) at the Berlin Observatory, also urging him to search for the planet in a particular region of the sky. That very evening, September 23, 1846, after receiving Leverrier's letter, Galle directed his telescope as Leverrier had suggested and almost immediately found the new planet within a degree of the predicted position. The director of the observatory called it the most brilliant of all planetary discoveries. The discovery of Neptune, as the planet came to be known, was another triumph for Newtonian physics, proving that Newton's laws are valid for the entire solar system.

Within a century, small perturbations were discovered in the orbit of Neptune itself. Percival Lowell (1855–1916) of the Lowell Observatory and, independently, William H. Pickering (1858–1938) at the Mt. Wilson Observatory ascribed them to yet another unknown planet. Lowell tried unsuccessfully for years to find what he called Planet X, and the search was continued after his death. In 1930, Clyde W. Tombaugh (b. 1906) at the Lowell Observatory detected an object that was identified as Planet X and named Pluto (Fig. 5–31).

Are there other planets in the solar system? In 1972, Joseph L. Brady at the University of California predicted the existence of a Planet Y based on a computer analysis of comet perturbations, which obey the same laws as planets. The new Planet Y is believed to be three times the size of Saturn and could account for perturbations in the orbit of Halley's Comet. Others maintain that a planet of the presumed size would have long since been discovered and that there are other reasons for the perturbations. Only time will tell which view is correct.

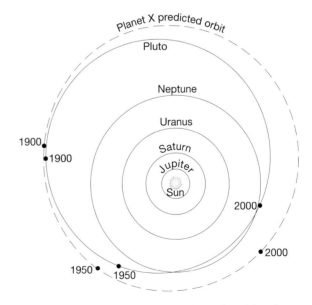

Figure 5–31　The orbit that Lowell predicted for Planet X in 1905 compared with that observed for the new planet, Pluto, discovered 25 years later. The relative positions for 1900 were very close. (After Federer and Tombaugh, *Sky and Telescope.* Sky Publishing Corp.)

5–11 EARTH SATELLITES

When launched at the proper angle and given enough speed, objects may be projected for long distances. A missile may be launched from one spot on earth and land at almost any other spot. Intercontinental ballistic missiles (ICBMs) are designed to do just this. Their trajectory may be measured by tracking with radar or other means.

Theoretically, if an apple were given a suitable horizontal speed from a high enough elevation, it could "fall around the earth," as does the moon. Actually, a projectile in an orbit this close to the surface of the earth would experience so much atmospheric drag that it would slow down and fall back to the ground. In principle, however, Newton was correct, and modern technology makes it possible to launch objects to altitudes so high that atmospheric resistance is negligible. Under these conditions, they go into stable orbit around the earth. In 1957, the earth satellite called *Sputnik* was successfully launched. Since then, scores of other satellites have been placed in orbits around the earth.

While in orbit, even a natural satellite such as our moon continually falls toward the earth. Its speed is great enough, however, that it can continue to move in orbit and not come any closer to the earth. The speed required by a satellite to remain in orbit depends on its height above the earth.

The gravitational attraction of the earth holds satellites in various orbits. For circular orbits, the gravitational force is equal to the centripetal force:

$$\mathbf{F} = \underbrace{\frac{Gm_1m_2}{r^2}}_{\substack{\text{gravitational} \\ \text{force}}} = \underbrace{\frac{m_1\mathbf{v}^2}{r}}_{\substack{\text{centripetal} \\ \text{force}}}$$

where m_1 is the mass of the satellite, m_2 is the mass of the earth, and r is the radius of the orbit measuring from the center of the earth. Solving for the velocity of the satellite in an orbit,

$$\mathbf{v}^2 = \frac{Gm_2}{r}$$

and

$$\mathbf{v} = \sqrt{\frac{Gm_2}{r}} \tag{5–7}$$

Observe that the mass of the satellite cancels out of the equation, meaning that all satellites, regardless of mass, travel with the same velocity in a particular orbit.

The closer a satellite is to the earth, the faster it needs to travel to stay in orbit. As a spaceship moves from one orbit to another, its orbital speed undergoes a change, decreasing as it moves from a lower to a higher orbit. Near the earth's surface, a satellite must travel at more than 17,000 miles per hour to remain in orbit, while one 4000 miles above the surface orbits at 11,500 miles per hour. At the distance of the moon, the orbital speed is 2000 miles per hour. The time required to revolve about the earth, called the **period of a satellite**, is obtained from kinematics: Time = distance/velocity, or $T = 2\pi r/\mathbf{v}$. Satellites traveling in larger orbits have longer periods.

Hundreds of satellites have been circling the earth since the first *Sputnik* and *Explorer* ushered in the space age. Satellites have revolutionized global communications. Most television watchers have seen an athletic event or news reports beamed "via satellite" from the far corners of the world. A three-minute telephone call from New York to London that cost about $12 before the launching of the first commercial communications satellite, *Early Bird,* in 1965, then cost about $3.20. Satellites hovering in **geosynchronous orbit**, in which they have the same period of revolution as the period of the earth's rotation and appear to be stationary over one location on the earth, relay messages over the Atlantic, Pacific, and Indian oceans to more than 100 countries.

Weather pictures taken by **meteorological satellites**, or **metsats**, are displayed on almost every televised news broadcast. The GOES metsat (geostationary operational environmental satellite) beams back to ground stations detailed photographs both day and night of the entire Western hemisphere from a fixed altitude of 22,300 miles. The eye of a hurricane only 50 miles across can easily be seen in the pictures. Other satellites monitor the unfolding panorama of the weather on the earth below.

Landsats, still another type of satellite, can make fast, accurate surveys of forests and crops. Their cameras go over the same point on earth every nine days and radio data and pictures back to earth. Landsats have located deposits of copper and manganese, as well as new sites for tuna and shrimp fishing.

Applied technology satellites, also known as **ATS**, beam educational and medical information. Others are used as navigational aids by ships and aircraft. Orbiting space laboratories and observatories are used to obtain knowledge about the universe that cannot be obtained in any other way.

SUMMARY OF EQUATIONS

$$\text{Acceleration} = \frac{\text{Force}}{\text{Mass}}$$

$$\mathbf{a} = \frac{\mathbf{F}}{m}$$

or

$$\mathbf{F} = m\mathbf{a}$$

Weight ($\mathbf{w}$) = mg

Centripetal acceleration

$$\mathbf{a} = \frac{\mathbf{v}^2}{r}$$

Centripetal force

$$\mathbf{F} = m\mathbf{a} = \frac{m\mathbf{v}^2}{r}$$

Gravitational force

$$\mathbf{F} = G\frac{m_1 m_2}{r^2}$$

KEY TERMS

Action-Reaction Pair   Newton's third law of motion states that to every action force there is an equal and opposite reaction force; the two forces are an action-reaction pair.

Applied Technology Satellites (ATS)   Earth satellites that are used in educational broadcasting, navigation, and as space laboratories and observatories.

Celestial Mechanics   The branch of astronomy that deals with the motions and gravitational influences of the members of the solar system.

Centrifugal Force   A force that an object moving along a circular path exerts on the body constraining the object.

Centripetal Force   A force that is necessary to keep an object in a circular path.

Dyne   The cgs unit of force; the force that accelerates a mass of 1 g by 1 cm/s^2 in the direction of the force.

Electromagnetic Force   One of the four basic forces in nature, operating between electrically charged particles.

Force   A push or a pull that imparts an acceleration to an object.

Friction   The force that resists relative motion between two bodies because of the interaction between them.

Geosynchronous Orbit   An orbit of an artificial satellite that travels above the equator and at the same speed as the earth rotates so that the satellite seems to remain in the same place.

Gravitational Force   One of the four fundamental forces of nature, the force by which two masses attract each other.

Inertia   The resistance of an object to changes in its state of motion.

Landsat   An artificial satellite that can survey forests and crops, and radio data and pictures back to earth.

Mass   A measure of an object's inertia.

Metsat   A meteorological satellite that monitors the weather on the earth below.

Newton (unit)   The mks unit of force; the force that accelerates a mass of 1 kg by 1 m/s^2 in the direction of the force.

Newton's First Law   An object will maintain its state of rest or motion unless acted on by a net external force.

Newton's Law of Universal Gravitation   The gravitational force between two objects is directly proportional to the product of their masses and inversely proportional to the square of the distance between their centers of mass.

Newton's Second Law   The acceleration of a body is directly proportional to the net force acting on it, in the same direction as the force, and inversely proportional to the mass of the body.

Newton's Third Law Whenever one body exerts a force on a second body, the second body exerts an equal and opposite force on the first.

Nuclear Force The strong nuclear force is a very short-range, attractive force that acts between all nuclear particles.

Period of a Satellite For an earth satellite, the time required for the satellite to revolve about the earth; the larger the orbit, the longer the period.

Perturbation A discrepancy between the predicted and actual orbits of a planet or satellite.

Pound The unit of force in the British system. The weight of 0.454 kilogram on earth.

Satellite An object that orbits around an astronomical body.

Weight The force of gravitational attraction on an object.

THINGS TO DO

1. (a) Quickly pull a card out from under a quarter. (b) Repeat with the card placed on a water glass. What does the coin do in each case? Explain.

2. Observe the effect on your body when the car in which you are riding (a) starts; (b) turns a sharp corner; (c) stops.

3. Turn on a rotary lawn sprinkler. Note its action and reaction.

4. Swing a tennis ball at the end of a strong cord in a circular path. Consider the forces produced.

5. Quickly pull a card out from under a tall metal cylinder.

6. Tighten the head on a hammer by pounding on the ground, handle first. Explain.

7. Swing a half-full pail of water rapidly in a vertical circle. Does the water spill when the pail is upside down? Explain.

EXERCISES

Inertia

5–1. Assume that you are standing in a bus. Explain why you tend to fall backward as the bus starts up.

5–2. Does an earth-orbiting space laboratory possess inertia? Explain.

5–3. A passenger sitting in the rear of a bus claims that he was injured when the driver slammed on the brakes, causing a suitcase to come flying toward the passenger from the front of the bus. If you were the judge in this case, what disposition would you make?

5–4. A magician pulls a tablecloth from underneath a setting of plates and glasses. Rather than falling to the floor and breaking, they remain on the table. Explain.

Force and Motion

5–5. Explain the statement that although your mass is the same everywhere, your weight may vary from place to place.

5–6. With a diagram, show the force, **F**, exerted by a child pulling a sled, and indicate the horizontal and vertical components.

5–7. (a) What is your weight? (b) What would your weight be on the moon, where the acceleration due to gravity is one sixth that on earth?

5–8. (a) A force of 16.0 newtons eastward is applied to a cart having a mass of 4.00 kg. What will the acceleration be? (b) If the cart started from rest, what is its velocity at the end of 5.00 s?

5–9. A net force of 2500 newtons southward acts on a car, accelerating it 2.50 m/s². What is the mass of the car?

5–10. An astronaut pulls a 1600-g camera across a spaceship. The camera is accelerated 25.0 cm/s². What force is exerted?

5–11. Once it has been in operation, why does it become easier to accelerate a rocket? (About 90% of the mass of a rocket at the beginning of a launch is fuel.)

5–12. A 1000-kg automobile accelerates westward from rest to 40.0 mi/h (18.0 m/s) in 5.00 sec. What force does the road exert on the car?

5–13. A sprinter of mass 70.0 kg accelerates uniformly to the right from a stationary start for the first 5.00 s of a race. His velocity is then 10.0 m/s. Calculate the force provided by his legs.

5–14. A body dropped from rest has a constant acceleration in free fall. Is the distance traveled by the body the same for each time interval? Explain.

5–15. Which is greater: 1 N, 1 dyne, or 1 lb?

5–16. Would you weigh more or less on top of Mt. Everest than you do at sea level?

5–17. A body having a mass of 50.0 kg moves with a constant acceleration of 2.00 m/s² to the right. What force must be acting on it?

The Action/Reaction Law

5–18. The earth exerts a gravitational force on you. Identify the "reaction" force.

5–19. Identify the action and reaction forces involved in kicking a football.

5–20. What causes a rotary lawn sprinkler to rotate?

Circular Motion

5–21. Explain the statement, "The moon is falling."

5–22. Account for the fact that the moon does not travel in a straight line.

5–23. Since the earth is attracted to the sun by gravitational force, what prevents the earth from falling into the sun?

The Gravitational Force

5–24. Does an earth satellite travel at constant speed in its orbit? Explain.

5–25. How does the speed of an earth satellite in orbit depend on its distance from the earth?

5–26. Calculate the orbital velocity of an earth satellite in an orbit 400 mi above the earth.

5–27. Explain why mass and weight are not expressed in the same units.

5–28. (a) How much does a 75.0-kg astronaut weigh on earth? (b) What would the astronaut's weight be on Venus, where the acceleration due to gravity is 8.70 m/s²?

5–29. (a) Would your mass be the same on the moon as on the earth? (b) Your weight?

5–30. Does the gravitational attraction of the sun on the planets act as a centripetal or as a centrifugal force? Explain.

5–31. *Multiple Choice*

A. Newton's first law is a statement of the law of
   (a) inertia.
   (b) force.
   (c) action-reaction.
   (d) gravitation.

B. A measure of the inertia of a body is its
   (a) speed.
   (b) friction.
   (c) mass.
   (d) weight.

C. In Newton's third law, the action and reaction forces
   (a) act on the same body.
   (b) act on different bodies.
   (c) are not necessarily equal.
   (d) are inversely proportional.

D. A satellite is held in orbit around the earth because of
   (a) a force pushing it around the orbit.
   (b) centrifugal force.
   (c) the sun's gravity.
   (d) the earth's gravity.

E. A baseball thrown by a pitcher
   (a) falls at a rate that depends on its horizontal speed.
   (b) begins to fall when it loses most of its speed.
   (c) falls just as fast as if it had no horizontal speed.
   (d) does not fall.

F. The weakest fundamental force is the
   (a) electromagnetic force.
   (b) gravitational force.
   (c) weak nuclear force.
   (d) strong nuclear force.

G. The constant G is
   (a) the acceleration due to gravity.
   (b) smaller on the moon than on earth.
   (c) the force of gravity.
   (d) a universal constant.

H. In free fall, which one of the following quantities increases with time?
   (a) acceleration
   (b) force
   (c) velocity
   (d) mass

I. Two equal and opposite forces of 3 N have a net force of
   (a) 9 N.
   (b) 6 N.
   (c) 3 N.
   (d) 0 N.

J. Which of the following is not constant for an object in uniform circular motion?
(a) distance with time

(b) speed
(c) velocity
(d) centripetal acceleration

SUGGESTIONS FOR FURTHER READING

Andrade, E. N. da C., *Sir Isaac Newton*. Garden City, N.Y.: Doubleday & Company, 1967.
Discusses Newton's discoveries and the changes they made in our view of the universe.

Boslough, John, "Searching for the Secrets of Gravity." *National Geographic*, May, 1989.
Discusses the possibility of a natural "fifth force" that counteracts gravity. If real, it would require major revisions in current theories.

Calkin, M. G., "The Motion of an Accelerating Automobile." *American Journal of Physics*, June, 1990.
Each month automotive magazines bring test results of accelerating automobiles to automobile enthusiasts. The kinematic data can be related to the coefficient of friction between the road and the drive wheels and the fraction of the engine power that goes to accelerate the automobile.

Clotfelter, B. E., "The Cavendish Experiment as Cavendish Knew It." *American Journal of Physics*, March, 1987.
Henry Cavendish did not report his work as a measurement of the universal gravitational constant G. In fact, that did not become the standard interpretation for more than 100 years. Cavendish himself thought that he had measured the mean density of the earth.

Cohen, I. Bernard, "Newton's Discovery of Gravity." *Scientific American*, March, 1981.
Isaac Newton's discovery of the law of universal gravitation marked the beginning of modern science. It was not the result of an isolated flash of genius, but the culmination of a series of exercises in problem solving.

Drake, Stillman. "Newton's Apple and Galileo's *Dialogue*." *Scientific American*, August, 1980.
If the familiar story of the apple is true, it still does not explain how Newton came to formulate his question about the moon in orbit. He may have been inspired by a diagram in Galileo's *Dialogue*.

Drake, Stillman, and Charles T. Kowal, "Galileo's Sighting of Neptune." *Scientific American*, December, 1980.
Although he did not identify it as a planet, Galileo first observed Neptune in December, 1612, 234 years before it was recognized as the eighth planet by Johann Gottfried Galle.

Franklin, Allan, "Principle of Inertia in the Middle Ages." *American Journal of Physics*, June, 1976.
Shows that the revolution in mechanics in the seventeenth century did not occur out of thin air.

Freedman, Daniel Z., and Peter van Kieuwenhuizen, "Supergravity and the Unification of the Laws of Physics." *Scientific American*, February, 1978.
A new theory of gravity that may lead to a unified theory of all the basic forces in nature.

Hardorp, Johannes, "An Experimental Test of Newton's Celestial Mechanics." *American Journal of Physics*, June, 1986.
Twelve years of pulsar observations prove that Newton was right: The sun does participate in the "dance of the planets" rather than just direct it.

Soderblom, Laurence A., "The Galilean Moons of Jupiter." *Scientific American*, January, 1980.
The *Voyager 1* and *Voyager 2* spacecraft flyby of Jupiter in 1979 returned close-up pictures of Jupiter's largest or Galilean moons: Io, Ganymede, Callisto, and Europa. It is now known that they belong to the family of objects designated as terrestrial, a family that includes Mercury, Venus, the earth, the earth's moon, and Mars.

Speers, Robert R., "Physics and Roller Coasters—the Blue Streak at Cedar Point." *American Journal of Physics*, June, 1991.
A roller coaster is used for studies of kinematics, forces, dynamics, and energy conservation. Experimental results and theoretical predictions are compared.

Thorne, Kip S., "Gravitational Collapse." *Scientific American*, November, 1967.
The weakest force known, gravity, becomes the dominant force on an astrophysical scale and plays the role of "midwife and undertaker" in the birth and death of stars.

Weber, Joseph, "The Detection of Gravitational Waves." *Scientific American*, May, 1971.
According to the general theory of relativity, accelerated masses should radiate gravitational waves, but the detection of these waves has been the subject of controversy.

ANSWERS TO NUMERICAL EXERCISES

5–7. (b) one-sixth that on earth

5–8. (a) 4.00 m/s²

    (b) 20.0 m/s

5–9. 1000 kg

5–10. 40,000 dynes

5–12. 3600 N

5–13. 140 N

5–15. 1 lb

5–17. 100 N

5–26. 7.54×10^3 m/s

5–28. (a) 735 N   (b) 653 N

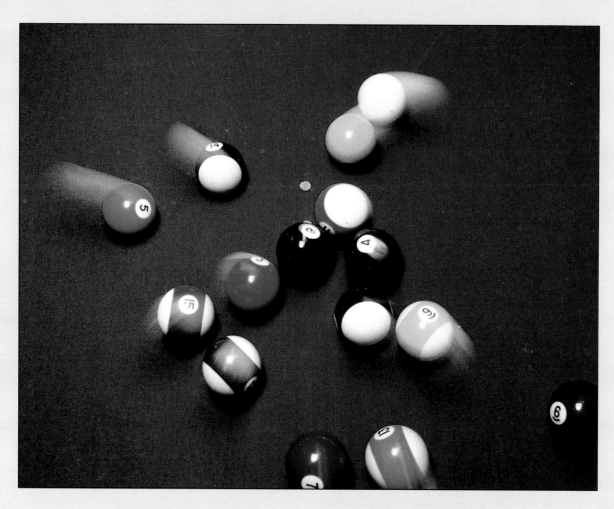

A game of billiards provides many examples of the conservation of momentum. (David Rodgers)

6

MOMENTUM, WORK, AND ENERGY

Imagine rolling a bowling ball against one end of a row of bowling balls in contact. Only the ball at the opposite end moves; it rebounds from the far end as though it had been struck in the original impact. If two bowling balls strike the row, then two rebound at the far end. (An entertaining device based on this principle is shown in Fig. 6–1.) Then imagine a bowling ball suspended like a pendulum and released from a position of rest. It rises no higher than its starting level when it completes a swing, as demonstrated in Figure 6–2. Two of the most powerful concepts in physical science are involved in such events: *momentum* and *energy*.

6–1 MOMENTUM

A bus is clearly harder to stop than a compact car moving at the same speed. We can express this fact by saying that the bus has greater **momentum** than the car. The linear momentum of an object of mass m moving with a velocity $\mathbf{v}$ is defined as the product of the mass and the velocity. The greater the mass or velocity, or both, the greater the momentum.

$$\text{Momentum} = \text{Mass} \times \text{Velocity} \qquad (6\text{–}1)$$
$$\mathbf{p} = m\mathbf{v}$$

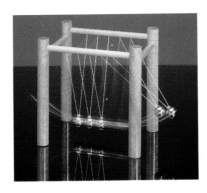

Figure 6–1 Collision spheres with a "memory." The release of one sphere on the left is followed by the ejection of one sphere on the right; the release of two on the left, by two on the right. The process then repeats from right to left. (Courtesy of Arbor Scientific, Inc.)

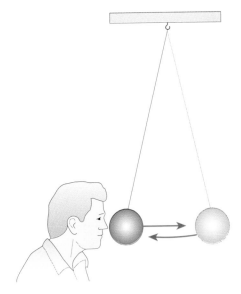

Figure 6–2 Bowling-ball pendulum. If released from the tip of one's nose with no initial velocity, the pendulum bob swings through an arc and returns just to the tip of the nose.

Table 6–1 UNITS FOR MOMENTUM

Momentum	=	Mass	×	Velocity		Unit
p	=	*m*		**v**		
SI System		1 kilogram	×	$1 \dfrac{\text{meter}}{\text{s}}$	=	$1 \dfrac{\text{kg} \cdot \text{m}}{\text{s}}$
cgs System		1 gram	×	$1 \dfrac{\text{centimeter}}{\text{s}}$	=	$1 \dfrac{\text{g} \cdot \text{cm}}{\text{s}}$
British Engineering		1 slug	×	$1 \dfrac{\text{foot}}{\text{s}}$	=	$1 \dfrac{\text{slug} \cdot \text{ft}}{\text{s}}$

Like velocity, momentum is a vector quantity, and its direction is the same as the direction of the velocity. It is often convenient to choose velocities in one direction as positive and those in the opposite direction as negative; thus, momentum may be either positive or negative, expressed as $+m\mathbf{v}$ and $-m\mathbf{v}$. Some appropriate units for expressing momentum are shown in Table 6–1. There are no special names for units of momentum; although some have been proposed, none have been accepted.

EXAMPLE 6–1

Compute the momentum of a baseball that has a mass of 0.200 kilogram and is moving at the rate of 30.0 m/s to the right.

SOLUTION

$m = 0.200$ kg $\mathbf{p} = m\mathbf{v}$

$\mathbf{v} = 30.0$ m/s $= (0.200 \text{ kg})(30.0 \text{ m/s})$

$\mathbf{p} = ?$

$= 6.00 \dfrac{\text{kg} \cdot \text{m}}{\text{s}}$

EXTENSION
A 10.0-ton freight car is moving at a speed of 10.0 ft/s. What is its momentum? (1 ton = 2000 lb.) (Answer: 6250 slug · ft/s)

∎

If the baseball in Example 6–1 were moving half as fast, at 15.0 m/s, its momentum would be only half as great: 3.00 kg · m/s. A fast-moving baseball has more momentum than a slow-moving baseball be-cause of its greater velocity. Similarly, a heavy bus has more momentum than a small car moving at the same speed because its mass is larger. A moving object, then, can have a large momentum if its mass is large, its speed is large, or both its mass and speed are large. A stationary object has zero momentum since it has zero velocity. It is the product of mass times velocity that is important.

The more momentum an object has, the greater will be its effect if it strikes a second object. For example, because of its greater momentum, a speeding car will cause a more serious accident in a collision than a slow-moving car. Similarly, a 225-pound football player running at top speed has considerable momentum and is more likely to stun an opposing player upon tackling him than would a 180-pound or slower-moving player. A player standing still, however, has no momentum at all.

6–2 IMPULSE

The product of mass and velocity, which we call momentum, Newton referred to as a "quantity of motion," and it is in those terms that he expressed his second law. Since

Force = Mass × Acceleration

and acceleration is equal to the rate of change of velocity, that is, the change of velocity per unit time, the second law may be stated as

$$\text{Force} = \text{Mass} \times \frac{\text{Change in velocity}}{\text{Time}}$$

or

$$\mathbf{F} = m\,\frac{(\mathbf{v} - \mathbf{v}_0)}{t}$$

Performing the indicated multiplication by m, we obtain

$$\mathbf{F} = \frac{m\mathbf{v} - m\mathbf{v}_0}{t}$$

Therefore,

$$\text{Force} = \frac{\text{Change in momentum}}{\text{Time}}$$

and, multiplying by time,

$$\text{Force} \times \text{Time} = \text{Change in momentum} \qquad (6\text{--}2)$$
$$\mathbf{F}t = m\mathbf{v} - m\mathbf{v}_0$$

A force, then, is required to change the momentum of a body, and the force must act in an interval of time. The product of the net force acting on the body and the time during which this force acts is known as **impulse**. Thus

$$\text{Impulse} = \text{Change in momentum} \qquad (6\text{--}3)$$

This result is often called the **impulse-momentum theorem**.

Figure 6–3 An example of an impulsive force. Jimmy Connors at the 1991 U.S. Open. A large impulse delivered by the tennis racket sets the tennis ball in motion with a large momentum. Note how the racket and ball are distorted during contact in this high-speed photograph. (Focus on Sports)

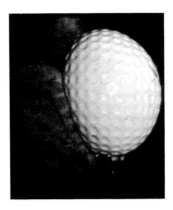

Figure 6–4 A golf club and ball in contact during delivery of impulse. The ball recovers its normal shape as it moves away from the club. (Courtesy of Dr. Harold Edgerton, Massachusetts Institute of Technology, Cambridge, Massachusetts)

The change in momentum of the body on which the force is acting is a measure of the impulse of the force and is equal to the impulse it receives. If the body starts from rest, the impulse of the force is just the momentum the body acquires. Some appropriate units for impulse are the newton-second, dyne-second, and pound-second.

In sports such as tennis, golf, or baseball, we often want to impart as large a momentum as possible to an object. Since a large impulse produces a large change in momentum, the impulse is made as large as possible by applying the greatest force we can and by extending the time of contact as much as possible with the tennis racket, baseball bat, or golf club (Figs. 6–3 and 6–4). A player who "follows through" when hitting a ball is thus more likely to deliver a maximum impulse to the ball.

EXAMPLE 6–2

A golf ball with a mass of 0.060 kg is subjected to an impulse that causes it to move with a velocity of 50.0 m/s. If the golf club and the ball are in contact for 5.00×10^{-3} seconds, what average force acts on the ball?

SOLUTION
Since the golf ball starts from rest, $\mathbf{v}_0 = 0$. Applying Equation 6–2, we have:

$m = 0.060 \text{ kg}$

$\mathbf{v} = 50.0 \text{ m/s}$

$\mathbf{v}_0 = 0$

$t = 5.00 \times 10^{-3} \text{ s}$

$\mathbf{F} = \text{?}$

$\mathbf{F}t = m\mathbf{v} - m\mathbf{v}_0$

$\therefore \mathbf{F} = \dfrac{m\mathbf{v} - m\mathbf{v}_0}{t}$

$\qquad = \dfrac{(0.060 \text{ kg}) \left(50 \dfrac{\text{m}}{\text{s}}\right) - 0}{5 \times 10^{-3} \text{ s}}$

$\qquad = 600 \dfrac{\text{kg} \cdot \text{m}}{\text{s}^2}$

$\qquad = 600 \text{ newtons}$

EXTENSION

An average force of 7.5×10^5 newtons stops a car in 0.10 s. Calculate the impulse. (Answer: 7.5×10^4 newton · seconds)

■

The impulse-momentum theorem can also be applied to the analysis of situations that involve a decrease of momentum. For example, a padded dash-board in a car is safer than a metal one. If your car came to a sudden stop and you were not wearing your seat belt or shoulder strap, your inertia would keep you in motion. If you hit the dashboard, your momentum would be reduced, but the padding would increase your contact time and reduce the impact force, possibly preventing injury.

The same principle underlies the use of the automobile air bag, designed to inflate automatically upon hard impact to prevent the driver and passengers from hitting the steering wheel, dashboard, or windshield. Similarly, catching a hard ball without having it "sting" your hands involves increasing the contact time with the ball by moving your hands in the direction of the motion of the ball (Fig. 6–5). The effect is to lessen the impulse force and, therefore, the sting. If you catch the ball with your hands rigidly extended, the small contact time makes the impulse force large, which stings your hands.

6–3 CONSERVATION OF MOMENTUM

Consider a system of two bowling balls, A and B, that have equal masses, each suspended to form a simple pendulum (Fig. 6–6). (A system is a collection of objects.) If we allow A to swing and collide head-on with B, then just before the collision, the momentum

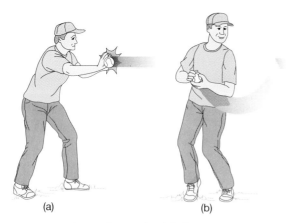

(a)   (b)

Figure 6–5 Catching a hard ball with no sting. The change in momentum in a ball being caught (stopped) is $m\mathbf{v}$, which is equal to the impulse. (a) If the contact time is small, the impact force is large and the ball "stings" the hands. (b) Increasing the contact time by moving the hands along with the ball reduces the impact force, and there is little or no sting.

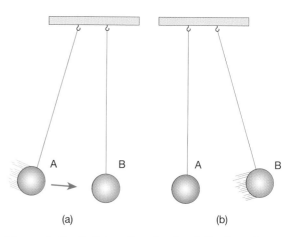

(a)   (b)

Figure 6–6 Collision of two bowling balls of equal mass, A and B, one moving and the other at rest, (a) before and (b) after impact. Ball B, which had been at rest, moves off with the same velocity that A had at the time of impact, while A comes to rest.

of the system is due to the motion of A and may be written:

$$\underbrace{\mathbf{p} = m\mathbf{v}_{\text{A}}}_{\text{before collision}}$$

This is true because the total momentum of a system is the vector sum of the momentum vectors of the individual objects. Since B is stationary, its momentum is zero. Following the collision, A comes to rest and B moves to the right with exactly the same velocity that A had just before the collision.

$$\underbrace{\mathbf{p} = m\mathbf{v}_{\text{B}}}_{\text{after collision}}$$

In a perfectly elastic collision (which occurs in nature only at the atomic level but is often closely approximated with hard objects such as steel balls and bowling balls), ball B would rise to exactly the same level from which ball A descended, there would be no lasting deformation, and heat would not be generated. In a real collision, B does not quite reach the same velocity and, hence, height.

During the impact between A and B, A exerts a force on B and B exerts a force on A. These two forces, by Newton's third law, are equal in magnitude but opposite in direction. Since they act for the same short time, they impart equal impulses to A and B. Equal impulses, accordingly, create equal changes in momentum. If one ball gains momentum, the other loses the same amount of momentum and the two changes, being vector quantities, cancel each other. The overall result is that the total momentum of the system remains constant. **A quantity whose total amount does not change in some process or is constant with time is said to be conserved.** Thus, momentum is conserved (Fig. 6–7). In all collisions that have ever been studied, the **law of conservation of momentum** has been found to apply:

> **The total momentum of a system (isolated from outside forces) after interaction is equal to the total momentum of the system before interaction.**

EXAMPLE 6–3

An astronaut and his equipment have a total mass of 100 kilograms in free space. If the astronaut removes his oxygen tank (mass = 8.00 kg) and throws it away

Figure 6–7   Momentum is transferred from a bowling ball to the pins. The momentum of the ball on impact is equal to the total momentum of the ball and pins after impact. (Focus on Sports)

with a velocity of 5.00 m/s, what velocity does he acquire?

SOLUTION

Since the mass of the oxygen tank is 8.00 kg, the mass of the astronaut and remaining equipment must be 100 kg − 8.00 kg = 92.0 kg. Knowing the velocity that the oxygen tank acquires, we can apply the law of conservation of momentum to determine the recoil velocity of the astronaut. Since the total momentum of the system (the astronaut and oxygen tank) is constant, there can be no net gain or loss of momentum. The momentum of the astronaut must be equal to the momentum of the oxygen tank, but opposite in direction, or sign; then the total momentum after the event will equal the total momentum before the event, and momentum will be conserved. Thus, using the following equation,

$$\begin{array}{ccc} \text{Momentum} & \text{Momentum} & \text{Total} \\ \text{of oxygen} + \text{of astronaut} = & \text{momentum} \\ \text{tank} & & \text{before event} \\ & & \text{and after} \\ & & \text{event} \\ m_1\mathbf{v}_1 \quad + \quad m_2\mathbf{v}_2 \quad = & 0 \end{array}$$

we can substitute the values given, and solve:

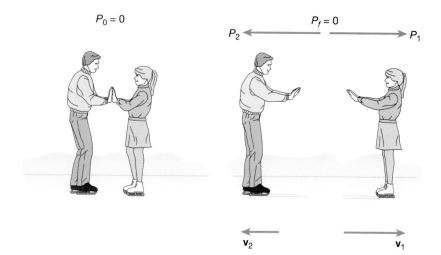

$P_0 = 0$

$P_f = 0$

$P_2 \longleftarrow \qquad \longrightarrow P_1$

$\mathbf{v}_2 \longleftarrow \qquad \longrightarrow \mathbf{v}_1$

Figure 6–8   Ice skaters "push off" each other. An example of conservation of momentum. At the start, the total momentum P_o of the skaters' system is zero. After "pushing off," the total momentum P_f remains zero. Total momentum is conserved.

$m_1 = 92.0$ kg

$m_2 = 8.00$ kg

$\mathbf{v}_2 = 5.00 \dfrac{\text{m}}{\text{s}}$

$\mathbf{v}_1 = ?$

$m_1\mathbf{v}_1 = -m_2\mathbf{v}_2$

$\mathbf{v}_1 = -\dfrac{m_2\mathbf{v}_2}{m_1}$

$= -\dfrac{(8.00\ \text{kg})\left(5.00\ \dfrac{\text{m}}{\text{s}}\right)}{92.0\ \text{kg}}$

$= -\dfrac{40.0\ \dfrac{\text{m}}{\text{s}}}{92.0}$

$= -0.435\ \dfrac{\text{m}}{\text{s}}$

The minus sign means that the astronaut acquires a velocity opposite in direction to that of the oxygen tank.

EXTENSION
Two ice skaters push off each other (Fig. 6–8). The mass of one is 75.0 kg, and his speed is 2 m/s. If the mass of the other is 30.0 kg, what speed does she acquire? (neglect friction.) (Answer: 5.00 m/s)

■

For a collision between two bodies, the sum of the momenta (plural of "momentum") of the two colliding bodies remains constant, that is,

$$\dfrac{\text{Total momentum}}{\text{before collision}} = \dfrac{\text{Total momentum}}{\text{after collision}}$$

or, in symbols,

$$m_1\mathbf{v}_1 + m_2\mathbf{v}_2 = m_1\mathbf{V}_1 + m_2\mathbf{V}_2 \qquad (6\text{–}4)$$

where m_1 and m_2 are the masses of the colliding bodies; $\mathbf{v}_1$ and $\mathbf{v}_2$ are their velocities before impact; and $\mathbf{V}_1$ and $\mathbf{V}_2$ are their velocities following impact. Although the momentum of each of the bodies may change, the sum of the two momenta is the same after the collision as it was before (Fig. 6–9).

Referring again to the two bowling balls, what would happen if we place some putty on bowling ball B at the point at which A strikes it and repeat the experiment? A and B will move off together with a lower velocity than A possessed at the time of impact

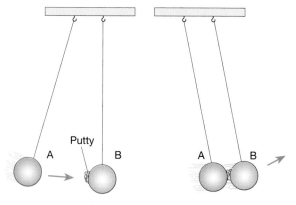

Putty

A   B

A   B

Figure 6–9   In this inelastic collision, following impact, A and B move off together with the same velocity (but lower than the velocity which A possessed at impact).

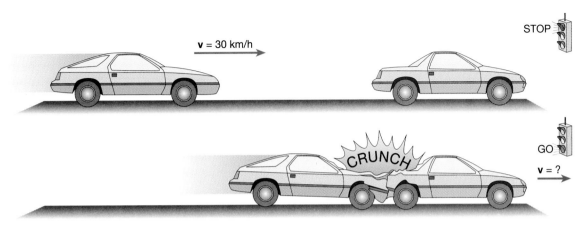

Figure 6–10   An inelastic collision. There is some deformation of the object(s), heat is generated during the collision, but momentum is conserved, provided there is no net external force.

(Fig. 6–10). In such **inelastic collisions**, there is often some deformation and heat is generated, but momentum is conserved provided there is no net external force. Knowing the velocity of A at the time of impact, we can calculate the velocity of the two bowling balls following impact from the law of conservation of momentum. For this system,

$$m_A\mathbf{v}_A + m_B\mathbf{v}_B = m_A\mathbf{V} + m_B\mathbf{V}$$

where $\mathbf{V}$ is the velocity of bowling balls A and B after they collide. Since B is stationary, $\mathbf{v}_B = 0$, and $m_B\mathbf{v}_B = 0$. Thus,

$$m_A\mathbf{v}_A = (m_A + m_B)\mathbf{V}$$

and

$$\mathbf{V} = \frac{m_A\mathbf{v}_A}{m_A + m_B}$$

EXAMPLE 6–4

A car traveling 30.0 km/h strikes a car of equal size that is stopped at a traffic light from the rear, and the two cars lock bumpers (Fig. 6–10). If the mass of each car is 500 kg, what is the common velocity of the two cars after the collision?

SOLUTION

The velocity of the parked car before the collision is 0. We can apply the law of conservation of momentum to find the common velocity after collision.

$$m_A = 500 \text{ kg} \qquad \mathbf{v}_A = 30.0 \text{ km/h}$$
$$m_B = 500 \text{ kg} \qquad \mathbf{v}_B = 0$$
$$\mathbf{V} = ?$$
$$m_A\mathbf{v}_A + m_B\mathbf{v}_B = m_A\mathbf{V} + m_B\mathbf{V}$$
$$m_A\mathbf{v}_A + m_B\mathbf{v}_B = (m_A + m_B)\mathbf{V}$$
$$\mathbf{V} = \frac{m_A\mathbf{v}_A + m_B\mathbf{v}_B}{m_A + m_B}$$
$$= \frac{\left(500 \text{ kg} \cdot 30.0 \, \dfrac{\text{km}}{\text{h}}\right) + (500 \text{ kg} \cdot 0)}{1000 \text{ kg}}$$
$$= \frac{15000 \, \dfrac{\text{km}}{\text{h}}}{1000}$$
$$= 15.0 \, \frac{\text{km}}{\text{h}}$$

∎

6–4 ANGULAR MOMENTUM

Anything that travels in a curved path, such as a planet around the sun, an electron around a nucleus, or a whirling rock at the end of a string, has **angular momentum**. Just as for linear or translational momentum, which applies to bodies moving in a straight line, angular momentum remains constant in a system in which no external forces are acting. That is, angular momentum is conserved.

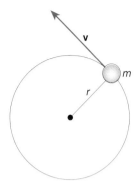

Figure 6–11 Angular momentum for circular motion. An object of mass m rotating in a circular path of radius r with speed **v** has angular momentum $L = m\mathbf{v}r$.

In Figure 6–11, an object of mass m rotates in a circular path of radius r with speed **v**. The magnitude of the angular momentum of the object is equal to the magnitude of its linear momentum, $m\mathbf{v}$, multiplied by the radius of the circular path, r.

$$\text{Angular momentum} = \text{Mass} \times \text{Radial distance} \times \text{Velocity} \tag{6-5}$$
$$\mathbf{L} = m\mathbf{v}r$$

Since the angular momentum for a given mass depends on both the *speed* and the *radius,* an object moving with a low speed on a large circle will have a large angular momentum.

Angular momentum is a vector quantity, like linear momentum. The direction of the vector can be determined using the right-hand rule. If you curl the fingers of your right hand in the direction that the object moves, your thumb will point in the direction of the angular momentum vector.

An interesting example of the conservation of angular momentum is the spinning ice skater in Figure 6–12. When spinning with her hands outstretched, most of the angular momentum is due to the mass in her hands and arms extended a distance r from the spin axis, the vertical line running down the center of the body. As the skater draws her hands and arms in closer to the body, the radius r decreases. This causes a faster spin, since angular momentum is conserved. That is, the angular velocity **v** of the skater increases as the radius r decreases to maintain the angular momentum **L** constant.

Another example is Kepler's law of equal areas for the planets in their orbits (Chapter 4). The planets

Figure 6–12 Ice skater Elizabeth Manley (Silver Medal winner at the 1988 Winter Olympics) spins faster when she raises her arms overhead, since angular momentum is conserved. (© duomo/Paul J. Sutton 1988)

speed up or slow down as they approach or recede from the sun, keeping their angular momentum constant.

6–5 WORK

There are two aspects of the effect of a force. We have previously discussed one of these, impulse (force × time). The second, the product of a force and the distance through which it acts (force × distance), is a measure of the *work* done. The concept of work is one of the most useful in physical science.

Work is defined as the product of the net force exerted on a body and the distance moved in the direction of the force. Work is a scalar quantity with no specific direction, even though force and displacement require both magnitude and direction.

$$\text{Work} = \text{Force applied} \times \text{Distance moved in the direction of the force.}$$

$$W = \mathbf{Fs} \tag{6-5}$$

Table 6–2 gives some units for work. Force units are summarized in Table 5–1.

The unit joule (J) (rhymes with "rule") is named after James Prescott Joule (1818–1889), a pioneer in the field of energy. Erg is from *ergon,* the Greek word for work.

Table 6–2   WORK UNITS

	Work W	$=$	Force $\mathbf{F}$	$\times$	Distance $\mathbf{s}$		Unit		
SI System			1 newton	$\times$	1 meter	$=$	1 newton-meter	$=$	1 joule
cgs System			1 dyne	$\times$	1 centimeter	$=$	1 dyne-centimeter	$=$	1 erg
British Engineering **System**			1 pound	$\times$	1 foot	$=$	1 foot-pound		

Since

$$1 \text{ joule} = 1 \text{ newton} \times 1 \text{ m}$$

then

$$1 \text{ joule} = 10^5 \text{ dynes} \times 10^2 \text{ cm}$$
$$= 10^7 \text{ dyne-cm}$$
$$= 10^7 \text{ ergs}$$

and

$$1 \text{ joule} = 0.738 \text{ ft-lb}$$

EXAMPLE 6–5

A boy pulls a sled to the right across a level snow slope. Assuming that the horizontal component of the force he applies is 15.0 newtons, how much work does he do while pulling the sled 50.0 meters?

SOLUTION
The work done on the sled depends on the net force exerted in the horizontal direction and the distance moved in the direction of the force.

$\mathbf{F} = 15.0$ N $W = \mathbf{Fs}$

$\mathbf{s} = 50.0$ m $= (15.0 \text{ N})(50.0 \text{ m})$

$W = ?$ $= 750$ N $\cdot$ m

          $= \boxed{750 \text{ joules}}$

EXTENSION
How much work is done in lifting a 10.0-kg trunk 1.00 m off the floor? (Answer: 98.0 joules)

■

As defined in the physical sense, work is associated with both force and motion. This meaning, unlike terms such as "velocity" or "force," is distinctly dif-ferent from its meaning in day-to-day situations, where it refers to any exertion maintained for some time. Whenever work is done, a net force is exerted and something is moved by that force. When you push on a crate and it moves, you are performing work on it (Fig. 6–13). An activity such as studying hard is not work in a scientific sense.

There are many cases in which you may exert a force and yet do no work. For example, you may push hard against a car, but if it does not move, you have done no work on the car: The distance is zero; there-fore, the product of force $\times$ distance is zero. When you lift a package of groceries or a flatiron, you do work against the earth's gravity (Fig. 6–14). But when you walk carrying the package or iron, you are doing no work against gravity. The force you exert on the iron is vertical and has nothing to do with the horizon-tal motion. A weightlifter holding a barbell over his head does no work (Fig. 6–15), although he did per-form work as he raised the barbell.

When you push a lawn mower you do work on it, even though you apply the force at an angle to the direction of motion (Fig. 6–16). The lawn mower moves horizontally because the applied force has a component parallel to the direction of motion. Only when the applied force is at right angles (90°) to the

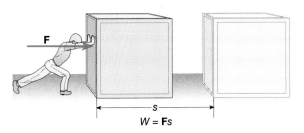

Figure 6–13   Work is done when a force $\mathbf{F}$ moves a crate a distance $\mathbf{s}$; the work $W = \mathbf{Fs}$.

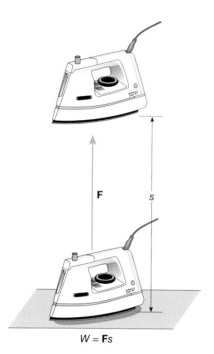

$$W = \mathbf{F}s$$

Figure 6–14 Work is done in lifting an object against gravity. The force **F** is applied vertically upward, moving the flatiron through a distance **s**. When the displacement is in the same direction as the applied force, the work done is $W = \mathbf{F}s$.

Figure 6–15 Work is done in lifting the barbell, but the weightlifter performs no physical work just holding it. (Focus on Sports)

Figure 6–16 Only the component of the applied force, **F**, in the direction of motion of the lawn mower, $\mathbf{F}_{\text{Horizontal}}$, is used to do work on the lawn mower.

direction of motion will it lack a component parallel to the direction of motion; in such cases, work, in the physical sense, is not done.

6–6 POWER

The total *work* done for a given purpose is often less important than the *rate* at which it is done. Given time, an automobile engine can do as much work as is done by the jet engine of an airliner during a New York–Chicago flight, but it could not do the work fast enough to get the airliner off the ground. A powerful engine is one that can perform a lot of work in a short time (Fig. 6–17). **Power** is defined as the rate of doing work:

$$\text{Power} = \frac{\text{Work}}{\text{Time}} \qquad (6\text{–}6)$$

$$P = \frac{W}{t}$$

The units for power were suggested by Thomas Savery (1650–1715), whose pumping engine was the first device to use steam power in industry. Since horses had been used to pump water in draining coal mines, Savery proposed as a standard of power the

Figure 6–17 These runners are doing the same amount of work, but they do not have the same power. The runner in front is using more power. Why? (Focus on Sports)

rate at which a horse could do work. James Watt (1736–1819), a mechanical engineer, improved the steam engine and in trying to sell his engines to mine owners was often asked, "If I buy one of your engines, how many horses will it replace?" To find out, Watt took Savery's suggestion and harnessed strong work horses to a load. He found that an average horse walking at the rate of 2½ miles per hour would steadily exert a 150-pound force for several hours. The rate at which a horse performed work became the **standard horsepower (hp)**. It has a value of 33,000 foot-pounds per minute, or 550 foot-pounds per second. In SI units, the unit of power is defined as that capable of performing work at a rate of 1 joule/second, or 1 **watt**:

$$1 \frac{\text{joule}}{\text{second}} = 1 \text{ watt (W)}$$

and 1 horsepower = 746 watts. A larger unit of power, the kilowatt (kW), is useful in rating engines and motors; it is equal to 1000 watts. A still larger unit, the megawatt (MW), 10^6 watts, is applied to the rating of power plants. Although they are basically mechani-

cal units, the watt and kilowatt can be used to express any power unit, such as electric power.

EXAMPLE 6–6

An 80.0-kilogram man runs up a flight of stairs 5.00 meters high in 10.0 seconds. What is the man's power output in watts and in horsepower?

SOLUTION

Since work = force × distance, the man's weight (force) must be considered. We can determine his weight, **w**, from the product of his mass and the acceleration due to gravity. Knowing the time, we can also determine his power and express it in appropriate units.

$m = 80.0$ kg

$\mathbf{s} = 5.00$ m

$\mathbf{g} = 9.80 \frac{\text{m}}{\text{s}^2}$

$t = 10$ s

$\mathbf{F} = \mathbf{w} = m\mathbf{g}$

$$P = \frac{W}{t}$$

$$= \frac{\mathbf{F}\mathbf{s}}{t}$$

$$= \frac{(m\mathbf{g})\mathbf{s}}{t}$$

$$= \frac{(80.0 \text{ kg}) \left(9.80 \frac{\text{m}}{\text{s}^2} \right) (5.00 \text{ m})}{10 \text{ s}}$$

$$= 392 \frac{\text{newton-meters}}{\text{s}}$$

$$= 392 \frac{\text{joules}}{\text{s}}$$

$$= \boxed{392 \text{ watts}} \times \frac{1 \text{ hp}}{746 \text{ watts}}$$

$$= \boxed{0.525 \text{ hp}}$$

EXTENSION

A pump lifts 30.0 kg of water a vertical distance of 20.0 m each second. What is its output power? (Answer: 5.90×10^3 watts)

6–7 KINETIC ENERGY

Whenever work is done, a net force is exerted on a body over a distance and something is moved by that force in the direction of the force. For example, when you throw a ball, your hand exerts a force on the ball, and this force acts through a distance. The work you do on the ball causes it to leave your hand with a certain velocity. Because of its motion, the ball then has the capacity to do work on another object it strikes; for example, it may shatter a window. **The ability to do work is energy.** The **energy of motion** is known as **kinetic energy** (KE), a term introduced by Lord Kelvin (1824–1907). Some other forms of energy are gravitational, thermal, chemical, electrical, and nuclear. We shall have more to say about these forms of energy in later chapters.

The connection between work and kinetic energy is illustrated as follows: When work is done on a body at rest, such as a ball, the body is accelerated to some velocity and acquires KE. If the body is already moving, the work goes into increasing the KE. Since work is going into increasing motion, the energy acquired can be expressed in terms of the amount of motion. The gain in the KE of the body is equal to the product of the force and the distance **s** through which the force acts, that is, the work done on the body, a relationship known as the **work-energy theorem**:

$$KE = \mathbf{F}\mathbf{s}$$

The energy that a body gains as a result of mechanical work being done on it is stored work, or **mechanical energy**. Starting from rest, the distance **s** for constant acceleration is $\mathbf{s} = \frac{1}{2}\mathbf{a}t^2$. Therefore,

$$KE = \mathbf{F}\left(\frac{1}{2}\mathbf{a}t^2\right)$$

From Newton's second law, $\mathbf{F} = m\mathbf{a}$; thus

$$KE = (m\mathbf{a})\left(\frac{1}{2}\mathbf{a}t^2\right) = \frac{1}{2}m\mathbf{a}^2t^2$$

Substituting $\mathbf{v} = \mathbf{a}t$ from kinematics, we get

$$KE = \frac{1}{2}m\mathbf{v}^2 \tag{6–7}$$

Kinetic energy = One half the mass times the velocity squared.

Kinetic energy is a scalar quantity. Some appropriate units are given in Table 6–3. So, kinetic energy is measured in the same units as those for work. The distinction between energy and work is that energy is something a body *has*, whereas work is something a body *does*. A body can do work if it has energy, work being a measure of how much energy is transferred from one system to another.

EXAMPLE 6-7

A girl on skis going 20.0 m/s, reaches the bottom of a hill. If her total mass, including equipment, is 60.0 kilograms, what is her kinetic energy?

SOLUTION

$$m = 60.0 \text{ kg} \qquad KE = \frac{1}{2}m\mathbf{v}^2$$
$$\mathbf{v} = 20.0 \text{ m/s}$$
$$KE = ? \qquad\qquad = \frac{1}{2}(60.0 \text{ kg})\left(20.0 \frac{\text{m}}{\text{s}}\right)^2$$
$$= 12{,}000 \frac{\text{kg} \cdot \text{m}}{\text{s}^2} \cdot \text{m}$$
$$= 12{,}000 \text{ newton-meters}$$
$$= \boxed{12{,}000 \text{ joules}}$$

Table 6–3   SOME UNITS FOR KINETIC ENERGY

	m	$\mathbf{v}^2$	$\frac{1}{2}m\mathbf{v}^2$			Unit	
SI System	kilograms $\times$	$\left(\dfrac{\text{meters}}{\text{second}}\right)^2$	$= \dfrac{\text{kg} \cdot m^2}{\text{s}^2}$	$\dfrac{\text{kg} \cdot m}{\text{s}^2} \times$ m	$=$ newton-meter	$=$ joule
cgs System	grams $\times$	$\left(\dfrac{\text{centimeters}}{\text{second}}\right)^2$	$= \dfrac{\text{g} \cdot \text{cm}^2}{\text{s}^2}$	$\dfrac{\text{g} \cdot \text{cm}}{\text{s}^2} \times$ cm	$=$ dyne-cm	$=$ erg
British Engineering System	slugs $\times$	$\left(\dfrac{\text{feet}}{\text{second}}\right)^2$	$= \dfrac{\text{slug} \cdot \text{ft}^2}{\text{s}^2}$	$\dfrac{\text{slug} \cdot \text{ft}}{\text{s}} \times$ ft	$=$ pound-ft	$=$ foot-pound

A sports car is moving at 4.00 m/s. If its mass is 800 kg, how much kinetic energy does it have? (Answer: 6400 joules)

■

6–8 POTENTIAL ENERGY

A stationary object may possess energy by reason of its position. In the process of raising an object—a pendulum, a sphere on an incline, a book—from one level to another, work is done on the body and its energy of position relative to the earth is increased (Fig. 6–18). In the process of falling, this energy is converted to energy of motion (**KE**). William Rankine's (1820–1872) suggestion that such "stored" energy be called **potential energy (PE)** found general favor, and this term has been used ever since. Two common types are gravitational potential energy and the potential energy of a spring.

We see this principle applied in a pile driver (Fig. 6-19). There, a ram acquires potential energy when raised to an elevated position, which is converted to kinetic energy in the process of falling and does work on a beam. When the ram strikes the beam, it is driven into the ground.

This energy, which is due to position, is called **gravitational potential energy** (PE) (Fig. 6–20).

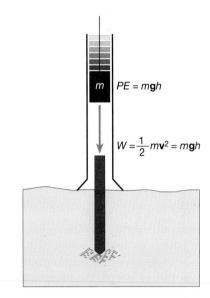

Figure 6–19   A pile driver does work. The potential energy of the block is transformed into kinetic energy by falling and drives the post into the ground.

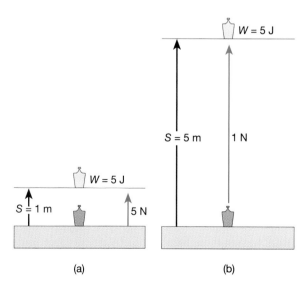

Figure 6–18   The work done in raising two different masses is the same in each case: 5 joules.

Figure 6–20   A skier descending a slope has both gravitational potential energy and kinetic energy, energy of motion. The downhill skiing record is 224 km/hr (139 mph). (Focus on Sports)

The body has PE by virtue of the earth's gravitational attraction. It takes work to separate it from the earth. The gravitational PE of a body is equal to the product of its weight (w) and its height (h) above the initial position. The reason for this is that you must exert a force on it at least equal to its weight and move it through a certain distance h.

$$\text{Gravitational potential energy} = \text{Weight} \times \text{Height} \qquad (6\text{–}8)$$

Since weight = $m\mathbf{g}$,

$$\text{PE} = m\mathbf{g}h$$

The work done against the gravitational force is stored by the body raised as potential energy. The body then has the ability to do work on something else. When a spring mechanism is wound, the compressed spring acquires the ability to do work on gears to run a clock, sound an alarm, close a screen door or mouse trap, or operate a staple gun as the spring returns to equilibrium.

6–9 CONSERVATION OF ENERGY

When a pendulum bob is raised to a certain height and released, its motion consists of a series of alternations between potential and kinetic energy (Fig. 6–21). The bob possesses maximum potential energy and zero kinetic energy at the two extremes of its swing. At the midpoint of its swing, the kinetic energy reaches its maximum value, and at that point the potential energy is at a minimum. As the bob travels up the arc, kinetic energy is converted into potential energy. In the absence of friction, the potential energy of the bob at the highest points in the arc is equal to the kinetic energy at the lowest point, and at all times the sum of the potential energy and kinetic energy is a constant. These energy relations are expressed in the statement of the principle of **conservation of mechanical energy**:

The sum of kinetic energy and potential energy is a constant if friction is not present.

KE + PE = A constant

or

KE + PE = E (total energy)

Here are some other applications of energy transformations (Figs. 6–22 and 6–23). The potential energy of water stored behind a dam, because of the elevated position of the water, is transformed into kinetic energy as the water falls (Fig. 6–24). An automobile perched at the top of a steep hill acquires kinetic energy if it rolls down the hill. A coiled spring contains potential energy owing to its configuration: When it is released, the spring can do work, that is, produce motion. In each of these examples, the potential energy is gravitational or mechanical in origin.

We have seen that an ideal pendulum would swing to precisely the same height when it reaches the end of its arc as that from which it was released initially. But a real pendulum does not; the swings become shorter and shorter and eventually stop. Where has the "missing" energy gone?

Whenever anything moves through a gas or a liquid, as a pendulum bob does in air, it encounters resistive or **frictional forces**. Whenever one surface slides or rolls over another, as a metal sphere does rolling down an incline, it also encounters resistive forces. When frictional forces work on a system, its total mechanical energy decreases. In performing work against these forces, the objects dissipate some of their mechanical energy. Some of the "missing" mechanical energy of the system is converted to an equivalent amount of energy in another form: heat energy.

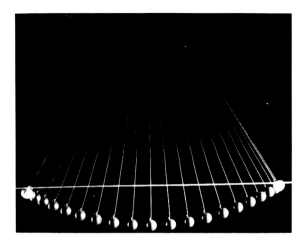

Figure 6–21   Stroboscopic photograph of a swinging pendulum. The bob has maximum potential energy at the extremes of each swing and maximum kinetic energy at the midpoint. (Courtesy of Fundamental Photographs, from the Granger Collection)

Figure 6–22 This photo of a pole-vaulter illustrates both kinetic energy and energy associated with position in space, gravitational potential energy. (© Harold E. Edgerton; courtesy of Palm Press, Inc.)

Figure 6–23 In archery, the potential energy of the bow is transformed into the kinetic energy of the arrow. (Focus on Sports)

Figure 6–24 The potential energy of water at the top of the Glen Canyon Dam is converted to kinetic energy as it falls, which then does work on an electric generator. (Stan Tess, The Stock Market)

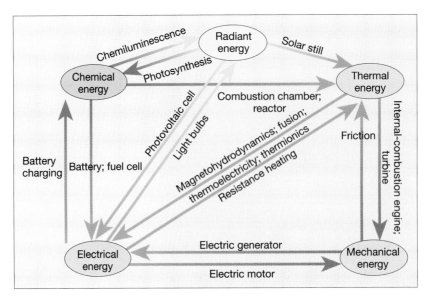

Figure 6–25   Some forms of energy and their conversion pathways.

Energy takes several forms, all of which will be discussed in this text. Some forms of energy are mechanical energy, heat, sound, electromagnetic energy, chemical energy, and nuclear energy (Fig. 6–25). These various forms of energy can be transformed from one form to another. The kinetic energy of falling water can be transformed into electrical energy, which in turn can be transformed into light or heat energy. The chemical energy of gasoline can be converted into the kinetic energy of an automobile. Whenever energy is transformed from one form to another, no energy is created or destroyed: That is,

the total amount of energy is conserved. It can very well be lost from doing useful work, but it is around someplace in the universe. The result of considerable work on energy transformations is the single most powerful generalization in physical science, the **law of conservation of energy**:

The total energy content of a closed system is constant. Energy is never created or destroyed. Energy may be transformed from one kind into another, but its total magnitude remains the same.

SUMMARY OF EQUATIONS

Momentum = mass × velocity

$$\mathbf{p} = m\mathbf{v}$$

Impulse = Change in momentum

$$\mathbf{F}t = m\mathbf{v} - m\mathbf{v}_0$$

Law of conservation of momentum:

Total momentum before collision
= Total momentum after collision

$$m_1\mathbf{v}_1 + m_2\mathbf{v}_2 = m_1\mathbf{V}_1 + m_2\mathbf{V}_2$$

Work = Force applied × Distance moved in the direction of the force

$$W = \mathbf{F}s$$

$$\text{Power} = \frac{\text{Work}}{\text{Time}} \qquad P = \frac{W}{t}$$

$$\text{Kinetic energy (KE)} = \frac{1}{2}m\mathbf{v}^2$$

$$\text{Potential energy (PE)} = m\mathbf{g}h$$

Conservation of mechanical energy:

$$\text{KE} + \text{PE} = \text{A constant}$$

KEY TERMS

Angular Momentum   An object rotating in a circular path has an angular momentum of mvr.

Conservation of Energy   Energy cannot be created or destroyed, but it can be transformed from one kind into another.

Conservation of Mechanical Energy   In the conversion of kinetic energy into potential energy, and vice versa, the sum of kinetic energy and potential energy is a constant, if friction is disregarded.

Energy   The capacity of an object to do work.

Erg   The cgs unit for work: 1 dyne-cm.

Frictional Forces   Forces that oppose the motion of a body moving through a gas or liquid, or the motion of one surface sliding or rolling over another.

Gravitational Potential Energy The energy that a body has by reason of its position in space. In the process of falling, it is converted to energy of motion, kinetic energy.

Horsepower   Work performed at the rate of 33,000 foot-pounds per minute, or 550 foot-pounds per second. 1 HP = 746 watts.

Impulse   The product of force and time. A measure of the change in momentum of an object.

Impulse Momentum Theorem Impulse is equal to change in momentum.

Inelastic Collision   A collision in which momentum is conserved but kinetic energy is not. In a perfectly inelastic collision, two objects collide, then stick together after the collision, so that their final velocities are the same.

Joule   The SI unit for work: 1 newton-meter.

Kinetic Energy   The energy a moving object has by virtue of its motion: KE = $\frac{1}{2}mv^2$.

Law of Conservation of Momentum   The total momentum of an isolated system of objects does not change when no external force acts on the system; the total momentum

of the system before a collision is equal to the total momentum of the system after the collision.

Mechanical Energy   The sum of the potential energy and kinetic energy of an object.

Momentum   The product of the mass of a moving object and its velocity.

Potential Energy   The energy possessed by a body by virtue of its position or condition.

Power   The rate at which work is done. Power = work/time.

Watt   The SI unit of power. Work performed at the rate of 1 joule/second.

Work   The product of force and the distance moving the direction of the force.

Work-Energy Theorem   The net work done on an object by the force or forces acting on it is equal to the change in the kinetic energy of the object.

THINGS TO DO

1. Calculate the power your body generates by running up a flight of stairs. Your weight, multiplied by the vertical height risen, gives the work done. That value divided by the time gives the power. Express in horsepower.

2. Experiment with a set of collision spheres similar to the one shown in Figure 6–1.

3. Set up two similar pendulums on the same support so that they touch. Set one in motion and allow it to collide with the other. Consider the momentum and energy changes that they undergo during "elastic" and inelastic collisions.

4. Aim spheres of the same size but different masses with the same velocity at a block of wood. Note the effect on the block.

EXERCISES

Momentum

6–1. What is the momentum of a 100-kilogram fullback moving at 4.00 m/s?

6–2. A heavy body and a light body have the same momentum. Which has the greater velocity?

6–3. Is it possible for a body to have energy without having momentum?

6–4. Calculate the momentum of a 75.0-kilogram sprinter running with a speed of 12.0 m/s.

6–5. A 1500-kilogram automobile travels east with a speed of 8.00 m/s. It slows to a speed of 3.00 m/s in 3.00 s. Find (a) the impulse delivered to the car and (b) the average force exerted on the car during this time.

6–6. A 100-gram lump of clay moving eastward at 10.0 m/s collides with a 150-gram lump of clay moving in the same direction at 5.00 m/s. The two coalesce after impact. With what velocity does the combined mass move?

6–7. As a tether ball winds around a pole, what happens to its angular velocity? Explain.

6–8. A boxer learns to move his head backwards when he sees he is going to receive a jab to the head. What does this head motion accomplish?

6–9. What does doubling the speed of a car do to the car's momentum?

6–10. Explain why follow-through is important in hitting a baseball.

6–11. Why do gymnasts use padded mats for landings?

6–12. As you jump one way from a rowboat, the boat moves in the opposite direction. Explain.

Work and Power

6–13. A man pushes a box across the floor by exerting a 10.0-newton force through a distance of 5.00 meters. How much work has he done?

6–14. For what quantities are the following units appropriate? (a) kg · m/s; (b) kg · m/s²; (c) kg · m²/s².

6–15. If you push vigorously against a brick wall and it does not move, how much work do you do on the wall? Explain.

6–16. How much power does a 70.0-kilogram person develop by running up a 5.00-meter stairway in 10.0 seconds?

6–17. A weightlifter 5 ft 2 in tall and another 6 ft 2 in both lift 300 lb. Who does more work?

6–18. Two movers raise identical boxes the same vertical distance h. Compare the work done if one lifts the box vertically and the other pushes it up a ramp.

6–19. A flight attendant pulls a wheeled suitcase, applying a 60-N force at a 35° angle to the horizontal. How much work does she do in pulling the suitcase 45 m?

6–20. A 1500-kg car drives up a mountain road a vertical rise of 1300 m from the base elevation in 30 min. What is the average power output?

6–21. A light bulb is rated 100-W. What is its horsepower rating?

Kinetic Energy

6–22. An electron is traveling toward the screen of a TV set with a velocity of 2.00×10^6 m/s. If its mass is 9.10×10^{-31} kilograms, what is its kinetic energy?

6–23. You roll a bowling ball down the alley. If its mass is 5.00 kilograms and it is moving with a velocity of 6.00 m/s, what is its kinetic energy?

6–24. A baseball pitcher throws a 140-gram ball at a speed of 50.0 m/s. How much energy does the catcher absorb when he catches the ball?

6–25. An athlete runs the 100-meter dash in 10.0 seconds. What is his kinetic energy while running if his mass is 75.0 kilograms?

6–26. A car (total mass = 1000 kilograms) is traveling at a speed of 50.0 m/s. What is the car's KE?

6–27. A 500-gram mass drops from a height of 10.0 meters. What is its kinetic energy just before it hits the ground?

6–28. A baseball (mass = 0.200 kilogram) is thrown with a speed of 15.0 m/s.
(a) What is its KE?
(b) If it is thrown at twice this speed, what is its KE?

6–29. What does doubling the speed of a car do to the car's kinetic energy?

Potential Energy

6–30. A helicopter lifts a load of supplies through a vertical distance of 500 meters. What potential energy does the load acquire if its mass is 100 kilograms?

6–31. Is it possible for a body to have energy without having momentum?

6–32. Explain how a sling shot can have potential energy.

6–33. Pulling the bow back farther causes an arrow to have a higher speed when released. Explain.

6–34. (a) How do you know that an archery bow has potential energy?
(b) What happens to this energy when an arrow is shot?

6–35. Does a biker ready to descend at the top of a hill have kinetic or potential energy? Explain.

Energy Transformations

6–36. A 60.0-gram ball is dropped 100 meters from rest.
(a) How much potential energy is lost in the fall?
(b) What speed does the ball have after it has fallen 100 meters?

6–37. Why does a swinging pendulum eventually come to rest?

6–38. Is the energy that any mechanical device "wastes" destroyed? Explain.

6–39. Explain what "conservation of energy" means in a physical sense.

6–40. (a) An apple hanging on a tree has potential energy. Why?
(b) When the apple falls, what happens to its potential energy just before it hits the ground?

6–41. Sound is considered to be a form of energy. Explain why.

6–42. Discuss the energy transformations that occur during the driving of an automobile.

6–43. *Multiple Choice*

A. Work is
(a) energy times distance.
(b) force times distance.
(c) force times time.
(d) momentum times distance.

B. You push against Plymouth Rock with a force of 100 newtons for 20.0 seconds. If the rock does not move, how much work have you done?
(a) 2000 joules
(b) 5.00 joules
(c) 1.00 joule
(d) 0 joule

C. When an automobile's speed is doubled, its kinetic energy is
(a) twice as large.
(b) four times as large.
(c) half as large.
(d) tripled.

D. The rate at which work is done is called
(a) momentum.
(b) potential energy.
(c) kinetic energy.
(d) power.

E. Energy is
(a) lost if heat is produced.
(b) a form of power.
(c) conserved in a closed system.
(d) equal to $m\mathbf{v}$.

F. Padded dashboards in automobiles reduce injury by
(a) increasing friction.
(b) increasing contact time.
(c) stopping a person more quickly.
(d) conserving momentum.

G. When objects stick together after colliding,
(a) momentum is not conserved.
(b) momentum is zero.
(c) the collision is inelastic.
(d) the collision is elastic.

H. Work is done
(a) by any applied force.
(b) by all components of force.
(c) when a force moves an object.
(d) by an applied force perpendicular to the direction of motion.

I. When a bird's speed is doubled, its kinetic energy is
(a) half as large.
(b) the same.
(c) twice as large.
(d) four times as large.

J. A heavy truck has more momentum than a passenger car moving at the same speed because the truck
(a) is not streamlined.
(b) has greater mass.
(c) has a large wheelbase.
(d) has greater velocity.

SUGGESTIONS FOR FURTHER READING

Brancazio, Peter J., "Physics of Basketball." *American Journal of Physics*, April, 1981.
Asserts that a knowledge of physics can make one a better basketball player.

Brancazio, Peter J., "Rigid-body Dynamics of a Football." *American Journal of Physics*, May, 1987.
According to the author, the reader will come to look at forward passes, kickoffs, and punts with a whole new perspective.

Brody, H., "Physics of the Tennis Racket." *American Journal of Physics*, June, 1979.
Theoretical and experimental studies suggest that it may be possible to design a better tennis racket.

Brody, H., "The Sweet Spot of a Baseball Bat." *American Journal of Physics*, July, 1986.
Discusses the three sweet spots of a baseball bat and determines the location of the ball impact point on the bat that leads to maximum "power" (greatest batted ball speed).

Damask, Arthur C., "Forensic Physics of Vehicle Accidents." *Physics Today*, March, 1987.
Surveys an emerging interdisciplinary field that is leading to safer vehicle design. The field focuses on the reconstruction of accidents and the analysis of the mechanisms of injury.

Dyson, Freeman J., "Energy in the Universe." *Scientific American*, September, 1971.
Shows that energy on the earth is part of the energy in the universe.

Feld, Michael S., Ronald E. McNair, and Stephen R. Wilk, "The Physics of Karate." *Scientific American*, March, 1979.
A karate expert can break wood and concrete blocks with bare hands. A large amount of momentum is applied to a small area, delivering several kilowatts of power over several milliseconds.

Frohlich, Cliff, "Aerodynamic Effects on Discus Flight." *American Journal of Physics*, December, 1981.
Calculates the effect on distance thrown caused by changes in wind velocity, altitude, air temperature, gravity, and release velocity.

Lawson, Ronald A., and Lillian C. McDermott, "Student Understanding of the Work-Energy and Impulse-Momentum Theorems." *American Journal of Physics*, September, 1987.
Students were asked to compare the changes in momentum and kinetic energy of two frictionless dry-ice pucks as they moved under the influence of the same constant force. Relating the theory learned in class to the simple motion that they observed proved to be a challenge to them.

McFarland, Ernie, "How Olympic Records Depend on Location." *American Journal of Physics*, June, 1986.
A number of anomalous records in track and field were set during the Summer Olympics at Mexico City in 1968. Some can be explained by the effects of low gravity at Mexico City.

Starr, Chauncey, "Energy and Power." *Scientific American*, September, 1971.
Discusses the role of energy in human life: past, present, and future.

Summers, Claude M., "The Conversion of Energy." *Scientific American*, September, 1971.
Light bulbs, automobile engines, and home furnaces are a few of the devices that convert one kind of energy into another.

Walker, Jearl, "The Essence of Ballet Maneuvers Is Physics." *Scientific American*, June, 1982.
Examines the movements of classical ballet as a blend of beauty and the physics of motion.

ANSWERS TO NUMERICAL EXERCISES

6–1. 400 kg m/s

6–4. 900 kg m/s

6–5. (a) -7.50×10^3 N · s
(b) -2.50×10^3 N

6–6. 7.00 m/s

6–13. 50.0 joules

6–16. 340 watts (0.460 hp)

6–19. 2200 J

6–20. 1.1×10^4 watts

6–21. 0.134 hp

6–22. 1.82×10^{-18} J

6–23. 90.0 joules

6–24. 175 J

6–25. 3750 J

6–26. 1.25×10^6 J

6–27. 49.0 J

6–28. (a) 22.5 J   (b) 90.0 J

6–30. 4.90×10^5 J

6–36. (a) 58.8 J   (b) 44.3 m/s

ESSAY

ALEGERNATIVE ENERGY SOURCES

ROBERT E. GABLER
Western Illinois University

ROBERT J. SAGER
Pierce College, Washington

DANIEL L. WISE
Western Illinois University

Ninety percent of the energy consumed in the United States is derived from the burning of fossil fuels. As those conventional energy sources dwindle and as we become more concerned with their role in escalating the greenhouse effect, alternative energy sources become more viable as new methods to heat and cool our homes and provide power to our factories. Nuclear power is a pollution-free alternative energy source. However, the possibility of the accidental release of radioactive material into the atmosphere through a nuclear plant accident, like that at Three Mile Island in 1979, is a major concern of environmentalists. While the 1979 incident did not release harmful levels of radiation into the atmosphere, it certainly illustrated the potential for such an event and greatly contributed to the public's reluctance to accept nuclear power as an energy source. Even though the probability of a catastrophic nuclear accident is remote, the environmental problem of how to dispose of the spent radioactive fuel is still another environmental stumbling block to its use.

Natural alternative energy sources, such as wind, water, and solar energy are much more attractive, since they are not only clean and safe sources of energy, but also renewable—indeed infinite—resources. Hydropower, power derived from the energy of moving water, is used in those regions where the damming of streams permits its use. Although the cost of generating electricity by this method is reasonable, the high cost of transmitting electrical power over long distances hinders widespread distribution of this energy source. Other problems are the shortage of good sites for development and opposition from environmentalists concerned about the flooding of cropland and wildlife habitats. In the near future, the two other natural alternative energy sources mentioned—wind and solar energy—might offer more acceptable solutions to our growing energy crisis.

WIND POWER

Wind power is an inexhaustible source of clean energy. Windmills, still visible on the landscape but seldom operational, were once used to pump water and to grind grain before the widespread availability of inexpensive electricity. While wind power, unlike solar power, is not limited to daytime use, it does have some problems. First, it is quite costly;

energy production by windmills costs more than twice that of conventional fuels. Some of the giant wind turbines used to generate power cost over a million dollars to build. A second factor is the need for a steady wind source. Because wind is so variable, a wind power system must be able to store the energy generated during windy periods for use during calm periods. This usually necessitates the use of expensive storage batteries to insure an adequate supply of power. To meet the electrical requirements of a typical home requires a wind power system that would cost about $10,000.

Rather than use individual wind turbines, it is much more economical to use a cluster of wind turbines, called *wind farms*. Wind farms consist of 50 or more wind generators, each producing at least one megawatt of electricity (Fig. A).

Most wind power systems can extract 30% to 40% of the wind's energy, with some experimental models reaching up to 60% efficiency. A typical wind power system requires that wind speeds must be at least 20 km (12 mi) per hour 40% of the time to operate economically. Since the power generated is proportional to the cube of the wind speed, a

doubling of the wind speed increases energy production eight times. Thus, we want persistent strong winds. However, if the winds are too strong or gusty, damage to the expensive turbines could result. Possible sites within the United States that hold the greatest potential for wind power are the western Great Plains from Wyoming to Oklahoma, the New England coast, the Pacific Northwest, and coastal California.

SOLAR ENERGY

The earth receives, in just two weeks' time, an amount of solar energy that would equal our entire known global supply of fossil fuels. Although the potential use of solar radiation for power production is enormous, various problems inhibit its use on a larger scale. Primarily, the intermittent nature of solar radiation (no radiation at night, cloud cover) coupled with the high cost of converting solar energy to electricity preclude anything but local application at this time. An economical means is needed of first collecting, then storing the energy for use during nonsunshine periods, and finally converting solar energy to electricity. Until such technology is available, solar energy will not be a major source of power production.

Small-scale use of solar energy, where cost is not a factor, is already with us. For many centuries, farmers and fishermen have used the power of the sun to evaporate their salt pans, dry fish, and preserve crops. In the space program and in sea buoys, road signs, and offshore oil rigs, solar power lights and heats small-scale operations. On the domestic scene, the use of solar space heating and hot water heating is a reality. One approach, the use of so-called **passive systems**, employs good architectural design and directional siting to warm interiors in winter and prevent overheating in summer.

From the beginning, Indian adobe structures of the Southwest were well-adapted to the desert sun, with their thick walls, small windows, and south-facing exposures set in overhangs on canyon walls to shelter inhabitants from the near-vertical summer rays, thus making maximum use of solar heating and cooling. The Chinese, very early on, placed their village homes with doors and windows facing south and thick adobe windowless walls on the north side facing into the cold winter winds. Architects in Arizona and California, as well as in Israel

Figure A   A wind farm in California. (Lowell Georgia/Science Source, Photo Researchers, Inc.)

and elsewhere, are experimenting with similar passive solar design, utilizing greenhouse-like attachments, ultrathick stone or concrete walls, double-glazed windows, and careful site placement. **Active systems** include flat-plate and collector panels that heat water to 70°C (158°F). The water is then circulated and/or stored for domestic heating arrangements. Obviously, initial installation costs are expensive, but estimates suggest that after about a year and a half, there are savings of 60 percent on the cost of heating water.

Large-scale solar energy operations are mainly in the experimental stage and, for reasons of cost, are generally not feasible at present. Two types of solar technology are being actively developed in various parts of the world, especially in the United States, France, Israel, and Australia. Photovoltaic cells that convert sunlight directly into electrical power (not unlike a camera's light meter) are already in use in Arizona and California and in remote communication systems (Fig. B). Prices for these photovoltaic cells have dropped dramatically over the last 30 years, since they are now mass-produced.

The other major type of solar technology involves solar thermal towers, where racks of tracking mirrors (a heliostat field) follow the sun and focus its heat on a steam boiler perched on a high tower. Temperatures in the boiler may be raised to over 500°C (900°F). In principle no different from our youthful experiments with a magnifying glass to set fire to paper, this device already is operative at experimental sites in France and elsewhere. In California, Solar One in the Mojave Desert is the world's largest solar thermal electric power plant of this kind. However, development is slow because

Figure B The world's largest solar electric generating system, located in Boron, California. (Ken Lucas, Biological Photo Service)

additional solar thermal towers must be considered in terms of current economic feasibility. Only as the costs of other energy sources rise and those of solar devices fall significantly will these become fully competitive.

When sites for possible future plant locations are considered, obviously the "sunbelt states" of the American Southwest are favored, where high percentages of possible annual sunshine amounts are assured. Flat ground is also necessary if large heliostat fields are to be constructed, and, as is the case with all new energy development, the impact of the new sites on the local environment must be carefully considered. Whether it be from earth-based solar energy stations or from insolation-collecting satellites beaming electricity to earth via microwave transmissions, or from some other as yet undeveloped technology, the sun will become increasingly important as humanity enters the twenty-first century.

It is apparent that the use of our climatic elements for power production is limited at present to local applications, and even then the cost factor may be prohibitive. As our conventional fuel supplies are depleted, however, these alternative sources will undoubtedly receive more attention. In the not-too-distant future, technological advances will be achieved, allowing these valuable energy sources to be utilized on a much larger scale than at present.

The above information illustrates some possible positive aspects of weather and climate, relative to energy production. Negative aspects also exist. Most power is transmitted or delivered by cables. Severe weather conditions (ice storms, strong winds, or electrical storms, for example) can reduce transmission or damage the system, causing increased needs for maintenance. Weather also causes erratic patterns in energy demands. Excessively hot summers increase the use of energy for cooling, and extremely cold winters require more energy for heating. Unfortunately, peak demands and increased maintenance requirements often occur at the same time, taxing the system to its limits. As a consequence of the influence of weather on a region, energy systems must be designed and constructed to accommodate the possible extreme climatic conditions even though these conditions are seldom encountered—a very costly, but necessary, precaution.

The eruption of a volcano, such as Hawaii's Mauna Loa, serves as a reminder of the considerable heat present in the interior of the earth. (Krafft/Photo Researchers Inc.)

7

HEAT—A FORM OF ENERGY

Although inventors have made ambitious claims, a perpetual motion machine—a device that would put out more energy than it takes in—has never been built. Whenever a machine performs work, a part of the energy put into the machine seems to disappear as friction converts it into heat. The useful work output, therefore, must be less than the original energy put into the machine. The efficiency of a machine, defined as the ratio of work output/work input, is thus always less than 100%. Since this frictional loss in the form of heat always occurs, it is important to understand the subject of heat in some detail. What is heat? How is it measured? How does it affect life?

7–1 TEMPERATURE MEASUREMENT

Our sense perceptions enable us to tell that an ice cube tray is cold and the sand on a sunny beach is hot. That such judgments can be ambiguous at times is indicated by a demonstration. Set up three pans of water—hot, cold, and lukewarm (Fig. 7–1). Place your right hand in the pan of hot water, a, and your left hand in the cold water, c, for the same length of time.

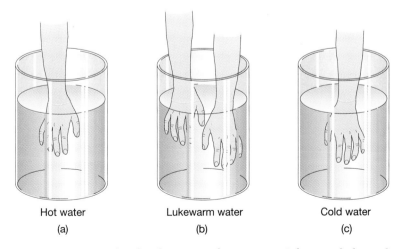

| Hot water | Lukewarm water | Cold water |
| (a) | (b) | (c) |

Figure 7–1 Your hand makes a poor thermometer. Subjective feelings of warmth and coldness can be misleading.

Then place both in the lukewarm water, b. To the hand that had been in a, b seems cool, whereas to the hand that had been in c, b seems warm. Since subjective feelings of warmth and coldness can be misleading, an objective instrument is needed to measure temperatures.

Thermometers depend on some property of matter that varies with temperature in a measurable way—a **thermometric property**—such as expansion (gases, liquids, or solids), color change, electrical resistance, or emission of infrared radiation (Fig. 7–2). Also, thermometers must have a reproducible scale, so that readings on different thermometers can be compared. For this purpose, two **fixed points** are usually calibrated, with the interval between them subdivided into equal degrees.

Probably the earliest thermometer was an instrument invented by Galileo, the air thermometer, or thermoscope. It has a glass bulb filled with air and a long stem with its end dipped into a reservoir of colored water (Fig. 7–3). When the bulb is warmed, the air expands and forces the water level to drop. Cooling the bulb causes the air to contract and the water level in the stem to rise. The instrument is surprisingly responsive to temperature changes, although Galileo apparently did not calibrate the stem. Santorio Santorio (1561–1636), a pupil of Galileo and a professor of medicine at the University of Padua, used a variation of the thermoscope to indicate changes in body temperature during illness and produced, in effect, the first clinical thermometer.

When Gabriel Daniel Fahrenheit (1686–1736) perfected a method of cleaning mercury, a substance that had been found to be particularly well suited to thermometers, he introduced the mercury thermometer into general use (Fig. 7–4). A quantity of mercury is enclosed within a bulb and extends into a narrow column in a glass tube. The volume of mercury responds uniformly to temperature changes. Even a small change in volume brought about by warming or cooling will cause a visible change in the length of the column of liquid in the tube. When the thermometer is placed in a cup of hot water, both the mercury and the glass expand, but the mercury ex-

Figure 7–2   An infrared thermometer (pyrometer). The instrument is used to detect the infrared radiation from an object and the radiation is then related to its temperature.

Figure 7–3   Galileo's thermoscope. When the bulb is warmed, the air expands and the water level drops; when the bulb cools, the air contracts and the water rises in the tube.

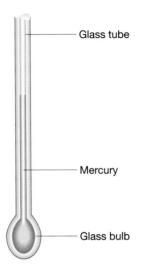

Figure 7–4 A mercury thermometer, a liquid-in-glass thermometer. As the mercury in the bulb expands or contracts with temperature changes, the level in the narrow column rises or falls.

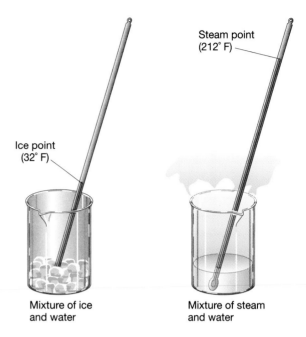

Mixture of ice and water

Mixture of steam and water

Figure 7–5 Calibrating a thermometer on the Fahrenheit scale. By definition, a mixture of ice and water is at 32°F, and a mixture of steam and water at 212°F.

pands more than the glass and rises in the stem of the thermometer. When the thermometer is cooled, the mercury contracts more than the glass and the level of mercury drops. Although a clinical thermometer contains mercury as the thermometric substance, other household thermometers are more likely to contain alcohol made more visible by a dye, usually red or blue.

7–2 TEMPERATURE SCALES

While experimenting with a number of systems as possible fixed points for the scale of his thermometers, Fahrenheit happened to visit the astronomer Olaf Römer (1644–1710) and found him calibrating thermometers. With some minor changes, Fahrenheit adopted the same principles in his own work. Römer had chosen the temperature of a mixture of ice, salt, and water as the lower fixed point and the boiling point of water as the higher fixed point. On **Fahrenheit's scale**, the **freezing point** of water, or **ice point**—the temperature of a mixture of pure ice and water at standard atmospheric pressure— became 32 degrees (Fig. 7–5). The **boiling point** of water, or **steam point**—the temperature of steam from pure boiling water at standard atmospheric

pressure—was assigned the value of 212 degrees. There are, therefore, 212 − 32, or 180 degrees between the steam point and the ice point.

Another temperature scale has been even more widely adopted. In 1742 Anders Celsius (1701–1744), an astronomer, proposed a scale based on 100 degrees between the steam point and the ice point. Although Celsius suggested that the boiling point of water be assigned the value 0 degrees, and the freezing point 100 (possibly to avoid negative temperatures below the melting point of ice), his colleague at the University of Uppsala, Märten Strömer, inverted the scale and identified 0 with the freezing point and 100 with the boiling point of water. The Celsius-Strömer scale is called simply the **Celsius scale** (Fig. 7–6).

Since there are 100 Celsius degrees between the steam point and the ice point, and 180 Fahrenheit degrees between these two points, each Celsius degree is nearly twice as large as a Fahrenheit degree or, more exactly, 180/100 = 9/5 as large. To convert temperatures from one scale to the other, this ratio is one factor we must consider. A second factor is the

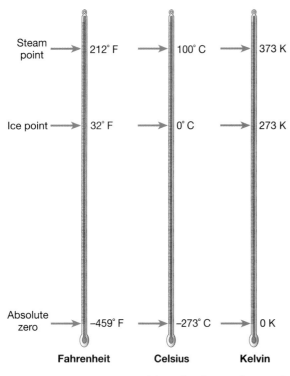

Figure 7–6   Comparison of the Fahrenheit, Celsius, and Kelvin scales.

difference in the calibration of the two scales. On the Celsius scale, the ice point is 0°, but on the Fahrenheit scale it is 32°. The number 32, therefore, also enters into the conversion factor.

Suppose we want to convert a reading of 10°C to its equivalent in °F. We are 10 degrees above the freezing point of water. On the Fahrenheit scale, we are 9/5 (10) or 18 degrees above the freezing point of water, 32°F. The reading of 10°C corresponds, therefore, to 32 + 18 = 50°F. On the other hand, suppose we want to convert a reading of 95°F to its equivalent in °C. On the Fahrenheit scale we are (95 − 32) = 63 degrees above the ice point. On the Celsius scale, we are 5/9 (63) or 35 degrees above the ice point, 0°C. Therefore, 95°F is equivalent to 35°C. Thus, by simple reasoning, we can convert from one scale to the other without any use of formulas. You are encouraged to practice this approach, or you may rely on the following conversion formulas:

$$°F = \frac{9}{5} °C + 32° \qquad (7–1)$$

$$°C = \frac{5}{9} (°F − 32°) \qquad (7–2)$$

EXAMPLE 7–1

A patient's temperature is 100°F, slightly above normal (98.6°F). What is the equivalent temperature on the Celsius scale?

SOLUTION

$$°C = \frac{5}{9} (°F − 32°)$$

$$= \frac{5}{9} (100° − 32°)$$

$$= \frac{5}{9} (68°)$$

$$= \boxed{37.8°C}$$

EXTENSION
One of the highest temperatures ever recorded in the shade was 136.4°F. What is the corresponding Celsius temperature? (Answer: 58.0°C)

■

EXAMPLE 7–2

The melting point of silver is 960.8°C. What is the equivalent Fahrenheit temperature?

SOLUTION

$$°F = \frac{9}{5} °C + 32°$$

$$= \frac{9}{5} (960.8°) + 32°$$

$$= 1729° + 32°$$

$$= \boxed{1761°F}$$

EXTENSION
One of the lowest temperatures recorded in the world was − 88.3°C. What is the Fahrenheit equivalent? (Answer: − 127°F)

■

Figure 7–7 Thermogram of a man standing with his hands on his hips. The color coding runs from white, the hottest regions, through yellow, red, blue, green, purple, and black, the coldest. (Adam Hart-Davis/Science Photo Library, Photo Researchers, Inc.)

Figure 7–8 A thermogram of a house shows varying degrees of heat loss through doors, windows, and walls; red and yellow indicate the greatest loss. How could the owners save on heating costs? (NASA/Science Source/Photo Researchers, Inc.)

7–3 THERMOGRAPHY

Although we think of normal human body temperature as about 98.6°F (37°C), that is only the oral temperature. The range of skin temperature is considerable. The temperature on the skin of the back may be 90°F, the legs 85°F, and the feet 50°F or even lower. The variations result from differences in the skin and in the tissues under it. Where disease or abnormal physiological states are present, there are often changes in skin temperature. More than 95% of all breast cancers, for example, are associated with a skin temperature of at least 1°C (1.8°F) higher than the uninvolved part of the same breast. Fractures, infections, and other conditions also yield their thermal "fingerprints."

A study of temperature data of the body can yield valuable information concerning the diagnosis and treatment of certain disorders. In **thermography**, variations in body temperature are detected and transformed into visible signals that can be recorded on photographic film. The result, called a **thermogram**, shows the parts of the body as white, the warmest, ranging to black, the coldest (Fig. 7–7). The progress of disease and the effectiveness of therapy can be monitored by studying the thermal patterns of the skin before and after treatment. Thermograms can be useful in aiding energy conservation practice. Figure 7–8 shows a thermogram of a house, with heat escaping from the windows, front door, and parts of the roof.

7–4 TEMPERATURE AND LIFE; THERMAL POLLUTION

Life is confined to a narrow range of temperatures, from about 0° to 50°C (32° to 122°F). Very few organisms can live long at temperatures above or below these limits. Among the exceptions to this rule are some bacteria that can survive for months at a temperature of −180°C (−292°F). At the other extreme, some bacteria thrive at 70°C (158°F) or above, and a blue-green alga can live in some of the pools of Yellowstone National Park at 85°C (185°F).

Life is sensitive to the temperature of the environment because of the effects of temperature on **enzymes**, those biochemical catalysts that control the rates of life-sustaining reactions. If the organism's temperature decreases, the rates of such reactions decrease. At high temperatures, the large, complex enzyme molecules become unstable, and enzyme-catalyzed reactions cease. Changes of temperature at both ends of the scale influence the rate of reactions, limiting the range within which life is sustained.

The temperature of the earth when life first appeared is not known, but it was probably higher than it is today. Life is so nicely attuned to present temperatures that it is estimated that if the average daily temperature were suddenly raised or lowered by only 10°C, most forms of life would perish. In spite of the large differences in environmental temperature—from the tundra to deserts and jungles, from season to season, and from day to night—the body temperature of most mammals remains close to 37°C (98.6°F).

The discharge of waste heat into natural bodies of water—**thermal pollution**—is, therefore, a matter of concern to environmentalists. The electric-power industry, which consumes more than 80% of the cooling water used by industry, is the principal source of this waste heat (Fig. 7–9). The production of waste heat is inevitable in generating plants, and water is the only practical medium for carrying the heat away. Water is pumped from a river or lake through steam condensers in generating plants. It then is discharged into the same body of water at a temperature that is from −12°C (10°F) to 10°C (50°F) higher than it was originally. The hot discharge water, being warmer and lighter than the main body of water, spreads over the surface and disperses its heat to the river or lake and to the atmosphere.

The increased temperature of the water has a number of effects, most of them harmful to the ecology of the system. It decreases the amount of the dissolved oxygen while at the same time raising the metabolic rate of fish and therefore their need for oxygen. Above some maximum temperature, from about 24° (75°F) to 33°C (91°F), thermal death follows from a breakdown of one or more vital processes: respiration, circulation, or nervous response.

The reproductive mechanisms of fish are stimulated by normal seasonal changes, such as the warm-

Figure 7–9   Cooling tower for an electric generating station. Such stations are normally used when a natural or artificial body of water is not available or to prevent thermal pollution. (Courtesy U.S. Department of Energy)

ing of waters in the spring, and can be upset by abnormal temperatures that simulate such changes. A desirable fish population may die out and be replaced by less desirable species as the waters warm up. The brook trout, for example, becomes more active as the temperature rises from 5° (41°F) to 9°C (48°F). Its swimming speed decreases as the temperature goes from 9° to 16°C (61°F), however; death occurs at 25°C (77°F).

The dispersion of waste heat may be approached from two directions: as the disposal of a waste product or as an opportunity for the recovery of a valuable by-product. As an example of the latter approach, heated water between 32° and 55°C (90° and 131°F) from a pulp and paper plant is used by farmers for such purposes as frost protection and extending the growing season by heating soil in spring and fall. Thermal energy has also been used to desalt seawater by evaporation, and the desalinated water, enriched with nutrients, is used to grow high-quality fruits and vegetables within controlled-environment greenhouses. As the economic feasibility of these and other uses is demonstrated, we may expect to see other productive applications of waste heat.

7–5 THE LOWEST TEMPERATURE

The highest temperatures in the universe probably occur in the interiors of certain stars, which may be as high as 4 billion degrees Celsius. There is no theoretical upper limit to temperature.

In contrast, there is a theoretical *lower* limit to temperature. If a gas at 0°C is cooled at constant volume, its pressure decreases by 1/273 for every decrease of temperature by 1°. (**Pressure** is the ratio of force/area; in SI units, N/m².) At −273°C, the gas pressure should drop to zero (Fig. 7–10). The temperature −273°C is known as the absolute zero of temperature. If no further reduction in pressure is possible, no further reduction in temperature is possible.

William Thomson, Lord Kelvin (1824–1907), a mathematician and professor of physics at the University of Glasgow, proposed a scale based on the **absolute zero of temperature**. The International System of Units (SI) uses the **Kelvin scale**, with the symbol K meaning "degrees Kelvin." The degree symbol (°) is not used. The Kelvin or absolute scale keeps the Celsius degree, but the zero point is intended to be the lowest attainable temperature, 273° below zero on the Celsius scale (459.69° below zero on the Fahrenheit scale). Temperatures within 0.000001 K have been reached. The freezing point of water is 0° + 273° = 273 K, and the boiling point of water is 100° + 273° = 373 K. As on the Celsius scale, there are 100 Kelvin degrees between the ice point and the steam point.

The Kelvin and Celsius scales are related in the following way:

$$K = °C + 273° \tag{7-3}$$

EXAMPLE 7–3

What is the equivalent of 113°F on the Kelvin or absolute scale?

SOLUTION
We first convert the Fahrenheit temperature to Celsius, then to Kelvin.

$$°C = \frac{5}{9}(°F - 32°)$$

$$= \frac{5}{9}(113° - 32°)$$

$$= \frac{5}{9}(81°)$$

$$= 45°C$$

$$K = °C + 273°$$
$$= 45° + 273°$$
$$= \boxed{318 \text{ K}}$$

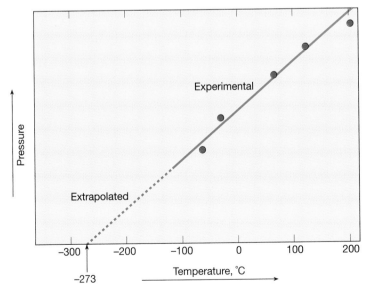

Figure 7–10 Plot of the variation of the pressure of a gas with temperature. When extrapolated downward, the pressure would become zero at −273°C, defined as the absolute zero of temperature.

EXTENSION

Dry ice, solid carbon dioxide, has a temperature of −80°C. Determine the equivalent Kelvin temperature. (Answer: 193 K)

■

7–6 KINETIC THEORY AND THE MOLECULAR INTERPRETATION OF TEMPERATURE

According to the **kinetic theory of matter**, matter is made up of tiny particles called molecules, and these molecules are in a constant state of motion. The molecules of a gas are thought to be rather far apart and to move in a constant, random, and rapid fashion (Fig. 7–11). Because of the wide separation of the molecules, a gas is mostly empty space and can be compressed easily, forcing the molecules closer together. A gas has no fixed shape or volume; the forces between the molecules are so weak that they do not stay together. A gas thus fills any container; the molecules move randomly in all directions, from top to bottom to all sides (Fig. 7–12).

Gas pressure results from collisions between molecules and the wall of the container. The collisions among gas molecules are assumed to be perfectly **elastic**, with both kinetic energy and momentum being conserved. Except for exchanges of kinetic energy during collisions, there are no forces acting between gas molecules or between them and the wall.

In a liquid, the molecules are much closer together than in a gas, so compression is more difficult. Al-

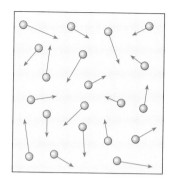

Figure 7–12 The molecules of a gas have velocities of random magnitudes and directions. The dots represent molecules, and the arrows their directions of motion.

though the forces between the molecules are not strong enough to hold them in fixed positions, they are strong enough to keep the molecules fairly close together. Thus, a liquid maintains its volume but assumes the shape of its container. For example, when a liter of water is poured from a pitcher into several glasses, the water assumes the shapes of the glasses but the volume remains 1 liter. Since liquids and gases both have the ability to flow, they are referred to as **fluids**.

In a solid, the molecules are arranged in an orderly, fixed array and are attracted to one another by relatively strong forces. Solids therefore have fixed shapes and sizes and are incompressible. The molecules cannot move very far but vibrate about nearly fixed positions. An increase in temperature causes the molecules of a solid to vibrate faster around these positions.

Figure 7–11 States of matter as described by the kinetic molecular theory.

Solid

Liquid

Gas

The most convincing direct evidence for the kinetic theory occurs in the motion of fine particles suspended in air or water. If a sample of air with fine smoke in it is observed with a microscope, the particles of smoke suspended in the air are seen to move constantly in a random zigzag path. The smaller the particle, the faster its motion. The effect was first discovered by Robert Brown (1773–1858), a botanist, while observing pollen granules suspended in water, and is called **Brownian motion** (Fig. 7–13).

Brownian motion was not really understood until Albert Einstein (1879–1955) explained it by the kinetic theory. Molecules of the medium in which the particles are suspended are in perpetual motion and collide with the particles. Air or water molecules are too small to be seen, but a particle's motion resulting from thee collisions can be followed under a microscope. Einstein used the kinetic theory and statistics to predict the average speed of a particle undergoing Brownian motion. Jean Perrin (1870–1942), noted for his quantitative studies of Brownian motion, experimentally determined the average velocity of a particle, measured the mass, and calculated its kinetic energy. He found the result to be in close agreement with the kinetic theory.

At the microscopic level, temperature is related to the average kinetic energy of molecules. The average kinetic energy of gas molecules is directly proportional to the absolute temperature of the gas. Kinetic theory tells us that the Kelvin temperature of a gas is a measure of the average kinetic energy of its molecules.

$$T = \frac{1}{2} m \overline{\mathbf{v}^2} \tag{7–4}$$

If there are N gas molecules in a box moving at various speeds in different directions and colliding with other gas molecules and with the molecules of the wall, the average kinetic energy of the gas molecules is given by the following expression:

$$\text{Average kinetic energy of molecules} = \frac{3}{2} kT$$

$$\overline{\text{KE}} = \frac{1}{2} m \overline{\mathbf{v}^2} = \frac{3}{2} kT \tag{7–5}$$

where T is the Kelvin temperature and k is $1.38 \times 10^{-23} \dfrac{\text{joule}}{\text{molecule} \cdot \text{K}}$. The square root of $\overline{\mathbf{v}^2}$ is called the root mean square (rms) speed of the molecules.

$$\mathbf{v}_{\text{rms}} = \sqrt{\frac{3kT}{m}} \tag{7–6}$$

The pressure on the walls of a container containing a gas results from collisions of the gas molecules against the walls. When the temperature of a gas is increased, the molecules have more kinetic energy and strike the walls with greater force. Molecules may well have speeds of hundreds or thousands of miles per hour. As can be seen in the preceding formulas, the greater the mass (m) of the molecule, the lower the rms speed for a given temperature. This concept can be extended to particles of enormous size compared to molecules. For example, in Brownian motion, the particle velocity is still appreciable, but it is negligible for more massive bodies.

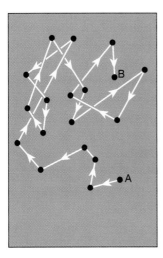

Figure 7–13 Brownian motion. Path of a pollen particle suspended in water. Positions were determined at two-minute intervals with a high-powered microscope.

EXAMPLE 7–4

What is the rms speed of oxygen molecules at 300 K? (Mass of an oxygen molecule = 5.32×10^{-26} kg.)

SOLUTION
Apply Equation 7–6. The constant k is known, and T and m are given. Solve for $\mathbf{v}_{\text{rms}}$.

$$T = 300 \text{ K}$$
$$m = 5.32 \times 10^{-26} \text{ kg}$$
$$\mathbf{v}_{rms} = ?$$

$$k = 1.38 \times 10^{-23} \frac{\text{joule}}{\text{molecule} \cdot \text{K}}$$

$$\mathbf{v}_{rms} = \sqrt{\frac{3kT}{m}}$$

$$= \sqrt{\frac{3(1.38 \times 10^{-23}) \dfrac{\text{joule} \cdot (300 \text{ K})}{\text{molecule} \cdot \text{K}}}{5.32 \times 10^{-26} \text{ kg}}}$$

$$= \sqrt{2.33 \times 10^{5} \frac{\text{m}^2}{\text{s}^2}}$$

$$= 483 \frac{\text{m}}{\text{s}}$$

EXTENSION

A virus (mass = 1.60×10^{-20} kg) is in a room at a temperature of 300 K. Calculate its root mean square speed. (Answer: 0.880 m/s)

∎

7–7 TEMPERATURE AND HEAT

When you boil a kilogram of water on a hot plate (Fig. 7–14), it takes a certain amount of time to raise the temperature of the water from room temperature to the boiling point. If you boil 5 kilograms of water on the same hot plate, the time required is five times as great. Something apparently passes from the hot plate

to the water, and more of this quantity is needed for the greater amount of water.

The first person to distinguish clearly between heat as a quantity distinct from, although related to, temperature was probably Joseph Black (1728–1799), a professor of medicine at the University of Glasgow. Heat is what is transferred from the hot plate to the water. It is a physical entity, and we can measure its amount just as we measure an amount of water. (The science of heat measurements is known as **calorimetry**.) **Temperature**, on the other hand, refers to the *intensity* of heat, its "hotness" or "coldness." At the boiling point, both samples of water have the same temperature, but the beaker with 5 kilograms of water has more of some entity.

According to the kinetic theory, **heat flow** is the transfer of molecular energy from one body to another because of a temperature difference (Fig. 7–15). The direction of flow is from a high temperature to a low temperature. For example, the fact that the plate is hot means that its molecules are moving very

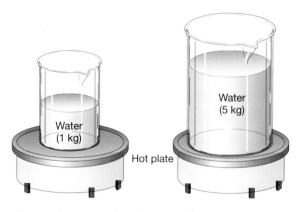

Figure 7–14 It takes five times as long to boil 5 kg of water as to boil 1 kg.

Figure 7–15 Thermal energy is transformed along the branding iron by conduction because the brand is hotter than the handle. (David Rogers)

rapidly. (Temperature is directly proportional to the average kinetic energy of the molecules.) When these high-speed molecules strike a beaker, they transfer some of their energy through collisions to the molecules of the beaker, thus raising its temperature. The faster-moving molecules of the beaker then collide with the cooler water molecules and increase their kinetic energy so that the temperature of the water increases as well. In this way, energy is transferred from the hot plate to the water.

Not all of the energy transferred to the water goes into increasing the **translational kinetic energy** of the water molecules, in which the molecules move from one place to another. Some of it increases the **rotational** and **vibrational motion** of the molecules (Fig. 7–16). The total energy of the water molecules —the sum of the kinetic energy, rotational energy, and vibrational energy—is referred to as **thermal energy**. The terms "temperature" and "thermal energy" thus refer to different properties. In 5 kilograms of water there is five times as much thermal energy as in 1 kilogram of water at the same temperature.

7–8 SPECIFIC HEAT

When a kilogram of warm water at 75°C is mixed with a kilogram of water at 25°C, the temperature of the mixture is found to be 50°C (Fig. 7–17). The temperature of the warm water is lowered 25°, and that of the cold water is raised 25°. It appears that the heat lost by the warm water is equal to the heat gained by the cold water.

Suppose that equal masses of two liquids are mixed, ethyl alcohol at 75°C and water at 25°C. The temperature of this mixture is only 43.6°C. The alcohol cools by 31.4°, whereas the water becomes warmer by only 18.6°. If we assume that **heat is conserved**, that is, that the heat lost by the warm alcohol is equal to the heat gained by the cold water, then a smaller quantity of heat is needed to increase the temperature of alcohol than is needed to increase the temperature of water by the same number of degrees. Alcohol has less **thermal capacity,** or capacity for heat, than does water. Every substance has its own thermal capacity, which can be determined by experiment.

The **specific heat** of a substance is defined as **the amount of heat required to raise the temperature of 1 gram of the substance 1 Celsius degree**. Let us take water as an example. The amount of heat required to raise the temperature of 1 gram of water 1 Celsius degree (from 14.5° to 15.5°C) is defined as

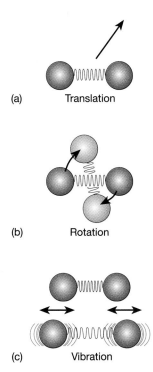

Figure 7–16 Molecules can have (a) translational kinetic energy, moving from one place to another; (b) rotational energy; and (c) vibrational energy—the atoms vibrate back and forth in the molecule.

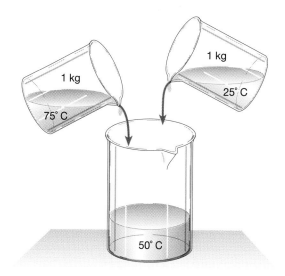

Figure 7–17 A kilogram of water at 75°C is mixed with 1 kg of water at 25°C. The mixture has a temperature of 50°C.

Table 7–1 SPECIFIC HEATS OF
VARIOUS SUBSTANCES

	Substance	Specific Heat (cal/g · C° or BTU/lb · F°)
Solids	Ice	0.49
	Wood (typical value)	0.40
	Aluminum	0.215
	Asbestos	0.195
	Glass (typical value)	0.19
	Copper	0.094
	Silver	0.056
	Tin	0.055
	Platinum	0.032
Liquids	Water	1.000
	Ethyl alcohol	0.60
	Benzene	0.41
	Mercury	0.033
Gases	Steam	0.48
	Air	0.17

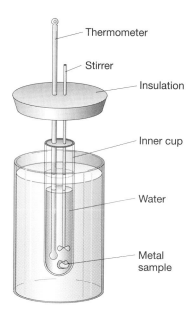

Figure 7–18 A simple, double-can calorimeter. The specific heat of a metal is determined by the method of mixtures. The metal sample, for example, copper, is preheated to 100°C and placed in the inner cup, which contains water at room temperature.

the **calorie** (lowercase "c"). Thus, the specific heat of water is 1 cal/g · C°. One **kilocalorie**, or "giant" calorie—**Calorie** (capital "C")—is the heat required to raise the temperature of 1 kilogram of water 1 Celsius degree. The kilocalorie equals 1000 "small" calories and is the unit used to rate the heat value of foods and fuels.

In engineering work, one **British thermal unit** (BTU) is defined as **the amount of heat required to raise the temperature of 1 pound of water 1 Fahrenheit degree.** For water, the specific heat is 1 BTU/lb · F°, and for other substances the numerical value of specific heat in BTUs is identical to the value of cal/g · C°. One British thermal unit is equal to 252 calories. Table 7–1 gives the specific heats of a number of substances.

7–9 CALORIMETRY

Measurements on the quantity of heat transferred from one substance to another are carried out in a specially designed vessel called a **calorimeter**, in which heat exchanges with the environment are reduced to a minimum (Fig. 7–18). In the **method of mixtures**, the heat given by a hot substance placed in the calorimeter is equal to the heat absorbed by the calorimeter and its contents (Fig. 7–19).

Since the specific heat of liquid water is 1.00 cal/g · C°, supplying 5.00 calories of heat to 1.00 gram of water increases the temperature 5.00 C°. The same quantity of heat supplied to 1.00 gram of alcohol raises its temperature 8.30 C°, since the specific heat of alcohol is 0.60 cal/g · C°. To raise the temperature of 1.00 gram of water from the freezing point (0°C) to the boiling point (100°C), 100 calories must be supplied (Fig. 7–20). The amount of heat that water releases to the environment upon cooling through the same temperature difference is equal to the amount of heat supplied in these processes intially. One gram of water thus furnishes 100 calories to its surroundings in cooling from its boiling point to its freezing point. Thus, the amount of heat a substance loses or gains depends on three factors: (a) the mass of the substance; (b) the nature of the substance; and (c) the temperature change. This relationship is expressed as follows:

$$\begin{pmatrix} \text{Heat lost} \\ \text{or gained} \end{pmatrix} = \begin{pmatrix} \text{Mass of} \\ \text{substance} \end{pmatrix} \cdot \begin{pmatrix} \text{Specific} \\ \text{heat} \end{pmatrix} \cdot \begin{pmatrix} \text{Change in} \\ \text{temperature} \end{pmatrix}$$

  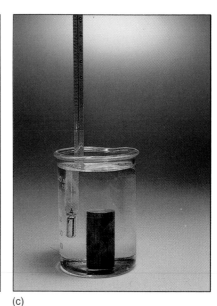

(a)   (b)   (c)

Figure 7–19   (a) An iron bar is heated in a flame. (b) When it is plunged into a beaker of cold water, heat is transferred from the hotter metal bar to the cooler water, causing the water immediately around the hot metal to boil. (c) Eventually the iron bar and the water reach the same temperature. (Charles D. Winters)

or, in symbols:

$$H = mc(T_2 - T_1) \tag{7-7}$$

in which H is the amount of heat lost or gained; m is the mass of the material; c is its specific heat; and $(T_2 - T_1)$ is the temperature change.

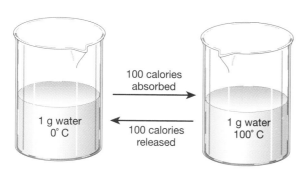

Figure 7–20   To raise the temperature of 1 g of water from 0°C to 100°C, 100 calories must be supplied. In turn, when the water cools from 100°C to 0°C, 100 calories are released.

EXAMPLE 7–5

How much heat must be supplied to 20.0 grams of tin, originally at 25.0°C, to raise its temperature to 100°C?

SOLUTION
The mass of tin, the initial temperature, and the final temperature are given. The specific heat of tin can be obtained from Table 7–1.

$m = 20.0$ g

$c_{tin} = 0.055$ cal/g · C°

$t_1 = 25.0$°C

$t_2 = 100$°C

$H = ?$

$H = m \cdot c \cdot (T_2 - T_1)$

$= (20.0\text{g}) \left(0.055 \dfrac{\text{cal}}{\text{g} \cdot \text{C}°} \right)$

$(100 - 25.0)°$ C

$= (20.0)(0.055)(75.0)$ cal

$= \boxed{82.5 \text{ cal}}$

How much heat is required to warm 150 kg of bathwater by 15 C°? (Answer: 2250 kilocalories)

∎

In a very real sense, foods are body fuels. The heat value of fuels and foods is determined from the **heat of combustion—the heat produced per unit mass of substance burned in oxygen**. With their heats of combustion known, it is possible to compare one kind of fuel with another or different grades of the same fuel with respect to their heat value. Kerosene, for example, yields 11,200 kcal/kg when burned, whereas alcohol furnishes 6400 kcal/kg. To get the same amount of heat from alcohol as from kerosene, the mass of alcohol burned must be nearly twice as great. For weight-watchers, the heat of combustion of foods is a factor in deciding on their daily diet; a scrambled egg yields more calories (2100 kcal/kg) than a boiled egg (1600 kcal/kg).

The heat of combustion of a fuel or food is measured in a **bomb calorimeter** (Fig. 7–21). The "bomb" is a strong steel cylinder fitted with a heavy cap that screws tightly into place. A crucible suspended from the cap contains a known mass of the material to be burned. A small platinum wire suspended from two insulated rods dips into the fuel. The cap is adjusted, and oxygen is admitted through a valve in the cap to a pressure of about 15 atmospheres. The bomb is then placed in a water calorimeter. An electric current sent through the wire heats it to incandescence and ignites the fuel, which burns violently in the oxygen atmosphere. The large amount of heat that is produced causes the temperature of the metal bomb and the surrounding water to increase. The heat of combustion is then computed by the method of mixtures:

$$\begin{array}{rl} \text{Heat produced} \\ \text{by fuel} \end{array} = \begin{array}{l} \text{Heat gained by} \\ \text{calorimeter} \end{array}$$
$$+ \begin{array}{l} \text{Heat gained} \\ \text{by water} \end{array}$$

Table 7–2 gives the heats of combustion of some fuels and foods.

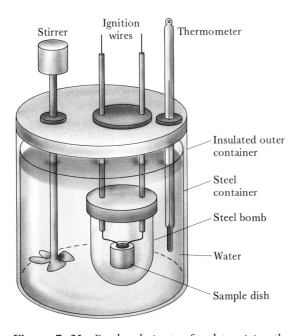

Figure 7–21 Bomb calorimeter for determining the heat of combustion of fuels or foods. (Source: William L. Masterton and Cecile N. Hurley, *Chemistry: Principles & Reactions*. Philadelphia: Saunders College Publishing, 1989, p. 216, Fig. 7.4.)

Stirrer
Ignition wires
Thermometer
Insulated outer container
Steel container
Steel bomb
Water
Sample dish

Table 7–2   HEATS OF COMBUSTION OF FUELS AND FOODS

	Heat of Combustion (kcal/kg)
Fuels	
Natural gas	8000–12,000
Gasoline	11,300
Diesel oil	10,500
Fuel oil	10,300
Anthracite coal	7000–8000
Alcohol	6400
Wood, pine	4500
Foods	
Bread, white	2600
Butter	7950
Eggs	
Boiled	1600
Scrambled	2100
Ice cream	2100
Meat, lean	1200
Milk	715
Peanuts	5640
Potatoes, white boiled	970
Rice, cooked	1120
Sugar, white	4000

7–10 CHANGE OF STATE

The sensation of coldness that you feel when you touch a piece of ice is an indication that heat is quickly drawn from your hand to the ice. Surprisingly, however, the temperature of the dripping water is no greater than the temperature of the ice. During the melting process, a large amount of heat—80 calories per gram—is absorbed by the ice *without* raising its temperature (Fig. 7–22). The heat that the ice absorbs in this way without increasing its temperature is known as the **latent heat of fusion**, L_f. The latent heat of fusion is defined as the **quantity of heat that must be added to 1 gram of solid at its melting point to change it to liquid at the same temperature and pressure**.

As viewed by kinetic theory, the molecules of ice vibrate faster and faster about their normal positions as the ice is heated. At 0°C (when the pressure is 1 atmosphere), the molecular vibrations are so vigorous that the molecules break out of their fixed positions in the crystal lattice and start rolling over one another; that is, the ice begins to melt. During the melting process, the temperature does not change: The heat of fusion does not increase the kinetic energy of the molecules. Instead, work must be done to overcome the attractive forces between the ice molecules, in order to separate them into the more random motion of the molecules of liquid water, and this raises their potential energy. The quantity of heat required to melt ice at 0°C and 1 atmosphere depends on the mass of ice and the heat of fusion of ice and may be calculated from the following equation:

$$H = mL_f \qquad (7-8)$$

in which H is the quantity of heat, m is the mass of ice, and L_f is the latent heat of fusion of ice.

EXAMPLE 7–6

How much heat is required to melt 50.0 grams of ice at 0°C?

SOLUTION

Use Equation 7–8 to determine the quantity of heat involved in the melting process of ice at 0°C, assuming that the temperature of the resulting water is 0°C. (See Table 7–3 for latent heat of fusion.)

$$m_{ice} = 50.0 \text{ g} \qquad H = mL_f$$
$$L_{f_{ice}} = 80.0 \text{ cal/g} \qquad = \left(50 \text{ g}\right)\left(80 \frac{\text{cal}}{\text{g}}\right)$$
$$H = ? \qquad\qquad = \boxed{4000 \text{ cal}}$$

EXTENSION

How much heat is required to melt 50.0 grams of copper at its melting point? (Answer: 2100 cal)

■

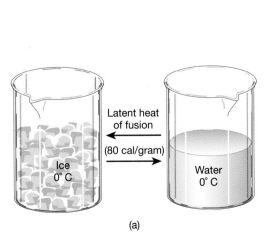

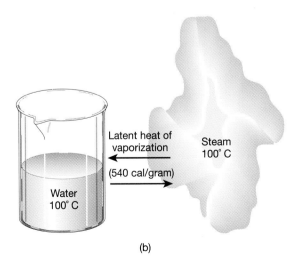

(a)   (b)

Figure 7–22   Phase changes. Ice absorbs 80 cal/g at 0°C to melt. When water freezes at 0°C, 80 cal/g are released. The heat of fusion of H_2O is 80 cal/g.

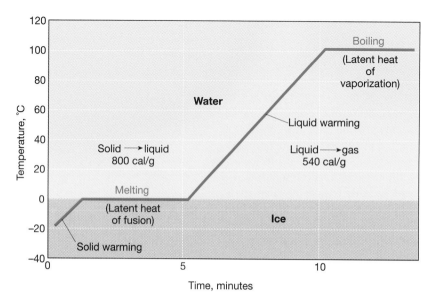

Figure 7–23 A graph showing the temperature of ice and water heated at a constant rate. Note that the temperature remains constant as the ice melts and again as the water boils.

Once the ice has melted, further heating increases the KE of the molecules and the temperature rises (Fig. 7–23). At the boiling point, the molecules are moving so rapidly that the attractive forces can no longer confine them to the limited volume of a liquid, and the molecules move out in all directions as a gas. The work done against the attractive forces raises the PE of the molecules. The energy required to do this work, called the **latent heat of vaporization**, L_v, is defined as **the quantity of heat that must be added to 1 gram of the liquid at its boiling point to change it to vapor at the same temperature and pressure**.

The value of L_v of water is 540 calories per gram, a value more than six times as large as its latent heat of fusion. The reason is that the energy needed to separate liquid molecules into the gas is greater than that needed for the smaller molecular separation required when a solid becomes a liquid. The corresponding equation for the process of vaporization is:

$$H = mL_v \qquad (7\text{–}9)$$

EXAMPLE 7–7

How much heat is required to vaporize 50.0 grams of water at 100°C?

SOLUTION
Apply Equation 7–9.

$$m_{water} = 50.0 \text{ g} \qquad H = mL_v$$
$$L_{v_{water}} = 540 \text{ cal/g}$$
$$H = ? \qquad = \left(50.0 \text{ g}\right)\left(540 \frac{\text{cal}}{\text{g}}\right)$$
$$= \boxed{27,000 \text{ cal}}$$

EXTENSION
During a body rub, 100 g of alcohol is converted from a liquid to a gas. How much body heat is removed? (Disregard the change in temperature before the alcohol changes state.) (Answer: 20,400 cal)

■

Table 7–3 gives the latent heats of fusion and vaporization of several substances.

The phase changes of a substance can be reversed. For example, water vapor condenses on the outside of a glass containing a cold drink. Water molecules in the air near the cold glass lose KE when they collide with the molecules of the glass. As more and more water molecules slow down and approach one another, the attractive forces between them cause them to come together and form a drop of water on the glass. In this

Table 7–3   LATENT HEATS OF FUSION AND VAPORIZATION (AT ATMOSPHERIC PRESSURE)

Substance	Melting Point (°C)	Latent Heat of Fusion, L_f (cal/g)	Boiling Point (°C)	Latent Heat of Vaporization, L_v (cal/g)
Alcohol, ethyl	−117.3	24.9	78.5	204
Ammonia	−75	108.1	−34	327
Copper	1083	42	2595	1760
Mercury	−38.7	2.8	357	70.6
Water	0	80	100	540

process, the latent heat of vaporization of water, 540 calories per gram, is released to the surroundings.

In a steam heating system, water is boiled in the furnace, absorbing 540 cal/g at the boiling point as it changes to steam. Entering a radiator at 100°C, steam is cooled and condenses to water, releasing the latent heat of vaporization to the surroundings. This is also the principle behind keeping tubs of water near vegetables or plants to protect them on a cold night. The freezing of the water is accompanied by the release of heat to the air—the latent heat of fusion—and this keeps the surrounding temperature from falling too low.

The cooling effect of a dab of rubbing alcohol on the skin comes from the withdrawal of surface heat, causing the alcohol to evaporate. For the same reason, a person feels cool after a swim, particularly if there is a breeze; the latent heat of vaporization of water is large, and as the water evaporates, it withdraws body heat. The cooling effect of evaporation may be used to keep liquids cool; evaporation from a moist, porous, canvas container helps to keep the contents cool. A burn from "live" steam is more harmful than one caused by an equal mass of hot water. As it comes into contact with the cooler skin, each gram of steam releases the latent heat of vaporization of steam—540 cal/g. This is a much greater value than the specific heat of water—1 cal/g · C°—released by each gram of water for each degree it is cooled.

When moth flakes are placed in a dresser drawer, they disappear eventually without leaving a trace. The moth flakes, a chemical substance called *p*-dichlorobenzene, **pass directly from the solid phase to the gas phase**, without passing through the liquid phase. This process is called **sublimation**. Be-

sides moth flakes, solid iodine and "dry ice" are familiar examples of substances that sublime (Fig. 7–24). Ordinary ice also sublimes to some extent. For example, when wet clothing is put out to dry in zero-degree weather, the water in it freezes, but after a period of time much of the ice sublimes.

We do not usually observe sublimation, because the more common vapors are usually colorless. Iodine crystals, however, give off a purple vapor when they sublime. If a few crystals are placed in a beaker covered with a watch glass, iodine vapor fills the beaker (Fig. 7–25). The presence of cold water in the watch glass causes iodine crystals to be deposited on the underside of the watch glass. The blackening inside a tungsten incandescent bulb is another example. When the tungsten filament is hot, some of the tungsten goes directly into the vapor state. As the vapor

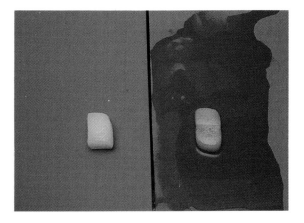

Figure 7–24   Dry ice (*left*) sublimes, leaving no puddle as the solid changes directly to a gas. Water ice (*right*) passes through the liquid state normally, then evaporates. (David Rogers)

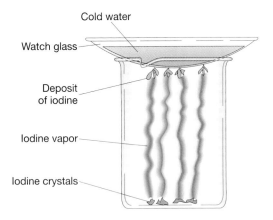

Figure 7–25 Sublimation of iodine crystals and their deposition on a watch glass.

Figure 7–27 A gas confined to a cylindrical container whose volume can be varied with a movable piston.

encounters the cooler glass of the bulb, tungsten is deposited on it and darkens it.

7–11 THERMAL EXPANSION

The volume of a gas at constant pressure decreases by 1/273 of its value at 0°C (273 K) for each degree decrease in temperature. The nature of the gas does not matter; all gases undergo approximately the same fractional change in volume with the same change in temperature (Fig. 7–26). This discovery was made independently by Jacques Charles (1746–1823) and Joseph Gay-Lussac (1778–1850) and is known as the Law of Charles and Gay-Lussac. Using absolute temperatures, it tells us:

the volume of a gas varies directly with its absolute temperature (if the pressure is held constant).

$$\frac{V}{V'} = \frac{T}{T'} \tag{7–10}$$

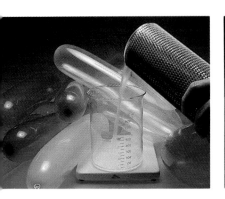

Figure 7–26 Air-filled balloons shrink when placed in liquid nitrogen at 77 K. When warmed back to room temperature, the balloons reinflate to their original volume. (Charles D. Winters)

Figure 7–28 A hot-air balloon rises because heating causes the air inside the balloon to expand and to become less dense than the surrounding air, providing the lift. (David Lissy/Focus on Sports)

in which V is the initial volume of the gas, V' the final volume, T the initial absolute temperature, and T' the final absolute temperature. This is an observation that led to the suggestion that molecules at higher temperatures are moving faster and on the average are farther apart from one another if the volume increases, one of the cornerstones of the kinetic theory. So, the gas will take up more space as its temperature is increased, provided the volume of the container can be varied (Figs. 7–27 and 7–28).

EXAMPLE 7–8

One liter of helium is heated from 25.0°C to 100°C. Assuming that the pressure is constant, what is the final volume?

SOLUTION

A key point in this example is that absolute temperature must be used, so that Celsius temperatures must be converted to their equivalents on the absolute scale. Since the temperature is increased, we predict that the volume of helium will increase.

$$V = 1.00 \text{ liter}$$
$$T = 25.0°C + 273° = 298K$$
$$T' = 100°C + 273° = 373K$$
$$V' = ?$$
$$\frac{V}{V'} = \frac{T}{T'}$$
$$V' = \frac{VT'}{T}$$
$$= \frac{(1 \text{ liter})(373K)}{298K)}$$
$$= \boxed{1.25 \text{ liters}}$$

(expansion)

EXTENSION

If a sample of 6.20 L of CO_2 gas at 810°C is cooled to 225°C at constant pressure, what is the new volume? (Answer: 2.85L)

■

When the temperature of a solid is increased, generally the solid expands. Unlike gases, however, different solids expand by different amounts for each degree change in temperature (Fig. 7–29). An aluminum rod, for example, expands in length twice as much as an iron rod and nearly six times as much as tungsten for the same change in temperature. Experiments show that the amount of expansion in length depends on three factors: (1) the increase in temperature—expansion is greater if the temperature is increased 50 degrees than if it is increased 10

Figure 7–29 The blade in the photograph was straight before being heated. It bends because it is a bimetallic strip, and each component metal expands to a different extent. (Courtesy of CENCO)

degrees; (2) the original length — a long rod expands more than a short one because of the greater number of crystals, each of which undergoes the same expansion; and (3) the nature of the metal. These facts are expressed in the equation for the **linear expansion** of a solid as follows:

$$\Delta L = \alpha L_0 (T_2 - T_1) \qquad (7\text{--}11)$$

in which ΔL (read "delta L") is the change in length; α (alpha) is the coefficient of linear expansion for the material (that is, the fractional expansion per degree change in temperature); L_0 is the original length; and $(T_2 - T_1)$ is the change in temperature. Table 7–4 gives the coefficients of linear expansion of several materials.

EXAMPLE 7–9

An aluminum rod is 50.0 centimeters long at 25.0°C. How long will the rod be at 100°C?

SOLUTION

We expect the rod to elongate with an increase in temperature. We solve for the expansion ΔL and add that amount to the original length to find the length at 100°C. (Find the coefficient of linear expansion of aluminum from Table 7–4.)

$$\alpha_{Al} = 24 \times 10^{-6}/°C$$
$$L_0 = 50.0 \text{ cm}$$
$$T_1 = 25.0°C$$
$$T_2 = 100°C$$
$$\Delta L = ?$$
$$\Delta L = \alpha L_0 (T_2 - T_1)$$
$$= \frac{(24 \times 10^{-6})}{°C} (50.0 \text{ cm})(100 - 25) °C$$
$$= 0.090 \text{ cm}$$

Length at 100°C = 50.0 cm + 0.090 cm

$$= \boxed{50.1 \text{ cm}}$$

EXTENSION

If concrete highway slabs are 30 m long, how much space should be allowed between slabs to prevent buckling if they are exposed to a range of temperature of 60 C°? (Answer: 0.022 cm)

■

Table 7–4   COEFFICIENTS OF LINEAR EXPANSION (CHANGE IN LENGTH PER UNIT LENGTH PER DEGREE CHANGE IN TEMPERATURE)

Material	$\alpha/°C$
Aluminum	24×10^{-6}
Copper	17×10^{-6}
Steel	13×10^{-6}
Iron	12×10^{-6}
Concrete	12×10^{-6}
Ordinary glass	8.5×10^{-6}
Tungsten	4.3×10^{-6}
Pyrex glass	3.3×10^{-6}
Invar (a nickel-steel alloy)	0.9×10^{-6}
Fused quartz	0.5×10^{-6}
Brass and bronze	19×10^{-6}

There are many familiar examples of expansion and contraction of solids caused by changes in temperature. A thick-walled glass tumbler may crack if hot water is poured into it, because stresses are created as the inner surface layers expand before the outer layers are heated. Concrete sidewalks and highways are laid with gaps between sections to accom-

Figure 7–30   Expansion slots allow bridges to expand and contract safely with temperature changes. (Robert Mathena/Fundamental Photographs)

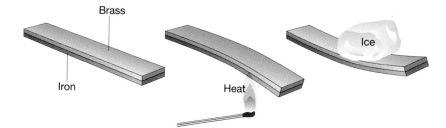

Figure 7-31 Differential expansion. A bimetallic strip of brass and iron bends, as shown, because brass expands (or contracts) more when heated (or cooled) than does iron.

modate expansion and prevent buckling. Expansion joints are provided between sections of a steel bridge to allow for daily and seasonal changes in temperature (Fig. 7–30). Power lines are more taut in winter than in summer. The design of certain types of thermostats is based on the thermal expansion of two metals, such as brass and iron, welded or riveted together (Fig. 7–31). When the temperature increases, the metals expand unequally, causing the bimetallic strip to bend. Many thermostats that control furnaces or air conditioners use bimetallic strips or coils to open and close electrical contacts (Fig. 7–32).

The behavior of liquid water with temperature changes is unusual. When a beaker of water at 4°C is heated, the water expands; but when it is cooled below this temperature, instead of contracting, the water expands again. A given mass of water thus has the smallest volume when it is at 4°C. Since **density** is the **ratio of mass to volume**, mass/volume, water has the greatest density at this temperature.

Aquatic life in colder climates is affected by the thermal behavior of water. In a body of fresh water, as the surface layer of water cools to 4°C, it becomes more dense than the water below and sinks, pushing up warmer water to the surface where it, in turn, is chilled. Considerable mixing occurs as successive layers at the surface cool to 4°C and sink, until the entire body of water is 4°C. When the surface layer is cooled below 4°C, its density becomes less than that of underlying layers and it remains at the surface while freezing (Fig. 7–33). Lakes and streams thus freeze from the top down, enabling plant and animal

Figure 7–32 The operation of a bimetallic thermostat depends upon the unequal expansion of the two metals. As electrical contact is made or broken, heating and cooling systems are turned on and off, respectively. (David Rogers)

Figure 7–33 Lakes freeze from the top down. But, since ice is a good thermal insulator, most lakes do not freeze to the bottom. Ice skating and ice fishing can coexist. (T. Rosenthal/Superstock, Inc.)

life to survive the winter in an above-freezing environment (0° to 4°C; 32° to 39°F). The lower density of the ice (ice is 10% less dense than water) keeps it on top, where it acts as an insulating blanket to retard the cooling of the underlying water. When water freezes in a confined space, it can exert a tremendous force and break even very strong metal. For this reason, antifreeze must be used in automobiles in cold climates and water pipes must be buried deep underground.

7–12 PRESSURE EFFECTS ON FREEZING AND BOILING POINTS OF WATER

The statement that water freezes at 0°C is valid only at normal atmospheric pressure. Since water expands upon freezing and cannot freeze without expanding, an increase in pressure lowers the freezing point of water by increasing the resistance to expansion.

One effect of the unusual behavior of water at the freezing point (most substances contract upon freezing) is that you can skate or ski on ice. When you do so, the ice or snow melts under two influences: (1) the pressure exerted by the skate blade or ski (a large force—weight—distributed over a small area) and (2) friction. A thin, slippery film of water is formed that acts as a lubricant between the blade and ice. As soon as the pressure returns to normal, the water refreezes.

Regelation—the process of ice melting at a temperature lower than 0°C because of increased pressure, then refreezing when the pressure is removed—can be demonstrated by suspending a heavy weight from a fine wire looped over a block of ice (Fig. 7–34). The increased pressure under the wire lowers the freezing point of the ice, causing some of it to melt and allowing the wire to drop down. The water refreezes above the wire, where the pressure is less and the freezing point is normal. In this way the wire passes through the ice, leaving a path of fine air bubbles, but the ice is still frozen solid.

The **boiling point** of a liquid is the **temperature at which its vapor pressure is equal to the surrounding atmospheric pressure**. As the boiling point is approached, the molecules are moving so rapidly that some of them may enter the gaseous state and form a bubble. With continued heating, the vapor pressure within the bubble builds up and the bubble

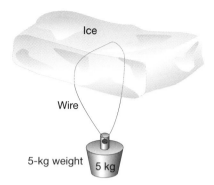

Figure 7–34 Regelation. A weighted wire loop cuts through a block of ice. The increased pressure under the wire lowers the freezing point of the ice. The water refreezes above the wire.

rises to the top (Fig. 7–35). At the boiling point, the vapor pressure within the bubble is equal to the pressure exerted on the liquid. If the external pressure is increased, it becomes more difficult for bubbles of vapor to form, and the boiling point is raised. In a **pressure cooker**, the confined steam increases the pressure on the surface of the liquid, thereby raising the boiling point of water from 100° to 110°C or more. This reduces the time required for cooking food. Table 7–5 gives the vapor pressure of water at various temperatures.

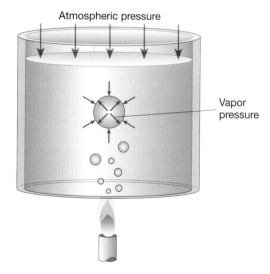

Figure 7–35 Boiling. When the vapor pressure within the bubble is equal to the pressure on the liquid, boiling starts. The bubbles rise and break through the surface, and the gas escapes.

Table 7–5 VAPOR PRESSURE OF WATER

Temperature (°C)	Pressure (torr = mm Hg)
0	4.58
20	17.53
40	55.32
60	149.4
80	355.1
100	760 (1 atm)
120	1489

Geysers are natural pressure cookers. Old Faithful, a geyser in Yellowstone National Park, is so named because on the average it erupts at regular intervals (about 65 minutes). One model of geyser action assumes that water from a hot spring collects at the bottom of a fissure—a vertical, pipe-like opening (Fig. 7–36). The water near the base of the fissure is under considerable pressure and may reach a temperature of 120° to 175°C—considerably above its normal boiling point and hotter than the water higher up in the fissure. When the water begins to boil and changes to steam, the expanding steam pushes some water out of the pipe in a brief, quiet fountain approximately 6 feet high. The pressure in the entire column is relieved by the overflow, but the water is now superheated at a temperature above its new boiling point. Within 60 seconds some of the water boils, and the expanding steam causes an eruption of steam and water 150 feet into the air that continues for about 4 minutes. The pressure and temperature at the base then return to normal, water seeps in again, and the cycle is repeated.

A decrease in the external pressure on the surface of a liquid has the opposite effect, that is, lowering the boiling point. Consider a round-bottomed flask in which water is boiled for a few minutes to expel the air. When steam comes from the mouth of the flask, it is tightly sealed with a stopper containing a thermometer (Fig. 7–37). Pouring cold water over the flask causes some of the steam above the surface of the liquid to condense to water, creating a partial vacuum, and the vapor pressure above the water is reduced. The boiling point of the water is thus lowered, and the

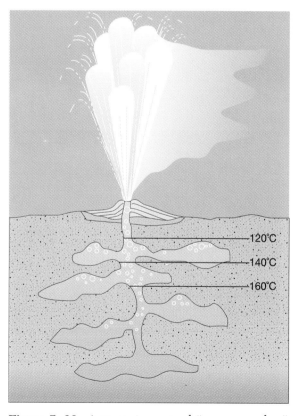

Figure 7–36 A geyser is a natural "pressure cooker." When superheated water under pressure boils, an eruption of steam and water soon follows.

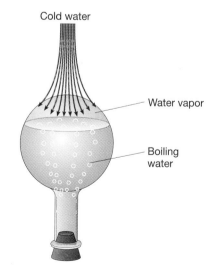

Figure 7–37 Water boiling at reduced pressure. Pouring cold water on a flask of boiling water causes steam to condense, reducing the pressure and lowering the boiling point.

water boils vigorously at a lower and lower temperature. Water can even be made to boil at room temperature in this way.

7–13 METHODS OF HEAT TRANSFER

When there is a temperature difference between two objects in close proximity or between two parts of the same object, the warmer region loses heat and the cooler region gains heat. The transfer of heat may occur by one or by a combination of processes: *conduction, convection,* and *radiation.*

Conduction.   Stirring boiling water with a silver spoon causes the spoon to heat up so quickly that it soon becomes too hot to handle. Silver conducts heat readily; the transfer of heat along the spoon is called **conduction**. Spoons made of other metals—aluminum, iron, lead, and brass—also conduct heat, but at different (and generally lower) rates. **Conduction involves the transfer of thermal energy through the interaction of neighboring molecules or atoms *without* the conducting material itself being moved.** With an increase in temperature, the molecules vibrate faster and faster and transmit this energy through collisions with adjacent molecules in the material.

Metals as a group are better conductors of heat than nonmetals, but metals differ in their conductivity. Silver is excellent, aluminum is good, and iron and lead are relatively poor metallic conductors. Glass is such a poor conductor of heat that one end of a 5-inch rod may be held comfortably in the hand while the other end is melting in a flame. Other nonmetals, such as wood, paper, and cotton, are poor conductors of heat and are therefore good insulators (Fig. 7–38). Air and gases generally are quite poor conductors of heat. The warmth-retaining quality of feathers, fur, cork, and other materials is frequently due to the insulating property of air trapped in small spaces between fibers or other structures (Fig. 7–39). Storm windows are effective heat retainers for this reason; two layers of glass are separated by a poorly conducting air space. The thermal conductivities of some materials are listed in Table 7–6.

Convection.   **Fluids** transfer heat by **convection, a process that causes mixing of the warmer with the cooler regions of the liquid or gas.** The

Figure 7–38   The left side of this ceramic dish is buried in ice (0°C) while its right side is very hot. This can occur because this ceramic material is a very poor conductor of thermal energy. (Courtesy of Corning Glass Works)

bottom-most layer of water in a kettle is heated by contact with the hot kettle and expands (Fig. 7–40). The increase in volume lowers the density of the warm water; being lighter, the warm water is pushed to the top by the buoyant force of the cooler, denser, surrounding water. The cooler water that moved in to replace it is itself heated and is then also pushed upward. The convection currents set up a circulation, and the mixing of the warm and cold water soon

Figure 7–39   The air trapped in the holes in fishnet underwear increases thermal insulation of clothing. (David Rogers)

Table 7–6 THERMAL CONDUCTIVITIES OF COMMON MATERIALS

Material	Thermal Conductivity (calories/cm · sec · C°)
Metals	
Silver	0.99
Copper	0.92
Gold	0.72
Aluminum	0.48
Iron and steel	0.11
Nonmetallic Solids	
Concrete	0.0022
Brick	0.0015
Asbestos sheet	0.0004
Wood	0.0003
Snow	0.0003
Glass wool	0.0001
Liquids	
Mercury	0.016
Water	0.0013
Gases	
Hydrogen	0.00041
Air	0.000057

Figure 7–41   Air convection around a flame. (FPG International/Bob Taylor)

establishes a more or less uniform temperature throughout.

Central heating systems in buildings operate by convection of hot air, hot water, or steam. Some systems rely on natural convection as just described. Other systems use supplementary blowers or pumps to circulate the heated fluid. When a room is warmed by a radiator, heat is transferred in part by convection currents that are set up in the air. Most of the heat in an open fireplace is lost by convection up the chimney (Fig. 7–41).

Radiation.   The sun is the source of most energy on the earth. Since the space between these two bodies is largely empty, it cannot transfer heat by conduction or convection, both of which require a material medium. Instead, energy is transferred by a third process—**electromagnetic radiation**.

Any body above absolute zero radiates heat to its surroundings. The hotter the body is, the greater the rate at which energy is radiated. **All bodies radiate energy as electromagnetic waves at a rate proportional to the fourth power of the absolute temperature of the body, T^4.** Radiant energy may pass through air or glass without warming it, because only a small part of it is absorbed. But when radiant energy falls on a rock, pavement, or leaf, which absorbs more of it, there is an increase in the internal energy of the object, followed by a rise in temperature. Experiments show that a rough, unpolished surface radiates heat better than a polished surface, and a black surface radiates heat better than a white surface at the same temperature. Good radiators of heat are also good absorbers of heat.

The methods of heat transfer have been considered in the design of the thermos bottle, which is intended to keep its contents at a given temperature: hot liquids hot or cold liquids cold (Fig. 7–42). The space between the walls of the double-walled thermos bottle is evacuated, minimizing heat transfer by conduction and convection, and the walls themselves are coated with silver to reduce radiation. Little heat

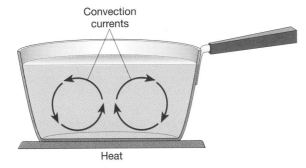

Figure 7–40   Convection currents circulate heated water.

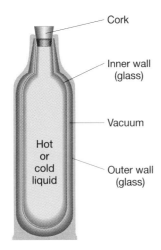

Figure 7–42 A thermos bottle keeps contents hot or cold by minimizing heat transfer by conduction, convection, and radiation.

is conducted along the glass or by the cork. Thus, there is a minimum amount of heat transferred into or out of the flask.

A relatively new development in cooking is heat transfer by radiation in the form of the microwave oven (Fig. 7–43). In a microwave oven, microwaves, a type of electromagnetic radiation, are reflected from a metal stirrer and the metal walls. Microwaves pass through plastic wrap and glass or ceramic cooking dishes but are absorbed by water molecules in the food. The outer layers of the food are heated; then heat is conducted to the interior, cooking the food throughout.

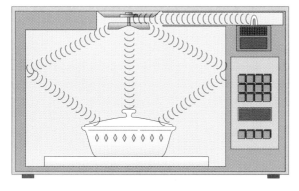

Figure 7–43 Heat transfer by radiation in a microwave oven. Microwaves are absorbed by water molecules in the food, thus heating it.

7–14 HEAT AND LIFE

Justus von Liebig (1803–1873) proved that heat is produced by the combustion of food in body tissues. Later it was shown that heat production and heat loss are controlled by the nervous system. The body reacts to cold with exercise, shivering, and the unconscious tensing of muscles. Heat loss, meanwhile, is minimized by cutting down the rate of circulation of blood to the skin. In hot weather, body heat is dissipated faster by increasing the circulation to the skin, by sweating, and by faster respiration.

In a very cold environment, the body's metabolic production of heat may not keep up with the loss of heat by radiation, conduction, and convection. Normally, the body's thermostat in the hypothalamus monitors and adjusts the internal temperature of the body from the inside. Even after a hot meal or a cold drink, the body's internal temperature is maintained at a practically constant value. Fever has long been recognized as a sign of disease, but why sickness is so often accompanied by a rise in body temperature is not yet known.

Animals living in cold environments have ways of conserving body heat. The fur of arctic animals is considerably thicker than that of tropical animals, and its insulating ability is many times greater. An arctic fox can be comfortable at −50°C without increasing its rate of metabolism, although the feet, legs, and nose are uncovered. If they allowed the escape of body heat, mammals and birds could not survive in cold climates. The warm feet of an animal standing on snow or ice would melt it and would soon freeze in place. A gull or duck swimming in icy water would lose heat through its webbed feet faster than it could produce it.

To reduce the loss of heat from unprotected extremities, the warm outgoing blood in the arteries heats the cool blood returning in the veins in a network of veins, the *rete mirabile*, near the junction between trunk and extremity. Therefore, the extremities can become much colder than the body without draining off the body heat. This mechanism also serves as a means of enabling thickly furred animals to release excess heat: The flow of blood to the extremities is increased and heat is radiated. Heat is also dissipated by evaporation from the mouth and tongue.

SUMMARY OF EQUATIONS

$$°F = \frac{9}{5}°C + 32°$$

$$°C = \frac{5}{9}(°F - 32°)$$

$$K = °C + 273°$$

Root mean square (rms) Molecular Speed

$$V_{rms} = \sqrt{\frac{3kT}{m}}$$

$$\begin{pmatrix} \text{Heat lost} \\ \text{or gained} \end{pmatrix} = \begin{pmatrix} \text{Mass of} \\ \text{substance} \end{pmatrix} \cdot \begin{pmatrix} \text{Specific} \\ \text{heat} \end{pmatrix} \cdot \begin{pmatrix} \text{Change in} \\ \text{temperature} \end{pmatrix}$$

$$H = mc(T_2 - T_1)$$

Heat of Fusion (Melting or Freezing)

$$H = mL_f$$

Heat of Vaporization or Condensation

$$H = mL_v$$

Law of Charles and Gay-Lussac

$$\frac{V}{V'} = \frac{T}{T'}$$

Linear Expansion of a Solid

$$\Delta L = \alpha L_0 (T_2 - T_1)$$

KEY TERMS

Absolute Zero of Temperature The theoretical lower limit temperature: $-273°C$.

Boiling Point The temperature at which the vapor pressure of a liquid is equal to atmospheric pressure.

Bomb Calorimeter An apparatus for measuring the heat of combustion of foods or fuels.

British Thermal Unit (BTU) The amount of heat that is required to raise the temperature of 1 pound of water 1 Fahrenheit degree.

Brownian Motion The zigzag motion of tiny particles suspended in a fluid due to the impacts of molecules of the fluid on these particles.

calorie ("small") The amount of heat that is required to raise the temperature of 1 gram of water 1 Celsius degree (from 14.5° to 15.5°C).

Calorie A kilocalorie or "giant" calorie is the heat required to raise the temperature of 1 kilogram of water 1 Celsius degree.

Calorimeter An apparatus for measuring the quantity of heat transferred from one substance to another.

Calorimetry The measurement of the quantity of heat transferred from one substance to another.

Celsius Scale A temperature scale on which the ice point is 0° and the steam point is 100°.

Conduction The transfer of heat through the collisions of neighboring molecules in a material without the conducting material itself moving.

Convection The transfer of heat through fluids by currents that cause the mixing of the warmer with the cooler regions of a liquid or gas.

Density The ratio of the mass of a substance to its volume: mass/volume.

Elastic Collision A collision between particles in which both kinetic energy and momentum are conserved.

Electromagnetic Radiation Radiation consisting of oscillating electric and magnetic fields and traveling at the speed of light.

Enzyme A biochemical catalyst.

Evaporation The change in state from a liquid to a vapor.

Fahrenheit Scale A temperature scale on which the ice point is 32° and the steam point is 212°.

Fixed Points Reference temperatures on a thermometer, usually the ice point and steam point of water.

Fluid A substance that has the ability to flow, such as a liquid or a gas.

Freezing Point The temperature at which a liquid changes to a solid at standard atmospheric pressure.

Geyser A natural spring that throws forth intermittent jets of heated water and steam.

Heat Flow The transfer of molecular energy from one body to another because of a temperature difference.

Heat of Combustion The heat produced per unit mass of substance burned in oxygen. Used to determine the heat value of fuels and foods.

Ice Point The freezing point of water.

Kelvin Scale A temperature scale based on the absolute zero of temperature. The size of a Kelvin degree is the same as that of a Celsius degree.

Kilocalorie The unit that is used for rating the heat value of a food or fuel; equal to 1000 "small" calories.

Kinetic Theory   The theory that matter is made up of tiny particles called molecules and that these molecules are in a state of constant motion.

Latent Heat of Fusion   The quantity of heat that must be added to 1 gram of solid at its melting point to change it to liquid at the same temperature and pressure.

Latent Heat of Vaporization   The quantity of heat that must be added to one gram of a liquid at its boiling point to change it to vapor at the same temperature and pressure.

Law of Charles and Gay-Lussac   The volume of a gas varies directly with its absolute temperature if the pressure is held constant.

Linear Expansion   The change in length of a solid upon heating.

Method of Mixtures   The heat given up by a hot substance placed in a calorimeter is equal to the heat absorbed by the calorimeter and its contents.

Pressure   Force per unit area.

Pressure Cooker   A utensil for quick cooking of foods by means of superheated steam under pressure.

Radiation   A method of heat transfer. Any object above absolute zero radiates energy to its surroundings as electromagnetic waves.

Regelation   The process of ice melting under pressure and then refreezing when the pressure is removed.

Specific Heat   The amount of heat that is required to raise the temperature of 1 gram of a substance 1 Celsius degree.

Steam Point   The boiling point of water at standard atmospheric pressure.

Sublimation   The change in state directly from the solid phase to the gas phase, without passing through the liquid phase.

Temperature   A measure of the intensity of heat, of the average kinetic energy of molecules.

Thermal Capacity   The capacity of a substance for heat.

Thermal Energy   The sum of the kinetic energy, rotational energy, and vibrational energy of molecules.

Thermal Pollution   The discharge of waste heat into natural bodies of water.

Thermogram   A photograph that shows variations in the heat emitted by a body or other structure.

Thermography   The detection and recording of variations in temperature on thermograms.

Thermometer   An instrument for determining temperature.

Thermometric Property   A property of matter that varies with temperature in a measurable way, such as expansion or electrical resistance.

Translational Kinetic Energy   The energy of an object due to its motion from one point to another.

THINGS TO DO

1. Construct a Galilean air thermometer. Note how a rising temperature is accompanied by a falling thermometer.

2. Obtain a ball-and-ring apparatus. When cold, the metal ball passes through a snugly fitting ring. Does the ball pass through when heated? What if the ring also is heated?

3. Add 25 ml of boiling water to 100 ml of cold tap water in a beaker. To a second beaker of similar size containing 100 ml of cold tap water, add 50 ml of boiling water. Stir well, then take the temperature of each. Note the difference between temperature and heat.

4. Sublime iodine crystals as in Figure 7–25.

5. Dab your arms with equal small quantities of water and alcohol. Note their relative cooling effects.

6. Place a small rubber balloon over the mouth of a long-stemmed flask. Heat the flask and note the effect on the balloon.

7. If you live where there is snow, make a snowball. Note how the pressure of squeezing melts the snow slightly and how it refreezes when the pressure is released.

8. (a) Heat a bimetallic strip (of brass and iron, for example). Explain what you observe. (b) Place ice on the strip. What is the effect?

9. If you have access to a convection apparatus, place a candle below one chimney, light the candle, and hold a smoking object above the other. What do you observe?

10. Fit a Brownian motion cell to the stage of a microscope. Draw smoke from an extinguished match into the cell and observe the motion of the smoke particles.

11. Press two ice cubes firmly together. What do you observe when you release them?

EXERCISES

Temperature Measurements

7–1. Determine the equivalent of 0°F on (a) the Celsius scale, (b) the Kelvin scale.

7–2. Express the following temperatures in °F: (a) −40.0°C; (b) 218 K; (c) 6000°C; (d) 360°C.

7–3. The material that has the lowest temperature most of us are likely to encounter is dry ice, −79.0°C. What would its temperature be on the Fahrenheit scale?

7–4. (a) What is the difference between 70.0°F and 70.0°C in Celsius degrees? (b) In Fahrenheit degrees?

7–5. Liquid oxygen rocket propellant is kept at its boiling point of 90.0 K. What would the temperature be on the Fahrenheit scale?

7–6. When a thermometer is placed in a beaker of hot water, the mercury level drops, then rises. Explain why.

7–7. Room temperature is often identified as 68.0°F. What temperature is this on (a) the Celsius scale; (b) the Kelvin scale?

7–8. Can you think of any advantage to using the Celsius scale over the Fahrenheit scale?

7–9. Explain the meaning of the absolute zero of temperature.

7–10. Why are high temperatures resulting from thermal pollution often fatal to fish?

Temperature, Heat, and Kinetic Theory

7–11. What is the basis for the belief that matter is made up of small particles in motion?

7–12. How is the concept of temperature interpreted by the kinetic theory?

7–13. How is the concept of gas pressure interpreted by the kinetic theory?

7–14. The gas pressure on a piston is 7.50×10^4 newtons per square meter. If the area is 2.00×10^{-3} square meters, what force does the gas exert?

7–15. A smoke particle undergoing Brownian motion in air has a mass of 2.00×10^{-5} kg.
  (a) What is its average velocity if the temperature is 300 K?
  (b) What is its average kinetic energy?

7–16. Distinguish between the temperature and the thermal energy of a cup of hot coffee.

Heat Measurements

7–17. How much heat is required to raise the temperature of 10.0 kilograms of aluminum by 80.0°C?

7–18. A hot water bottle containing 1 kilogram of water cools from 70.0°C to 20.0°C. How much heat was given off by the water?

7–19. A kilogram of copper gains 1 kilocalorie of heat. By how many degrees is the temperature of the copper changed?

7–20. If 500 kilograms of water at 200°F is poured into a lake that is at 50.0°F, how much heat is added to the lake?

7–21. To make iced coffee, 1.00 kilogram of ice at 0°C is added to 2.00 kilograms of coffee at 95.0°C. Assuming that the specific heat of coffee is the same as that of water, what is the temperature of the mixture?

7–22. A 65-kilogram woman is on a 2500-kcal/day diet. If a corresponding amount of heat were added to 65 kilograms of water at 37.0°C, to what temperature would the water be raised?

Changes of State

7–23. Why does the temperature of boiling water remain the same as long as the boiling continues?

7–24. How much more heat is required to vaporize 100 grams of water than to melt 100 grams of ice?

7–25. Explain why evaporation is a cooling process.

7–26. Explain why water condenses on the inside of a window on a cold day.

7–27. In a steam heating system, water leaves a radiator at the same temperature as steam entered, 100°C. How is the room heated?

7–28. What characteristic of water makes it useful as an automobile engine coolant?

7–29. How much heat does 50.0 grams of steam at 125°C release in the process of condensing to water at 100°C?

7–30. Why is an ice cube at 0°C more effective in cooling a drink than the same quantity of water at 0°C?

7–31. A 20.0-gram ice cube at 0°C melts, and the temperature of the water rises to 25.0°C. How much heat is involved?

7–32. Explain why perspiration occurs when the body is hot, but produces a cooling effect.

Thermal Expansion and Contraction

7–33. Do all liquids expand with an increase in temperature? Explain.

7–34. What water temperature would you expect to find near the bottom of an ice-covered lake?

7–35. What is the seasonal variation in the height of a 300-foot steel-frame building if the temperature range is $-20.0°$ to $95.0°F$?

7–36. Explain why metal lids to glass fruit jars can often be loosened by heating them under hot water.

7–37. When you fill your gasoline tank and then leave the car in a hot parking lot, some of the gasoline may overflow. Explain.

Pressure Effects

7–38. When the pressure on water is increased, will its boiling point be higher or lower than 100°C?

7–39. Why might an astronaut's blood boil if he left his capsule without his pressurized spacesuit?

7–40. Why is it possible to make a snowball?

Methods of Heat Transfer

7–41. Why does an oak floor feel colder to the bare foot than a rug at the same temperature?

7–42. If a house is insulated to reduce heat loss in winter, how comfortable will it be in summer?

7–43. Why are thermopane windows, which have double panes of glass separated by a small air space, better for insulation than single-pane windows?

7–44. A fireplace is only about 10% efficient in heating a room. Why?

7–45. Why does it cool more quickly at night when the sky is clear than when it is cloudy?

7–46. *Multiple Choice*
   A. Temperature is a measure of
      (a) the amount of heat in a body.
      (b) the average total energy of molecules.
      (c) the average kinetic energy of molecules.
      (d) all of these.
   B. As water freezes, its temperature
      (a) increases.
      (b) decreases.
   (c) first increases, then decreases.
   (d) does not change.
   C. The motion of small particles due to collisions with molecules in a fluid is called
      (a) evaporation.
      (b) sublimation.
      (c) Brownian motion.
      (d) pressure.
   D. The lowest possible temperature is
      (a) unknown.
      (b) absolute zero.
      (c) infinitely low.
      (d) 273 K.
   E. If the temperature of a gas is increased, its volume is
      (a) reduced by half.
      (b) reduced by one fourth.
      (c) doubled.
      (d) increased.
   F. Heat is
      (a) a fluid called caloric.
      (b) the average kinetic energy of molecules.
      (c) a nonmaterial substance.
      (d) a transfer of energy because of a difference in temperature.
   G. When 1 kg of ice at 0°C absorbs 80 kilocalories of heat, the ice undergoes
      (a) a change of state.
      (b) a loss of energy.
      (c) a rise in temperature.
      (d) an increase in volume.
   H. We obtain energy from the sun by
      (a) conduction.
      (b) convection.
      (c) radiation.
      (d) molecular travel.
   I. Forced air home heating is an example of heat transfer primarily by
      (a) conduction.
      (b) convection.
      (c) radiation.
      (d) none of these.
   J. A solid-gas phase change is called
      (a) melting.
      (b) boiling.
      (c) sublimation.
      (d) evaporation.

SUGGESTIONS FOR FURTHER READING

Barker, J. A., and Douglas Henderson, "The Fluid Phases of Matter." *Scientific American*, November, 1981.
   The structure of liquids and gases has been under scrutiny for more than a century. A successful theory has been based on one of the earliest and simplest models.

Bartels, Richard A., "Do Darker Objects Really Cool Faster?" *American Journal of Physics*, March, 1990.
Illustrates that one should be careful in using the statement, "a good absorber is a good emitter," or misconceptions can occur. For example, there is essentially no difference in the cooling rates of two water-filled metal cans, one painted white and the other painted black.

Bent, Henry A., "Energy and Exercise," *Journal of Chemical Education*, August, 1978.
Discusses the caloric costs of jogging, bicycling, and sweating.

Berry, R. Stephen, "When the Melting and Freezing Points Are Not the Same." *Scientific American*, August, 1990.
Hardly anything could appear simpler than to say that almost every solid has a melting point, and almost every liquid has a freezing point, and these two points are one and the same. But appearances can be deceiving. Clusters of atoms or molecules, numbering from 4 or 5 to 100 or 200, can have distinctly different melting and freezing points.

Delorenzo, Ronald, "Fire Walking, Temperature, and Heat." *Journal of Chemical Education*, November, 1986.
Fire walkers can walk across beds of glowing coals without apparent harm. How do they do it?

Donnelly, Russell J., "Leo Dana: Cryogenic Science and Technology." *Physics Today*, April, 1987.
Low-temperature physics is experiencing a period of rapid progress. This article reviews basic research on liquid helium and a range of activities in cryogenic technology.

Gershon-Cohen, Jacob, "Medical Thermography." *Scientific American*, February, 1967.
The measurement of heat emitted through the skin is useful in the diagnosis and treatment of various disorders.

Proctor, Warren G., "Negative Absolute Temperatures." *Scientific American*, August, 1978.
No substance can be cooled below absolute zero. But some physical systems can have a negative temperature on such a scale.

Schmidt-Nielsen, Knut, "Countercurrent Systems in Animals." *Scientific American*, May, 1981.
Many species of animals survive comfortably in an inhospitable environment. A countercurrent arrangement of arteries and veins is the basis for a variety of survival mechanisms.

Sproull, Robert L., "The Conduction of Heat in Solids," *Scientific American*, December, 1962.
The measurement of heat conduction by metals and nonmetals relates it to the conduction of both sound and electricity.

Storey, Kenneth B., and Janet M. Storey, "Frozen and Alive." *Scientific American*, December, 1990.
How do cold-blooded animals—frogs, turtles, beetles, spiders—endure when environmental temperatures fall below the freezing point of their body fluids? Some species avoid freezing, but many others freeze solid during the winter months and thaw in the spring. Some secrets of freeze tolerance are known; more may be unraveled.

Velarde, Manuel G., and Christiane Normand, "Convection." *Scientific American*, July, 1980.
Convection is such a complex process that an exact solution is still lacking.

Walker, Jearl, "The Secret of a Microwave Oven's Rapid Cooking Action Is Disclosed." *Scientific American*, February, 1987.

Walker, Jearl, "What Happens When Water Boils . . .?" *Scientific American*, December, 1982.
More is going on when water boils than meets the eye or ear.

ANSWERS TO NUMERICAL EXERCISES

7–1. (a) $-17.7°C$
     (b) 255 K

7–2. (a) $-40°F$
     (b) $-67°F$
     (c) 10,832°F
     (d) 680°F

7–3. $-110°F$

7–4. (a) 48.9 C°
     (b) 88 F°

7–5. $-297°F$

7–7. (a) 20°C
     (b) 293 K

7–14. 150 N

7–15. (a) 2.49×10^{-3} m/s
      (b) 6.21×10^{-21} J

7–17. 1.68×10^5 cal

7–18. 50.0 kcal

7–19. 10.6°C

7–20. 41,700 kcal

7–21. 36.7°C

7–22. 75.5°C

7–24. 46,000 cal

7–29. 27,600 cal

7–31. 2100 cal

7–35. 0.249 ft

THE ENERGY-EFFICIENT HOUSE OF THE FUTURE

KAREN ARMS

Life would be very different for us all if we lived in houses with utility bills of only a few hundred dollars a year. Such houses have been built.

The house's construction emphasizes insulation, because much of the energy we use to heat and cool the air in an ordinary house is lost through the windows and walls. The insulating effect of a substance is described by a number known as its R value. A substance with a high R value is a good insulator. New insulating material is thin so that the roof of our energy-efficient house is insulated to R-100 and the walls to R-40 without being thick. In an ordinary house, large amounts of heat seep through the wood studs between the outer and inner walls. Our energy-efficient house avoids this problem. It is built of metal bars thinner than the usual wooden timbers and studs, and the outer and inner walls are held together by cross beams (Fig. A).

Heat loss through double-glazed windows is half that in a conventional house, partly because the gap between the two panes is filled with a chemically unreactive gas such as argon or xenon instead of with air. In addition, one of the inner surfaces of the window is coated with tin oxide, which reduces the transmission of infrared radiation (heat) through the window while letting visible light pass through (see Fig. A). These "low-emissivity" windows will be put into all new houses shortly after 2000. Compared with today's windows, they will save the equivalent of one sixth of all the energy in the Alaska oil fields.

The way the sun strikes the windows of our new house is carefully controlled. Several windows are angled toward the winter sun and backed by a heat-absorbing wall (Fig. B). Fans draw the hot air into the house. A roof overhang lets the winter sun shine in but excludes the higher summer sun. When the temperature in the house rises, insulated louvers (like those of a venetian blind) automatically turn to exclude the sun. Like all the heating and cooling devices in the house, the louvers are controlled by a microprocessor. Other microprocessors are programmed to control the temperature by opening and closing windows and by turning fans on and off.

In the ground underneath the house is a water tank connected to a heat pump. This is the heart of the heat exchange system. It acts as a central heating and air conditioning unit. The water tank acts as a heat (or cold) store. The heat pump passes heat to it in summer and cold to it in winter. The heat pump is powered by electricity from a small cogeneration unit. This unit produces hot water and steam as needed and can also use the steam to generate electricity. It is about the size of a refrig-

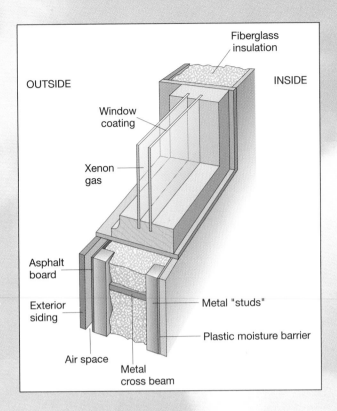

Figure A Part of a wall and window of a superinsulated house shows the structures that make the house so efficient at retaining hot and cold air.

erator and runs on natural gas or synthetic gas (made from coal). The house is not even connected to the utility company's electrical supply, although it contains all the latest energy-efficient appliances.

This house costs perhaps 10% more than an energy-efficient house built in 1988, but it will save at least ten times that amount over a 40-year period by approximately halving lifetime heating and cooling bills.

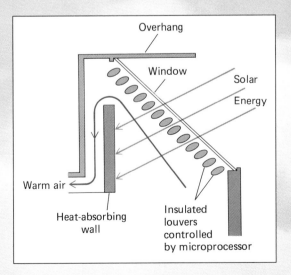

Figure B A window that collects solar heat, with insulated blinds to slow heat transfer at night and during the summer.

Fireworks over the San Francisco skyline exemplify energy conversion. (© David J. Cross/Peter Arnold, Inc.)

8

ENERGY CONVERSION

The fact that you can produce fire from the friction of rubbing together two sticks of wood, and from flint and iron, suggests a connection between motion and heat. Benjamin Thompson, Count Rumford (1753–1814), an American scientist and soldier who had immigrated to Europe and had become a count of the Holy Roman Empire, demonstrated through cannon-boring experiments that mechanical friction could produce large quantities of heat, enough to boil a massive amount of water without any fire. On the basis of his experiments, he asserted that work must be equivalent to heat. Rumford's experiments were not quantitative, however. That is, he could not say *how much* work was equivalent to a given amount of heat, and he admitted that he did not keep track of the heat absorbed in other ways than boiling the water. We now look at the connection between heat and work and at the conversion of the one into the other, called *thermodynamics*.

8–1 MECHANICAL EQUIVALENT OF HEAT

James Prescott Joule (1818–1889), a brewery owner and amateur scientist, established the equivalence of work and heat through experimentation. Although he was not a trained scientist in the usual sense, Joule did scientific research throughout his life as an avocation. Realizing the need for precise instruments, he and a Manchester instrument maker developed the first ac-

curate and reliable thermometers in England. For about 40 years, Joule experimented on the mechanical equivalent of heat, an interest he came to from studies with electric motors.

The electric motor, based on Michael Faraday's discoveries in electromagnetism, had recently been invented. Joule saw in the electric motor the possibility of converting the family brewery from steam to electric power, and he set out to improve it. But when he compared the work done by the motor to the cost of the zinc consumed in the batteries that operated the motor, he became convinced that the electric motor would be an expensive substitute for the steam power produced by burning coal.

In his investigations, Joule measured the heat developed by an electric current conducted in a wire. When he turned his attention to Faraday's dynamo —a machine in which a spool of wire is mechanically rotated between the poles of a magnet and an electric current is generated—he found that the current produced heat in the wire also. Thinking that the source of this heat was the work done in turning the handle that rotated the spool, he measured the work done on the handle and related it to the heat in the wire.

It then occurred to Joule that mechanical work might be converted directly to heat, avoiding the electrical step. For many years (from 1842 to 1878) he did experiments on such conversions: friction between two cast-iron plates moving over one another; liquids heated by the rotation of paddle wheels; fluid friction when liquid was forced through small tubes; and compression of air with a hand pump. His goal was to determine the quantitative conversion factor between work and heat for every conceivable situation that lent itself to experimental study. In some of the experiments, the temperature changes in his cal-

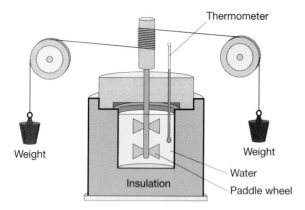

Figure 8–1 Joule's apparatus for measuring the mechanical equivalent of heat. As the weights descend, they cause a paddle wheel to turn in water. The potential energy of the weights is converted into kinetic energy of the paddle wheel, then into heat energy of the water.

orimeter were less than 1°F, but the sensitive thermometers that he designed enabled him to obtain readings to 1/200°F.

In Joule's paddle-wheel experiments, a vessel filled with water contained a rotating shaft with several stirring paddles attached to it (Fig. 8–1). Vanes attached to the wall of the vessel prevented the water from rotating freely with the paddles, increasing the internal friction. The shaft and paddles were driven by a falling weight suspended from a pulley. The mechanical work, the product of the weight and its distance, was converted into heat through the friction of the rotating water, and the temperature of the water rose. The effect on the water was the same as if the water had been heated. Knowing the amount of water, its specific heat, and the rise in temperature, Joule could calculate the amount of heat produced $[mc(T_2 - T_1)]$ and relate it to the work done.

Joule found that the same amount of work, however it was done, yielded the same quantity of heat. He concluded, therefore, that heat was a form of energy. The equivalence between mechanical units of energy and heat is called the **mechanical equivalent of heat** and is a constant designated by the symbol J (in honor of Joule).

$$\text{Mechanical equivalent of heat} = \frac{\text{Work}}{\text{Heat}} \qquad (8\text{–}1)$$

$$J = \frac{W}{H}$$

The value of J, as determined from the results of many experiments, is

$$J = 4.18 \frac{\text{joules}}{\text{calorie}}$$

Since the mechanical equivalent of heat is a constant, calories and joules must be different units of measurement for the same quantity. In other words, 1 calorie is the same quantity of energy as 4.18 joules but is just expressed in different units. Further, all of the energy in the systems measured is conserved: The input energy (measured in joules) ultimately appears as heat (measured in calories).

EXAMPLE 8–1

An automobile with a mass of 1000 kilograms (including passengers) is traveling at 36.0 km/h. How many calories of heat are produced in bringing it to a stop?

SOLUTION
The kinetic energy of the automobile is equivalent to the work necessary to stop it. From the information given, we calculate the KE, then relate it by Equation 8–1 to the heat evolved in the brakes.

$$m = 1000 \text{ kg}$$

$$\mathbf{v} = 36.0 \frac{\text{km}}{\text{h}} = \frac{3.60 \times 10^1 \times 10^3 \text{ m}}{3600 \text{ s}}$$

$$= \frac{3.60 \times 10^4 \text{ m}}{3.6 \times 10^3 \text{ s}} = \frac{10 \text{ m}}{\text{s}}$$

$$J = \frac{4.18 \text{ joules}}{\text{calorie}}$$

$$H = ?$$

$$\text{KE} = \frac{1}{2} m\mathbf{v}^2$$

$$= \frac{1}{2} (1000 \text{ kg}) \left(10 \frac{\text{m}}{\text{s}}\right)^2$$

$$= \frac{1}{2} (1000 \text{ kg}) \left(100 \frac{\text{m}^2}{\text{s}^2}\right)$$

$$= 50\,000 \frac{\text{kg} \cdot \text{m}}{\text{s}^2} \cdot \text{m}$$

$$= 5.00 \times 10^4 \text{ joules}$$

$$J = \frac{W}{H}$$

$$\therefore H = \frac{W}{J}$$

$$= \frac{5.00 \times 10^4 \text{ joules}}{4.18 \frac{\text{joules}}{\text{calorie}}}$$

$$= \boxed{1.20 \times 10^4 \text{ calories}}$$

EXTENSION
How much work must a person do to offset the eating of a piece of cake that furnishes 500 kcal? (Answer: 2.09×10^6 joules)

■

Like Joule, the physician Julius Robert Mayer (1814–1878) acquired an interest in the idea of the conservation of energy. Neither man knew that the other had simultaneously arrived at the same ideas. After taking his MD degree, Mayer served for a time as ship's physician on a Dutch East Indies freighter, the *Java*. In 1840, he sailed for the East Indies and arrived there after a four-month voyage.

Mayer had taken along Antoine Laurent Lavoisier's (1743–1794) treatise on chemistry, the first modern chemistry textbook. He soon became fascinated by Lavoisier's suggestion that animal heat is generated by the slow combustion of food. This suggestion set him to thinking about the relation between oxidation and the production of energy.

When the *Java* reached the East Indies, many of the crewmen were ill with fever. Mayer resorted to the accepted treatment of the day—bleeding the patient—and made an interesting observation. In his medical studies, Mayer had learned that blood in the arteries is bright red and blood in the veins dark red. But he observed that the crewmen's venous blood was bright red, almost as red as blood in the arteries.

Lavoisier had stated in his treatise that in warm surroundings, less internal combustion is needed to keep the body warm than in cold surroundings. Arterial blood is bright red because it has a high oxygen content. Mayer concluded, therefore, that the venous blood of his patients was bright red like arterial blood because it, too, had a high oxygen content. This meant that in the tropical East Indies the body did not consume as much oxygen as in cooler places. Mayer went beyond Lavoisier and related the heat of metabolism to the heat the body loses to its surroundings and to

the work the body performs. Although he guessed that there was an equivalence between heat and work and that they were two manifestations of energy, he did not have any experimental proof. Both Joule and Mayer received fame and recognition, however, following one of the most intense disputes over priority in the history of science.

8–2 THE FIRST LAW OF THERMODYNAMICS

The equivalence of heat and work is incorporated in the **first law of thermodynamics**, which is the same as the **law of conservation of energy**:

> **Energy cannot be created or destroyed but may be converted from one form to another, as from work into heat. However, the total energy in the universe is constant.**

The statement, "You can't get something for nothing" that is, it is impossible to create energy—characterizes the first law. This truism is based on human experience down through the ages and means that it is futile to try to invent a perpetual motion machine that would perform work without absorbing energy. Although claims for such machines have been offered, none has ever been substantiated.

The first law of thermodynamics means that the total energy, including heat, of a closed system isolated from its environment remains constant (Fig. 8–2). Hermann von Helmholtz (1821–1894), who was trained as a physician but made inquiries into physics, physiology, and meteorology as well, convinced the scientific community of the validity of the first law. Declaring that perpetual motion machines were impossible, he showed mathematically and experimentally that this led to the conclusion that energy is always conserved. Both heat and work, he pointed out, are forms of energy; what is conserved is the total of the two forms rather than either taken separately.

8–3 THE SECOND LAW OF THERMODYNAMICS

Although mechanical work can be converted entirely into an equivalent amount of heat, by the first law of thermodynamics the reverse is not true: **Heat cannot**

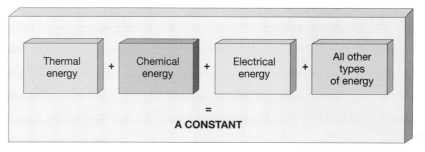

Figure 8–2   The total of all types of energy in an isolated system remains a constant. Transformations of one type into another may occur, as the first law of thermodynamics states.

be completely converted into work. In any operating heat engine, such as a steam engine, in which heat is transformed into work, some of the initial energy is wasted. Of the unavailable waste energy in a real engine, some is lost through overcoming friction, some through warming the engine and the atmosphere, and some through leakage and other ways. The **second law of thermodynamics** means that "You can't break even," because in energy-exchange processes heat cannot be completely converted into work.

Nicolas Léonard Sadi Carnot (1796–1832), an army engineer, imagined an ideal engine that was completely insulated, frictionless, and leakproof. He then invented the idea of a cycle, a series of events that brings a system back to its original condition. For example, water is heated until it is vaporized, the expanding steam forces a piston to move, the steam cools and condenses into water, and the piston returns to its starting position. In the process of cooling and condensation, however, there is an unavoidable loss of thermal energy, so that a complete conversion of heat into work is impossible. In any type of heat engine, a certain fraction of heat must be conducted to the surroundings at a lower temperature. This fact is the basis of another expression of the second law:

The natural direction of heat flow is from a reservoir of heat at a high temperature to a reservoir of heat at a low temperature.

The second law imposes a limit on the efficiency with which heat can be converted to other forms of energy. Every heat engine uses a hot working fluid such as steam or combustion gases in converting heat to mechanical work (Fig. 8–3). The converted energy is obtained from the hot working fluid at a temperature T_1. At the lowest temperature of the cycle, however, T_2, there is heat remaining in the fluid that is not converted to work but is dissipated—either discarded with the working fluid or carried away by the condenser cooling water. In a jet engine, heat is lost in the hot gases of the jet; in a steam turbine, heat is lost in the condenser cooling water.

Suppose Q_1 is the heat that flows into the engine, Q_2 is the heat that flows out of the engine, and W is the work that the engine does (Fig. 8–4). Since heat is equivalent to energy, Q_1 also represents the total energy input. The heat expelled and the work, $Q_2 + W$,

Figure 8–3   The "iron horse," or steam locomotive, is a heat engine. (Ed Bohon/The Stock Market)

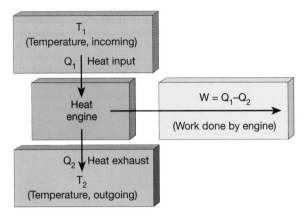

Figure 8–4 Principle of a heat engine. Heat (Q_1) flows into the engine at temperature T_1 and is exhausted (Q_2 at temperature T_2. The work (W) done by the engine = $Q_1 - Q_2$.

represent the total energy output. According to the first law of thermodynamics, energy is conserved; therefore,

$$\text{Heat input} = \text{Heat output} + \text{Work done} \quad (8\text{--}2)$$
$$Q_1 = Q_2 + W$$

from which

$$W = Q_1 - Q_2 \quad (8\text{--}3)$$

From Equation 8–3, we see that the maximum amount of work is obtained from the quantity of heat Q_1 when Q_2 is zero. How can Q_2 be made equal to zero? For the ideal or Carnot engine, it can be shown that

$$\frac{Q_1}{Q_2} = \frac{T_1}{T_2} \quad (8\text{--}4)$$

from which

$$Q_2 = Q_1 \frac{T_2}{T_1} \quad (8\text{--}5)$$

in which T_1 and T_2 stand for the initial and final absolute temperatures, respectively, between which the engine is operating. Only when $T_2 = 0$ (absolute zero) is the heat expelled, Q_2, equal to zero:

$$Q_2 = \frac{0}{T_1} Q_1 = 0$$

If no heat is exhausted, no energy is lost. Thus, all the heat input in an engine can be converted to useful work only if the final temperature is absolute zero.

The maximum possible **thermal efficiency** of a **Carnot engine**, that is, the fraction of heat that can be transformed into work, is given by

$$\begin{aligned}\text{Efficiency} &= \frac{\text{Useful work}}{\text{Energy input}} \\ &= \frac{W}{Q_1} \\ &= \frac{Q_1 - Q_2}{Q_1}\end{aligned}$$

and by Equation 8–6

$$\text{Efficiency} = \frac{T_1 - T_2}{T_1} \quad (8\text{--}6)$$

The maximum efficiency of a steam engine depends, Carnot discovered, only on the temperature difference between the boiler and condenser and the amount of heat that passes between them. All heat engines working between the same temperature levels would have the same maximum efficiency, whether they are turbine or cylinder engines, or whether they use steam, air, or any other working substance.

EXAMPLE 8–2

A highly efficient engine operates between 2000 K (T_1) and 700 K (T_2). What is its maximum possible efficiency?

SOLUTION
We can determine the efficiency of an ideal (Carnot) engine operating between these temperatures. Since a real engine can do no better than this, the efficiency represents the maximum possible.

$$T_1 = 2000 \text{ K}$$
$$T_2 = 700 \text{ K}$$
$$\text{Efficiency} = ?$$

$$\begin{aligned}\text{Efficiency} &= \frac{T_1 - T_2}{T_1} \\ &= \frac{2000 \text{ K} - 700 \text{ K}}{2000 \text{ K}} \\ &= \frac{1300 \text{ K}}{2000 \text{ K}} \\ &= 0.650, \text{ or } 65.0\%\end{aligned}$$

EXTENSION

A heat engine takes in superheated steam at 150°C. If the efficiency is 11.8%, at what temperature is the steam exhausted? (Answer: 100°C)

■

Since we cannot make $Q_2 = 0$ or avoid friction, some of the original energy Q_1 is not put to work and becomes unavailable, though not destroyed. Consequently, our supply of energy must be decreasing. This principle of the degradation of energy leads to another statement of the second law of thermodynamics:

As a result of natural processes, the energy in our world available for work is continually decreasing.

Rudolph Clausius (1822–1888), a mathematical physicist, combined the work of Joule, Mayer, and Helmholtz on conservation of energy with the work of Carnot and gave the second law still another formulation:

As a result of natural processes, the entropy of our world is continually increasing.

Entropy is the quantitative measure of the **disorder of a system**. Whenever there is a process of mixing, entropy is created (Fig. 8–5). Entropy, S, is defined as the amount of heat received or lost by the body divided by the body's absolute temperature:

$$\text{Entropy} = \frac{\text{Heat for process}}{\text{Absolute temperature}} \qquad (8\text{–}7)$$

$$S = \frac{Q}{T}$$

Figure 8–5 If you shake red and black marbles with each other, the random distribution of red and black marbles at the left is a much more probable state than the highly ordered distribution at the right. (Charles Steele)

The second law of thermodynamics says that the total entropy of any closed system must either remain constant or increase with time; it can never decrease. A mixture, for example, will not unmix itself spontaneously; this would reduce its entropy. Mixing is an irreversible process.

8–4 "HEAT DEATH," "TIME'S ARROW," AND MAXWELL'S "DEMON"

Clausius predicted that a time will come when the entire universe will be at the same temperature, if the laws of thermodynamics apply to the universe as a whole. This end is called the **"heat death"** of the universe, when there will be no available energy. Although the total amount of energy will be the same as always, there will be no way to use it. With no differences in temperature—no "hills" of heat—there can be no work. The universe will be incapable of further change and will be in a state of final chaos. The entropy of the universe will have reached its maximum value.

There is no evidence to either confirm or contradict the applicability of thermodynamics to the universe as a whole. We know that conservation laws that are believed to be universally valid must be reexamined and tested constantly. But even if the "heat death" prediction is valid, that event will not come to pass for billions of years.

An interesting aspect of the second law of thermodynamics is that it differentiates between forward and backward in time. There is a preferred direction of natural processes called the **"arrow of time."** Imagine a motion picture of any scene of ordinary life; then run it backward in your mind. In the time-reversed view of diving off a board into a pool, the diver leaps backward from the pool to the diving board. The backward-in-time views are amusing because things do not happen that way, not because they are forbidden by energy considerations—that is, the first law—but because they violate the second law of thermodynamics.

We know that a sequence of events can occur in one order and not another, because a spontaneous process is associated with an entropy increase. We recognize the impossibility of a natural process in which entropy is decreasing. Of course, there are many examples of a seeming decrease in entropy—a

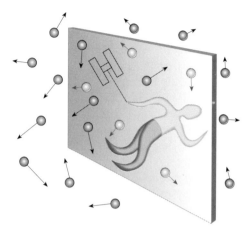

Figure 8–6 Maxwell's "demon" can allegedly establish a temperature difference by separating fast from slow molecules.

Figure 8–7 In this model of Hero of Alexandria's aeolipile, the escape of steam through the nozzles causes the sphere to rotate. (Courtesy of CENCO)

picture is painted, a letter is written, a symphony is played, a baseball game is held—in each case, something becomes more highly organized than it was before. But to accomplish each such organization of matter, energy from food is being inefficiently converted—into sound waves, muscular motion, and so on—so that the total entropy of the universe is still increased.

James Clerk Maxwell (1831–1879), a professor of physics who supervised the building of the Cavendish Laboratory at Cambridge University, imagined a **"demon"** that could violate the second law and create what Maxwell called a "perpetual motion machine of the second kind." Work would be obtained from a source at a uniform temperature by separating it into a region with molecules having greater than average energy (hotter) and a region with less than average velocities (colder) without expending any energy (Fig. 8–6). Once a temperature difference was established, it could be used to drive a heat engine that would perform useful work. However, if the energy expended by the demon is taken into account, as it must be, then there is no conceivable way a demon could be constructed that could violate the second law, and none has been.

8–5 THE STEAM TURBINE

There are practical limits to the sizes of different types of engines. Watt's reciprocating steam engine, which inaugurated the "Age of Steam" and com-

pletely revolutionized world industry, was capable of developments on a larger scale than Thomas Newcomen's (1663–1729) atmospheric engine. The turbine can be built on a still larger scale.

At about the time when reciprocating steam engines had been pushed to their limit, Charles A. Parsons (1854–1931), an engineer, demonstrated the reaction steam turbine. In principle, this turbine is a descendant of Hero of Alexandria's reaction sphere (Fig. 8–7). Some of the early Parsons turbines were built as small as 10 horsepower, but the large sizes were more common. The high speeds of the turbine (10,000 to 30,000 revolutions per minute for the primary shaft) are achieved with a minimum of vibration.

The steam turbine differs from the reciprocating steam engine in that the work expended by the steam in expanding and cooling is done on a set of rotating blades. In principle, a steam turbine is the essence of simplicity; it is a pinwheel driven by high-pressure steam rather than by air. It is subject to the same limitation on thermal efficiency, however, as the steam engine.

In the **steam turbine engine**, jets of steam are directed against the blade or "buckets" of the turbine, exert forces on the blades, and keep the wheel in motion (Fig. 8–8). Fixed blades are mounted between adjacent moving wheels. The steam jet strikes against the first set of moving buckets and rebounds, giving up some of its energy. Then the jet, reversed by the fixed blades, strikes the next revolving wheel. The steam gradually gives up momentum and energy as it does work. The turbine, connected to an electric generator, does work in turning the shaft. The heat energy of the steam is converted in this way to the ki-

Figure 8–8   One of the largest steam turbines built with a single shaft. Its capacity is 400,000 horsepower. (Courtesy of Brown, Boveri & Company, Ltd., Baden, Switzerland)

netic energy of rotary motion. Steam turbines account for more than 75% of the electric power generated in the world (most of the rest is hydroelectric). When large unit output and maximum efficiency are essential, as for the propulsion of big ships and power generation, the steam turbine is unchallenged.

8–6 INTERNAL COMBUSTION ENGINES

The essential concept of **internal combustion** is that combustion can take place inside the working fluid without any need for a firebox and a boiler. Carnot had suggested that air in a heat engine can serve as the medium of energy conversion and at the same time, because of its oxygen content, can participate in the

generation of heat. By successfully working out Carnot's idea, Nicolaus August Otto (1832–1891), an inventor, and Rudolf Diesel (1858–1913), a professor of thermodynamics and mechanical engineer, achieved a revolution in energy conversion.

More than 10 million reciprocating internal combustion engines are built in the United States each year to provide power for automobiles, trucks, boats, locomotives, and small airplanes. More than 100 million such engines are in service. Nearly all are based on a principle first demonstrated in 1876 by Otto, who had found a way to burn a compressed mixture of illuminating gas and air in the cylinder of an engine without producing destructive explosions.

In an **"Otto cycle,"** four functions must be performed (although not necessarily in four distinct strokes): fuel-and-air intake, compression, power, and exhaust (Fig. 8–9). Two, four, or six strokes may be used to accomplish the four functions. An engine of the Otto type takes in a controlled mixture of fuel and air, compresses it, and ignites it by a spark plug. The original **Otto engine** had an efficiency three to four times that of a steam engine.

Although automobiles are usually powered by an Otto engine, the diesel has revolutionized heavy transportation—heavy trucking, earth-moving equipment, and ships under 20,000 horsepower. Rudolf Diesel considered the engine that he patented in 1892 to be not just an improved heat engine but a revolutionary one. He insisted that it was an engine built on scientific principles, a "rational" engine, and he expected it to displace the steam engine within a few years and to drive everything from battleships to sewing machines. He thought of it as a Carnot engine that would have a thermal efficiency of 73% (compared with 7% for the best steam engines of the time). As it turned out, the **diesel engine** we know today is not a Carnot engine at all and is quite different from Diesel's original idea.

The diesel evolved over 25 years. In World War I, most submarines and some ships were powered by diesels. The modern diesel appeared in the 1920s, when suitable fuel injectors were perfected (Fig. 8–10). It is a high-compression engine, with a compression ratio of about 16:1 (compared with Otto ratios of about 7 or 8 to 1). Air is the working fluid. The fuel is injected near the end of the compression stroke and is ignited by the heat of compression. The thermal efficiency is high—up to 40% for the best diesels.

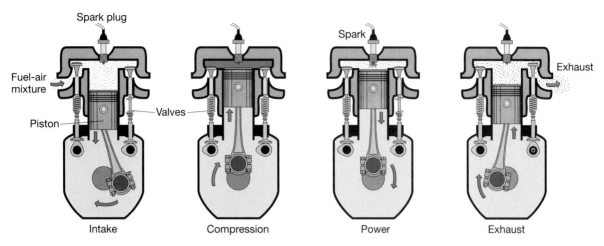

Figure 8–9 The conventional piston engine employs the four-stroke cycle first demonstrated by Nicolaus August Otto: intake, compression, power, and exhaust. After compression, the mixture of fuel and air is ignited by a spark plug. The expanding gases produce the power stroke.

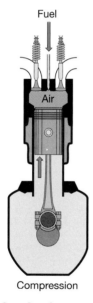

Figure 8–10 Modern diesel engine. Near the end of the compression stroke, fuel is injected and ignites by the heat of compression.

8–7 THE ROTARY ENGINE

There are strong incentives to replacing the piston engine. Foremost among them are new laws requiring a reduction in the pollutants contained in the exhaust gases and a reduction in manufacturing costs. Several alternatives to the reciprocating piston engine are being explored: the gas turbine, electric propulsion, and a number of steam or vapor engines.

A proposed successor to the reciprocating piston engine is one invented by Felix Wankel (b. 1902) and known as the Wankel engine, the **rotary engine**, or the RC (rotating combustion) engine (Fig. 8–11). The apexes of the triangular rotor of a Wankel are always in contact with the housing and create three separate variable volumes. The Wankel has no valves; the rotor itself opens and closes the fuel-air inlet port and the exhaust port at appropriate times in the combustion cycle. With only about half as many moving parts as a conventional engine, a Wankel is only about half the size and weight and is almost completely vibrationless. It may be more reliable and serviceable than piston engines. Its fuel-octane requirement is less, and it can run well with unleaded fuel.

8–8 GAS TURBINE OR TURBOJET ENGINE

For powering high-speed aircraft, the **gas turbine** (also known as the **turbojet**, or **jet engine**) has replaced the piston engine. It is a nonreciprocating internal combustion engine that transforms heat energy directly into work, eliminating the intermediate steps

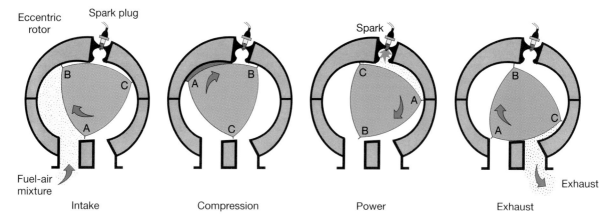

Figure 8–11   The Wankel engine, a rotary engine, undergoes a four-stroke Otto cycle in each of three small, variable chambers in one revolution of the rotor: intake, compression, ignition by a spark plug, and exhaust.

of the piston, shaft, wheel, and propeller. The gas turbine liberates a large amount of energy in a short time by operating continuously and at higher speeds than a piston engine, since it does not have reciprocating parts.

The gas turbine works as follows: Air is drawn into a compressor in which its pressure is raised, then into a combustion chamber (Fig. 8–12). There it is mixed with fuel and produces an intense flame. The combustion gas passes through a turbine, its velocity increasing, and then through an exhaust nozzle and out to the atmosphere. A powerful thrust (net force) is developed at high speeds from the change in momentum of the gases between the inlet and the exhaust. The turbine drives the compressor.

The efficiency of jet engines, as with other heat engines, depends on the temperature limits between which they operate. The most serious efficiency limitation is the temperature of the combustion gas, which must be kept low enough to prevent destruction of the turbine blades, usually below 1000°C. The efficiency of an aircraft gas turbine is about 20%.

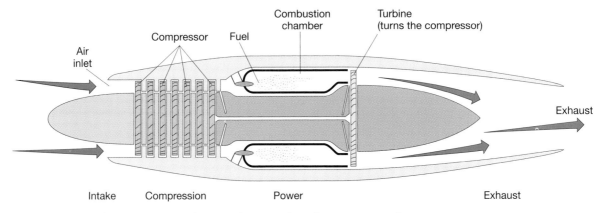

Figure 8–12   Turbojet engine. In relation to the internal combustion engine, the basic four-stroke functions are carried out simultaneously in different parts of the turbojet engine.

8–9 ROCKETS

The simplest type of jet propulsion, in principle, is the **rocket**. A gas is generated at high temperature and high pressure as it flows through the exhaust nozzle. In the combustion chamber, chemical energy is converted into heat with an efficiency on the order of 95%, and in the exhaust nozzle, heat is converted into the kinetic energy of the jet with an efficiency of about 50%. The unconverted heat is expelled with the hot gas. Unlike the turbojet, the rocket engine carries its own oxidizer, usually liquid oxygen or concentrated nitric acid, and does not take in air. Liquid hydrogen and liquid oxygen powered the 200,000-lb thrust J-2

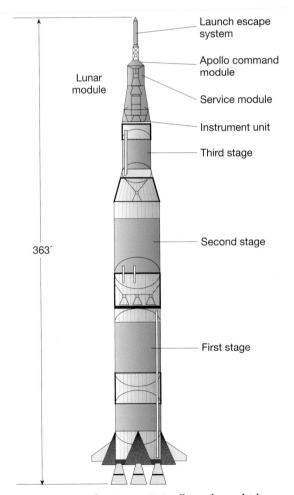

Figure 8–13 The Saturn V-Apollo rocket, which propelled astronauts to the moon. (Courtesy of NASA)

Labels on figure:
- Launch escape system
- Apollo command module
- Service module
- Instrument unit
- Third stage
- Second stage
- First stage
- Lunar module
- 363′

engines in the second and third stages of the Saturn V-Apollo system that sent astronauts to the moon.

The rocket principle itself is centuries old. The Chinese may have used rockets in the thirteenth century; Arabs of the period referred to rockets as "Chinese arrows." These rockets were fueled with black powder, a mixture of sodium nitrate, charcoal, and sulfur. For fireworks rockets, fine metal filings might be added to produce a shower of sparks. Their tendency to explode during manufacture led Alfred Nobel (1833–1896), a chemist, to develop smokeless powders from nitrocellulose, nitroglycerine, and stabilizers, which became the basis of modern **solid-fuel rockets**.

With a solid-fuel rocket, nothing can be done to control its flight after it has been ignited. In 1903, Konstantin Tsiolkovsky (1857–1935), an inventor and a rocket expert, suggested a new approach — that rockets be designed for **liquid fuels**. With liquid fuels, the rate of burning can be adjusted by means of valves or injection nozzles, either by a radio signal or directly in a manned rocket.

Robert Hutchins Goddard (1882–1945), a physicist at Clark University in Worcester, Massachusetts, in 1919 presented the first modern paper on rocket theory, "A Method of Reaching Extreme Altitudes," in which he included a reference to liquid fuels. In his book, *The Rocket into Interplanetary Space* (1923), Hermann Oberth strongly recommended the use of liquid fuels, with liquid oxygen as the oxidizer. On March 16, 1926, Goddard launched the first **liquid-fuel rocket**, containing gasoline and liquid oxygen, on a farm in Auburn, Massachusetts. The rocket landed 184 feet away, the precursor to the flights to the moon of the 1960s and 1970s (Fig. 8–13).

8–10 THE REFRIGERATOR

The **refrigerator** acts as a heat engine in reverse, a **heat pump**. It extracts heat from food and water at a low temperature, does work, and delivers heat to the air surrounding the refrigerator (the kitchen).

In the electric refrigerator, an electric motor operates a compressor (Fig. 8–14). The working substance, called the refrigerant, is a gas that is easily condensed by pressure. Ammonia (NH_3), sulfur dioxide (SO_2), or Freon (CCl_2F_2) is commonly used. The refrigerant is taken through a cycle that is repeated

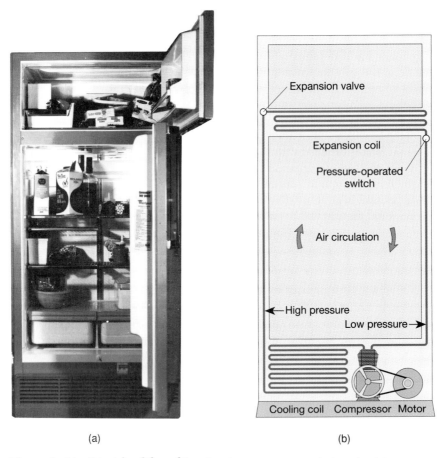

(a) (b)

Figure 8–14 Principle of the refrigerator. (Source: G. L. Lee *et al.*, *General and Organic Chemistry.* Philadelphia: W.B. Saunders Company, 1971, p.400.)

until the temperature is reduced to the point at which a thermostat opens a switch and cuts off the motor. When the interior warms up enough, the thermostat closes the switch and the cooling cycle begins again.

A typical cycle may start with the refrigerant as a vapor. The vapor is taken to the compressor, where a pump compresses it and heats it to a temperature slightly above room temperature. In the pipes of the condenser, the compressed and liquefied fluid gives up heat to the circulating air and is cooled. Then it goes to the cooling unit, where it evaporates and reduces the temperature of the surroundings. In this process, heat is extracted from the contents of the cooling unit of the refrigerator and is used to vaporize the refrigerant. The vapor is returned to the compressor, and the cycle is repeated.

8–11 FUEL CELLS

Most **electrical energy** produced in the world is generated through some form of heat engine fueled by the combustion of coal, petroleum, or natural gas. The source of that energy is the **chemical bonds** in these **fossil fuels**. In the process of obtaining electrical energy in this way, most of the chemical energy is wasted in the three-step conversion process: The chemical energy is first converted by combustion of the fuel into heat; the heat is then transformed into mechanical energy in a heat engine; and, finally, the mechanical energy is converted into electrical energy in a generator.

The overall efficiency of this process suffers from incomplete combustion, heat losses, and the limitations of mechanical devices. Even under ideal conditions, the thermodynamics of heat engines limits the efficiency of converting thermal energy into mechanical work. At the operating temperatures of heat engines, the second law of thermodynamics decrees that more than half of the original chemical energy must be lost as wasted heat. (Nuclear power plants are subject to this same limitation and are even less efficient than fossil-fuel plants because of their lower operating temperature.)

In a **fuel cell**, the heat cycle is bypassed and the **chemical energy** of fuels is **converted directly into electricity**. This idea is not new. Sir William Grove (1811–1896), a lawyer whose hobby was physics, in 1839 constructed a battery in which hydrogen reacted with oxygen to form water and an electric current was generated. The basic fuel is used directly in a true fuel cell, unlike zinc or lead in batteries, which must be refined at great expense. With fuel cells, efficiencies of 75%, more than twice that of the average steam-power station, are possible. Considerable development work will be required to realize such possibilities on a wide commercial scale, although fuel cells have already proved feasible in the Gemini and Apollo spacecrafts (Fig. 8–15).

To send astronauts to the moon and back, efficient and lightweight energy sources had to be developed. Since batteries and solar cells were judged inadequate for such space flights, the fuel cell appeared to

Figure 8–15   A hydrogen/oxygen fuel cell is used in spacecraft. (United Technologies)

be the best solution. A fuel cell can produce up to six times as much electricity as the best alkaline battery. Moreover, a fuel cell operating on hydrogen and oxygen produces pure water as a by-product, which can be used by the astronauts for drinking, for cooking, and for cooling equipment. Fuel cells provided onboard electrical power for Apollo spacecraft, each vehicle using three power plants of 31 cells each, connected in series. Many believe that the fuel cell has a bright future, although commercial fuel cells may rely on the conversion of hydrocarbons and air, rather than hydrogen and oxygen, into electricity. If they are right, cars and homes could eventually be powered by fuel cells.

SUMMARY OF EQUATIONS

Mechanical equivalent of heat $= \dfrac{\text{Work}}{\text{Heat}}$

$$J = \frac{W}{H}$$

$$J = 4.18 \, \frac{\text{joules}}{\text{calorie}}$$

Efficiency $= \dfrac{\text{Useful work}}{\text{Energy input}}$

$$\begin{matrix}\text{Efficiency}\\ \text{of a}\\ \text{Carnot Engine}\end{matrix} = \frac{T_1 - T_2}{T_1}$$

KEY TERMS

Arrow of Time   A preferred direction of natural processes.

Carnot Engine   An ideal engine —completely insulated, frictionless, and leakproof.

Chemical Bond   A linkage between two atoms.

Diesel Engine   A high-compression engine in which air is the working fluid. The fuel is in-

jected near the end of the compression stroke and is ignited by the heat of compression.

Electrical Energy   The energy associated with charged particles.

Entropy A measure of the disorder of a system.

First Law of Thermodynamics The law of conservation of energy: Energy cannot be created or destroyed but may be converted from one form to another.

Fossil Fuel A fuel that is formed in the earth from plant or animal remains, such as coal, petroleum, or natural gas.

Fuel Cell A device in which the chemical energy of fuel is converted directly into electricity.

Heat Death of the Universe A prediction of the second law of thermodynamics that as time goes on the universe will approach a state of maximum disorder and all the energy of the universe will have been degraded to thermal energy.

Heat Pump A device that transfers heat from a low-temperature reservoir to a high-temperature reservoir.

Internal Combustion Combustion takes place inside the working fluid without any need for a firebox and a boiler.

Liquid Fuel A liquid such as kerosene, gasoline, or methanol that can be used as fuel.

Maxwell's "Demon" An imaginary being that could establish a temperature difference from a source of uniform temperature by separating fast from slow molecules that could then drive a heat engine.

Mechanical Equivalent of Heat Mechanical work is converted into heat by the ratio Work/Heat = 4.18 joules/calorie.

Otto Engine A reciprocating, piston, internal combustion engine that employs a four-stroke cycle first demonstrated by Nicolaus August Otto: intake, compression, power, and exhaust.

Refrigerator An appliance or room for keeping food or other items cool.

Rocket Engine An engine that consists essentially of a combustion chamber and an exhaust nozzle and is used for the propulsion of a missile or a vehicle such as a space shuttle.

Rotary Engine A rotating combustion engine invented by Felix Wankel in which there are no valves; the rotor itself opens and closes the fuel-air input port and the exhaust port.

Second Law of Thermodynamics One formulation of this law is that the energy in the world available for work is continually decreasing. Another formulation is that the entropy of the world is continually increasing.

Steam Turbine Engine A rotary engine usually made with a series of curved vanes on a central rotary spindle that is driven by the pressure of steam discharged at high velocity against the vanes.

Thermal Efficiency The efficiency of an ideal (Carnot) engine operating between temperatures T_1 and T_2 is given by $(T_1 - T_2)/T_1$.

Thermodynamics The branch of science that deals with the relationship of heat and mechanical energy.

Turbojet Engine A gas turbine or "jet engine"; a nonreciprocating internal combustion engine that transforms heat energy directly into work.

THINGS TO DO

1. Bend a piece of wire back and forth. What does this do to its temperature?

2. Pull a nail from a board with a hammer. How does the nail feel to the touch?

3. Place a bowl containing some water under an electric mixer, and note the temperature of the water. Run the mixer for several minutes; then take the temperature again.

4. Quickly stretch a thick rubber band. What effect has this on its temperature?

5. Make a working model of Hero's engine, or examine one in a laboratory. Escaping steam makes the system rotate. How do you suppose the temperature of the exhaust steam compares with that of the steam in the engine?

6. Examine a working model of an Otto engine.

EXERCISES

Mechanical Equivalent of Heat

8–1. What did Joule's experiments prove about energy?

8–2. Water at Niagara Falls undergoes a 160-foot drop. Would you expect a difference in the temperature of the water between the top and the bottom of the Falls? Explain.

8–3. A 75.0-kilogram man climbs a staircase 7.00 meters high. How much energy (in calories) does he expend?

8–4. If the energy in the food that a man consumes in one day, 3000 kcal, were all converted into mechanical energy, how high would it lift a 75.0-kilogram mass?

The First Law of Thermodynamics

8–5. Explain what is meant by the conservation of energy.

8–6. Give an example of a device that can perform the conversion of (a) electrical energy to thermal energy; (b) mechanical energy to thermal energy; (c) chemical energy to thermal energy.

8–7. In highly developed countries, a person consumes approximately 3000 kilocalories per day as food. Explain what happens to this energy.

8–8. Which law of thermodynamics can be paraphrased as follows: "You can't win; the best you can do is break even"? Discuss.

8–9. Apply the first law of thermodynamics to the motion of an automobile.

The Second Law of Thermodynamics

8–10. How are the concepts of entropy and "time's arrow" related?

8–11. The first law says that we cannot get more out of a process than we put in, but the second law says that we cannot break even. Explain.

8–12. Which do you think has the greater entropy, 1 kg of water or 1 kg of ice? Why?

8–13. Give an example of a natural process in which order goes to disorder.

8–14. Does straightening out a room after a party violate the second law of thermodynamics? Explain.

Heat Engines

8–15. The input of a heat engine is 5000 calories per cycle, and the work output is 2000 joules per cycle. What is the efficiency of the engine?

8–16. Could a heat engine be designed with a thermal efficiency of 100%? Explain.

8–17. The efficiency of an engine that obtains heat at 260°C and exhausts heat at 60.0°C is 16.0%. What is its maximum possible efficiency?

8–18. A steam turbine operates between the temperatures of 800°C and 20.0°C with an efficiency of 33.0%. How does its efficiency compare with the efficiency of a Carnot engine operating between the same temperatures?

8–19. What is the Carnot efficiency of a steam turbine working between 300°C and 40.0°C?

8–20. How did James Watt influence the modern world?

8–21. Describe what happens to the waste heat in an automobile engine that has an efficiency of 25%.

8–22. State some advantages and disadvantages of rotary engines over reciprocating engines.

8–23. How does a rocket engine differ from a turbojet engine?

8–24. List some favorable and adverse effects of the internal combustion engine on society.

8–25. A major component of an electric power plant is a steam turbine. What is the advantage of raising the temperature of the steam as much as possible?

8–26. What is the purpose of an automobile radiator?

8–27. *Multiple Choice*

A. A heat engine is designed to
   (a) heat houses.
   (b) cool houses.
   (c) change work into heat.
   (d) change heat into work.

B. The law of conservation of energy is expressed by the
   (a) first law of thermodynamics.
   (b) second law of thermodynamics.
   (c) first law of motion.
   (d) second law of motion.

C. A mixture of salt and pepper separating into different layers when shaken violates the
   (a) conservation of energy.
   (b) conservation of momentum.
   (c) law of entropy.
   (d) principle of efficiency.

D. The mechanical equivalent of heat is expressed by
   (a) J.
   (b) W.
   (c) H.
   (d) C.

E. The Carnot efficiency of a diesel engine is
   (a) easy to attain.
   (b) its maximum possible efficiency.
   (c) 100%.
   (d) a measure of its thermal energy.

F. For every natural process, the entropy of the universe
   (a) decreases.
   (b) is destroyed in part.
   (c) increases.
   (d) remains constant.

G. The larger the thermal efficiency of a heat engine,
  (a) the greater the heat input.
  (b) the more heat output for a given heat input.
  (c) the greater the work output for a given heat input.
  (d) the larger the Carnot efficiency of the engine.

H. A diesel engine
  (a) is a type of rotary engine.
  (b) operates in a four-stroke cycle.
  (c) is a Carnot engine.
  (d) has no spark plug.

I. The theoretical limit for the efficiency of a cyclic heat engine is given by
  (a) the first law of thermodynamics.
  (b) entropy.
  (c) Carnot efficiency.
  (d) the work output.

J. In a thermodynamic process,
  (a) the entropy always increases.
  (b) energy is conserved.
  (c) heat is always transferred.
  (d) heat flows spontaneously from a colder body to a warmer body.

SUGGESTIONS FOR FURTHER READING

Angrist, Stanley W., "Perpetual Motion Machines." *Scientific American,* January, 1968.
Proposals of getting something for nothing have foundered on either the first or second law of thermodynamics.

Bryant, Lynwood, "Rudolf Diesel and His Rational Engine." *Scientific American,* August, 1969.
Although the diesel engine failed to attain the ideal of the Carnot cycle, it has revolutionized heavy transportation.

Dinga, Gustav P., "Hydrogen: The Ultimate Fuel and Energy Carrier." *Journal of Chemical Education,* August, 1988.
Hydrogen has much in its favor as a fuel: high efficiency, compatibility with the biosphere, cost-effectiveness, and resource-conserving. It is being applied to transportation and energy systems around the world.

Fickett, Arnold P., "Fuel-Cell Power Plants." *Scientific American,* December, 1978.
These devices convert the energy of fuel directly into electricity. They are being scaled up for possible use by electric utilities.

Gibbons, John H., Peter D. Blair, and Holly L. Gwin, "Strategies for Energy Use." *Scientific American,* September, 1989.
Energy efficiency can stretch energy supplies and buy time to develop alternative energy resources.

Hafemeister, David, "Science and Society Test X: Energy Conservation." *American Journal of Physics,* April, 1987.
Since the oil embargo of 1973–74, energy consumption in the United States has remained constant while the GNP has increased by about 30%. The physics behind some of the improvements in efficiency in such areas as buildings, solar energy, appliances, and lighting is discussed.

Leff, Harvey S., and Andrew F. Rex, "Resource Letter MD-1: Maxwell's Demon." *American Journal of Physics,* March, 1990.
A guide to the literature on Maxwell's demon, which Maxwell introduced over 120 years ago. The issues continue to inspire research and debate.

Pierce, John R., "The Fuel Consumption of Automobiles." *Scientific American,* January, 1975.
Can the efficiency of American cars be increased by at least 40%? The author suggests ways to do this.

Scaife, W. Garrett, "The Parsons Steam Turbine." *Scientific American,* April, 1985.
Parsons' invention revolutionized shipping and the generation of electric power. Many of his innovations are still in use.

Smith, Norman, "The Origins of the Water Turbine." *Scientific American,* January, 1980.
The water turbine evolved from the water wheel and was an alternative to the steam engine for driving machinery. Today it produces about one fourth of the world's electric power and will continue to be an important component of energy production.

Wilson, David Gordon, "Alternative Automobile Engines." *Scientific American,* July, 1978.
The search for an alternative to the conventional Otto engine is stimulated by requirements for cleaner and more efficient engines.

Wilson, S. S., "Sadi Carnot." *Scientific American,* August, 1981.

Carnot's main interest was not theoretical physics, but the improvement and the wider use of steam engines and other heat engines in France.

Zweibel, Kenneth, "Photovoltaic Cells." *Chemical and Engineering News,* July 7, 1986.

Photovoltaic cells may one day supplement or even replace conventional fuels for generating electricity in large-scale power plants.

ANSWERS TO NUMERICAL EXERCISES

8–3. 1230 cal	8–15. 9.60%	8–18. 72.7%
8–4. 1.70×10^4 m	8–17. 37.5%	8–19. 45.4%

A circular waveform generated by water drops. (Richard Megra/Fundamental Photographs)

9

WAVE MOTION AND SOUND

As we have seen, a particle is an entity that has a definite mass associated with it. A baseball moving through the air may be viewed as a particle that by virtue of its motion carries energy and momentum. Although a wave, too, can carry energy and momentum, it does not necessarily carry any mass with it. Yet a wave can cause dishes to rattle, or the eardrum to oscillate, or a glass to shatter.

Waves are a direct connecting link between our senses and our physical environment. The colors of a sunset, the sounds of children at play, the surface of the sea, the flapping of a flag, the shock waves emitted by a jet plane flying at supersonic speeds—these are all wave phenomena. Waves of all kinds—mechanical, acoustical, electrical, and optical—behave fundamentally alike in most respects. They propagate, reflect, refract, superpose, interfere, and diffract.

Figure 9–1 A surfer riding an ocean wave. (Sylvain Cazenave/Agence Vandystadt/Photo Researchers, Inc.)

9–1 THE WAVE CONCEPT

A wave is a disturbance that is propagated through a system. The disturbance—a change in the condition of the system—may be a displacement of a material medium, as in the case of water waves and sound waves (Fig. 9–1). Or it may be a change in an electromagnetic field in empty space, as in the case of light and radio waves. In all waves, energy is transferred from one point in space to a neighboring point.

When you drop a stone into a quiet pool of water, circular ripples spread out from the point of impact and travel over the surface of the pool (Fig. 9–2). If the wave passes a toy boat, the boat bobs up and down but does not move forward with the advancing wavefront (Fig. 9–3). Although there is no large-scale movement of water from the point of impact of the stone to any other point, the wave causes physical effects some distance away. The features of mechanical wave motion, then, are a source of energy—the splashing stone; a medium—the water; and a moving pattern of displacement that we call the wave.

If the disturbance is a single event of short duration, it is called a **pulse** (Fig. 9–4). Supersonic jet aircraft set up pulses called **shock waves** in air, speedboats produce them in water, and earthquakes produce them in rocks (Fig. 9–5). When the distur-

Figure 9–2 Circular ripples spreading out from two sources in a body of water. (Martin Dohrn/Science Photo Library)

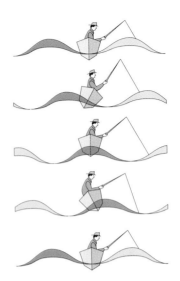

Figure 9–3 As the wavefront proceeds from left to right, the boat bobs up and down in place.

bance is repeated at a regular interval, called the **period**, T, a **periodic wave** is propagated.

Some wave properties can be demonstrated with a ripple tank. A shallow glass-bottomed tank containing water is illuminated from above. Ripples are produced by a rod that pushes up and down on the surface of the water, and their shadow is cast on a screen of white paper on the table under the glass. The highest points of waves are called **crests**, the lowest points **troughs**. The crests and troughs of the waves appear on the screen as alternating bright and dark circular bands moving outward from the rod (Fig. 9–6). The **amplitude**, representing the maximum displacement, is the height of a crest or the depth of a trough, as measured from the water level when there is no disturbance (the equilibrium position).

As long as the rod oscillates at a constant rate, the disturbance is repeated at a regular interval, the period, T. The number of waves produced per unit time is the **frequency**, f (Fig. 9–7). If the frequency is 10 waves/second, the period is 0.1 second/wave, and therefore

$$\text{Period} = \frac{1}{\text{frequency}} \qquad (9\text{–}1)$$

$$T = \frac{1}{f}$$

from which

$$f = \frac{1}{T} \qquad (9\text{–}2)$$

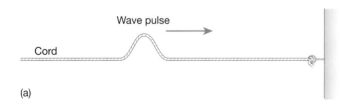

Wave pulse

Cord

(a)

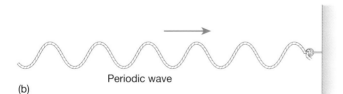

Periodic wave

(b)

Figure 9–4 (a) A single wave pulse sent out along a cord. (b) A periodic wave.

Figure 9–6 A ripple tank for studying water waves. (E.D.C. Distribution Center)

Figure 9–5 The V-shaped bow wave is formed because the boat's speed is greater than the speed of water waves. A jet plane travelling faster than the speed of sound forms shock waves in air. (Mike Sheldon/FPG International)

We see then that T and f are two ways of expressing the same idea. If the period is short, the frequency is high. The SI unit for frequency, the cycle/second (cps), is named the hertz (Hz), after Heinrich Hertz (1857–1894), the physicist:

$$1 \text{ hertz (Hz)} = 1 \, \frac{\text{cycle}}{\text{second}}$$

The **wavelength**, λ (Greek letter lambda), is the distance between any two points in the medium that, at any instant, are in precisely the same state of disturbance (Fig. 9–8). The distance from crest to crest on two successive waves, therefore, or from trough to trough, is the wavelength. Points on a wavetrain that are a wavelength apart are said to be **"in phase,"** since they are points of equal disturbance; they are also in phase if they are 2, 3, 4 . . . wavelengths apart. On the other hand, points on a crest and a trough, and similar points that differ by half a wavelength, are completely **"out of phase."**

If we watch a particular crest of a wave, we will see it traveling with a constant velocity (Fig. 9–9). Initially at point P, the crest moves to points P_1 and P_2 in

Figure 9–7 Frequency of a wave, measured in cycles per second (hertz), is the number of crests or troughs that pass a given point in one second. The frequency of wave (b) is greater than that of wave (a).

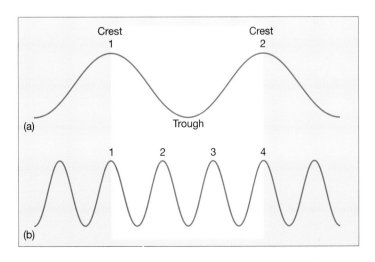

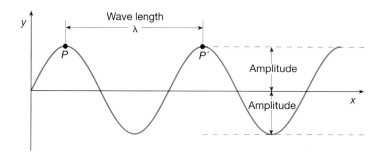

Figure 9–8 Wavelength and amplitude of a periodic wave. Points P and P′ are "in phase."

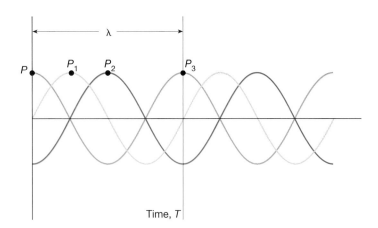

Figure 9–9 A periodic wave propagating to the right, shown at different time intervals. The crest at P moves forward a whole wavelength in time, T. The velocity, **v**, therefore equals λ/T.

successive small time intervals. During the time T, the crest has moved forward a whole wavelength to the position occupied by the preceding crest at the start of this time interval. The velocity **v** with which a periodic wave is propagated is, therefore, given by

$$\mathbf{v} = \frac{s}{t}$$

$$= \frac{\lambda}{T} \quad \text{or} \quad \frac{1}{T} \times \lambda$$

Since $1/T = f$, we have a useful relationship called the **wave equation: velocity = frequency × wavelength**, or

$$\mathbf{v} = f\lambda \tag{9–3}$$

For many types of waves, the velocity is well known, so that we may use the wave equation to calculate the wavelength when the frequency is known, or the frequency when the wavelength is known. For a wave of a particular velocity, if we generate waves at a faster rate, their crests will be closer together. So, the higher the frequency, the shorter the wavelength.

<hr>

EXAMPLE 9–1

In a ripple tank, waves are generated at a rate of ten per second. If the wavelength is 2.50 centimeters, (a) what is the wave velocity? (b) what is the period?

SOLUTION
Since the frequency and wavelength are known, the wave equation may be applied to find the velocity. We can get the period from Equation 9–1.

$f = 10 \, \dfrac{\text{waves}}{\text{s}}$ (a) $\mathbf{v} = f\lambda$

$\lambda = 2.50 \text{ cm}$ $= \left(10 \, \dfrac{\text{waves}}{\text{s}} \right)\left(\dfrac{2.50 \text{ cm}}{\text{wave}} \right)$

$\mathbf{v} = ?$

$T = ?$ $= \boxed{25.0 \, \dfrac{\text{cm}}{\text{s}}}$

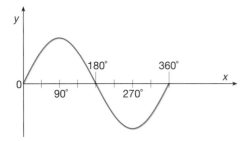

Figure 9–10   One wavelength of a sine wave divided into 360°. Waves of this type are referred to as sinusoidal.

(b) Period $= \dfrac{1}{\text{frequency}}$

$$T = \dfrac{1}{10\,\dfrac{\text{waves}}{\text{s}}}$$

$$= 0.10\,\dfrac{\text{s}}{\text{wave}}$$

Figure 9–11   The Foucault pendulum at the Smithsonian Institution in Washington, D.C. (Courtesy of the Smithsonian Institution)

EXTENSION

A water wave has a wavelength of 10.0 ft and is traveling at a speed of 60.0 ft/s. What is its frequency? (Answer: 6.00 Hz)

■

The simplest type of regular wavetrain is called **sinusoidal** (since it has the same shape as a graph of the sine function in trigonometry), or a **sine wave** (Fig. 9–10). A small cork on water undergoes sinusoidal motion as it is displaced to a maximum from its equilibrium position, first in one direction, then in the opposite direction, by a progression of wave crests and wave troughs. The motion of a swing and the oscillations of a pendulum are also sinusoidal, alternating between a maximum displacement in one direction, a return to the equilibrium or starting position, and a maximum displacement in the opposite direction (Fig. 9–11).

9–2 TRANSVERSE WAVES

Imagine a long horizontal spring with one end attached to a wall and the other end held in the hand and pulled taut. By moving the near end up and down rapidly, you can generate a wave that travels along the spring in a horizontal direction. Each particle in the spring, however, oscillates up and down in a vertical plane at right angles to the direction in which the wave is traveling (Fig. 9–12). This kind of wave is called a **transverse wave**.

Light waves are transverse waves. They differ from transverse waves on a spring or on the surface of water; the latter are **mechanical waves**, in which material particles are displaced. Light waves are **electromagnetic waves** in which the disturbance involves changing electric and magnetic fields traveling with a velocity of 3.00×10^8 m/s (about 186,000 miles/s) in a vacuum, which is equivalent to traveling around the earth at the equator seven times in one second! Visible light ranges in wavelength from about 4000 Å (angstrom units) at the violet end to about 8000 Å at the red end. Although our eyes cannot see beyond this range on either side, we know from measurements that there are electromagnetic waves both longer and shorter; radio waves are among the longest, and gamma rays the shortest (Table 9–1).

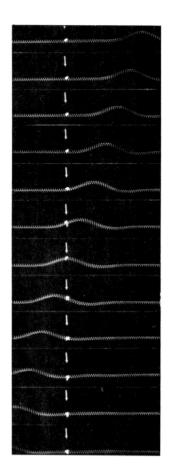

Figure 9-12 Transverse wave traveling along a coiled spring from right to left. The ribbon vibrates at right angles to the direction of motion of the wave. (Educational Development Center, Newton, MA)

EXAMPLE 9-2

A television channel broadcasts waves with a frequency of 55.0 MHz (megahertz). What is their wavelength?

SOLUTION

Since we know the frequency and velocity of the waves, we apply the wave equation to get the wavelength.

$$\mathbf{v} = 3.00 \times 10^8 \text{ m/s}$$
$$f = 55.0 \text{ MHz}$$
$$= 55.0 \times 10^6 \frac{\text{cycles}}{\text{s}}$$
$$= 5.50 \times 10^7 \frac{\text{cycles}}{\text{s}}$$
$$\lambda = ?$$

$$\mathbf{v} = f\lambda$$

$$\therefore \lambda = \frac{\mathbf{v}}{f}$$

$$= \frac{3.00 \times 10^8 \frac{\text{m}}{\text{s}}}{5.50 \times 10^7 \frac{\text{cycles}}{\text{s}}}$$

$$= 0.550 \times 10^1 \frac{\text{m}}{\text{cycle}}$$

$$= \boxed{5.50 \frac{\text{m}}{\text{cycle}}}$$

EXTENSION

A radar wave has a wavelength of 2.00 cm. Calculate its frequency. (Answer: 1.50×10^{10} Hz)

■

Table 9-1 WAVELENGTHS AND FREQUENCIES OF TYPICAL ELECTROMAGNETIC RADIATIONS

Type of Wave	Wavelength (cm)	Range (Å)°	Frequency Range (Hz)
AM radio	6×10^4 to 1.5×10^3	6×10^{12} to 1.5×10^{11}	0.5×10^6 to 2×10^7
TV and FM radio	7.5×10^2 to 1.5×10^2	7.5×10^{10} to 1.5×10^{10}	4×10^7 to 2×10^8
Microwaves	30 to 0.1	3×10^9 to 10^7	10^9 to 3×10^{11}
Near infrared	0.03 to 1.0×10^{-4}	3×10^6 to 8000	10^{11} to 3.0×10^{14}
Visible light			
Red	7.6×10^{-5}	6300–8000	3.9×10^{14}
Orange	6.1×10^{-5}	5900–6300	4.9×10^{14}
Yellow	5.9×10^{-5}	5600–5900	5.1×10^{14}
Green	5.4×10^{-5}	4900–5600	5.6×10^{14}
Blue	4.6×10^{-5}	4000–4900	6.5×10^{14}
Ultraviolet	4×10^{-5} to 3×10^{-6}	4×10^3 to 3×10^2	7.5×10^{14} to 10^{16}
X-rays	3×10^{-6} to 10^{-10}	3×10^2 to 10^{-2}	10^{16} to 3×10^{20}
Gamma rays	3×10^{-8} to 10^{-13}	3 to 10^{-5}	10^{18} to 3×10^{23}

° 1 Å = angstrom = 10^{-10} m.

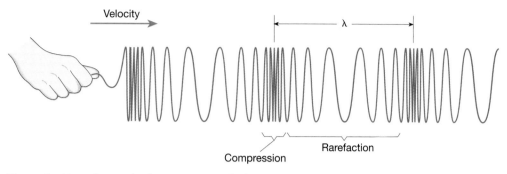

Figure 9–13 A longitudinal wave on a stretched spring. The particles of the spring oscillate in the direction of the wave. A wavelength is the distance between two adjacent points that are in phase.

9–3 LONGITUDINAL WAVES

The spring described at the beginning of Section 9–2 can oscillate in another mode—forward and backward in a horizontal direction. The spring remains straight, but a wave of alternating compressions and rarefactions travels along its length (Fig. 9–13). (A **compression** is a region in which the coils are closest together; a **rarefaction** is one in which they are spread out.) A particular coil in the spring undergoes periodic motion in a horizontal direction parallel to the direction in which the wave is traveling. Wave motion of this kind is called **longitudinal** (or **compressional**). Sound waves are our most familiar example of longitudinal waves: The disturbance of the medium occurs parallel to the direction in which the wave is proceeding.

The wavelength of a longitudinal wave is defined as the distance from a point on one compression to a point in the same phase of oscillation on an adjacent compression. Each wave is composed of a compression and a rarefaction. The frequency is the number of compressions or rarefactions that are produced in one second.

The energy transfer that can take place with compressional waves can be illustrated as follows. A number of metal or glass spheres are laid side by side along a grooved ruler (Fig. 9–14). When a sphere is rolled along the groove and strikes one end of the row, a sphere at the other end is kicked away some distance. The spheres in the middle of the row are not perceptibly disturbed, although they transmit energy, as is evident from the behavior of the sphere at the end of the row. These spheres transmit the disturbance while they themselves oscillate imperceptibly about their equilibrium positions in the direction in which the wave is traveling.

A **sound wave** in air consists of a train of compressions (high-pressure regions) following one another in rapid succession with the frequency of the sound and separated by rarefactions (regions of lower pressure). For the sake of convenience, we represent

Figure 9–14 Illustration of energy transfer with compressional waves. The spheres in the middle transmit energy, although they are not perceptibly disturbed. (Source: Jerry D. Wilson, *Practical Physics,* Philadelphia: Saunders College Publishing, 1986, p. 47, Fig. 3.12)

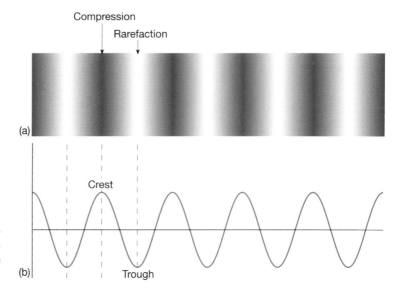

Figure 9–15 Compressions and rarefactions of longitudinal waves (a) compared to crests and troughs of transverse waves (b).

sound waves graphically as sinusoidal waves (Fig. 9–15). A crest on a sine wave representing sound corresponds to a compression, and a trough a rarefaction. If a light pith ball is held by a thread close to a vibrating tuning fork, the ball is thrown violently outward and comes to rest only as the sound ceases. The air is compressed when the tine of the fork moves outward and is rarefied when it moves inward (Fig. 9–16). The vibrating surfaces of the tuning fork thus become the

center of compressions and rarefactions that spread out in all directions with the speed of sound. The individual molecules of air through which sound is being transmitted oscillate backward and forward in a direction parallel to the disturbance.

That sound can be heard only if there is a medium — solid, liquid, or gaseous — between the source and the ear to transmit it, is shown graphically by the **bell-in-vacuum** demonstration (Fig. 9–17). A ring-

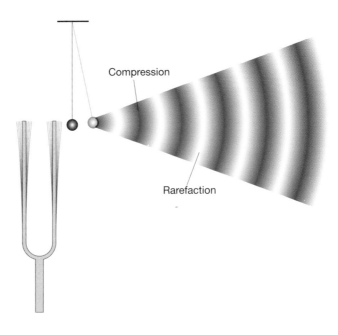

Figure 9–16 A vibrating tuning fork is a center of compressions and rarefactions that travel outward with the speed of sound.

Vacuum

To
vacuum
pump

Figure 9–17 The sound of the bell is not transmitted through a jar from which the air has been pumped out.

ing alarm clock cannot be heard in a jar from which the air has been exhausted but becomes audible when air is reintroduced. The ear is sensitive to an amplitude in air of around a thousand-millionth of a centimeter (even less amplitude in liquids and solids). The maximum displacement of the molecules in air from their original positions is seldom more than 0.5 millimeter.

9–4 VELOCITY OF SOUND WAVES

It is a common observation that sound has a definite velocity in air. We hear thunder moments after seeing a flash of lightning. Sitting in a grandstand in a baseball park, we see a fly ball before hearing the crack of the bat. This means that sound takes a measurable period of time to reach us.

The speed of sound can be determined directly by time-interval measurements. A gun or cannon is fired while observers stationed at various distances measure the time interval between the flash of the gun and sound of the firing. Dayton C. Miller (1866–1941), a physicist at what is now Case Western Reserve University, applied this method with coastal defense guns and made some of the most accurate measurements in modern times. The development of modern electronics has made possible even more precise measurements of small time intervals. In air at 0°C and 1 atmosphere of pressure, the velocity of

sound is 331 m/s (1087 ft/s). The velocity varies with the temperature, increasing approximately 0.60 meter per second per Celsius degree (2.00 ft/s/C°). Therefore, over a wide range of temperature, T, the velocity of sound in air is given by the following expressions:

$$\mathbf{v}_T(\text{m/s}) = \mathbf{v}_0 + 0.60\ T \tag{9–4}$$
$$\mathbf{v}_T(\text{ft/s}) = \mathbf{v}_0 + 2.00\ T \tag{9–5}$$

in which $\mathbf{v}_0$ is the velocity of sound at 0°C, and $\mathbf{v}_T$ the velocity of sound at temperature T. As the temperature increases, so does the speed of the molecules. As a result, the molecules collide more often and a disturbance is transmitted more quickly through the air.

EXAMPLE 9–3

Calculate the velocity of sound (m/s) in air when the temperature is 35.0°C.

SOLUTION
Use Equation 9–4 for determining the velocity of sound in m/s in air at various temperatures.

$$T = 35.0°C \qquad \mathbf{v}_{35.0°C} = \mathbf{v}_0 + \frac{0.60\ \dfrac{\text{m}}{\text{s}}}{°C}(35.0°C)$$

$$\mathbf{v}_0 = 331\ \text{m/s} \qquad\qquad = 331\ \frac{\text{m}}{\text{s}} + 21.0\ \frac{\text{m}}{\text{s}}$$

$$\mathbf{v}_{35.0°C} = ? \qquad\qquad = 352\ \frac{\text{m}}{\text{s}}$$

EXTENSION
The velocity of sound on a warm day was 1157 ft/s. How warm was it? (Answer: 35.0°C, or 95.0°F)

■

The speed of sound waves is different for each gas. In liquids and solids, the speed of sound is considerably greater, as Table 9–2 shows. The molecules in a solid are closer together than in a gas and hence respond more rapidly to a disturbance. The difference is apparent if a steel rail is struck by a hammer. A distant observer will hear two sounds, the first coming through the rail, which transmits the sound readily; the second through the more slowly transmitting air. In general, sound travels more slowly in liquids than in solids because liquids are more compressible.

Table 9–2   VELOCITY OF SOUND IN VARIOUS MEDIA

	Medium	m/s	ft/s
Gases	Air (0°C)	331.4	1087
	Carbon dioxide (0°C)	258	846
	Helium (20°C)	927	3040
	Hydrogen (0°C)	1270	4165
	Oxygen (0°C)	317	1041
	Water vapor (35°C)	402	1320
Liquids	Mercury	1450	4760
	Sea water	1520	4990
	Water (20°C)	1461	4794
Solids	Glass	5500	18,033
	Granite	3950	12,600
	Lead	1230	4030
	Pine	3320	10,900
	Steel	5000	16,400

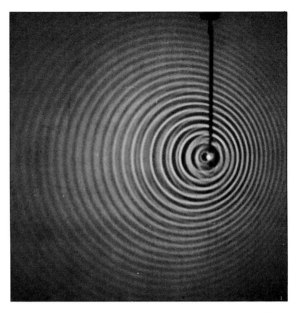

Figure 9–18   Doppler effect. Waves are produced by a source moving to the right. The waves bunch up on the right, the wavelength is reduced, and the frequency is higher than if the source were at rest. The waves behind the source are spread out and have a lower frequency. (Courtesy of Educational Development Center, Newton, MA)

9–5 THE DOPPLER EFFECT

The word "frequency," as we have seen, is an objective property; that is, it is the number of cycles per second. When we hear a high-frequency sound, we say, subjectively, that the "pitch" is high. Likewise, for low-frequency sounds, we say that we hear a low pitch. When there is relative motion between a source of waves and an observer, an apparent change occurs in the frequency of the sound; that is, there is a change in pitch as perceived by the observer (Fig. 9–18). A few familiar examples illustrate this. The pitch of the siren on an ambulance moving in traffic seems to be higher than normal as it approaches a pedestrian on a sidewalk, then suddenly drops to lower than normal as it passes. The pitch of an automobile horn or a jet plane undergoes a similar change as it approaches or recedes from us.

Christian Doppler (1803–1853), a physicist and mathematician, worked out the theory of such frequency shifts, now called the **Doppler effect**. Experimental tests confirmed his equations. The sound was provided by several trumpeters, and musically trained observers evaluated it by ear. First, the trumpeters were positioned on a moving railroad flat car while the observers were stationed on the ground. Then they exchanged places, and the observers moved past the stationary trumpeters.

Although the pitch of the trumpets is unchanged as played, to an observer in relative motion with the source it seems to increase or decrease. When the source is moving toward the observer, with each new wave the wavetrain is closer to the observer than when the preceding wave started. The distance that each new wave has to travel is therefore less than that traveled by the preceding one, and more waves reach the observer per second than when the train is stationary. The effect is to raise the apparent pitch of the trumpet. When the source is receding, the pitch is lowered, since fewer waves reach the observer per second.

The Doppler effect applies to light waves as well as to sound waves. Armand Fizeau (1819–1896) explained the application of the Doppler effect using the light coming from a star. If a star is approaching the earth, its light seems more violet, indicating higher frequencies; its radiation is said to **shift toward the violet**. If it is receding, the star's radiation becomes redder, indicating lower frequencies, and there is a **"red shift."**

9–6 REFLECTION OF WAVES

When waves encounter a barrier or an abrupt change in the nature of the medium, some **reflection** occurs at the surface in a manner similar to that in which a ball bounces after striking a wall. The waves turn back on themselves; what is not reflected is absorbed or transmitted by the medium in which they are traveling. Light is reflected from mirrors, sound from buildings or canyons, certain radio waves from the ionosphere and satellites, and laser beams from the moon (Fig. 9–19). It is found experimentally for all waves that the **law of reflection** applies (Fig. 9–20):

> **The angle of incidence is equal to the angle of reflection.**

The incident and reflected waves are measured from a line perpendicular to the plane of the reflecting surface called the normal. The two waves and the normal lie in the same plane.

A reflected sound that can be easily distinguished from the original is called an **echo**. To hear an echo, it is necessary to stand 60.0 feet or more away from a large reflecting barrier and to produce a loud pulse of sound. The distance is critical, because the sensation of sound persists for about one tenth of a second. If the speed of sound is 1100 feet/second, sound travels

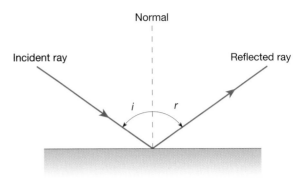

Figure 9–20 *Law of reflection.* The incident ray and the reflected ray make equal angles with the normal to the surface: $i = r$.

110 feet in 0.10 second. Therefore, if the distance between source and reflector is 60.0 feet or more, the pulse has time to travel to the barrier and back (120 feet) and to be heard as a sound distinct from the original. By then the sensation of the original sound will have decayed.

Sound can undergo rapid multiple reflections that cannot be analyzed into distinct echoes, an effect called **reverberation**. Such sound may persist for several seconds after the original sound has ceased. Wallace C. Sabine (1868–1919), who established the basic principles of modern architectural acoustics, found that reverberation time, an index of the "liveness" of sound, is one of the factors determining the acoustical qualities of auditoriums and concert halls. By using appropriate materials on interior surfaces —fabrics, acoustic tiles and panels, stone, brick, wood, and plaster—architects attempt to design music rooms and lecture halls that have optimum reverberation times (Fig. 9–21). Boston Symphony Hall, the first structure designed on the basis of Sabine's work, is still regarded as one of the excellent concert halls in the world. But, as we have learned to our chagrin in the past 25 years, the job of designing an opera house or a concert hall is too complex to be an exact science.

9–7 REFRACTION OF WAVES

When the speeds of the waves differ in two adjacent media, such as in air and water, then in addition to reflection, a wavetrain undergoes a change in direc-

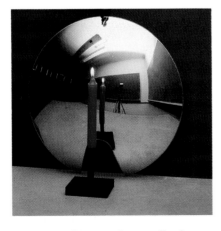

Figure 9–19 Reflection of a candle from a convex spherical mirror. (David Rogers)

Figure 9–21   Davies Symphony Hall in San Francisco exemplifies the synthesis of architecture and acoustics. (Mark E. Gibson/The Stock Market)

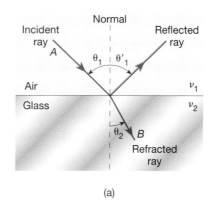

(a)

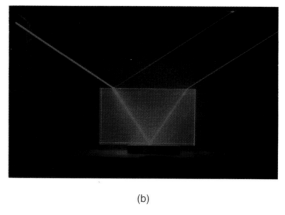

(b)

Figure 9–23   Some of the light incident on a lucite block is reflected and some is refracted. The refracted ray is bent toward the normal when it enters the block and away from the normal when it leaves the block. (Source: Serway and Faughn, *College Physics*, 3rd edition. Philadelphia: Saunders College Publishing, p. 756.)

tion and bends as it passes at an angle from one medium into the other. This bending of the waves upon entering a new medium is called **refraction** (Fig. 9–22). Investigation shows that an increase in the angle of incidence is followed by an increase in the angle of refraction. It has been determined that the ratio of the speeds of the waves in the two media is a constant, *n*, called the **index of refraction** of the second medium relative to the first. Values of *n* for various media relative to air are listed in tables.

The bending of the wavefront is toward the normal as the waves enter a medium in which their velocity is

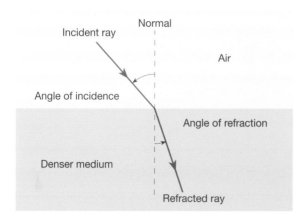

Figure 9–22   Refraction of waves. As they pass from one medium into another at an angle, waves change direction.

slower (Fig. 9–23). As part of the wavefront enters the second medium in advance of the rest of the front, its velocity changes, and there is a change in the direction of the front (Fig. 9–24). If the wavefront is found to bend away from the normal, the waves must be traveling with a greater velocity in the second medium than in the first.

A layer of warm air acts as if it were acoustically different from a layer of cold air, owing to the difference in the velocities of sound waves in the two layers. Sound refraction occurs when sound passes obliquely from cold air into warm air. This explains why sound travels great distances over a frozen lake. As the sound waves travel upward from the ice, their speed is

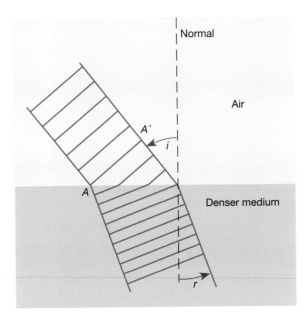

Figure 9–24 Refraction explained by a difference in the wave velocity in the two media. Point A enters the denser medium first and slows down, while A′ is still traveling in the less dense medium. (Photography by Henry Leap)

greater at the higher altitude, where the air is warmer (Fig. 9–25). The waves are refracted by the warmer air and strike the surface of the lake again. There they are reflected, and the cycle of refraction and reflection is continued, with the result that sound travels great distances and can be heard quite distinctly.

9–8 DIFFRACTION

Although we cannot see around corners, we can hear a person talking around a corner or in the next room. Low-pitched tones, in particular, carry well. In similar fashion, ocean waves bend around a breakwater and cause boats behind the breakwater to bob up and down. The bending of waves around barriers is not explained by reflection or refraction, since the medium is the same and we may assume that the temperature is the same. The bending can be explained, however, by diffraction.

Diffraction is the property that enables waves to bend around an obstacle or spread through an opening and not cast a completely sharp shadow (Fig. 9–26). To shut out noise from a room, a door or window must be closed completely; leaving it open even slightly may admit nearly as much sound as when it is wide open, because of the tendency of the waves to spread. Sound cannot be blocked out by holding a book between yourself and the source; the sound waves will bend around and behind the book. The waves then unite in various combinations and yield diffraction patterns.

Diffraction spreading is explained by **Huygens' theory of wave propagation** (Fig. 9–27). Each point on a wavefront may be regarded as a source of secondary wavelets that spread out with the speed **v** of the wave at that point. After a time t, these secondary wavelets have radii **v**t, and the envelope of these wavelets will be the new wavefront at the end of the time interval. The wavelets combine to form a replica of the original wavefront, and the wave has advanced.

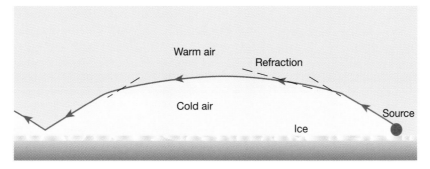

Figure 9–25 Refraction of sound waves over a frozen lake. Sound waves are bent —*refracted*—downward by a layer of warm air above cold, then reflected by the ice. Distant sounds thus can be heard quite distinctly that ordinarily would be too faint to be heard.

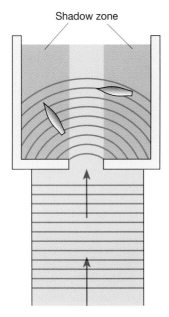

Figure 9–26 Diffraction of waterwaves. The waves spread when they bend around obstacles or pass through openings.

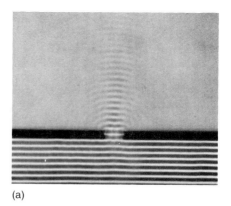

(a)

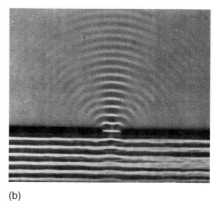

(b)

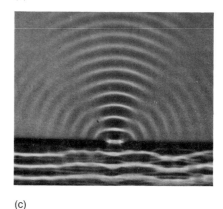

(c)

Figure 9–28 The diffraction of waves through a narrow opening. The longer the wavelength, the greater the amount of diffraction. (Courtesy of Educational Development Center, Newton, MA)

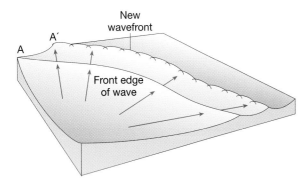

Figure 9–27 Huygens' theory of wave propagation. Every point on an advancing wavefront is considered a source of secondary wavelets. The envelope around the secondary wavelets marks the new position of the wave.

When waves pass around an obstacle or through a small opening, they can thus change direction and spread. The greater the wavelength and the smaller the opening, the more pronounced the diffraction (Fig. 9–28).

9–9 SUPERPOSITION OF WAVES; INTERFERENCE

When two or more waves move through a medium at the same time, each wave proceeds as if it were the only one present, except where they meet. In those

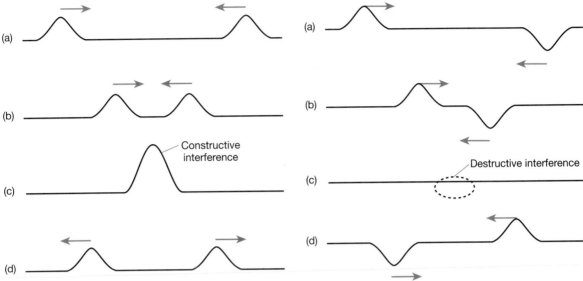

Figure 9–29 *Superposition.* Two waves traveling in opposite directions on a cord add where they meet and then pass through each other.

Figure 9–30 *Destructive interference.* A crest and a trough traveling in opposite directions cancel each other when they meet.

regions, the resultant displacement of the medium is the vector sum of the displacements of the individual waves that would occur if each wave alone were present (Fig. 9–29). Then each wave passes on in its original direction without change, as though the other waves were not present. This behavior, unique to waves, is known as the **superposition principle**.

If two waves having the same amplitude arrive at a given point where each wave is at a crest, the two amplitudes combined will be exactly twice the amplitude of a single wave alone. This effect is known as **constructive interference**. But if one wave is at crest and the other is at a trough, the two waves give a combined amplitude of zero, and there is **destructive interference** (Fig. 9–30). For other combinations, the resultant amplitude will be greater than zero but less than the maximum.

Thomas Young (1773–1829), a physician and physicist, performed experiments that did much to establish the **wave nature of light**. When a panel in which two thin slits had been cut was illuminated by sunlight, two light beams emerged from the other side of the panel, each slit itself acting as a source. On a screen some distance away, Young saw alternating areas of brightness and darkness (Fig. 9–31). The results can be explained only if light is considered to have a wave nature as opposed to a particle nature.

Bright fringes occur where light waves from the two sources superpose in similar phases, as crest superposed on crest, and the dark areas result from destructive interference, as when a trough superposes with a crest. The rainbow of colors of soap films and oil slicks on a wet pavement are due to interference of light waves reflected from the two surfaces of the film. The combination of constructive and destructive interference causes some wavelengths to be reinforced and others to be canceled.

Two wavetrains of sound of slightly different frequencies but equal amplitudes interfere to produce a series of **beats**: points of maximum amplitude alternating with points of minimum amplitude in the vibration (Fig. 9–32). The beat frequency is given by the difference of the two wave frequencies. Thus, when two tones of slightly different frequency are sounded together, say 128 Hz and 136 Hz, the loudness fluctuates at a beat frequency of eight beats per second.

9–10 SOUND AS VIBRATION

When you strike a tuning fork with a rubber mallet, it emits its tone. Although the vibrations of the tuning fork are too rapid to be seen, the edges present a fuzzy

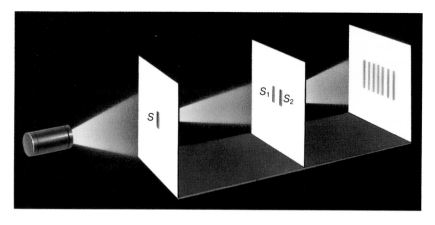

(a)

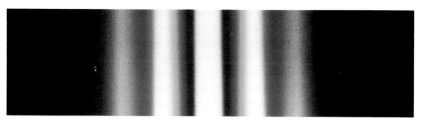

(b)

Figure 9–31 (a) Young's double-slit experiment that demonstrates the interference of light. (b) The interference pattern produced by white light. (Kodansha Publications, Japan)

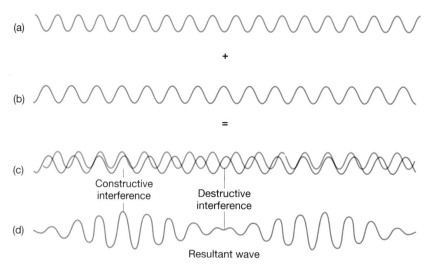

Figure 9–32 Beats result from interference of two waves having slightly different frequencies. A periodic change in amplitude results in fluctuation in loudness of the sound waves.

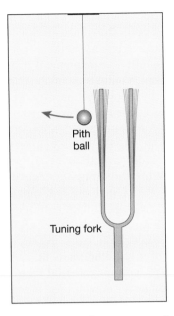

Figure 9–33 A vibrating tuning fork.

outline while sound is emitted, and the vibrations can be detected by touching the fork with the finger (Fig. 9–33). The vibrating surface becomes the center of compressions and rarefactions in the air, which spread out in all directions. Any vibrating object will produce a compressional wave in a medium—the buzzing wings of a fly, the vibration of the vocal cords, the speaker of a radio or stereo system.

Although a tuning fork emits a pure tone, the sound cannot be heard more than a short distance away. If the handle is pressed against a tabletop, though, the sound can be heard throughout the room. The tuning fork forces the tabletop into vibration, and the large vibrating area of the table radiates sound

well. A vibrating string by itself also produces very little sound, since it is too narrow to disturb much air. But the sound of the string can be amplified by passing the string over a support, which rests on a wide plate called the soundboard (Fig. 9–34). The vibrations of the string are conveyed to the soundboard, which in turn communicates them to a much greater mass of air. Without amplification, many sounds would be inaudible. The three stages in the production of sound are the initial disturbance, the amplification of the disturbance, and the radiation of sound.

The use of a stretched string or wire as a source of sound is the basis of such instruments as the violin, guitar, and piano. The tone it emits when twanged, bowed, or struck depends on the length, mass, and tension of the string. The vibration frequencies of strings are inversely proportional to their lengths, directly proportional to the square roots of their tensions, and inversely proportional to the square roots of their linear densities.

A string fixed at each end is set into vibration by plucking, bowing, or striking. In the violin and guitar, varying the length of a string by fingering causes the frequency to vary. Since the four strings of a violin are all of the same length but of different thickness, their **fundamental frequencies—the lowest frequency the string can emit**—differ. The wooden bridge communicates the vibrations of the strings to the body of the violin and the air inside it, which are then set into vibration and amplify the sound (Fig. 9–35). In the piano, the length, tension, and thickness of a wire vary from note to note. Like the body of the violin, the soundboard of the piano moves a large mass of air that radiates sound to the surroundings.

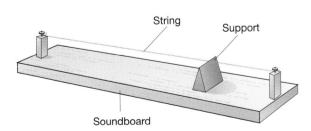

Figure 9–34 String and soundboard, which amplifies the sound produced by the vibrating string. Many sounds would be inaudible without amplification.

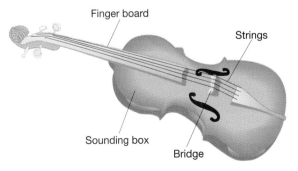

Figure 9–35 The four strings of a violin are tuned to different frequencies, which can be compared to certain notes on a piano keyboard.

9–11 STANDING WAVES

If the frequencies are properly selected, a string may be made to vibrate in several natural vibrational modes. Each direct wave is reflected at the end of the string and upon its return meets an ongoing wave. Regions of constructive and destructive interference occur at fixed positions, and a **standing wave** is produced (Fig. 9–36). The stationary pattern is called a "standing" wave because it involves no motion along the length of the string; unlike a traveling wave, a crest does not move from one point to another. Each section appears, instead, to move transversely to the length. Vibrating at its fundamental frequency, a string forms a simple pattern with a region of maximum constructive interference, the **antinode**, in the middle, and a stationary point, or **node**, at each end, the result of complete destructive interference.

Standing waves are useful because of the fixed wave pattern they form. To get a specific note from a musical instrument, a fixed wave pattern—that is, one that does not change its shape too rapidly—is necessary. Otherwise, a mixture of sounds is produced that is not identifiable as a note. The sound that comes from a violin string, an organ pipe, or air columns of wind instruments is the result of standing waves. Plates, rods, and diaphragms that are parts of percussion instruments also produce standing waves.

As the frequency of vibration is increased, standing wave patterns consisting of two, three, or more segments can be developed with nodes alternating with antinodes (Fig. 9–37). The distance between two consecutive nodes or antinodes is one-half wavelength, $\frac{1}{2}\lambda$. When the string vibrates as one segment, with a node at each end and only one antinode between them, the length L of the string equals one-half wavelength, $\frac{1}{2}\lambda$; the wavelength, therefore, is twice the length of the string: $\lambda = 2L$. This is the maximum wavelength of a standing wave for the string and corresponds to the **fundamental frequency**, f_0— the lowest frequency of which the string is capable.

When the string is made to vibrate in two segments, the length of the string contains two half-wavelengths: $L = 2\frac{\lambda}{2}$. The wavelength thus equals the length of the string: $\lambda = L$. Since the frequency and wavelength are inversely proportional, according

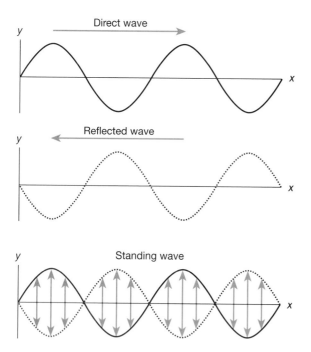

Figure 9–36 When two waves of the same frequency and amplitude travel in opposite directions, a standing wave is produced.

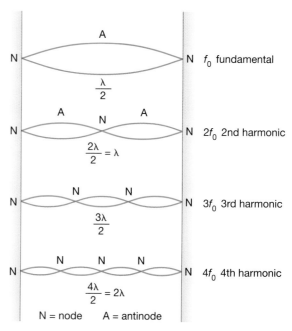

Figure 9–37 Standing wave patterns in a stretched string. The first four harmonic modes of vibration are shown.

Figure 9–38   A guitarist shortens the vibrating portion of the strings to change notes by pushing them against the frets of the fingerboard. (R. Heinzen/Superstock, Inc.)

to the wave equation ($\mathbf{v} = f\lambda$), the corresponding frequency is twice that of the fundamental, or $2f_0$. This frequency is called the **first octave**, or **first overtone**. Three segments in the standing wave pattern $\left(L = 3\dfrac{\lambda}{2}\right)$ produce a frequency three times the fundamental, $3f_0$, called the **second overtone**. Four segments produce the frequency $4f_0$, the **third overtone**, and so on. The frequencies of these modes of vibration form the series f_0, $2f_0$, $3f_0$, $4f_0$, in which the frequencies are integral multiples of the frequency of the fundamental. The frequencies in such a series are called **harmonics**. The fundamental is called the **first harmonic**, the higher frequencies the **second harmonic**, **third harmonic**, and so on. Although it is difficult to visualize, a string vibrates in multiple modes at the same time, producing the fundamental frequency and various overtones (Fig. 9–38).

EXAMPLE 9–4

Give three possible (a) wavelengths and (b) corresponding frequencies of standing waves in a wire stretched between two supports that are 1.00 meter apart. (Assume the velocity of sound in the wire to be 5000 m/s.)

SOLUTION

For the fundamental frequency, the length L of the wire equals $\dfrac{\lambda}{2}$, since there will be a node at each end. Therefore, $\lambda = 2L$. For the first overtone, $\lambda_2 = L$; and for the second overtone, $\lambda_3 = \dfrac{2L}{3}$. The frequencies are in the ratio $1:2:3$.

$L = 1.00$ m

$\mathbf{v} = 5000$ m/s

$\lambda_0, \lambda_1, \lambda_2 = ?$

$f_0, f_1, f_2 = ?$

(a) $\lambda_0 = 2L$
$ = 2(1.00 \text{ m})$
$ = \boxed{2.00 \text{ m}}$

$\lambda_1 = L = \boxed{1.00 \text{ m}}$

$\lambda_2 = \dfrac{2L}{3} = \dfrac{2 \text{ m}}{3} = \boxed{0.670 \text{ m}}$

(b) $\mathbf{v} = f_0 \cdot \lambda_0$

$f_0 = \dfrac{\mathbf{v}}{\lambda_0}$

$ = \dfrac{5000 \dfrac{\text{m}}{\text{s}}}{2.00 \text{ m}}$

$ = \boxed{2500 \text{ Hz}}$

$f_1 = 2f_0$
$ = 2(2500 \text{ Hz})$
$ = \boxed{5000 \text{ Hz}}$

$f_2 = 3f_0$
$ = 3(2500 \text{ Hz})$
$ = \boxed{7500 \text{ Hz}}$

■

9–12 VIBRATING AIR COLUMNS

Since wind instruments are silent unless someone blows on them, the vibrations are driven by energy put in from the outside. The standing wave pattern set up in an air column is like the standing wave on a string except that the air has a longitudinal motion in a direction parallel to the wave. The simplest standing wave pattern has an antinode at the open end, where

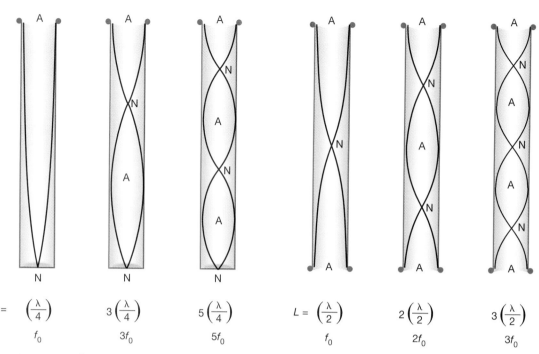

$$L = \left(\frac{\lambda}{4}\right) \qquad 3\left(\frac{\lambda}{4}\right) \qquad 5\left(\frac{\lambda}{4}\right)$$

$$f_0 \qquad\qquad 3f_0 \qquad\qquad 5f_0$$

Figure 9–39 Standing wave patterns of an air column closed at one end. There is always a node at the closed end and an antinode at the open end.

$$L = \left(\frac{\lambda}{2}\right) \qquad 2\left(\frac{\lambda}{2}\right) \qquad 3\left(\frac{\lambda}{2}\right)$$

$$f_0 \qquad\qquad 2f_0 \qquad\qquad 3f_0$$

Figure 9–40 Standing wave patterns of an air column open at both ends.

the air has maximum possible motion, and a node at the closed end, where the air cannot move. Many other standing wave patterns are possible, but in all there will always be an antinode at the open end and a node at the closed end.

For the simplest case mentioned, the length (L) of the tube is very close to one-quarter wavelength, $\frac{1}{4}\lambda$ (Fig. 9–39). The maximum wavelength is, therefore, $4L$, corresponding to the fundamental, the lowest frequency of sound possible for the tube to emit. The ratio of the set of frequencies for such a tube is $1f_0 : 3f_0 : 5f_0 : \ldots$, or **odd multiples** of the fundamental. The frequency $3f_0$ corresponds to the **first overtone** or **second harmonic**; $5f_0$ to the **second overtone** or **third harmonic**; and so on.

In a **tube open at both ends**, the simplest standing wave has an antinode at each end with one node between them (Fig. 9–40). The length L of the tube is very close to one-half wavelength; therefore, the wavelength of the fundamental is approximately twice the length of the tube, or $2L$. The possible frequencies emitted by an open tube are in the ratio

$1f : 2f : 3f : \ldots$, or both odd and even multiples of the fundamental.

The same tone, say with a fundamental frequency of 100 Hz, sounds different, depending on whether it is produced by a closed tube or an open tube, since the number and strength of overtones present differ. With the closed tube, only odd multiple overtones may be present in the tone: 300 Hz, 500 Hz, and so on. With the open tube, both the odd and even multiples are possible: 200 Hz, 300 Hz, 400 Hz, 500 Hz, and so on. While the lowest tone generally establishes the pitch, the remaining harmonics determine the timbre or quality of the sound.

9–13 TIMBRE OR TONE QUALITY

A **musical note** emitted by a vibrating string or air column is a combination of pure tones. The mathematician Jean Baptiste Joseph Fourier (1758–1830) showed that any **wave motion** can be represented as a **sum of sinusoidal motions** of different frequencies and amplitudes, each of which is known as a

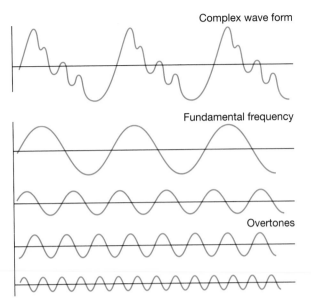

Complex wave form

Fundamental frequency

Overtones

Figure 9–41 Complex wave associated with a violin note. The sound is analyzed into four prominent harmonics, each of which is a pure sine wave. The musical note is a combination of pure tones.

component (Fig. 9–41). One sine curve represents the fundamental, and there is a sine curve corresponding to each of the overtones, which are heard simultaneously. A sound may be regarded as a combination of many independent sinusoidal sound waves, each behaving as if it were the only one present.

The "vibration recipe" of a sound—that is, the presence in it of certain frequencies—explains our ability to perceive sounds as being different even when they appear to have the same pitch and loudness. Each of us speaks with a complex vibrating system that has a recognizable vibration recipe. A sound produced by a musical instrument is a complex of tones of definite pitch, detectable by the ear and called **partial tones** or **partials**. The partials give the particles of the medium a motion that is the resultant of a group of simple periodic motions.

The distinctive **tone quality** of a sound depends on the particular mixture of component frequencies that compose it. Most instruments give both the fundamental and higher harmonics. Without the latter, all instruments would produce rather lifeless sounds of the same quality, indistinguishable by the ear. The ear tends to hear each harmonic as a separate tone,

and the brain combines them in some way. The fundamental normally has a greater amplitude than any of the overtones and is the frequency that we generally hear as the pitch. Different instruments emphasize different overtones. The difference in quality between a violin and a cello when sounding the same note depends on the relative intensities of the partials or harmonics produced by each instrument. The fifth partial of a violin is given particular emphasis. The sound of a tuning fork consists almost entirely of the fundamental and is musically uninteresting. The presence of harmonics in a sound gives it brilliance and brightness.

9–14 INTENSITY AND LOUDNESS

The **intensity** of a sound wave refers to the amount of **energy** the wave is carrying. It is defined as the **amount of sound energy crossing a unit area in unit time** and is commonly measured in watts per square meter. The intensity is directly related to the amplitude of the wave. The least intense sound detectable by the human ear is about 10^{-12} watt/m^2, corresponding to amplitudes in air that may be less than one millionth of a centimeter. The amplitude in air produced by rather intense sounds (the largest intensity that can be detected by the ear is 1 watt/m^2) may be about one hundredth of a centimeter.

Loudness is a physiological sensation that is experienced when sound waves strike the eardrum. It is related to the physical intensity of the compressions and rarefactions of the air. Loudness is a highly subjective perception, and individual differences can be quite large.

The smallest change in intensity to which the average ear is sensitive is an increased intensity 1.26 times the original intensity. The change in intensity of a sound resulting from ten such increases in the original intensity, $(1.26)^{10}$ times, is called a **bel**, after Alexander Graham Bell (1847–1922), a teacher of speech to the deaf and inventor of the telephone. Each of the ten equal steps, representing an increase by a factor of $10^{0.1}$ or 1.26, is one tenth of a bel, or a **decibel** (dB). Thus, 1 bel = 10 decibels (10 dB). The average individual can distinguish about 130 steps in intensity, from the threshold of hearing to the level just below that at which the ear interprets the intensity as pain rather than sound.

Table 9-3   INTENSITIES OF TYPICAL SOUNDS

Source of Sound	Intensity Level in Decibels (dB)
Rustle of leaves (faint)	10
Soft whisper at 3 feet	20
Library	30
Mosquito buzzing	40
Ordinary conversation	50
Noisy store, busy office	60
Vacuum cleaner/average street traffic	70
Garbage disposal/diesel truck/ tabulating room	80
Motorcycle/symphony music	90
Subway train/jet fly-over	100
Hard-rock music/outboard motor	110
Thunderclap/construction noise	120
Jet aircraft taking off	140
Rifle	160
Moon rocket	200

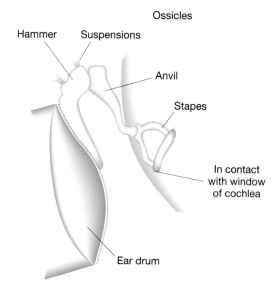

Figure 9-42   Structure of the ear, showing the three tiny bones (ossicles) that connect the eardrum to the oval window of the cochlea.

The relative intensity of two sound waves is measured in decibels, a unit that expresses a ratio between two sound intensities, I/I_0: the reference intensity I_0 and the intensity I of the source being measured, such as a dishwasher or a truck. A particular intensity is selected to represent 0 decibels, the standard reference level: 10^{-12} watt/m^2 at a frequency of 1000 Hz. This is close to the faintest sound that can be heard by a young adult in a quiet location. A wave 10 times as intense is at 10 decibels; a wave 100 times as intense, representing another tenfold increase, is at 20 decibels; and so on. Table 9-3 lists some typical sound levels.

9-15 THE EAR AND HEARING

A sound wave striking the ear penetrates the outer opening and sets the eardrum into vibration (Fig. 9-42). Three small bones (**ossicles**) in the middle ear transmit these vibrations to a fluid in the cochlea of the inner ear, forcing the thin basilar membrane into vibration. The latter transmits the stimulus to the organ of Corti, where the endings of the auditory nerves are located, and the organ of Corti converts the vibrations into nerve impulses. This organ can identify the component frequencies in a complicated sound. The combination of ear and brain distinguishes pitch, loudness, and quality.

To be audible, a sound must have a minimum intensity. This intensity, called the **threshold of hearing**, is different for different frequencies. A low-frequency sound at 30 Hz cannot normally be heard until its sound intensity reaches 60 decibels, whereas a 4000-Hz tone might be heard at a level of 5 decibels. An **audiogram** gives a profile of an individual's threshold of hearing for a range of frequencies (Fig. 9-43). Pure tones at various frequencies are presented, and the intensity that is necessary for audibility at each frequency is noted.

The ear is least sensitive to low frequencies and most sensitive to the range of 500 to 4000 Hz. The insensitivity to the lower frequencies is fortunate, for otherwise we would hear the bodily movements, muscular contractions, and vibrations of our own bodies. With advancing age, the sensitivity of the ear to high-frequency sound normally deteriorates. A person at age 40 may lose about 80 Hz sensitivity at the higher frequencies every six months.

Hearing also occurs via **bone conduction** through the skull. The sound of clicking teeth is heard mainly through vibrations of the bones in the skull; some of the vibrations bypass the middle ear and are transmitted directly to the inner ear. When we speak, some of the sound reaches the ears via the air and some

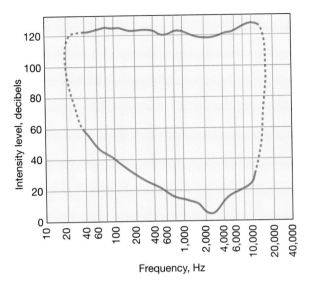

Figure 9–43   An audiogram. The ear is most sensitive to frequencies near 2000 Hz. Above or below this frequency, the intensity must be increased considerably for the sound to be audible.

through bone conduction. Humming with closed lips is transmitted almost entirely by bone conduction. Speaking involves two different sounds to the speaker, but only one—through air conduction—to another listener. We normally do not hear ourselves as others hear us because the low-frequency vibrations we hear through bone conduction are not heard by others. A recording of one's own voice may seem disappointingly thin for this reason.

9–16 NOISE POLLUTION

The range of sound quality extends from pure sine waves, through musical tones consisting of a set of harmonically related frequencies, to the chaos of unrelated sounds called **noise**. Beyond ordinary noise there is a pure confusion of sound called *gaussian noise,* or *white noise,* containing sound of *all* frequencies, just as white light is made up of all frequencies of light.

The British Committee on the Problems of Noise defines noise as **"unwanted sound."** In a noisy environment—near elevated railroads, trucking routes, airplane flight paths—about 25% of the people say they are not disturbed by the noisy activities. Another 10% seem to be annoyed by any noise, re-

gardless of how faint, that they themselves do not make.

Noise is a by-product of the conversion of energy. The basic problem is essentially incurable; it is a price we must pay for a machine civilization. Although the noise of modern technology cannot be eliminated, it can be controlled to minimize its adverse effects. There is legislation establishing industrial standards, building codes, and transportation requirements to limit the transmission of noise. The Noise Control Act of 1972 authorized the federal government to study controls and set standards to protect the public health and welfare. The Walsh-Healy Act allows an overall sound intensity level of 90 dB for eight hours' exposure of employees in a plant selling a product or providing a service to the government. Deafness, partial or total, has become a major category of industrial injury, and workers are now protected with compensation for work-associated hearing disabilities. If the noise intensity is too high, special ear muffs or ear plugs must be worn (Fig. 9–44).

Noise is one of the most annoying and pervasive of all types of pollution. Certain types of noise and vibration can cause disease, dizziness, and nausea. The psychological effects are more difficult to measure, but through the persistence of annoyance, fatigue, and the inability to concentrate, efficiency may be reduced and accidents caused.

Every problem having to do with excessive noise or vibration originates in the periodic motion of a structure. The most widely used means of controlling vibrations is to isolate the mechanical energy of a vi-

Figure 9–44   Special ear protectors must be worn in some occupations to protect workers from hearing loss. (Superstock)

brating structure from surrounding areas by installing soft, elastic separators along the transmission route. An example is the isolation of automobile-engine mount vibrations from the passenger compartment by means of coil springs and pads of elastomer.

Another approach to controlling noise and vibration is **acoustical absorption**, used mainly in architectural applications to make buildings quieter. Acoustic tiles and baffles constructed from porous materials are used; the air penetrates the structure and expends its energy through frictional flow. This interception of acoustical energy in the air protects the ears from exposure to intense pressure waves. Acoustical building codes are applied to new construction.

Basic research is needed in many areas if much noise control is to be realized. Noise production by jet-engine exhausts is not well understood. The impact of one body on another is perhaps the least understood of all mechanical sources of noise. Yet the noise of industrial machinery is produced largely by impacts: Pneumatic drills, hammers, saws, engine blocks, milling machines—all make high-speed impacts. Sustained public concern for a quieter environment can lead to results. For example, consumer demand for less noise in appliances resulted in reduction of the noise level of a portable dishwasher from 85 to 70 dB through a combination of vibration control mechanisms, and the more annoying high-frequency components were decreased by 40 dB.

An especially promising approach to noise control is **active noise control (ANC)**. The idea behind this new technology was first studied in the 1870s by Lord Rayleigh. When sound waves of equal amplitude and frequency but opposite phase collide, they cancel each other and leave behind nothing but silence. Lord Rayleigh experimented with canceling sound from two organ pipes using tones generated by an electric tuning fork. Although ANC is simple in principle, it is a very complex problem in application. The sound frequencies making up noise must be analyzed, and a precise mirror image of each component created.

The microelectronics revolution is making it possible to meet this challenge. Powerful digital signal processors on one chip can convert analog sounds into digital signals and then analyze those signals. In a matter of microseconds, the computer generates the appropriate **"antinoise"** to cause noise cancellation.

Since 1987, ANC systems have been installed in dozens of factories, office buildings, hospitals, and recording studios to quiet air conditioning, heating systems, and ventilating fans. In the 1990s, grain-loading equipment, refrigerators, and magnetic resonance imagers (MRIs) will be silenced. Even luxury cars will be equipped with ANC mufflers. This technology, it is felt, could easily become a multibillion-dollar industry because of the growing demand for better noise control in the home, on the street, and on the job. A variant of ANC may even nip noise in the bud by generating **antivibrations** directly on the noise-producing object. Not only would this cut noise; it could improve durability, since vibration is an important cause of wear and tear.

9–17 ULTRASONICS

Sound waves with frequencies ranging from 20,000 Hz, the upper limit of human hearing, to a billion or more Hz are described by the term **ultrasonics**. Waves of these frequencies can be generated by the vibration of thin pieces of quartz or other crystals, set into motion by electrical means. Frequencies up to 20 billion Hz (20 gigahertz) have been achieved. The wavelengths of these ultrahigh-frequency waves are about the same as those of visible light.

Ultrasonic waves travel through liquids and solids in the same way that audible sound waves do. They are used for such purposes as underwater echo-ranging by **sonar** (*so*und *n*avigation *a*nd *r*anging)—locating the object and measuring the distance between it and the source; detecting flaws in the interior of solids; destroying microorganisms; and mapping underground structures for oil and mineral deposits. They are also applied widely in medical diagnosis and treatment. In sounding out the abdomen, as an example, the sound waves pass through the different tissues at speeds that depend on the elasticity and density of the tissue. As they collide with different structures, they send back echoes, which are picked up by sensitive microphones and turned into electrical signals on a TV screen. From the pattern of the echoes, tumors, abscesses, lesions, and other abnormalities can be picked up within the liver, pancreas, kidneys, heart, and other organs. **Medical ultrasonography** is nearly ideal for use in human beings (Fig. 9–45).

More efficient and more sensitive than any sonar contrived by man, however, is the sonar system of certain bats that, like dogs, crickets, and other animals, perceive acoustic frequencies inaudible to us.

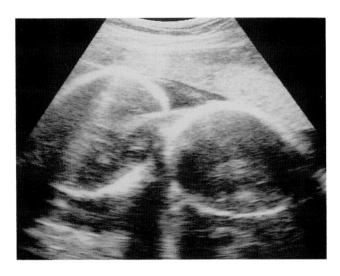

Figure 9–45 An ultrasonic scan of the womb, showing the heads of twins at 25 weeks. (St. Bartholomew's Hospital/Science Photo Library/Photo Researchers, Inc.)

Not all bats use sonar; the largest bats have prominent eyes and depend on their vision instead. Small bats rely largely on echolocation; they emit and hear sounds of frequencies at least up to 100,000 Hz, and the precision of their sonar echolocation is so great that they can get along perfectly well without eyesight (Fig. 9–46). Insect-eating bats can easily catch hundreds of moths in an hour in the dark of night. Far from being a crude collision-warning device, the echolocation system of bats is very sharp and precise.

Example 9–5 illustrates a typical echo-ranging situation.

EXAMPLE 9–5

An acoustical signal is transmitted from the bottom of a ship and is reflected from the bottom of a channel in 0.034 s. Assuming the velocity of sound in seawater to be 1520 m/s, how deep is the channel?

SOLUTION

If the elapsed time is 0.034 s, the time required for the signal to travel from the ship to the bottom is one half of this, or 0.017 s. From kinematics, the depth can be determined.

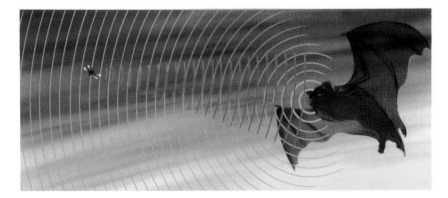

Figure 9–46 The bat uses sonar. It emits sound waves that reflect off the insect and return to the bat, giving away the insect's position. (Source: Kirkpatrick/Wheeler, *Physics: A World View*. Philadelphia: Saunders College Publishing, 1992, Fig. 15.3, p. 325.)

$$\bar{\mathbf{v}} = 1520 \text{ m/s} \qquad \mathbf{s} = \bar{\mathbf{v}}t$$

$$t = 0.017 \text{ sec} \qquad = 1520 \frac{\text{m}}{\text{s}}(0.017 \text{ s})$$

$$\mathbf{s} = ? \qquad = \boxed{26.0 \text{ m}}$$

EXTENSION

A bat receives a reflected sound from a moth 0.06 s after it is emitted. How far is the moth if the velocity of sound is 350 m/s? (Answer: 10.5 m)

■

9–18 RESONANCE

Resonance occurs when a system that can vibrate with a certain natural frequency is acted upon by a periodic disturbance that has the same frequency (Fig. 9–47).

A child's swing illustrates this principle. The swing is essentially a pendulum with a natural frequency that depends on the length. A small effort applied at the right moment produces the best effect, either by the child's own "pumping" of the swing or by gentle pushes from another. A suspension bridge, also, can swing to and fro like a pendulum; if excited into its natural resonant vibration by a wind, the amplitude can lead to its collapse. This happened to the Tacoma Narrows Bridge in 1940, which was set into such vio-

lent vibration during a gale that it collapsed only four months after its completion. A musical instrument may set windows or other objects into vibration (Fig. 9–48).

Many sources of sound, such as the vocal cords or a violin string, produce such faint sounds that they would be almost inaudible if there were no way to increase the sound. By means of resonance, a relatively small amount of energy can be amplified by the larger mass of air set into vibration. The resonant cavities in the head reinforce and amplify the sound produced by the vocal cords.

When you hold a seashell to your ear, you hear sounds of certain frequencies that are reinforced by the shell and stand out from other sounds. If you speak into a bottle, certain frequencies present in your voice are emphasized by the resonance of the bottle. The air is set strongly in motion by a sound wave that has very nearly the same frequency as one of its resonant frequencies.

Figure 9–48   A wine glass shattered by resonance. The frequency of a singer's sound wave matches one of the natural frequencies of vibration of the wine glass and sets up resonant vibrations in the glass, causing the glass to shatter. (Ben Rose/The IMAGE Bank)

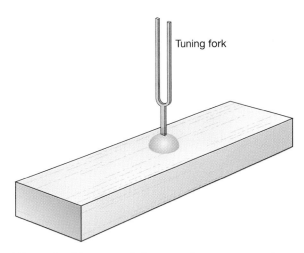

Figure 9–47   Tuning fork mounted on a resonating box. The sound of the vibrating tuning fork is amplified by the resonance of the air in the box.

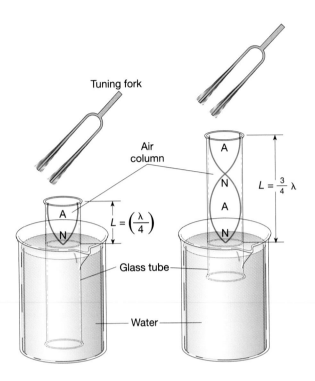

Tuning fork

Air
column

$L = \left(\dfrac{\lambda}{4}\right)$

$L = \dfrac{3}{4}\,\lambda$

Glass tube

Water

Figure 9–49 Resonance between a tuning fork and an air column. The tube is moved up and down in the water to tune it to the frequency of the fork. When the length of the column L, equals $\lambda/4$, the sound increases greatly in loudness, as it does again at $L = \frac{3}{4}\,\lambda$.

A striking demonstration of resonance involves a tuning fork struck and held over an adjustable column of air (Fig. 9–49). Several positions of the column can be found that reinforce the sound of the tuning fork. The shortest length L of the tube in which the air can be in resonance with a wavelength λ is $\lambda/4$, since there will be a node at the closed end and an antinode at the open end. With a tube that is long enough, resonance occurs again when the length of the tube is $3\,\lambda/4$, since there will again be a node at the closed end and an antinode at the open end. Musical instruments are essentially resonators; the mouthpiece, string, or reed is the source of excitation, and certain frequencies are emphasized through resonance by the instrument.

SUMMARY OF EQUATIONS

Period $= \dfrac{1}{\text{frequency}}$

$$T = \dfrac{1}{f}$$

Wave Equation

Velocity = Frequency × Wavelength

$$\mathbf{v} = f\lambda$$

Velocity of Sound in Air at Various Temperatures

$$\mathbf{v}_T(\text{m/s}) = \mathbf{v}_0 + 0.60\,T$$

$$\mathbf{v}_T(\text{ft/s}) = \mathbf{v}_0 + 2.00\,T$$

KEY TERMS

Acoustical Absorption   The interception of acoustical energy; in architectural applications, the use of tiles and baffles to make buildings quieter.

Active Noise Control (ANC)   A new technology that applies microelectronics to noise control to quiet air conditioning, heating systems, ventilating fans, and other generators of noise.

Amplitude   Maximum displacement of a medium through which a wave is passing from its equilibrium condition.

Antinode   On a standing wave, a region of maximum constructive interference.

Antinoise   A signal generated by a computer that causes the cancellation of a noise signal.

Antivibration   A technology that would reduce noise and improve the durability of a noise-producing object by generating antivibrations directly on the object and nipping noise in the bud.

Audiogram   A profile of an individual's threshold of hearing for a range of sound frequencies.

Beats   A periodic fluctuation in loudness caused when two wavetrains of sound of slightly different frequencies interfere.

Bel   Ten decibels.

Bell-in-Vacuum   A demonstration that a ringing bell cannot be heard in a jar from which air has been exhausted but becomes audible when air is reintroduced.

Bone Conduction   Hearing through vibrations of bones in the skull that bypass the middle ear and are transmitted directly to the inner ear.

Compression   The region on an oscillating spring in which the coils are closest together.

Compressional Wave   A wave such as a sound wave in which the disturbance of the medium occurs parallel to the direction in which the wave is proceeding.

Constructive Interference   An effect caused when the crests of two waves having the same amplitude arrive at a given point and the two amplitudes combine.

Crest   The highest point of a wave.

Decibel (dB)   A unit of sound intensity.

Destructive Interference   A crest of one wave and a trough of a second wave encounter and cancel each other, giving a combined amplitude of zero.

Diffraction   The bending of waves around obstacles or through openings.

Doppler Effect   The change in the frequency (or wavelength) of waves as sensed by an observer when there is relative motion between a source of waves and the observer.

Echo   A reflected sound that can be distinguished from the original.

Electromagnetic Wave   A transverse wave involving changing electric and magnetic fields and traveling with the speed of light.

Frequency   The number of waves passing a given point per unit time.

Fundamental   The lowest frequency of which a vibrating string or an air column is capable.

Harmonic   An overtone whose vibration frequency is an integral multiple of that of the fundamental.

Huygens' Theory of Wave Propagation   Each point on a wavefront is a source of secondary wavelets that spread out with the speed of the wave at that point and combine to form a replica of the original wavefront.

Index of Refraction   The ratio of the speed of light in a vacuum to that in the substance.

In Phase   Term applied to two or more waves whose crests (and troughs) arrive at a point at the same time, so that their effects reinforce each other.

Intensity   The amount of energy carried by a wave.

Interference   The superposition of waves, producing regions of reinforcement and regions of cancellation.

Law of Reflection   For all waves, the angle of incidence is equal to the angle of reflection.

Law of Refraction   The ratio of the sine of the angle of incidence to the sine of the angle of refraction is a constant.

Longitudinal Wave   A wave consisting of alternating compressions and rarefactions, such as a sound wave, in which the particles of the medium oscillate in the direction of the disturbance.

Loudness   A physiological sensation that is experienced when sound waves strike the eardrum.

Mechanical Wave   A wave such as a wave on a spring or on the surface of water in which material particles are displaced.

Medical Ultrasonography   The application of ultrasonic waves in medical diagnosis and treatment.

Musical Note   A combination of pure tones emitted by a vibrating string or air column.

Node   A position in a standing wave pattern where the displacement is at a minimum.

Noise   Sounds consisting of unrelated frequencies; any sound regarded as a nuisance.

Octave   The interval between two frequencies having a ratio of 2 to 1.

Ossicles   Three small bones in the middle ear.

Out of Phase   Two waves in which the crest of one wave arrives at a point at the same time as a trough of the second wave arrives so that their effects cancel each other.

Overtones   Multiples of the fundamental frequency of a vibrating string or an air column.

Partial   Component tones of definite pitch present in a sound produced by a musical instrument.

Period   The time to complete one cycle of a wave.

Quality   Timbre or tone color of a sound.

Rarefaction   On an oscillating spring, a region in which the coils are spread out.

Red Shift   A decrease in the measured frequency of radiation from a receding source, because the decrease is toward the low-frequency, or red, end of the color spectrum.

Reflection   The change in direction that waves undergo when they encounter a barrier.

Refraction   The bending of waves when they pass from one medium into another.

Resonance   The effect produced when a system that can vibrate at a certain frequency is acted upon by a periodic disturbance that has the same frequency.

Reverberation   A multiple reflection of sound that cannot be analyzed into distinct echoes.

Shock Wave   A pulse set up by a jet aircraft or speedboat that travels faster than the speed of the waves in the surrounding medium.

Sine Wave   The path traced by the simplest type of regular wavetrain.

Sinusoidal   A wavetrain that has the same shape as a graph of the sine function in trigonometry.

Sonar   Underwater echo-ranging using ultrasonic waves.

Sound Wave   A disturbance in a medium consisting of a train of compressions separated by rarefactions.

Standing Wave   The interference pattern produced by two waves of equal amplitude and frequency traveling in opposite directions and characterized by alternating nodes and antinodes.

Threshold of Hearing   The lowest intensity at which the sound of a given frequency can be heard by an individual.

Tone Quality   The timbre of a sound that enables us to distinguish it from another with the same intensity and frequency.

Transverse Wave   A wave in which the disturbance travels in one direction while the particles of the medium oscillate in a direction perpendicular to it.

Trough   The lowest point of a wave.

Ultrasonics   Sound waves with frequencies above the upper limit of human hearing.

"Unwanted Sound"   Noise.

Wave Equation   The velocity of a wave is equal to its frequency multiplied by its wavelength.

Wavelength   The distance between successive crests or troughs of the same wave.

THINGS TO DO

1. Strike a tuning fork with a mallet. Note the source of the sound in the vibrations of the fork by holding a pith ball near one of its prongs.

2. Examine a variety of musical instruments—piano, violin, xylophone, clarinet, trumpet.

3. Set up a slinky or screendoor spring and place it in longitudinal vibration.

4. Experience the Doppler effect through the change in the pitch of a horn of a passing car.

5. Use a ripple tank or bathtub to study the production of waves, frequency, wavelength, velocity, reflection, refraction, diffraction, and interference.

6. Study the transverse nature of water waves by the motion of a cork stopper or toy boat floating on the water.

7. Record your voice on tape and play it back. Does the sound of your voice surprise you?

8. Study resonance in an air column as discussed in the chapter. Use tuning forks of various frequencies.

9. Set the stem of a vibrating tuning fork on a desk or table. What effect has this on the amplitude of the sound?

10. Make a simple sonometer by stretching wire over a board with clamps.

11. Investigate the relation between wave form and tone quality by using a microphone with a cathode-ray oscilloscope.

12. Determine the velocity of sound in air by means of a resonance tube, as illustrated in the text.

13. Set up standing waves in a cord with a mechanical vibrator.

14. Carry out the bell-in-vacuum demonstration to convince yourself that a material medium is necessary for the transmission of sound.

15. Blow across the mouth of a bottle three-fourths filled with water. Pour out some water, then repeat. Does the pitch change? Empty the bottle and repeat again.

16. Tune a set of open softdrink bottles to various musical notes by adding water and blowing across the tops of the bottles.

17. Pluck a string of a sonometer at different distances from the end. Do the tones change?

EXERCISES

Wave Properties

9–1. Discuss (a) some ways in which sound waves resemble light waves; (b) some ways in which they differ.

9–2. Why are sound waves described as longitudinal?

9–3. What evidence can you give that sound is a form of energy?

9–4. What evidence is there that light is a form of wave motion?

9–5. When you are watching a distant batter, why does the crack of the bat striking the ball seem to occur moments after the event?

9–6. There may be a problem for the marchers in a long column to keep in step with a band up front. Explain.

9–7. Six seconds after a lightning flash was seen, thunder was heard. How many feet away was the lightning if the temperature was 30.0°C?

9–8. An echo returns from a building in 0.500 second. How many feet away is the building? (Assume the temperature of the air to be 20.0°C.)

9–9. Assuming that your range of hearing is from 16 Hz to 20,000 Hz, what are the corresponding wavelengths in feet (a) in air at 20.0°C? (b) in water at 20.0°C?

9–10. Dolphins emit very short sound waves with a frequency of 2×10^5 Hz. What is their wavelength in meters in water? (Consider the velocity of sound in seawater to be 1520 m/s.)

9–11. An ocean wave with a wavelength of 100 meters has a speed of 16.3 m/s. How many waves would pass a given point in 1.00 second?

9–12. Waves with a wavelength of 3.00 centimeters are generated in a ripple tank when the frequency is set for 10 Hz. What is the speed of propagation?

9–13. A particular wavelength of orange light is 6.06×10^{-5} cm. (a) What is its frequency? (b) What is its period?

9–14. A broadcasting station transmits radio waves at a frequency of 10^6 Hz. What is their wavelength?

9–15. Arrange the following in order of their decreasing frequency: (a) gamma rays; (b) orange light; (c) radio waves; (d) infrared rays.

9–16. Distinguish between reflection and refraction.

9–17. Waves traveling in air at 3.00×10^8 m/s enter a medium in which their velocity is reduced to 2.00×10^8 m/s. If the waves strike the interface at an angle of incidence of 35.0°, in which direction will they move in the second medium?

9–18. A tone with a frequency of 512 Hz is sounded at the same time as a 520-Hz tone. How many beats per second are produced?

9–19. Explain why sound travels well on a quiet night in summer.

9–20. Why is a long reverberation time undesirable in a concert hall?

9–21. Assume that you are in a moving car. Apply the Doppler effect to an ambulance siren approaching you from the opposite direction and passing you.

Standing Waves

9–22. An air column closed at one end resonates with a tuning fork that has a frequency of 128 Hz. What is the length of the air column?

9–23. Two pipes, one open at both ends and the other closed at one end, are 0.500 meter in length. What is the fundamental frequency of each? (The velocity of sound is 340 m/s.)

9–24. What length of open organ pipe would give a frequency of 440 Hz at 20.0°C?

9–25. Are the long pipes or short pipes of an organ the high-frequency pipes? Explain.

9–26. Explain how a wire of a given length is capable of emitting notes of different frequencies.

9–27. What is "stationary" about standing waves on vibrating strings?

9–28. Use diagrams to show the fundamental and first two overtones of (a) a vibrating string; (b) a vibrating air column in a tube closed at one end.

9–29. Assuming that the frequency of the fundamental wave in Exercise 9–28 is 100 Hz, what are the frequencies of the overtones in each case?

9–30. A string 2.00 meters long is fixed at both ends. What are the three longest wavelengths of standing waves established in the string? (Assume the velocity of sound to be 340 m/s.)

9–31. What conditions must be satisfied for resonance to occur between two mechanical systems?

9–32. What are some possible wavelengths of standing waves in a cord stretched between two supports 50.0 centimeters apart?

Music and Noise

9–33. Determine the fifth overtone of middle A (440 Hz) on a piano.

9–34. The frequency of a tuning fork is 256 Hz. What is the frequency of a tuning fork one octave higher?

9–35. Why does a recording of one's own voice often seem surprising when heard for the first time?

9–36. What is considered musical in one country may be considered noise in another. Discuss.

9–37. Why do identical notes emitted by a trumpet and a trombone sound different?

9–38. A quieter environment will not come cheaply. Discuss whether you would be willing to pay your share toward it in the form of higher prices for goods and services.

9–39. Discuss whether, in your opinion, the quality of life in terms of noise pollution has improved during the past five years.

9–40. *Multiple Choice*

A. A wave transmits
   (a) matter.
   (b) momentum.
   (c) energy.
   (d) velocity.

B. The distance from crest to crest of a wave is called its
   (a) frequency.
   (b) wavelength.
   (c) amplitude.
   (d) velocity.

C. A sound wave is
   (a) longitudinal.
   (b) transverse.
   (c) resonant.
   (d) harmonious.

D. The bending of a wave around a barrier is called
   (a) reflection.
   (b) refraction.
   (c) interference.
   (d) diffraction.

E. The frequency of a sound wave is related to its
   (a) pitch.
   (b) amplitude.
   (c) quality.
   (d) loudness.

F. The difference between a transverse and a longitudinal wave is
   (a) a longitudinal wave carries more energy.
   (b) the oscillations are perpendicular to the direction of travel in a transverse wave and parallel in a longitudinal wave.
   (c) the reverse of (b).
   (d) a transverse wave has greater amplitude.

G. A vibrating object completes one cycle in 0.500 s. Its frequency is
   (a) 0.500 Hz.
   (b) 2.00 Hz.
   (c) 1.50 Hz.
   (d) 5.00 Hz.

H. The amplitude of a sound corresponds to
   (a) pitch.
   (b) quality.
   (c) loudness.
   (d) frequency.

I. The lowest note that a given string can produce is called
   (a) an overtone.
   (b) the fundamental.
   (c) an octave.
   (d) the pitch.

J. A tone with a frequency of 512 Hz is sounded at the same time as a 520-Hz tone. How many beats per second are produced?
   (a) 1032
   (b) 512
   (c) 520
   (d) 8

SUGGESTIONS FOR FURTHER READING

Aljishi, Samer, and Jakub Tatarkiewicz, "Why Does Heating Water in a Kettle Produce Sound?" *American Journal of Physics*, July, 1991.
Much of the sound emitted in the boiling process can be attributed to a coupling between the generation and collapse of vapor bubbles with the natural acoustic modes of the kettle.

Beranek, Leo L., "Wallace Clement Sabine and Acoustics." *Physics Today*, February, 1985.
Sabine was the world's first acoustical scientist. He changed architectural acoustics from an obscure body of knowledge to an experimental science.

Devey, Gilbert B., and Peter N. T. Wells, "Ultrasound in Medical Diagnosis." *Scientific American*, May, 1978.
It is possible to explore structures within the human body painlessly, safely, and at relatively low cost with sound waves.

Fletcher, Neville H., and Suzanne Thwaites, "The Physics of Organ Pipes." *Scientific American*, January, 1983.
Oscilloscopes and other modern devices have made it possible to develop an understanding of the mechanism of how organ pipes generate their sheer majesty of sound.

Kottick, Edward L., Kenneth D. Marshall, and Thomas J. Hendrickson, "The Acoustics of the Harpsichord." *Scientific American*, February, 1991.
Despite five centuries of existence, how the harpsichord produces its resonating sound remained poorly understood until recently.

Matthews, Max V., and John R. Pierce, "The Computer as a Musical Instrument." *Scientific American*, February, 1987.
Electronic instruments and computer programs can produce any sound, including some never before heard. Such digital sound synthesis has found a place in popular music, as in the production of sound tracks for film and television, and in the future may be a part of every symphony orchestra.

Quate, Calvin F., "The Acoustic Microscope." *Scientific American*, October, 1979.
Previously inaccessible properties of microscopic objects are coming into view in the acoustic images. The acoustic microscope is expected to find its place as one of a complementary set of tools with the optical microscope and the electron microscope.

Reiser, Stanley Joel, "The Medical Influence of the Stethoscope." *Scientific American*, February, 1979.
The stethoscope was the first instrument generally used for diagnosis and transformed the practice of medicine.

Rossing, Thomas D., "Resource Letter MA-2: Musical Acoustics." *American Journal of Physics*, July, 1987.
A guide to the literature on musical acoustics.

Schelleng, John C., "The Physics of the Bowed String." *Scientific American*, January, 1974.
The physics of the behavior of strings can be of considerable importance to the player of a stringed instrument.

Shaw, Edgar A. G., "Noise Pollution—What Can Be Done?" *Physics Today*, January, 1974.
Noise is associated with a vast increase in the use of energy for transportation and labor-saving machinery. The control of noise is likely to be of major concern for years to come.

Shen, Sinyan, "Acoustics of Ancient Chinese Bells." *Scientific American*, April, 1987.
A set of Chinese bronze chime bells was unearthed in China in 1978. Archeologists have revealed the sophisticated acoustical design of these important orchestral instruments, which vanished from history 2000 years ago.

Suga, Nobuo, "Biosonar and Neural Computation in Bats." *Scientific American*, June, 1990.
An echolocating bat can pursue and capture a fleeing moth with a facility and success rate that would be the envy of any military aerospace engineer.

Suslick, Kenneth S., "The Chemical Effects of Ultrasound." *Scientific American*, February, 1989.
Chemistry is the interaction of energy and matter. Although chemical applications of ultrasound are still in the early stages of development, the next few years promise rapid progress in sonochemistry.

Walker, Jearl, "What Makes You Sound So Good When You Sing in the Shower?" *Scientific American*, May, 1982.
The secret lies in the acoustic resonance of the shower stall.

ANSWERS TO NUMERICAL EXERCISES

9–7. 6880 ft

9–8. 282 ft

9–9. (a) 74.0 ft; 0.056 ft
(b) 300 ft; 0.24 ft

9–10. 0.0076 m

9–11. 0.163 wave/s

9–12. 30.0 cm/s

9–13. (a) 4.95×10^{14} Hz
(b) 2.02×10^{-15} s

9–14. 300 m

9–18. 8 beats/s

9–22. $l = 0.670$ m

9–23. Open pipe: 340 Hz
Closed pipe: 170 Hz

9–24. 0.78 m

9–29. (a) 100 Hz; 200 Hz; 300 Hz
(b) 100 Hz; 300 Hz; 500 Hz

9–30. 1.33 m; 2.00 m; 4.00 m

9–32. 100 cm; 50.0 cm; 33.3 cm

9–33. 2640 Hz

9–34. 512 Hz

ESSAY

ULTRASOUND AND ITS APPLICATIONS

RAYMOND A. SERWAY
James Madison University

JERRY S. FAUGHN
Eastern Kentucky University

Ultrasonic waves are sound waves whose frequencies are in the range of 20 kHz to 100 kHz, which is beyond the audible range. Because of their high frequency, and corresponding short wavelengths, ultrasonic waves can be used to produce images of small objects and are currently in wide use in medical applications, both as a diagnostic tool and in certain treatments. Various internal organs in the body can be examined through the images produced by the reflection and absorption of ultrasonic waves. Although ultrasonic waves are far safer than X-rays, their images do not always provide as much detail. On the other hand, certain organs, such as the liver and the spleen, are invisible to X-rays but can be diagnosed with ultrasonic waves.

It is possible to measure the speed of blood flow in the body using a device called an ultrasonic flow meter. The technique makes use of the Doppler effect. By comparing the frequency of the waves scattered by the blood vessels with the incident frequency, one can obtain the speed of blood flow.

The technique used to produce ultrasonic waves for clinical use is illustrated in Figure A. Electrical contacts are made to the opposite faces of a crystal, such as quartz or strontium titanate. If an alternating voltage of very high frequency is applied to these contacts, the crystal will vibrate at the same frequency as the applied voltage. As the crystal vibrates, it emits a beam of ultrasonic waves. At one time, almost all of the headphones used in radio reception produced their sound in this manner. This method of transforming electrical energy into mechanical energy is called the **piezoelectric effect**. This effect is also reversible. That is, if some external source causes the crystal to vibrate, an alternating voltage is produced across the crystal. Hence, a single crystal can be used to both transmit and receive ultrasonic waves.

The production of electric voltages by a vibrating crystal is a technique that has been used for years in stereo and hi-fi equipment. In this application, a phonograph needle is attached to the crystal,

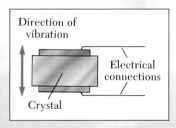

Figure A An alternating voltage applied to the faces of a piezoelectric crystal causes the crystal to vibrate.

and the vibrations of the needle as it rides in the groove of the record are translated by the crystal into an alternating voltage. This voltage is then amplified and used to drive the system's speakers.

The primary physical principle that makes ultrasound imaging possible is the fact that a sound wave is partially reflected whenever it is incident on a boundary between two materials having different densities. It is found that, if a sound wave is traveling in a material of density p_i and strikes a material of density p_t, the percentage of the incident sound wave reflected, PR, is given by

$$PR = \left(\frac{p_i - p_t}{p_i + p_t}\right)^2 \times 100$$

This equation assumes that the incident sound wave travels perpendicular to the boundary and that the speed of sound is approximately the same in both materials. This latter assumption holds very well for the human body since the speed of sound does not vary much in the various organs of the body.

Physicians commonly use ultrasonic waves to observe a fetus. This technique offers far less risk than X-rays, which can be genetically dangerous to the fetus and can produce birth defects. First the abdomen of the mother is coated with a liquid, such as mineral oil. If this is not done, most of the incident ultrasonic waves from the piezoelectric source will be reflected at the boundary between the air and the skin of the mother. Mineral oil has a density similar to that of skin, and as the equation indicates, a very small fraction of the incident ultrasonic wave is reflected when $p_i \approx p_t$. The ultrasound energy is emitted as pulses rather than as a continuous wave so that the same crystal can be used as a detector as well as a transmitter. The source-receiver is then passed over a particular line along the mother's abdomen. The reflected sound waves picked up by the receiver are converted to an electric signal, which forms an image along a line on a fluorescent screen. The sound source is then moved a few centimeters on the mother's

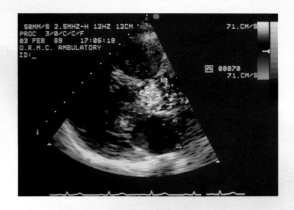

Figure B   An ultrasound image of the heart. (Schleichkorn, Custom Medical Stock Photo)

body, and the process is repeated. The reflected signal produces a second line on the fluorescent screen. In this fashion a complete scan of the fetus can be made. Difficulties with the pregnancy, such as the likelihood of abortion or of breech birth, are easily detected with this technique. Also, such fetal abnormalities as spina bifida and water on the brain are readily observable.

Another interesting application of ultrasound is the ultrasonic ranging unit designed by the Polaroid Corporation. This device is used in some of their cameras to provide an almost instantaneous measurement of the distance between the camera and the object to be photographed. The principal component of this device is a crystal that acts as both a loudspeaker and a microphone. A pulse of ultrasonic waves is transmitted from the transducer to the object to be photographed. The object reflects part of the signal, producing an echo that is detected by the device. The time interval between the outgoing pulse and the detected echo is then electronically converted to a distance value, since the speed of sound is a known quantity.

A tunnel of light marks the entrance to the Kodak pavilion at Epcot Center, Florida. (Ann Purcell/Photo Researchers, Inc.)

10

LIGHT AND OTHER ELECTROMAGNETIC WAVES

It is difficult to imagine anything "purer" or more elementary than a beam of sunlight. The discovery of its complex nature, therefore, came as a surprise. Through a small hole in a window shutter, Newton admitted a beam of sunlight to a darkened room and directed the beam through a prism. A band of colors—*a spectrum*—appeared like a rainbow on the opposite wall (Fig. 10–1). Newton then used a second prism, inverted with respect to the first, and recombined the colors into white light (Fig. 10–2).

Dispersing sunlight through a prism as Newton had done, the astronomer William Herschel (1738–1822) placed several thermometers in the region just beyond the red end of the spectrum and discovered invisible but heat-carrying radiation (Fig. 10–3). This region of the spectrum is now called the *infrared* (*infra*, meaning "below"). Later, the physicists Wilhelm Ritter (1776–1810) and William Hyde Wollaston (1766–1828) detected *ultraviolet* rays (*ultra*, meaning "beyond") at the opposite end of the visible spectrum through the blackening effect of these rays on crystals of silver iodide. Gradually, other radiations from various sources were discovered. Although the concept of "light" is applied to stimuli that affect the sense of vision, a more inclusive definition includes radiation which shares with visible light many of the same properties, such as reflection, refraction, polarization, and the same velocity.

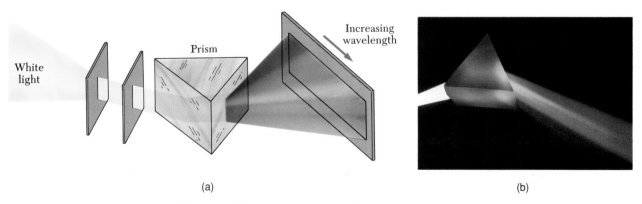

(a)   (b)

Figure 10–1 Isaac Newton discovered that a prism separates a beam of sunlight into its component colors and produces a spectrum. (Photograph courtesy of Bausch and Lomb)

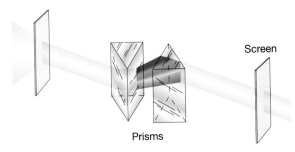

Figure 10–2 A second prism recombines the colors; a white spot appears on the screen.

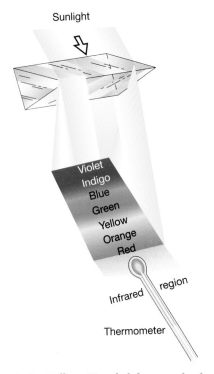

Figure 10–3 William Herschel discovered infrared radiation while measuring the temperature of the spectrum. The temperature increased from the violet end to the red end and, surprisingly, remained high beyond the visible red.

10–1 THE VELOCITY OF LIGHT

While studying the eclipses of Io, the innermost of the satellites of the planet Jupiter, Olaus Roemer (1644–1710), the astronomer, came upon an apparent discrepancy in the satellite's behavior. After determining that the average period of Io was about 42½ hours, Roemer thought he could predict the times when Io emerged from Jupiter's shadow, and he prepared a long-term table of Io's appearances. Io appeared further and further behind schedule, however, as the earth receded more and more from Jupiter in its annual orbit around the sun. As the distance between the earth and Jupiter decreased on the earth's return, the discrepancy between the predicted and observed times also decreased until Io's appearances were back on schedule.

Roemer reasoned that Io's appearances were on schedule and that the delay represented the transit time of light across the earth's orbit (Fig. 10–4). If the difference between the maximum and minimum separation between the earth and Jupiter is the diameter of the earth's orbit around the sun, 186,000,000 miles, and the observed delay in Io's appearance is 1000 seconds, then the velocity of light is 186,000,000 miles/1000 seconds = 186,000 miles/second, or 300,000 km/s. (The data used here are more accurate than those that were available to Roemer.)

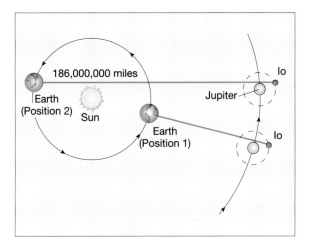

Figure 10–4 Olaus Roemer's method of determining the speed of light. With the earth at position 2, light from Jupiter's satellite, Io, must travel an added distance, the diameter of the earth's orbit.

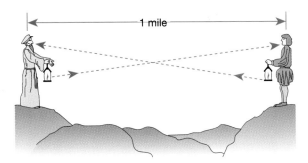

Figure 10–5 With an assistant stationed on a hill about a mile away, Galileo attempted to measure the speed of light by flashing his lantern and recording the time elapsed to receive a flash of light from the assistant. The time was too short to detect.

Galileo also had believed that light traveled with a finite velocity and attempted to measure it by timing the travel of a pulse of light over a measured distance (Fig. 10–5). He used a hand-operated shutter to transmit flashes from a lantern to an assistant stationed on a hill about a mile away. When the light beam reached the assistant, he was to uncover his lantern; Galileo, upon seeing this light, would note the elapsed time.

Galileo failed in his purpose, although he did demonstrate that the velocity of light must be very great. One weakness of Galileo's method was the reaction time of the assistant. The physicists Hippolyte-Louis Fizeau (1819–1896) and Léon Foucault (1819–1868) made refinements in the method that paved the way for modern precise measurements. They replaced the assistant with a mirror that returned the light to its point of origin (Fig. 10–6). They improved upon another drawback, the manually operated shutter, first by replacing it with a toothed wheel in front of a light beam that transmitted the light in short pulses, and later by using a rotating mirror. Knowing the distance of the round trip and the time of travel of light to the mirror and back, Foucault determined the velocity of light to be 298,000 km/s.

Albert A. Michelson (1852–1931), a physicist at the University of Chicago and the first American to receive a Nobel Prize in physics (1907), applied the Galileo-Fizeau-Foucault method with an unprecedented degree of precision. He positioned a rotating mirror on Mt. Wilson in California; 22 miles away he set up a fixed mirror on Mt. San Antonio (Fig. 10–7). The distance between the two mirrors, measured by the U.S. Coast and Geodetic Survey, was based on a primary baseline that may have been the most accu-

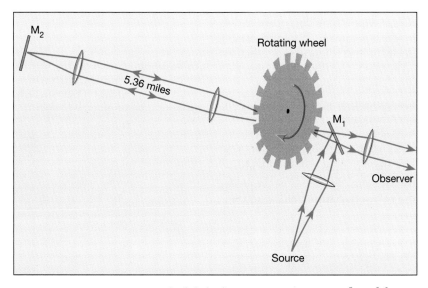

Figure 10–6 In Fizeau's method, light from an arc source was reflected from a mirror (M_1) toward a rotating wheel. Bursts of light passed through the wheel and were reflected to the eye of the observer from a mirror (M_2) 5.36 miles away.

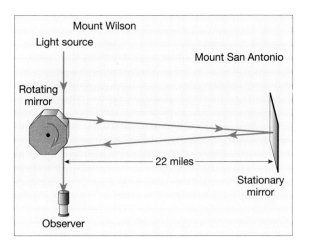

Figure 10–7 Albert A. Michelson used an octagonal mirror in rapid rotation to reflect light from an intense source to a stationary mirror 22 miles away. The speed of rotation of the mirror was adjusted so that the mirror moved one eighth of a revolution while the light made the 44-mile round trip. The time for the mirror to make one eighth of a revolution was then equal to the time for the light to travel 44 miles.

Figure 10–8 Light refraction. A pencil appears to be bent at the surface of the water. (Leonard Lessin/Peter Arnold, Inc.)

rately measured line ever marked on the surface of the earth. Michelson determined the velocity of light to be 299,796 km/s. The currently accepted value, based on an average of several determinations, is 299,792 km/s, or 186,282 mi/s.

The velocity of light depends on the nature of the medium through which it travels. Foucault established that the velocity of light is less in water than in air, and a group led by Michelson determined that it is slightly greater in a vacuum than in air.

The change in the velocity when light passes from one medium into another causes refraction (Fig. 10–8). The **index of refraction** is the ratio of the velocities of light in the two media. If the first medium is a vacuum, the ratio of the velocities defines the **absolute index of refraction**, usually called just the index of refraction or the **refractive index** of the second medium, such as water.

$$\text{Absolute index of refraction (water)} = \frac{\text{Velocity of light in a vacuum}}{\text{Velocity of light in water}}$$

$$n = \frac{c}{v} \qquad (10\text{–}1)$$

Since v is always less than c, the index of refraction is always greater than 1. For example, in water n is 1.333. Table 10–1 lists the absolute indexes of refraction for some selected materials.

Table 10–1   ABSOLUTE INDEXES OF REFRACTION OF SELECTED MATERIALS

Materials	n
Air	1.0003
Water (20°C)	1.333
Alcohol (ethyl)	1.362
Glycerine	1.473
Benzene	1.501
Glass (crown)	1.517
Quartz (crystalline)	1.544
Mica	1.561
Carbon disulfide	1.63
Glass (flint)	1.65
Diamond	2.417

EXAMPLE 10-1

Calculate the velocity of light in benzene.

SOLUTION

Knowing the value of c, and obtaining the refractive index of benzene from Table 10-1, we determine the velocity of light in benzene from Equation 10-1.

n (benzene) = 1.501

$c = 3.00 \times 10^8$ m/s

Velocity of light in benzene = ?

$$n \text{ (benzene)} = \frac{\text{Velocity of light in a vacuum}}{\text{Velocity of light in benzene}}$$

$$n = \frac{c}{v}$$

$$v = \frac{c}{n}$$

$$v = \frac{3.00 \times 10^8 \, \frac{m}{s}}{1.501}$$

$$v = \boxed{2.00 \times 10^8 \, \frac{m}{s}}$$

EXTENSION

Ice has an index of refraction of 1.305. What is the velocity of light in ice? (Answer: $2.29 \times 10^8 \, \frac{m}{s}$)

■

10-2 POLARIZED LIGHT

While examining a crystal of calcium carbonate brought from Iceland to Copenhagen and known as calcite or Iceland spar, Erasmus Bartholinus (1625–1692) discovered a puzzling phenomenon. Looking through the crystal at an object, he observed two images rather than one (Fig. 10-9). He also found that if the crystal was rotated, one of the images, which he called the **extraordinary one**, rotated with the crystal, but the other, which he called the **ordinary one**, remained fixed. He called this phenomenon **double refraction**, because an incident beam of light seemed to be split into two beams (Fig. 10-10).

Figure 10-9 Photograph of printed material viewed through an Iceland spar crystal (calcite). The letters appear double.

A satisfactory explanation of the phenomenon had to wait until other properties of double-refracting crystals had been discovered. The index of refraction of the ordinary ray in calcite is always 1.658; for the extraordinary ray it varies with the angle of incidence. In some double-refracting crystals, one of the beams is nearly completely absorbed on its way through the crystal, and only the other is transmitted undiminished. Such a crystal is tourmaline, in which sections only 1.00 mm thick completely absorb one of the rays. Thomas Young suggested that these effects could best be explained if light is regarded as having a **wave nature** and, more specifically, as having a **transverse** rather than a longitudinal wave motion.

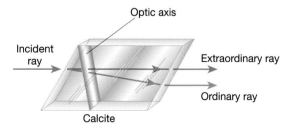

Figure 10-10 Double refraction by a calcite crystal. The crystal separates the E and O rays from incident light and transmits them in different directions and at different speeds.

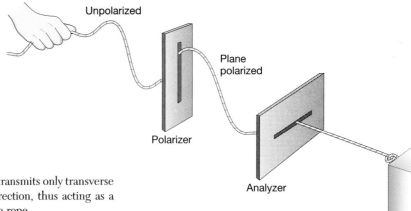

Figure 10–11   The vertical slit transmits only transverse waves vibrating in the vertical direction, thus acting as a vertical polarizer for waves along a rope.

Ordinarily, light waves oscillate in every possible direction at right angles to the beam's path. If the vibrations are confined to just one of these directions, the light is **plane polarized** just as transverse waves along a rope are plane polarized when the vibrations are restricted to a single plane (Fig. 10–11). If light is a transverse wave motion, it should be possible to polarize it. Longitudinal waves cannot be polarized. Double-refracting crystals are best understood, according to Young, by assuming that the ordinary and extraordinary rays are polarized at right angles to each other.

Polarized light became a useful scientific tool when William Nicol (1768–1851), a physicist, prepared crystals in such a way that one of the rays, the ordinary ray, was absorbed by the crystal, and only the extraordinary, polarized ray emerged (Fig. 10–12). A

Nicol prism alone constitutes a **polarizer**; a second crystal, an **analyzer**, transmits more or less of the polarized light, depending on how it is aligned with reference to the polarizer. An arrangement consisting of a polarizer and an analyzer is called a **polariscope** (Fig. 10–13).

Glass, Lucite®, and many other transparent materials become double-refracting under stress. When placed between the polarizer and analyzer of a polariscope, a colored pattern resulting from the **interference effects** of polarized light is observed that can be related to strains in the material (Fig. 10–14). By building models of structures with transparent materials, engineers can study stresses with polarized light.

Edwin H. Land (b. 1909), an inventor and physicist, succeeded in 1929, while a freshman in college, in orienting large numbers of the minute, needle-

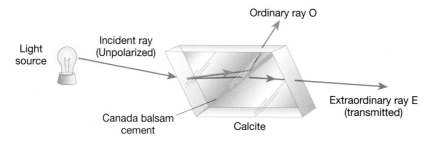

Figure 10–12   In a Nicol prism, a calcite crystal has been cut at a particular angle and re-cemented with Canada balsam. The refractive indices of the polarized E and O rays are such that only the E ray passes through the cement and then through the crystal. A beam of plane-polarized light is thus obtained.

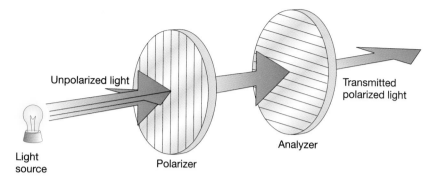

Figure 10–13   Principle of a polariscope. The amount of light transmitted depends upon the alignment of the polarizer and analyzer and on the medium between them.

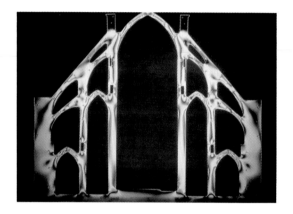

Figure 10–14   Strain pattern in a sample of plastic under stress. The plastic is placed between crossed polarizers. Strain distributions in transparent samples can be analyzed in this manner. The photograph shows a model of the flying buttress of the Notre Dame Cathedral. (Peter Aprahamian/Sharples Stress Engineers Ltd./Science Photo Library)

shaped crystals of the organic compound quinine iodosulfate. He did this by suspending them in a plastic film and then stretching the film in one direction. The polarizing action of **Polaroid**, as the material is called, is based on the uniform alignment of the double-refracting crystals, which absorb all of one ray and transmit the other.

A sheet of Polaroid acts very much like an optical picket fence, transmitting plane-polarized light in which the vibrations of light are parallel to the crystals. For optical use, films are mounted between glass plates (Fig. 10–15). Used in sunglasses, Polaroid reduces glare from roads and bodies of water. Light reflected from such surfaces is strongly polarized in a horizontal plane and is absorbed by the vertically oriented crystals in the Polaroid (Fig. 10–16).

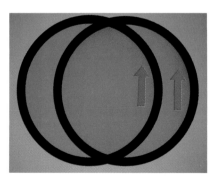 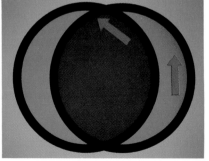 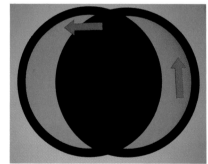

Figure 10–15   Transmission of light through Polaroid filters. A maximum amount of light is obtained when the transmitted Polaroids are parallel, and a minimum amount when they are crossed. (Courtesy of Henry Leap)

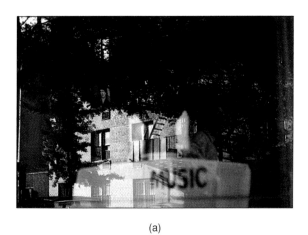

(a)

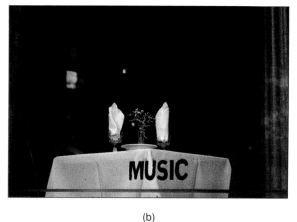

(b)

Figure 10–16   (a) Window reflection before polarization. (b) Window reflection after polarization. (Leonard Lessin/Peter Arnold, Inc.)

10–3 ELECTROMAGNETIC WAVES

The optical phenomena of diffraction and interference support the hypothesis that light has a wave nature, and polarization the hypothesis that light waves are transverse. But what is it that actually waves, and what is the medium that carries this wave? Huygens and Young agreed that light was transmitted by mechanical waves in an elastic medium called the *luminiferous* ("light-bearing") *ether,* which was assumed to pervade all space.

James Clerk Maxwell gave a different answer to the first question. His masterpiece, *A Treatise on Electricity and Magnetism* (1873), contains his famous equations of the electromagnetic field. From his equations, Maxwell deduced the existence of **electromagnetic waves** in which changing fields produce other fields in their vicinity and the disturbance travels through space at the speed of 3.00×10^8 m/s, the speed of light. Electromagnetic waves would have all the known properties of light waves. Light is thus just a form of electromagnetic radiation of a certain wavelength. The propagation of the waves involves a changing magnetic field that generates a changing electric field; the electric field in turn generates a changing magnetic field. In an electromagnetic wave, it is the electric and magnetic fields that propagate through space. Both fields are perpendicular to the direction of propagation, so the wave is transverse (Fig. 10–17).

For years there appeared to be no way to test experimentally Maxwell's theory of the electromagnetic nature of light. The problem was to prove the existence of a relation between light and electricity. Then Heinrich Rudolf Hertz (1857–1894) succeeded in producing and detecting other electromagnetic waves. Already convinced that every kind of light — sunlight, candlelight, the light from a glowworm — was basically electrical, he set out to prove it on the basis of Maxwell's equations.

Hertz's design was to generate waves electrically and to demonstrate that they traveled through the "ether" in the same way and with the same speed as light. He reasoned that such oscillations should be detected by a resonant receiver at some distance from the transmitting source. The receiver would pick up the oscillations, and the induced current would produce a spark across a gap.

Hertz's transmitter consisted of two polished metal spheres separated by a small air gap and connected to an induction coil, which built up a large voltage across the gap and caused sparks (Fig. 10–18). The sparks produced current oscillations that, according to Maxwell's theory, generate electromagnetic waves that then propagate through space. The receiver — a piece of circular wire with a tiny spark gap formed by a little knob and a sharp point — was set up at the other end of a darkened physics lecture hall. The instant the transmitter was turned on and

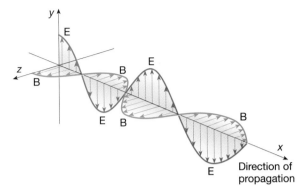

Figure 10–17 In electromagnetic waves, electric and magnetic fields propagate through space. The electric field, E, is vertical here; the magnetic field, B is horizontal; and both are at right angles to the direction of propagation.

sparks jumped across it, tiny sparks—only a few thousandths of an inch long and lasting not more than a microsecond—jumped across the receiving "antenna."

To prove that the electromagnetic waves were like light waves, Hertz measured the wavelength and frequency of the radiation and determined that the speed was approximately 3.00×10^8 m/s—the speed of light, as Maxwell had predicted. He showed further that the electromagnetic waves behaved like light

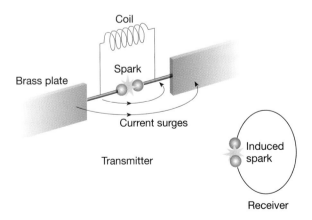

Figure 10–18 Hertz's experiment in which electromagnetic waves were generated and detected. Sparks in the transmitter cause waves that travel through space and induce sparks at the receiver.

waves in other respects. He reflected them from polished metal sheets and mirrors, refracted them with large prisms, made them interfere, and polarized them. In every case the new radiation behaved just as light did, except that it was not visible. Hertz proved that light and electromagnetic radiation were identical and that they both are explained by the electromagnetic theory of light.

10–4 THE ELECTROMAGNETIC SPECTRUM: RADIO, TV, MICROWAVES

The waves in the various regions of the electromagnetic spectrum share similar properties but differ in wavelength, frequency, and method of production. There is no sharp dividing line between one kind of wave and the next. Radio and TV waves and microwaves are generated by oscillating electrical circuits and range in wavelength from 20,000 meters down to a fraction of a centimeter. Infrared waves originate in molecular systems; their wavelengths extend from a fraction of a centimeter to approximately 7000 Å. Visible light, from 7000 Å to about 4000 Å; ultraviolet radiation, as short as 100 Å; and X-rays, to 1 Å, arise from disturbances in the electronic structure of atoms. Gamma rays are as short as 10^{-5} Å and originate in atomic nuclei (Fig. 10–19).

The transmission of a radio or television wave begins with a high-frequency (possibly 1 MHz for radio) oscillating current in a long metal rod, a transmitting antenna. Each electron is accelerating since its position and velocity are changing. Because of the continually changing electric and magnetic fields, due to the oscillating electrons, electromagnetic waves are radiated. These changing fields induce other fields in space, the disturbance traveling outward with the speed of light (Fig. 10–20).

When the wave reaches a receiving antenna, the oscillating electric field exerts an oscillating force on the electrons in the antenna, causing them to oscillate in the same way as the electrons in the transmitting antenna. Communication thus becomes possible through the intervening space as energy is carried by the electromagnetic waves. In **amplitude modulation (AM) radio**, an audible signal frequency of several hundred or thousand hertz is impressed upon a carrier wave of megahertz frequency, modulating its

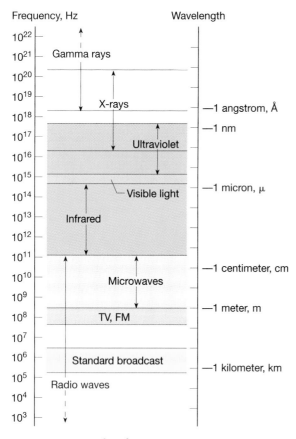

Figure 10–19 The electromagnetic spectrum. (Source: Raymond A. Serway and Jerry S. Faughn, *College Physics,* 3rd edition. Philadelphia: Saunders College Publishing, 1992.)

amplitude but having no effect on its frequency (Fig. 10–21). The signal is recovered at the receiver end, amplified, and converted to a sound wave. In **frequency modulation (FM)**, only the frequency of the carrier wave is modulated, increasing and decreasing at the same rate as the impressed audible signal frequency. Since atmospheric static affects only the amplitude and not the frequency of the modulated wave, FM transmission is essentially noise free. Television combines a video AM signal with an audio FM signal. In other regions of the spectrum, energy transfer is manifested in various ways — for example, in the visible range, the stimulation of appropriate nerve endings in the eye and the initiation of chemical reactions through suntanning or photosynthesis.

Microwaves are so named because they are very short radio waves compared with AM waves, having wavelengths between about 1.00 mm and 30.0 cm. They, too, are used in communication, such as long-distance telephone transmission. In this process, the information is encoded on microwaves by either amplitude or frequency modulation and carried via relay towers from point to point.

We have already referred to the domestic application of these waves in microwave ovens. Water molecules in food resonate at frequencies in the microwave region, heating the foot internally and resulting in a very efficient method of cooking. **Radar** (*ra*dio *de*tecting *a*nd *r*anging) systems also use microwaves

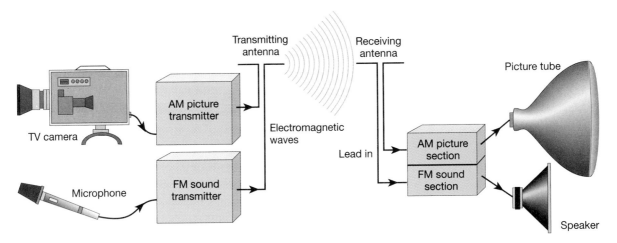

Figure 10–20 Television transmission and reception use both an AM picture signal and an FM audio signal. (Adapted from Jerry Wilson, *Technical College Physics,* 3rd edition. Philadelphia: Saunders College Publishing, 1992.)

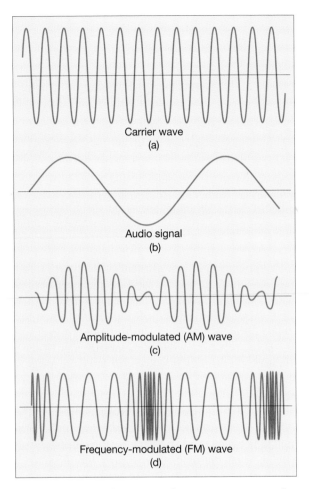

Figure 10–21 The carrier frequency in AM radio broadcasting ranges from 550 to 1700 KHz. FM stations broadcast in the range from 88 to 108 MHz. Represented in this figure are an unmodulated "carrier" radio wave (a) an audio-frequency wave (b); an AM (amplitude modulation) wave (c); and an FM (frequency modulation) wave (d).

to determine distances to objects by measuring microwave echo time, as sonar does with ultrasonic waves. The principles of radar are similar to those used in radio broadcasting, except that the electromagnetic wave consists of short bursts, or pulses, rather than of a continuous wave. The antenna is then used as a receiver to detect any reflected pulses, and the distance to the reflecting object is determined. Police officers can catch speeders by observing the Doppler shift of reflected microwaves. A computer quickly calculates the speed from the magnitude of

the frequency shift and displays it for the officer to see.

10–5 SIMPLE LENSES

The action of optical instruments such as cameras, telescopes, and microscopes depends on the refraction of light by lenses made of transparent glass or plastic. A **convex** or **converging lens** is thicker at the center than at the edges. Its structure is like that of two prisms placed base to base (Fig. 10–22). Parallel rays of light incident upon such a prism arrangement obey the law of refraction. Each ray is bent toward the normal as it enters a prism and away from the normal as it emerges into the air. The net effect is that the rays of light are bent toward the thicker part of the prism; that is, they converge.

Similarly, parallel rays are refracted toward the thick part of a double convex lens and converge at a point called the **principal focus**, *F* (Fig. 10–23). (There is a principal focus on each side of such a lens.) The distance between *F* and the optical center, *C*, of the lens is called the **focal length**, *f*. This is the primary characteristic of the lens. The optical center and the two principal foci are situated on an imaginary line through *C*, perpendicular to the plane of the lens. This imaginary line is known as the **optic axis**. Incident light rays that are parallel to the optic axis are all refracted through the point *F* (Fig. 10–24).

The position of the image formed by a double convex lens of a small object standing on the optic axis on one side of the lens at a distance beyond 2*f* may be

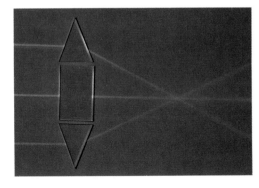

Figure 10–22 Parallel light rays are bent toward the thicker section of the two prisms placed. The action resembles that of a converging or convex lens. (David Rogers)

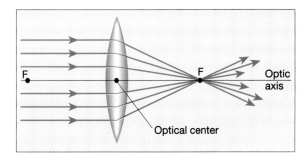

Figure 10–23   Action of a converging lens on light rays parallel to the optic axis. The rays are refracted and pass through the principal focus (*F*) on the optic axis.

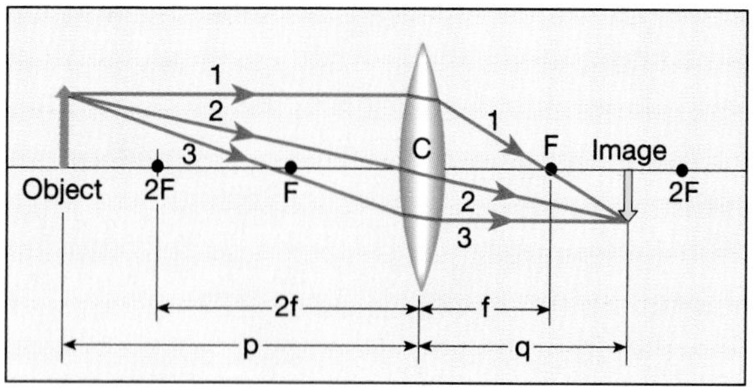

Figure 10–25   Ray diagram of image formation by a converging lens. With the object beyond 2*f*, the image is real, inverted, and smaller than the object.

determined by selecting three rays from each point on the object. This is shown for a single point in Figure 10–25. A ray parallel to the optic axis (ray 1) passes through the focal point on the opposite side of the lens after refraction by the lens. A second ray (ray 2) from the same point on the object is selected to pass through the optical center of the lens; it passes through essentially undeviated. A third ray (ray 3) passing through the principal focus on the same side of the lens emerges from the lens parallel to the optic axis. The point at which the three rays converge locates the image of the point. Three rays are similarly selected for a point at the opposite end of the object, and the image of that point is located where the three rays intersect. The image formed by the lens is made up of millions of such point images. Since light rays actually go to this image, it is called a **real image**. In this case the image is real, inverted, and smaller than the object.

When the object is at a distance from the converging lens greater than the focal length, the image is always real and inverted. For distances greater than 2*f*, the image is smaller than the object. When the object distance is 2*f*, the image distance is also 2*f* and the image is the same size as the object (Fig. 10–26). If the object distance is between 2*f* and *f*, the image is enlarged. Try to draw the principal rays for this case yourself.

In a single lens **motion picture projector**, the film is placed just beyond the focal length of the lens, that is, between *f* and 2*f*. By tracing the ray optics, you will see that the image is real, inverted, and much larger than the object. The image on the screen is certainly much larger than the film. To have a picture that is right side up, though, the film must be inverted when placed in the projector. The illusion of motion is obtained by quickly advancing the film at a given speed so that one slightly different picture frame after another is projected on the screen.

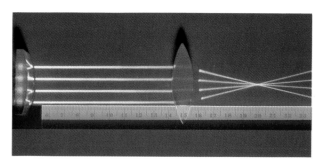

Figure 10–24   Photograph of the effect of a converging lens on parallel rays. (Courtesy of Jim Lehman, James Madison University)

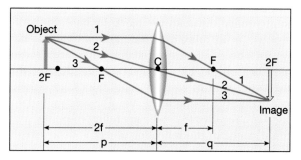

Figure 10–26   When the object is at 2*f*, the image is at 2*f* and is the same size as the object.

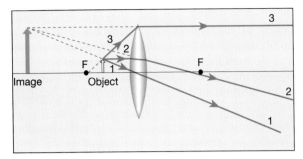

Figure 10–27 With a magnifying glass, the object is between *F* and the lens. A virtual image is formed that is right side up and enlarged.

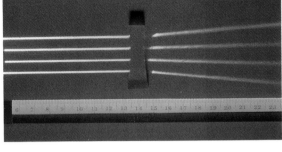

Figure 10–29 Effect of a diverging lens on parallel rays.
(Courtesy of Jim Lehman, James Madison University)

A converging lens is unable to converge rays to a real focus when the object is nearer the lens than the principal focus. Since the rays from the object are so highly diverging when they are incident on the lens, they are still diverging after they pass through the lens, so that no real image is formed (Fig. 10–27). The rays appear to diverge from a point on an image

that is virtual, erect, and enlarged. The image is **virtual** because there is no light at the image position and no image would form on a screen placed there. This is how a magnifying glass works. You hold the magnifying glass close to an object and see an erect, magnified image (Fig. 10–28).

A **diverging lens** is thinner at the center than at the edge (Fig. 10–29). A ray from a point on the object that is parallel to the optic axis is refracted in such a way that it appears to come from the principal focus. A second ray passes through the center of the lens with essentially no deviation. After passing through the lens, these rays appear to come from a single point on an image on the same side as the object (Fig. 10–30). The image is virtual, upright as long as the object distance is greater than *f*, and inside the focal length regardless of the object distance. A diverging lens can never form a real image, because the rays passing through the lens can never be brought to a real focus.

Figure 10–28 A magnifying glass is a converging lens. By holding it close to an object, such as this map, you see an enlarged image. (Richard Megna/Fundamental Photographs)

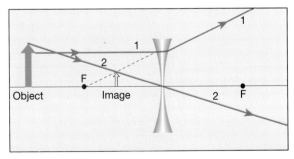

Figure 10–30 A diverging lens forms a virtual, upright image that is reduced in size.

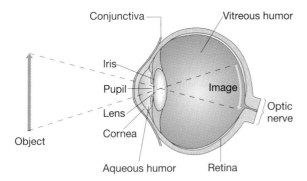

Figure 10–31   Structure and optics of the human eye.

10–6 THE OPTICS OF THE EYE

Light enters the eye at the curved front surface of the cornea, then passes through the aqueous humor, crystalline lens, and vitreous humor, before reaching the light-sensitive retina, which contains cone cells and rod cells (Fig. 10–31). These cells change light energy into electrical signals that travel along the nerves. A sharp image must focus on the retina for clear vision to occur. Although the image is inverted,

the brain interprets it as right side up. The distance between the lens and retina is fixed, but focusing can be accomplished by changing the curvature of the lens through the action of the muscles of the ciliary body on the ligaments around the rim of the lens. The lens is flexible and by such accommodation can adapt itself to objects at various distances. The amount of light entering the eye is controlled by the iris diaphragm; its opening, the pupil, automatically contracts in bright light and enlarges in dim light.

Most of the ability of the eye to form an image is concentrated in the cornea and the aqueous humor, their combined power being approximately twice as great as that of the lens itself. Even if the lens must be removed, vision is still possible, although an auxiliary spectacle lens is necessary to compensate for its loss, and no accommodation can occur. **Myopia**, or **nearsightedness**, is a condition in which the eyeball is so lengthened that a sharp image of a distant object cannot be focused on the retina without the aid of diverging lenses (Fig. 10–32). The rays that should converge on the retina actually converge in front of the retina, so that the image on the retina is blurred.

The weakening of the eye muscles with age and a hardening of the lens reduce the power of the lens,

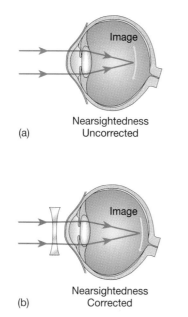

(a)   Nearsightedness Uncorrected

(b)   Nearsightedness Corrected

Figure 10–32   Nearsightedness (myopia) is corrected with a diverging lens.

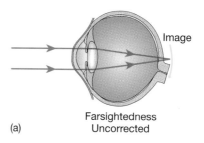

(a)   Farsightedness Uncorrected

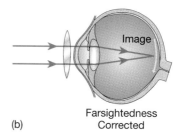

(b)   Farsightedness Corrected

Figure 10–33   Farsightedness (hyperopia) is corrected with a converging lens.

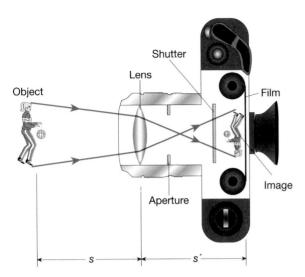

Figure 10–34   A simple camera. Note its similarity to the structure of the eye. (Source: Raymond A. Serway and Jerry S. Faughn, *College Physics,* 3rd edition. Philadelphia: Saunders College Publishing, 1992.)

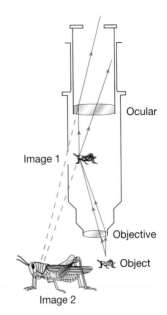

Figure 10–35   Principle of the compound microscope. The objective forms a real, enlarged image of the specimen in front of the ocular, and the ocular magnifies this image.

and **farsightedness (hyperopia)** becomes common. Rays converge behind the retina in this case. Distant objects are seen clearly, but near objects are blurred. This condition is corrected with auxiliary converging lenses (Fig. 10–33). In middle age, bifocal lenses, invented by Benjamin Franklin, are often fitted. The lower part is convex and assists in converging light for near vision; the upper part is used in distant vision. A single-lens photographic camera has parts that correlate with those of the eye (Fig. 10–34).

10–7 THE MICROSCOPE AND TELESCOPE

The principles of image formation by lenses are also applied in the microscope and telescope. The **microscope** is essentially a combination of two converging lenses (Fig. 10–35). The **objective**, the lens at the lower end of the microscope tube, acts as a projector, forming a magnified real image of the specimen (object) on the slide. The **ocular**, or **eyepiece**, the lens at the upper end of the tube, acts as a magnifying lens; it forms a magnified virtual image of the image formed by the objective. In more complex microscopes, the objective and the ocular are themselves composed of several lenses to eliminate distortions, but their function is the same as in the simpler case.

The **refracting astronomical telescope** is similar in principle to the microscope. It is a two-lens instrument, consisting of an objective and an ocular, designed to view objects at a distance. The objective in the telescope forms a real inverted image of a distant object in front of the ocular, the image being far smaller than the object. The ocular forms an enlarged virtual image that remains inverted. Telescopes are so long because large magnification requires an objective with a large diameter, hence a very long focal length, to gather more light. Since for viewing things on earth we prefer to have the image upright, an additional convex lens is inserted between the objective and the ocular to invert the image. These are often called terrestrial telescopes.

In very large telescopes, a concave mirror is used in place of the objective lens (Fig. 10–36). The purpose of the mirror, like that of the objective, is to gather as much light as possible and form an image that can be magnified by the ocular. These are called **reflecting telescopes**. The objective lens in the refracting tele-

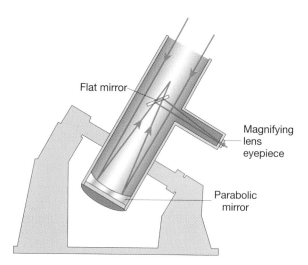

Figure 10–36   Reflecting telescope. A parabolic mirror reflects light to a focal point near the top of the tube. There a small flat mirror deflects the light out the side of the tube toward the magnifying lens eyepiece.

scope at the Yerkes Observatory, the largest in use, is 40 inches in diameter (Fig. 10–37). The mirror in the reflecting Hale telescope of Mt. Palomar Observatory is 200 inches in diameter (Fig. 10–38).

10–8 COLOR

A **rainbow** in the sky, a garden spray, or a fountain is caused by droplets of water dispersing white light into its colors, each droplet acting like a **prism** (Fig. 10–39). A beam of light is refracted as it enters a droplet, dispersed into its component colors, reflected internally, then refracted a second time as it re-emerges (Fig. 10–40). With the sun, observer, and droplets positioned at an appropriate angle, a bow is formed with red at the outer edge (42°2′), since red is refracted least, and violet at the inner edge (40°17′). The rainbow that passengers in an airplane at high altitude see forms a great circle, since they can see light from drops above and below the plane.

The colorful display on the surface of soap bubbles, oil slicks, and similar **thin films**, as well as iri-

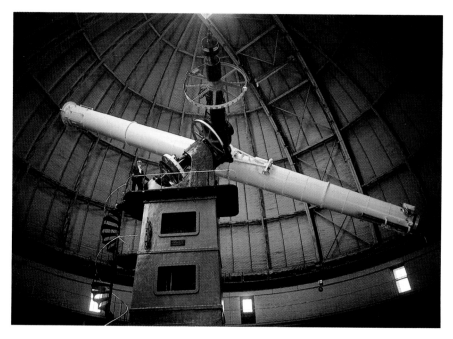

Figure 10–37   The 40-inch refracting telescope at Yerkes Observatory, Wisconsin, is the largest telescope of its kind in the world. It has a 1-m-diameter lens. (Dennis Milon)

Figure 10–38 View of the 200-inch reflecting Hale telescope at Mt. Palomar, California. The mirror, with a diameter of 200 in. and 27 in. thick, weighs 15 tons. The dome was rotated during this long-time exposure, making it seem transparent. (©1987 Roger Ressmeyer/Starlight)

Figure 10–39 Photograph of a primary rainbow. Under certain conditions, a fainter secondary rainbow can also be seen. (D. Davis/FPG International)

descence of the wings and bodies of many insects, is due to **interference** effects (Fig. 10–41). Light reflected from the two surfaces of a thin film travels different path lengths (Fig. 10–42). As they recombine, the light waves undergo constructive and destructive interference, depending on whether they are in or out of phase.

The process of **light scattering—the reflection of light in many directions**—accounts for the blueness of the sky (Fig. 10–43). Particles that are small compared with the wavelength of light, such as the nitrogen and oxygen molecules and other small particles in the air, scatter light inversely as the fourth

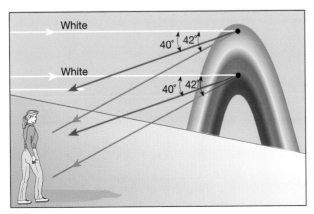

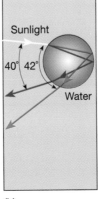

(a) (b)

Figure 10–40 A primary rainbow formed from sunlight refracted and internally reflected by water droplets.

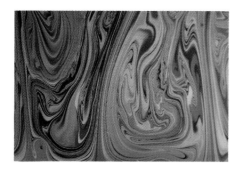

Figure 10–41 Photograph of the pattern of colors displayed by a thin film of oil on water when white light is incident on the film. (© Tom Branch 1984/Photo Researchers, Inc.)

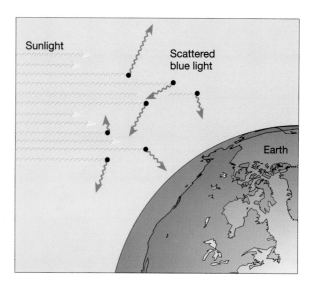

Figure 10–43 The blue color of the sky is produced by the greater amount of scattering of the shorter wavelengths of sunlight by air molecules. Thus, the light from the sky is predominantly scattered blue light.

power of the wavelength; the shorter, blue waves are scattered far more than the longer, red waves. Large particles scatter light of all wavelengths about equally and therefore appear white. That is why snow, salt, sugar, powders, soapsuds, clouds, and seafoam are white.

Most of the colors we see—in leaves, flowers, birds, fabrics, paints, dyes, and minerals—are due to **selective absorption**. This is a subtractive process in which certain wavelengths are removed from the incident light (Figs. 10–44, 10–45). The chlorophyll in green leaves and grass absorbs much of the light that

enters the cells, allowing only green light to escape. A green leaf illuminated by light lacking green wavelengths appears black, since all of the incident light is absorbed and none is reflected. The petals of red flowers absorb most of the wavelengths except red. The color of cloth, therefore, depends on whether it is

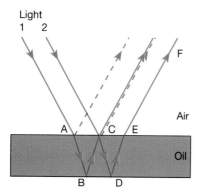

Figure 10–42 Iridescent colors of an oil slick arise from the constructive and destructive interference of light waves traveling along different path lengths. The difference is twice the thickness of the oil film. If rays 1 and 2 meet in phase at C, a bright spot appears. If there is destructive interference for a particular wavelength, the color associated with that wavelength will be absent from the reflected light.

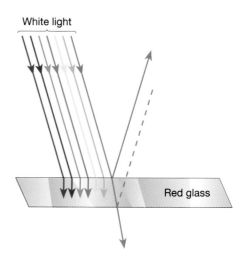

Figure 10–44 Colors of most objects are due to selective reflection and absorption of light. Red glass appears red because it transmits the red component present in white light while absorbing all other colors and reflects some of the incident light from the back surface.

Figure 10–45 Photograph of a red rose *(left)* and a white rose *(right)* taken under blue light. Some of the incident light is absorbed and some is reflected, causing the red rose to appear magenta and the white rose blue. (Leonard Lessin/ Peter Arnold, Inc.)

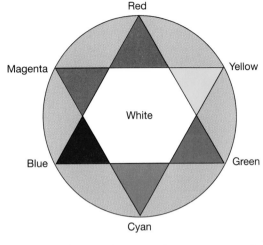

Figure 10–47 Complementary hues are any two hues that produce white when blended in some proportion, for example, blue and yellow. On this color triangle, the complementary light colors are opposite each other.

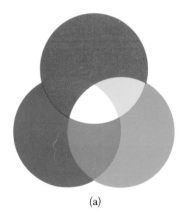

(a)

illuminated by natural or various forms of artificial illumination.

Three **spectral hues** or **pure colored lights**— red, green, and blue—combine to form white. They are called the **primary light colors** (Fig. 10–46). All the other colors in the visible spectrum and many that are not can be produced by combinations of these three. When light beams of these colors are projected on a screen so that the beams overlap, **additive mixtures** of colors and white are produced. **Complementary colors** are any two colored lights that blend to give white. On a color triangle they are opposite each other: blue and yellow, green and magenta, red and cyan (Fig. 10–47).

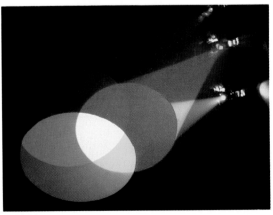

(b)

Figure 10–46 Additive mixing of colored lights from more than one source. Mixing of the additive primary colors, red, blue, and green, produces the colors illustrated in the regions of overlapping light beams, yellow, cyan, and magenta. (Fritz Goro, *Life*)

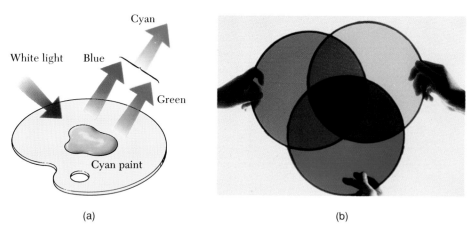

(a)   (b)

Figure 10–48   Subtractive mixing occurs when white light passes through colored filters—yellow, magenta, and cyan—that allow certain wavelengths to pass. (Fritz Goro, *Life*)

Mixing pigments is different from mixing spectral hues. The **primary pigments** or **subtractive primaries** of artists are yellow, cyan (blue-green), and magenta (blue-red), the complementaries of the primary light colors (Fig. 10–48). Yellow reflects the red and green components of incident white light, while absorbing the blue. Cyan reflects blue and green and absorbs red. If equal amounts of yellow and cyan are mixed, the result is green. The reason is that both yellow and cyan reflect green. Although cyan also reflects blue, the blue is absorbed by the red in the yellow pigment; similarly, the red in the yellow is absorbed by the blue in cyan. In this example of the **subtractive method** of color production or mixing, certain colors are subtracted and the eye sees the color that is not subtracted or absorbed. When all three primary pigments are combined, the result is black, if the pigments are pure.

According to the **Young-Helmholtz theory of color vision**, three light-sensitive pigments are distributed among three kinds of receptor cells in the retina, each sensitive in a different region of the spectrum. One is primarily responsible for sensing red light, one for sensing green, and one for blue: the R-cones, G-cones, and B-cones (Fig. 10–49). The three-receptor hypothesis is supported by recent spectrophotometric measurements of individual cone cells in the retinas of goldfish, rhesus monkeys, and human beings. But how this information is coded in the retina, transmitted to the brain, and decoded there are questions that remain to be answered.

When pure spectral yellow enters the eye, the R- and G-cones respond equally, and the sensation is yellow. A similar sensation can be produced when no spectral yellow is presented. Pure red and green cause both the R- and G-cones to respond equally, and the sensation is again yellow. When blue and

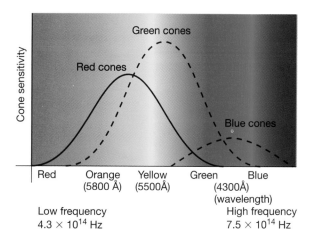

Figure 10–49   Color sensitivity of different cones of the eye: red cones, green cones, and blue cones. Each type is sensitive in a particular region of the spectrum.

green enter the eye, the stimulation of B- and G-cones produces the senation of cyan. The absence of R-cones in the retina leads to a type of **color-blindness** called protanopia. Only two primary colors, blue and green, are at the disposal of these individuals, who can see only to 6800 Å instead of the normal 7600 Å. The number of hues that the color-blind person can see is therefore reduced. Most animals live in a color-blind world; only a few species see the world in full color.

SUMMARY OF EQUATIONS

$$\text{Index of refraction} = \frac{\text{Velocity of light in a vacuum}}{\text{Velocity of light in medium}} \qquad n = \frac{c}{v}$$

KEY TERMS

Absolute Index of Refraction   The ratio of the velocity of light in a vacuum to the velocity of light in a medium (such as water).

Additive Primaries or Primary Colors   Lights of the colors red, green, and blue can be mixed or "added" to give any color of the visible spectrum or white.

Amplitude Modulation (AM) Radio   An audible signal frequency is impressed upon a carrier wave, modulating the amplitude.

Analyzer   In a polariscope, a Nicol prism that transmits more or less polarized light.

Color-blindness   Partial or total inability to distinguish one or more colors.

Complementary Colors   Any two colored lights that give white when combined, such as blue and yellow.

Converging Lens   A lens that is thicker at the center than at the edge and focuses rays parallel to the optic axis at a focal point, the principal focus.

Diverging Lens   A lens that is thinner at the center than at the edge causes rays parallel to the optic axis to diverge so that they appear to come from a point on an image on the same side as the object.

Double Refraction   A property of calcite crystals and other materials of splitting an incident beam of light into two beams and causing a double image.

Electromagnetic Wave   Electric and magnetic fields that propagate through space at the speed of light at right angles to each other and to the direction of propagation.

Extraordinary Image   With a double-refracting crystal, an image that rotates with the crystal.

Farsightedness   A condition in which distant objects are seen clearly, but near objects are blurred.

Focal Length   The distance between the principal focus of a lens and the optical center.

Frequency Modulation (FM)   An audible signal frequency is impressed on a carrier wave, modulating only the frequency.

Index of Refraction   The ratio of the speed of light in the two media when light passes from one medium into another.

Infrared   The region of the electromagnetic spectrum with wavelengths from 8000 Å to 1.00 mm.

Interference   The superposition of two or more waves.

Light Scattering   The absorption and re-emission of light in many directions by particles.

Microscope   An optical instrument consisting of a lens or combination of lenses for making enlarged images of minute objects.

Microwave   Electromagnetic waves with a frequency range of 10^9 to 10^{11} Hz used in communication and radar applications and in microwave ovens.

Nearsightedness   A condition in which a person can see near objects clearly, but no distant objects.

Objective Lens   A lens at the lower end of a microscope tube that forms a magnified real image of the specimen (object) on the slide.

Ocular Lens   A lens at the upper end of a microscope tube that forms a magnified virtual image of the image formed by the objective.

Optic Axis   An imaginary line drawn through the optical center and the two principal foci of a lens, perpendicular to the plane of the lens.

Polariscope   An instrument that enables an engineer to infer the stresses in structural materials from colored patterns produced by the interference effects of polarized light.

Polarized Light Light waves that oscillate in just one direction at right angles to the beam's path.

Polarizer A crystal or other object that polarizes light.

Polaroid A material invented by Edwin H. Land that polarizes light.

Primary Colors Three pure colored lights—red, green, and blue—that form white when combined.

Primary Pigments Yellow, cyan (blue-green), and magenta (blue-red), which can be mixed to give paints or dyes that appear as any color or black through selective absorption of light.

Principal Focus The location at which a lens focuses rays parallel to the optic axis or from which such rays appear to diverge.

Prism A triangular piece of glass that separates a beam of light into its component colors.

Radar *Ra*dio *d*etecting *a*nd *rang*ing systems that use microwaves to determine distances to objects.

Real Image An image formed by the convergence of light.

Reflecting Telescope A type of telescope that uses a mirror as the objective.

Refracting Telescope A type of telescope that uses a lens as the objective.

Selective Absorption A subtractive process in which certain wavelengths of light are removed from the incident light and absorbed, while the remaining wavelengths are reflected.

Spectrum The spread of colors seen when light is passed through a prism.

Subtractive Method of Color Production In this method of color mixing, certain colors are subtracted and the eye sees the color that is not subtracted or absorbed.

Subtractive Primaries See primary pigments.

Ultraviolet The region of the electromagnetic spectrum consisting of wavelengths 100 to 4000 Å.

Young-Helmholtz Theory The three-receptor theory of color vision that certain cells in the retina sense red light, others green, and still others blue.

THINGS TO DO

1. Determine the approximate focal length of a double-convex lens by focusing a well-illuminated object several hundred feet away on a paper screen until the image is sharp. The distance between the lens and screen is the focal length.

2. Note how light is produced in a (a) candle; (b) incandescent bulb; (c) fluorescent bulb; (d) neon light.

3. Illuminate fabrics of various colors in a dark room with light from (a) an incandescent bulb, (b) a fluorescent bulb, and note their appearances.

4. Observe the double images produced by Iceland spar when rotated on printed material.

5. Looking toward a window, rotate a sheet or disk of Polaroid completely around a second sheet or disk. Note the positions of maximum and minimum transmission of light.

6. Examine strains (a) in a piece of stretched plastic wrap and (b) in a plastic comb when placed between two Polaroid disks and one of the disks is rotated.

7. Note how sunglasses with polarizing filters reduce the glare from reflecting surfaces—water, glass, ice, and cars.

8. Place an automobile license plate in front of a plane mirror and observe the properties of the image: size, distance, right-to-left, and uprightness.

9. Show that a bottle or flask filled with water acts like a lens.

10. Determine the focal length of a magnifying glass. Then place it near a book or newspaper until the print is focused as large as possible. Compare the distance from the printed material with the focal length.

11. Examine a compound microscope.

12. Rotate a prism in the path of a beam of light from the sun or from a strong incandescent bulb until a good spectrum shows on a wall or screen. Note the order of the colors.

13. With your back to the morning or afternoon sun, adjust a water sprinkler or hose to give a fine spray. Observe the rainbow or rainbows produced and the order of the colors.

14. Place a red bulb and a blue bulb about 2 feet apart, and put a piece of white paper between them. Note the colors on the paper (a) near each bulb and (b) near the center where the colors mix. Repeat, with other light combinations.

15. Mark a card with a black circle and black cross about 3 inches apart. Close one eye, hold the card about a foot from the other eye, and move the card with your attention focused on the circle until you find a spot where the cross disappears, the blind spot.

16. Observe the colors produced on the surface of (a) soap bubbles; (b) oil slicks.

17. With an optical disc, if available, verify the (a) law of reflection; (b) law of refraction.

18. Place a stick in a glass of water. Note how refraction makes it appear to be broken at the surface.

19. Illuminate fluorescent material such as rocks and teeth with an ultraviolet "black light" lamp.

20. Illuminate a bottle of corn syrup with a bright light. Place a piece of Polaroid between the two, and view the light through a second Polaroid that you rotate. Note the colored patterns.

21. Make a pinhole camera. Replace one end of a small cardboard box with onion-skin paper. Make a pinhole at the other end. (a) Place a candle in front of the camera and look at its image on the paper. (b) Aim the camera at a bright scene and repeat.

EXERCISES

Velocity of Light

10–1. How long does it take light to travel from the sun to the earth, 93.0 million miles away?

10–2. A light-year is a unit of length equal to the distance light travels in a year. What is the length of a light-year in (a) kilometers? (b) miles?

10–3. The star Arcturus is 40.0 light-years from the earth. What is this distance in (a) kilometers? (b) miles?

10–4. How far away is an object if the reflection from a radar pulse is received in 93.0 microseconds?

10–5. Determine the frequency of light that has a wavelength of 5890 angstrom units.

10–6. How long does it take light to cross a room 30.0 m long?

10–7. If electromagnetic waves on transmission lines travel at the speed of light, how long would it take your voice to travel 5000 km on a telephone call?

10–8. Imagine that you are 200 m from home plate watching a baseball player hit a ball.
(a) How long does it take the light wave to reach you?
(b) How long does it take the sound wave from the bat striking the ball to arrive, if the speed of sound in air is 345 m/s?

10–9. A radar pulse returns to the receiver after a total travel time of 5×10^{-4} s. How far away is the object that reflected the wave?

Properties of Electromagnetic Waves

10–10. How do Polaroid sunglasses reduce glare from a road or a body of water?

10–11. Discuss the evidence for the transverse-wave nature of light.

10–12. What experimental evidence strengthened Maxwell's hypothesis that light and electromagnetic waves are similar?

10–13. List the forms of behavior in which light manifests wave properties.

10–14. Which color has the higher frequency, red or violet?

10–15. Which radiation has the shorter wavelength, radio waves or X-rays?

10–16. The speed of light in a clear plastic material is 1.50×10^8 m/s. What is the index of refraction?

10–17. If they are visible, what colors would you assign the following wavelengths: (a) 400 Å; (b) 4000 Å; (c) 6000 Å; (d) 40,000 Å?

10–18. Microwave ovens use frequencies of 900 and 2560 MHz. What is the wavelength associated with each frequency?

10–19. What type of electromagnetic wave in free space would have a wavelength of (a) 1.00 m? (b) 1.00 mm?

10–20. How does a microwave oven cook food?

10–21. When electromagnetic radiation such as light travels across a region, what is it that moves?

Seeing and Color

10–22. Explain the formation of a rainbow.

10–23. What spectral colors are complementary to (a) blue; (b) yellow; (c) magenta?

10–24. Why is the index of refraction always greater than or equal to 1.00?

10–25. (a) What color would the sky be if there were no atmosphere? (b) What proof is there?

10–26. Compare the number of "octaves" to which the eye responds with the number to which the ear responds.

10–27. Will light bend toward or away from the normal (perpendicular) when it goes from (a) air to glass? (b) water to glass?

10–28. Explain why the mixing of red and blue lights gives a different color than the mixing of equal amounts of red and blue pigments.

10–29. What part of a camera plays a role similar to that of the retina of the eye?

10–30. *Multiple Choice*

A. Compared with the velocity of light, the velocity of radio waves is
(a) less.
(b) greater.
(c) the same.
(d) infinite.

B. Polarized light is explained by the hypothesis that light consists of
(a) transverse waves.
(b) longitudinal waves.
(c) particles.
(d) an ether.

C. Color is related to
(a) amplitude.
(b) frequency.
(c) velocity.
(d) quality.

D. The bending of a ray of light as it passes at an angle from air into water is an example of
(a) reflection.
(b) diffraction.
(c) refraction.
(d) interference.

E. The pupil of the eye adjusts for
(a) long-distance vision.
(b) color.
(c) size of object.
(d) amount of light.

F. Parallel rays of light falling on a converging lens pass through the lens and
(a) spread out.
(b) are brought to a focus.
(c) are reflected back in the original direction.
(d) remain undeflected.

G. A piece of green cloth
(a) reflects all colors but green.
(b) absorbs all colors but green.
(c) absorbs only red light.
(d) reflects only red light.

H. Some of the most precise determinations of the velocity of light were made by
(a) Land.
(b) Maxwell.
(c) Michelson.
(d) Helmholtz.

I. The region of the electromagnetic spectrum with wavelengths longer than 8000 Å is known as the
(a) ultraviolet.
(b) infrared.
(c) X-ray.
(d) gamma ray.

J. Light in a rainbow is
(a) refracted only.
(b) reflected only.
(c) both refracted and reflected.
(d) neither refracted nor reflected.

SUGGESTIONS FOR FURTHER READING

Bates, Harry E., "Resource Letter RMSL-1: Recent Measurements of the Speed of Light and the Redefinition of the Meter." *American Journal of Physics*, August, 1988.
A guide to the literature on recent measurements of the speed of light and the redefinition of the meter.

Babović, V. M., D. M. Davidović, and B. A. Aničin, "The Doppler Interpretation of Rømer's Method." *American Journal of Physics*, June, 1991.
Demonstrates that Rømer's method for determining the speed of light can be considered an application of the Doppler effect.

Brittain, James E., "The Magnetron and the Beginnings of the Microwave Age." *Physics Today*, July, 1985.
A device developed in 1916 as an alternative to grid control in vacuum tubes became in 1940 the key to the successful development of radar by the Allied forces during World War II.

Brou, Philippe, Thomas R. Sciascia, Lynette Linden, and Jerome Y. Lettvin, "The Colors of Things." *Scientific American*, September, 1986.
Proposes the unconventional notion that the perceived colors of things do not depend on the light from each object, sensed independently, but on a comparison of the lights from an object and its surroundings.

Ennis, John L., "Photography at Its Genesis." *Chemical and Engineering News*, December 18, 1989.

The year 1989 marked the one hundred and fiftieth anniversary of the daguerreotype—generally thought of as the progenitor of photography. But the daguerreotype process was only one of several photographic processes disclosed during that single eventful year.

Land, Edwin H., "The Retinex Theory of Color Vision." *Scientific American*, December, 1977.
The eye has evolved to see the world in unchanging colors. How does it achieve this feat?

Levine, Joseph S., and Edward F. MacNichol, Jr., "Color Vision in the Fishes." *Scientific American*, February, 1982.
Fishes are exposed to visual environments that are blue, green, or near-infrared. The pigments in the retina that diverse species acquire in adapting to these environments are a clue to the evolution of the eye.

MacAdam, David L., "Color Science and Color Photography." *Physics Today*, January, 1967.
James Clerk Maxwell invented three-color photography and explained the principle on which modern color photography, color printing, and color television are based. Modern workers in these fields still have much to learn from him.

Mulligan, Joseph F., "The Influence of Hermann von Helmholtz on Heinrich Hertz's Contributions to Physics."*American Journal of Physics*, August, 1987.
Heinrich Hertz made many contributions to physics during his brief life (1857–1894). A former student of Hermann von Helmholtz, he maintained a very close personal and scientific relationship with his mentor for the rest of his life.

Nassau, Kurt, "The Causes of Color." *Scientific American*, October, 1980.   Practically all the mechanisms of color have an element in common: The colors come about through the interaction of light waves with electrons.

Nathans, Jeremy, "The Genes for Color Vision." *Scientific American*, February, 1989.

The author and his colleagues have now identified the genes that code for the pigments in color discrimination, deciphered their structures, and deduced the amino acid sequences of the encoded proteins. Their studies have provided clues to the evolution of normal color perception and "color-blindness."

Nijhout, H. Frederick, "The Color Patterns of Butterflies and Moths." *Scientific American*, November, 1981.
A few simple rules guide the development of more than 100,000 different wing patterns.

Rowell, J. M., "Photonic Materials." *Scientific American*, October, 1986.
Photonics is a new technology, analogous to electronics, for generating, transmitting, receiving, and processing signals made up of light photons instead of electrons. Photonic materials are revolutionizing communications.

Shankland, R. S., "Michelson: America's First Nobel Prize Winner in Science." *The Physics Teacher*, January, 1977.
Discusses the famous experiment that disproved the hypothesis of a "luminiferous ether."

Swenson, Loyd S., "Michelson and Measurement." *Physics Today*, May, 1987.
Albert Abraham Michelson became a preeminent master of measurement by light waves largely through his playful work with interferometers. His name became practically synonymous with extremely precise measurement.

Walker, Jearl, "Mysteries of Rainbows, Notably Their Rare Supernumerary Arcs." *Scientific American*, June, 1980.
Extra bands of color, called supernumerary arcs, sometimes accompany a natural rainbow. Such arcs can be investigated in a variety of ways.

Winston, Roland, "Nonimaging Optics." *Scientific American*, March, 1991.
Lenses and mirrors are not the best tools for the job of concentrating light. Within the past 20 years or so, devices called nonimaging concentrators have been developed that concentrate light to the highest possible limit.

ANSWERS TO NUMERICAL EXERCISES

10–1. 500 s

10–2. (a) 9.45×10^{12} km
(b) 5.90×10^{12} m

10–3. (a) 3.84×10^{14} km
(b) 2.40×10^{14} mi

10–4. 13.95 km

10–5. 5.09×10^{14} Hz

10–6. 10^{-7}

10–7. 1.67×10^{-2} s

10–8. (a) 0.667×10^{-6} s
(b) .580 s

10–9. 7.5×10^{4} m

10–16. 2.00

10–18. 30.0 cm; 11.7 cm

10–26. Eye: 1 octave   Ear: 10 octaves

Electrostatics. A glass rod charged by rubbing with a silk scarf attracts bits of paper. (©Paul Silverman/Fundamental Photographs)

11

ELECTRICITY AND MAGNETISM

On the muggy evening of July 13, 1977, two bolts of lightning struck two transmission towers, causing New York City to experience its second major blackout in 12 years. Eight million persons lost their electricity, and subways and elevators went dead. The great blackout that descended upon the populous northeastern sector of the United States on the evening of November 9, 1965, had given convincing proof of the electrical basis of modern civilization (Fig. 11–1). Normal living patterns ground to a sudden halt, indicating how vulnerable that basis is and how dependent we have become on a technology that was developed for the benefit of humanity only about a century ago. There is no assurance that such an event will not happen again, as the increasing demand for electrical energy taxes our ability to provide it.

Four natural manifestations of electricity have been known for centuries: lightning; the ability of the resinous substance amber to attract objects when rubbed; "animal electricity" as it occurs in the discharge of the electric eel and other species; and St. Elmo's fire, a discharge of light seen near pointed objects in storms.

Our electric technology, which has superseded the steam age, is based on magnetism. Electric power generators, motors, radio and television, the telephone—all depend on magnetism and its relation to electricity.

Figure 11–1   The fragility of our electrical age was dramatically demonstrated on the evening of November 9, 1965, when a faulty electrical relay caused a power failure and plunged the northeastern part of the United States into total darkness. (*Left,* Jake Rajs/The IMAGE Bank; *right,* Douglas Kirkland/The IMAGE Bank)

11–1 THE AMBER PHENOMENON

You are probably acquainted with some of the simplest facts of electrical force. A balloon rubbed with wool, flannel, or dry hair will seemingly defy gravity and stick to a wall or ceiling (Fig. 11–2). Moreover, a balloon "electrified" in this way attracts bits of paper and other small objects and is attracted to the material that was used to electrify it (Fig. 11–3). You can easily demonstrate that an electrified balloon repels another balloon treated in the same way and that these effects are more pronounced on a dry day than on a humid day (Fig. 11–4).

The basic discoveries of frictional electricity were made over a period of more than 2000 years. The philosopher and scientist Thales of Miletus (640?–546 BC), one of the Seven Wise Men of Greece, is credited with the discovery (about 600 BC) that when the resin **amber** is rubbed with wool or fur, it temporarily attracts small bits of matter, such as straw, leaves, and hair. The Greek word for amber is "elektron." But neither the Greeks nor others did much more with electricity than to make this and similar observations. A basic difficulty is that frictional electricity is not usually produced in great amounts and does not lend itself to many kinds of experiments. Moreover, it is quickly lost in damp surroundings.

The modern development of **electrostatics**—**electric charges at rest**—dates from the publication in 1600 of the book *De Magnete* by Sir William Gilbert (1544–1603), chief physician to Queen Eliz-

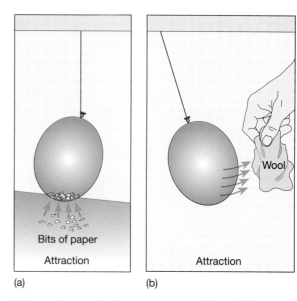

Figure 11–3 The basic experimental facts of electricity are illustrated. (a) A balloon rubbed with wool attracts bits of paper and (b) is attracted to the wool.

abeth I. In this classic of scientific literature, Gilbert brought together everything that was known about magetism and electricity, described many of his own experiments, and set down his conclusions and speculations. He demonstrated that the amber phenomenon was not limited to amber and a few other materials but that many types of bodies could be electrified,

Figure 11–2 Two balloons electrified by rubbing with a cloth stick to a wall. This method is commonly used to decorate the walls of a room at a party. (Courtesy of Henry Leap and Jim Lehman)

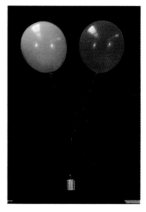

Figure 11–4 These two helium-filled balloons have been charged by rubbing them with a cloth. Both carry the same kind of charge and hence repel each other. (Courtesy of Henry Leap and Jim Lehman)

Figure 11–5 A stream of water is deflected to the right by a charged glass rod because of an attractive force between the water and the rod. (Courtesy of Henry Leap and Jim Lehman)

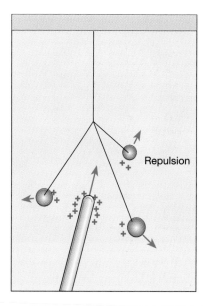

Figure 11–6 If the charge on a glass rod is shared by several pith balls, the similar charges cause mutual repulsion.

among them diamonds, sapphires, glass, sulfur, sealing wax, and mica. Moreover, electrified bodies attracted not only chaff and twigs, but metals, wood, stones, even water and oil. You can demonstrate one of these effects by rubbing a ball-point pen on wool and placing it near a gentle stream of water from a faucet (Fig. 11–5). The pen diverts the direction of the stream.

Electrified bodies can be divided into two classes. Materials such as glass and porcelain are included in one class. The second class includes amber, silk, and other materials. In the terminology proposed by Benjamin Franklin (1706–1790), statesman, scientist, and philosopher, the electrical condition glass acquires when it has been rubbed with silk is called **positive electric charge.** The kind of charge that hard rubber acquires when it has been rubbed with fur is called **negative electric charge.** The **law of electric charges** (Fig. 11–6) is based on the experimental finding:

> **Like charges repel each other; unlike charges attract.**

Our present view of electricity is based on the **electron theory of matter.** Matter is composed of atoms with a positive nucleus surrounded by a cloud of orbiting negative electrons. Charges are associated with particles and are properties of those particles just as their masses are. Ordinarily, an atom has equal numbers of positively charged protons and negatively charged electrons and is electrically neutral. When

two different materials, such as rubber and fur, are rubbed together, however, the one with a greater affinity for electrons (rubber) attracts some from the other (fur). The body that gains electrons acquires an excess negative charge, and the body that loses electrons acquires a net positive charge (Fig. 11–7). Whenever one body acquires a negative charge, a second body will acquire an equal positive charge.

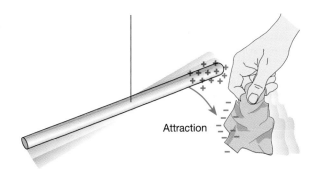

Figure 11–7 When a glass rod is rubbed with silk, electrons pass from the glass to the silk. The glass rod is left with a net positive charge, the silk with an equal negative charge. Charges of opposite sign attract.

Figure 11–8 A nylon comb can acquire an electrical charge and attract light objects, such as bits of paper. (Fundamental Photographs)

This observation is embodied in the principle of the **conservation of charge**:

> **Charge cannot be created or destroyed; the total charge in the universe is constant.**

Although rubbing causes a charge transfer, the mere contact between two unlike materials causes them to become electrically charged to some extent. Rubbing, by bringing more areas into contact, increases the effect by making many points of contact instead of the relatively few formed when two rough surfaces are placed in contact. Your shoes scuffing over the pile of a rug, the tires of a moving car, a comb passing through the hair—all acquire electrical charges in this way (Fig. 11–8).

In Table 11–1, various materials are listed on the basis of the electrostatic charges they acquire upon contact with other materials. Each material becomes more positive than the one below it. If glass is in

Table 11–1   ELECTROSTATIC SERIES

Glass	Cotton
Mica	Wood
Wool	Cork
Quartz	Amber
Cat's fur	Graphite
Silk	Rubber
Felt	Sulfur

contact with silk, glass acquires a positive charge and silk a negative charge, showing that it is easier to remove electrons from glass than from silk. But if sulfur is in contact with silk, silk becomes positive with respect to sulfur; that is, it is easier to remove electrons from silk than from sulfur.

11–2 CONDUCTORS, SEMICONDUCTORS, AND INSULATORS

Since many ordinary metals can transmit electricity readily, they are classified as good electrical **conductors.** On the other hand, many other materials— glass, rubber, amber—do not conduct electricity well and are called **insulators**. When placed on an insulator, a charge does not migrate but remains approximately where it is placed. Because they are good conductors, metals are used in electric wires to take electricity where we want it to go, whereas glass, fabrics, and plastics are used to insulate wires, that is, to prevent electricity from going where it is not wanted.

Materials do not always fall neatly into these two classes. On a dry winter day, for example, air behaves as an insulator. As we walk across a rug and accumulate charge on our bodies, the charge remains fixed and we may experience a shock as we reach for an electric wall switch or a metal doorknob. On a damp day this is less likely to happen; then air is more conducting, and the moisture in the air drains off charge from our bodies. Some typical conductors and insulators are listed in Table 11–2.

A third class of materials, called **semiconductors**, includes materials such as carbon, silicon, and germanium, which ordinarily are slightly conducting. Semi-

Table 11–2   ELECTRICAL CONDUCTORS AND INSULATORS

Conductors	Insulators
Aluminum	Amber
Copper	Fur
Gold	Glass
Iron	Mica
Mercury	Rubber
Seawater	Sealing wax
Silver	Silk
	Wool

conductors are not good insulators, but at the same time they do not conduct electricity well enough to be considered good conductors. When light falls on them or when their temperature is raised, however, they conduct well. This property allows them to play an important technological role.

11–3 THE ELECTROSCOPE

In the early days of experiments with electrical charges, the amount of the charge was estimated roughly by the size of the sparks produced or by the size of the shock the experimenter received upon touching the electrified body. Today, a gold-leaf electroscope may be used to tell whether there is a charge, whether it is positive or negative, and, more or less, the amount.

An **electroscope** has two thin gold or aluminum leaves mounted at the end of a metal rod with a metal knob at the top (Fig. 11–9). The rod and metal leaves are enclosed in a container made either of glass or of metal with glass windows, to avoid the effects of air currents and to prevent any electrical effects on the leaves other than those produced on the knob. The rod is mounted in a plug of rubber or some other good insulator to separate it from the box.

If a negative charge from a rubber rod that has been rubbed with fur is placed directly on the knob of the electroscope—a process called **charging by conduction** or contact—it spreads over the knob, rod, and leaves. This is because the negative charge consists of electrons that are free to move through metallic parts. The electrons get as far away from the rod as possible. In this case the leaves are the farthest point. Each leaf acquires a net negative charge. Since like charges repel, the leaves repel each other and diverge. The amount of divergence depends on the quantity of charge on each leaf and the mass of each leaf. The electroscope is now charged negatively. It may be charged positively in a similar way with a charging body such as a glass rod rubbed with silk. In this case, electrons migrate from the leaves, rod, and knob to the glass rod, and each leaf acquires a positive charge. The leaves again diverge.

The electroscope is a sensitive instrument. The amount of charge required to produce a deflection of the leaves is small, owing to the small mass of the thin leaves. It is therefore often preferable to charge it by another method—**induction**—to prevent damage

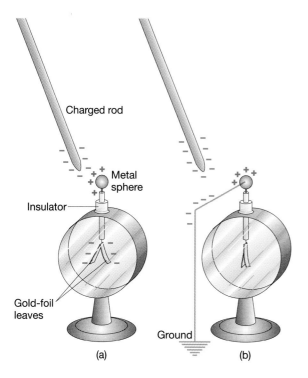

(a)   (b)   (c)

Figure 11–9 Charging an electroscope by induction. (a) A negative rod brought near the electroscope repels electrons to the leaves, which diverge. (b) Electrons are removed by grounding when the knob is touched; the leaves collapse. (c) With the ground connection broken first and then the inducing charge removed, positive charges on the leaves cause them to diverge. The electroscope is now charged positively.

to the leaves. In this process there is no direct contact between the charging body and the electroscope (Fig. 11–9). If a negatively charged rubber rod is brought near the knob of the electroscope, some of the electrons in the knob are repelled to the leaves, which then diverge. You will note that no electrons are added to the electroscope in this process and that the electroscope is still neutral.

With the rubber rod still in position, the knob is momentarily touched with a finger. Since there is now a connection between the electroscope and the ground through another conductor (the body), this step is called **grounding**. When grounded, electrons can get even farther away. The earth can be considered a reservoir for electrons; it can accept or supply an unlimited number of electrons. Some of the electrons in the electroscope are repelled from the leaves to the rod, knob, finger, and ground. Having lost their excess electrons, the leaves collapse. When the finger is removed, followed by the removal of the rubber rod, the leaves diverge again. The reason is that the electroscope has lost some electrons to the ground and, because it is deficient in electrons, has acquired a positive charge. Although the charging body, the rubber rod, is negative, the electroscope has acquired the opposite charge by induction; it becomes positive. To charge an electroscope negatively by induction, the charging body must be positive.

With a charged electroscope, we can determine the presence of a charge on any body brought near it and also the sign of the electric charge. Suppose the electroscope is charged negatively (Fig. 11–10). If a body is brought near the knob and the leaves diverge still farther, we can be sure that the body has a negative charge; more electrons are repelled from the knob onto the leaves, giving the leaves a greater negative charge and hence greater repulsion. On the other hand, if the leaves collapse to some extent, we conclude that the body has a positive charge, since some of the electrons are withdrawn from the leaves and the force of repulsion between them is reduced. Once charged, an electroscope may retain its charge for some hours, depending on the moisture content of the air.

Electroscopes are used with other devices to study radiation from X-rays, radioactivity, and cosmic rays. Such radiations charge particles in the air, and the charged particles, in turn, cause the discharge of an electroscope. The rate of discharge is related to the amount of radiation.

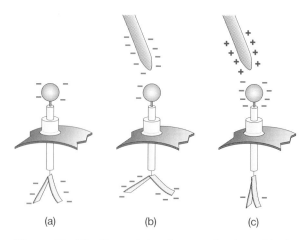

Figure 11–10 Determining the sign of a charged body with a negatively charged electroscope (a). If the leaves diverge farther, more electrons are driven from the ball to the leaves, so the rod must be negative (b). If the leaves draw together, we conclude that electrons are attracted from the leaves to the rod, which must be positive (c).

11–4 FORCES BETWEEN ELECTRIC CHARGES

Electrical repulsions and attractions are pushes and pulls and are therefore forces. Using a torsion balance, the physicist Charles-Augustin de Coulomb (1736–1806) discovered how electrical forces varied with the distance between the charges.

The torsion balance was set up as follows: By means of a fine thread, Coulomb suspended a light rod that had a metal sphere at each end (Fig. 11–11). At the top, he put a support that he could turn and a means to measure the amount of turn. When he charged two spheres in the same way, the repulsive forces between them caused the one on the rod to move. Since the rod was attached to the thread, the thread twisted; but it could be twisted back by turning it at the top until the two charged spheres returned to their original distance apart. The amount of twist became a measure of the force of repulsion. Later, using another arrangement, Coulomb measured the force of attraction between two charges.

Coulomb found that the electrical force of attraction or repulsion between two bodies varies inversely with the square of the distance between them. He also found that the electrical force is proportional to the amount of electric charge on each of the bodies.

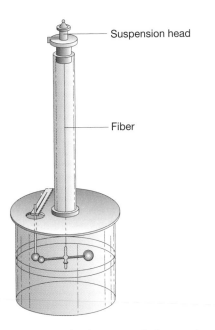

Suspension head

Fiber

Figure 11–11   Coulomb's torsion balance. The repulsion of two like charges causes the suspending wire to twist. The angle of twist is related to the force of repulsion. (Drawing from Coulomb's memoir to the French Academy of Sciences, 1785)

Coulomb's law of electrostatic force is expressed as follows:

> **The force between two electric charges, q_1 and q_2, is proportional to the magnitude of each of the charges and inversely proportional to the square of the distance between them.**

In equation form:

$$F = k\frac{q_1 q_2}{r^2} \qquad (11\text{–}1)$$

where, in SI units,

$$F = \text{the force of attraction}$$
$$\text{or repulsion in newtons (N)}$$
$$q_1 \text{ and } q_2 = \text{the electric charges in coulombs (C)}$$
$$r = \text{the distance separating the}$$
$$\text{charges in meters (m)}$$
$$k = 9.00 \times 10^9 \frac{\text{N} \cdot \text{m}^2}{\text{C}^2}$$
$$= \text{the Coulomb constant}$$

When q_1 and q_2 have like signs, the force F is positive and is interpreted as a repulsion; when q_1 and q_2 have unlike signs, F is negative and is interpreted as an attraction.

You will note the mathematical similarity between Coulomb's force law and the law of gravitational force. Both are inverse-square laws. However, there are differences between them. Electric forces can be either attractive or repulsive, whereas gravitational force is always attractive. Furthermore, electric forces are considerably stronger than the gravitational force, approximately 10^{40} times as strong (10,000 trillion, trillion, trillion times as strong). Being a vector quantity, electric force has direction as well as magnitude; the direction of the force is given by the law of charges. When the charges are of the same sign, the force is repulsive. When the charges are of opposite sign, the force is attractive.

The coulomb is a very large amount of charge. A body with 1 coulomb of charge 1 meter away from a similar body would exert a force on it of 9.00×10^9 newtons (about 2 billion pounds). A lightning discharge may transfer 200 coulombs of charge. Therefore, charges are often expressed in terms of the submultiples, the micro- or nanocoulomb. The charge of a comb rubbed on the sleeve is less than 1 microcoulomb.

Electrical charges cannot be subdivided indefinitely. Robert A. Millikan (1868–1953), a physicist at the University of Chicago and the California Institute of Technology, and others proved that the minimum possible electric charge e is equal to 1.60×10^{-19} coulomb. This is the charge associated with the electron (negative) and proton (positive). All charges possessed by the material particles are integral multiples of this minimum charge. (If particles called *quarks* exist, as some theories have proposed, they would have charges of $e/3$ and $2e/3$; however, they have not been detected conclusively.)

EXAMPLE 11–1

What is the force between two electrons separated by 10^{-10} m (1 angstrom unit)?

SOLUTION

Knowing the values of the charges and the distance separating them, we apply Equation 11–1 to calculate the force.

$q_1 = -1.60 \times 10^{-19}$ coul

$q_2 = -1.60 \times 10^{-19}$ coul

$r = 1.00 \times 10^{-10}$ m

$F = ?$

$$F = \frac{9.00 \times 10^9 \text{ N} \cdot \text{m}^2}{\text{coul}^2} \frac{q_1 q_2}{r^2}$$

$$= \frac{9.00 \times 10^9 \text{ N} \cdot \text{m}^2}{\text{coul}^2}$$

$$\times \frac{(-1.60 \times 10^{-19} \text{ coul})(-1.60 \times 10^{-19} \text{ coul})}{(1.00 \times 10^{-10} \text{ m})^2}$$

$$= \frac{9.00 \times 10^9 \text{ N} \cdot \text{m}^2}{\text{coul}^2} \times \frac{(2.56 \times 10^{-38}) \text{ coul}^2}{1.00 \times 10^{-20} \text{ m}^2}$$

$$= \frac{23.0 \times 10^9 \times 10^{-38}}{10^{-20}} \text{ N}$$

$$= \boxed{2.30 \times 10^{-8} \text{ newton (repulsion)}}$$

EXTENSION

The force between two electrons is 1.00×10^{-12} newton. How far apart are they? (Answer: 1.50×10^{-8} m)

∎

11-5 LIGHTNING

The nature of lightning was in considerable doubt until Benjamin Franklin demonstrated that a lightning stroke was an electric discharge and, therefore, a natural phenomenon that could be described by scientific laws. Franklin carried out his famous kite experiment in the summer of 1752. He constructed a kite from a large silk handkerchief and cedarwood, attached to it a length of twine, held it by a silk ribbon tied to the lower end of the twine, and attached a door key where the ribbon and twine were joined. Standing under a shed to keep himself dry, he raised the kite and after a while saw the loose fibers of the twine stand erect; this meant that they were electrified and that mutual repulsion existed between them. When he placed a knuckle to the key, he received a spark and a shock. Franklin concluded that lightning, too, is an electric spark (Fig. 11-12). (That his experiment was a dangerous one to perform was proved a year later when Georg Richmann [1711–1753] was struck dead in St. Petersburg while he experimented with lightning.)

It occurred to Franklin that buildings and ships could be protected from lightning by placing on the

Figure 11-12   Multiple lightning bolts illuminate Kitt Peak National Observatory in Arizona in this dramatic one-minute exposure. (©Gary Ladd, 1972)

highest parts of those structures upright sharp-pointed iron rods with a wire running from the base of the rod down the side of the structure into the subsoil or water. He thought that such rods would safely draw electricity from a cloud and thus prevent a damaging lightning stroke. Or, if a stroke did occur, it would be conducted away harmlessly. Lightning rods soon appeared on many buildings here and abroad. Seldom has an application of a scientific discovery moved so swiftly from the realm of speculation to the arena of practice.

The way in which lightning discharges are produced has been difficult to explain. The basic mechanism—how water droplets in a thundercloud become electrified—is still not clear. It is known, however, that a weather system that produces updrafts of warm, moist air reaching a velocity of 160 miles/hour or more and rising above the freezing levels in the atmosphere becomes a huge generator of electricity. The atomizing action of the wind currents on water droplets probably produces a fine, negatively charged spray, leaving the larger droplets positively charged. Within a cloud, there is a separation of large masses of positively and negatively charged water droplets.

As a charged thunderhead approaches, trees, buildings, and other objects on the ground become oppositely charged by induction. A corona discharge, popularly known as St. Elmo's fire, sometimes issues from a church steeple or an airplane wing during such times. The electric charge within a cloud is much greater than a charge on the ground, and lightning discharges occur between cloud and earth. Many more discharges occur within a cloud, however, than between a cloud and the earth. Such discharges are often concealed by the cloud and account for the so-called sheet or heat lightning visible in the distance on dark nights. Tall structures such as skyscrapers, chimneys, and bridges are struck scores of times, though usually without damage.

11–6 THE ELECTRIC BATTERY

Aside from the amber phenomenon, lightning, and St. Elmo's fire, natural electricity has long been known to be associated with some animals. Certain fishes of the Nile and other waters can deliver paralyzing electric shocks. In South American waters, there is an eel that shocks handlers when the top of the head and the underside of its body are touched with both hands. In Senegal there is an electric catfish, and in the Mediterranean the electric ray, or torpedo.

Luigi Galvani (1737–1798), a physician and professor of anatomy whose interests lay in the actions of nerves and muscles, studied the effects of electric discharges on the nervous system of the frog. He found that the amputated hind legs of a frog would kick convulsively if they were made part of an electric circuit. Galvani established that pairs of different metals caused the legs to twitch. He found, for example, that a fork with one iron and one copper tong caused the leg to contract each time contact was made with the nerve and muscle. Galvani explained the effect by assuming that electricity was a special animal property; that animal tissue contained a vital force, which he named "animal electricity"; and that to produce the currents, living tissue must touch the metals.

Galvani sent a copy of his results to his physicist friend Alessandro Volta (1745–1827), a pioneer in electricity. Volta repeated Galvani's experiments and constructed the most sensitive electroscope of his time. He became convinced that the contraction of the frog's legs was an inorganic phenomenon and that he could produce an electric current with two wires of different metals and a salt-water solution. In honor of his friend, Volta called the phenomenon **galvanism**. He proposed that the electricity came from the two dissimilar metals, not from the frog.

Volta proved his point by inventing the first electric battery, or "voltaic pile," consisting of a series of alternating copper and zinc disks separated by pads moistened with salt water (Fig. 11–13). A stack with 20 or so pairs produced a vigorous deflection of the leaves of his electroscope, as well as other effects. A wire, for example, connected across the opposite plates became hot, as did the plates themselves. With the development of the battery, electricity became available for study in the laboratory. One of the many discoveries that followed was the electrolysis of water.

Galvani and Volta were both partly right in their views of animal and metallic electricity. Electrical forces do arise at the junctions of dissimilar metals and a solution, and living cells do produce electricity (Fig. 11–14). Michael Faraday (1791–1867), a chemist and physicist, proved that the electricity from electric fishes and voltaic piles is the same. Many examples of animal electricity have been measured, such as that produced by the heart and brain. The

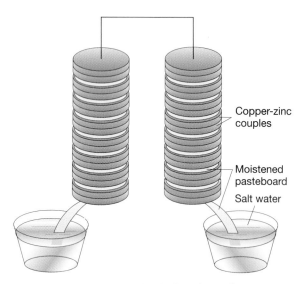

Figure 11–13   Voltaic pile, the first electric battery, consisted of disks of two dissimilar metals and absorbent material immersed in salt water.

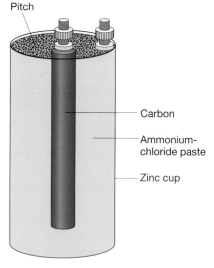

Figure 11–15   A "dry" cell generates electricity from the chemical action of an electrolyte and two dissimilar electrodes. A battery consists of a group of cells.

electrocardiograph and the electroencephalograph have become powerful tools in medical diagnosis and research.

The ordinary **"dry" cell** used in flashlights is the best-known type of voltaic cell. It is not really dry (Fig. 11–15). The cell consists of two dissimilar materials, called **electrodes**, and a conducting solution, the **electrolyte**. The electrodes can be zinc in the form of a cylindrical can and a solid carbon rod in the center of the cell but not touching the cylinder. In place of the salt water, a moist paste of ammonium chloride and zinc chloride often serves as the electrolyte. It is embedded in a mixture of powdered carbon, sawdust, and manganese dioxide. The cell is sealed at the top

with pitch or sealing wax, to prevent evaporation of the moisture. When the cell is operating, a chemical reaction between the zinc and ammonium chloride converts chemical energy into electrical energy. Some metallic zinc dissolves while the ammonium chloride solution decomposes, and an excess of electrons accumulates on the zinc. If a wire connects the zinc and carbon terminals, an electric current flows through this external circuit and operates a lamp or other devices in the circuit (Fig. 11–16). Many combinations of chemical substances are used in the design of cells for different purposes. A group of connected cells is called a **battery** (Fig. 11–17).

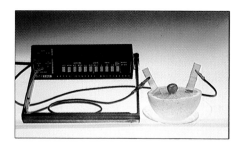

Figure 11–14   A cell consisting of zinc and copper rods inserted into a grapefruit is a modern version of Volta's "pile." (Charles D. Winters)

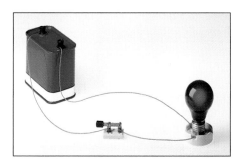

Figure 11–16   An electric current flowing through this current operates a light bulb when the switch is closed. (Richard Megna/Fundamental Photographs)

Figure 11–17   Photograph of an assortment of batteries.
(Courtesy of Henry Leap and Jim Lehman)

11–7 ELECTROSTATIC GENERATORS

Benjamin Franklin and his predecessors had no other way to produce electric charges than by rubbing dissimilar materials. For the continuous generation of charge, a circular glass plate with a pad of silk pressing against it was mounted so that it could be rotated. Later, some **electrostatic generators** were constructed that depended on the induction principle. The **electrophorus**, invented by Volta, generates large electrostatic charges in this manner.

The electrophorus consists of a flat disk made of sealing wax, hard rubber, or some other insulator and a metal plate with an insulating handle (Fig. 11–18). When the wax is rubbed with fur or wool, it acquires a negative charge. Placing the metal plate in contact with the charged disk causes electrons in the plate to be repelled to the upper surface by electrons in the disk, leaving the lower surface of the plate positively charged. The plate is then grounded by touching with a finger or grounded wire; electrons escape from the plate to the ground, leaving the plate with a net positive charge. When the plate is lifted from the disk, it becomes a source of strong positive charge that may be used for a variety of purposes, such as producing sparks or discharging a neon tube. The process may be repeated many times. During this process, the mechanical energy of rubbing the disk is converted into electrical energy.

One type of electrostatic machine is based on the principle that when an electric charge is introduced into the interior of a hollow metal sphere, the charge becomes distributed along the outer surface of the sphere. The reason is that the mutual repulsion between electrons drives them as far away as possible. Called the **Van de Graaff generator**, after the inventor–physicist Robert J. Van de Graaff (1902–1967), it consists of a hollow metal sphere mounted on an insulated hollow cylinder, the cylinder containing a running belt that is continuously charged at the bottom and discharged after entering the sphere (Fig. 11–19). The transferred positive charge accumulates

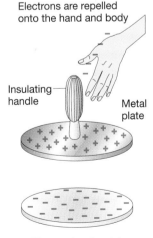

Figure 11–18   The electrophorus is a means of generating electrostatic charge.

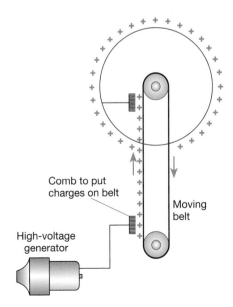

Figure 11–19   Principle of a Van de Graaff generator. Charges from a small generator are continuously carried upward by the moving belt and transferred to the metal sphere at the top of the insulated column.

Figure 11–20   Standing on a pedestal insulated from ground, the young lady is holding on to the sphere of a Van de Graaff generator. Notice her hair standing on end. (Courtesy of Henry Leap and Jim Lehman)

on the outside of the sphere (Fig. 11–20). Van de Graaff electrostatic generators are adapted to such varied uses as cancer therapy, industrial radiography, and research with electron and X-ray beams.

11–8 ELECTROSTATICS TODAY

The scope of modern electrostatics goes far beyond the literal definition of electric charges at rest. A main application of electrostatics has been in the prevention of air pollution by the electrostatic precipitation of particulate industrial wastes, including fly ash, dust, and fumes. Although many communities continue to produce a great deal of water pollution by dumping sewage into our streams and rivers, electrostatic precipitators can keep the air clean and thus

make life livable near cement mills and plants that process mineral ores (Fig. 11–21). Evidence of their effectiveness in keeping our canopy of air breathable is the fact that, in the United States alone, electrostatic precipitators remove fly ash from coal-burning power plants at a rate of approximately 20 million tons per year.

Frederick G. Cottrell (1877–1948), a physical chemist, developed the first effective **electrostatic precipitator** in 1905. It is based on the principle that unlike charges attract each other. A grounded duct, the passive electrode, carries the flue or other gas loaded with waste particles. Gas ions (charged particles such as oxygen molecules that have captured an electron) are produced around a central wire, the active electrode (Fig. 11–22). The ions in turn charge the waste particles to which they may become at-

Figure 11–21 Electrostatic precipitators prevent air pollution by removing particulates from a smokestack. These pictures indicate how effective they can be in operation. (Courtesy of Eastman Kodak Company)

tached, and the composites move across the gas stream and collect on the walls of the duct. Periodically, the duct is rapped to shake the residue into a hopper. In practice, an industrial precipitator consists

High-voltage wire
(30,000 volts or more)

Clean
air out

Dirty
air in

Dust particles
out

Electric ground

Figure 11–22 Simplified diagram of an electrostatic precipitator. An industrial precipitator consists of a large assembly of units.

of an assembly of such units. Efficiencies can exceed 99% by weight. If the particles are liquid, as they are in most fumes, the residue runs down the duct walls.

Perhaps the best-known application of modern electrostatics is **xerography**, the dry-copy imaging process invented by Chester Carlson (1906–1968) in the 1930s and developed into automated duplicating machines by the Xerox Corporation in the 1940s. In a typical copying machine, a rotating drum coated with a semiconductor such as selenium is charged in the dark (Fig. 11–23). Then an optical system focuses an image of the page to be copied on the drum. The semiconductor has a **photoelectric property**—it releases electrons when light strikes it; the light removes all the charge except where the images of black areas appear. These images then attract a black dust called the toner. Precharged paper is fed in, makes contact with the drum, attracts the toner from the drum, and moves through a heating stage that fuses the toner to the paper to make a permanent copy.

Electrostatic effects can be annoying and dangerous as well as useful. It is impossible to move sugar, flour, or any similar dry powder through ducts without charging the powder. Belts running over pulleys often generate sparks. If this occurs with flammable materials, such as paper or plastic films, fires and explosions can be set off. Excess charge can some-

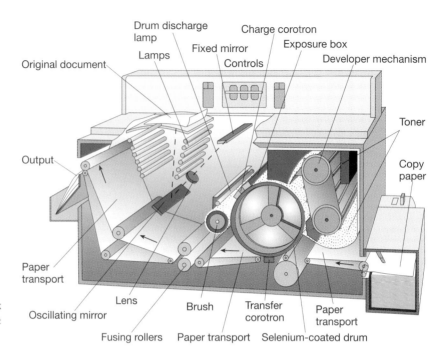

Figure 11–23 Typical Xerox dry-copying machine. (Copyright ©1965 by Ziff-Davis Publishing Co.)

times be removed by simply raising the humidity, thereby increasing the conductivity of the air. Airplanes can become highly charged flying through dust, sleet, or highly charged clouds, and their communications and control systems can be damaged. Solutions to this problem have yet to be found.

11–9 MAGNETS

Just as amber was involved in so many of the early experiments in electricity, so has the **lodestone** (*lode*, meaning to lead or attract) been associated with magnetism. One story tells of a shepherd named Magnes who on the slopes of Mt. Ida on the island of Crete found his iron-tipped crook attracted to the ground. Digging to find why, he discovered the magnetic stones called lodestones.

Lodestones are now recognized as a fairly common magnetic iron ore called **magnetite** and are found in many parts of the world (Fig. 11–24). Their property of attracting iron was noted by various civilizations early in their Iron Age. Until 1820, the lodestone was the only source of magnetism available.

A stone possessing such an almost magical property became prime material for spinners of yarns, so it

was sometimes difficult to separate fact from fancy. The belief arose that there were mountains of lodestones that attracted or repelled iron. The mountains were then placed on a magnetic island that attracted ships put together with iron nails and held them

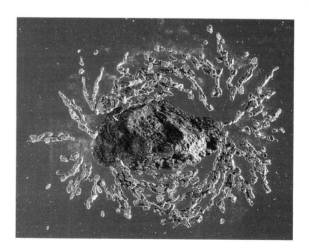

Figure 11–24 The magnetic field of a lodestone, a natural magnet consisting of the iron ore magnetite. (Fundamental Photographs)

Figure 11-25 A magnetized needle, supported so that it can swing freely horizontally, turns so that it is aligned with the earth's magnetic field. One end then points north.

there. Magnets were also said to pull nails out of ships, and Archimedes (287–212 BC), a mathematician and inventor, is supposed to have sunk enemy ships in this way.

Iron becomes magnetic when rubbed with lodestone. Both lodestone and iron rubbed in this way

turn to the north–south orientation when suspended and allowed to swing freely horizontally (Fig. 11–25). The tendency of a magnet to orient itself in this way is also exhibited when it is fastened to a piece of wood and floated on water. Put to navigational use in the mariner's compass, this property enabled seafarers to venture confidently far out to sea.

A magnet dipped into a pile of iron filings will collect the filings around regions called **magnetic poles**. Poles come in pairs, north and south. **Like poles repel, and unlike poles attract** (Fig. 11–26). Whenever a magnet is cut, opposite poles appear on either side of the cut and each segment has two dissimilar poles (Fig. 11–27). Sprinkling iron filings on a piece of paper or glass placed over a magnet causes the filings to arrange themselves in curved paths (Fig. 11–28).

Magnetism seemed at first to be limited to lodestone and iron. But the elements cobalt and nickel also exhibit magnetic attraction, and certain mixtures of elements (alloys) are magnetic. Even nonmagnetic elements, such as copper, tin, and manganese, are present in a group of magnetic alloys known as Heusler alloys.

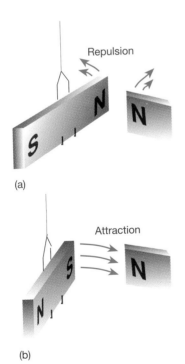

Figure 11-26 Like magnetic poles (a) repel; unlike poles (b) attract.

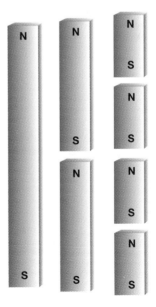

Figure 11-27 Cutting a magnet produces new magnets with a pair of north and south poles at each break. The poles do not become isolated.

Figure 11–28 Iron-filing pattern produced by a bar magnet. (Richard Megna/Fundamental Photographs)

Permanent magnets are made in many shapes: bar, circular, U, V, and the familiar horseshoe (Fig. 11–29). A rod of magnetic material can be magnetized with a permanent magnet by stroking the rod from one end to the other with one pole of the magnet, always starting at the same end. Another method is to hammer the rod while it is aligned in a north–south direction. Water pipes and steel girders in buildings become magnetized merely by standing in the earth's magnetic field. You can verify this fact with a magnetic compass. A magnetic substance can also become magnetized by heating, then cooling, in the presence of a magnetic field or by being placed in a coil or wire through which a current is passed.

11–10 THE EARTH AS A MAGNET

The end of a magnetic needle that points north is called a north pole, and the opposite end is called a south pole. Since a north pole is attracted by a south pole, there must be a south (magnetic) pole in the northern hemisphere of the earth. This south magnetic pole of the earth and the north magnetic pole (in the southern hemisphere) form a **magnetic axis** that does not coincide with the **geographical axis**. Thus, except in very restricted parts of the earth, compasses do not point to true geographical north (Fig. 11–30). At present, the magnetic pole in the northern hemisphere lies considerably below the surface of the earth at a point above the latitude of 70° in the islands to the north of Canada; the magnetic pole in the southern hemisphere is near the Ross Sea in Antarctica.

Since the magnetic poles and the geographical poles do not coincide (they are separated by about 1000 miles), a compass needle in most places points to

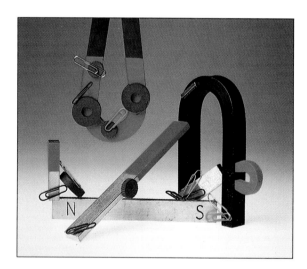

Figure 11–29 An assortment of magnets. Many are made of iron, nickel, and cobalt alloy steels. (Richard Megna/Fundamental Photographs)

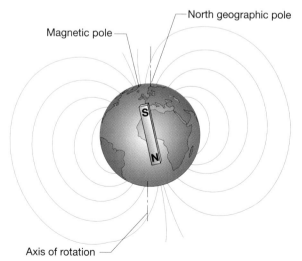

Figure 11–30 The earth's magnetism is like that of a bar magnet at the center of the earth, with the south pole near the north geographic pole. The magnetic poles are more than 1000 miles from the geographic poles.

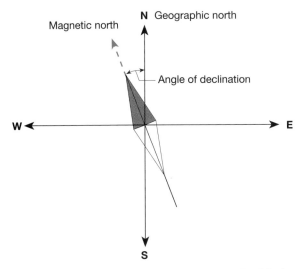

Figure 11–31 Magnetic declination. The angle of declination is measured in degrees east or west of geographic north. It is 11°50′W for New York City, 2°56′E for Chicago.

the east or west of true north. (There is a story that Columbus, by shifting the compass card at night, averted a mutiny of his sailors when they saw the compass needle deviate 10 degrees from true north. Another tale is that he persuaded the sailors that the North Star had moved, not the compass.) The angle between magnetic north and geographical north for

any locality on the earth's surface is called the **angle of declination** for that locality (Fig. 11–31).

Through surveys of the magnetic conditions of the earth, carried out over the past several hundred years, it has been learned that the angle of declination at a given locality is not constant but changes with time, in some places several degrees in a century. Since 1600, the magnetic declination in London has varied from 11°E to 24°W. On the extreme east and west coasts of the United States, the declination may be 15° or more. For Boston, the declination currently is about 15°W; for Los Angeles, 15°E (Fig. 11–32).

A magnetized needle not only points approximately north and south, but if suspended at its center so as to be free to swing in a vertical north–south plane, it sets itself at an angle (Fig. 11–33). At the earth's magnetic poles, the needle dips straight down; the **angle of dip** or inclination is 90° at those two locations. At points along the magnetic equator the angle of dip is 0°, and elsewhere it is between 0° and 90° measured from the horizontal. Magnetic surveys have shown that the angle of dip for a given locality is not constant but may change as much as several degrees in a century. The angle of dip is currently about 72° for New York City and 55° for Los Angeles.

Sir William Gilbert shaped a lodestone in the form of a sphere, called a **terrella** ("little earth"), to simulate the earth and found that a magnetized needle placed on the surface acts in the same way as a compass needle does at different places on the earth's

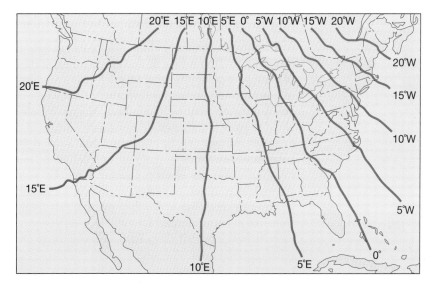

Figure 11–32 A map of the United States showing the declination of a compass from true north. (Raymond A. Serway and Jerry S. Faughn, *College Physics,* 3rd edition. Philadelphia: Saunders College Publishing, 1992, Fig. 20.3, p. 630.)

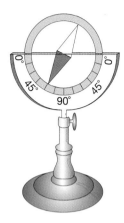

Figure 11–33 A magnetic dipping needle determines the inclination, or angle of dip.

surface. He demonstrated inclination with his terrellas and tiny magnets and predicted that the angle of dip would increase as one traveled north, a prediction soon verified by explorers (Fig. 11–34). From this he deduced that the earth itself is a giant lodestone with a covering of rocks, soil, and water. Terrestrial magnetism, according to this view, arises within the earth and is not caused by the sun, the North Star, or any other source outside the earth.

Magnetized needle

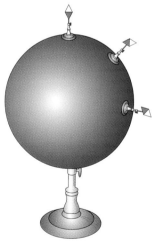

Figure 11–34 A magnetized needle placed on the surface of Gilbert's "terella" behaves like a compass needle on the earth's surface.

This view of terrestrial magnetism was novel at the time, but it has since been amply substantiated. Karl Friedrich Gauss (1777–1855), a mathematician and astronomer, derived a mathematical expression for the strength and shape of the earth's magnetic field that closely fit the observed values. If the earth's magnetism were caused by some external agency, the field would have a very different shape from the one it has.

11–11 MAGNETIC POLES AND FORCES

Magnetic poles behave differently from electric charges. Magnetic poles always occur in pairs, while electric charges can be separated. In theory there is no reason for **monopoles** not to exist, but a search has thus far failed to reveal them. In a long, thin magnet, the north and south poles may be well separated from each other at opposite ends of the magnet, although they can never be completely separated and isolated.

Using the same torsion balance that he used to study electrostatic forces. Coulomb suspended such a magnet from the thread and placed a similar magnet near it. He proved that the law that applies to electrostatic forces also holds for magnetic forces and is also analogous to Newton's law of gravitational force—all three are inverse-square laws. According to **Coulomb's law** of force between magnetic poles:

> **Like poles repel, unlike poles attract, and the force between magnetic poles varies directly with the pole strengths and inversely with the square of the distance between them.**

Despite the similarity of the laws of magnetic and electric force, the magnetic and electrical properties of matter are quite different. Magnetism is a property of iron and only a relatively few other substances, but *any* substance may be electrified. Magnetic poles come in pairs that cannot be separated; positive and negative charges can be separated and isolated. A substance may retain its magnetism indefinitely, but it rapidly loses its electric charge. Magnetic pole strength is confined to certain regions in a magnet, whereas electric charge can flow more or less freely from one place to another. Despite these differences, there is evidence that magnetism and electricity are related phenomena.

11–12 MAGNETIC EFFECT OF A CURRENT

Having heard of Volta's work with electricity, Hans Christian Oersted (1777–1851), a professor at the University of Copenhagen, constructed an electric pile of his own. At the time, some felt that there was a connection between electricity and magnetism. In a classic case of **serendipity**, Oersted discovered a missing link between the two.

During a lecture in his course on electricity and magnetism, Oersted placed his voltaic pile on the lecture table. There was a compass needle nearby. Upon connecting a platinum wire to the battery, thus causing a current to flow, and holding the wire above the compass needle, Oersted observed the compass needle swing around. Instead of orienting itself in the north–south direction, the needle, to his surprise, moved and came to rest in a direction perpendicular to the wire (Fig. 11–35).

Oersted followed up the experiment after class to make certain that he had really found something. When he placed the compass needle above the wire, the needle moved in the opposite direction. Reversing the direction of the current 180 degrees caused the compass needle to turn 180 degrees. Oersted was then certain that he had discovered a connection between electricity and magnetism: Electric current flowing through a wire exerted a force on a magnet—the compass needle—and the direction in which the compass needle was oriented depended on the direction of current in the wire (Fig. 11–36). By Newton's

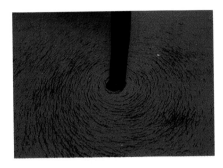

Figure 11–36 Circular magnetic pattern of iron filings surrounding a wire conducting an electric current. (Courtesy of Henry Leap and Jim Lehman)

third law, the magnet should exert an equal and opposite force on the wire carrying the current, and Oersted discovered this effect as well.

Oersted was surprised by the fact that in this interaction the needle was neither attracted nor repelled directly but set itself at right angles to the direction of the current. He repeated the experiment with many variations, then wrote a report based on his discovery.

André Marie Ampère (1775–1836), a professor of physics in Paris, was among those who were inspired by Oersted's findings to investigate the interactions of electricity and magnetism. Within a few weeks, Ampère established not only that an electric current acts on a magnetic needle, but also that two electric currents interact with each other. If the currents in two parallel wires are in the same direction, the wires attract each other, and if the currents travel in opposite directions, the wires repel one another (Fig. 11–37). Ampère attributed these interactions to magnetic force.

Ampère also established that two copper coils conducting an electric current interact in the same way as two bar magnets. He coined the word **"solenoid"** ("canal") for a wire bent into the form of a helix that showed magnetic properties, since it acted to channel and thus to intensify the magnetism (Fig. 11–38). Ampère's work led him to the hypothesis that natural magnetism arises from electric currents in the atoms and molecules of magnetized bodies. In modern terms, atoms contain electrons that move rapidly in orbits around the nuclei of the atoms and spin on their own axes as well. Since the electrons are charged, each of these motions constitutes an electric current. Each atom becomes, in effect, a tiny magnet with a

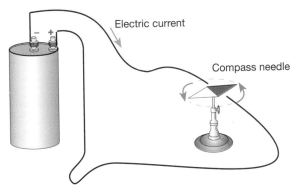

Figure 11–35 Oersted's discovery. Electric current in the wire caused the compass needle to swing around at a right angle to the wire.

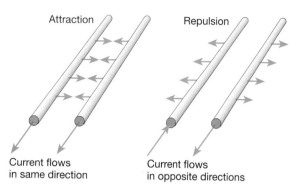

Figure 11–37 Ampère's discovery of the magnetic force of one electric current on another. (a) Parallel currents in a common direction attract each other; (b) currents having opposite directions repel.

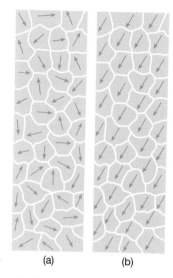

Figure 11–39 Magnetic domains. (a) In unmagnetized iron, the domains point in different directions. (b) In magnetized iron, they form a parallel array.

north and a south pole. In most materials, however, the magnetic effects of these atomic currents cancel out, because the electron orbits are oriented at random.

From experiments, we learn that the individual atomic magnets of iron act not separately but in groups called **magnetic domains**, in which whole blocks of atoms have a common orientation. The domains are small, with about a million domains per cubic millimeter, but each domain may have 10^{12} to 10^{15} atoms. In an unmagnetized sample of such material, the domains are randomly oriented (Fig. 11–39). To make the material magnetic, an external force

must overcome thermal agitation and line up the domains in the same direction.

There is visual evidence for the existence of domains in microphotographs of magnetite powder in colloidal suspension applied to the polished surface of a magnetic crystal. When observed through a powerful microscope, the powder is found to have collected along well-defined lines, the **domain boundaries**. With this technique, the process of magnetization can be monitored. When an external magnetic field is applied, there is a shift in the wall or boundary between one domain and the next. Domains with favorable orientation grow at the expense of those less favorably oriented, and the latter shrink.

11–13 FIELDS

Many scientists have felt uncomfortable with the concept of **"action-at-a-distance"** between pairs of practically isolated bodies (for example, between the sun and the earth) between which various kinds of forces are exerted and yet there is no apparent contact. It has seemed better to say that each body sets up a "field" of some sort and that each body is disturbed by the field of the other.

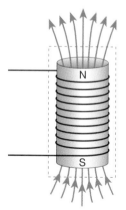

Figure 11–38 A solenoid has the properties of a bar magnet when it is carrying current.

A field is a region of space in which certain physical effects exist and can be detected. There are gravitational fields, electrostatic fields, and magnetic fields. A **gravitational field** is a region in which gravitational forces act on bodies. The gravitational field at any point is defined as the force per unit mass acting on a body at that point. If the acceleration due to gravity, **g**, is 9.80 m/s², then a force of 9.80 newtons acts on a 1-kilogram body at that point. The strength of the earth's gravitational field is, therefore, 9.80 newtons/kilogram at that point. Here we have the characteristic property of a field. A field is described by sets of numbers, each set denoting the field strength and direction at a point in space.

An **electric field** exists in any region of space in which an electric charge would experience an electric force. The strength of the electric field, or **electric intensity**, at a point is defined as the electric force **F** that would be exerted on a very small positive test charge $+q$ placed at that point. The force per unit test charge is known as the electric field intensity **E**, a vector quantity. The direction of the electric field at a point is defined as the direction of the force on a positive charge placed at that point and always points away from the positive charge.

$$\text{Electric field intensity} = \frac{\text{Force}}{\text{Quantity of charge}}$$

$$\mathbf{E} = \frac{\mathbf{F}}{q} \qquad (11-2)$$

where

$\mathbf{F}$ = force in newtons,

q = electric charge in coulombs, and

$\mathbf{E}$ = electric field intensity in newtons/coulomb

The **magnetic field** at any point in space is the force that a magnet would exert on a unit north pole placed at that point if one existed. The direction of the magnetic field is given by the direction of a compass needle placed at that point.

Michael Faraday proposed a useful approach for describing an electric field. It considers an electric charge as having associated with it a number of **lines of force** that extend outward from the charge in every direction (Fig. 11–40). **The lines of force start at a positive charge and end on a negative one.** At

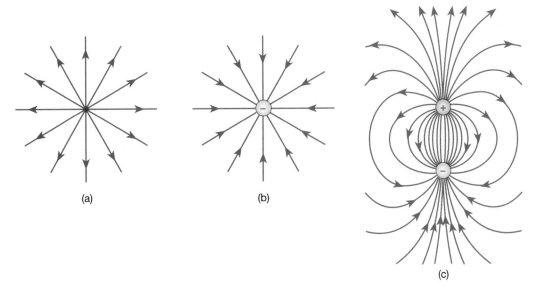

Figure 11–40 Electric field patterns. The lines of force for (a) a single-point positive charge; (b) a single-point negative charge; (c) a positive charge and a negative charge of equal magnitude.

every point these lines give the direction of the force that would be exerted on a small positive test charge if placed at that point. The arrowheads always point away from a positive charge and toward a negative charge. When the lines are closely packed, the electric force is large and the field is strong. In the same way, magnetic fields exist near magnets (Fig. 11–41). **The lines of force of a magnet originate at a north pole and enter at a south pole.** The field strength decreases as the square of the distance from a charge or pole.

The classical gravitational and electromagnetic field theories satisfactorily explain all large-scale physical phenomena. They are correct as long as they are applied to objects much bigger and heavier than a single atom, but they fail to describe the behavior of individual atoms. For the small-scale arena of physics, other theories give more accurate explanations.

11–14 ELECTRICITY FROM MAGNETISM

Although he is regarded as the father of the modern electrical age, Michael Faraday was interested in science for its own sake and not in its practical applications. He no doubt realized the practical importance of his discoveries, but he left the development of the generator and motor to others. The prime minister of Great Britain, perhaps impatient with pure science, upon seeing the wires, coils, and magnets and learning about electromagnetic induction from Faraday, remarked: "That's interesting, Mr. Faraday, but of what possible use is it?" Faraday replied: "Well, one day, Mr. Prime Minister, you will be able to tax it."

Oersted's discovery that an electric current produced a magnetic field motivated Faraday to search for the reverse effect: the production of electricity by magnetism. After several years of trying, the breakthrough came in 1831 with his discovery of **electromagnetic induction**. Faraday connected a coil, consisting of many turns of wire, to a galvanometer, an instrument for measuring electric current. When he

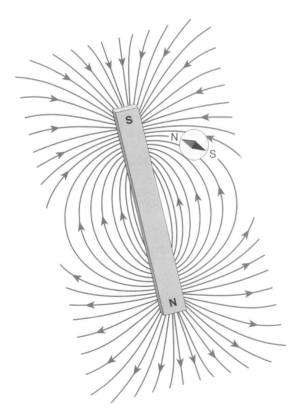

Figure 11–41   The field in the neighborhood of a magnet represented by lines of force. Arrows indicate the direction a north pole would move. The north pole of a compass points in the direction of the field.

Figure 11–42   Faraday's discovery of electromagnetic induction proved that magnetism could produce electricity. By either thrusting a magnet into a coil of wire or withdrawing it, an electric current was generated. (Richard Megna/Fundamental Photographs)

inserted a magnet into the coil, the needle of the galvanometer moved over and back momentarily, indicating that a current flowed. Then, as he withdrew the magnet, the needle moved in the opposite direction (Fig. 11–42). Faraday repeated this process many times and established that the key to the production of an electric current by magnetism was the relative motion between the coil and the magnet. The result was surprising, for in Oersted's discovery, a steady current produces a steady magnetic field.

Faraday's discovery of electromagnetic induction showed that it was feasible to produce electricity by mechanical means. Several days after making his discovery, Faraday built the first **dynamo**, or electric generator, which produced a continuous flow of electricity. A 12-inch copper disk mounted on an axle was made to rotate between the poles of a powerful magnet, and an electric current was induced in the disk (Fig. 11–43). The energy of mechanical motion is converted into electric current in the disk. Fifty years were to elapse, however, before Faraday's discovery was applied to commercial generators. The production of electricity on a large scale then became possible from such sources as the potential energy of water at high elevations and the mechanical motion produced by a steam engine.

It does not detract from Faraday's greatness in any way to point out that he was not quite the first to produce an electric current by magnetism. The same discovery was made by Joseph Henry (1799–1878) at nearly the same time. Then a teacher at Albany Academy, in New York, Henry made his discovery during a one-month summer vacation, but he did not publish

his work until a year later. By then, Faraday had published his results following an exhaustive investigation into many aspects of the subject.

11–15 SOURCE OF THE EARTH'S MAGNETISM

After centuries of research, the earth's magnetic field is one of the best described but least understood of all planetary phenomena. Its origin and existence are still unexplained, although Gilbert and Gauss did establish that it arises within the earth and not outside it. Gilbert's idea that the field might be produced by a large body of permanently magnetic material inside the earth had to be dropped, however, when it was realized that the temperature of the earth's interior is too high to allow any material to retain its magnetism. Although no theory of the earth's magnetism has yet been generally accepted as correct, the **dynamo theory** proposed by Walter M. Elsasser and Sir Edward C. Bullard accounts for many of the observed facts.

An electric current produces a magnetic field not only in a bar of iron but in a gas or other fluid as well. The earth's iron-and-nickel core is analogous to the copper disk of Faraday's dynamo and acts like a huge natural dynamo.

According to the dynamo theory, the radioactive content of the earth's core supplies heat to produce motion, or convection currents, in the liquid of the core. The convection currents cause the inner part of the core to rotate at a faster rate than the outer part. This difference in rotation can produce electric currents if a magnetic field like the earth's is already present; the electric currents then cause and maintain the earth's field. The process may have begun with a stray magnetic field early in the earth's formative period, or with small currents produced by some kind of battery-like action. Support for the dynamo theory comes from the strong magnetic fields of various stars. The stars rotate; their gaseous matter conducts electricity well; their magnetic fields change rapidly —all of these factors point to a dynamic rather than a static explanation of their fields.

Any theory of the earth's magnetism must explain the curious **reversal of its magnetic field** that has occurred from time to time in the past. When molten volcanic rocks cool and solidify, the magnetic materi-

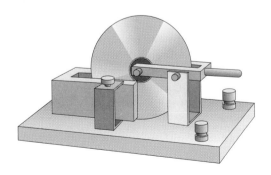

Figure 11–43 The first dynamo, or electric generator. Faraday rotated a copper disk between the poles of a magnet. A steady electric current was induced across the disk.

als in them are magnetized in the direction of the earth's magnetic field. Bernard Brunhes found some volcanic rocks that were magnetized not in the direction of the earth's present field but in exactly the opposite direction, and he concluded that the field must have reversed. It has been established that the earth's magnetic field has two stable states: It can point either toward the North Pole, as it does today, or toward the South Pole, and it has alternated between the two orientations. It is estimated that there have been at least nine reversals of the earth's field in the past 3.6 million years.

11–16 ELECTRIC CURRENT

Any motion of an electric charge constitutes an electric current. In this sense, electric currents include charged water droplets rising or falling in a cloud, the charged moving belt of a Van de Graaff generator, the migration of ions (charged particles) in an electric cell, or a lightning stroke. Usually, however, we think of an electric current as a flow of electrons in a metal wire. This is the kind of current that operates our radios, toasters, and light bulbs.

The flow of electrons in a conductor is comparable to the flow of water in a pipe. We can express the rate of flow of electric charge quantitatively, just as we do the rate of flow of water. The rate of flow of electrons is a measure of the electric current. When the rate of flow past a given point is 1 coulomb per second (1 coulomb = 6.24×10^{18} electrons), 1 **ampere** of current is flowing.

$$\text{Current} = \frac{\substack{\text{Quantity of charge} \\ \text{flowing by a point} \\ \text{in a conductor}}}{\text{Time required for the flow}} \qquad (11\text{–}3)$$

$$I = \frac{Q}{t}$$

$$1 \text{ ampere} = \frac{1 \text{ coulomb}}{1 \text{ second}}$$

EXAMPLE 11–2

What quantity of charge passes through the wire filament in a light bulb in 30.0 minutes if the current is 0.500 ampere?

SOLUTION

Since the time unit in the definition of the ampere is the second, convert 30.0 minutes to seconds. The current is given, leaving one unknown, the quantity of charge.

$$I = 0.500 \text{ ampere}$$

$$t = 30.0 \text{ min} \times 60 \, \frac{\text{s}}{\text{min}} = 1800 \text{ s}$$

$$Q = ?$$

$$I = \frac{Q}{t}$$

$$Q = It$$

$$= 0.5 \text{ ampere} \cdot 1800 \text{ s}$$

$$\boxed{900 \text{ coulombs}}$$

EXTENSION

A battery is charged for 2.00 hours with a current of 30.0 amperes. How much charge passes through it? (Answer: 2.16×10^5 coulombs)

■

Just as a force must be exerted on water to move it through a pipe, so a force must be exerted on the electrons to cause them to flow through a conductor. This means that work is done on the electrons. The driving force behind the flow of electrons is the difference in electrical condition between one point and another in a conductor. Electrons flow from a region of high potential, where they have more energy per unit charge, to a region of low potential, where they have less. (Compare this with water, which flows from a region of high pressure to one of low pressure.) The electrical potential, or **voltage, is defined as the work done in moving a charge between two points, divided by the quantity of charge and is measured in joules/coulomb, or volts.** An automobile battery supplies either 6 or 12 volts, and an electric power company typically provides a potential difference of 110 volts in an electrical outlet.

$$\text{Electric potential} = \frac{\text{Work}}{\text{Charge}} \qquad (11\text{–}4)$$

$$V = \frac{W}{Q}$$

$$1 \text{ volt} = \frac{1 \text{ joule}}{1 \text{ coulomb}}$$

EXAMPLE 11-3

An electron is moved from one terminal of an accelerator to another. The two terminals are separated by a potential difference of 3.00 million volts. How much work is expended?

SOLUTION

The charge on an electron is 1.60×10^{-19} coulomb. Solve for the work done on the electron.

$$V = 3.00 \times 10^6 \text{ volts}$$
$$Q = 1.60 \times 10^{-19} \text{ coulomb}$$
$$W = ?$$
$$V = \frac{W}{Q}$$
$$\therefore W = V \cdot Q$$
$$= 3.00 \times 10^6 \text{ volts} \cdot 1.60 \times 10^{-19} \text{ coulomb}$$
$$= \boxed{4.80 \times 10^{-13} \text{ joule}}$$

EXTENSION

A lightning flash transfers 15.0 coulombs of charge to earth from a potential difference of 6.00×10^7 volts. How much energy is released? (Answer: 9.00×10^8 joules)

■

Georg Simon Ohm (1787–1854), a physicist and mathematician, studied electric currents in circuits. With voltaic piles and a galvanometer to measure the strength of the electric current, he conducted experiments with wires made of different metals and of different lengths and thicknesses. He found that the current was greater for short wires than for long wires and greater for thick wires than for thin wires. The current also depended on the nature of the material in the wire and the potential difference between the ends of the wire. He found a relationship that is called **Ohm's law**:

At a given temperature the current through a conductor is directly proportional to the potential difference between the ends of the conductor and inversely proportional to a property of the wire that is called its electrical resistance.

Ohm's law is expressed in equation form as follows:

$$\text{Current} = \frac{\text{Potential difference}}{\text{Resistance}} \quad (11\text{--}5)$$

$$I = \frac{V}{R}$$
$$1 \text{ ampere} = \frac{1 \text{ volt}}{1 \text{ ohm}}$$

The unit of resistance, the **ohm**, is defined as the resistance of a conductor such that when the current through it is exactly 1 ampere, the potential difference between its ends is exactly 1 volt.

EXAMPLE 11-4

A current of 7.50 amperes flows through an air conditioner. If the voltage in the wall outlet is 110 volts, what is the resistance of the air conditioner?

SOLUTION

Knowing the current I and the voltage V, we use Ohm's law to determine the resistance R.

$$I = 7.50 \text{ amperes}$$
$$V = 110 \text{ volts}$$
$$R = ?$$

$$I = \frac{V}{R}$$
$$\therefore R = \frac{V}{I}$$
$$R = \frac{110 \text{ volts}}{7.50 \text{ amperes}}$$
$$= \boxed{14.7 \text{ ohms}}$$

EXTENSION

The resistance of the filament in a bulb is 5.00 ohms. If the bulb is connected to two cells with a potential difference of 3.00 volts, how much current flows through the bulb? (Answer: 0.600 ampere)

■

Every conductor opposes the flow of current through it to an extent that depends on the length, thickness, material, and temperature of the wire. The moving electrons undergo collisions with the vibrating atoms in the conductor. Doubling the length of the wire doubles the number of collisions and halves the current for a given potential difference. When the temperature is raised, the atoms vibrate with a greater amplitude and make a greater number of collisions with the "current-carrying" electrons, thereby impeding the forward motion of the electrons. Therefore, the current is reduced in most substances. There are many situations in which high-resistance wires are desired since the heat generated from collisions between the electrons and atoms can be put to practical use. Such is the case with electric heaters, stoves, toasters, and flatirons; for these purposes, a wire made of the alloy nichrome is commonly used.

11–17 ELECTRIC CIRCUITS

When the same current flows through several conductors, the conductors are said to be connected in **series** (Fig. 11–44). The resistance of the combination is the sum of the individual resistances, since each device in the circuit resists the flow of current (Fig. 11–45).

$$R_{total} = R_1 + R_2 + R_3 + \cdots \qquad (11–6)$$

EXAMPLE 11–5

Three coils with resistances of 5.00, 10.0, and 15.0 ohms are connected in series. If a voltage of 120 volts is maintained across the group, how much current will flow?

SOLUTION
We find the total resistance first, then apply Ohm's law to determine the current.

Figure 11–44   A series circuit consisting of a 12-volt battery and two light bulbs. The same current flows through each bulb. (Richard Megna/Fundamental Photographs)

$$R_1 = 5.00 \text{ ohms}$$
$$R_2 = 10.0 \text{ ohms}$$
$$R_3 = 15.0 \text{ ohms}$$
$$V = 120 \text{ volts}$$
$$R_{total} = ?$$
$$I = ?$$
$$R_{total} = R_1 + R_2 + R_3$$
$$= 5.00 \text{ ohms} + 10.0 \text{ ohms} + 15.0 \text{ ohms}$$
$$= 30.0 \text{ ohms}$$
$$I = \frac{V}{R}$$
$$= \frac{120 \text{ volts}}{30.0 \text{ ohms}}$$
$$= \boxed{4.00 \text{ amperes}}$$

EXTENSION
Five 100-ohm bulbs are connected in series. What is their total resistance? (Answer: 500 ohms)

■

In a divided or **parallel circuit**, more than one path is provided for the current, which divides and passes through each resistance independently (Fig. 11–46). Each additional path makes it easier for current to flow, just as a multiple-lane highway makes it easier for traffic to flow (Fig. 11–47). The total resistance in a parallel circuit is less than any single resistance and is given by the expression

$$\frac{1}{R_{total}} = \frac{1}{R_1} + \frac{1}{R_2} + \frac{1}{R_3} + \cdots \qquad (11–7)$$

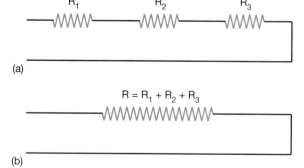

(a)

(b)

Figure 11–45   (a) Resistors connected in series. (b) An equivalent circuit consists of one resistor equal to the sum of the three.

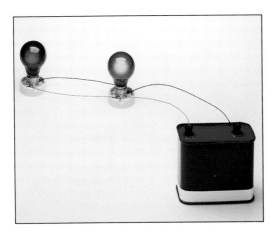

Figure 11–46 A 12-volt battery and two light bulbs connected in parallel. The current divides and passes through each bulb independently. (Richard Megna/Fundamental Photographs)

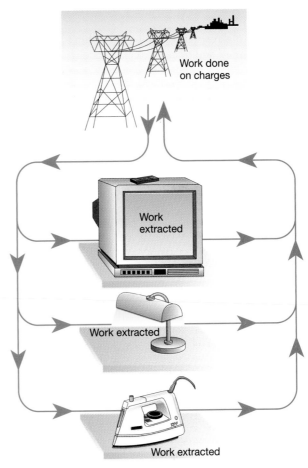

Figure 11–48 In a parallel circuit, each appliance provides a separate path for the flow of electric current. The combined resistance is less than any individual resistance in the circuit.

Consider what would happen if light bulbs, toasters, electric stoves, and the many other appliances in a home were connected in series. The combined resistance would be so great that there would probably not be current enough to operate the appliance that had the lowest resistance. Moreover, in a series circuit, to operate any one appliance, all would have to be used; otherwise, the circuit would be opened and current could not flow. In a parallel circuit, on the other hand, each unit is connected separately from the others to the two main supply lines by two wires and can be operated without using any of the other appliances (Fig. 11–48). The connections in wiring a home, therefore, are parallel circuits, each circuit being connected separately between the main supply lines.

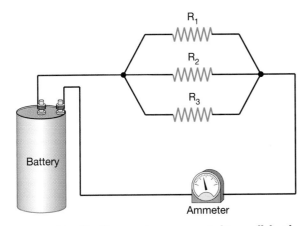

Figure 11–47 Three resistors connected in parallel with each other and then in series with a battery and ammeter.

EXAMPLE 11–6

Assume that the three coils in Example 11–5 are connected in parallel and that there is a potential difference of 120 volts across their common terminals. (a) What is the combined resistance of the three coils? (b) What is the current in the overall circuit? (c) What is the current through each coil?

SOLUTION

We find total resistance. Then apply Ohm's law to the overall circuit and to each coil separately.

(a)
$$\frac{1}{R_{total}} = \frac{1}{R_1} + \frac{1}{R_2} + \frac{1}{R_3}$$

$$\frac{1}{R_{total}} = \frac{1}{5.00 \text{ ohms}} + \frac{1}{10.0 \text{ ohms}} + \frac{1}{15.0 \text{ ohms}}$$

$$\frac{1}{R_{total}} = \frac{6 + 3 + 2}{30.0 \text{ ohms}} = \frac{11.0}{30.0 \text{ ohms}}$$

$$11.0 \, R = 30.0 \text{ ohms}$$

$$R = \frac{30.0 \text{ ohms}}{11.0} = \boxed{2.72 \text{ ohms}}$$

(b)
$$I = \frac{V}{R}$$

$$= \frac{120 \text{ volts}}{2.72 \text{ ohms}}$$

$$= \boxed{44.0 \text{ amperes}}$$

(c)
$$I_1 = \frac{V}{R_1} \qquad\qquad I_2 = \frac{V}{R_2}$$

$$= \frac{120 \text{ volts}}{5.00 \text{ ohms}} \qquad = \frac{120 \text{ volts}}{10.0 \text{ ohms}}$$

$$= \boxed{24.0 \text{ amperes}} \qquad = \boxed{12.0 \text{ amperes}}$$

$$I_3 = \frac{V}{R_3}$$

$$= \frac{120 \text{ volts}}{15.0 \text{ ohms}}$$

$$= \boxed{8.00 \text{ amperes}}$$

EXTENSION

Calculate the equivalent resistance of five 100-ohm bulbs connected in parallel. (Answer: 20.0 ohms)

■

If the current passing through home wiring reaches a level too high for these wires to carry, the wires may overheat to the point of causing a fire. That is why fuses are connected in series with the supply line. A **fuse** consists of a strip of metal that has a low melting point, housed in a porcelain or glass receptacle (Fig. 11–49). When the current reaches a predetermined value, say 20 amperes, the wire melts; the

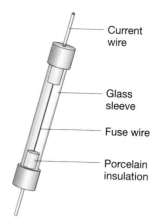

Figure 11–49 A fuse. The low melting point of the metallic element causes it to melt and break the circuit if the current should become excessive.

circuit is thus opened, and current stops flowing. The fuse in this way protects the circuit from an electrical fire. Most houses have several circuits, each individually fused. In new homes, a **circuit breaker** is generally used instead of a fuse to accomplish the same thing (Fig. 11–50). When the overload is removed, the circuit breaker can be reset by pressing a lever, whereas a fuse must be replaced.

Figure 11–50 A bank of circuit breakers protects household circuits from drawing too much current and starting a fire. A breaker is reset by pressing a lever. (Paul Silverman/Fundamental Photographs)

11–18 ELECTRIC POWER AND ENERGY

The production of heat is one of the best-known effects of electric current. It is the basis of such appliances as the toaster, hair dryer, steam iron, and electric blanket (Figs. 11–51, 11–52). In a power transmission line, however, heat represents a waste of electrical energy, and in a faulty section of wire in the home, it is a potential source of fire.

In Joule's experiment involving the conversion of electrical energy into heat (discussed in Chapter 8), the heating element was a coil of wire immersed in a known mass of water in a calorimeter. A current I was maintained by a known potential difference V for a time t; the **electrical energy *VIt*** was measured and compared to the heat energy absorbed by the water and calorimeter. Joule found in experiments of this kind that the rate at which heat was developed, or power delivered to a conductor, is proportional to the square of the current flowing through the conductor. Known as **Joule's law**, this relationship is expressed as follows:

$$\text{Power} = (\text{Current})^2 \times \text{Resistance} \qquad (11\text{–}8)$$

$$= \text{Voltage} \times \text{Current}$$

$$P = I^2R = VI$$

$$1 \text{ watt} = 1 \text{ volt} \times 1 \text{ ampere}$$

In SI units, when V is in volts and I is in amperes, P is in joules/s or watts (W). (In mechanical units, recall that $1 \text{ W} = 1 \text{ J/s}$.) Table 11–3 lists the power re-

Figure 11–52   A steam iron uses the heating effect of an electric current. Water is turned to steam by heat from the heating coil. The steam is sprayed from jets in the bottom of the iron.

quirements of some common appliances and power tools.

A **watt-meter** is a device that measures V and I simultaneously and automatically performs the multiplication to get P. **Watt-hour-meters** in house electrical connections go one step beyond and accumulate readings of power for an elapsed time (Fig. 11–53). These values correspond to the **total electrical energy** consumed. This is the quantity—energy—that is reflected in the electric bill. Electricity is not consumed in a circuit; we pay the utility company for the energy of the electric current we use

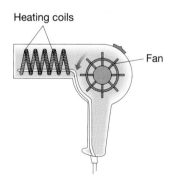

Figure 11–51   In a hair dryer, a fan blows air past the heating coils. The warm air can be used to dry hair. The same principle is used to dry clothes and to heat buildings.

Figure 11–53   A watt-hour meter. (Courtesy of Westinghouse Electric Corporation)

Table 11–3 POWER REQUIREMENTS OF SOME
COMMON APPLIANCES AND POWER TOOLS*

Appliance	Power Requirement	Tool	Power Requirement
Air conditioner			
Room	1500 W	Arc welder, portable	68,000 W
Central	4500 W	Belt sander	600 W
Blanket, electric	180 W	Bench grinder	650 W
Blender	800 W	Drill press	1250 W
Coffee maker,		Motor, ½ hp	500 W
automatic	1625 W	Saw	
Cooker, slow	300 W	Circular, portable	550 W
Dishwasher	1200 W	Radial-arm	1325 W
Food processor	330 W	Table	1500 W
Heater			
Portable	1400 W		
Water	4500 W		
Iron	1100 W		
Microwave oven	625 W		
Refrigerator			
Regular	400 W		
Frost-free	500 W		
Stove			
Range-top	6000 W		
Oven	4500 W		
Television			
Black and white	50 W		
Color	100 W		
Toaster	950 W		

*From Wilson, Jerry D., *Technical College Physics,* 2nd edition. Philadelphia: Saunders College Publishing, 1987, p. 398, Table 21.2.

that is forced through appliances yielding heat, light, and so on (Table 11–4). The filament of an ordinary light bulb is designed to be heated to incandescence at 2500 to 3000°C by the conversion of electrical energy into heat and light (Fig. 11–54). The process is inefficient, however. More than 95% of the electrical energy is lost in the forms of nonvisible radiation and thermal energy, an example of an unwanted heating effect.

EXAMPLE 11–7

An electric light bulb is marked 60 watts, 120 volts. (a) What current flows through the bulb? (b) What is the resistance of the filament?

SOLUTION

We calculate the current from Equation 11–8, then apply Ohm's law to obtain the resistance.

Figure 11–54 A carbon-filament incandescent lamp converts electrical energy into heat and light. (Courtesy of CENCO)

Table 11–4 AVERAGE HOUSEHOLD APPLIANCE USAGE AND COSTS

	Estimated Kilowatt-Hour Usage in 2 Months	Estimated Cost For 2-Months' Usage (in Dollars)
Food Preparation		
Blender	2.5	0.09
Broiler	16.7	0.60
Carving knife	1.3	0.05
Coffee maker	17.7	0.60
Deep fryer	13.8	0.50
Dishwasher	60.5	2.18
Frying pan	31.0	1.12
Hot plate	15.0	0.54
Mixer	2.2	0.08
Oven, self-cleaning	191.0	6.88
Range	195.8	7.05
Roaster	34.2	1.23
Toaster	5.5	0.20
Waste disposer	5.0	0.18
Food Preservation		
15-cu-ft freezer	199.2	7.17
frostless	293.5	10.57
12-cu-ft refrigerator	121.3	4.37
frostless	202.8	7.30
14-cu-ft refrigerator/freezer	189.5	6.82
frostless	304.8	10.97
Laundry		
Clothes dryer	165.5	5.96
Iron (hand)	24.0	0.86
Washing machine (automatic)	17.1	0.62
Water heater	870.0	31.32
Comfort and Health		
Air conditioner (room)	231.5	8.33
Dehumidifier	62.8	2.26
Fan (attic)	48.5	1.75
Fan (circulating)	7.2	0.26
Fan (window)	28.3	1.02
Hair dryer	2.3	0.08
Humidifier	27.2	0.98
Shaver	0.3	0.01
Lights (equivalent of five 150-watt bulbs burning 5 hours a day)	225.0	8.10
Home Entertainment		
Radio	14.3	0.51
Radio/record player	18.2	0.66
Television (black and white)	60.3	2.17
Television (color)	83.7	3.01
Housewares		
Clock	2.8	0.10
Sewing machine	1.8	0.06
Vacuum cleaner	7.7	0.28

$P = 60$ watts

$V = 120$ volts

$I = ?$

$R = ?$

(a) $P = VI$

$\therefore I = \dfrac{P}{V}$

$= \dfrac{60 \text{ watts}}{120 \text{ volts}}$

$= \boxed{0.50 \text{ ampere}}$

(b) $I = \dfrac{V}{R}$

$R = \dfrac{V}{I}$

$= \dfrac{120 \text{ volts}}{0.50 \text{ ampere}}$

$= \boxed{240 \text{ ohms}}$

EXAMPLE 11–8

An electric clothes dryer is connected to a 110-volt outlet. How much current does it use if it requires 2000 watts of electric power?

SOLUTION

The power and voltage are given. We calculate the current from Equation 11–8.

$P = 2000$ watts

$V = 110$ volts

$I = ?$

$P = VI$

$\therefore I = \dfrac{P}{V}$

$= \dfrac{2000 \text{ watts}}{110 \text{ volts}}$

$= \boxed{18.2 \text{ amperes}}$

EXTENSION

A 9.0-volt transistor radio draws a current of 0.50 ampere. How much electrical power is consumed? (Answer: 4.5 watts)

■

11–19 SUPERCONDUCTIVITY

In the normal conduction of an electric current by a metal, electrons are pumped through the crystal lattices of the metal under the influence of an electromotive force supplied by a cell, battery, generator, or the like. The motion of the electrons is impeded by collisions with the atoms in the lattice; as we have seen, this impedance is responsible for the conductor's electrical resistance. The resistance increases as the temperature increases, because the more energetic vibrations of the atoms interfere with the electrons' motion more strongly.

It seems reasonable to assume that if the atoms' vibrations are completely stopped by reducing the temperature to absolute zero, electrical resistance might vanish. To test the idea, Heike Kamerlingh-Onnes (1853–1926), a physicist at the University of Leiden, passed an electric current through some frozen mercury in 1911 and discovered that all detectable resistance to the flow of current disappeared at 4.15 K. This phenomenon is called **superconductivity**.

Superconductivity is not a rare phenomenon; at least 26 elements and several hundred compounds and alloys are already known to be superconductors. In each case the electrical resistance disappears at a particular temperature, for example, 20.7 K for an alloy of aluminum, germanium, and niobium. It is not necessary to go to absolute zero.

In 1986, 75 years after the discovery of superconductivity, Karl Alex Müller and Johannes Georg Bednorz of IBM Zurich observed that copper oxide ceramic materials might be superconducting at temperatures up to 35 K (Fig. 11–55). In one of the

Figure 11–55   A permanent magnet is levitated above a disk of the superconductor yttrium barium copper oxide, $YBa_2Cu_3O_{7-\delta}$, which is maintained at 77 K using liquid nitrogen as a coolant. The superconductor has a critical temperature of 92 K, below which its electrical resistance is zero. (Courtesy of Argonne National Laboratory)

fastest awards on record, the two received the 1987 Nobel Prize in physics. Meanwhile, several groups around the world have reported seeing superconductivity above 90 K in a number of oxides containing rare earth elements and are exploring applications for these materials. Why materials are superconducting at such high temperatures is one of the many questions that theorists are trying to answer.

In the superconducting state, a stream of electrons can flow without encountering any resistance. Such currents, once started, persist for long periods with little or no energy. In one experiment, an electric current induced in a small ring of metal was still circulating years later and had not noticeably diminished in strength.

Practical applications of superconductivity are being made. Superconducting magnets are the largest class of devices in use, and large electric motors using superconductors have been built. They have been proposed for trains, which could travel 300 miles per hour by eliminating the problem of friction; a train equipped with such a motor would, in effect, be lifted off the tracks by using the force between current elements (Fig. 11–56). Superconductors have also been proposed for amplifiers, particle accelerators, and computers.

Figure 11–56 Superconducting magnets levitate this prototype train a few inches above the track and propel it smoothly at speeds of 300 miles per hour or more. (T. Matsumoto/SYGMA)

SUMMARY OF EQUATIONS

Coulomb's Law of Electrostatic Force

$$F = k\,\frac{q_1 q_2}{r^2}$$

Electric potential $= \dfrac{\text{Work}}{\text{Charge}}$

$$V = \frac{W}{Q} \qquad 1\ \text{volt} = \frac{1\ \text{joule}}{1\ \text{coulomb}}$$

Joule's Law

Power $=$ (Current)^2 $\times$ Resistance $=$ Voltage $\times$ Current

$$P = I^2 R = VI$$

$$1\ \text{watt} = 1\ \text{volt} \times 1\ \text{ampere}$$

or

$$1\ \text{joule/s}$$

Current $= \dfrac{\text{Quantity of charge flowing by a point in a conductor}}{\text{Time required for the flow}}$

$$I = \frac{Q}{t} \qquad 1\ \text{ampere} = \frac{1\ \text{coulomb}}{1\ \text{second}}$$

Ohm's Law

Current $= \dfrac{\text{Potential difference}}{\text{Resistance}}$

$$I = \frac{V}{R} \qquad 1\ \text{ampere} = \frac{1\ \text{volt}}{1\ \text{ohm}}$$

Resistance of conductors connected in series

$$R_{\text{total}} = R_1 + R_2 + R_3 + \cdots$$

Resistance of conductors connected in parallel

$$\frac{1}{R_{\text{total}}} = \frac{1}{R_1} + \frac{1}{R_2} + \frac{1}{R_3} + \cdots$$

KEY TERMS

Action-at-a-Distance   The action of forces—gravitational, electric, and magnetic—between pairs of bodies that are not in contact.

Amber   A fossil resin that becomes strongly electrified when rubbed. The Greek word for amber is "elecktron."

Ampere   The unit of electric current equal to the rate of flow past a given point of 1 coulomb per second.

Angle of Declination   The angle between magnetic north and geographical north for any locality.

Angle of Dip   The angle made by a magnetized needle that is suspended at its center and swings freely in a vertical north-south plane.

Animal Electricity   An assumption of Galvani's that electricity was a special animal property.

Battery   A group of electric cells connected together to furnish electric current.

Charging by Conduction   Placing a positive or negative electric charge directly on the knob of an electroscope.

Circuit Breaker   A switch that automatically interrupts an electric circuit under an abnormal condition.

Conductor   Any material that can readily transmit electricity or heat.

Conservation of Charge   In an isolated system the total charge is conserved.

Coulomb's Law   The force between two electric charges is proportional to the magnitude of each of the charges and inversely proportional to the square of the distance between them.

Domain Boundaries   Lines of separation between magnetic domains.

Dry Cell   A voltaic cell whose contents are not spillable.

Dynamo   An electric generator.

Dynamo Theory   A theory of the source of the earth's magnetic field that relies upon convection currents and differential rotation of the inner and outer parts of the core to produce electric currents and a magnetic field.

Electric Battery   See Battery.

Electric Current   A flow of electric charge; measured in amperes.

Electric Field   A region of space in which an electric charge would experience an electric force.

Electric Intensity   The strength of an electric field at a point, defined as the electric force that would be exerted on a very small positive test charge placed at that point.

Electrode   A conductor used to establish electrical contact with a nonmetallic part of a circuit.

Electrolyte   A material that dissolves in water to give a solution that conducts an electric current.

Electromagnetic Induction   The production of electricity by magnetism, discovered by Faraday, which depends upon the relative motion between a coil and a magnet.

Electron Theory of Matter   Matter is composed of atoms that have equal numbers of positively charged protons and negatively charged electrons.

Electrophorus   A device invented by Volta, consisting of a disk and a metal plate, that generates electrostatic charges by induction.

Electroscope   A device that may be used to determine the presence and kind of electric charge on an object and, to some extent, the quantity of charge.

Electrostatic Generator   Any device that produces electric charges.

Electrostatic Precipitator   A device that removes particulate industrial wastes—fly ash, dust, and fumes —from air.

Electrostatics   Deals with attractions or repulsions of electric charges at rest.

Field   A region of space in which certain physical effects exist and can be detected, such as gravitational fields, electrostatic fields, and magnetic fields.

Fuse   An electric safety device consisting of a wire or strip of metal that melts and interrupts the circuit when the current exceeds a predetermined amperage.

Galvanism   Direct current of electricity, especially when produced by chemical action. Term coined by Volta in honor of his friend Galvani.

Gravitational Field   A region in which gravitational forces act on bodies.

Grounding   Establishing an electrical connection to the earth in order to neutralize an object.

Induction   A method of producing electrostatic charges in which there is no direct contact between the charging body and the object.

Insulator   Any material through which electricity or heat does not readily flow.

Joule's Law   The rate at which heat is developed by an electrical conductor is proportional to the square of the current flowing through the conductor.

Law of Electric Charges   Like charges repel each other; unlike charges attract.

Lines of Force   Lines that are used to represent the direction of an electrical or magnetic field.

Lodestone   A natural magnetic stone.

Magnetic Axis   A line connecting the south magnetic pole of the earth and the north magnetic pole that does not coincide with the geographical axis.

Magnetic Domain   Clusters of atoms that have a common orientation. If aligned, the substance is magnetic.

Magnetic Field   The force that a magnet would exert on a unit north pole placed at any point in space.

Magnetic Field Reversal   The earth's magnetic field has two stable states and has alternated between them over the ages.

Magnetic Poles   The north and south ends of a magnet.

Magnetite   A magnetic iron ore (Fe_3O_4).

Monopole   A hypothetically isolated magnetic pole.

Negative Electric Charge   The kind of electric charge that hard rubber acquires when rubbed with fur.

Ohm   The SI unit of electrical resistance.

Ohm's Law   The current through a conductor is directly proportional to the potential difference across the conductor and inversely proportional to its resistance.

Parallel Circuit   A divided electrical circuit that provides more than one path for the current. The current passes through each resistance independently.

Photoelectric Property   An electrical effect due to the interaction of radiation (as light) with matter.

Positive Electric Charge   The kind of electric charge that glass acquires when rubbed with silk.

Semiconductor   Materials such as silicon and germanium, which ordinarily are not good conductors or good insulators, can be made to conduct well when made with certain impurities or when their temperature is raised.

Serendipity   The faculty of finding valuable or agreeable things not sought for.

Series Circuit   An electrical circuit in which the same current flows through several conductors in turn.

Solenoid   A coil of wire in the form of a long cylinder that when carrying a current resembles a bar magnet.

Superconductivity   The disappearance of all resistance to the flow of electric current.

Terrella   A lodestone shaped in the form of a sphere to simulate the earth.

Van de Graaff Generator   An electrostatic machine consisting of a hollow metal sphere mounted on an insulated hollow cylinder containing a running belt that is continuously charged at the bottom and discharged in the sphere.

Volt   The unit of electrical potential. 1 volt = 1 joule/coulomb.

Voltage   Electric potential expressed in volts; the work done in moving a charge between two points, divided by the quantity of charge.

Voltaic Pile   The first electric battery. A series of alternating copper and zinc disks separated by pads moistened with salt water.

Watt-Hour-Meter   An instrument that measures the electrical energy consumed by in-house electrical connections.

Watt-Meter   An instrument for measuring electric power in watts.

Xerography   The dry-copy imaging process underlying the Xerox and other copying machines.

THINGS TO DO

1. Inflate two rubber balloons and stroke with wool or fur. What is the effect when they are brought close to one another?

2. Rub a ball-point pen, hard rubber rod, or comb with wool or fur. Bring it near small pieces of paper, sawdust, and other objects.

3. Charge a glass or Lucite® rod with silk. Test for the presence of charge on small objects.

4. Place a charged ball-point pen or rubber rod near a stream of water coming from a faucet.

5. Charge an electroscope negatively by conduction. Test its ability to detect (a) a source of positive and (b) a source of negative charge by the collapsing or diverging of the leaves.

6. Suspend a charged rubber or ebonite rod from a support by a thread. (a) Bring another rod similarly charged to it. (b) Repeat, but with a glass rod that has been rubbed with silk.

7. Dip a charged rubber rod into a mass of small paper pieces. What happens to the paper attracted to the rod after a short interval of time?

8. Suspend two pith balls from a common point on a horizontal rod. Charge them by contact (a) positively, (b) negatively, and observe.

9. Hold a sheet of paper against the wall and rub it with the hand or with a ball-point pen.

10. Coat two Ping-Pong balls with silver paint and suspend them from separate supports. Charge them by contact with a (a) rubber rod; (b) glass rod.

11. Suspend two bar magnets on stands that rotate about vertical axes. Observe the forces between (a) like poles; (b) unlike poles.

12. Place a piece of glass or paper over a bar magnet. Sprinkle iron filings on the glass, tap gently, and observe the field lines.

13. With a dip needle placed in a N–S direction, determine the inclination of the earth's magnetic field in your locality.

14. Magnetize a steel knitting needle by stroking it with a strong permanent magnet. Begin at the center, proceed to the end, and lift for the return stroke.

15. Dip a lodestone into a mass of iron filings.

16. With iron filings and a glass plate, study the magnetic field patterns around (a) two bar magnets 1 inch apart with like poles together; (b) the same magnets in (a) with unlike poles together; (c) a horseshoe magnet.

17. Map the magnetic field around a bar magnet using a small compass and a sheet of paper. Move the compass about and draw a line in the direction of the needle at each position.

18. Make slits in a lemon or orange. Insert strips of two metals, such as copper and zinc. Connect the strips by wires to a galvanometer to detect a current.

19. Examine the parts of a flashlight cell that has been disassembled.

20. Examine an incandescent light bulb.

21. Examine various kinds of (a) electrical wire; (b) insulation.

22. Examine various types of fuses.

23. Examine a watt-hour-meter. The speed of the rotating aluminum disk is directly proportional to the power being consumed.

24. With a compass, check a home iron radiator, refrigerator, and stove for magnetism by the earth's field (a) near the top; (b) near the bottom of each object.

25. Repeat Oersted's experiment using a circuit consisting of a compass, a dry cell, a flashlight bulb in series with it, a switch, and a long piece of electrical wire.

26. Connect a dry cell, a switch, and three flashlight bulbs (a) in series; (b) in parallel.

27. Examine your utility bill. How much does electrical energy cost in your area?

EXERCISES

Electricity at Rest

11–1. If a positively charged glass rod attracts a pith ball suspended by a silk thread, must the pith ball be negatively charged? Explain.

11–2. Although rubbing a ball-point pen with a piece of flannel is not essential for electrifying the pen, it acquires a stronger charge than when it is merely in contact with the flannel. Explain.

11–3. Why is repulsion rather than attraction a conclusive test of electrification?

11–4. An unknown charge is brought near the knob of a positively charged electroscope, and the leaves converge. What is the nature of the unknown charge?

11–5. Assuming that an object is charged, explain how you would decide whether the charge is positive or negative.

11–6. Why is it more difficult to maintain an electric charge on an electroscope on a damp day than on a dry day?

11–7. Explain how you would charge an electroscope negatively by induction.

11–8. Explain how a charged body attracts an uncharged one.

11–9. Why do small bits of dry paper that are attracted to a charged comb often fly away when they touch the comb?

11–10. Explain why you may receive a shock from a metal door handle when you slide across a seat in an automobile.

11–11. The atoms of conductors and insulators are composed of the same constituents: electrons, protons, and neutrons. In what way, then, do conductors and insulators differ?

11–12. Name (a) three good conductors of electricity, and (b) three good insulators.

Forces Between Electric Charges

11–13. (a) In what ways are electrical forces and gravitational forces similar? (b) In what ways do they differ?

11–14. In hydrogen, the simplest atom, the electron and the proton (nucleus) are separated by 0.530 angstrom unit. If the charge on each is 1.60×10^{-19}

coulomb (− for the electron, + for the proton), what is the force between them?

11–15. Calculate the force between two electrons that are separated by 10^{-9} meter. (Charge on the electron is -1.60×10^{-19} coulomb.)

Lightning

11–16. Explain the discharge of a lightning stroke between a negatively charged cloud and the earth.

11–17. How does an automobile that has a metal top and body shield its occupants from lightning?

11–18. Why is standing in an open field during a thunderstorm a dangerous practice?

Magnetic Poles and Forces

11–19. (a) In what ways are electric charges and magnetic poles similar? (b) Different?

11–20. Why may heating reduce the strength of a magnet?

11–21. Discuss the significance of Faraday's and Henry's discovery of the production of an electric current by means of magnetism.

11–22. Can you create monopoles by breaking a bar magnet in half? Explain.

11–23. Like poles repel, yet the north-seeking pole of a magnet is attracted to the geographic North Pole of the earth. Explain.

11–24. Explain why a nail is attracted to either pole of a magnet, but another magnet is attracted to only one of the poles.

11–25. A magnet can be weakened by dropping it on a hard floor. Explain.

11–26. The atoms in a piece of wood are tiny magnets, yet the piece of wood is not. Explain.

11–27. Why will an unmagnetized nail that is in contact with a magnet attract an unmagnetized paper clip?

Fields

11–28. Discuss the meaning of the term "electric field."

11–29. Calculate the magnitude and direction of the electric field 3.00 meters from a positive charge of 10^{-5} coulomb.

11–30. Sketch the electric field pattern between two point charges of opposite sign separated by a small distance.

11–31. (a) Two bar magnets are placed with their south poles facing each other. Represent the magnetic field between them in a diagram.

(b) Repeat with the north pole of one magnet opposite the south pole of the second magnet.

Electric Currents

11–32. A transistor radio draws 0.500 ampere of current. If it uses a 9.00-volt battery, what is the resistance of the radio?

11–33. What is the current through a 60.0-watt lamp that operates at 120 volts?

11–34. How many electrons flow through the filament of a 120-volt, 60.0-watt light bulb per second?

11–35. When people say that they left lights in the house "burning," is this a correct statement? How might it have originated?

11–36. An electric clock uses 5.00 milliamperes of current on 120 volts. Find its resistance.

11–37. Why is it dangerous to replace a 15-ampere fuse that blows repeatedly with one having a larger current rating?

11–38. How much current flows through a 60.0-watt headlight of an automobile that has a 12-volt electrical system?

11–39. Calculate the resistance of a vacuum cleaner that uses a current of 14.0 amperes and operates on 120 volts.

Electric Circuits

11–40. Three incandescent lamps with resistances of 220, 440, and 660 ohms are connected in parallel across a 110-volt line.
(a) What is their combined resistance?
(b) What total current is drawn by the three lamps?
(c) How much current does each lamp draw?

11–41. Explain why operating too many electrical devices at one time may cause a fuse to blow.

11–42. Are the currents in resistors that are connected in parallel generally the same? Explain.

11–43. (a) What is the total resistance of five 20-ohm resistors connected in series? (b) Parallel?

11–44. Three resistors of 5, 10, and 15 ohms each are connected in series in a circuit with a 9.0-volt battery. What is the current through each resistor?

Electric Power and Energy

11–45. A 100-watt lamp draws 0.800 ampere in operation. What is its resistance?

11–46. Which uses more energy: a 250-watt TV set in 1 hour or a 1200-watt toaster in 10 minutes?

11–47. A 100-watt lamp was accidentally left burning in a storeroom for one week. How much did it cost at 10 cents per kilowatt-hour?

11–48. An electric range draws 20 amperes at 240 volts. What does it cost per hour to operate at 10 cents per kilowatt-hour?

11–49. Does a 5000-watt heater use twice the energy of a 2500-watt heater? Explain.

11–50. *Multiple Choice*

A. The force of attraction between two charged spheres can be calculated from
   (a) Newton's second law.
   (b) Coulomb's law.
   (c) Joule's law.
   (d) Ohm's law.

B. An instrument for determining the magnitude of a charge is the
   (a) Van de Graaff generator.
   (b) electrostatic precipitator.
   (c) electroscope.
   (d) electrophorus.

C. An electric current in a wire is a flow of
   (a) atoms.
   (b) protons.
   (c) ions.
   (d) electrons.

D. Regions where the magnetic fields of atoms are lined up in the same direction are known as
   (a) superconductors.
   (b) magnetic domains.
   (c) currents of spinning electrons.
   (d) dynamos.

E. A magnetic field is produced by
   (a) static negative charges.
   (b) an electron current.
   (c) static positive charges.
   (d) falling water.

F. Which of the following substances differs from the others with respect to electricity?
   (a) plastic
   (b) glass
   (c) copper
   (d) cotton

G. A natural magnet is
   (a) iron.
   (b) amber.
   (c) aluminum.
   (d) lodestone.

H. Ohm's law is correctly expressed by
   (a) $I = \dfrac{V}{R}$
   (b) $V = \dfrac{I}{R}$.
   (c) $R = VI$.
   (d) $I = VR$.

I. A light bulb glows because
   (a) there is burning inside.
   (b) an electric current glows.
   (c) the filament is hot.
   (d) a very small current is flowing.

J. Which of the following does not produce a magnetic field?
   (a) a current-carrying wire
   (b) a stationary electric charge
   (c) a moving electric charge
   (d) an electromagnet

SUGGESTIONS FOR FURTHER READING

Bloxham, Jeremy, and David Gubbins, "The Evolution of the Earth's Magnetic Field." *Scientific American*, December, 1989.
   The magnetic field at the boundary between the earth's core and mantle has been mapped. These maps reveal remarkable features of the earth's magnetic field and elucidate the geodynamo, the process that maintains the field.

Burrill, E. Alfred, "Van de Graaff, The Man and His Accelerators." *Physics Today*, February, 1967.
   A sketch of Robert J. Van de Graaff and a class of electrostatic generators that have become associated with his name.

Carrigan, Richard A., and W. Peter Trower, "Superheavy Magnetic Monopoles." *Scientific American*, April, 1982.
   Although they are predicted to exist, isolated north and south magnetic poles have never been observed. Their discovery would rank as one of the finds of the century.

Devons, Samuel, "Benjamin Franklin as Experimental Philosopher." *American Journal of Physics*, December, 1977.
   Presents Franklin's contributions to science and to electricity in particular.

Goldhaber, Alfred S., "Resource Letter MM-1: Magnetic Monopoles." *American Journal of Physics*, May, 1990.

Isolated magnetic poles continue to challenge experimenters and theorists. Their existence has yet to be established.

Moore, A. D., "Electrostatics." *Scientific American*, March, 1972.
Modern electrostatics goes far beyond the limited domain of nonmoving electric charges implied by the literal definition of the word.

Müller, K. Alex, and J. Georg Bednorz, "The Discovery of a Class of High-Temperature Superconductors." *Science*, Vol. 237, 4 September, 1987, pp. 1133–1138.
Discussion of a new class of oxide superconductors and of the concepts that led to their discovery. The authors were the recipients of the 1987 Nobel Prize in physics for their discovery.

Mulligan, Joseph F., "Hermann von Helmholtz and His Students." *American Journal of Physics*, January, 1989.
Hermann von Helmholtz, one of the most versatile scientists who ever lived, set the tone for scientific research in Germany during the nineteenth century. He served as a role model for young physicists from many nations, including the United States, who flocked to Berlin to study and do research with him.

Parfit, Michael, "Coping with Blackout: What Happens when the Lights Go Out?" *Smithsonian*, February, 1987.
Electricity has become so central to American life that when it disappears the darkness seems alien and dangerous.

Uman, Martin A., and E. Philip Kreider, "Natural and Artificially Initiated Lightning." *Science*, 27 October, 1989.
Modern research on the physics of lightning began in the early twentieth century with the work of C. T. R. Wilson, of cloud chamber frame. Scientists are trying to understand better the processes that produce electric fields in all types of clouds and how these fields initiate a discharge.

Van Vleck, J. H., "Quantum Mechanics: The Key to Understanding Magnetism." *Science*, Vol. 201, 14 July, 1978.
Quantum mechanics is the master key that unlocks each door to understanding magnetism.

Walsh, William J., "Advanced Batteries for Electric Vehicles—A Look at the Future." *Physics Today*, June, 1980.
By the year 2000, millions of electric vehicles may be in use. The actual number will depend in part on research on advanced batteries.

Williams, Earle R., "The Electrification of Thunderstorms." *Scientific American*, November, 1988.
Two centuries ago, Benjamin Franklin demonstrated that lightning is a form of electricity. Today, there are still many unanswered questions concerning the exact origins and the mechanism by which rain clouds are electrified.

Williams, L. Pearce, "André-Marie Ampère." *Scientific American*, January, 1989.
Ampère carried out the first systematic investigations of the magnetic fields produced by electric currents. He is less well known for his achievements as a philosopher of science.

Williams, L. Pearce, "Why Ampère Did Not Discover Electromagnetic Induction." *American Journal of Physics*, April, 1986.
Michael Faraday announced his discovery of electromagnetic induction in 1832. André Marie Ampère claimed that he had actually discovered it in 1822. In fact, he had, but he did not really publish the fact at that time. Why did Ampère fail to lay claim to a discovery that would have guaranteed him scientific immortality?

Wolsky, Alan M., Robert F. Giese, and Edward J. Daniels, "The New Superconductors: Prospects for Applications." *Scientific American*, February, 1989.
Identifying new applications of superconductors, as yet unforeseen, may have a greater impact than achieving the applications now being envisioned.

ANSWERS TO NUMERICAL EXERCISES

11–14. -8.20×10^{-8} N (attraction)

11–15. 2.30×10^{-10} N (repulsion)

11–29. 10^4 N/coul away from the positive charge

11–32. 18.0 ohms

11–33. 0.500 ampere

11–34. 3.12×10^{18} electrons

11–36. 24,000 ohms

11–38. 5 amperes

11–39. 8.57 ohms

11–40. (a) 120 ohms
(b) 0.920 ampere
(c) 0.500 ampere
0.250 ampere
0.170 ampere

11–45. 156 ohms

11–46. TV set

11–47. $1.68

11–48. $0.48

In fluorescence, a substance absorbs ultraviolet light and re-emits it as visible light. The key is the absorption and re-emission of quanta of radiation (Paul Silverman, Fundamental Photographs)

12

THE QUANTUM THEORY OF RADIATION AND MATTER

Classical physics, the physics of Newton and his successors to the year 1900, seemed so magnificent that it left many scientists with the feeling that the essential task of physical science had been completed. There were no new worlds to conquer, only "mopping up" operations, such as extending the accuracy of measurement to the fourth, fifth, or sixth decimal place.

At the same time, however, cracks appeared in the structure of physics. Although the transmission of light, with its associated phenomena such as interference, diffraction, polarization, and its finite velocity, was satisfactorily accounted for by the wave theory, the emission and absorption of radiation were not. The wave theory also provided no satisfactory explanation for either atomic spectra or the photoelectric effect. Then a series of discoveries in the five-year period preceding 1900—X-rays, radioactivity, the electron—and theoretical breakthroughs produced an upheaval in thought that is not resolved to this day. Chief among them was the quantum theory.

12–1 SPECTROSCOPY

It is through the analysis of light from the sun, planets, stars, and galaxies that we know their chemical composition. This has to be considered an achievement of the first rank. There were many who believed that such knowledge would be forever denied to us. No elements have been found to occur in these bodies that do not also occur on the earth. An apparent exception—the discovery on the sun of a hitherto unknown element, helium (*helios*, meaning "sun") —was eventually discovered to exist on earth as well.

Such useful applications of spectroscopy flowed from two developments: (1) the introduction of refinements that led to an increase in the resolving power of **spectroscopes**, instruments that spread light out by wavelength and made finer analysis possible, and (2) the interpretation of the spectra (Fig. 12–1). Joseph von Fraunhofer (1787–1826), an optician and a physicist whose skillful use of lenses and slits indicated the presence in the solar spectrum of many dark lines, called **Fraunhofer lines**, mapped hundreds of them. The interpretation of these lines by Robert Bunsen (1811–1899), a chemist (also known for the Bunsen burner), and Gustav Kirchhoff (1824–1887), a physicist, pointed out the potential value of spectroscopy. They found that each chemical element gives rise to a characteristic pattern of lines— a fingerprint, as it were—of the element (Fig. 12–2). As the sun's rays pass through its outer atmosphere, certain elements absorb rays at sharp and well-defined wavelengths (Fig. 12–3). For example, a pair of closely spaced dark lines in the yellow region of the spectrum, the Fraunhofer D lines, are due to the absorption of sunlight by sodium atoms in the sun's atmosphere.

Fraunhofer's invention of the **diffraction grating** gave scientists a tool that far exceeded the prism in its ability to disperse light (Fig. 12–4). In transmission gratings, a large number of parallel rulings are made on glass, plastic, or other transparent material, and

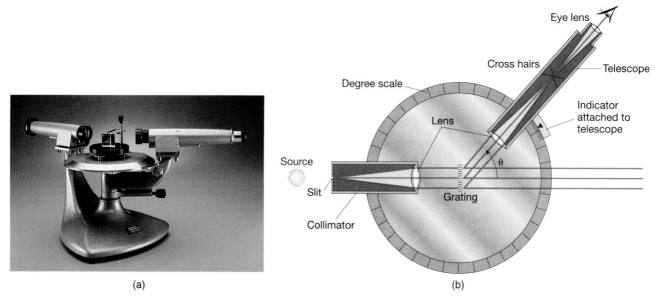

(a)   (b)

Figure 12–1   (a) A spectrometer that uses a prism. (Courtesy CENCO) (b) the essential parts of a type of spectrometer that uses a diffraction grating.

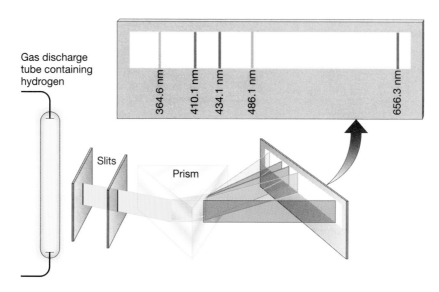

Figure 12–2   The line spectrum of excited hydrogen atoms. Light from glowing hydrogen gas is passed through narrow slits in a spectroscope, forming a narrow beam of light. The beam is then passed through a prism or a diffraction grating that separates the light into its component wavelengths as individual lines that we observe as a line spectrum. (Source: John C. Kotz and Keith F. Purcell, *Chemistry and Chemical Reactivity,* 2nd edition. Philadelphia: Saunders College Publishing, 1991, p. 288.)

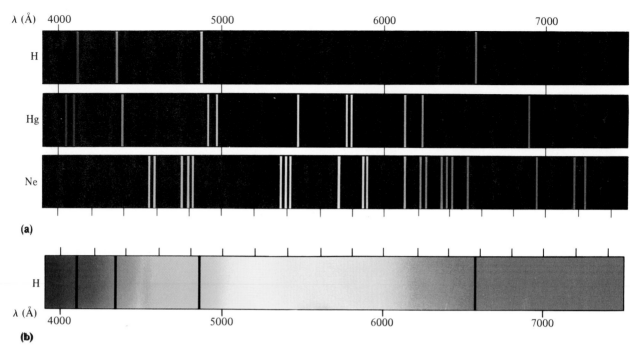

Figure 12–3 (a) The line emission spectra for various elements. Note that the spectra consist of well-defined colors and differ from one another. (b) Absorption spectrum for hydrogen. When a beam of white light is passed through unexcited hydrogen and then through a slit and a prism, the transmitted light is lacking in intensity at the same wavelengths as the emission spectra in (a). (Source: Kenneth W. Whitten, Kenneth D. Gailey, and Raymond E. Davis, *General Chemistry*, 4th edition. Philadelphia: Saunders College Publishing, 1992, p. 185.)

light goes through the unscratched area. A reflection grating employs polished metal, with reflection coming off the unscratched areas. Typical gratings have from 10,000 to 30,000 lines or slits per inch and may be as much as 6 inches in width. In either case, the constituent waves of a mixture constructively interfere in different directions depending on their wavelength and are thereby separated. In place of the subjective concept of color, we get wavelength, a precisely measurable quantity (Table 12–1).

Figure 12–4 A transmission diffraction grating that can be used with a spectrometer. It has 600 lines/mm. (Courtesy CENCO)

Table 12–1   WAVELENGTH RANGES OF COLOR

Color	Wavelength (Å)
Violet	3800–4500
Blue	4500–4900
Green	4900–5600
Yellow	5600–5900
Orange	5900–6300
Red	6300–7600

Table 12–2   WAVELENGTHS OF CERTAIN SPECTRAL LINES

Element	Region of Spectrum	Wavelength (Å)
Hydrogen	Red	6562.8
	Blue	4861.3
	Violet	4340.5
	Violet	4101.7
Sodium	Yellow	5895.9
	Yellow	5890.0
Mercury	Yellow	5790.7
	Green	5460.7
	Blue	4916.0
	Violet	4358.3

Table 12–2 gives the wavelengths of some selected lines.

12–2 BALMER'S FORMULA AND THE HYDROGEN SPECTRUM

The light given off by an element when it is made to burn is found to contain only certain colors in the form of bright lines. The apparent simplicity of the four prominent lines in the visible region of the hydrogen spectrum aroused the curiosity of Johann Jakob Balmer (1825–1898), a teacher of mathematics. Following years of attempting to discover a numerical relationship among the lines, Balmer in 1885 published a paper with the title "Notice Concerning the Spectral Lines of Hydrogen." It contained a formula that yielded the observed lines to the most re-

markable degree of accuracy. In modern notation, **Balmer's formula** is written:

$$\frac{1}{\lambda} = R_{\mathrm{H}}\left(\frac{1}{2^2} - \frac{1}{n^2}\right) \tag{12–1}$$

in which λ is the wavelength in centimeters; R_{H} is a constant called the Rydberg constant for hydrogen, equal to $109{,}678 \text{ cm}^{-1}$, and n is a running integer that takes the values $3, 4, 5, 6, \ldots$. A comparison of the wavelengths as given by Balmer's formula and as measured by the astronomer and physicist Anders Ångström (1814–1874) is given in Table 12–3.

Balmer predicted the existence at the extreme violet end of the hydrogen spectrum of a fifth line corresponding to $n = 7$ and a wavelength of 2969.65 Å. This line had not been seen before because the human eye is not very sensitive in that range. Its prompt discovery by the astronomer Sir William Huggins (1824–1910) and the chemist Hermann Vogel (1834–1898), and the discovery of other lines for values of n greater than 7, gave strong support to his formula. Balmer further speculated that the term $1/2^2$ in his formula might be replaced by $1/1^2$ or $1/3^2$ and predicted the existence of lines in the ultraviolet and infrared regions of the spectrum. In time, these regions were explored with apparatus sensitive in these regions, and other series of lines were discovered as predicted. They are shown in Table 12–4 and correspond to the general formula

$$\frac{1}{\lambda} = R_{\mathrm{H}}\left(\frac{1}{m^2} - \frac{1}{n^2}\right) \tag{12–2}$$

in which m is 1 for the Lyman series, 2 for the Balmer series, 3 for the Paschen series, and so on, and n is a running integer for each respective series.

Table 12–3   WAVELENGTHS OF LINES IN VISIBLE REGION OF HYDROGEN SPECTRUM

Line	n	According to Balmer's Formula (Å)	Measured By Ångström (Å)	Difference
Hα	3	6562.08	6562.10	+ 0.02
Hβ	4	4860.80	4860.74	− 0.06
Hγ	5	4340.0	4340.1	+ 0.1
Hδ	6	4101.3	4101.2	− 0.1

Table 12–4 COMPLETE HYDROGEN SPECTRUM $\dfrac{1}{\lambda} = R_{\text{H}}\left(\dfrac{1}{m^2} - \dfrac{1}{n^2}\right)$

Series	Year Discovered	Region	Formula	m	n
Lyman	1906–14	Ultraviolet	$\dfrac{1}{\lambda} = R_{\text{H}}\left(\dfrac{1}{1^2} - \dfrac{1}{n^2}\right)$	1	2, 3, 4, . . .
Balmer	1885	Visible	$\left(\dfrac{1}{2^2} - \dfrac{1}{n^2}\right)$	2	3, 4, 5, . . .
Paschen	1908	Infrared	$\left(\dfrac{1}{3^2} - \dfrac{1}{n^2}\right)$	3	4, 5, 6, . . .
Brackett	1922	Infrared	$\left(\dfrac{1}{4^2} - \dfrac{1}{n^2}\right)$	4	5, 6, 7, . . .
Pfund	1924	Infrared	$\left(\dfrac{1}{5^2} - \dfrac{1}{n^2}\right)$	5	6, 7, 8, . . .

Although the Balmer formula applies only to the hydrogen spectrum, Johannes Robert Rydberg (1854–1919) developed a general formula that applies to a wide variety of line spectra. The fact that each element has its own distinct spectrum is striking enough (Fig. 12–5). Since the line spectra of elements fit into a general pattern, speculation grew that they were produced by some common mechanism. The fundamental and harmonic frequencies of a mu-sical note suggested a similar regularity in the spacings of the line spectra, but no such relationship could be found.

EXAMPLE 12–1

Calculate the wavelength of the first spectrum line of the Lyman series of hydrogen.

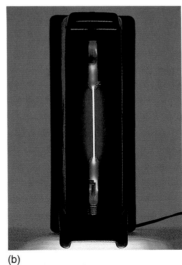
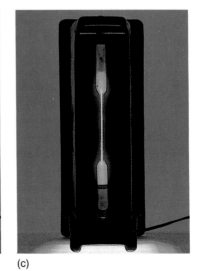

(a)   (b)   (c)

Figure 12–5 A gas at low pressure emits light of characteristic frequencies when excited. Neon lights used in advertising signs are an example. (a) Neon, (b) argon, (c) mercury. (Photographs by Charles D. Winters)

SOLUTION

We apply Balmer's formula. From Table 12–4 we see that $m = 1$ for the Lyman series, and $n = 2$ for the first line. Knowing that for hydrogen, $R_H = 109\ 678$ cm^{-1} = 1.10×10^5 cm^{-1} (rounded off for convenience), we solve for the wavelength λ.

$$m = 1$$
$$n = 2$$
$$R_H = 1.10 \times 10^5 \text{ cm}^{-1}$$
$$\lambda = ?$$

$$\frac{1}{\lambda} = R_H\left(\frac{1}{m^2} - \frac{1}{n^2}\right)$$

$$\frac{1}{\lambda} = \frac{109\ 678}{\text{cm}}\left(\frac{1}{1^2} - \frac{1}{2^2}\right)$$

$$\frac{1}{\lambda} = \frac{1.10 \times 10^5}{\text{cm}}\left(1 - \frac{1}{4}\right)$$

$$\frac{1}{\lambda} = \frac{1.10 \times 10^5}{\text{cm}}\left(\frac{3}{4}\right)$$

$$\frac{1}{\lambda} = \frac{3.30 \times 10^5}{4 \text{ cm}}$$

$$\lambda = \frac{4 \text{ cm}}{3.30 \times 10^5}$$

$$\lambda = 1.20 \times 10^{-5} \text{ cm} \times \frac{10^8 \text{ Å}}{\text{cm}}$$

$$\lambda = \boxed{1200 \text{ Å (ultraviolet)}}$$

EXTENSION

Calculate the wavelength of the third line of the Brackett series of hydrogen. (Answer: 21,700 Å)

■

12–3 THE ELECTRON

When Heinrich Geissler (1814–1879), a glassblower and mechanic, invented the mercury diffusion pump —the first major improvement in the vacuum pump in 200 years—he opened up a new field of study. Geissler was able to remove enough gas from a strong glass tube to reduce the pressure to 1/10,000 of standard atmospheric pressure. He prepared the tubes with such great skill that evacuated tubes became known as **Geissler tubes**. By sealing a wire into each end of a tube, the wire ending in a metal plate called an **electrode,** and running the wires to a battery, the

mathematician and physicist Julius Plücker (1801–1868) found, to his surprise, that electricity flowed through the tube (Fig. 12–6). It was the forerunner of present-day fluorescent and neon tubes and television tubes.

When a Geissler tube is connected to an induction coil or electrostatic machine and an electric discharge passes between the **negative terminal (cathode)** and the **positive terminal (anode)**, a dark region appears around the cathode that gradually spreads through the tube. The glass itself at the opposite end glows with a color that depends on the nature of the glass making up the tube.

Sir William Crookes (1832–1919), a physicist and chemist, studied these effects with tubes of his own design. He introduced into one a metal barrier in the form of a Maltese cross and observed that instead of an intense glow, a shadow appeared at the end of the tube. Emanating from the cathode were what Eugen Goldstein (1850–1930), a physicist, called **cathode rays**, which caused the glow to appear when they reached the glass (Fig. 12–7). From the shadows formed, Crookes concluded that cathode rays traveled in straight lines, like light rays. However, he found that unlike light rays, a magnetic field deflects the path of cathode rays. Further, he found that cathode rays produced from many different materials are similar.

There was intense speculation concerning the nature of cathode rays in the last quarter of the nineteenth century. Were they waves or particles? In

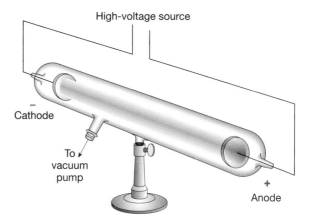

Figure 12–6 Discharge of electricity through a gas at low pressure is accompanied by various effects as the vacuum is increased.

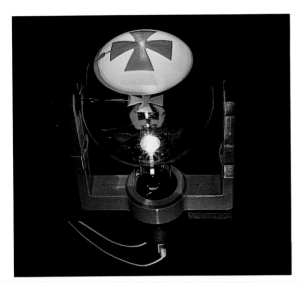

Figure 12-7 A Crookes Maltese Cross tube shows that cathode rays travel in straight lines and cast sharp shadows. (Courtesy CENCO)

1895, Jean Perrin (1870–1942) collected some on an insulated conductor and with an electroscope proved that they carry a negative charge. Then, in 1897, Sir Joseph John ("J. J.") Thomson (1856–1940), director of the Cavendish Research Laboratory at Cambridge University, made a quantitative study of cathode rays.

Thomson found that by applying an electric field across two plates, or a magnetic field around a **Crookes tube**, he could **deflect the cathode rays**. The deflection showed that they act like charged particles rather than light (Fig. 12–8). Further, the deflection is in the direction to be expected for particles carrying a negative charge. He was able to calculate the **ratio of the charge of a particle to its mass**, **e/m**, and found that the rays from cathodes made of different materials all had the same value: 1.76×10^{11} coulombs/kilogram. This showed conclusively that particles could be split off from atoms and that the atom was, therefore, not the ultimate subdivision of matter as had been believed. Thomson identified the cathode rays as a building block of atoms and called them **electrons**.

Between 1909 and 1916, Robert A. Millikan (1868–1953), then at the University of Chicago, measured the **quantity of charge** on the electron in his famous **oil-drop experiment**. When oil is sprayed in a chamber, the atomizer effect on the droplets gives them an electrical charge (Fig. 12–9), some positive and others negative. The force of air resistance on the minute droplets causes them soon to acquire a very low terminal speed of a fraction of a centimeter per second. In a properly illuminated chamber, the droplets appear as bright stars of light, and their motion can be followed with a telescope for hours. By applying an electric field to the drops, Mil-

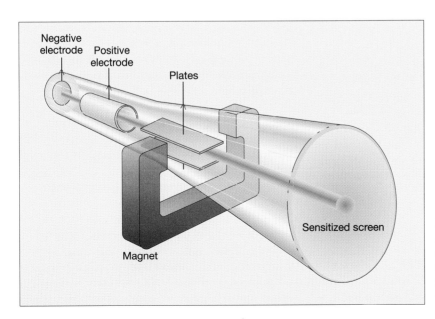

Figure 12-8 Diagram of the apparatus with which J. J. Thomson measured the deflection of cathode rays by electric and magnetic fields, their velocity, and their ratio of charge to mass (e/m).

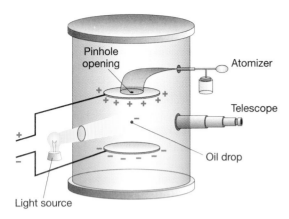

Figure 12-9 Millikan's oil-drop experiment for determining the charge on an electron. The pull of gravity on the drop is balanced by the upward electrical force.

likan could alter their motion, balancing it against the force of gravity and even causing the droplets to rise. When the electric field is adjusted to balance the downward force of gravity,

$$\mathbf{F}_{\text{electric}} = \mathbf{F}_{\text{gravitational}}$$
$$e\mathbf{E} = m\mathbf{g} \qquad \text{and}$$
$$e = \frac{m\mathbf{g}}{\mathbf{E}}$$

where e is the electric charge, $\mathbf{E}$ is the electric field strength, m is the mass of the droplet, and $\mathbf{g}$ is the acceleration due to gravity.

Millikan found that the **minimum electric charge** that an oil drop can acquire is 1.60×10^{-19} coulomb, which he assumed to be the charge on the electron. Electric charge comes in multiples of this unit: 3.20×10^{-19} coulomb, 4.80×10^{-19} coulomb, and so on. By combining Thomson's charge/mass value for the electron, e/m, with Millikan's value for the charge, e, the mass of the electron is 9.11×10^{-31} kilogram, about 1840 times smaller than the mass of hydrogen, the lightest atom. For this work, Millikan became the second American to receive a Nobel Prize in physics (1923).

12-4 X-RAYS

Wilhelm Konrad Roentgen (1845–1923), a physicist, was also curious about the cathode rays produced in a Crookes tube. Testing the opacity of black paper for experiments he was planning, he happened, on November 8, 1895, to cover a Crookes tube with pieces of it, then darkened the laboratory and turned on the current. There was darkness except for a glimmer originating on something on a little bench that was about a meter away. Roentgen lit a match and discovered that the light was coming from a screen coated with barium platinocyanide, a fluorescent material, lying on the bench.

What could have caused the fluorescence? Ordinarily, barium platinocyanide fluoresces when illuminated with ultraviolet light, but there was no source of ultraviolet light in Roentgen's apparatus. Might cathode rays have caused it? Cathode rays had not been known to travel more than a few centimeters through air, and the screen was well beyond their range. Roentgen was forced to conclude that the fluorescence was due to a new kind of ray, which he named **X-ray**, X standing for unknown. During the following weeks and months he studied the properties of X-rays intensively, reporting his first results to a medical society seven weeks later in a paper entitled, "On a New Kind of Ray," and publishing two additional papers describing the properties of this new radiation the following year.

Roentgen established that X-rays originate in the glass walls of a Crookes tube where the cathode rays strike. Since they are not deflected by a magnetic field or an electric field, they cannot be charged particles. The rays travel in straight lines, darken a photographic plate, and have a remarkable penetrating ability, passing easily through a 1000-page book, a double pack of cards, tin foil, blocks of wood, and many other materials. X-rays differ from visible light only in that their wavelength is considerably shorter, like ultra-ultraviolet rays, and their frequency is very great. Like light waves, X-rays can be reflected, refracted, polarized, and diffracted. Roentgen found that if he held his hand between a Crookes tube and the screen, the bones appeared as a dark shadow within the slightly dark shadow image of the hand itself (Fig. 12-10).

Few scientific discoveries have had such an immediate impact on the public. Early in January, 1896, some of Roentgen's X-ray photographs were exhibited at a meeting of the Berlin Physical Society. The first press reports appeared on the following day, and the news was carried around the world in newspapers and periodicals. Roentgen's experiments were repeated and extended in many laboratories, since

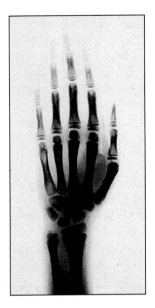

Figure 12–10   X-ray photograph of a human hand.

undergo a rapid deceleration and radiate X-rays, a form of electromagnetic radiation.

X-rays are used in medical diagnosis and treatment; in industrial diagnosis for possible defects; in the detection of artificial gems; in the analysis of crystal structure; and in many other ways. One of the more unusual applications is in the field of art, to confirm the age and integrity of paintings and to detect forgeries and fakes. With X-rays, information may be obtained about the structure of the painting that lies below the surface. The old masters used white leaded paint to shape the underlying structure of their paintings. Since lead absorbs X-rays very well, this underlying structure can be explored. Modern paints differ in chemical content from older paints, leading to differences in X-ray absorption. Zinc-oxide white was not used before 1870, and titanium-oxide white not until the twentieth century; therefore, they should not appear in a painting that is supposedly 300 years old. On the basis of X-rays, Hans Van Meegeren, an art dealer, was discovered in 1945 to have forged 300-year-old paintings by the artist Vermeer that were sold for more than $2 million.

cathode-ray tubes were readily available and X-rays are produced easily. Within weeks, X-rays were being used to photograph broken bones and to locate bullets in wounds.

In a modern X-ray tube, X-rays are produced when a beam of electrons is accelerated across a voltage of 30,000 to 40,000 volts (Fig. 12–11). Stopped suddenly by a metal target—the anode—the electrons

12–5   RADIOACTIVITY

Having heard a report of Roentgen's discovery of X-rays, the physicist Antoine Henri Becquerel (1852–1909) wondered whether they could be associated with fluorescence and **phosphorescence—**

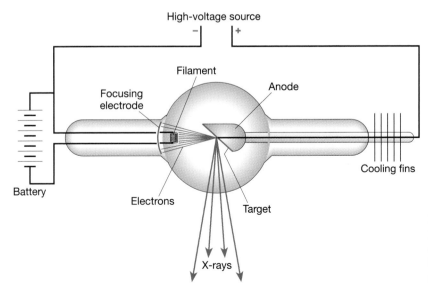

Figure 12–11 X-ray tube. Electrons emitted by the heated cathode are accelerated and strike the metal target at high speeds. X-rays are emitted from the target.

the ability of certain crystals to glow in the dark. In particular, he was aware of the pronounced phosphorescence of potassium uranyl sulfate, a uranium salt, when excited by ultraviolet light.

Becquerel knew that uranium salts, when exposed to sunlight, emit radiations that pass through aluminum, cardboard, and other materials, and blacken a photographic plate, just as X-rays do. During a spell of cloudy weather in February, 1896, he wrapped his photographic plates in black paper and stored them along with some uranium salts in a cabinet drawer. Believing that sunlight or ultraviolet light was necessary to excite the radiation. Becquerel did not expect to find the photographic plates exposed when he developed them a few days later. To his great surprise, he found that they were intensely exposed.

Becquerel continued to experiment for several months and obtained other remarkable results. He found that the salts continue to emit radiations when kept in complete darkness for months. They radiate whether they are in crystal form or in solution and with an intensity proportional to the uranium content. He established that the element uranium radiates even more than its salts and attributed this new radiation to the presence of uranium.

Coming as it did on the heels of the more spectacular discovery of X-rays, Becquerel's discovery evoked little interest, with one important exception.

The husband-and-wife team of physical chemists, Pierre Curie (1859–1906) and Marie Sklodowska Curie (1867–1934), pursued the matter further and found that the element thorium displayed a radiation similar to that of uranium. The Curies introduced the name **radioactivity** to describe it. Finding that pitchblende, an ore rich in uranium, was far more radioactive than its uranium content suggested, from a ton of pitchblende the Curies in 1898 isolated minute amounts of two new elements, which they named polonium and radium. They found polonium to be 400 times as radioactive as uranium itself, and radium a million times. They also showed that radioactivity is a spontaneous process in certain elements and is unaffected by pressure, heat, or chemical combination.

Some scientists at first believed that the Becquerel rays were probably weak X-rays. But Ernest Rutherford (1871–1937), then at McGill University, using magnetic and electric fields, found that radioactivity consisted of two components. Rutherford designated them alpha particles and beta particles, from the first two letters of the Greek alphabet. **Alpha particles** are heavy, carry a positive charge, can be stopped by a thick sheet of paper, and bear a resemblance to helium. **Beta particles** are light, are negatively charged, and can be stopped by a thin piece of aluminum; they were later shown to be high-energy elec-

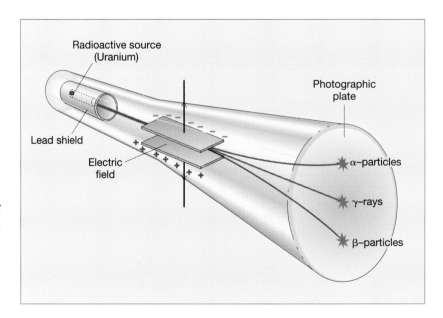

Figure 12–12 Separation of alpha, beta, and gamma rays from a radioactive source by an electric field. Alpha and beta rays are deflected in opposite directions; gamma rays are undeflected.

trons. Paul Villard (1860–1934) discovered a third component, **gamma rays**; they have no observable charge or mass and are highly penetrating resembling high-energy X-rays (Fig. 12–12).

12–6 RUTHERFORD'S MODEL OF THE ATOM

In a **scattering experiment**, a narrow beam of particles or rays is aimed at a target usually consisting of a thin foil or film of some material. As the beam strikes the target, some of the projectiles are deflected or scattered from their original direction. The scattering is the result of interactions between the projectile and the atoms of the material. From a knowledge of the mass, velocity, and direction of the projectiles, and the scattering angles, the properties of the atoms that scattered them can be deduced.

The alpha particle joined the growing arsenal of projectiles—electrons and X-rays—that could be used to probe the structure of matter. No one put alpha particles to more effective use than Ernest Rutherford. Under his direction, Hans Geiger (1882–1945) and Ernest Marsden (1889–1970) carried out studies of the **scattering of alpha particles** by thin metal foils (Fig. 12–13). Most of the particles went right through with little or no deflection, as observed with a scintillation screen and microscope. The screen contained a phosphor, zinc sulfide, that glowed momentarily when struck by an alpha particle. Surprisingly, however, a small fraction of the alpha particles were scattered through very large angles, a few even bounding straight back. To Rutherford, "It was quite the most incredible event that has ever happened to me in my life. It was almost as incredible as if you fired a 15-inch shell at a piece of tissue paper and it came back and hit you." Strong forces are needed to turn back or deflect an alpha particle moving with a speed greater than 10^9 cm/s and having a mass thousands of times as great as an electron.

The results of the alpha-scattering experiments could not be explained by the widely accepted atomic theory of J. J. Thomson. In this model, the positive charge is distributed uniformly over the atom, and negative electrons are imbedded in it like raisins in a cake (Fig. 12–14). The positive charge is so diffuse, however, that an alpha particle could never be close to more than a small part of it, and any deflection of a positive alpha particle would be minimal. The experimental results called for a concentration of positive charge small enough to permit a close approach of an alpha particle.

Rutherford was convinced by the experimental evidence that a new model of the atom was necessary. He proposed one in which the positive charge is confined to a small sphere, the **nucleus**, which has a radius no larger than 10^{-14} m and which also contains

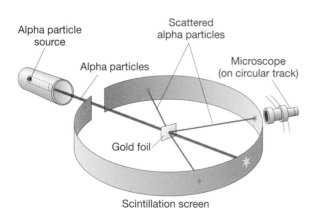

Figure 12–13 Experiment of Rutherford, Geiger, and Marsden, showing that alpha particles are scattered by a metal foil in all directions.

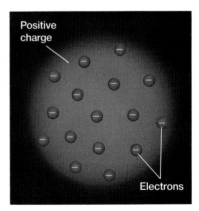

Figure 12–14 J. J. Thomson's model of the atom. Negative electrons are imbedded in a diffuse positive charge much like plums in a pudding. (Source: Larry D. Kirkpatrick and Gerald F. Wheeler, *Physics: A World View*. Philadelphia: Saunders College Publishing, 1992, p. 497.)

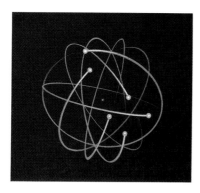

Figure 12–15   Rutherford's model of the atom. The positive charge is concentrated in a tiny spot at the center, the nucleus. The negative electrons orbit the nucleus as planets orbit the sun. (Source: Larry D. Kirkpatrick and Gerald F. Wheeler, *Physics: A World View.* Philadelphia: Saunders College Publishing, 1992, p. 500.)

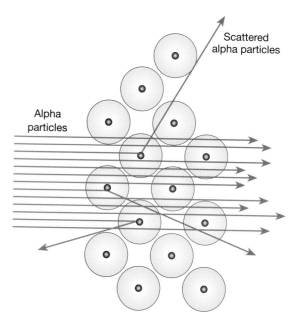

Figure 12–16   Alpha particle scattering according to Rutherford's nuclear model of the atom. Most of the particles pass through the foil undeflected. Those particles that approach a positive nucleus closely experience Coulomb repulsive forces and are scattered through various angles.

most of the mass of the atom (Fig. 12–15). The radius of an atom is 10^{-10} m, approximately 10,000 times as great as that of the nucleus. An atom, therefore, is almost entirely empty space. Light, negative electrons in a number that will balance the positive charge on the nucleus are distributed over a region within 10^{-10} m from the center. Thus, hydrogen has a positive charge of 1 unit on its nucleus and one extranuclear electron; helium has a charge of 2+ on the nucleus and two electrons; gold has a nuclear charge of 79+ and 79 electrons. Most of the alpha particles pass through the largely empty atom undeflected; those few that encounter a tiny nucleus account for the observed large-angle scattering (Fig. 12–16).

Every atomic model must account for the stability of the atom. How is the hydrogen atom held together? Gravitational forces are far too small. If we assume only electrical forces, a possibility is that the electron is stationary outside the nucleus. But this arrangement is unstable: The electrostatic attraction between the unlike charges of the nucleus and the electron would cause them to merge, destroying the atom.

What if the electron could travel in orbit around the nucleus, like a charged solar system with electrical rather than gravitational forces? The planetary model, alas, had a fatal flaw. It comes from Maxwell's theory predicting that an electron orbiting an atom would radiate energy in the form of electromagnetic radiation with a frequency equal to its frequency of revolution, because it is undergoing acceleration. As

the electron loses energy by radiation, it should spiral closer and closer to the nucleus, radiating more and more energy until it reached the nucleus within billionths of a second. During this process, radiation should be given off in a continuous range of frequencies, a band spectrum, contrary to the observed sharp and discrete line spectrum. The Rutherford nuclear atom was, therefore, inherently unstable whether the electron was stationary or in orbital motion around the nucleus.

12–7 PLANCK'S QUANTUM HYPOTHESIS

The failure of classical radiation theory to explain the stability of the atom was but one of the shortcomings of that theory. Another that troubled physicists in the last quarter of the nineteenth century concerned a prediction that it made about radiation.

A hot piece of iron emits electromagnetic radiation. If hot enough, much of this radiation is in the visible part of the spectrum. Thus, as the iron is heated, it glows first a dull red, then a brighter red, and, if heated to a high enough temperature, white like the tungsten filament in a light bulb. The actual spectrum of colors emitted from a very hot object differs completely from the prediction of classical physics. According to the theory, the radiation should always be in the form of very short-wavelength or high-frequency electromagnetic waves, such as ultraviolet light waves and X-rays. Lord Rayleigh (John William Strutt, 1842–1919) and Sir James Jeans (1877–1946) gave the name "**ultraviolet catastrophe**" to this prediction. If borne out by experience, it would mean that if you opened the door of a hot oven, you would be bombarded with deadly, high-energy X-rays. The oven actually emits mostly red light.

The solution to the radiation dilemma came in 1900. Max Planck (1858–1947), a physicist at the University of Berlin, discovered a formula that correctly describes the distribution of intensity with respect to wavelength of a "black body"—a perfect absorber or emitter of radiation (Fig. 12–17). To explain why the formula worked, Planck had to assert that radiant energy comes in discrete amounts, or "**quanta**," and that the energy content of each quantum was directly proportional to the frequency.

Energy of a quantum =

    A constant × Frequency of quantum   (12–3)

$$E = hf$$

or, since

$$f = \frac{c}{\lambda}$$

$$E = \frac{hc}{\lambda}$$

The constant, h, known as **Planck's constant**, has since proved to be a fundamental constant of nature. By matching the theory with observations, it was determined to be 6.63×10^{-34} joule-second. Planck's theory avoided the ultraviolet catastrophe by limiting the energy of an incandescent body to a finite number of sources. The high-frequency quanta require more energy; hence, fewer of them would be radiated.

Planck's hypothesis had an important virtue: It worked; that is, it agreed with experiment. The theoretical energy-distribution curve for radiation matched the experimental curve. The price that was

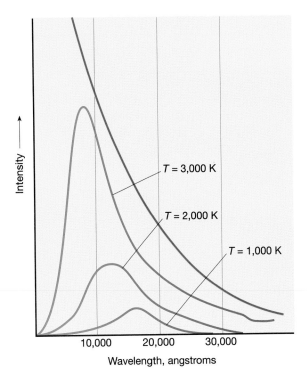

Figure 12–17 Radiation curves of a "black body." At a particular temperature, the radiation has a characteristic distribution of wavelengths (solid curves). The hotter the body, the more it radiates at shorter wavelengths. However, the classical theory predicted infinite radiation at short wavelengths (red line).

paid for this success was a revolution in the way people thought about electromagnetic radiation: not as waves, but as discrete bundles or quanta of energy.

To see how great a departure Planck's quantum hypothesis is from classical theory, recall that in classical theory, the energy of a wave is related to the amplitude. Large waves, such as ocean waves, have a large energy. There is no necessary relation between the energy and the frequency. One can have waves of low energy and high frequency, or of high energy and low frequency. According to Planck, however, each electromagnetic wave carries a minimum energy that depends on its frequency. It took a quarter-century for this strange idea to be fully assimilated into the mainstream of scientific thought. Planck himself tried for years to reconcile the quantum hypothesis with classical theory but did not succeed. He had created a revolution, almost to his dismay.

EXAMPLE 12–2

Calculate the energy of an X-ray quantum of 1.00 Å.

SOLUTION

We convert the wavelength, given in angstrom units, to meters. Knowing Planck's constant, h, and the velocity of light, c, we apply Equation 12–3.

$$h = 6.63 \times 10^{-34} \text{ joule} \cdot \text{s}$$

$$\lambda = 1.00 \text{ Å} = 10^{-10} \text{ m}$$

$$c = 3 \times 10^8 \text{ m/s (velocity of light)}$$

$$c = f \cdot \lambda \text{ (wave equation)}$$

$$\therefore f = \frac{c}{\lambda}$$

$$E = ?$$

$$E = hf = \frac{hc}{\lambda}$$

$$= \frac{(6.63 \times 10^{-34} \text{ joule} \cdot \text{s})\left(3 \times 10^8 \, \frac{\text{m}}{\text{s}}\right)}{10^{-10} \text{ m}}$$

$$= 19.89 \times 10^{-34 + 8 + 10} \text{ joule}$$

$$= 19.89 \times 10^{-16} \text{ joule}$$

$$= \boxed{1.99 \times 10^{-15} \text{ joule}}$$

EXTENSION

Calculate the energy of a quantum that has a wavelength of 4000 Å. (Answer: 5.00×10^{-19} joule)

■

12–8 EINSTEIN'S PHOTOELECTRIC EQUATION

Planck's quantum hypothesis was soon applied successfully to something that had long been mystifying, the **photoelectric effect** (*photo*, "light") (Fig. 12–18). Discovered by Heinrich Hertz and Wilhelm Hallwachs (1859–1922), the photoelectric effect has these features:

1. Electrons are emitted from the surfaces of certain metals when exposed to visible or ultraviolet light.
2. Increasing the intensity of the light increases the number of photoelectrons, but not the velocity with which they leave the surface of the metal.
3. There is a threshold frequency for each substance below which the effect does not occur.
4. The higher the frequency of the light, the greater the kinetic energy of the photoelectrons.

The mystifying aspects of the photoelectric effect, in terms of classical theory, come from the lack of a relation between the intensity of the incident light and the velocity of the photoelectrons. No matter how weak the illumination, provided that it exceeds the threshold frequency, the emission of photoelectrons takes place instantly. This fact defied explanation.

From the point of view of classical theory, there should be no threshold frequency. Given time, electrons might soak up enough energy from the electric field vector of the incident light to escape from the metal surface at any applied frequency. According to classical wave theory, the velocity of photoelectrons should depend on the amplitude of the electric field vector in the incident light, and therefore on the intensity rather than the frequency, just as a cork on the surface of a pond bobs weakly if the height—amplitude—of the wave is small. Classical theory utterly failed to explain the facts.

Albert Einstein (1879–1955), the theoretical physicist, resolved those difficulties when he ex-

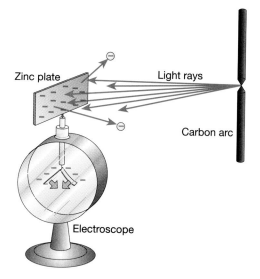

Figure 12–18 The photoelectric effect. Light striking the negatively charged zinc plate causes it to lose charge by emitting electrons. The negatively charged electroscope connected to the zinc therefore is discharged.

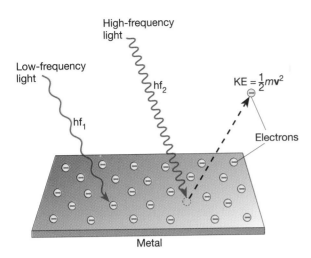

Figure 12–19 Light below the threshold frequency for a given substance does not eject photoelectrons; high-frequency radiation does.

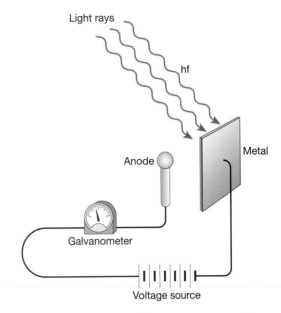

Figure 12–20 Diagram showing arrangement for measuring the kinetic energy of photoelectrons. The retarding voltage is adjusted until the current in the galvanometer just becomes zero, indicating that photoelectrons are not reaching the anode. This retarding voltage is a measure of the maximum kinetic energy of the photoelectrons.

tended Planck's quantum hypothesis to light. He proposed that light is not only emitted in discrete packets of energy, hf, but is also absorbed as such. (The quanta of light were later given the name **photons** by Gilbert N. Lewis.) A bright red light is regarded as a stream of many relatively weak photons; a faint blue light as relatively few energetic photons.

The energy of a light wave, according to Einstein, is not spread out all along the wave front but is concentrated in photons. The energy transferred from a photon to an electron in the photoelectric effect is, therefore, related only to the frequency, not the intensity. No matter how intense the illumination, if the frequency is too low—below the threshold—a photon cannot transfer enough energy to an electron to eject it from the surface (Fig. 12–19). At a given frequency of incident light, the more intense the light, the greater the number of photoelectrons emitted with the same energy. These ideas are expressed in Einstein's photoelectric equation:

$$\begin{pmatrix} \text{Energy of} \\ \text{incident} \\ \text{photon} \end{pmatrix} =$$

$$\begin{pmatrix} \text{Energy required} \\ \text{to free electron} \\ \text{from metal} \end{pmatrix} + \begin{pmatrix} \text{Kinetic energy of} \\ \text{electron as it leaves} \\ \text{the metal surface} \end{pmatrix}$$

$$E = \quad W \quad + KE \qquad (12\text{--}4)$$

$$\text{Work function}$$

Robert A. Millikan tested Einstein's theory in a series of experiments between 1912 and 1917. Using sodium and other metals as targets, he illuminated a target with light of various frequencies and studied the maximum kinetic energy of photoelectrons as a function of the frequency of the incident light (Fig. 12–20). His results were in excellent agreement with Einstein's theory. Millikan also determined the value of Planck's constant from his measurements and found it to be in close agreement with values derived by other methods. By showing that the straight lines obtained for different metals all had the same slope (Fig. 12–21), Millikan proved that Planck's constant is the same in all cases. Einstein's theory thus passed the quantitative test with flying colors, as did the quantum hypothesis upon which it rested.

The photoelectric effect is applied in **photocells**, which are used to open store doors when patrons cut off a light beam, to sound burglar alarms, to turn on street lights, and to produce pictures in television cameras and the soundtracks of motion pictures. In one type of photocell, the cathode is coated with ce-

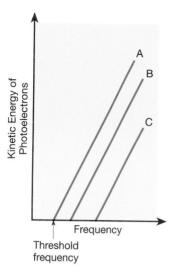

Figure 12–21   Plots of maximum kinetic energy of photoelectrons against frequency of incident light for different metals are parallel, although threshold frequencies vary.

sium, a light-sensitive material that emits electrons when illuminated. The electrons collect at a second electrode, the anode, the current being proportional to the illumination. The sound information is stored on the film in the form of spots of varying widths. When a film is run between a light source and the photoelectric cell, variations in the light intensity reaching the cell cause pulsations in the current that, after being amplified, activates a loudspeaker and reproduces the sound (Fig. 12–22).

12–9 BOHR'S ENERGY-LEVEL ATOM

The success of Planck and Einstein with the quantum theory was extended by the physicist Niels Bohr (1885–1962) in 1913 to the dilemma of the stability of the hydrogen atom. Bohr's model of the atom combined classical physics, the quantum theory of light, and some new ideas.

Bohr assumed **Rutherford's nuclear model** of the hydrogen atom of an electron orbiting around a

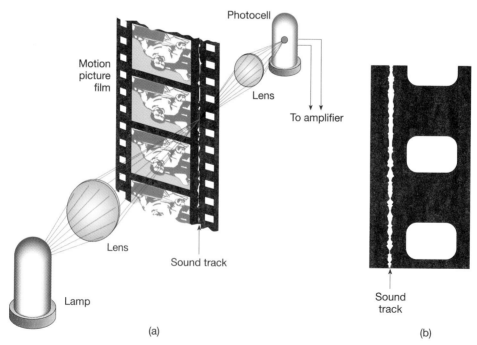

Figure 12–22   Use of a photocell in a sound motion picture projector. The varying current from the sound track is converted to sound by an amplifier and speaker.

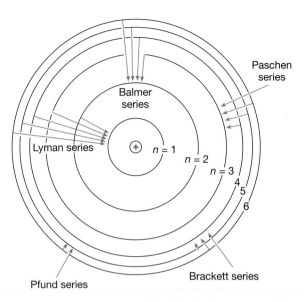

Figure 12–23 Bohr model of the hydrogen atom. The circular electron orbits have radii in the ratios of 1:4:9:16. An electron emits light of a characteristic frequency (indicated by arrows) when it falls from an orbit of higher energy to one of lower energy. A group of energy transitions gives rise to a spectral series. (Not drawn to scale)

tiny nucleus and held there by Coulomb's law. Reasoning that there must be definite energy levels to account for the discrete spectrum observed, Bohr proposed that:

1. Of all the possible circular orbits, only orbits with certain radii are allowed. (Bohr could not say why; the idea just "worked.")
2. When the electron is in one of the allowed orbits, it does not radiate energy despite the fact that it is undergoing acceleration.
3. When an electron makes a transition from a large orbit to a smaller one, the energy lost by the electron in falling to a lower level appears as a photon of light.

To obtain photons with frequencies that were in agreement with those observed. Bohr discovered that the allowed radii corresponded to specific values of **angular momentum**:

$$mvr = \frac{nh}{2\pi}$$

where n is any integer: 1, 2, 3, ... The Balmer formula now had meaning. For the Lyman series,

$$\frac{1}{\lambda} = R_{\mathrm{H}}\left(\frac{1}{1^2} - \frac{1}{n^2}\right)$$

"1" represents the "**ground state**" for the series, and the values of $n = 2, 3, 4, ...$ represent higher or "**excited**" **energy levels** (Fig. 12–23). Ordinarily, the electron occupies the lowest or ground state, but by absorbing a quantum of energy, it is promoted to a higher level, where it remains briefly (on the order of 10^{-8} second) before falling back to a lower level. When an electron returns from the second orbit to the first, the difference in energy between the two orbits is emitted as a photon with a wavelength of 1200 Å (in the ultraviolet). Bohr's theory accurately explained the hydrogen spectrum and provided a physical model for a stable atom.

12–10 THE COMPTON EFFECT

In the 1920s, Arthur Holly Compton (1892–1962), then at Washington University, carried out scattering experiments using a beam of X-rays having a single, well-defined frequency. Since earlier experiments had established the wave nature of X-rays, it could be assumed that the scattered X-rays would have the same wavelength as the rays of the incident beam. Ordinary light, for example, is scattered by air particles in all directions with no change in wavelength.

Compton did find such scattering, the "**primary**" **scattering**, but he also detected a different kind in which the scattered X-rays had a longer wavelength than the rays of the original beam. Compton could not explain the presence of "**secondary**" **scattering** by classical theory. He could interpret the results by treating X-rays as photons having energy and momentum and the scattering as the result of collisions between a photon and an electron in the target. In effect, in scattering experiments, an X-ray beam behaves like a stream of particles.

Before a collision, an electron is assumed to be stationary in a target atom; the energy that an X-ray photon brings to the collision sets the electron in motion. The collision causes the X-ray to be deflected with reduced energy, the energy difference appearing as the recoil energy of the electron (Fig. 12–24). The laws of the conservation of momentum and en-

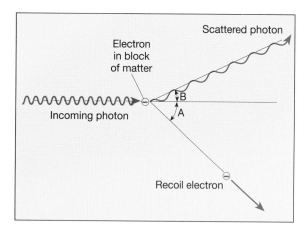

Figure 12–24   The Compton effect explained how X-rays passing through matter may increase in wavelength, producing "secondary" scattering. An X-ray photon, like a particle, collides with an electron, is deflected, and loses energy to the electron.

ergy apply in every photon-electron collision. In a typical result, with incident X-rays having a wavelength of 0.708 Å, Compton found an increase of 0.022 Å in wavelength for secondary X-rays scattered at 90 degrees, corresponding to a decrease in photon energy. The difference in energy was imparted to an electron.

The **Compton effect** indicated that radiation has a **wave-particle duality**. Light behaves in some cases—interference, polarization—like a wave, and in other cases, as in the photoelectric effect and the Compton effect, like a particle. It is wave-like in interactions with radiation and particle-like in its interactions with matter.

12–11 MATTER WAVES

Compton's proof that X-rays have a particle as well as a wave character suggested to Louis de Broglie (1892–1987) that duality might extend to electrons and other atomic particles, since nature is in general symmetrical. Now that it had been shown that radiation had wave and particle properties, perhaps matter did also. De Broglie introduced the hypothesis that there is a **wavelength associated with a moving particle** and that the wavelength is governed by the following relation:

$$\text{Wavelength} = \frac{\text{Planck's constant}}{\text{Momentum}} \qquad (12\text{–}5)$$

$$\lambda = \frac{h}{m\mathbf{v}}$$

EXAMPLE 12–3

What is the de Broglie wavelength associated with an electron traveling at 3.00×10^7 m/s (1/10 the speed of light)?

SOLUTION
Since Planck's constant and the mass and velocity of the electron are known, we apply Equation 12–5 to determine the wavelength of the electron. We reduce joules to fundamental units in order to express the result in a proper unit of length.

$$h = 6.63 \times 10^{-34} \text{ joule} \cdot \text{s}$$
$$m = 9.11 \times 10^{-31} \text{ kg}$$
$$\mathbf{v} = 3.00 \times 10^7 \text{ m/s}$$
$$\lambda = ?$$
$$\text{joule} = \mathbf{F}d$$
$$\mathbf{F} = m\mathbf{a}$$
$$\therefore \text{joule} = m\mathbf{a}d$$
$$= \frac{\text{kg} \cdot \text{m} \cdot \text{m}}{\text{s}^2}$$
$$= \frac{\text{kg} \cdot \text{m}^2}{\text{s}^2}$$

$$\lambda = \frac{h}{m\mathbf{v}}$$

$$= \frac{6.63 \times 10^{-34} \text{ joule} \cdot \text{s}}{(9.11 \times 10^{-31} \text{ kg}) \left(3.60 \times 10^7 \dfrac{\text{m}}{\text{s}} \right)}$$

$$= \frac{0.224 \times 10^{-10} \dfrac{\text{kg} \cdot \text{m}^2}{\text{s}^2}}{\text{kg} \dfrac{\text{m}}{\text{s}}}$$

$$= .242 \times 10^{-11} \text{ m} \times 10^{10} \frac{\text{Å}}{\text{m}}$$

$$= \boxed{0.242 \text{ Å}}$$

(which is in the X-ray range)

EXTENSION

A proton (mass $= 1.67 \times 10^{-27}$ kg) is traveling at 3.00×10^7 m/s. What is the de Broglie wavelength? (Answer: 1.32×10^{-14} Å)

■

George P. Thomson (son of J. J., 1892–1975), set out to test de Broglie's hypothesis of the wave nature of particles. He designed an experiment to observe interference effects for an electron beam. His approach was to direct a beam of high-energy electrons through a thin gold foil. Since electrons would have about the same wavelength as X-rays, he expected to get an electron diffraction picture similar to X-ray diffraction pictures (Fig. 12–25).

Thomson demonstrated the **wave nature of the electron** in 1927, just 30 years after his father had established its particle nature and several months after Clinton J. Davisson (1881–1958) and Lester H. Germer (1896–1971) at the Bell Telephone Laboratory obtained an electron diffraction pattern from a

beam of electrons incident on a nickel crystal. The experiments of Davisson, Germer, and Thomson showed that electrons do have wave properties and that their wavelengths are correctly given by de Broglie's formula. Matter, like radiation, was thus shown to have a dual nature: particle and wave.

Wave properties have since been discovered for other particles—protons, neutrons, alpha particles, even whole atoms and molecules. They are all accurately described by de Broglie's formula. Interference effects have not been observed with such things as Ping-Pong balls or baseballs, simply because their wavelengths are so short as to be unobservable. The de Broglie wavelength of a 0.250-kilogram baseball traveling at 20.0 m/s is 1.32×10^{-34} meters, or only about 10^{-27} as large as a light wave.

De Broglie's hypothesis changed Bohr's picture of the atom (in which the electron is a locatable particle orbiting the nucleus at a radius r with a specified velocity **v**) to one in which the electron becomes a wave along that orbit still located at distance r from the nucleus. De Broglie drew a comparison between

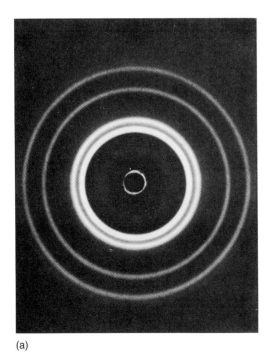

(a)

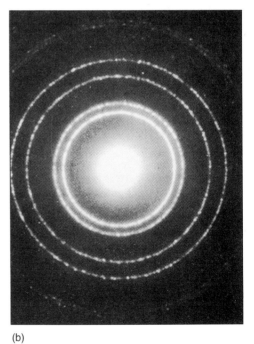

(b)

Figure 12–25 (a) Diffraction pattern made by a beam of X-rays passing through thin aluminum foil. (b) Diffraction pattern made by a beam of electrons passing through the same foil. (Source: Holton, G., et al., *The Project Physics Course Text.* New York: Holt, Rinehart & Winston, Inc., 1970, Unit 5, Chapter 20, p. 94. Photographs from the P.S.S.C. film *Matter Waves.*)

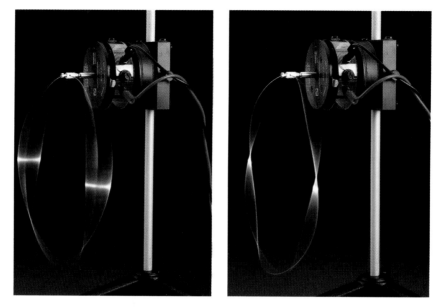

Figure 12–26   Standing waves on a circular wire can form when the circumference is equal to a whole number of wavelengths. (Courtesy of PASCO Scientific)

Figure 12–27 De Broglie electron waves in an atom. Each Bohr orbit, circumference = $2\pi r$, accommodates an electron standing wave pattern that exactly fits the equation $n\lambda = 2\pi r$, for which $n = 1, 2, 3$, etc. If $n = 2r$, the wave interferes destructively with itself and cannot exist.

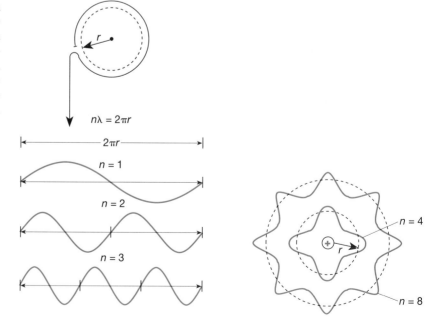

the set of **discrete energy levels** of an atom and the set of **discrete standing waves** observed in vibrating strings and air columns. Wave motion that is confined in any way, as on a string fixed at both ends, results in standing waves, for which only certain wavelengths are possible (Fig. 12–26). If an electron is confined around a nucleus, then it too must be present as a standing wave that closes on itself; otherwise, destructive interference will cause the amplitude of the wave to drop to zero and the wave will die out.

In the atom, the possible modes correspond to Bohr's orbits, and only those orbits for which standing electron waves are possible. The condition is that the circumference of an orbit, $2\pi r$, must be equal to a whole number of wavelengths λ: $\lambda, 2\lambda, 3\lambda, 4\lambda, \ldots$ The length of the whole circumference of an orbit at radius r must be a whole wavelength, or two shorter wavelengths, or three still shorter wavelengths, and so on (Fig. 12–27).

The wave properties of electrons are applied in the **electron microscope** (Fig. 12–28). To see an object, an agent, such as light, that is smaller in wavelength than the object is required; otherwise, the light will pass by the object as though it did not exist. Since electron wavelengths are shorter than light wavelengths, electrons can reveal a wealth of detail about the structure of such specimens as bacteria and viruses, which are beyond the capability of even the best optical microscopes. For a 50,000-volt accelerat-

ing voltage, common in electron microscopy, the electron wavelength is about 5×10^{-11} meter, roughly 1/10,000 the wavelength of visible light. An electron microscope has a resolving power about a thousand times better than that of an ordinary microscope and can resolve objects down to about a hundred-millionth of an inch. An ordinary microscope can reveal details down to only about a hundred-thousandth of an inch.

In an electron microscope, a beam of high-speed electrons is passed through a sample thin enough to transmit the beam. Instead of optical lenses, electric fields or magnetic fields are used to refract the beam. A greatly magnified image can be photographed or made visible on a fluorescent screen. Ernst Ruska built the first electron microscope in 1931, for which he received half of the 1986 Nobel Prize in physics.

12–12 QUANTUM THEORY AND LUMINESCENCE

A fluorescent lamp and a television picture tube have this property in common: Light is produced by exciting a **phosphor**, a substance that absorbs electromagnetic energy and then re-emits it as visible light. Wurtzite (zinc sulfide) and fluorite (calcium fluoride) are two of the many natural phosphors. The difference between the devices lies in the agent used for excitation. In a fluorescent lamp, the phosphor is excited by ultraviolet rays and in a television picture tube by a beam of high-speed electrons. The key to the emission of light is the ability of electrons, upon absorbing a quantum of energy, to jump to an "excited" state. On returning to a lower state, they re-emit the absorbed energy. In a phosphor, part of this energy is emitted as light.

The **fluorescent lamp** was first exhibited in the United States at the New York World's Fair in 1939 and now challenges the incandescent lamp in popularity. Essentially, it is a mercury-vapor lamp in a long glass tube with a phosphor coated on the inside surface of the tube (Fig. 12–29). When the two filaments built into the ends of the tube are heated, they emit electrons that move through the tube, colliding with mercury atoms. The primary source of light is the excitation of the mercury atoms. The collisions also

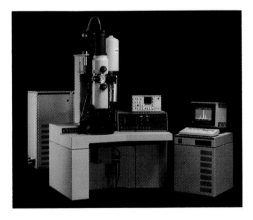

Figure 12–28   The electron microscope depends on the wave properties of electrons. (Philips Electronic Instruments, Inc.)

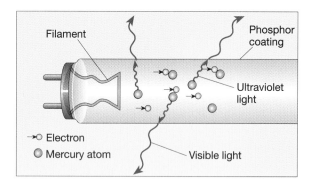

Figure 12–29   Fluorescent lamp.

yield ultraviolet light, part of which is absorbed by the phosphor on the wall of the tube and reradiated as visible radiation. By a suitable selection of phosphors, light of various wavelengths can be added to the primary source from the mercury to produce any desired color.

Fluorescence and phosphorescence differ in the persistence of the emission. In **fluorescence**, the emission of light persists for but a few hundred-millionths of a second after the source is removed.

This is the time required for an electron to make a transition in an atom and emit a photon. If the emission persists for minutes or even hours, it is described as **phosphorescence**. The picture tube of a black-and-white television receiver is covered with a fluorescent phosphor that glows with a slightly bluish light when sprayed with a jet of electrons (Fig. 12–30). The more intense the jet, the brighter the glow. Phosphors that fluoresce with different colors are used in color television.

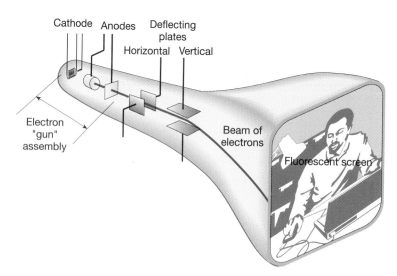

Figure 12–30   Diagram of a picture tube in a television receiver. High-speed electrons produce fluorescence.

12–13 LASER LIGHT

"**Laser**" is an acronym for "*l*ight *a*mplification by *s*timulated *e*mission of *r*adiation." Laser radiation has three main characteristics: The waves are coherent—all in step; are highly **monochromatic** —all with the same wavelength; and can be propagated over long distances in the form of well-collimated beams. The light emitted by an incandescent lamp, in contrast, consists of uncoordinated waves of many different wavelengths. The first operating laser was constructed in 1960 by Theodore H. Maiman (b. 1927) of the Hughes Research Laboratory.

Stimulated emission lies at the heart of the laser principle. By absorbing a photon, an atom is raised to an excited quantum state. As Albert Einstein suggested in 1917, the atom, while still in the excited state, can be stimulated to emit a photon if it is struck by an outside photon that has precisely the energy of the photon that would otherwise be emitted spontaneously (Fig. 12–31). The incoming photon is thereby augmented by the one given up by the excited atom. What is remarkable is that the wavelengths of the two photons are precisely in phase.

With an active medium, most of the atoms can be placed in an excited state so that an electromagnetic wave of the right frequency passing through them will stimulate a **cascade of photons**, thereby forming an intense light wave. Through the activating process called pumping, an excess of excited atoms—a population inversion—is created. In a gas laser or a solid-state laser, this is done by passing an electric current through it or by illuminating it. In a typical laser, a ruby crystal (aluminum oxide) to which 0.05% of the active ingredient, chromium atoms, is added, optical pumping is achieved from a xenon flash lamp that emits white radiation. Reflection of the coherent light back and forth from a pair of mirrors through the excited atoms also augments the cascade process. After the laser light has built up, it emerges at one end.

Lasers have many applications, although their total potential remains to be tapped in the years ahead. Surveying, communications, holography, surgery, photography, spectroscopy, and Doppler-shift mea-

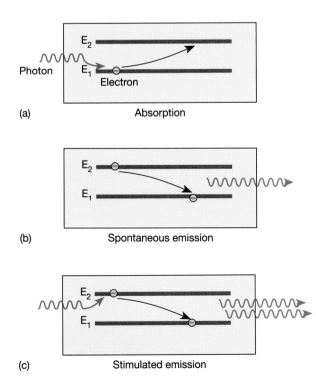

(a)   Absorption

(b)   Spontaneous emission

(c)   Stimulated emission

Figure 12–31 Comparison of (a) absorption, (b) spontaneous emission, and (c) stimulated emission of photons. Stimulated emission is the basis of the laser principle.

surements are some of the areas in which they have been applied (Fig. 12–32). The lunar laser reflector placed on the moon in 1969 by the astronauts of *Apollo 11*, Neil A. Armstrong and Edwin E. Aldrin, Jr., made it possible to measure the distance between the earth and the moon within about 6 inches. The distance was determined by aiming the intense light from a laser at the moon and measuring the time required for a brief pulse to travel to the lunar laser reflector on the moon and back. Studies over a period of years may reveal important information about changes in the location of the North Pole and whether the gravitational constant, G, is constant or is weakening.

Figure 12–32   A laser reads a bar code at a supermarket checkout. (Shambroom/Photo Researchers, Inc.)

SUMMARY OF EQUATIONS

Balmer's Formula for the Hydrogen Spectrum

$$\frac{1}{\lambda} = R_{\mathrm{H}}\left(\frac{1}{m^2} - \frac{1}{n^2}\right)$$

Einstein's Photoelectric Equation

$$\begin{pmatrix}\text{Energy of}\\ \text{incident}\\ \text{photon}\end{pmatrix} = \begin{pmatrix}\text{Energy required}\\ \text{to free electron}\\ \text{from metal}\end{pmatrix} + \begin{pmatrix}\text{Kinetic energy of}\\ \text{electron as it leaves}\\ \text{the metal surface}\end{pmatrix}$$

$$E = W + \mathrm{KE}$$
$$\text{(Work function)}$$

De Broglie Wavelength of a Moving Particle

$$\text{Wavelength} = \frac{\text{Planck's constant}}{\text{Momentum}}$$

$$\lambda = \frac{h}{m\mathbf{v}}$$

Energy of a quantum = Planck's constant × Frequency of quantum

$$E = hf = \frac{hc}{\lambda}$$

KEY TERMS

Alpha Particle   A positively charged particle composed of two protons and two neutrons (helium nucleus) emitted by certain radioactive substances.

Angular Momentum   For an object orbiting a point, the product of the linear momentum and the radius of the path: $m\mathbf{v}r$.

Anode   In a Crookes tube, the positive terminal.

Balmer's Formula   A formula that yields the prominent lines in the visible region of the hydrogen spectrum with an amazing degree of accuracy.

Beta Particle   A high-energy electron emitted by a radioactive nucleus.

Black-Body Radiation   The radiation distribution of a perfect emitter of radiation.

Cathode   In a Crookes tube, the negative terminal.

Cathode Ray   An electron emitted from the negative electrode in an evacuated tube.

Compton Effect   X-ray bombardment of various targets produces primary scattering and secondary scattering, the latter from the collision of an X-ray photon with an electron of the target material, as

Arthur Holly Compton demonstrated.

Crookes Tube   Cathode-ray tube designed by Sir William Crookes that showed that cathode rays travel in straight lines.

De Broglie Wavelength   The wavelength associated with a moving particle, such as an electron or proton.

Deflect   To turn or bend from a straight course.

Diffraction Grating   A system of close parallel lines ruled on a polished surface to produce spectra by diffraction.

Discrete Energy Level   In an atom, according to Bohr's model, the location of an electron in an allowed orbit.

Discrete Standing Wave   The interference pattern produced by two waves of equal amplitude and frequency traveling in opposite directions.

Electrode   An anode or a cathode.

Electron   A subatomic particle that has a very low mass and carries a single negative electric charge.

Electron Microscope   An instrument in which a beam of electrons focused by means of an electron lens is used to produce an enlarged image of a minute object.

Excited Energy Level   A state that has a higher energy than the ground state of an atom.

Fluorescence   The property of producing light by exciting a phosphor, a substance that absorbs electromagnetic energy and then re-emits it as visible light.

Fluorescent Lamp   A tubular electric lamp having phosphor on its inner surface and containing mercury vapor.

Fraunhofer Lines   Dark lines on a spectrum that represent absorption lines.

Gamma Ray   High-energy and short-wavelength radiation that frequently accompanies alpha- and beta-emission.

Geissler Tube   A vacuum tube first made by Heinrich Geissler.

Ground State   The lowest possible energy state of an atom or molecule.

Laser   An acronym for "*l*ight *a*mplification by *s*timulated *e*mission of *r*adiation." An optical device that emits coherent waves that are highly monochromatic and can be propagated in the form of well-collimated beams.

Monochromatic   A beam of radiation having a single, well-defined frequency.

Nuclear Atom   In Rutherford's model of the atom, the positive charge of an atom and most of its mass are confined to a small sphere, the nucleus, and the negative electrons are distributed at some distance from the nucleus.

Nucleus   The positively charged central core of an atom, which contains the protons and neutrons.

Oil-Drop Experiment   An experiment designed by Robert A. Millikan that enabled him to measure the charge on the electron.

Phosphor   A substance that emits light when excited by radiation.

Phosphorescence   Similar to fluorescence, except that the emission of light may persist for minutes or even hours.

Photocell   A cell whose electrical properties are modified by the action of light.

Photoelectric Effect   The emission of electrons from a metal surface when light shines on it.

Photon   A packet of electromagnetic energy, hf. A light quantum.

Planck's Constant   The proportionality constant h in the Planck relationship between energy and the frequency of radiation.

Primary Scattering   In the Compton effect, scattered X-rays that have the same wavelength as the rays of the incident beam.

Quantum   The unit of radiant energy. According to Max Planck, radiant energy comes in discrete amounts, or packets, called quanta.

Radioactivity   A process in which an element is transmuted into another element through the emission of nuclear particles and radiations.

Scattering Experiment   A beam of particles or rays is directed at a target. Interactions between the projectiles and the atoms of the target can lead to a better understanding of the atoms.

Secondary Scattering   In the Compton effect, scattered X-rays that have a longer wavelength than the rays of the incident beam.

Spectroscope   An instrument that spreads light out by wavelength.

Stimulated Emission   The basis of the laser principle. Atoms in the excited state are stimulated to emit photons with wavelengths that are precisely in phase with light of the right frequency passing through them.

Ultraviolet Catastrophe   Classical theory predicts that most of the radiation from a very hot body will be in the ultraviolet of X-ray range. This prediction fails.

Wave-Particle Duality   The idea that matter and radiation manifest both wave and particle properties.

X-Rays   High-energy photons with wavelengths from approximately 0.1 Å to 50 Å.

THINGS TO DO

1. (a) Dip a loop of moistened wire in a little table salt and hold it over a nearly colorless flame. The color characteristic of sodium is produced. (b) Repeat with calcium chloride.

2. Observe the fluorescence of teeth, nails, fluorescent light bulbs, and various minerals in the presence of ultraviolet light.

3. Take readings with a light meter of the light intensity in various parts of a room.

4. If you have access to a Crookes deflection-ray tube, hold one pole of a strong magnet near the top of the tube, then the other pole, and observe the deflection of the fluorescent line.

5. If a spectroscope is available, observe the spectrum emitted by various light sources such as a fluorescent bulb.

6. Examine closely the screen of a color television tube, possibly with a magnifying glass. Note the pattern of colored dots or streaks.

7. Watch demonstrations of the excitation of various gases in Geissler tubes.

8. Certain museums offer laser shows. Is there one in your area that does?

9. Interview an ophthalmologist on the practice of laser surgery.

10. Invite a radiologist to discuss developments in the field.

EXERCISES

Atomic Spectra and Balmer's Formula

12–1. Discuss the basis for a claim that there is gold on the sun.

12–2. Calculate the wavelength of the second line of the Balmer series for hydrogen.

12–3. Calculate the wavelength of the third line of the Paschen series for hydrogen.

12–4. A photon is emitted from a hydrogen atom that undergoes a transition from the $n = 4$ state to the $n = 2$ state. Calculate the wavelength of the emitted photon.

12–5. Why does the "fingerprint" of an atom have more than one characteristic frequency of light?

12–6. Why was the Balmer series detected before the Lyman series?

12–7. Does a neon sign emit light of a continuous spectrum or only a few discrete wavelengths? Explain.

12–8. What do the lines in the spectrum of an atom represent?

The Rutherford-Bohr Model of the Atom

12–9. How did Bohr's model of the atom account for the Balmer and other spectral series of the hydrogen atom?

12–10. Compare radioactivity and X-rays.

12–11. What was the experimental basis of Rutherford's concept of a nuclear atom?

12–12. Although hydrogen has but a single electron, the complete hydrogen spectrum consists of numerous lines. How do you account for this?

Quantum Physics

12–13. Discuss two instances in which classical electromagnetic theory failed to account for the experimental observations.

12–14. In what way are quanta of radiation a return to the corpuscular theory of light?

12–15. In the photoelectric effect, the energy of a photoelectron is less than that of the incident photon. Explain.

12–16. Yellow light can induce the photoelectric effect in a certain metal, but orange light cannot. Would blue light? Explain.

12–17. The human eye is most sensitive to light of approximately 5500 Å in wavelength. What is the energy in joules of such a photon?

12–18. Which photon has the greater energy—infrared or ultraviolet?

12–19. How does the frequency of an X-ray photon that is scattered by an electron compare with the frequency of the incident photon?

12–20. If the frequency of a photon is doubled, how is its energy affected?

12–21. If the wavelength of a photon is doubled, what happens to its energy?

12–22. Explain why ultraviolet light causes sunburn, but visible light does not.

12–23. Which colors of the visible spectrum have photons of higher energy?

12–24. Why was the Compton effect considered important for verifying Einstein's idea of light quanta?

Wave-Particle Duality

12–25. How did de Broglie's hypothesis help to explain the stability of the atom?

12–26. What is the de Broglie wavelength of a 0.500-kilogram football traveling at 15.0 m/s?

12–27. Calculate the de Broglie wavelength associated with the earth, if its mass is 6.00×10^{24} kilograms and it revolves around the sun with a velocity of 3.00×10^4 m/s.

12–28. How is the wave-particle duality related to the concept of symmetry in nature?

12–29. Discuss the meaning of the dual nature of radiation and matter.

12–30. Account for the considerably greater detail seen in a bacterial specimen revealed by an electron microscope as compared with the same specimen revealed by a light microscope.

12–31. Is light a wave or a particle? Explain.

12–32. Is an electron a particle or a wave? Explain.

12–33. Calculate the momentum of a photon of wavelength 5000 Å.

12–34. Calculate the de Broglie wavelength in meters for a proton moving with a speed of 10^6 m/s.

12–35. What is the de Broglie wavelength in meters of a 75-kg person who is jogging at 5 m/s?

12–36. If matter has a wave nature, why is this not observable in our daily experiences?

12–37. *Multiple Choice*
A. When light shines on a metal surface, the energy of the photoelectrons depends on the
   (a) intensity of the light.
   (b) frequency of the light.
   (c) velocity of the light.
   (d) all of these.

B. A bright-line spectrum of an element
   (a) is concentrated at the red end.
   (b) is evenly distributed.
   (c) is concentrated at the blue end.
   (d) is characteristic of the element.

C. Radioactivity is a property of
   (a) excited electrons.
   (b) ultraviolet light.
   (c) atomic nuclei.
   (d) X-rays.

D. The quantum theory does not explain
   (a) interference.
   (b) the photoelectric effect.
   (c) the Compton effect.
   (d) line spectra.

E. Compared with a light microscope, an electron microscope
   (a) uses less energy.
   (b) works faster.
   (c) can magnify in greater detail.
   (d) uses longer wavelengths.

F. Energy equals Planck's constant times
   (a) $\dfrac{c}{\lambda}$.
   (b) mc.
   (c) mc^2.
   (d) $c\lambda$.

G. Rutherford's gold-foil experiment was instrumental in establishing the concept of
   (a) an atom.
   (b) a molecule.
   (c) subatomic particles.
   (d) a nucleus in an atom.

H. An alpha particle is
   (a) negatively charged.
   (b) electrically neutral.
   (c) a helium nucleus.
   (d) approximately the same as an electron.

I. Experimental support for the arrangement of electrons in distinct energy levels is based primarily on
   (a) atomic spectra.
   (b) scattering experiments.
   (c) the photoelectric effect.
   (d) radioactivity.

J. Which of the following forms of electromagnetic radiation is the most energetic?
   (a) red
   (b) ultraviolet
   (c) infrared
   (d) green

SUGGESTIONS FOR FURTHER READING

Andrade, E. N. da C., *Rutherford and the Nature of the Atom.* Garden City, N.Y.: Doubleday & Co., 1964.
An account of the life and work of the foremost experimental nuclear scientist of the first third of the twentieth century.

Banet, Leo, "Balmer's Manuscripts and the Construction of His Series." *American Journal of Physics*, July, 1970.
Analyzes the development of the Balmer formula and shows its role in Bohr's atomic theory.

Bent, Henry A., "Einstein and Chemical Thought." *Journal of Chemical Education*, June, 1980.
Einstein, the first chemical physicist, showed that the physical atom and the chemical atom were one. For advancing the doctrine of atomism, Einstein deserved another Nobel Prize, in chemistry.

Berns, Michael W., "Laser Surgery." *Scientific American*, June, 1991.
Lasers are increasingly important medical tools. They can be used to unclog obstructed arteries, break up kidney stones, clear cataracts, and even alter genetic material.

Bernstein, J., and S. S. Shaik, "The Wave-Particle Duality." *Journal of Chemical Education*, April, 1988.
A series of pictures that are based on some well-known optical illusions dramatically conveys the concept of the wave-particle duality.

Bragg, Sir Lawrence, "The Start of X-Ray Analysis." *Chemistry*, December, 1967.
Sir William Bragg and Sir Lawrence Bragg, father and son, determined the structure of many substances and started the science of X-ray crystallography.

Bromberg, Joan Lisa, "The Birth of the Laser." *Physics Today*, October, 1988.
The laser was born between September, 1957, when Charles H. Townes conceived the idea of a maser at optical frequencies, and December, 1960, when Ali Javan, William Bennett, and Donald Herriott operated the first continuous laser.

Compton, A. H., "The Scattering of X-Rays as Particles." *American Journal of Physics*, December, 1961.
Reviews the experiments and theory that led to the discovery and interpretation of the Compton effect.

Deltete, Robert, "Einstein's Opposition to the Quantum Theory." *American Journal of Physics*, July, 1990.
Einstein regarded quantum theory as incomplete and inadequate. He hoped and expected that a better theory would be developed, from which the quantum theory could be recovered as an approximation.

Fletcher, Harvey, "My Work with Millikan on the Oil-Drop Experiment." *Physics Today*, June, 1982.
Relates his experiences as a graduate student and his contribution to the determination of the charge of the electron.

Gehrenbeck, Richard K., "Electron Diffraction: Fifty Years Ago." *Physics Today*, January, 1978.
A look at the experiment that established the wave nature of the electron, and at the investigators, Clinton Davisson and Lester Germer.

Hielbron, J. L., "Rutherford-Bohr Atom." *American Journal of Physics*, March, 1981.
The Rutherford-Bohr atom occupies a position in the middle of an evolution initiated by J. J. Thomson and concluded by the invention of quantum mechanics.

Kevles, Daniel J., "Robert A. Millikan." *Scientific American*, January, 1979.
The most famous American scientist of his day. Millikan was also a notable public figure with a penchant for controversy in both science and public policy.

Kidd, Richard, James Ardini, and Anatol Anton, "Evolution of the Modern Photon." *American Journal of Physics*, January, 1989.
Traces the history and evolution of the photon concept, which represents at least four distinct models. The authors favor replacing the corpuscular and wave packet models with a semiclassical approach.

Li, Tingye, "Lightwave Telecommunication." *Physics Today*, May, 1985.
High-speed lasers transmit billions of bits of data per second to sensitive solid-state detectors through glass fibers more than a hundred kilometers long.

Medicus, Heinrich A., "Fifty Years of Matter Waves." *Physics Today*, February, 1974.
Louis de Broglie's theory of matter waves inaugurated the era of modern quantum mechanics.

Mermin, N. David, "Is the Moon There When Nobody Looks? Reality and the Quantum Theory." *Physics Today*, April, 1985.
Albert Einstein, Boris Podolsky, and Nathan Rosen in 1935 published an argument that quantum theory fails to provide a complete description of physical reality. Today, 50 years later, experiments have shown that what bothered Einstein is the observed behavior of the real world.

Schawlow, Arthur L., "Laser Light." *Scientific American*, September, 1968.
An authoritative account of lasers by one who was intimately involved in their development.

Thomson, George, *J. J. Thomson, Discoverer of the Electron*. Garden City, N.Y.: Doubleday & Co., 1966.
An eminent physicist in his own right describes his fa-

ther's work and life at the Cavendish Laboratory that "J. J." headed for many years.

ANSWERS TO NUMERICAL EXERCISES

12–2. 4850 Å

12–3. 11,000 Å

12–4. 4850 Å

12–17. 3.60×10^{-19} joule

12–26. 8.82×10^{-35} m

12–27. 3.68×10^{-68} m

12–33. $1.33 \times 10^{-27} \dfrac{\text{kg m}}{\text{s}}$

12–34. 3.97×10^{-13} m

12–35. 1.77×10^{-36} m

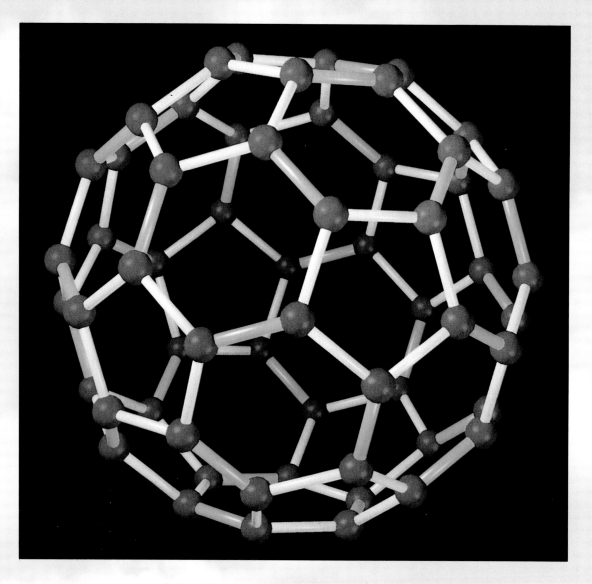

A supercomputer simulation of the atomic structure of a single molecule of buckminsterfullerene
(C_{60}). Carbon atoms appear as spheres. The 5- and 6-membered rings are arranged in the shape of
a soccer ball or one of Buckminster Fuller's geodesic domes. Families of fullerenes of various sizes
and shapes are known. Interesting applications may be expected. (Courtesy J. Bernhole *et al*, North Carolina
State University/Science Photo Library)

II

CHEMISTRY

PERSPECTIVE ON . . . CHEMISTRY

When you were young, you may have played with Lego blocks, Tinkertoys, or similar interconnecting pieces of wood or plastic. You stuck them together and tore them apart again, giggling at your ever-changing inventions. From a simple collection of perhaps ten or 12 shapes and sizes, the possibilities for creation seemed endless.

Your game, it turns out, was a child's version of how the material world works. Everything around you is assembled from atoms, which are the building blocks of the elements—substances that cannot be broken down into something simpler. But instead of ten or 12 elements, nature boasts 89. (Scientists have made another 20 in the laboratory.) Imagine a box of Tinkertoys with 89 varieties of pieces, and you will begin to appreciate the flexibility nature has at its disposal to construct our universe.

Learning how those elements interact is the essence of chemistry. There are many rules elements must follow once the actions and reactions begin. For instance, "noble gases" such as helium and neon avoid other elements entirely, just as nobility shuns the common elements of society. Conversely, fluorine and chlorine are highly reactive, grabbing other atoms with abandon. With practice, it is not hard to figure out which elements can form a proper molecule together, how many atoms each element must contribute to do so, and what the properties of that compound will be. The answers lie in the periodic table of the elements, which hangs in every science classroom and appears in every chemistry textbook. This table is not there because it is great art. Rather, its pattern of rows and columns—that is, its periodicity—nicely illustrates the physical causes of chemical behaviors that otherwise might seem random.

The periodic table is such an icon of science, in fact, that a panel of researchers and artists recently found a most unusual use for it. The panel was tackling a challenging problem: At a site of buried nuclear waste, how might one devise markers that would warn future societies—no matter what their culture or language—against digging it up? Such a warning would have to last 10,000 years, a testament to the staying power of radioactivity. The panelists proposed intimidating earthworks, monoliths, and symbols of danger. In addition, they suggested placing a reproduction of the periodic table in an underground chamber. Prominent marks would highlight the squares representing uranium, plutonium, and the other hazardous elements buried below. The panelists reasoned that the periodic table would survive any conceivable cultural change because it is so recognizable. Even intelligent extraterrestrial beings would understand its significance because elements have the same properties throughout the universe.

Chemistry is far more interesting than simply mastering the periodic table and balancing chemical equations, however. To a greater extent than in any other physical science, in chemistry the results apply to the "real world." Most materials around you owe their origins to basic chemical research, for instance. Every cell in your body carries out hundreds of chemical reactions each second to keep you alive. Physicians rely on chemists to develop new drugs for treating diseases. And chemicals from industry and other human sources, paradoxically, have caused but also may help cure many of earth's environmental problems.

One well-known example of this double-edged sword is the "ozone hole," the alarming thinning of

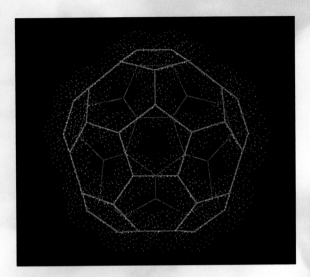

The molecular structure of C_{60}/buckminsterfullerene, or "Buckyball." (Ken Eward/Science Source/Photo Researchers)

tubes. Compounds that kill those cells move on to tests in tissue cultures, then in mice and rats. Fewer than one in a thousand compounds ever advance to trials in people with cancer. This laborious process can last a decade or more from initial discovery to hospital shelf.

Bacteria, fungi, and land plants are the most common sources of medicinal drugs. Recently, scientists have also turned to more exotic supplies — the potent secretions of tropical frogs, for instance, or self-defense poisons from marine sponges. But many of these promising compounds are rare in nature, so chemists must try to synthesize them in the laboratory. One massive such effort is under way for taxol, an extract of the bark of yew trees that may become the best treatment yet for ovarian cancer. Unfortunately, yew trees are not plentiful; they mature very slowly, and they grow mainly in ecologically sensitive forests. Moreover, obtaining just one pound of taxol requires stripping the bark from 5000 yew trees. If chemists can surmount the challenge of making the complex taxol molecule from scratch, they may save the lives of thousands of cancer patients. Until then, the trees will remain rooted in an environmental and ethical tug of war.

Many other current topics of research in chemistry could also lead to profound changes in society. One is the drive to create new superconductors — materials that carry electricity without losing energy. Today's superconducting substances operate only in extreme cold, but many physicists and chemists feel that room-temperature superconductors are inevitable. Such materials would make power and transportation vastly less expensive and more efficient. Other research areas in chemistry are more theoretical but equally intriguing, such as the attempt to identify the molecules responsible for the origin of life on earth.

Despite these exciting advances, chemists occupy a frustrating niche in the public's perception of science. They often are blamed for the planet's ills, and their contributions to our quality of life certainly are underappreciated. As with physics and the atomic bomb, and as with the exploration versus the exploitation of space, society must weigh the benefits of chemical research against its risks and then render a judgment. But whatever the future of chemistry holds, one thing is certain: We have come a long way since earth, air, fire, and water were the only elements in town.

the ozone layer each year over Antarctica. The culprits are chlorofluorocarbons (CFCs), artificial chemicals that work wonders in refrigerators and spray cans. However, these substances have had unforeseen and disastrous side effects. After a CFC molecule drifts high into the atmosphere, sunlight breaks it apart and frees an extremely reactive chlorine atom. Over time, that single atom can destroy thousands of molecules of ozone, a type of oxygen that absorbs ultraviolet light from the sun.

Media reports are full of predictions about the consequences of a weakened ozone shield: more skin cancer, damage to human immune systems, and crop failures. An international agreement will ban CFCs by 2000, but chemists are not sitting idle in the meantime. Some are creating substitute chemicals that might do the same jobs without the side effects. Others are proposing schemes to stop chlorine's wicked chain reactions, such as flying planes into the upper atmosphere and spraying propane or ethane gas to mop up the chlorine. Such an approach, although fraught with potential new hazards, may be necessary to reverse a problem that will not fade quietly on its own.

One of the hottest pursuits in chemistry today is just as important but more down-to-earth: the search for new drugs to fight cancer. In the most successful approach so far, scientists search for toxic compounds made by microorganisms, plants, or animals, and mix them with cancer cells in test

Periods in the periodic table and the number of seats in a stadium's rows increase in capacity as the shells of the elements are located further from the nucleus and as the rows are further from the field of play. (Jerry Cooke, Photo Researchers, Inc.)

13

ATOMIC STRUCTURE AND THE PERIODIC TABLE OF ELEMENTS

It had long been suspected that electricity and chemistry were closely connected. Years before the discovery of the electron, for example, Michael Faraday had experimentally shown interactions between electricity and chemical activity. Ernest Rutherford's nuclear model of the atom with its electrical basis also turned out to be very successful. Then when Niels Bohr applied the quantum theory to the periodic table of the elements and showed that atomic structure is the basis of the periodic table, he cleared a path toward understanding how atoms combine to form molecules and other aggregates of matter. This is the concern of chemistry.

13-1 THE CHEMICAL ELEMENTS— ACTINIUM TO ZIRCONIUM

The idea that a few elementary substances combine to form the many thousands of substances known has always been attractive in attempting to answer the question, "What is the world made of?" Of course, there was not universal agreement as to which substances were "elements," although Empedocles' (about 500–430 BC) theory of the four elements— earth, air, fire, and water—was widely adopted.

The modern concept of chemical elements stems from the work of Robert Boyle (1627–1691). In his book, *The Sceptical Chemist* (1661), Boyle proposed the operational definition that an **element is a pure substance that cannot be decomposed into** simpler substances by ordinary chemical or physical means. Since neither hydrogen nor oxygen is decomposable into simpler substances by ordinary methods, they are classified as elements. (Water can be decomposed into hydrogen and oxygen, so water is not considered to be an element.) On this basis, there are today 100-odd chemical elements, each with its set of characteristics and of which everything else in the universe is believed to be composed. Eighty-eight of the elements occur naturally; the rest can be produced in a laboratory. (The list of elements will be increased if experiments leading to the synthesis of a number of "superheavy" elements in nuclear accelerators are successful.) The 20 most abundant elements on earth are listed in Table 13–1, some familiar elements in Table 13–2, and the elements required for life in Table 13–3.

13-2 SYMBOLS FOR THE ELEMENTS

The chemical elements are represented by a set of symbols that is accepted throughout the world. Compared to a time when as many as 14 different symbols were used for lead and 20 for mercury, this is indeed fortunate. You need only learn, for example, that the symbol for lead is Pb and the symbol for mercury is Hg. We owe the existence of the present system to Jöns Jakob Berzelius (1779–1848), a chemist at the University of Stockholm. Alchemical signs such as the ones shown in Table 13–4—circles, squares, dots, arrows, and various combinations of these—had all been tried before the modern symbols were adopted.

Table 13–1 THE 20 MOST ABUNDANT ELEMENTS ON EARTH

Element	Percent by Weight	Element	Percent by Weight
Oxygen	49.5	Chlorine	0.2
Silicon	25.7	Phosphorus	0.1
Aluminum	7.5	Manganese	0.09
Iron	4.7	Carbon	0.08
Calcium	3.4	Sulfur	0.06
Sodium	2.6	Barium	0.04
Potassium	2.4	Nitrogen	0.03
Magnesium	1.9	Fluorine	0.03
Hydrogen	0.9	Nickel	0.02
Titanium	0.6	Strontium	0.02

Table 13–2 SOME COMMON ELEMENTS AND THEIR USES

Element	Use	Element	Use
Aluminum	House siding; boats, foil; cans	Lead	Engine antiknock additive; batteries
Chlorine	Swimming pools; household bleach	Magnesium	Ladders; flares; airplanes
Chromium	Automobile trim; furniture	Mercury	Thermometers
Copper	Electric wire; plumbing	Neon	Advertising signs
Fluorine	Water supplies; toothpaste	Oxygen	Diving; flying; rockets; welding
Gold	Jewelry; coins; foil	Silver	Silverware; coins; medallions
Helium	Balloons; diving	Sulfur	Matches; sulfuric acid
Iodine	Antiseptic; iodized salt	Tungsten	Incandescent light bulb filaments; X-ray tubes
Iron	Structural steel; automobiles; farm machinery	Zinc	Brass; batteries; plating

Table 13–3 RELATIVE ABUNDANCE OF ELEMENTS IN THE HUMAN BODY

Element	Percent (Atoms)	Element	Percent (Atoms)
Hydrogen	60.3	Potassium	0.036
Oxygen	25.5	Chlorine	0.032
Carbon	10.5	Magnesium	0.010
Nitrogen	2.42	Manganese	trace
Sodium	0.730	Iron	"
Calcium	0.226	Copper	"
Phosphorus	0.134	Zinc	"
Sulfur	0.132	Cobalt	"

Table 13–4 SOME SYMBOLS OF ELEMENTS USED BY ALCHEMISTS

Antimony	Lead (Saturn)	Gold (Sun)	Sulfur
Arsenic	Quicksilver (Mercury)	Iron (Mars)	Tin (Jupiter)
Copper (Venus)	Silver (Moon)		

Table 13–5   ORIGINS OF SOME CHEMICAL SYMBOLS

Element	Symbol	Origin of Symbol	Element	Symbol	Origin of Symbol
Antimony	Sb	Stibium ("mark")	Potassium	K	Kalium ("potash")
Copper	Cu	Cuprum (Cyprus, source of copper)	Silver	Ag	Argentum
Gold	Au	Aurum ("shining dawn")	Sodium	Na	Natrium
Iron	Fe	Ferrum	Tin	Sn	Stannum
Lead	Pb	Plumbum ("heavy")	Tungsten	W	Wolfram (wolframite, a mineral)
Mercury	Hg	Hydrargyrum ("liquid silver")			

In place of the signs that were then in vogue, Berzelius proposed that letters be used for chemical symbols, since they are simple and easy to write. A **symbol** may be the capitalized first letter of the name of the element, such as H for hydrogen. If two or more elements share the same first letter, a second letter in lowercase is added; thus, He is the symbol for helium. Elements for which there were Latin names often have symbols derived from them. For example, the Latin name for mercury was "hydrargyrum" (meaning "quicksilver" or "liquid silver"), and the chemical symbol for mercury is Hg. The symbols of some elements that are unrelated to their common English names are given in Table 13–5. Nearly all the symbols suggested by Berzelius are in use today. Even though most of the elements have been discovered since Berzelius' time, their symbols are nonetheless based on his suggestions.

Anyone who discovers an element has the right to propose a name for it, but the name has to be ac-

cepted by a body representing the scientists of the world. That body is the Commission on the Nomenclature of Inorganic Chemistry (CNIC) of the International Union of Pure and Applied Chemistry (IUPAC). Elements have been named for the sun, moon, and planets: helium (He), sun; selenium (Se), moon; tellurium (Te), earth; neptunium (Np); plutonium (Pu); and uranium (U). Various qualities or properties have also served as the basis of the names of elements: argon (Ar), inactive; beryllium (Be), sweet; bromine (Br), stench; chlorine (Cl), light green; hydrogen (H), water-former; neon (Ne), new; oxygen (O), acid-former; phosphorus (P), light-bearing; technetium (Tc), artificial; and xenon (Xe), stranger. A recent trend is to name the elements for a person or for their place of discovery, as shown in Table 13–6.

Table 13–6   RECENTLY DISCOVERED ELEMENTS

Element	Symbol	Year Discovered	Derivation
Americium	Am	1944	The Americas
Curium	Cm	1944	Pierre and Marie Curie
Berkelium	Bk	1949	Berkeley (California)
Californium	Cf	1950	California
Einsteinium	Es	1952	Albert Einstein
Fermium	Fm	1952	Enrico Fermi
Mendelevium	Md	1955	Dmitri Mendeléev
Nobelium	No	1958	Alfred Nobel
Lawrencium	Lr	1961	Ernest Lawrence
Kurchatovium (proposed)	Ku	1964	Igor Kurchatov
Hahnium (proposed)	Ha	1971	Otto Hahn

13–3 THE ATOMIC MASS SCALE; ISOTOPES; ATOMIC NUMBER

An **atom** is the smallest particle of an element that shows the chemical behavior of that element. The lightest atom has a mass of about 1.66×10^{-24} gram; the heaviest, only about 250 times as much. Since it is impossible to measure such masses directly, a relative scale is used to compare the masses of individual atoms. The atomic masses of the elements are their relative masses compared with an arbitrary standard. Since 1961, by international agreement, the **unified atomic mass unit** (abbreviated u or amu) or dalton (after John Dalton, 1766–1844) has been defined as 1/12 the mass of a certain variety of carbon atom, carbon-12, taken as exactly 12.00000. On this scale, the lightest atom, hydrogen, has a mass of about 1 atomic mass unit or 1 dalton.

The relative masses of atoms are precisely determined with the **mass spectrometer** (Fig. 13–1). Using this instrument, Francis William Aston (1877–1945) at the Cavendish Laboratory made an exhaustive study of the elements. The mass spectrometer is based on a principle that J. J. Thomson employed, that a moving charged particle is deflected into a curved trajectory by a magnetic field perpendicular to the direction of motion. A magnetic field exerts a force on any moving charged particle (Fig. 13–2).

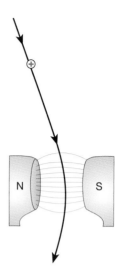

Figure 13–2   A magnetic field exerts a force upon a moving electrically charged particle, changing its direction of motion.

The force is perpendicular to the direction of motion of the electric charge and to the magnetic field (not attractive or repulsive to either pole of the magnet). The magnitude of the force **F** on the charged particle is directly proportional to the charge q on the particle, to the velocity **v** of the particle, and to the magnetic field strength **B**.

A heated filament emits electrons that bombard the gas sample (Fig. 13–3). The high-speed electrons knock electrons off the atoms or molecules of the gas, creating charged particles called **positive ions.** When accelerated by a voltage and subjected to a magnetic field, the positive ions are bent in a path governed by their charge-to-mass (q/m) ratio. Ions of different q/m values are curved into different paths and separated at the collector. The more massive the particle, the less it is deflected. The largest q/m value for positive ions is obtained when hydrogen is the gas and is approximately 1/1837 of the q/m value for electrons.

With neon in a mass spectrometer, the particles have three different masses relative to carbon-12: 20, 21, and 22 (Fig. 13–4). We have to conclude that three varieties of neon atoms are present. The chemist Frederick Soddy (1877–1956) proposed the name **isotopes** (*iso*, "same"; *topos*, "place") for the different varieties of the atoms of an element. Knowing the relative abundances of the isotopes of an element and their masses compared with carbon-12, we can calculate the **atomic mass** of the element—a weighted

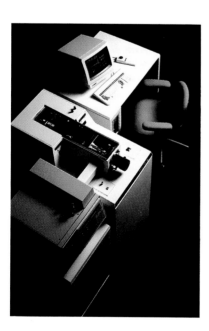

Figure 13–1   A modern mass spectrometer. (Courtesy of Finnigan Corporation)

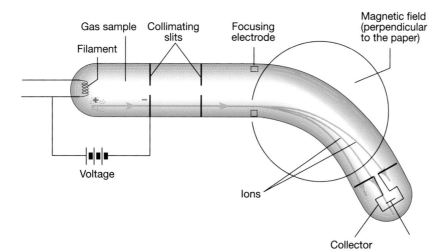

Figure 13-3 A mass spectrometer. A gas sample bombarded by electrons becomes a source of positively charged ions that are curved into different paths by a magnetic field and separated, depending upon their charge-to-mass (q/m) ratio.

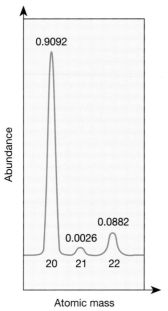

Figure 13-4 The mass spectrum of neon. The higher the peak, the more abundant is the particular isotope.

nucleus. The charge and mass of these particles are listed in Table 13-8.

The number of **protons** in the nucleus is called the **atomic number** of the element, and is designated Z. All atoms of a given element have the same number of protons and, thus, the same atomic number, ranging from 1 for hydrogen to 109 for Une; for neon, Z = 10. The nuclear charge, or number of protons, determines which element an atom represents. In a neutral atom, electrons are distributed outside the nucleus in a number to balance the positive charge on the nucleus. Each neon isotope thus has ten protons in the nucleus and ten extranuclear electrons. An element is a substance, therefore, in which all the atoms have the same atomic number.

The neon isotopes differ in mass, owing to the presence in the nucleus of a varying number of uncharged particles called **neutrons** (Fig. 13-5). These particles were discovered by James Chadwick (1891-1974) at the Cavendish Laboratory in 1932, and they have practically the same mass as protons. The term **"nucleon"** refers to a nuclear particle, either a proton or a neutron. The sum of the protons and neutrons

average of the isotopes as they occur naturally. A sample calculation is shown in Table 13-7.

An atom is composed of a nucleus consisting of protons and neutrons and of electrons outside the

Table 13-7   CALCULATING THE ATOMIC MASS OF NEON

Isotope	Mass Compared with Carbon-12 as 12.00 u		Proportion of all Neon Atoms		Total Mass Contributed (u)
Neon-20	19.99	×	0.9092	=	18.17
Neon-21	20.99	×	0.0026	=	0.05
Neon-22	21.99	×	0.0882	=	1.94
			Atomic mass of neon:		20.16 u

Table 13–8 SUBATOMIC PARTICLES

Particle	Charge	Mass in Atomic Mass Units (u)	Relative Mass
Electron	−1	0.00055	1/1837
Proton	+1	1.00728	1
Neutron	0	1.00866	1

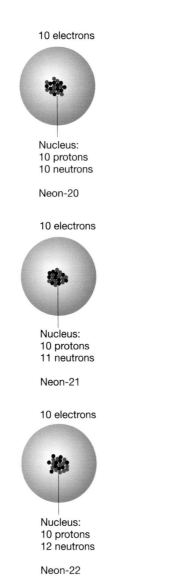

10 electrons

Nucleus:
10 protons
10 neutrons

Neon-20

10 electrons

Nucleus:
10 protons
11 neutrons

Neon-21

10 electrons

Nucleus:
10 protons
12 neutrons

Neon-22

Figure 13–5 Structures of the neon isotopes. Protons and neutrons comprise the nucleus; electrons are extranuclear. (Not drawn to scale.)

in a nucleus is called the **mass number,** A. Isotopes are denoted by their symbols with a subscript to the left to represent the atomic number and a superscript to the left to denote the mass number, as in:

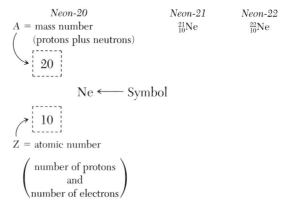

Neon-20

A = mass number
(protons plus neutrons)

$\boxed{20}$

Ne $\longleftarrow$ Symbol

$\boxed{10}$

Z = atomic number

$\left(\begin{array}{c}\text{number of protons}\\\text{and}\\\text{number of electrons}\end{array}\right)$

Neon-21 $^{21}_{10}\text{Ne}$

Neon-22 $^{22}_{10}\text{Ne}$

Although the number of elements at present is somewhat over 100, about 1400 isotopes are known. Even hydrogen occurs in three isotopic forms: $^{1}_{1}\text{H}$, $^{2}_{1}\text{H}$, and $^{3}_{1}\text{H}$. Since the atomic number and the chemical symbol both denote the element, the subscript can be omitted. Thus, one of the varieties of hydrogen can be referred to as "hydrogen-2." The atomic number, atomic mass, and isotopes of several elements are given in Table 13–9.

EXAMPLE 13–1

Give the values of the atomic number Z and the mass number A, and the numbers of electrons, protons, and neutrons, for $^{39}_{19}\text{K}$.

SOLUTION
The subscript 19 denotes the atomic number Z, which gives the number of electrons and the number of protons; therefore, there are 19 electrons and 19 protons. The superscript 39 denotes the mass number A, the sum of protons and neutrons. Having determined that 19 protons are present, we know that the number of neutrons must be A minus Z, or $39 - 19 = 20$ neutrons.

EXTENSION
Give Z, A, and the atomic composition of $^{184}_{74}\text{W}$. (Answer: $Z = 74$; $A = 184$; 74 protons, 74 electrons, 110 neutrons)

■

The atomic masses of some environmentally important elements are listed in Table 13–10.

Table 13–9 ATOMIC MASSES AND ISOTOPES OF SEVERAL ELEMENTS

Element	Atomic Number	Atomic Mass	Isotopes, Mass Number
Helium	2	4.0026	^{3}He ^{4}He ^{5}He ^{6}He
Carbon	6	12.01115	^{10}C ^{11}C ^{12}C ^{13}C ^{14}C ^{15}C
Nitrogen	7	14.0067	^{12}N ^{13}N ^{14}N ^{15}N ^{16}N ^{17}N
Oxygen	8	15.9994	^{14}O ^{15}O ^{16}O ^{17}O ^{18}O ^{19}O
Sodium	11	22.9898	^{20}Na ^{21}Na ^{22}Na ^{23}Na ^{24}Na ^{25}Na

13–4 THE PERIODIC TABLE

Julius Lothar Meyer (1830–1895), a chemist at the University of Tübingen, and Dmitri Ivanovitch Mendeléev (1834–1907), a chemist at the University of St. Petersburg, discovered independently that when the elements are arranged in a table ordered from light to heavy, the chemical and physical properties recur at definite intervals. The **periodicity** involves such properties as density, solubility, melting point, boiling point, ionization energy, and hardness. Meyer plotted curves that graphically illustrate the periodicity of each property, as in Figure 13–6). When the physicist H. G. J. Moseley (1887–1915),

Table 13–10 ATOMIC MASSES OF SOME ELEMENTS IMPORTANT IN THE ENVIRONMENT

Element	Symbol	Atomic Number	Atomic Mass	Element	Symbol	Atomic Number	Atomic Mass
Hydrogen	H	1	1.008	Chlorine	Cl	17	35.453
Helium	He	2	4.0026	Potassium	K	19	39.102
Carbon	C	6	12.011	Calcium	Ca	20	40.08
Nitrogen	N	7	14.0067	Iron	Fe	26	55.847
Oxygen	O	8	15.9994	Copper	Cu	29	63.54
Fluorine	F	9	18.9984	Arsenic	As	33	74.9216
Sodium	Na	11	22.9898	Strontium	Sr	38	87.62
Magnesium	Mg	12	24.305	Cadmium	Cd	48	112.40
Aluminum	Al	13	26.9815	Iodine	I	53	126.9045
Silicon	Si	14	28.086	Mercury	Hg	80	200.59
Phosphorus	P	15	30.9738	Lead	Pb	82	207.2
Sulfur	S	16	32.06	Uranium	U	92	238.03

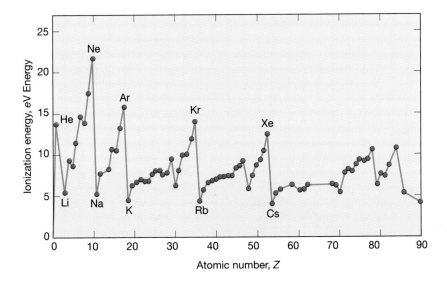

Figure 13–6 Variation of first ionization energy of the elements with atomic number. Symbols of elements with very high and very low ionization energy are shown.

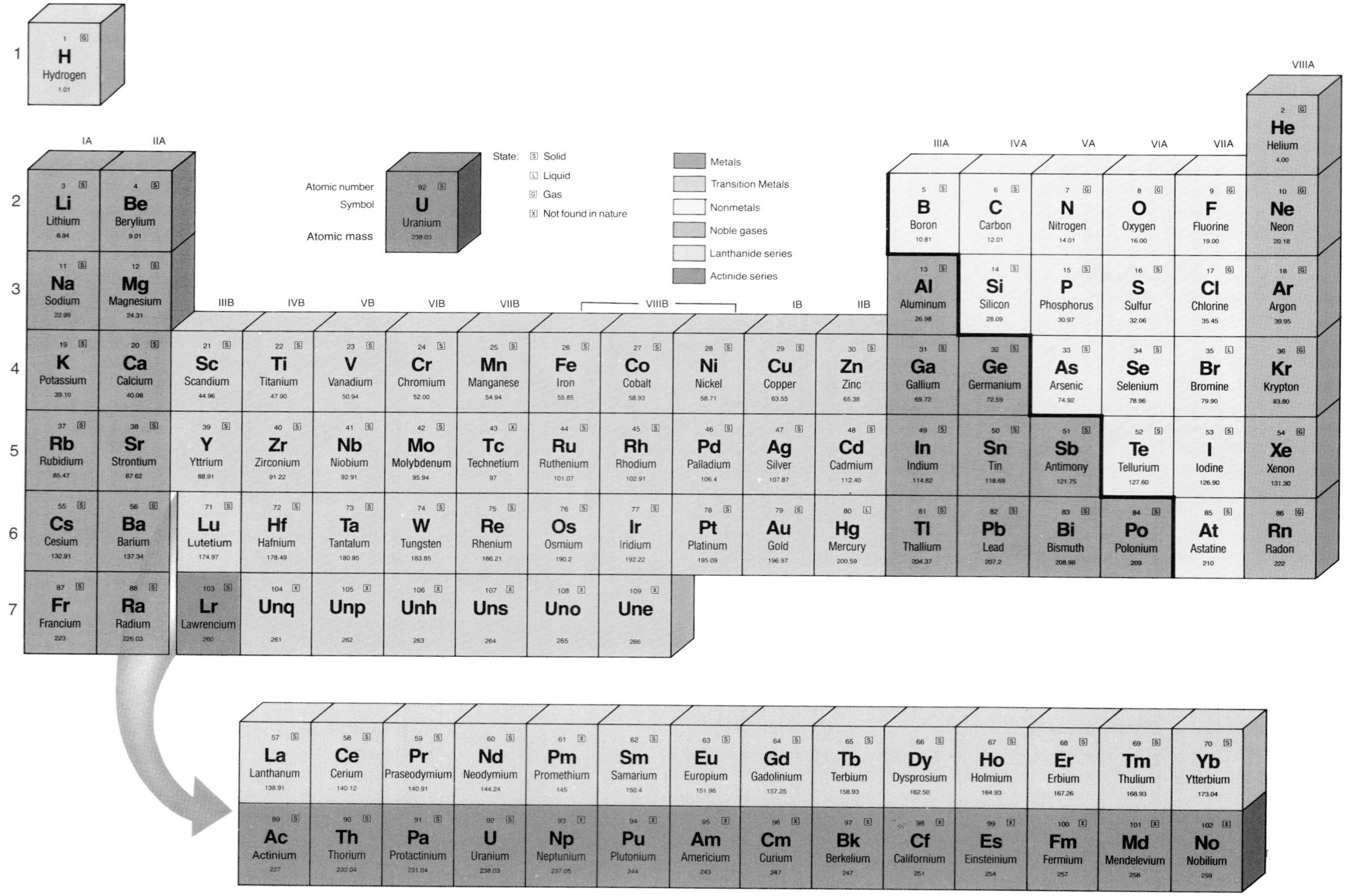

Figure 13–7 The modern periodic table of the elements.

Table 13–11 SOME PROPERTIES OF GERMANIUM

Property	Predicted for Ekasilicon (Mendeléev, 1871)	Observed for Germanium (Winkler, 1886)
Atomic mass	72	72.3
Atomic volume	13 cm^3	13.22 cm^3
Boiling point of chloride	100°C	86°C
Boiling point of ethyl derivative	160°C	160°C
Color	Dark gray	Grayish white
Specific gravity	5.5	5.469
Specific gravity of chloride	1.9	1.887
Specific gravity of oxide	4.7	4.703

working with Rutherford, found that the atomic number of an element is a fundamental property more important than the atomic mass, he formulated the new periodic law, or **Moseley's law,** which states:

The properties of the elements are a periodic function of their atomic numbers.

A **periodic table** is useful to the extent that it correlates the physical and chemical properties of groups of elements. In his 1871 table, the forerunner of modern classification, Mendeléev left some of the spaces vacant rather than insert elements that did not fit properly into a family. He made the bold prediction that new elements would be discovered that would fit. He also predicted some of the properties of three such elements, which he designated "ekaboron," "ekaaluminum," and "ekasilicon." Within 15 years, all three were discovered and were named scandium, gallium, and germanium. A comparison of

the properties Mendeléev predicted for ekasilicon and those observed for the new element germanium is given in Table 13–11. The periodic table can still be used to show where undiscovered elements are to be expected and provides a means for predicting their chemical and physical properties.

The usual version of the modern periodic table is divided into **periods**—horizontal rows running from left to right across the table (Fig. 13–7). There are seven periods, the first consisting of only 2 elements, hydrogen and helium. The second and third periods contain 8 elements each. The fourth and fifth periods have 18 elements each. The sixth period has 32 elements, 14 of which are usually shown in a separate block at the bottom of the table. The seventh and last period appears incomplete.

Periodicity in chemical properties is illustrated by the fact that, except for the first period, each period begins with a very reactive metal (Fig. 13–8). Succes-

Figure 13–8 The elements of period 3. Properties progress *(left to right)* from solids (Na, Mg, Al, Si, P, S) to gases (Cl, Ar) and from the most metallic (Na) to the most nonmetallic (Ar). (Charles Steele)

Figure 13–9 The elements of group IVA: carbon (C) *(bottom),* silicon (Si) *(left middle),* tin (Sn) *(right middle),* lead (Pb) *(top).* (Charles D. Winters)

Figure 13–10 Three of the elements of group VIIA: *(left to right)* chlorine (Cl_2), bromine (Br_2), and iodine (I_2). The deep red color of bromine and the violet of iodine are those of the vapors in equilibrium with liquid Br_2 and solid I_2. (Leon Lewandowski)

sive elements within a period are less and less reactive and increasingly nonmetallic, until a very reactive nonmetal is encountered. Each period finally closes with a noble gas.

The vertical columns are occupied by elements that have related chemical and physical properties and make up a chemical family or **group** (Fig. 13–9). The members of a group designated IA to VIIA are called **main-group** or **representative elements.** Those in a column headed by a Roman numeral and the letter B, IB to VIIB, are known as **transition metals.** The elements within a given subgroup—for example, Li, Na, K, Rb, and Cs in group IA— resemble one another more closely than they resemble the members of the other family subgroup—Cu, Ag, and Au in group IB (Fig. 13–10). The elements in group IA (except H) are very reactive metals that never occur uncombined in nature. On the other hand, Cu, Ag, and Au, the elements in group IB, are relatively inert and can be recovered from their compounds with such ease that they are among the oldest metals known. The gradation in physical properties in passing successively from one element to the next in a group is typified by the alkali metals in Table 13–12.

Table 13-12 PHYSICAL PROPERTIES OF ALKALI METALS

Property	Li	Na	K	Rb	Cs
Melting point, °C	180	98	63.4	38.8	28.7
Boiling point, °C	1336	889	757	679	690
Density, g/mL at 20°C	0.535	0.971	0.862	1.53	1.90
Electronegativity	1.0	0.9	0.8	0.8	0.7
Atomic radius, Å	1.52	1.86	2.31	2.44	2.62
Ionic radius, Å	0.60	0.95	1.33	1.48	1.69

13-5 WAVE MECHANICS

Why do the elements in a given subgroup in the periodic table exhibit similar properties? The answer lies in the *electronic structures* of their atoms. The chemical and physical properties of an atom depend on the arrangement of its electrons. Elements whose atoms have similar electronic structures in their outer energy levels have many properties in common.

Many of the lines in the emission spectrum of an element are not individual lines but closely spaced *groups* of lines. The spectral lines are split into additional lines when the light-emitting source is placed in a strong magnetic field. Although Bohr's theory of the atom correctly predicts the wavelengths of the principal lines of the hydrogen spectrum, it does not account for the "fine structure" and "hyperfine structure" of many of the lines. The existence of this structure implies that the principal energy levels consist of groups of sublevels that differ slightly in energy.

Erwin Schrödinger's (1887–1961) wave equation, developed while he taught at the University of Zurich, resolves some of these difficulties. Building upon de Broglie's discovery that matter has a wavelike character, Schrödinger assumed the wave nature of electrons and described electrons as standing waves in the atom. This is the basis of **wave mechanics,** or **quantum mechanics.** The Schrödinger equation can be solved exactly for the electron in the hydrogen atom and approximately for the electrons in other atoms.

It is not necessary here to know the Schrödinger equation or how to solve it, only that the solutions are used to describe the arrangement of electrons in atoms. A solution is expressed as a set of permitted values of **four quantum numbers**—n, l, m, and s—that express the energy of the electrons and their most probable location in the atom. The first three are mathematically related to one another. Electrons are located in energy levels around the nucleus and their energies are **quantized;** that is, they can take on only certain values. Each principal energy level consists of one or more sublevels. These sublevels, in turn, have one or more orbitals. The arrangements of electrons in atoms can be predicted from the quantum numbers.

The **principal quantum number,** n, determines the main energy level in which the electron is located. Corresponding to Bohr's numbers for orbits, n can take on any positive integral value; for the elements known at present, the values range from $n = 1$ to $n = 7$. The energy level for which $n = 1$ is closest to the nucleus and has the lowest energy.

The **angular momentum quantum number,** l, designates the sublevel within a main energy level. It may have any integral value from $l = 0$ to $l = (n - 1)$. The sublevels for which $l = 0$, 1, 2, and 3 are denoted by the letters, s, p, d, and f (from spectral lines that were originally designated *s*harp, *p*rincipal, *d*iffuse, and *f*undamental). Thus, an energy level for which $n = 1$, and $l = 0$ is designated 1s; for $n = 3$ and $l = 1$, 3p; and so on.

Each sublevel consists of one or more sublevels, or **orbitals,** defined by a set of permitted values of n, l, and the magnetic quantum number, m. According to the **Heisenberg uncertainty principle,** if we can measure the energy of an electron precisely, we cannot simultaneously determine the exact position of the electron. Since the energy of an electron in a given orbital is specified precisely by its quantum numbers, the position of the electron is uncertain. We can therefore talk only about the **probability** of the electron being in a certain region. An **orbital** is defined as a **three-dimensional region in space around the**

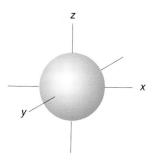

Figure 13–11   An s orbital is spherically symmetrical, with the nucleus located at the center of the orbital.

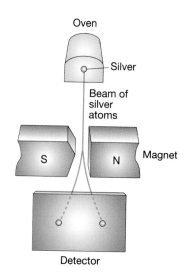

Figure 13–13   The Stern-Gerlach experiment. A beam of silver atoms is split into two beams by a magnetic field.

nucleus where there is the greatest probability of locating a particular electron. The probability may be expressed diagrammatically as an **"electron cloud"** that has a shape and density (Fig. 13–11).

The **magnetic quantum number** (m) describes the possible orientations of an electron cloud in space in the presence of an external magnetic field. The permitted values of m range from $-l$ to $+l$, including 0. For p orbitals, therefore, where $l = 1$, m can have three values: -1, 0, and 1; thus, there are three p orbitals, designated p_x, p_y, and p_z (Fig. 13–12).

The **spin quantum number,** s, was formulated not from the wave equation but from a hypothesis put forth to explain certain experiments. Otto Stern (1887–1969) and Walther Gerlach (1889–1979) had shown that if a beam of silver atoms is allowed to pass by the poles of a magnet, the beam is split into two beams (Fig. 13–13). Similar results are obtained with H, Na, and other elements. Something in the atom is

apparently affected by the magnet. George E. Uhlenback (1900–1988) and Samuel A. Goudsmit (1902–1978) proposed that an electron spins about its own axis like a top as it moves about the nucleus, generating a magnetic field and behaving like a tiny magnet (Fig. 13–14). The spin is quantized in one of two equivalent directions, clockwise and counterclockwise, accounting for the two beams produced (Fig. 13–15). An external magnet therefore interacts with a spinning electron. For each value of m, there are two values of s: $+ 1/2$ and $- 1/2$. The two spin orientations will be designated by arrows pointing up ($\uparrow$) and

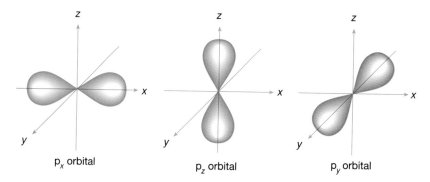

Figure 13–12   Electron clouds corresponding to three p orbitals are directed at 90° angles to each other. The electron density in the p_x orbital is symmetrical about the x-axis; the p_y orbital about the y-axis; and the p_z orbital about the z-axis.

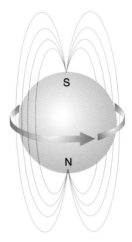

Figure 13–14 Electron spin. An electron in motion in an atom generates a magnetic field.

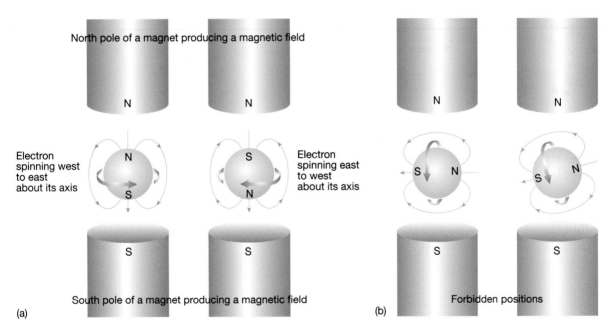

North pole of a magnet producing a magnetic field

Electron spinning west to east about its axis

Electron spinning east to west about its axis

South pole of a magnet producing a magnetic field

Forbidden positions

(a)   (b)

Figure 13–15 (a) The two assumed spins of the electron relative to a magnetic field: one spinning west to east, with the north pole of the electron facing the north pole of the external magnet; the second spinning in the opposite direction, with the north pole of the electron facing the south pole of the external magnet. Any other position, such as those illustrated in (b), is forbidden. Therefore, we can say that the spin of an electron is quantized.

down ($\downarrow$). Table 13–13 summarizes the properties of the four electron quantum numbers.

EXAMPLE 13–2

For the principal quantum number $n = 2$, how many orbitals are possible?

SOLUTION

In the main energy level ($n = 2$), the number of possible sublevels depends on the permitted values of l: $0(s)$, $1(p)$. In the s sublevel, $l = 0$; the only possible value for m is 0; therefore, the s sublevel consists of just one orbital. In the p sublevel, $l = 1$, and m may take the values -1, 0, and $+1$; therefore, three p

Table 13–13 CHARACTERISTICS OF THE ELECTRON QUANTUM NUMBERS

Symbol	Name	Description	Permitted Values
n	Principal quantum number	Main energy level	1,2,3, . . .
l	Angular momentum quantum number	Sublevel	0,1,2,3, . . . $(n - 1)$
			$\underbrace{}$
			s p d f
m	Magnetic quantum number	Orbital	$-l, 0, +l$
s	Spin quantum number	Electron spin	$+ 1/2, - 1/2$

orbitals are possible. In all, $1 + 3 = 4$ orbitals are possible when $n = 2$.

EXTENSION

How many orbitals are possible in the third energy level, $n = 3$? (Answer: 9)

■

The relations among the quantum numbers n, l, and m are shown in Table 13–14 for the first three energy levels.

13–6 BUILDING UP THE PERIODIC TABLE

In the **ground state** of an atom, the electrons occupy the lowest energy levels available and the atom is stable. Any other arrangement corresponds to an **excited state** of the atom. Starting with a nucleus, which has a positive charge equal to the atomic number of the element, we can get the electronic configuration of the ground state of an atom by adding electrons equal in number to the atomic number. Niels

Table 13–14 DERIVATION OF ORBITALS FROM QUANTUM NUMBERS

n	l	m	Orbital
1	0	0	1s
2	0	0	2s
	1	-1	$2p_x$
		0	$2p_y$
		$+1$	$2p_z$
3	0	0	3s
	1	-1	$3p_x$
		0	$3p_y$
		$+1$	$3p_z$
	2	-2	$3d_{xy}$
		-1	$3d_{xz}$
		0	$3d_{yz}$
		$+1$	$3d_{z^2y^2}$
		$+2$	$3d_{z^2}$

Increasing energy (↓)

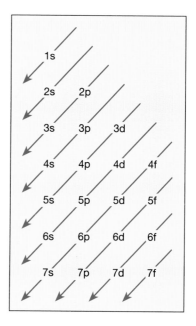

Figure 13–16 Electron configuration sequence. Proceed from the head of one arrow to the tail of the next.

Table 13–15 ELECTRON CAPACITIES OF SUBLEVELS

Sublevel	Angular Momentum Quantum Number, l	Magnetic Quantum Number, m	Orbitals	Electron Capacity
s	0	0	1	2
p	1	$-1, 0, +1$	3	6
d	2	$-2, -1, 0, +1, +2$	5	10
f	3	$-3, -2, -1, 0, +1, +2, +3$	7	14

Bohr referred to this process as the **Aufbau principle,** that is, the building-up principle of the periodic system of the elements.

Why are not all of the electrons in an atom found in the lowest energy orbital? It is a law of nature that **no two electrons in the same atom can have the same set of four quantum numbers.** The statement of this law is known as the **Pauli exclusion principle,** after Wolfgang Pauli (1900–1958). Every electron must have a separate set of quantum numbers. Since the quantum numbers n, l, and m specify an orbital, an orbital can accommodate only two electrons, and they must have opposite spins, clockwise and counterclockwise. It is impossible for any two electrons in an atom to exist in the same state, that is, to have identical sets of quantum numbers.

The order in which electrons enter the energy levels of an atom is determined by their energies. The electrons of lowest energy enter levels closest to the nucleus. The order of increasing energy for atomic orbitals is given by following the arrows in the scheme in Figure 13–16. Following this sequence, we see that when the 3p energy level has been filled, the next electron enters the 4s level. Only when this level has been filled do electrons enter the 3d level. The sequence, therefore, is: 1s 2s 2p 3s 3p 4s 3d 4p 5s 4d. . . . On this basis we can predict the location of electrons in atoms provided that we also know the **electron capacity** of each sublevel. The capacity depends on the number of orbitals, as given in Table 13–15. The electron capacities of levels 1 to 4 are shown in Table 13–16.

Table 13–16 ELECTRON CAPACITIES OF ENERGY LEVELS 1 TO 4

Main Level n	Sublevel l	Number of Electrons	Total Number of Electrons in Energy Level
$n = 1$	s	2	2
$n = 2$	s	2	
	p	6	8
$n = 3$	s	2	
	p	6	
	d	10	18
$n = 4$	s	2	
	p	6	
	d	10	
	f	14	32

To write the **electron configuration** for an element, we show the number of electrons that occupy each sublevel, listed in order of increasing energy. A superscript denotes the number of electrons occupying a sublevel. The sum of the superscripts is the number of electrons in the atom. The electron configuration of hydrogen, which has one electron, is expressed as follows:

symbol and atomic number of hydrogen $= {}_1H$

$1s^1 \leftarrow$ one electron

principal quantum number
$n = 1$

s sublevel
$l = 0$

This notation means that the hydrogen electron is in the s sublevel of the first energy level. For sodium, the electron configuration is:

$${}_{11}Na \qquad 1s^2 2s^2 2p^6 3s^1$$

It is often convenient to group the electrons by main energy level, so that we may express the configuration of sodium as 2,8,1 a total of 11 electrons.

Another way of representing the electron configuration of sodium is $[Ne] 3s^1$. Neon, atomic number 10, is a noble gas that precedes sodium in the periodic table and has the configuration 2,8. With the exception of helium, all of the **noble gases** have an **octet** of electrons in their outermost energy level, a particularly stable configuration. According to the Pauli exclusion principle, the first ten electrons of sodium have the same quantum numbers and configuration as the electrons of neon, that is, $1s^2 2s^2 2p^6$. The eleventh electron is characteristic of sodium. Therefore, the symbol [Ne], the **"core,"** designates the ten electrons of sodium that have the same quantum numbers as the ten electrons of neon. Representing the core of a nearby noble gas in the electron configuration of an element is particularly useful with elements of higher atomic number, since this saves both space and time. Table 13–17 gives the electron configurations of the first 20 elements, some with abbreviated notations.

EXAMPLE 13–3

Identify the element that has the following electronic structure:

$$1s^2 2s^2 2p^6 3s^2 3p^6 4s^2 3d^{10} 4p^6 5s^2$$

Table 13–17   ELECTRON CONFIGURATIONS OF THE FIRST 20 ELEMENTS

${}_1H$	$1s^1$	${}_{11}Na$	$[Ne] 3s^1$
${}_2He$	$1s^2$	${}_{12}Mg$	$[Ne] 3s^2$
${}_3Li$	$1s^2 2s^1$	${}_{13}Al$	$[Ne] 3s^2 3p^1$
${}_4Be$	$1s^2 2s^2$	${}_{14}Si$	$[Ne] 3s^2 3p^2$
${}_5B$	$1s^2 2s^2 2p^1$	${}_{15}P$	$[Ne] 3s^2 3p^3$
${}_6C$	$1s^2 2s^2 2p^2$	${}_{16}S$	$[Ne] 3s^2 3p^4$
${}_7N$	$1s^2 2s^2 2p^3$	${}_{17}Cl$	$[Ne] 3s^2 3p^5$
${}_8O$	$1s^2 2s^2 2p^4$	${}_{18}Ar$	$[Ne] 3s^2 3p^6$
${}_9F$	$1s^2 2s^2 2p^5$	${}_{19}K$	$[Ar] 4s^1$
${}_{10}Ne$	$1s^2 2s^2 2p^6$	${}_{20}Ca$	$[Ar] 4s^2$

SOLUTION

From the sum of the superscripts we get the atomic number of the element, 38. The element is strontium.

EXTENSION

Which element has the configuration $[Ar] 4s^1$? (Answer: K)

◼

EXAMPLE 13–4

Show the electron configuration of ${}_{53}I$.

SOLUTION

We have to assign the 53 electrons of iodine to their respective energy levels and sublevels. Following the sequence shown in Figure 13–9, the configuration is

$$1s^2 2s^2 2p^6 3s^2 3p^6 4s^2 3d^{10} 4p^6 5s^2 4d^{10} 5p^5$$

EXTENSION

Give the configuration of ${}_{21}Sc$. (Answer: $1s^2 2s^2 2p^6 3s^2 3p^6 4s^2 3d^1$)

◼

The Aufbau principle explains the **periodicity of chemical and physical properties.** Families of elements have similar electron configurations in their highest energy levels, for example, the alkali metals in Group IA (Fig. 13–17). Each member of the family has an unfilled level of one electron surrounding the nucleus and one or more filled levels of electrons.

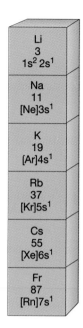

Figure 13–17 Electron configurations of group IA elements, the alkali metals.

EXAMPLE 13–5

The elements neon, argon, and krypton are members of the noble gas family. Do they share any similarity in their electron configurations?

SOLUTION
We write the spdf structures of these elements directly from their atomic numbers.

$_{10}Ne$ $1s^2 2s^2 2p^6$ $= 2,8$
$_{18}Ar$ $1s^2 2s^2 2p^6 3s^2 3p^6$ $= 2,8,8$
$_{36}Kr$ $1s^2 2s^2 2p^6 3s^2 3p^6 4s^2 3d^{10} 4p^6 = 2,8,18,8$

Grouping the electrons by main energy level reveals the structural similarity of eight electrons in the outermost level surrounding inner, filled energy levels of electrons.

EXTENSION
Why are $_6C$ and $_{14}Si$ placed in the same group in the periodic table? (Answer: $_6C$: $1s^2 2s^2 2p^2$; $_{14}Si$: $1s^2 2s^2 2p^6 3s^2 3p^2$. The configurations of the outer energy levels are similar.)

■

The electron configuration of an atom can also be expressed in an **energy-level diagram** (Fig. 13–18). A circle or box is used to represent an orbital; one for an s orbital, three for p orbitals, five for d orbitals, and seven for f orbitals. An arrow represents an electron, pointing up for one spin orientation and down for the other spin orientation. Any orbital has a capacity for two oppositely spinning electrons. For sodium, $_{11}Na$, the 11 electrons are arranged as follows: $1s^2 2s^2 2p^6 3s^1$. Two electrons of opposite spin are accommodated in the 1s orbital; two in 2s; six in 2p—two each in $2p_x$, $2p_y$, and $2p_z$; and the eleventh electron in the 3s orbital.

An energy-level diagram provides a pictorial method of applying another important principle of electron configurations. Consider the arrangement of the electrons in carbon, $_6C$: $1s^2 2s^2 2p^2$. Although it seems reasonable to expect the fifth and sixth electrons to fill the $2p_x$ orbital, the evidence is that this does not happen and that the sixth electron enters a different 2p orbital. The rule, known as **Hund's rule of maximum multiplicity,** states (Fig. 13–19):

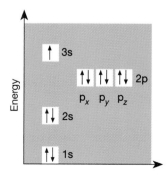

Figure 13–18 Energy-level diagram of the ground state of sodium.

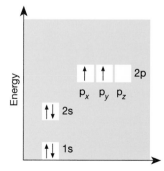

Figure 13–19 Hund's rule as applied to carbon. The $2p_x$ and $2p_y$ orbitals are half occupied.

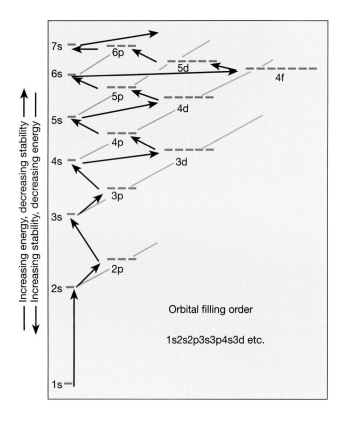

Figure 13–20 Order of energy of atomic orbitals.

Electrons do not enter into joint occupancy of an orbital until all of the available orbitals of the same type in that shell are half occupied.

When several orbitals of the same type are available, a single electron is assigned to each orbital before any electrons are allowed to pair. Figure 13–20 gives the order of the energy of atomic orbitals.

KEY TERMS

Angular Momentum Quantum Number Designates a sublevel within a main energy level: s, p, d, f.

Atom The basic unit of an element that can enter into chemical combination.

Atomic Number The number of protons in the nucleus of an atom.

Atomic Mass A weighted average of the naturally occurring isotopes of an element compared with carbon-12.

Aufbau Principle The building-up principle of the periodic system of the elements, based on four

quantum numbers assigned to every electron in an atom.

Core In the electronic configuration of an element, the symbol in brackets of the noble gas preceding the element in the periodic table.

Electron Capacity The number of electrons that can be accommodated by each electron shell, subshell, and orbital.

Electron Cloud The electron probability density at each point in an atom; the probability of finding an electron at that point.

Electron Configuration The distribution of electrons in the quantum shells of an atom.

Element A pure substance that cannot be decomposed into simpler substances by ordinary chemical or physical means.

Energy-level Diagram The representation of the electron configuration of an atom using circles or boxes for orbitals and arrows for electrons.

Excited State Any state other than the ground state of an atom.

Ground State   In the ground state of an atom, the electrons occupy the lowest energy levels available and the atom is stable.

Group   A vertical column in the periodic table occupied by elements with related chemical and physical properties that make up a chemical family.

Heisenberg Uncertainty Principle   According to this principle, if we can measure the energy of an electron precisely, we cannot simultaneously determine the exact position of the electron.

Hund's Rule   Electrons do not enter into joint occupancy of an orbital until all of the available orbitals of a given subshell are half occupied.

Isotopes   Different varieties of the atoms of an element. Their nuclei have the same atomic number but different mass numbers.

Magnetic Quantum Number   Designates a sublevel or orbital.

Main-group Elements   The elements in a vertical column in the periodic table designated IA to VIIA.

Mass Number   The total number of protons and neutrons in the nucleus of an atom.

Mass Spectrometer   An instrument that separates the isotopes of an element.

Moseley's Law   The properties of the elements are a periodic function of their atomic numbers.

Neutron   A nuclear particle with approximately the same mass as a proton but lacking an electrical charge.

Noble Gas   One of the family of elements helium, neon, argon, krypton, xenon, and radon. All except helium have eight electrons in their outermost energy level.

Nucleon   A proton or a neutron.

Octet   The presence of eight electrons in the outermost energy level of an atom.

Orbital   A sub-sublevel of an atom that can accommodate two electrons of opposite spin.

Pauli Exclusion Principle   No two electrons of the same atom may have the same set of four quantum numbers.

Period   A horizontal row running from left to right across the periodic table.

Periodic Table   A chart showing all the elements arranged in columns in such a way that the elements in a given column exhibit similar chemical properties.

Periodicity   The recurrence at definite intervals of the chemical and physical properties of the elements when arranged in a table according to increasing atomic number.

Positive Ion   An atom that has lost one or more electrons and carries a positive charge.

Principal Quantum Number   Determines the main energy level in which an electron is located.

Probability   The position of an electron in an atom is uncertain, so that we can only refer to the probability of an electron being in a certain region.

Proton   A nuclear particle of an atom carrying a unit positive charge.

Quantized   The energies of electrons located in energy levels around the nucleus can take on only certain values specified by four quantum numbers.

Quantum Mechanics   A mathematical method of treating particles on the basis of quantum theory.

Representative Elements   See Main-Group Elements.

Shell   An orbit or main energy level of an atom.

Spin Quantum Number   One of two equivalent electron spin orientations.

Symbol   A representation of an element.

Transition Metals   Elements in the B columns in the periodic table in which d or f orbitals are being filled.

Unified Atomic Mass Unit   An arbitrary mass unit defined to be exactly one-twelfth the mass of the carbon-12 isotope.

Wave Mechanics   A system based on Erwin Schrödinger's wave equation, in which the wave nature of electrons is assumed and the solutions describe the arrangement of electrons in atoms.

THINGS TO DO

1. Examine models of atoms, nuclei, and crystals.

2. Construct a mobile of an atom to portray its three-dimensional structure.

3. Make an energy-level chart of the atomic structures of the first 20 elements in the periodic table.

4. Examine a sample of a representative element from group I to group VII.

5. Construct models of the isotopes of an element.

6. Make a chart of the Longuet-Higgins periodic table.

7. Pick out the elements in the periodic table that are named after (a) people; (b) places; (c) planets.

EXERCISES

Symbols for the Elements

13–1. Give the name of the element for each atomic symbol.
(a) Pb
(b) Sn
(c) Hg
(d) Ag
(e) Si

13–2. Give the atomic symbol for each of the following elements.
(a) potassium
(b) sodium
(c) phosphorus
(d) aluminum
(e) fluorine

13–3. Give the symbols and names of the elements whose symbols consist of only one letter.

13–4. Name five elements whose symbols appear to be unrelated to the names we use for the elements.

Atomic Number; Mass Number; Atomic Mass; Isotopes

13–5. Give the number of electrons, protons, and neutrons for these environmentally important elements: (a) fluorine-19; (b) sulfur-32; (c) arsenic-75; (d) strontium-88; (e) mercury-201.

13–6. Determine the nuclear structure of the following isotopes: (a) $^{35}_{17}Cl$; (b) $^{36}_{17}Cl$; (c) $^{37}_{17}Cl$.

13–7. The abundances of the isotopes of magnesium (based on the mass of ^{12}C as 12.00000) are as follows:

Isotope	Mass	% Abundance
^{24}Mg	23.9847	78.6
^{25}Mg	24.9858	10.1
^{26}Mg	25.9814	11.3

Determine the atomic mass of magnesium.

13–8. When Mt. St. Helens erupted in 1980, a considerable amount of a radioactive element was produced. The atomic number of the element is 86. What are the name, symbol, and atomic mass of this element?

13–9. Give the mass of an atom with
(a) 49 protons, 49 electrons, 64 neutrons.
(b) 31 protons, 31 electrons, 38 neutrons.
(c) 91 protons, 91 electrons, 140 neutrons.

13–10. Determine the number of neutrons in each of the following.

(a) a nitrogen atom of mass number 13
(b) a potassium atom of mass number 41
(c) a lead atom of mass number 207

13–11. Where in the atom are these particles located?
(a) protons
(b) electrons
(c) neutrons

13–12. How many nucleons are there in each of the following?
(a) ^{40}Ca
(b) ^{56}Fe
(c) zinc-65
(d) iodine-127

13–13. Which of the following atoms are isotopes of each other?
(a) $^{15}_{7}A$
(b) $^{14}_{7}B$
(c) $^{14}_{8}C$

13–14. Do the symbols $^{40}_{20}Ca$ and $_{20}Ca$ have the same meaning? Explain.

13–15. Cobalt-60 is used in radiation therapy for certain cancers. How many
(a) protons are in its nucleus?
(b) neutrons are in its nucleus?
(c) electrons are in a cobalt atom?

13–16. In what way do the isotopes copper-63 and copper-65 differ from each other?

13–17. Are there any atoms that have no neutrons? Explain.

13–18. Arrange the following in order of increasing mass.
(a) 2 Na atoms
(b) 3 Mg atoms
(c) a K atom

Electron Configurations and Quantum Numbers

13–19. Name the elements associated with the following electron configurations.
(a) $1s^2 2s^2$
(b) [Ne] $3s^1$
(c) [Ar] $4s^2 3d^7$
(d) $1s^2 2s^2 2p^6 3s^2 3p^6 4s^2 3d^{10} 4p^5$
(e) [Kr] $5s^2 4d^{10} 5p^2$

13–20. What experimental evidence is there for the idea that electronic spin is "quantized" in two possible directions?

13–21. In the energy level $n = 3$,
(a) how many orbitals are possible?
(b) what is the electron capacity of this level?

13–22. Write the electron configurations for (a) $_5$B; (b) $_9$F; (c) $_{21}$Sc; (d) $_{33}$As; (e) $_{56}$Ba.

13–23. The highest-energy electron of a chlorine atom is designated $3p^5$. (a) Explain why. (b) Deduce from this notation the values of the four quantum numbers (n, l, m, s).

13–24. (a) Construct an "electron-in-the-box" energy-level diagram for phosphorus, $_{15}$P. (b) What principle dictates the arrangement of the outermost electrons?

13–25. Relate the Heisenberg uncertainty principle to the idea of an electron orbital.

13–26. Explain why the following is not a possible set of quantum numbers for an electron in an atom: $n = 2, l = 2, m = 0$.

13–27. How many electrons in an atom can have the following sets of quantum numbers?
(a) $n = 2$
(b) $n = 2, l = 0$
(c) $n = 2, l = 0, m = 0$

13–28. Emission spectra can be used to confirm the presence of an element in a material. How is this possible?

Periodic Properties

13–29. On the basis of one property that you select, discuss the meaning of the term "periodic" in the expression "periodic table."

13–30. Experiments are under way to synthesize element 114. In which group of the periodic table would you expect it to be located?

13–31. Sodium initiates the third period, and argon closes it. What feature of atomic structure do the elements in this period share?

13–32. What similarity in electronic arrangement is shared by the members of the nitrogen family?

13–33. Which of the following elements have fairly similar properties: Ne, Sr, Kr, Br, Rb, Ca, He, Cl, F, Li, K?

13–34. Why does the periodic table work?

13–35. *Multiple Choice*
A. When the electron of a hydrogen atom is in the energy level $n = 2$, we say that
(a) it emits light.
(b) it is in the ground state.
(c) it is in an excited state.
(d) it ionizes the atom.
B. The atomic number of gold is 79 and the mass number of one of its isotopes is 197. The number of neutrons is
(a) 79.
(b) 197.
(c) 118.
(d) 158.
C. Elements in the same group in the periodic table
(a) have similar chemical properties.
(b) are called isotopes.
(c) have consecutive atomic numbers.
(d) make up a period of elements.
D. The principle that a maximum of two electrons is allowed in any orbital is known as
(a) Bohr's Aufbau principle.
(b) Heisenberg's uncertainty principle.
(c) Hund's rule.
(d) the Pauli exclusion principle.
E. The total electron capacity of the energy level $n = 4$ is
(a) 8.
(b) 16.
(c) 18.
(d) 32.
F. All of the following are located in the nucleus of atoms, *except*
(a) protons.
(b) electrons.
(c) neutrons.
(d) nucleons.
G. An element has eight protons and seven neutrons. The element is
(a) phosphorus.
(b) nitrogen.
(c) oxygen.
(d) chlorine.
H. Neutral atoms of ^{117}Sn and ^{119}Sn differ in
(a) the number of protons in the nucleus.
(b) the number of neutrons in the nucleus.
(c) their atomic numbers.
(d) the number of extranuclear electrons.
I. The atomic weight of potassium is 39.1. We can conclude that
(a) an atom of potassium has a mass of 39.1 g.
(b) potassium is a mixture of different masses, the average of which is 39.1 u.
(c) every atom of potassium has a mass of 39.1 u.
(d) potassium is an isotope of $_{39}$Y.
J. Experimental support for the arrangement of electrons in distinct energy levels is based primarily on
(a) scattering experiments.
(b) the photoelectric effect.
(c) atomic spectra.
(d) radioactivity.

SUGGESTIONS FOR FURTHER READING

Ball, David, W., "Elemental Etymology: What's in a Name." *Journal of Chemical Education*, September, 1985.
A close look at the origin of the names, or etymologies, of the chemical elements.

Belloni, Lanfranco, "Pauli's 1924 Note on Hyperfine Structure." *American Journal of Physics*, May, 1982.
Goudsmit and Uhlenbeck were influenced by Pauli's introduction of a fourth quantum number for the electron. But Pauli had doubted the concept of electron spin on first hearing.

Bloch, Felix, "Heisenberg and the Early Days of Quantum Mechanics." *Physics Today*, December, 1976.
Heisenberg's first doctorate student, later a Nobel laureate, reminisces about the formative years of the new mechanics.

Fernelius, W. Conrad, "Some Reflections on the Periodic Table and Its Use." *Journal of Chemical Education*, March, 1986.
The periodic table can be the most useful means of generalization in relating, retrieving, and predicting properties and reactions. But the choice of one form of the periodic table as the "official" or "best" is a mistake, according to the writer.

Fowler, William A., "The Quest for the Origin of the Elements." *Science*, Vol. 226, 23 November, 1984, pp. 922–935.
All life on earth depends on the energy in sunlight. But the sun did not produce the chemical elements that are found in the earth and in our bodies.

Gale, Noel, H., and Zofia Stos-Gale, "Lead and Silver in the Ancient Aegean." *Scientific American*, June, 1981.
Lead and silver were smelted from the same ores. The pattern of the isotopes of lead shows that most of the ore came from only two sources.

Goudsmit, Samuel A., "It Might as Well Be Spin." *Physics Today*, June, 1976.
Co-discoverer with George Uhlenbeck of electron spin, the author recalls the atmosphere in the "springtime of modern atomic physics."

Hanle, Paul A., "The Schrödinger-Einstein Correspondence and the Sources of Wave Mechanics." *American Journal of Physics*, July, 1979.
Letters between Einstein and Schrödinger in 1925 and de Broglie's doctoral dissertation reveal sources of Schrödinger's inspiration to invent wave mechanics.

Jacob, Maurice, and Peter Landshoff, "The Inner Structure of the Proton." *Scientific American*, March, 1980.
There is evidence that protons and neutrons are not elementary objects. Like Rutherford's atoms, they are mostly empty space. But embedded within them are small, hard constituents called either partons or quarks.

Jaffe, Bernard, *Moseley and the Numbering of the Elements*. Garden City, N.Y.: Doubleday & Co., 1971.
An absorbing account of the discoverer of the new periodic law.

Kragh, Helge, "Chemical Aspects of Bohr's 1913 Theory." *Journal of Chemical Education*, April, 1977.
Discusses the chemical content of the 1913 theory that revolutionized our concept of matter.

Longuet-Higgins, H. C., "A Periodic Table: The Aufbauprinzip as a Basis for Classification of the Elements." *Journal of Chemical Education*, January, 1957.
Presents a periodic table that shows how the position of an element in the table is related to the electronic structure of the atom.

Penzias, Arno A., "The Origin of the Elements." *Science*, 10 August, 1979.
Reviews the modern understanding of the origin of the chemical elements.

Perrino, Charles T., and Donald L. Peterson, "Another Quantum Number?" *Journal of Chemical Education*, August, 1989.
The authors propose that presenting a fifth quantum number, m_s, for spin magnetic quantum number (z-axis component), would promote consistency and clarity in the understanding of the quantum numbers and spectroscopy.

Rich, Ronald L., and Robert W. Suter, "Periodicity and Some Graphical Insights on the Tendency *Toward* Empty, Half-full, and Full Subshells." *Journal of Chemical Education*, August, 1988.
Understanding physical properties and certain aspects of reactivity requires a knowledge of the energy levels available to electrons and the extent to which these levels are populated.

Ringnes, Vivi, "Origin of the Names of Chemical Elements." *Journal of Chemical Education*, September, 1989.
Traces the etymology of the names of the elements and the reasons scientists coined a specific name for a newly discovered element.

Ruby, Lawrence, "Modern Architects of the Periodic System." *American Journal of Physics*, January, 1984.
During the nineteenth century, Mendeléev and others revised the ordering of the elements many times. In the

twentieth century, the entire basis of classification has been changed.

Solov'ev, Yu. I., "D. I. Mendeléev and the English Chemists." *Journal of Chemical Education*, December, 1984.
Mendeléev recognized that the cooperation of scholars of different countries was required in order to explain the periodic law.

Spronsen, Jan W. von, "The Priority Conflict Between Mendeléev and Meyer." *Journal of Chemical Education*, March, 1969.

Discusses one of the classical priority disputes in science.

Szokefalvi-Nagy, Zoltan, "How and Why of Chemical Symbols." *Chemistry*, February, 1967.
Although chemical symbols have been in use for thousands of years, their full potential has been realized only recently.

ANSWERS TO NUMERICAL EXERCISES

13–5. (a) 9 p, 9 e, 10 n
        (b) 16 p, 16 e, 16 n
        (c) 33 p, 33 e, 42 n
        (d) 38 p, 38 e, 50 n
        (e) 80 p, 80 e, 121 n

13–6. (a) 17 p, 18 n
        (b) 17 p, 19 n
        (c) 17 p, 20 n

13–7. 24.310 amu

13–9. (a) 113
        (b) 69
        (c) 231

13–10. (a) 6
         (b) 22
         (c) 125

13–12. (a) 40
         (b) 56
         (c) 65
         (d) 127

13–21. (a) 9 orbitals
         (b) 18 electrons

13–27. (a) 8
         (b) 2
         (c) 2

13–30. group IV

ESSAY

THE ORIGIN OF THE ELEMENTS

JOHN C. KOTZ

State University of New York College at Oneonta

KEITH F. PURCELL

Kansas State University

By now you have seen some of the chemistry of the elements and their compounds, and you have begun to connect these to chemical processes in the world around you. But where did the elements originate? Why are there so few of them? Why are carbon-based compounds so numerous? Why does the hemoglobin of your blood contain iron and not ruthenium or uranium? Are there elements somewhere in the universe that are not found on earth? These are all very interesting questions, and we are most fortunate in now having some answers!

The composition of a number of stars in the universe and of the planets and moons of our solar system has been determined. In one sense it is a great relief, and in another a disappointment, to find that there is no evidence for an element found in some distant star that does not exist on earth. Further, although there are variations from star to star and planet to planet, the relative abundances of the elements throughout the universe are approximately the same. These are very important observations, since they mean that the element-forming processes are general throughout the universe.

The relative abundances of a few of the elements in the universe are illustrated in Figure A. The most striking features of this figure are listed here and are the facts that must be taken into account in any theory of the origin of the elements.

(a) 90% to 95% of the atoms in the universe are hydrogen atoms.

(b) 4% to 9% of all atoms are helium.

(c) All of the other elements taken together make up only about 1% of the universe, even on a weight basis.

(d) Lithium, beryllium, and boron are mysteriously rare.

(e) Elements of even atomic number are more abundant than those with odd atomic number.

(f) There is a general decline in abundance from oxygen to lead. However, there is a very pronounced maximum in relative abundance around iron.

(g) There are no stable elements with mass numbers greater than about 210.

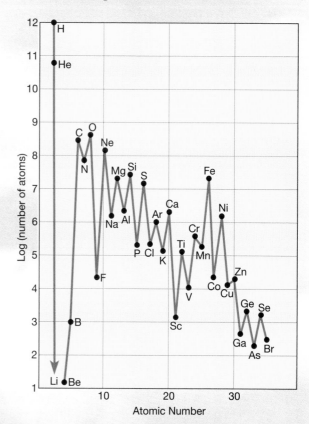

Figure A   The cosmic abundances of the lighter elements as a function of atomic number. Abundances are expressed as numbers of atoms per 10^{12} atoms of H and are plotted on a (base-10) logarithmic scale. (Data taken from G. O. Abell, *Exploration of the Universe*, 4th edition. Philadelphia: Saunders College Publishing, 1982, p. 706.)

The universe is thought to have begun about 20 billion years ago as a subatomic particle soup (electrons, positrons or positive electrons, protons, neutrons, and massless particles such as neutrinos and antineutrinos and photons in a ball recently estimated to be only about 10^{-28} cm in diameter). The "big bang" theory hypothesizes that an explosion of unimaginable fury began the series of events that has led ultimately to the universe and to life on earth.

Within seconds of the big bang—the explosion of the super-dense ball of particles—the synthesis of the elements began. Free neutrons are known to decay to give protons, electrons, and a large amount of energy.

$$\text{Neutron} \rightarrow \text{Proton} + \text{Electron} + \text{Energy}$$

Since a hydrogen atom has a nucleus consisting of a single proton, this is the origin of the hydrogen in our universe. To continue the process of element formation, hydrogen nuclei fused in the high-temperature furnace of the big bang and continue to do so in stars, to give helium nuclei, the next most important building block of the elements.

In the seconds, minutes, and hours following the moment of Creation, the gas-filled, chaotic universe rapidly expanded and began to cool. After perhaps a billion years, it had cooled to the point that the hydrogen and helium began to coalesce into galaxies and, within the galaxies, into stars—a process that continues even today. The primordial stars contracted under the force of gravitational attraction, and the temperature began to climb, finally reaching about 10 million degrees, the critical temperature for "hydrogen burning" or fusion to form more helium. This exothermic reaction raises the temperature of the star, and helium nuclei can begin to coalesce at a hundred million degrees or so into nuclei of still heavier elements. In the "helium-burning" cycle, it is the carbon nucleus that is first formed (Fig. B). The direct formation of carbon from helium means that the process has circumvented lithium, beryllium, and boron and explains their very low abundance in the universe. The fact that these elements exist, however, must mean that they are formed in a different way than by the fusion of helium nuclei.

Our sun is a relatively small star, since ones of much larger mass, say 10 to 30 solar masses, have been observed. Because of their very great mass, the core temperatures of large stars must become very great, and fusion reactions to form heavier nuclei than carbon are possible. Heavier and heavier elements are built up, in a process involving he-

lium, ^4_2He, similar to that in Figure B. Since helium is an element of "even mass," this helps explain the observation that elements of even mass are more abundant than those of odd mass.

Energy is evolved when light nuclei coalesce or fuse to form nuclei lighter than iron. However, iron formation is the end of this exothermic series! To continue the fusion process requires energy. When a massive star has essentially consumed its fuel and iron has been formed, the star can no longer support its outer layers against gravity, and the core collapses to a super-dense state *in less than a second.* The shock waves from this collapse expand outward through the envelope of gases surrounding the core, heating the gases and providing the energy to create heavy nuclei. This outward blast is seen as a gigantic explosion called a supernova, many of which have been observed. The explosion sends debris rocketing into space, where over additional millions of years it coalesces with hydrogen, helium, and other elements to form new stars. Our sun is just such a "second-generation" star, and our earth and the other planets of the solar system were formed from the debris of the explosions of massive stars. It is for this reason that there is a wide variety of elements found on the earth and other planets.

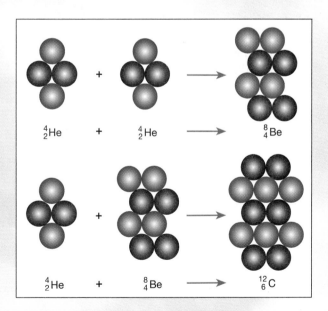

Figure B   Hydrogen fusion continues for 99% of the lifetime of a star. When the hydrogen is essentially depleted, the star collapses until its core reaches a temperature in excess of 100 million degrees Celsius. At this temperature, helium nuclei begin to fuse to produce carbon nuclei. The process is thought to proceed through an unstable Be isotope as an intermediate.

Yellowish needle-like crystals of rutile (titanium oxide) embedded in a crystal of colorless quartz (silicon dioxide). The molecular architecture is reflected in the geometry of a crystal. This specimen was found in Bahia, Brazil. (Roberto De Gugliemo/Science Photo Library/Photo Researchers, Inc.)

14

CHEMICAL BONDING AND MOLECULAR GEOMETRY

We usually encounter the 100-odd elements not as pure substances but as constituents of compounds. Bulk matter occurs in combination rather than in atomic form and is held together by forces called *chemical bonds*. Simple or complex, all compounds —water, sugar, salt—are formed of aggregates of atoms in a definite ratio represented by a chemical formula, as shown in Table 14–1. Chemical bonds are the "glue" that holds the atoms together and with molecular geometry determines the properties of the compound.

condition that apparently confers unusual stability upon the atoms involved. In all of the noble gases except helium there are eight electrons in the outermost energy level; in helium there are two electrons. Before an atom can accept one or more electrons, there must be orbital vacancies available to accommodate them. Since all of the orbitals in noble gas atoms are filled, there are no orbitals available for bond formation.

The electrons of the noble gas atoms are held tenaciously. A measure of the tightness with which an atom holds on to its electrons is the **ionization energy**—the energy necessary to remove an elec-

14–1 THE NOBLE GASES

Although most of the **noble gases** were discovered in the 1890s, until 1962 it was believed that they were chemically inert. They were not known to take part in the formation of any compounds whatsoever and appeared to be models of chemical stability. Then Neil Bartlett of the University of British Columbia and others prepared compounds of the heavier noble gases, Kr, Xe, and Rn, with fluorine, and found certain of these compounds, such as XeF_2, XeF_4, and XeF_6, and KrF_2, also to be quite stable (Fig. 14–1). This work, a startling discovery, caused a sensation.

Their *almost* complete lack of chemical reactivity suggests that the noble gases have little inclination to lose, gain, or share electrons in order to form chemical bonds. As their electron configurations in Table 14–2 indicate, their orbitals are filled to capacity, a

Figure 14–1 Crystals of xenon tetrafluoride, XeF_4. Before 1962, the rare gases had been thought to be chemically inert, so that the discovery that they did form compounds came as a revolutionary development. (Argonne National Laboratory)

335

Table 14–1   SOME FAMILIAR COMPOUNDS

Name	Formula	Name	Formula
Alcohol (ethyl)	C_2H_5OH	Milk of magnesia	$Mg(OH)_2$
Aspirin	$C_9H_8O_4$	Photographer's hypo	$Na_2S_2O_3$
Baking soda	$NaHCO_3$	Plaster of Paris	$CaSO_4 \cdot \frac{1}{2}H_2O$
Borax	$Na_2B_4O_7 \cdot 10H_2O$	Sugar (sucrose)	$C_{12}H_{22}O_{11}$
Dry ice	CO_2	Table salt	$NaCl$
Epsom salt	$MgSO_4 \cdot 7H_2O$	TNT	$C_7H_5N_3O_6$
Lime	CaO	Vinegar	$C_2H_4O_2$
Lye	$NaOH$	Water	H_2O

Table 14–2   ELECTRON CONFIGURATIONS OF THE NOBLE GASES

Element	Symbol and Atomic Number	Electronic Configuration	Number of Electrons in Shell					
			1	2	3	4	5	6
Helium	$_2He$	$1s^2$	2					
Neon	$_{10}Ne$	$[He]\ 2s^2 2p^6$	2	8				
Argon	$_{18}Ar$	$[Ne]\ 3s^2 3p^6$	2	8	8			
Krypton	$_{36}Kr$	$[Ar]\ 4s^2 3d^{10} 4p^6$	2	8	18	8		
Xenon	$_{54}Xe$	$[Kr]\ 5s^2 4d^{10} 5p^6$	2	8	18	18	8	
Radon	$_{86}Rn$	$[Xe]\ 6s^2 4f^{14} 5d^{10} 6p^6$	2	8	18	32	18	8

Table 14–3   IONIZATION ENERGIES°

Atomic Number	Atom	Ionization Energies (eV)							
		1st	2nd	3rd	4th	5th	6th	7th	8th
1	H	13.6							
2	He	24.6	54.4						
3	Li	5.4	75.6	122.4					
4	Be	9.3	18.2	153.9	217.7				
5	B	8.3	25.1	37.9	259.3	340.1			
6	C	11.3	24.4	47.9	64.5	392.0	489.8		
7	N	14.5	29.6	47.4	77.5	97.9	551.9	666.8	
8	O	13.6	35.1	54.9	77.4	113.9	138.1	739.1	871.1
9	F	17.4	35.0	62.6	87.2	114.2	157.1	185.1	953.6
10	Ne	21.6	41.1	64	97.2	126.4	157.9	—	—
11	Na	5.1	47.3	71.7	98.9	138.6	172.4	208.4	264.2
12	Mg	7.6	15.0	80.1	109.3	141.2	186.9	225.3	266.0
13	Al	6.0	18.8	28.4	120.0	153.8	190.4	241.9	285.1
14	Si	8.1	16.3	33.5	45.1	166.7	205.1	246.4	303.9
15	P	10.6	19.7	30.2	51.4	65.0	220.4	263.3	309.3
16	S	10.4	23.4	35.0	47.3	72.5	88.0	281.0	328.8
17	Cl	13.0	23.8	39.9	53.5	67.8	96.7	114.3	348.3
18	Ar	15.8	27.6	40.9	59.8	75.0	91.3	124.0	421
19	K	4.3	31.8	46	60.9	—	99.7	118	155
20	Ca	6.1	11.9	51.2	67	84.4	—	128	147

°From Johnston et al., *Chemistry and the Environment*. Philadelphia: W.B. Saunders Company, 1973, p. 60.

Figure 14–2 Crystals of rock salt, naturally occurring NaCl. (Harold L. Levin)

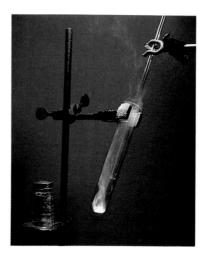

Figure 14–3 Sodium reacting with chlorine to produce table salt, sodium chloride. (Charles Steele)

tron from an atom in the gaseous state. Ionization energy is usually expressed in units of electron volts/atom or kilojoules/mol (1 eV/atom = 96.49 kJ/mol).

The first electron of an atom is removed with less difficulty than a second or third, since the latter are removed from positively charged ions. The energies are referred to as the first ionization energy, the second ionization energy, and so on, as diagrammed in Table 14–3). Note the trend in the first ionization energy for the first 20 elements. In each period, a maximum is reached with the noble gas that completes that period.

14–2 THE IONIC BOND

Although the noble gases are quite inert, they are flanked on both sides in the periodic table by very active groups of elements. The halogens, a group of nonmetals, and the alkali metals readily enter into reactions with one another, forming compounds called salts (*halogen*, salt-former) of which ordinary table salt, sodium chloride, is an example. These salts are crystalline solids; they are hard and brittle, have relatively high melting (600°–1000°C) and boiling points, are generally soluble in water, and are good conductors of electricity when molten or in water solution (but not in the crystalline state) (Fig. 14–2). Table 14–4 shows the electronic structures of atoms of the halogen and alkali metal families.

Atoms of the halogens and the alkali metals can acquire a stable noble gas structure by gaining or losing one electron, respectively. Thus, a sodium atom can acquire the neon electronic structure by losing an electron and forming an ion with a single positive charge, Na^+. By gaining an electron, a chlorine atom requires an octet of electrons in its outermost energy level and becomes the negatively charged chloride ion, Cl^-, an **anion**, with the electron configuration of argon. When the elements sodium and chlorine are brought together (Fig. 14–3), elec-

Table 14–4 ELECTRONIC STRUCTURES OF THE HALOGENS AND ALKALI METALS

Halogens		Noble Gases	Alkali Metals	
		$_2$He	$_3$Li	[He] $2s^1$
$_9$F	[He] $2s^2 2p^5$	$_{10}$Ne	$_{11}$Na	[Ne] $3s^1$
$_{17}$Cl	[Ne] $3s^2 3p^5$	$_{18}$Ar	$_{19}$K	[Ar] $4s^1$
$_{35}$Br	[Ar] $4s^2 4p^5$	$_{36}$Kr	$_{37}$Rb	[Kr] $5s^1$
$_{53}$I	[Kr] $5s^2 4d^{10} 5p^5$	$_{54}$Xe	$_{55}$Cs	[Xe] $6s^1$
$_{85}$At	[Xe]$6s^2 4f^{14} 5d^{10} 6p^5$	$_{86}$Rn	$_{87}$Fr	[Rn] $7s^1$

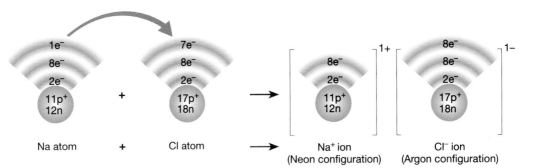

Figure 14–4   Ionic bonding. Formation of sodium chloride by the transfer of electrons between atoms.

trons are transferred to form Na^+ and Cl^- ions (Fig. 14–4). The stable ions in NaCl are held together by the Coulomb force of attraction between oppositely charged ions and form an **ionic** or **electrovalent bond** (Fig. 14–5). From such observations the "rule of eight" or **octet rule** has been developed: **Atoms enter into combination to obtain a structure that has eight electrons in the outermost energy level.**

Bond formation usually involves only those electrons that occupy the highest-energy orbitals of the atom, called the **valence electrons**. Electrons in fully occupied lower-lying energy orbitals are ordinarily held too tightly by the nucleus to participate in ionic bonding. The highest-energy s and p orbitals are valence orbitals, containing from one to eight valence electrons. The number of charges on the ion defines its **electrovalence**, or **oxidation number**, and represents the combining power of the element. Thus,

sodium has a valence of $+1$, and chlorine has a valence of -1.

In acquiring the electronic structures of neon and argon, sodium and chlorine are not changed into the noble gas atoms. The sodium atom, sodium ion, and neon atom are related, as shown in Figure 14–6. Since the nuclear charge determines the nature of the element, the Na^+ with 11 positive charges on the nucleus is different from neon, which has only 10 positive charges.

The charge of $+1$ on the sodium ion comes from the imbalance between the number of positively charged protons in the nucleus and the number of negatively charged electrons:

Electric charges in the sodium ion

$11 +$ from 11 protons
$\underline{10 -}$ from 10 electrons
 $1 +$ net charge on sodium ion

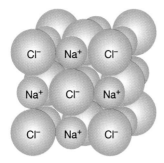

Figure 14–5   Structure of a crystal of sodium chloride. Each sodium ion is surrounded by chloride ions, and each chloride ion is surrounded by sodium ions.

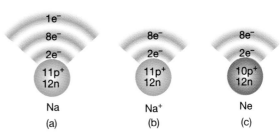

Figure 14–6   Comparison of a sodium atom (a), a sodium ion (b), and a neon atom (c).

Table 14-5 SELECTED SETS OF ISOELECTRONIC IONS AND NOBLE GASES

Noble Gas	Number of Electrons	Ion
Neon, Ne	10	Oxide, O^{2-} Fluoride, F^- Sodium, Na^+ Magnesium, Mg^{2+} Aluminum, Al^{3+}
Argon, Ar	18	Sulfide, S^{2-} Chloride, Cl^- Potassium, K^+ Calcium, Ca^{2+}
Krypton, Kr	36	Selenide, Se^{2-} Bromide, Br^- Rubidium, Rb^+ Strontium, Sr^{2+}

Figure 14-7 Sodium, a silvery metal, reacts with chlorine, a green gas, to form sodium chloride, a white, crystalline solid. (Chip Clark)

Since the sodium ion and the neon atom have the same number of electrons, ten, they are said to be **isoelectronic** (*iso*, same) with each other.

There is a limit to the number of electrons that an atom can lose or gain. Positive ions (cations) commonly have charges of $+1$, $+2$, or $+3$ at most. On the other hand, a nonmetallic atom may gain an extra one, two, or perhaps three electrons, but electrovalences lower than -3 are unknown. The noble gas structures of various ions are shown in Table 14–5.

The often remarkable differences in the properties of the uncombined elements and the compounds that they form are evident in sodium chloride, as Table 14–6 and Figure 14–7 show.

Table 14–7 gives the names and chemical formulas of some common ionic compounds. Figure 14–8 shows some common colored ionic compounds.

14-3 THE COVALENT BOND

Although ionic bonding accounts for the properties of many important compounds, there are many thousands more for which there is no evidence to indicate that the elements exist as ions. These compounds may be gaseous, liquid, or solid at room temperature. They usually have rather low melting and boiling points. They do not conduct an electric current, and they are not very soluble in water as a rule. The experimental evidence indicates that they are made up of discrete molecules, a **molecule** being the **smallest particle of a substance that can have a stable independent existence** (Fig. 14–9).

Gilbert Newton Lewis (1875–1946), at the University of California, studied the electronic structure of molecular compounds. He proposed that a stable

Table 14-6 PROPERTIES OF SODIUM AND CHLORINE IN THE FREE AND COMBINED STATES

Substance	State at Room Temperature	Melting Point, °C	Boiling Point, °C
Sodium	Silvery, soft solid	98	892
Chlorine	Greenish, irritating gas	-101	-35
Sodium chloride	Colorless, crystalline solid	801	1,413

Table 14–7   REPRESENTATIVE IONIC COMPOUNDS

Common Name	Formula	Chemical Name
Baking soda	$NaHCO_3$	Sodium bicarbonate
Bleaching powder	$Ca(ClO)_2$	Calcium hypochlorite
Limestone	$CaCO_3$	Calcium carbonate
Rock salt	$NaCl$	Sodium chloride
Silver bromide	$AgBr$	Silver bromide

Table 14–8   LEWIS STRUCTURES OF HYDROGEN THROUGH ARGON

I	II	III	IV	V	VI	VII	VIII
H ·							He :
Li ·	· Be ·	· Ḃ ·	· Ċ ·	· N̈ ·	: Ö ·	: F̈ ·	: N̈e :
Na ·	· Mg ·	· Al ·	· Si ·	· P̈ ·	: S̈ ·	: C̈l ·	: Är :

Figure 14–8   Examples of ionic compounds: nickel(II) nitrate (green); potassium dichromate (orange); and copper(II) sulfate, blue. (Charles D. Winters)

Figure 14–9   Molecular solids such as violet-black iodine and white, solid carbon dioxide (dry ice), consist of molecules. (Charles D. Winters)

noble gas structure could be achieved by the **sharing of electrons** between atoms. Irving Langmuir (1881–1957), a chemist at the General Electric Research Laboratory, elaborated on Lewis' theory and introduced the term **covalent bond** to describe the Lewis electron-pair bond.

Lewis introduced a useful notation to represent atoms, ions, and molecules. In **electron-dot symbols**, called **Lewis structures**, the valence electrons are shown as dots surrounding the symbol of the element. The symbol itself represents the **kernel**, consisting of the atom's nucleus and inner, or core, electrons. Except for helium, the number of valence electrons of the elements in the first three periods is equal to the number of their periodic group. For example, aluminum in Group III has three valence electrons. The Lewis structures of the first 18 elements are shown in Table 14–8.

To apply the Lewis-Langmuir approach to the structure of the hydrogen molecule, we see that an individual H atom has one electron and is, therefore, one electron short of the noble gas configuration of helium. The stable diatomic hydrogen molecule (H_2) results from the sharing of two electrons of opposite spin by the two hydrogen atoms (Fig. 14–10). In this way, each H atom is surrounded by two electrons and

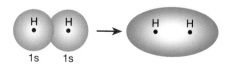

Figure 14–10   The s orbitals from two H atoms overlap to form an H_2 molecule. The electron density is concentrated in the region between the two bonded atoms.

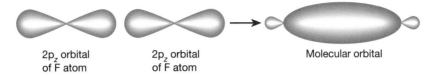

2p$_z$ orbital   2p$_z$ orbital   Molecular orbital
of F atom   of F atom

Figure 14–11 When two F atoms form a molecule of F_2, their two 2p$_z$ orbitals overlap head-to-head to form a sigma covalent bond or molecular orbital. (The other orbitals are not affected and are not shown in this figure.)

acquires the stable helium structure. The shared pair of electrons, represented by two dots or a dash, is called a covalent bond.

H : H   H — H

covalent bond

The Lewis-Langmuir model of the covalent bond accounts for the diatomic nature of fluorine (F_2) and other halogen molecules (Cl_2, Br_2, I_2) as well. The valence shell of a fluorine atom has seven electrons ($2s^2 2p^5$), so it is one electron short of the noble gas configuration of neon ($2s^2 2p^6$). A stable diatomic molecule (F_2) results from the sharing of a pair of electrons by two fluorine atoms (Fig. 14–11). In this way, each fluorine atom acquires an octet of electrons in its valence shell. The formation of Cl_2, Br_2, and I_2 is accounted for in the same way.

: F : F :   F — F

covalent bond

Some atoms acquire a noble gas configuration by sharing more than one pair of electrons. In these cases, the atoms in the molecules are linked by **multiple bonds**. Two nitrogen atoms, for example, become bonded in a nitrogen molecule by the sharing of three pairs of electrons. Each N atom has the structure $1s^2 2s^2 2p_x^1 2p_y^1 2p_z^1$. The three p electrons of one N atom pair with three electrons of opposite spin from the second N atom.

: N ⦂⦂ N :   : N ≡ N :

nitrogen (N_2)

The sharing of two pairs of electrons leads to a **double bond**; three pairs, a **triple bond**. A molecule having only single bonds is said to be **saturated**, whereas one with double or triple bonds is **unsaturated**.

Several covalent molecules are listed in Table 14–9 with their Lewis structures.

Table 14–9 LEWIS STRUCTURES OF SOME COVALENT MOLECULES

Substance	Lewis Structure	Dash Formula	Substance	Lewis Structure	Dash Formula
Bromine	: Br : Br :	Br — Br	Carbon dioxide	· Ö ⦂⦂ C ⦂⦂ Ö ·	O=C=O
Chlorine	: Cl : Cl :	Cl — Cl	Ammonia	H : N : H H	H — N — H | H
Fluorine	: F : F :	F — F	Methane	H H : C : H H	H | H — C — H | H
Iodine	: I : I :	I — I			
Hydrogen	H : H	H — H			
Oxygen	Ö ⦂⦂ Ö	O=O	Carbon tetrachloride	: Cl : : Cl : C : Cl : : Cl :	Cl | Cl — C — Cl | Cl
Nitrogen	: N ⦂⦂ N :	N ≡ N			

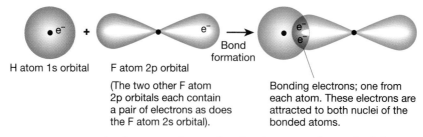

H atom 1s orbital   F atom 2p orbital

(The two other F atom
2p orbitals each contain
a pair of electrons as does
the F atom 2s orbital).

Bond
formation

Bonding electrons; one from
each atom. These electrons are
attracted to both nuclei of the
bonded atoms.

Figure 14–12   The formation of a covalent bond in HF. The s orbital from a hydrogen atom overlaps the p_z orbital from a fluorine atom.

EXAMPLE 14–1

Draw an electron-dot diagram (Lewis structure) for a molecule of hydrogen fluoride, HF.

SOLUTION

Since Lewis structures involve only valence electrons and kernels, we write the electronic configurations of H and F to determine the number of electrons in the outermost energy level of each.

$_1$H $1s^1$ H ·

$_9$F $1s^2 2s^2 2p^5$ · F̈ :

Hydrogen has one valence electron, and fluorine has $(5 + 2) = 7$ valence electrons. Fluorine acquires an octet by sharing the electron of hydrogen (Fig. 14–12). The Lewis structure for hydrogen fluoride is, therefore,

H : F̈ :

EXTENSION

Write a Lewis structure for water, H_2O. (Answer: H : Ö : H)

■

In the hydrogen molecule (H_2) and chlorine molecule (Cl_2), two identical atoms are joined by a covalent bond,

H : H   : C̈l : C̈l :

and both atoms share equally in the electrons of the bond. The covalent bond in each molecule is **nonpolar**, because the region of the greatest electron density is exactly midway between the nuclei. Nonpolar substances do not conduct electricity, since they do not have charged particles to respond to an electric field. Although the atoms in the molecules are held together by strong covalent bonds, the forces between the molecules are very weak, and the molecules are easily separated. Nonpolar covalent compounds are, therefore, often gases or liquids, or solids that sublime easily, and tend to have rather low melting points (under 300°C) and boiling points (under 500°C) (Fig. 14–13).

Figure 14–13   Naphthalene, $C_{10}H_8$, a white, crystalline, nonpolar covalent solid that you may recognize as mothballs. (Paul Silverman/Fundamental Photographs)

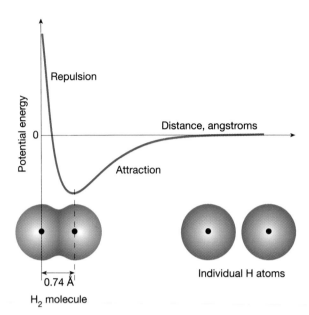

Figure 14–14 Potential energy diagram for two hydrogen atoms. At a distance of 0.74 Å, the potential energy is at a minimum and stability is at a maximum. The H_2 molecule forms at this position.

14–4 QUANTUM MECHANICS AND THE COVALENT BOND

Quantum mechanics was first applied to the problem of the structure of the hydrogen molecule in the 1920s by Walter Heitler and Fritz W. London, who calculated the energy and thus the strength of the covalent bond. Imagine two H atoms approaching one another (Fig. 14–14). When they are far apart, the attractive force exerted by each upon the other is slight. At shorter distances, this force increases while the energy of the system decreases.

When they are close, each electron is attracted not only by its own nucleus but also by that of the other atom. At distances closer than 0.74 Å, however, the repulsive forces between the two electrons and between the two protons increase greatly, and the energy rises. At that distance there is a minimum in potential energy; the attractive force between the electron of one atom and the proton of the other far exceeds the repulsive forces, and the system reaches a state of maximum stability, the hydrogen molecule,

H_2. The orbitals merge or overlap, provided that the electron spins are opposed, and there is then a chance of finding each electron in the vicinity of either nucleus. The electrons now travel about both nuclei, the resulting mutual attractions of the two electrons and the two nuclei binding the atoms together. The pair of electrons is shared by the two atoms, and the result is a covalent bond.

14–5 POLAR MOLECULES

In a molecule of hydrogen fluoride (HF), the fluorine nucleus exerts a greater attraction for the shared pair of electrons than the hydrogen nucleus (Fig. 14–15). The region of maximum electron density therefore lies closer to fluorine:

$$H^{\delta+} : \overset{..}{\underset{..}{F}} : ^{\delta-}$$

Since one end of the bond is negative and the other end positive, owing to the partial separation of electronic charge, the bond is called a **polar covalent bond**. Bond polarity can be indicated by the symbols $\delta+$ and $\delta-$, meaning partially positive and partially negative, respectively.

A measure of the ability of a bonded atom in a molecule to attract electrons is given by a positive number called the **electronegativity**. On the electronegativity scale proposed by Linus Pauling (b. 1901) at the California Institute of Technology, fluorine, the element with the greatest electron attracting power, is assigned the highest value, 4.0, followed by oxygen, nitrogen, and chlorine. Cesium, on the other hand, gives up an electron most readily and is as-

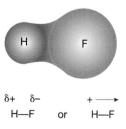

Figure 14–15 In the HF molecule, the electron density is shifted in the direction of the F atom, leaving H somewhat positive.

Table 14-10   ELECTRONEGATIVITY VALUES (PAULING'S SCALE)

1 **H** 2.1																	
3 **Li** 1.0	4 **Be** 1.5											5 **B** 2.0	6 **C** 2.5	7 **N** 3.0	8 **O** 3.5	9 **F** 4.0	
11 **Na** 1.0	12 **Mg** 1.2											13 **Al** 1.5	14 **Si** 1.8	15 **P** 2.1	16 **S** 2.5	17 **Cl** 3.0	
19 **K** 0.9	20 **Ca** 1.0	21 **Sc** 1.3	22 **Ti** 1.4	23 **V** 1.5	24 **Cr** 1.6	25 **Mn** 1.6	26 **Fe** 1.7	27 **Co** 1.7	28 **Ni** 1.8	29 **Cu** 1.8	30 **Zn** 1.6	31 **Ga** 1.7	32 **Ge** 1.9	33 **As** 2.1	34 **Se** 2.4	35 **Br** 2.8	
37 **Rb** 0.9	38 **Sr** 1.0	39 **Y** 1.2	40 **Zr** 1.3	41 **Nb** 1.5	42 **Mo** 1.6	43 **Tc** 1.7	44 **Ru** 1.8	45 **Rh** 1.8	46 **Pd** 1.8	47 **Ag** 1.6	48 **Cd** 1.6	49 **In** 1.6	50 **Sn** 1.8	51 **Sb** 1.9	52 **Te** 2.1	53 **I** 2.5	
55 **Cs** 0.8	56 **Ba** 1.0	57 **La** 1.1	72 **Hf** 1.3	73 **Ta** 1.4	74 **W** 1.5	75 **Re** 1.7	76 **Os** 1.9	77 **Ir** 1.9	78 **Pt** 1.8	79 **Au** 1.9	80 **Hg** 1.7	81 **Tl** 1.6	82 **Pb** 1.7	83 **Bi** 1.8	84 **Po** 1.9	85 **At** 2.1	
87 **Fr** 0.8	88 **Ra** 1.0	89 **Ac** 1.1															

Legend:
- < 1.0
- $1.0-1.4$
- $1.5-1.9$
- $2.0-2.4$
- $2.5-2.9$
- $3.0-4.0$

signed the lowest value, 0.7, as shown in Table 14–10. Electronegativities increase from left to right across a period and decrease from the top to the bottom of a group. Thus, the most electronegative elements are located in the upper right corner of the periodic table; the least electronegative in the lower left corner.

With the electronegativity scale we can predict the **polarity** of molecules. The greater the electronegativity difference between two bonding atoms, the greater the polarity of the bond. In the hydrogen molecule or chlorine molecule, the electronegativity difference between the two atoms is zero; hence, the bond is nonpolar, with equal sharing of electrons.

	H—H	Cl—Cl
Electronegativity:	2.1 2.1	3.0 3.0
Difference:	0	0
	(nonpolar covalent)	(nonpolar covalent)

With an electronegativity difference of 0.9 between H and Cl, the hydrogen chloride molecule is **polar covalent**.

	$\delta +$ $\delta -$
	H—Cl
Electronegativity:	2.1 3.0
Difference:	0.9 (polar covalent)

When the electronegativity difference is greater than 1.7, the shift in electron density is great enough to describe the bond as **ionic**, rather than covalent.

	+	−
	Na	Cl
Electronegativity:	0.9	3.0
Difference:	2.1	(ionic)

The relation between electronegativity difference and bond type is given in Table 14–11.

Electronegativity values can also be used to estimate the **ionic character** of a bond. The bond between most atoms is neither completely ionic nor completely covalent. A percentage describes the ionic or covalent character of a bond, as in Table 14–12.

Table 14-11   BOND TYPE PREDICTED FROM ELECTRONEGATIVITY DIFFERENCE

Electronegativity Difference	Bond Type
0	Nonpolar covalent
0 to 1.7	Polar covalent
Greater than 1.7	Ionic

Table 14–12   ELECTRONEGATIVITY DIFFERENCE AND PERCENT
IONIC CHARACTER OF A CHEMICAL BOND

Electronegativity Difference	% Ionic Character	Electronegativity Difference	% Ionic Character
0.2	1	1.8	55
0.4	4	2.0	63
0.6	9	2.2	70
0.8	15	2.4	76
1.0	22	2.6	82
1.2	30	2.8	86
1.4	39	3.0	89
1.6	47	3.2	92
		3.4	96

When the electronegativity difference is greater than 1.7, the bond is more than 50% ionic.

EXAMPLE 14–2

Predict (a) whether a carbon–chlorine bond (C—Cl) is relatively ionic or covalent and (b) which atom is relatively positive and which negative.

SOLUTION

First, we determine the electronegativity difference between the atoms from Table 14–10. Then we refer to Table 14–12 for the percent ionic character of the bond.

	C—Cl	
Electronegativity:	2.5	3.0
Difference:		0.5

(a) An electronegativity difference of 0.5 corresponds to approximately 7% ionic character (or 93%

covalent character). A C—Cl bond is, therefore, primarily covalent.

(b) The greater attraction that chlorine has for electrons creates an unbalanced electron distribution and makes the chlorine side of the molecule more negative than the carbon side:

$$C^{\delta+}—Cl^{\delta-}$$
$$\overset{\longrightarrow}{\vdash}$$

EXTENSION

Would you expect a bond between calcium and chlorine to be ionic or covalent? (Answer: Ionic)

■

The presence of polar bonds increases the tendency of molecules to stick to or associate with one another in the liquid state. The positive end of one molecule attracts the negative end of another to form clumps of several molecules (Fig. 14–16). Polar covalent compounds do not conduct electricity, how-

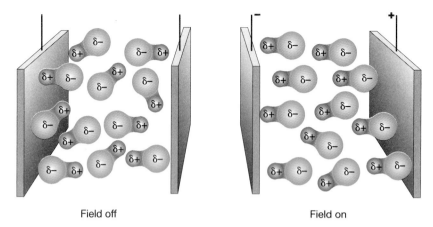

Field off   Field on

Figure 14–17 Polar molecules such as HF tend to line up very slightly in a direction opposite to that of the field. Nonpolar molecules are not oriented in an electric field.

ever. The negative and positive charges of the dipole are contained in the same particle, and the particle is attracted by the same amount in opposite directions by the electric field (Fig. 14–17). Polar covalent compounds such as hydrogen chloride (HCl) have higher melting points and boiling points than comparable nonpolar molecules such as silane (SiH_4). In the solid state, the crystals of polar molecules are harder than nonpolar molecular crystals.

14–6 DIPOLE MOMENT

In a polar molecule such as HCl, the centers of positive and negative charges do not coincide. Such a molecule is called a **dipole,** represented by the symbol $\leftrightarrow$ with the arrow pointing from the positive toward the negative pole. Quantitatively, the polar character of a molecule is expressed by its dipole moment, μ ("mu"), defined as the product of the effective charge (either positive or negative) and the distance between the centers of opposite charges.

$$+Q \qquad -Q$$

$$\longleftarrow r \longrightarrow$$

$$\text{Dipole moment} = \text{Charge} \times \text{Distance} \qquad (14-1)$$
$$\mu = Qr$$

The dipole moments of various molecules are given in Table 14–13 in Debye units (D), after Peter Debye (1884–1966), a pioneer in the field. A Debye is 3.33×10^{-30} C · m (C · m stands for coulomb · meter.)

The dipole moments of H_2 and O_2 are zero, since the electronic charge is symmetrically distributed about two similar atoms in these molecules. The zero dipole moment of CH_4 is surprising, however. The electronegativities of C and H are 2.5 and 2.1. Each

Table 14–13   DIPOLE MOMENTS OF SOME MOLECULES (DEBYE UNITS) (1 D = 3.33×10^{-30} coulomb · meters, C · m)

H_2	0	HF	1.91	H_2O	1.84	NH_3	1.46
O_2	0	HCl	1.03	H_2S	0.92	PH_3	0.55
CH_4	0	HBr	0.80	CH_3Cl	1.86	PCl_3	0.78
CCl_4	0	HI	0.38	CH_2Cl_2	1.59	PF_3	1.025

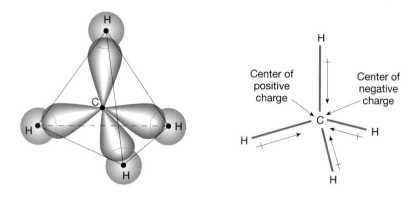

Figure 14–18. A nonpolar molecule, CH_4, $\mu = 0$. The vector sum of the dipole moments is zero, because the four C—H bonds are symmetrically distributed in space. All of the bond dipole moments are cancelled, and the center of positive charge coincides with the center of negative charge.

C—H bond, therefore, has an electronegativity difference of 0.4, corresponding to 4% ionic character. Although each C—H bond is slightly polar, in the molecular architecture of CH_4 the centers of negative and positive charges of the molecule as a whole coincide; hence, there is no net polarity. In other words, the vector sum of the four dipole moments is zero (Fig. 14–18).

For the same reason, the dipole moment of CCl_4 is zero. But if one, two, or three of the hydrogens in methane are replaced by chlorine, the polar nature of the molecules is rather pronounced. Thus, the dipole moment of methyl chloride (CH_3Cl) is 1.86 D; dichloromethane (CH_2Cl_2), 1.59 D; chloroform ($CHCl_3$), 1.15 D. The dipole moment of water,

1.84 D, leads us to conclude that the molecule is bent rather than linear, for if it were linear, the dipole moment would be zero (Fig. 14–19).

14–7 MOLECULAR GEOMETRY

Oxygen is bonded covalently to two hydrogen atoms in the water (H_2O) molecule. The fact that water has a permanent dipole moment (1.84 D), as well as other evidence, leads us to believe that the water molecule is bent and that the O—H bonds form a bond angle of approximately 105° with respect to one another. The **bond angles** in molecules determine how the atoms are arranged in space—that is, the

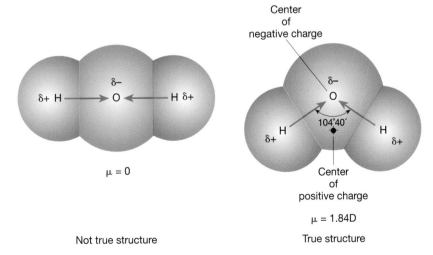

Figure 14–19 The water molecule. If the structure were linear, the dipole moment μ would be 0. Since $\mu = 1.84$ D, the molecule must be bent; the centers of positive and negative charge do not coincide.

geometry of the compound. Since molecular structure has a decided influence on the properties of the substance, it is important to know that structure. Only a few geometrical structures are necessary to account for most of molecular chemistry.

The **valence shell electron pair repulsion (VSEPR) theory**, proposed by R. J. Gillespie at McMaster University, allows one to predict the molecular geometry of a wide variety of compounds. The VSEPR (pronounced "vesper") theory says that the number of valence shell electron pairs surrounding an atom determines the arrangement of the bonds around the atom. The electron pairs may be bound or unbound. The lowest energy, and therefore preferred, arrangement of the electron pairs is assumed to be the molecular geometry in which there is a minimum of electron-electron repulsion. This condition is satisfied when the electron pairs are kept as far apart as possible.

Consider the structure of beryllium hydride, BeH_2. The two valence electrons of beryllium form electron pairs with the electrons from two hydrogen atoms, in what may be considered two positions on a sphere representing the beryllium atom, the central atom. The bond angle is at a maximum 180° when the distance between the two electron pairs is greatest, since this is as far apart as two locations on a sphere can be—at opposite ends of an axis. The bonds, therefore, are in the same straight line, and the molecule is linear: $H—Be—H$ (Fig. 14–20).

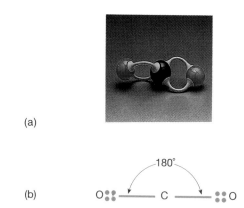

(a)

(b)

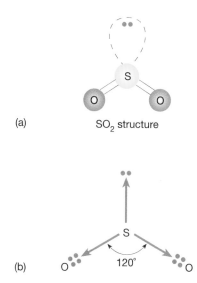

Figure 14–21 Structure of carbon dioxide, CO_2. The two electron locations are as far apart as possible: 180°. (Charles D. Winters)

The carbon atom, the central atom of carbon dioxide (CO_2), is linked to each oxygen atom by a double bond. All four valence electrons of carbon are used—two for each bond. For the two electron positions on the carbon atom in CO_2 to be as far apart as possible, the oxygen atoms must be at opposite sides of the carbon. The result, again, is a linear molecule, $O=C=O$, with a bond angle of 180° (Fig. 14–21).

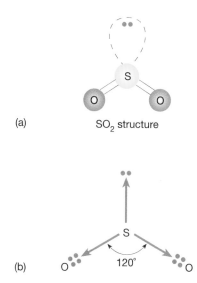

(a) SO_2 structure

(b)

(a)

(b)

Figure 14–20 Structure of beryllium hydride, BeH_2. All three atoms lie in the same straight line. The bond angle is 180°. (Charles Steele)

Figure 14–22 Structure of sulfur dioxide, SO_2. The three electron locations define the corners of an equilateral triangle. The atoms lie in the same plane, at a bond angle of 120°.

Although two double bonds are also present in a molecule of sulfur dioxide (SO_2), these bonds are not linear, but bent. In this case, a lone pair of electrons is also present, since sulfur has six valence electrons, but the lone pair is uninvolved in the bonding. There are, then, three electron locations around the central atom, sulfur. The repulsion between the bonding electrons and the **lone-pair electrons** forces the three locations as far apart as possible on the surface of a sphere, namely, to the corners of an equilateral triangle, with the center of the sphere in the plane of the triangle. The angles are each 120° (Fig. 14–22).

Boron forms three covalent bonds with chlorine in boron trichloride (BCl_3). Since all three valence shell electrons of boron are involved in the bonds, the electron pairs are located at only three positions on the atomic surface of the central atom, boron. The farthest apart these three positions can be is a trigonal planar arrangement with bond angles of 120°, as for SO_2 (Fig. 14–23).

In ammonia (NH_3), nitrogen forms three covalent bonds with hydrogen. The nitrogen atom also has a lone pair of electrons, however, that is not involved in the three covalent bonds. The mutual repulsion among these four electron locations directs the electron pairs to the four corners of a tetrahedron. The observed bond angle in ammonia is 107°, close to that for a tetrahedral arrangement (109°28′). The hydrogen atoms are located at three corners and a lone pair of electrons at the fourth; in effect, this causes the ammonia structure to be a low pyramid, with the nitrogen atom over a triangle of hydrogen atoms (Fig. 14–24). Table 14–14 and Figure 14–25 summarize the geometries of some simple molecules.

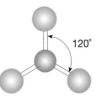

Figure 14–23 Boron and chlorine form a trigonal planar BCl_3 molecule.

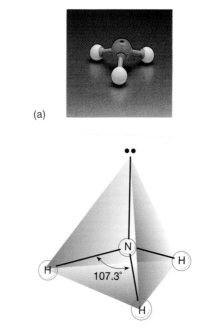

(a)

(b)

Figure 14–24 Structure of the ammonia molecule, NH_3. Four electron locations are involved. The observed bond angle is 107.3°, close to that for a tetrahedron (109°28′). (Charles D. Winters)

Table 14–14 MOLECULAR STRUCTURES

Number of Bonds	Geometry of Molecule	Bond Angle	Examples
2	Linear	180°	BeF_2, $MgCl_2$
2	Angular	90°–120°	H_2O, SO_2
3	Trigonal planar (equilateral triangle)	120°	BF_3, SO_3
3	Pyramidal	90°–114°	NH_3, PBr_3
4	Tetrahedral	109°28′	CH_4, SiF_4

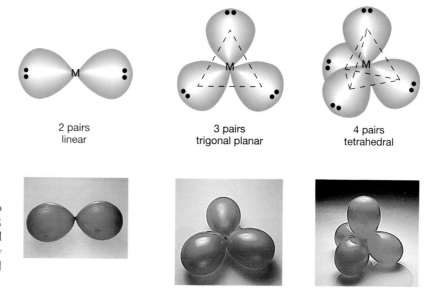

Figure 14–25 Geometries to be expected from the VSEPR model for molecules with 2, 3, and 4 electron pairs, represented by balloons, surrounding a central atom. (Charles D. Winters)

2 pairs
linear

3 pairs
trigonal planar

4 pairs
tetrahedral

EXAMPLE 14–3

Predict the molecular structure of PCl_3.

SOLUTION

The electronic structure of phosphorus can be determined from its atomic number, 15: $1s^22s^22p^63s^23p^3$. The five valence electrons are represented as follows in the Lewis structure:

$$\cdot \overset{\displaystyle \cdot\cdot}{\underset{\displaystyle \cdot}{P}} \cdot$$

There are thus three electrons and a lone pair. Chlorine, a halogen, has the Lewis structure

$$: \overset{\displaystyle \cdot\cdot}{\underset{\displaystyle \cdot\cdot}{Cl}} \cdot$$

Phosphorus forms three covalent bonds with chlorine. We have, therefore, three bonds and a lone pair, for a total of four electron-pair locations. The best separation in this case gives a pyramidal structure, with the phosphorus atom over a triangle of chlorine atoms.

EXTENSION

Predict the molecular structure of CF_4. (Answer: Tetrahedral)

14–8 COORDINATE COVALENT BOND

Soon after the nature of the covalent bond was explained. Langmuir and others, including Nevil V. Sidgwick (1873–1952) at Oxford University, realized that there was nothing to prevent both electrons of an electron-pair bond to come from the same atom. For example, the nitrogen atom in an ammonia molecule (NH_3) has a lone pair of electrons that it can share with another atom that is short of electrons, such as the boron atom of boron trifluoride (BF_3):

$$
\begin{array}{ccc}
\text{H} & :\overset{\cdot\cdot}{\text{F}}: & \\
\text{H}:\overset{\cdot\cdot}{\text{N}}: + & \overset{\cdot}{\underset{\cdot\cdot}{\text{B}}}:\overset{\cdot\cdot}{\text{F}}: & \longrightarrow \quad \text{H}:\overset{\cdot\cdot}{\text{N}}:\overset{\cdot\cdot}{\text{B}}:\overset{\cdot\cdot}{\text{F}}: \\
\text{H} & :\overset{\cdot\cdot}{\underset{\cdot\cdot}{\text{F}}}: &
\end{array}
$$

A bond of this type is called a **coordinate covalent bond,** or **dative bond.** In some molecules, a polarity is created, the donor atom becoming slightly positive and the acceptor atom slightly negative. This fact is represented by an arrow pointing away from the atom that has furnished the bonding pair of electrons, as in $H_3N \rightarrow BF_3$. The nitrogen atom is the donor, the boron atom the acceptor; the arrow indicates the relationship of donor to acceptor.

Once formed, however, a coordinate covalent bond is similar to any other covalent bond. The "donation" is a special case of electron sharing and does not involve a complete transfer of electrons. Nevertheless, the donor atom, nitrogen, no longer has full possession of the pair of electrons and acquires a **formal positive charge**. The acceptor atom, boron, now has a share in more electrons than it did before the bond was formed and acquires a **formal negative charge**.

14–9 THE METALLIC BOND

The structure and properties of a great majority of the elements in the periodic table (about 75%) are due to the **metallic bond**. These elements exist as metallic solids. A typical metal, such as copper, is opaque, except in very thin layers, and **lustrous**; that is, it reflects light of any wavelength. It is relatively dense and strong. It can be rolled or hammered into shape —it is **malleable**—and it can be drawn into a wire —it is **ductile** (Fig. 14–26). It is a good conductor of heat and electricity and has a high melting point. Metals are used primarily for these mechanical, electrical, and thermal properties.

Because of **close packing**, the mass per unit volume of metals—the **density**—is high (Fig. 14–27). Metals typically crystallize in either a face-centered cubic or hexagonal close-packed lattice, in which each metal atom is surrounded by 12 nearest neighbors, or in a slightly more open structure, the body-centered cubic lattice, in which each atom is surrounded by 14 others: 8 nearest neighbors and 6 others slightly far-

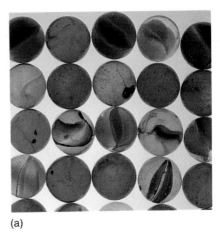

(a)

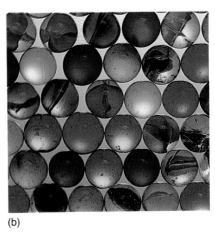

(b)

Figure 14–27 Packing of spheres in one layer. (a) Each marble contacts four others. (b) Each marble contacts six others at the corners of a hexagon—a more efficient packing arrangement. (Charles D. Winters)

Figure 14–26 Samples of copper in different forms. Because of their malleability and ductility, metals can be formed into many shapes. (Charles D. Winters)

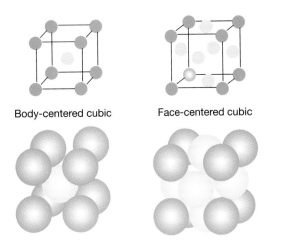

Body-centered cubic   Face-centered cubic

Figure 14–28 Body-centered and face-centered cubic lattices are efficient for filling space.

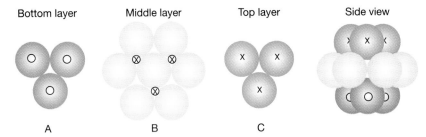

(a) Hexagonal close-packing

Bottom layer   Middle layer   Top layer   Side view

A   B   C

(b) Cubic close-packing = face centered cubic

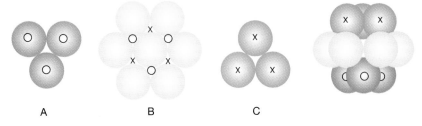

A   B   C

Figure 14–29 Metals generally crystallize in either a (a) hexagonal close-packed lattice or (b) a face-centered cubic lattice.

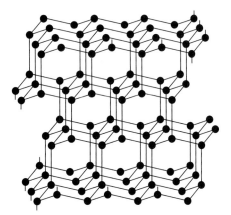

Figure 14–30 The structure of diamond. Each carbon atom is surrounded by four other carbon atoms at the corners of a tetrahedron.

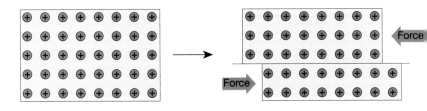

Figure 14–31 Simple diagram of a metallic crystal: An array of cations in a "sea of electrons."

Table 14–15 RELATION OF BONDING FORCE AND PHYSICAL PROPERTIES

	Ionic	Covalent	Metallic
Structural Unit	Ions	Atoms	Positive ions
Attractive Force	Strong; ionic or electrostatic	Strong; sharing of electron pairs	Moderate to strong; electrostatic attraction between positive ions and "sea of electrons"
Melting Point	High	Low to very high	Medium to high
Boiling Point	High	Low to high	Medium to high
Hardness	Hard and brittle	Soft to very hard	Soft to medium
Electrical Conductivity	Very low	Very low	Good
Examples	Calcium fluoride (CaF_2)	Carborundum (SiC)	Copper (Cu)
	Silver bromide (AgBr)	Diamond (C)	Iron (Fe)
	Sodium chloride (NaCl)	Quartz (SiO_2)	Mercury (Hg)

ther away (Fig. 14–28). Such arrangements are efficient for filling space, far more so than in nonmetals (Fig. 14–29). For example, in a nonmetallic covalent crystal like the diamond, each carbon atom is bonded to only four other carbon atoms (Fig. 14–30).

The atoms of metals must be bonded quite differently from those in covalent substances, since the one, two, or three valence electrons in metals are not nearly enough to form electron-pair bonds with 12 or 14 adjacent atoms. With only one valence electron per atom, the metal lithium crystallizes in a body-centered cubic lattice in which each atom is bound to 14 other atoms tightly enough to give it a melting point of 186°C.

A simple picture of a metal crystal visualizes it as an array of positive metal ions—**cations**—suspended in a "**sea of electrons**" (Fig. 14–31). The valence electrons are considered to be **delocalized**; that is, they no longer belong to individual atoms but to the crystal as a whole. The sea of negatively charged, mobile electrons overcomes the repulsive forces between the cations and acts as a kind of "glue" that holds the crystal together by electrostatic attraction.

The easy mobility of the electrons in metals accounts for their high thermal and electrical conductivities. Metals conduct electricity because the electrons are free to move in an electric field. Metals are opaque and lustrous because free electrons absorb and radiate back most of the light energy that falls on them. Metals conduct heat effectively because electrons can transfer thermal energy.

The **electron-sea model** also explains the mechanical properties of malleability and ductility. The nondirectional bonding in metals enables the metallic ions to give up old loyalties and form new bonds easily. A group of ions in a metal may move past another group without changing the environment of the positive ions. The shape of the crystal can, therefore, be changed without destroying its integrity. Metals can be rolled or hammered into shape and welded and alloyed because of the unselectivity of the metallic bond (Fig. 14–32). A comparison of physical properties and the type of bonding force is given in Table 14–15.

Figure 14–32 Metals can be hammered or rolled into shape.

14–10 BAND THEORY OF SOLIDS

The **band theory** provides a good working model for solids in general, including **conductors (metals)**, **semiconductors**, and **insulators (nonmetals)**. It assumes that isolated electron energy levels of atoms are broadened into energy bands that belong to the crystal as a whole and not to individual atoms and that the Pauli exclusion principle applies to each energy level. The lower bands are completely filled by electrons, whereas the bands of higher energy—the valence band and the conduction band—are separated by an energy gap (Fig. 14–33).

The electron bands in metals are incompletely filled. In $_3$Li, for example, (electronic configuration $1s^2 2s^1$), the band originating from the broadening of the 1s atomic orbital is fully occupied (Fig. 14–34). The 2s band, however, is only half filled, since each lithium atom has one electron in 2s level, but this level has a capacity for two electrons; this is the **conduction band** of lithium. When an electric field is applied to the metal, electrons at the top of the partially filled 2s band gain small amounts of energy and migrate into the available states lying immediately above. Electrons in the unfilled valence band of a metal can also absorb wavelengths of light over a broad range and be promoted to the unfilled energy states immediately above the filled states, accounting for their opaqueness and luster. That is why most metals have no color of their own; they absorb and re-emit all colors of light.

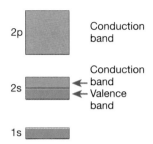

Figure 14–34 Band model of lithium, showing partly filled conduction band.

In metals that have a filled s band, the close packing causes the p band to combine with it or overlap so that there is no "forbidden" energy zone between them. Electrons are then easily promoted from the s band to the empty p band. The large energy barrier between the valence and the conduction bands of an insulator such as diamond (about 6 eV) effectively prevents the promotion of electrons into the conduction band. The forbidden zone separating the bands in silicon, a semiconductor, is small enough (about 1.1 eV) to permit bridging by electrons under the influence of an electric field, at high temperatures, or even by light. The electrical conductivity of semiconductors thus increases as the temperature increases. Germanium and silicon are two of the important semiconductors used in transistors and other "solid-state" devices (Fig. 14–35).

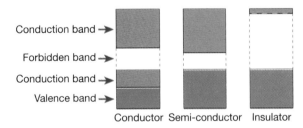

Figure 14–33 Electron energy bands in metals, semiconductors, and insulators.

Figure 14–35 Silicon is used in solar cells to collect energy from the sun. (Standard Oil Company, Ohio)

KEY TERMS

Anion   A negatively charged ion.

Band Theory   The band theory of solids assumes that isolated energy levels of atoms are broadened into energy bands that belong to the crystal as a whole.

Bond Angle   The angle formed between three atoms in a molecule.

Cation   A positively charged ion.

Chemical Bond   The force that holds two atoms together in a compound.

Close Packing   A space-filling arrangement in metals in which as many atoms as possible are packed into a given volume.

Conduction Band   The higher energy levels in a solid where electrical conduction can occur.

Conductor   A substance that can transmit electricity or heat.

Coordinate Covalent Bond   A covalent bond in which both electrons of the bond come from the same atom.

Covalent Bond   A chemical bond formed by the sharing of one or more electron pairs between two atoms.

Dative Bond   A coordinate covalent bond.

Delocalized Electrons   Bonding electrons distributed among more than two atoms that are bonded together, as in a crystal.

Density   A property of matter representing the mass per unit volume, such as grams per milliliter.

Dipole   A molecule in which the centers of positive and negative charge do not coincide.

Dipole Moment   A measure of the electrical character of a molecule; it is defined as the product of the effective charge and the distance between the centers of opposite charge.

Double Bond   Occurs when two atoms are joined by sharing two pairs of electrons.

Ductile   The ability to be drawn into a wire; a characteristic of metals.

Electron-dot Symbol   A Lewis structure.

Electronegativity   A measure of the ability of a bonded atom in a molecule to attract electrons.

Electron-Sea Model   The picture of a metal crystal as an array of positive metal ions suspended in a "sea of electrons," delocalized valence electrons.

Electrovalence   The combining capacity of an atom involved in ionic bonding.

Electrovalent Bond   A chemical bond formed by the transfer of electrons between atoms.

Formal Negative Charge   The negative charge on an atom in a coordinate covalent compound, calculated on the assumption that the electrons in each bond are shared equally between the two atoms involved.

Formal Positive Charge   The positive charge on an atom in a coordinate covalent bond.

Insulator   A substance incapable of conducting electricity.

Ionic Bond   See Electrovalent Bond.

Ionic Character   A percentage that describes the ionic nature of a chemical bond.

Ionization Energy   The energy necessary to remove an electron from an atom in the gaseous state.

Isoelectronic   Atoms or ions that possess the same number of electrons and hence the same ground-state electron configuration.

Kernel   The nucleus of an atom and all of the electrons in the inner filled shells of the atom.

Lewis Structure   A representation of atoms, ions, and molecules in which the valence electrons are shown as dots surrounding the symbol of an element.

Lone-Pair Electrons   A pair of electrons not involved in a chemical bond.

Lustrous   The ability of a metal to reflect light of any wavelength.

Malleable   The ability to be pounded into a thin sheet or rolled; a characteristic property of metals.

Metallic Bond   The chemical bond that joins the atoms of metals.

Molecular Geometry   The arrangement of the atoms of a molecule in space.

Molecule   The smallest particle of a substance that can have a stable independent existence.

Multiple Bond   A double or triple covalent chemical bond.

Noble Gas   One of the family of elements helium, neon, argon, krypton, xenon, and radon. All except helium have eight electrons in their outermost energy level.

Nonpolar Bond   A covalent bond in which both atoms share equally in the electrons of the bond.

Octet Rule   Atoms enter into combination to obtain a structure that has eight electrons in the outermost energy level.

Oxidation Number   For single-atom ions, the charge on the ion.

Polar Covalent   A covalent bond in which the electron pair is not shared equally by the joined atoms.

Polarity   The distribution of electric charge between bonded atoms in a molecule.

Saturated Molecule   A molecule that has only single covalent bonds.

Semiconductor   Materials such as silicon and germanium, which ordinarily are not good conductors or good insulators, can be made to conduct well when made with certain impurities or when the temperature is raised.

Sharing of Electrons   Characteristic of a covalent or electron-pair bond.

Triple Bond   A covalent bond resulting from the sharing of three pairs of electrons between two atoms.

Unsaturated Bond   A double or triple covalent bond.

Valence Electron   An electron located in the outermost electron shell of an atom.

Valence Shell   Electrons in the valence shell of an atom occupy the highest-energy orbitals of the atom.

VSEPR Theory   The valence shell electron pair repulsion theory, proposed by R. J. Gillespie, says that the number of valence shell electron pairs surrounding an atom determines the arrangement of the bonds around the atom.

THINGS TO DO

1. Study a model of the crystal structure of sodium chloride (NaCl), a representative ionic compound.

2. Construct a model of ethyl alcohol (C_2H_5OH), a covalent substance, with a molecular model kit.

3. With styrofoam balls held together by sticks, show close-packing that is characteristic of metals.

4. Test dilute solutions of (a) table salt, (b) table sugar, and (c) vinegar for electrical conductivity with a single bulb conductivity apparatus.

5. Examine samples of (a) zinc, (b) sulfur, (c) zinc sulfide.

6. Compare crystals of (a) sodium fluoride, (b) sodium chloride, (c) sodium bromide, (d) sodium iodide.

7. Examine a model of a tetrahedron.

8. Compare the properties of iodine crystals and moth flakes with those of table salt.

9. Observe various forms of copper-sheet, wire, turnings, shot, and cylinder.

10. Expose aluminum foil to various colored light sources.

EXERCISES

The Noble Gases

14–1. List the noble gas elements.

14–2. What is the relation between the octet rule and a noble gas structure?

14–3. Explain why the chemical activity of the noble gases went undiscovered for so long.

Lewis Structures

14–4. Draw the Lewis dot structures for the following atoms
(a) N
(b) Na
(c) Cl
(d) Ne
(e) Br

14–5. The following substances are isoelectronic with one another. Write a Lewis structure for each.
(a) CH_4
(b) NH_3

(c) H_2O
(d) HF
(e) Ne

14–6. Write Lewis structures for the following compounds.
(a) sodium fluoride: NaF
(b) carbon tetrachloride: CCl_4
(c) boron trifluoride: BF_3
(d) molecular nitrogen: N_2
(e) carbon monoxide: CO

14–7. Differentiate between the molecular formula and Lewis formula of water.

14–8. Give the total number of outer-shell electrons in each of the following.
(a) NH_3
(b) $CaCl_2$
(c) Al_2S_3
(d) NO_3^-
(e) PO_4^{3-}

14–9. Write the Lewis structure for O_2. What kind of bond has to be formed between the two oxygens to satisfy the octet rule?

14–10. Give the number of the periodic group and the number of valence electrons for each of the following atoms.
(a) S
(b) C
(c) Mg
(d) Ne
(e) B

Ionic Bonding

14–11. Does the Ca^{2+} ion possess a noble gas configuration? Explain.

14–12. Is the formula MgF_3 likely to represent a stable compound? Explain.

14–13. Using Lewis structures, diagram the reaction that occurs between Ca atoms and F atoms.

14–14. Account for the fact that the second ionization energy of lithium (75.6 eV) is greater than the first (5.4 eV).

14–15. (a) List four ions that are isoelectronic with the noble gas xenon, and (b) give the charge of each.

14–16. Write Lewis symbols to represent the electron transfer between K and S to form ions with noble gas configurations.

14–17. Explain why argon does not form ionic bonds.

14–18. What are the charges on the ions in Al_2S_3?

14–19. Why would it be misleading to refer to "molecules" of sodium chloride (NaCl) in the crystalline state?

Covalent Bonding

14–20. Explain why the halogens (fluorine, chlorine, bromine, and iodine) occur as diatomic molecules.

14–21. How do a bromine atom, a bromine molecule, and a bromide ion differ? Write the formula for each.

14–22. Why does neon not form covalent bonds?

14–23. What is (a) a double bond; (b) a triple bond?

14–24. How many covalent bonds are normally formed by
(a) F?
(b) O?
(c) N?
(d) C?

14–25. Which electron pairs in the Lewis formula of H_2S are bonding and which are lone pairs?

14–26. How does a coordinate covalent bond differ from an ordinary covalent bond?

Electronegativity and Bond Type

14–27. Predict (a) whether the bonds between the following pairs of elements would be ionic or covalent, and (b) if covalent, the percent ionic character of the bond.
(a) Ba, O
(b) Al, S
(c) N, Cl
(d) C, S
(e) Si, C

14–28. Although fluorine is the most electronegative element, silicon tetrafluoride (SiF_4) does not have a dipole moment. Explain.

14–29. Based on the electronegativities of the atoms, arrange the following bonds in order of increasing ionic character.

$$B—Cl \qquad P—F \qquad H—S$$
$$Mg—O \qquad C—N$$

14–30. Why is an HCl molecule polar whereas a Cl_2 molecule is nonpolar?

14–31. Predict whether a bond will form and, if so, whether it will be ionic or covalent, between
(a) Br and F.
(b) Li and Na.
(c) K and S.
(d) N and Ne.

14–32. Which atom in ClF is more negatively charged?

14–33. How do the electronegativities of the elements vary from (a) left to right across a period and (b) from the top to the bottom of a group?

Molecular Shape

14–34. What geometric structure would you predict for each of the following substances?
(a) CF_4
(b) $BeCl_2$
(c) BBr_3
(d) AsH_3
(e) SF_6

14–35. (a) What bond angle would you predict when all of the valence electrons of an atom are involved in three single bonds? (b) Would the presence of a lone pair of electrons in addition to three bonds change your prediction? If so, how?

14–36. On the basis of molecular structure, explain why NH_3 has a dipole moment but BF_3 does not.

14–37. The H_2S molecule has a geometry similar to that of water. Is the H_2S molecule a dipole? If so, where are the partial charges located?

14–38. How many atoms are directly bonded to the central atom in a tetrahedral molecule?

14–39. What are some properties of water that are determined by the presence of a dipole moment?

Metallic Bonding

14–40. Explain why copper wire is a good conductor of electricity.

14–41. (a) List three types of solids based on the nature of the bonding forces, and (b) give an example of each.

14–42. Account for the characteristic luster of silver metal.

14–43. Why are metals good conductors of heat and electricity?

14–44. Account for the relatively high density of metals.

14–45. How are the atoms in a metal held together?

14–46. *Multiple Choice*

    A. In the Lewis structure for fluorine, the number of dots surrounding the symbol for fluorine is
      (a) one.
      (b) four.
      (c) five.
      (d) seven.

    B. The most electronegative element on the Pauling electronegativity scale is
      (a) oxygen.
      (b) neon.
      (c) fluorine.
      (d) cesium.

    C. The kind of bonding in sodium chloride is
      (a) ionic.
      (b) covalent.
      (c) coordinate covalent.
      (d) metallic.

    D. The compound that has the greatest covalent character is

      (a) H_2O.
      (b) NaCl.
      (c) Br_2.
      (d) Fe_2O_3.

    E. The bond between two nitrogen atoms in nitrogen gas is
      (a) single.
      (b) double.
      (c) triple.
      (d) electrovalent.

    F. The chemical formula for the compound formed when $_{14}Si$ reacts with $_{17}Cl$ is
      (a) SiCl.
      (b) $SiCl_2$.
      (c) $SiCl_3$.
      (d) $SiCl_4$.

    G. Which of the following molecules could not possibly be polar?
      (a) H_2O
      (b) HCl
      (c) I_2
      (d) NH_3

    H. Sodium chloride has
      (a) ionic bonds.
      (b) covalent bonds.
      (c) nonpolar bonds.
      (d) coordinate covalent bonds.

    I. When K and O react to form the ionic compound K_2O,
      (a) each potassium atom loses one electron.
      (b) each potassium atom loses two electrons.
      (c) each oxygen atom loses one electron.
      (d) each oxygen atom loses two electrons.

    J. A covalent bond is formed
      (a) when electrons are transferred.
      (b) when electrons are shared.
      (c) when a cation and anion come together.
      (d) only when shared electrons come from the same atom.

SUGGESTIONS FOR FURTHER READING

Benfey, Theodor, "Geometry and Chemical Bonding." *Chemistry,* May, 1967.
    Emphasizes the continuum of bond types from nonpolar covalent to polar covalent to ionic and from metallic to both ionic and covalent.

Cohen, Marvin, Volker Heine, and James C. Phillips, "The Quantum Mechanics of Materials." *Scientific American,* June, 1982.
    Recent applications of quantum mechanics to crystalline solids—covalent, ionic, or metallic—are making materials science one of the better developed parts of human knowledge.

Cotton, F. Albert, and Malcolm H. Chisholm, "Bonds Between Metal Atoms." *Chemical and Engineering News,* 28 June, 1982.
    Coordination chemistry is presently one of the great growth areas of chemical science. Yet only the tip of the iceberg has been seen so far in the chemistry of the solid state.

Cottrell, A. H., "The Nature of Metals." *Scientific American,* September, 1967.

Correlates the close-packed crystal structure of metals with their mechanical properties.

DeKock, Roger L., "The Chemical Bond." *Journal of Chemical Education*, November, 1987.
Discusses ionic bonds, covalent bonds, hydrogen bonds, bonds in the solid state, and variation in bond strengths.

Gillespie, R. J., "The Electron-Pair Repulsion Model for Molecular Geometry." *Journal of Chemical Education*, January, 1970.
Applies the valence shell electron pair repulsion (VSEPR) theory to the problem of molecular shape.

Gillespie, R. J., "A Defense of the Valence Shell Electron Pair Repulsion (VSEPR) Model." *Journal of Chemical Education*, June, 1974.
Asserts that the VSEPR theory is more reliable for predicting molecular shape than any other theory we have at present.

Hyman, Herbert H., "The Chemistry of the Noble Gases." *Journal of Chemical Education*, April, 1964.
Discusses the experimental background of noble gas chemistry and theoretical approaches to explain it.

Jensen, William B., "Abegg, Lewis, Langmuir, and the Octet Rule." *Journal of Chemical Education*, March, 1984.
Surveys the history of the significance of the number eight in valence relationships. That history extends back to the work of Mendeléev on the periodic table.

Klein, Douglas J., and Nenad Trinajstić, "Valence Bond Theory and Chemical Structure." *Journal of Chemical Education*, August, 1990.
Valence bond (VB) has emerged as one of the two most influential quantum-mechanical theories of chemical structure and the nature of the chemical bond. The other is molecular-orbital (MO) theory.

Mickey, Charles D., "Molecular Geometry." *Journal of Chemical Education*, March, 1980.
Discusses the main experimental methods for determining molecular geometry—diffraction, spectroscopy,

resonance—and the VSEPR theory to account for the molecular geometry of compounds.

Ogilvie, J. F., "The Nature of the Chemical Bond–1990." *Journal of Chemical Education*, April, 1990.
Asserts that quantum chemistry is unnecessary in the undergraduate curriculum in chemistry, at least in the compulsory component. The time saved could be spent on chemical reactions, chemical substances, and mixtures as materials.

Packer, John E., and Sheila D. Woodgate, "Lewis Structures, Formal Charge, and Oxidation Numbers." *Journal of Chemical Education*, June, 1991.
Outlines a "user-friendly" approach for writing Lewis structures, requiring only the ability to add and subtract, count, and know the number of valence electrons of neutral atoms.

Pauling, Linus, "G. N. Lewis and the Chemical Bond." Journal of Chemical Education, March, 1984.
Personal reflections on a great chemist by another great chemist.

Sanderson, R. T., "Principles of Electronegativity: Part I. General Nature; Part II. Applications." *Journal of Chemical Education*, February and March, 1988.
Electronegativity is a link to understanding the cause-and-effect relationship between atomic structure and the properties of compounds, the origin of bond energy, and the causes and direction of chemical reactions.

Stranges, Anthony N., "Reflection on the Electron Theory of the Chemical Bond: 1900–1925." *Journal of Chemical Education*, March, 1984.
Asserts that Lewis' theory of the shared electron pair remains to this day the foundation of modern chemical bond theory.

Tykodi, R. J., "Identifying Polar and Nonpolar Molecules." *Journal of Chemical Education*, December, 1989.
Develops a scheme, based on the most elementary ideas of molecular symmetry, for determining the polar/nonpolar character of simple molecules.

ANSWERS TO NUMERICAL EXERCISES

14–15. Te^{2-}, I^{1-}, Xe, Cs^{1+}, Ba^{2+}

14–27. (a) ionic
(b) polar covalent; 22% ionic character
(c) nonpolar covalent
(d) nonpolar covalent
(e) polar covalent; 12% ionic character

14–29. H—S
C—N
B—Cl
P—F
Mg—O
⎫ increasing ionic character ⎭

14–35. (a) trigonal planar, 120°
(b) yes, pyramidal (90°–114°)

The reaction of potassium permanganate (KMnO₄) with sodium chloride (NaCl) forms the light pink solution manganese(II) chloride. (Richard Megna/Fundamental Photographs)

15

CHEMICAL FORMULAS
AND EQUATIONS

If the elements are comparable to the letters of the alphabet as structural units of words—they are the structural units of bulk matter—then chemical formulas make up the vocabulary of chemistry, the words themselves. To understand the matter around us and within us, we should become familiar with that vocabulary. The common names for many materials are carryovers from an earlier time when knowledge of their chemical composition was limited. In this chapter we take up some principles of systematic chemical nomenclature. Then we look at the quantitative significance of formulas, and finally we apply this vocabulary to the writing of statements about chemical reactions called *chemical equations.*

15–1 CHEMICAL COMPOUNDS AND FORMULAS

We have seen (in Chapter 13) that every element is assigned a unique name that identifies it. In systematic chemical nomenclature the chemist aims to give each *compound* a unique name, too, one that indicates its chemical composition. The names run well into the hundreds of thousands.

A compound is a pure substance composed of two or more elements that are chemically combined in a definite proportion by mass. Table salt, for example, is composed of 39.3% by mass of sodium

and 60.7% by mass of chlorine in chemical combination (Fig. 15–1). A compound is represented by a **formula**—a combination of chemical symbols that tells what elements are present and how many atoms of each are involved in the compound. The formula NaCl for table salt shows that sodium and chlorine are present and that they are present in a 1 : 1 ratio. The **systematic name**, sodium chloride, indicates the

Figure 15–1 Table salt is sodium chloride, NaCl. This systematic name indicates its chemical composition. (Charles Steele)

Figure 15–2   The gemstone amethyst is quartz (silicon dioxide, SiO_2) that contains traces of Fe^{3+} ions. Its color can range from pale lilac to royal purple depending on the amount of Fe^{3+} present. (Charles D. Winters)

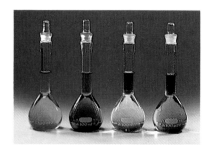

Figure 15–3   The aqueous solution of the binary compound chromium(II) chloride ($CrCl_2$) is blue; chromium(III) chloride ($CrCl_3$) is green. The same two elements can form different compounds. The yellow solution is potassium chromate ($K_2Cr_2O_4$); the orange solution, potassium dichromate ($K_2Cr_2O_7$). (Charles Steele)

chemical composition of the substance and identifies it uniquely.

Sodium chloride is an example of a **binary compound**, one that consists of two different elements. Many binary compounds contain a metal or hydrogen united with a nonmetal. Both in writing the formulas of binary compounds and in naming them, we place the more metallic element first (Fig. 15–2). The suffix -*ide* is added to the stem of the name of the more nonmetallic element, as in

KBr = potassium bromide

ZnO = zinc oxide

HF = hydrogen fluoride

$CaCl_2$ = calcium chloride

EXAMPLE 15–1

Name the compound BaS.

SOLUTION
We note that BaS is a binary compound. We write the name of the first element, barium, and modify the name of the second, sulfur, in order to add the suffix -*ide*. The name is barium sulfide.

EXTENSION
Name the compound Al_2O_3. (Answer: aluminum oxide)

15–2 OXIDATION NUMBERS

The same two elements can often form two or more binary compounds (Fig. 15–3). Iron and chlorine, for example, are combined in iron(II) chloride ($FeCl_2$) and iron(III) chloride ($FeCl_3$) (Fig. 15–4). A knowledge of oxidation numbers enables us to write and name such chemical formulas.

The **oxidation number of an element is the positive or negative charge of its atoms in a compound**. In ionic compounds, the charge comes from the transfer of electrons from one atom to another,

Figure 15–4   A reaction of iron with chlorine, forming iron(III) chloride ($FeCl_3$). (Charles Steele)

while in covalent compounds it comes from the unequal sharing of electrons between atoms. The oxidation number, therefore, is a measure of the combining capacity of an element and indicates the number of electrons that one of its atoms has gained, lost, or shared in forming chemical bonds with other atoms.

Since an iron atom transfers an electron to each of two chlorine atoms in $FeCl_2$, iron is assigned an oxidation number of $+2$ in this compound, and chlorine an oxidation number of -1. In $FeCl_3$, the oxidation number of chlorine is again -1, but that of iron is $+3$. The oxidation number of an atom in an ionic compound is thus equal to the charge of the ion. For covalent compounds, a negative oxidation number is assigned to the more electronegative atom of a bonded pair, and a positive charge to the other atom. Thus, in HF, H is assigned an oxidation number of $+1$ and F an oxidation number of -1. Since a compound is electrically neutral, the sum of the oxidation numbers of all the atoms must be zero. These rules apply to oxidation numbers:

1. The oxidation number of a free, uncombined element is zero.
2. For a neutral substance, the sum of the oxidation numbers of all the atoms in it is zero.
3. The oxidation number of a simple ion (a single atom that has lost or gained electrons) is equal to the charge of the ion.
4. The sum of the oxidation numbers of all the atoms in a polyatomic ion (one consisting of two or more atoms) is equal to the charge on the ion.
5. In most compounds, hydrogen is assigned an oxidation number of $+1$.
6. In most compounds, oxygen is assigned an oxidation number of -2.

A **polyatomic ion** is a group of atoms having either a positive or negative charge that occurs as a unit in compounds but is not stable by itself. Many polyatomic anions contain oxygen and are referred to as **oxyanions**. An example is the sulfate ion, SO_4^{2-}, that is found in compounds such as Na_2SO_4, sodium sulfate. The sulfur and oxygen atoms within the ion are bonded covalently and behave as a single unit in many chemical reactions:

$$Na^+ \begin{bmatrix} & \overset{..}{\underset{..}{O}} : & \\ : \overset{..}{\underset{..}{O}} : & S & : \overset{..}{\underset{..}{O}} : \\ & : \overset{..}{\underset{..}{O}} : & \end{bmatrix}^{2-} Na^+$$

Although one sulfur and four oxygen atoms have (5×6) or 30 outer-shell or valence electrons, the sulfate ion has 32 valence electrons and carries a charge of -2. The two excess electrons have come from the two sodium atoms that form ionic bonds with the SO_4^{2-} ion.

Certain metals have two different oxidation numbers. The classic method of distinguishing between the two is to add the suffix *-ous* to the base of the name of the metal for the lower oxidation number, and the suffix *-ic* for the higher, as in the following examples:

Fe^{2+}	Ferrous
Fe^{3+}	Ferric
Sn^{2+}	Stannous
Sn^{4+}	Stannic

For example, $FeCl_2$ is ferrous chloride, and $FeCl_3$ is ferric chloride. In the system proposed by the inorganic chemist Alfred Stock (1876–1946) and widely adopted, the oxidation number of the metal is indicated by Roman numerals enclosed in parentheses following the name of the metal, and suffixes for the metals are not used. The **Stock system** name for $FeCl_2$ is iron(II) chloride; $FeCl_3$, iron(III) chloride. Table 15–1 gives the names, formulas, and oxidation numbers of some common ions.

15–3 WRITING CHEMICAL FORMULAS FROM OXIDATION NUMBERS

To write a formula for a compound of aluminum and sulfur, refer to Table 15–1 for oxidation numbers. The oxidation number of Al is $+3$, and the oxidation number of S is -2.

$$Al^{3+}S^{2-}$$

Applying the principle of **electroneutrality**—that a chemical compound is electrically neutral—we see that two Al^{3+} ions are needed to balance three S^{2-} ions for the sum of the charges to equal zero. Write the number of each ion as a subscript to the right of the symbol for the simplest formula of aluminum sulfide:

$$Al_2S_3$$

We can get the correct formula by simply crossing over the numbers (but not the plus or minus signs) and writing them as subscripts.

Table 15–1   OXIDATION NUMBERS OF SOME ELEMENTS
AND POLYATOMIC IONS

	+ 1		**− 1**
H	Hydrogen	F	Fluoride
Au	Gold(I) or aurous	Cl	Chloride
Ag	Silver	Br	Bromide
Cu	Copper(I) or cuprous	I	Iodide
Hg	Mercury(I) or mercurous	ClO	Hypochlorite
K	Potassium	ClO_2	Chlorite
Li	Lithium	ClO_3	Chlorate
Na	Sodium	ClO_4	Perchlorate
NH_4	Ammonium	NO_2	Nitrite
		NO_3	Nitrate
		CN	Cyanide
		$C_2H_3O_2$	Acetate
		HCO_3	Bicarbonate
		MnO_4	Permanganate
		OH	Hydroxide

	+ 2		**− 2**
Ba	Barium	O	Oxide
Ca	Calcium	S	Sulfide
Cd	Cadmium	CO_3	Carbonate
Mg	Magnesium	C_2O_4	Oxalate
Sr	Strontium	CrO_4	Chromate
Zn	Zinc	Cr_2O_7	Dichromate
Co	Cobalt(II) or cobaltous	SO_3	Sulfite
Cu	Copper(II) or cupric	SO_4	Sulfate
Fe	Iron(II) or ferrous	SiO_3	Silicate
Pb	Lead(II) or plumbous		
Mn	Manganese(II) or manganous		
Hg	Mercury(II) or mercuric		
Ni	Nickel(II) or nickelous		
Sn	Tin(II) or stannous		

	+ 3		**− 3**
Al	Aluminum	N	Nitride
Bi	Bismuth	P	Phosphide
B	Boron	AsO_4	Arsenate
As	Arsenic(III) or arsenious	PO_3	Phosphite
Sb	Antimony(III) or antimonous	PO_4	Phosphate
Cr	Chromium(III) or chromic		
Co	Cobalt(III) or cobaltic		
Au	Gold(III) or auric		
Fe	Iron(III) or ferric		

	+ 4
Pb	Lead(IV) or plumbic
Mn	Manganese(IV) or manganic
Sn	Tin(IV) or stannic

Such formulas usually are reduced to the lowest ratio and represent the *ratio* of the ions.

EXAMPLE 15–2

Write the chemical formula for iron(III) nitrate.

SOLUTION

We know from the Roman numeral III in the Stock name that the oxidation number of iron in this compound is $+3$. In Table 15–1 the nitrate oxyanion has an oxidation number of -1 that applies to the entire ion and not to the oxygen or nitrogen alone: NO_3^{1-}. We enclose the nitrate ion within parentheses:

$$Fe^{3+}(NO_3)^{1-}$$

Next, we apply the electroneutrality principle: One Fe^{3+} ion will balance three NO_3^{1-} ions. If more than one ion is needed, we write a subscript to the right of the ion or we merely cross over the number. Either way, the formula is

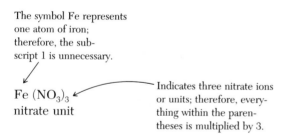

The symbol Fe represents one atom of iron; therefore, the subscript 1 is unnecessary.

$$Fe(NO_3)_3$$
nitrate unit

Indicates three nitrate ions or units; therefore, everything within the parentheses is multiplied by 3.

EXTENSION

Write the chemical formula for lead(II) phosphate. (Answer: $Pb_3(PO_4)_2$)

■

Since iron(III) nitrate contains three different elements—Fe, N, and O—it is called a **ternary compound**. Many ternary compounds are composed of a metal in combination with an oxyanion consisting of another element and oxygen (Fig. 15–5). If the ion occurs in the formula of a compound more than once, as does the NO_3^{1-} in $Fe(NO_3)_3$, it is enclosed in parentheses. A subscript following the parentheses indicates the number of times the ion occurs in the compound. Since the subscript outside the parenthe-

Figure 15–5 Two ternary compounds. On the left, solid potassium permanganate, $KMnO_4(s)$, and an aqueous solution of it, $KMnO_4(aq)$; on the right, solid potassium dichromate, $K_2Cr_2O_7(s)$, and an aqueous solution of it, $K_2Cr_2O_7(aq)$. (Charles D. Winters)

ses acts as a multiplier of every atom within the parentheses, we could have written the formula FeN_3O_9. Chemists prefer $Fe(NO_3)_3$, however, since it preserves the identity of the polyatomic ion, which behaves as a unit within the compound.

EXAMPLE 15–3

Give an appropriate name for $SnSO_3$.

SOLUTION

The formula signifies a ternary compound: tin combined with the SO_3^{2-} oxyanion. In Table 15–1 the oxidation number of SO_3 is -2. According to the principle of electroneutrality, the oxidation number of Sn in this compound must be $+2$:

$$Sn^{2+}SO_3^{2-}$$

We can call the Sn^{2+} ion either tin(II) or stannous. So two appropriate names for the compound are

Tin(II) sulfite
and
Stannous sulfite

EXTENSION

Name the compound $Cu_3(PO_3)_2$. (Answer: Copper(II) phosphite or cupric phosphite)

∎

15–4 MOLECULAR MASS AND FORMULA MASS

The **molecular mass** is the sum of the atomic masses of all the atoms in a molecule. For a diatomic molecule such as H_2, N_2, or O_2, the molecular mass is twice the atomic mass. For phosphorus, which forms P_4 molecules, the molecular mass is four times the atomic mass of phosphorus (Fig. 15–6).

The concept of molecular mass does not apply to ionic compounds, since they exist as ions rather than molecules. Instead, the formula mass is computed. The **formula mass** of a compound is the sum of the atomic masses of all the atoms in its formula. Since molecular and formula masses are computed in the same way, the more inclusive term, formula mass, applies to a substance whether it is molecular or ionic.

EXAMPLE 15–4

What is the formula mass of aspirin, $C_9H_8O_4$?

SOLUTION

The formula mass of aspirin represents the mass of an aspirin molecule relative to carbon-12. Simply add the atomic masses of each atom in the molecule.

$$\text{Atomic mass of C} = 12.0 \times 9$$
$$= 108.0 \text{ amu}$$
(contributed by C)

$$\text{Atomic mass of H} = 1.0 \times 8$$
$$= 8.0 \text{ amu}$$
(contributed by H)

$$\text{Atomic mass of O} = 16.0 \times 4$$
$$= 64.0 \text{ amu}$$
(contributed by O)

Formula mass $= \text{total} = 180.0$ amu

The formula mass may be expressed with or without atomic mass units (amu).

Figure 15–6 A model of the P_4 molecule, white phosphorus. (Charles Steele)

EXTENSION

Calculate the molecular mass of cholesterol, $C_{22}H_{45}OH$. (Answer: 326 amu)

∎

15–5 PERCENT COMPOSITION OF COMPOUNDS

The formula mass of a compound represents its total mass, or 100% of the compound. The mass of each element, therefore, represents a percentage of the total mass, just as a piece of pie represents a percentage of the whole pie. The **percent composition** of a compound is the mass-percent of each element present, that is, the percentage by mass contributed by each element in the compound. It is determined by dividing the total mass of each element by the formula mass of the compound and multiplying by 100%.

EXAMPLE 15–5

Calculate the percent composition of sugar (sucrose), $C_{12}H_{22}O_{11}$.

SOLUTION

First, we determine the formula mass of $C_{12}H_{22}O_{11}$. Then, we divide the total mass of each element by the formula mass.

12 C = 12 × 12.0	= 144.0 amu
22 H = 12 × 1.0	= 22.0 amu
11 O = 11 × 16.0	= <u>176.0</u> amu
Formula weight = total	= 342.0 amu

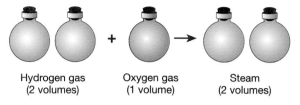

Hydrogen gas
(2 volumes)

Oxygen gas
(1 volume)

Steam
(2 volumes)

Figure 15–7 Gay-Lussac's law of combining volumes is illustrated by the reaction of two volumes of hydrogen gas with one volume of oxygen gas to form two volumes of steam; the volumes of the reactants and product are in the ratio of small whole numbers.

$$\%C = \frac{144 \text{ amu}}{342 \text{ amu}} \times 100\% = 42.1\% \text{ C}$$

$$\% \text{ H} = \frac{22 \text{ amu}}{342 \text{ amu}} \times 100\% = 6.4\% \text{ H}$$

$$\% \text{ O} = \frac{176 \text{ amu}}{342 \text{ amu}} \times 100\% = \underline{51.5 \% \text{ O}}$$

$$100.0\%$$

EXTENSION

Determine the percentage composition of acetylene, C_2H_2. (Answer: 92.3% C, 77.7% H)

■

15–6 GAY-LUSSAC'S LAW OF COMBINING VOLUMES

Joseph Louis Gay-Lussac (1778–1850), a chemist and physicist, found that two volumes of hydrogen gas react with one volume of oxygen gas to produce two volumes of steam when the gases are measured under the same conditions of temperature and pressure: 100°C and 1 atmosphere pressure (Fig. 15–7).

He also found that two volumes of carbon monoxide gas combine with one volume of oxygen gas to form two volumes of carbon dioxide gas. From these and other experiments, he proposed a generalization known as **Gay-Lussac's law of combining volumes**:

> **When gases combine to form new compounds, the volume of the reactants and products are in the ratio of small whole numbers.**

15–7 AVOGADRO'S HYPOTHESIS

Amadeo Avogadro (1776–1856), a chemist and physicist, saw in Gay-Lussac's law of combining volumes a clue to molecular formulas. He assumed that:

1. Equal volumes of all gases under the same conditions of temperature and pressure contain the same number of molecules.
2. Molecules of certain elements consist of two identical atoms that separate when the element reacts to form a compound.

If two hydrogen molecules each contain two identical hydrogen atoms and produce two water molecules, then each water molecule must contain two hydrogen atoms. If each oxygen molecule contains two oxygen atoms and produces two water molecules, then each water molecule must contain one oxygen atom (Fig. 15–8). The molecular formula of water must therefore be H_2O (not HO as John Dalton had proposed).

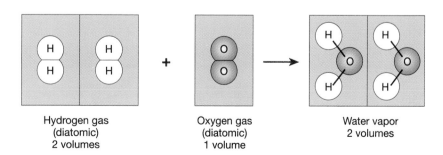

Hydrogen gas
(diatomic)
2 volumes

Oxygen gas
(diatomic)
1 volume

Water vapor
2 volumes

Figure 15–8 Avogadro's hypothesis explained Gay-Lussac's observations. The hydrogen molecule is diatomic, as is the oxygen molecule.

Figure 15–9   One gram-atom of any element contains one mole of atoms. The photograph shows one mole of atoms of some common elements. Back row (*left to right*): bromine, aluminum, mercury, copper. Front row (*left to right*): sulfur, zinc, iron. (Charles Steele)

15–8 THE MOLE: AVOGADRO'S NUMBER

Atomic, molecular, and formula masses can all be expressed in atomic mass units (amu). The conversion of these units to grams gives a useful unit of both mass and number.

One **gram-atomic mass** (gram-atom) is the mass of the element in grams that is numerically equal to its atomic mass (Fig. 15–9). One gram-atom of any element contains the same number of *atoms* as one gram-atom of any other element (Table 15–2).

One **gram-molecular mass** of a substance is the mass of the substance in grams that is numerically equal to the molecular mass of the substance. One gram-molecular mass of any substance contains the same number of *molecules* as one gram-molecular mass of any other substance (Table 15–3).

One **gram-formula mass** of a substance is the mass in grams numerically equal to the sum of the atomic masses appearing in the formula (Fig. 15–10). One gram-formula mass of any substance contains the same number of *formula units* as one gram-formula mass of any other substance (Table 15–4).

Table 15–2   GRAM-ATOMIC MASSES

Element	Atomic Mass	Value of One Gram-Atom
Carbon	12.01	12.01 g
Gold	196.97	196.97 g
Silver	107.87	107.87 g
Titanium	47.9	47.9 g

Table 15–3   GRAM-MOLECULAR MASSES

Substance	Molecular Mass	Value of One Gram-Molecular Mass
H_2O	18.0	18.0 g
O_2	32.0	32.0 g
CCl_4	154.0	154.0 g
S_8	256.5	256.5 g

Figure 15–10   One-mole quantities of compounds. The white compound is NaCl (58.44 g/mol); blue, $CuSO_4 \cdot 5H_2O$ (249.68 g/mol); deep red, $CoCl_2 \cdot 6H_2O$ (165.87 g/mol); green, $NiCl_2 \cdot 6H_2O$ (237.70 g/mol); and orange, $K_2Cr_2O_7$ (294.19 g/mol). (Richard Roese)

The unit, **mole**, abbreviated mol, is used interchangeably with gram-atomic mass, gram-molecular mass, and gram-formula mass. Thus, one mole, one gram-atom of any element, one gram-molecular mass of any substance, and one gram-formula mass of any compound, whether in the solid, liquid, or gaseous state, all contain the same number of atoms, molecules, or formula units. This number, called **Avogadro's number** (symbol N_A) has been evaluated experimentally in various ways and has the value 6.02×10^{23} (Fig. 15–11).

For example, one gram-atom of oxygen, or **molar mass** of oxygen, the mass of one mol or 16.0 g, is the mass of 6.02×10^{23} oxygen atoms. One mole of HCl,

molecular mass 36.5, has a mass of 36.5 g and contains 6.02×10^{23} molecules of HCl. One mole of KCl, an ionic compound, contains N_A formula units, or 6.02×10^{23} K^+ ions and 6.02×10^{23} Cl^- ions. A chemical formula, therefore, may refer to one atom, molecule, or formula unit or to a mole of atoms, molecules, or formula units, just as we may refer to a pair of gloves, a six-pack, or a dozen eggs. Avogadro's number is also the number of molecules of a gas confined in 22.4 liters at 0°C and 1 atmosphere pressure, a volume called the **gram-molecular volume** of a gas. Finally, since they represent relative numbers of atoms, molecules, or other units, quantities expressed in moles mean more than quantities expressed in grams. Thus, chemical symbols and formulas have both qualitative and quantitative significance, as Table 15–5 shows.

EXAMPLE 15–6

How many molecules are there in 40.0 g of I_2?

SOLUTION
Knowing that a mole of iodine (I_2) contains N_A iodine molecules, we can use dimensional analysis to deter-

Table 15–4   GRAM-FORMULA MASSES

Substance	Formula Mass	Value of One Gram-Formula Mass
$CaCl_2$	111.1	111.1 g
Na_2CO_3	106.0	106.0 g
KOH	56.1	56.1 g
$BaSO_4$	233.4	233.4 g

12 eggs
or
1 dozen eggs
or
24 ounces of eggs

6.022×10^{23} Fe atoms
or
1 mole of Fe atoms
or
55.847 grams of iron

Figure 15–11   Three different ways of representing amounts. (Charles Steele)

mine the number of molecules in 40.0 g. The atomic mass of iodine is 127, but remembering that iodine is diatomic, we use (127×2) for the molecular mass.

Table 15–5   INFORMATION CARRIED BY CHEMICAL SYMBOLS AND FORMULAS

The Symbol	Signifies
Ni	1. The element nickel
	2. One atom of nickel
	3. One atomic mass of nickel: 58.7 amu
	4. One gram-atom of nickel: 58.7 g
	5. One mole of nickel atoms: 6.02×10^{23} atoms (Avogadro's number)

The Formula	Signifies
CO_2	1. Carbon dioxide
	2. One molecule of carbon dioxide
	3. One molar mass of carbon dioxide: 44.0 amu
	4. One gram-molecular mass of carbon dioxide: 44.0 g
	5. One mole of carbon dioxide molecules: 6.02×10^{23} molecules (Avogadro's number)

$$40.0 \text{ g } I_2 \times \frac{6.02 \times 10^{23} \text{ molecules}}{(127 \times 2)\text{g } I_2}$$

$$= 40.0 \text{ g } I_2 \times \frac{6.02 \times 10^{23} \text{ molecules}}{254 \text{ g } I_2}$$

$$= 0.158 \, (6.02 \times 10^{23}) \text{ molecules}$$

$$= 0.950 \times 10^{23} \text{ molecules}$$

$$= 9.50 \times 10^{22} \text{ molecules}$$

EXTENSION
How many molecules are there in 5.00 g of HF? (Answer: 1.50×10^{23} molecules)

■

EXAMPLE 15–7

How many moles are there in 798 g of octane, C_8H_{18}?

SOLUTION
We evaluate the molecular mass of C_8H_{18} and apply dimensional analysis to get the number of moles of octane.

C_8H_{18}

$$8C = \quad 8 \times 12 = 96 \text{ amu}$$
$$18H = 18 \times \quad 1 = \underline{18} \text{ amu}$$
$$\text{Molar Mass} = 114 \text{ amu}$$

$$1 \text{ mol } C_8H_{18} = 114 \text{ g}$$

$$798 \text{ g } C_8H_{18} \times \frac{1 \text{ mole}}{114 \text{ g } C_8H_{18}} = \boxed{7.00 \text{ moles}}$$

EXTENSION

How many moles are there in 96.0 g of O_2 molecules?
(Answer: 3.00 moles)

■

15–9 EMPIRICAL AND MOLECULAR FORMULAS

The **empirical formula** is the simplest formula that we can write for a compound and gives the relative numbers of each kind of atom in the compound. The **molecular** or **true formula** represents the total number of atoms of each element present in one molecule or one formula unit of the compound, and it may or may not be the same as the empirical formula. The empirical and molecular formulas of H_2O are exactly the same: A molecule of water is composed of two atoms of hydrogen and one atom of oxygen. If the molecular formula is not the same, it will be an integral multiple of the empirical formula.

For example, CH (molar mass 13.0) is an empirical formula derived from a chemical analysis of a compound that showed 92.3% C and 7.7% H. The compounds acetylene and benzene both have this empirical formula. Other information will be necessary to decide which compound we have. Thus, the molar mass of acetylene is known to be 26.0 (Fig. 15–12). This is equal to 2×13.0; the molecular formula is therefore $2 \times$ CH, or C_2H_2. From the molar mass of benzene, 78.0 (6×13.0), we conclude that the molecular formula is $6 \times$ CH, or C_6H_6.

$$\text{CH} = \text{Empirical formula}$$
$$2 \times \text{CH} = C_2H_2 = \text{Acetylene (molecular formula)}$$
$$6 \times \text{CH} = C_6H_6 = \text{Benzene (molecular formula)}$$

The empirical formula of a compound is calculated from its percentage composition. By assuming that we have 100 g of material, we can express the percentage of each element in grams and convert to gram-atoms. We then divide the relative number of gram-atoms by the smallest relative number to give whole numbers. These small whole numbers become the subscripts in the formula.

Figure 15–12   Acetylene gas, C_2H_2, is used in the oxyacetylene torch for welding. (Randy Duchaine/The Stock Market)

EXAMPLE 15–8

Calculate the empirical (simplest) formula of the refrigerant Freon, which analysis shows contains 9.92% C, 58.68% Cl, and 31.40% F.

SOLUTION

Step 1: We assume that we have 100 g of Freon and express the percentage of each element in grams.

$$C = 9.92\% = 9.92 \text{ g}$$
$$Cl = 58.68\% = 58.68 \text{ g}$$
$$F = 31.40\% = 31.40 \text{ g}$$

Step 2: We then convert each mass to gram-atoms through dimensional analysis.

$$9.92 \text{ g C} \times \frac{1 \text{ gram-atom C}}{12.0 \text{ g C}} = 0.83 \text{ g-atom C}$$

$$58.68 \text{ g Cl} \times \frac{1 \text{ gram-atom Cl}}{35.5 \text{ g Cl}} = 1.65 \text{ g-atoms Cl}$$

$$31.40 \text{ g F} \times \frac{1 \text{ gram-atom F}}{19.0 \text{ g F}} = 1.65 \text{ g-atoms F}$$

Step 3: We express the relative numbers of gram-atoms as whole numbers by dividing each by the smallest, 0.83.

$$C = \frac{0.83 \text{ g-atom}}{0.83} = 1.00 \text{ g-atom C} = 1$$

$$Cl = \frac{1.65 \text{ g-atoms}}{0.83} = 1.99 \text{ g-atoms Cl} = 2$$

$$F = \frac{1.65 \text{ g-atoms}}{0.83} = 1.99 \text{ g-atoms F} = 2$$

The empirical formula is CCl_2Fe_2 .

EXTENSION

Analysis shows that naphthalene consists of 93.7% C and 6.3% H. What is its empirical formula? (Answer: C_5H_4)

■

EXAMPLE 15–9

The molar mass of Freon is 121.0. What is its molecular formula?

SOLUTION

In Example 15–8, the empirical formula of Freon was found to be CCl_2F_2. We calculate the molar mass corresponding to the empirical formula as follows:

$$1 \text{ C} = 1 \times 12.0 = 12.0 \text{ amu}$$
$$2 \text{ Cl} = 2 \times 35.5 = 71.0 \text{ amu}$$
$$2 \text{ F} = 2 \times 19.0 = \underline{38.0 \text{ amu}}$$
$$\text{Molar mass} = 121.0 \text{ amu}$$

Therefore, the empirical and molecular formulas of Freon are the same: CCl_2F_2 .

EXTENSION

What is the molecular formula of naphthalene (Ext. 15–8) if its molecular mass is 128? (Answer: $C_{10}H_8$)

■

15–10 CHEMICAL EQUATIONS

A **chemical equation** is a symbolic expression of a chemical reaction that shows what has taken place qualitatively and quantitatively. It tells what sub-

stances react, what substances are produced, and in what relative amounts. The reactants and products are represented by symbols and formulas. The **reactants** are on the left and the **products** on the right; the two sides of the equation are separated by a single arrow, →, a double arrow, ⇌, or an equal sign, =. The physical states of substances taking part in the reaction may be designated by subscripts following the formula or symbol: (*s*), (*l*), and (*g*), representing solid, liquid, and gas. The escape of a gaseous product may be shown by an upward arrow (↑), and a solid product precipitating out of a solution during the reaction by an arrow pointing downward (↓) or by underscoring. Symbols commonly used in equations are summarized in Table 15–6.

In chemical reactions, bonds are made and broken between atoms. Atoms are not created or destroyed by the reaction process; they are rearranged. This is implied by the **law of conservation of mass**, first stated by Antoine Lavoisier (1743–1794). Each atom that enters into a chemical reaction must, therefore, be accounted for in the end.

An equation is **balanced** by equalizing the number of atoms of each element participating in the reaction with those appearing in the products. The atom inventory of the reactants and products must be identical when the process is completed. The required number of each kind of atom is obtained by writing numbers called **coefficients** in front of formulas or symbols. The subscripts of a formula are never changed, since this would change the nature of the substance. The *coefficient* is a multiplier for the formula; for example, $3CaSO_4$ indicates 3 Ca atoms,

Table 15–6 SYMBOLS USED IN CHEMICAL EQUATIONS

Symbol	Interpretation
+	Plus
→	Yields; produces
⇌	The reaction is reversible.
=	Equilibrium between reactants and products
△	Heat
↑	Gaseous product
↓	Precipitate; solid product
(aq)	Aqueous solution
(*s*)	Solid
(*l*)	Liquid
(*g*)	Gas

3 S atoms, and 12 O atoms. The coefficients of an equation are reduced to their lowest possible integral values. Consider the following reaction:

Aluminum + Iron (III) oxide $\longrightarrow$
Aluminum oxide + Iron

$2Al + Fe_2O_3 \longrightarrow Al_2O_3 + Fe$ (not balanced)

Our atom inventory shows the following:

Atom Inventory:	*Reactants*	*Products*
	1 Al	2 Al
	2 Fe	1 Fe
	3 O	3 O

The number of oxygen atoms is three on the left and three on the right; therefore, oxygen is balanced. There is one Al atom on the left, but two on the right; by writing the coefficient 2 before the symbol Al on the left, we balance aluminum. Finally, there are two Fe atoms on the left but only one on the right; we write the coefficient 2 in front of Fe, and this balances iron. The balanced equation is

$2Al + Fe_2O_3 \longrightarrow Al_2O_3 + 2Fe$ (balanced)

Atom Inventory:	*Reactants*	*Products*
	2 Al	2 Al
	2 Fe	2 Fe
	3 O	3 O

Most chemical reactions can be classified as belonging to one of four types:

1. **Direct Combination** or **Synthesis Reactions** (Figs. 15–13 and 15–14). In a synthesis reaction, two elements, two compounds, or an element and a compound react to form a single compound. For example,

$2Al + N_2 \longrightarrow 2AlN$

$SO_3 + H_2O \longrightarrow H_2SO_4$

2. **Simple Decomposition** or **Analysis Reactions** (Fig. 15–15). A compound is broken down into simpler substances, usually when heated. The products may be the elements composing the compound, simpler compounds, or some of each. Examples are

$2HgO \longrightarrow 2Hg + O_2 \uparrow$

$2KNO_3 \longrightarrow 2KNO_2 + O_2 \uparrow$

3. **Simple Replacement** or **Substitution Reactions** (Figs. 15–16 and 15–17). An atom or

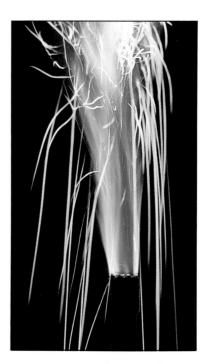

Figure 15–13   White phosphorus reacts with chlorine to form PCl_5. (Charles Steele)

Figure 15–14   Zinc powder sprayed into a Bunsen burner flame burns to give zinc oxide, ZnO. (Charles D. Winters)

Figure 15–15   Red mercury(II) oxide decomposes into mercury and oxygen when heated. When inserted into the test tube, a glowing splint will burst into flame as it reacts with the oxygen. (Charles Steele)

(a)

(b)

Figure 15–17   (a) Copper wire and a silver nitrate solution. (b) Finely divided silver deposited on the wire. The solution is blue because it contains copper(II) nitrate. (Charles Steele)

Figure 15–16   Potassium reacts vigorously with water, displacing hydrogen and generating considerable light. (Charles Steele)

polyatomic ion replaces another atom or ion in a compound in a simple replacement reaction. In the first example, aluminum replaces chlorine in HCl and releases hydrogen gas. In the second, zinc replaces silver in $AgNO_3$ and combines with the nitrate oxyanion.

$$2Al + 6HCl \longrightarrow 2AlCl_3 + 3H_2 \uparrow$$
$$Zn + 2AgNO_3 \longrightarrow Zn(NO_3)_2 + 2Ag$$

Figure 15–18 Ammonium sulfide solution is poured into a solution of cadmium nitrate, giving a precipitate of cadmium sulfide.

$$(NH_4)_2S + Cd(NO_3)_2 \longrightarrow CdS(s) + 2NH_4NO_3$$

In this metathesis reaction, one of the products is a solid.
(J. Morgenthaler)

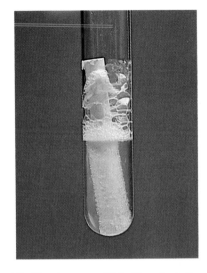

Figure 15–19 A piece of blackboard chalk, which is mostly $CaCO_3$, reacts with HCl to give $CaCl_2$, water, and CO_2 gas.

$$CaCO_3 + 2HCl \longrightarrow CaCl_2 + H_2CO_3$$
$$\hookrightarrow H_2O + CO_2(g)$$

(Charles D. Winters)

4. **Double Replacement** or **Metathesis Reactions** (Figs. 15–18 and 15–19). A double exchange is involved between two compounds, the two exchanging their positive and negative parts. Thus, the sodium from NaCl combines with the SO_4^{2-} oxyanion from H_2SO_4, and hy-drogen combines with chlorine. One of the products is usually a gas or a solid.

$$2NaCl + H_2SO_4 \longrightarrow Na_2SO_4 + 2HCl \uparrow$$
$$Fe_2(SO_4)_3 + 3Ca(OH)_2 \longrightarrow$$
$$2Fe(OH)_3 \downarrow + 3CaSO_4$$

KEY TERMS

Analysis Reaction A chemical reaction in which a compound is broken down into simpler substances.

Avogadro's Number The number of units in one gram-atom, one gram-molecular mass, and one gram-formula mass: 6.02×10^{23}.

Balanced Equation An equation containing the same number of atoms of each of the elements for both the reactants and the products.

Binary Compound A substance consisting of two different elements in chemical combination.

Chemical Formula A combination of chemical symbols and numbers that tells what elements and how many atoms of each are present in a compound.

Coefficient A number placed in front of a symbol or formula to equalize the number of atoms of each element in a balanced equation.

Compound A pure substance composed of two or more elements that are chemically combined in a definite proportion by mass.

Direct Combination Reaction A reaction in which two elements, two compounds, or an element and a compound react to form a single compound.

Double Replacement Reaction A reaction in which two compounds exchange their positive and negative parts.

Electroneutrality The principle that a chemical compound is electrically neutral.

Empirical Formula The simplest formula for a compound, giving the relative numbers of each kind of atom present in the compound.

Formula Mass The sum of the atomic masses of all the atoms in the chemical formula of a substance.

Gram-Atomic Mass The amount of an element in grams that is numerically equal to its atomic mass.

Gram-Formula Mass The amount of a substance in grams that is numerically equal to the sum of the atomic masses appearing in its formula.

Gram-Molecular Mass The amount of a substance in grams that is numerically equal to the molecular mass of the substance.

Gram-Molecular Volume The volume occupied by one mole of a gas at 0°C and 1 atmosphere pressure; 22.4 liters.

Law of Combining Volumes When gases combine to form new gaseous compounds, the volumes of the reactants and products are in the ratio of small whole numbers.

Law of Conservation of Mass There is no detectable gain or loss of mass during a chemical reaction. This indicates that atoms are neither created nor destroyed in chemical reactions; they are merely rearranged.

Metathesis Reaction A chemical reaction between two compounds in which there is a double replacement of atoms or polyatomic ions in the compounds.

Molar Mass The mass in grams of one mole of a substance; numerically equal to the formula mass of the substance.

Mole Abbreviated mol. A unit numerically equal to Avogadro's number: 6.02×10^{23}. This unit is used interchangeably with the terms "gram-atom," "gram-molecular mass," and "gram-formula mass," or "molar mass."

Molecular Formula The molecular or true formula represents the total number of atoms of each element present in one molecule of a compound.

Molecular Mass The sum of the atomic masses of all the atoms in a molecule.

Oxidation Number The positive or negative charge assigned to the atoms of an element in the formula of a compound. It is a measure of the combining capacity of an element.

Oxyanion A polyatomic ion that contains oxygen.

Percent Composition The mass-percent of each element present in a compound.

Polyatomic Ion A group of atoms having either a positive or negative charge that occurs as a unit in compounds but is not stable by itself.

Product A substance produced in a chemical reaction; written on the right side of a chemical equation.

Reactant A starting substance in a chemical reaction, it appears on the left side in a chemical equation.

Simple Decomposition Reaction A reaction in which a compound is broken down into simpler substances.

Simple Replacement Reaction A chemical reaction in which an atom or polyatomic ion replaces another atom or ion in a compound.

Stock System A system of chemical nomenclature in which the oxidation number of a metal is indicated by Roman numerals enclosed in parentheses following the English name of the metal; for example, Tin(II) chloride.

Substitution Reaction A simple replacement reaction.

Synthesis Reaction A chemical reaction in which atoms or molecules combine to form larger molecules.

Systematic Name A name that indicates the chemical composition of a substance and identifies it uniquely.

Ternary Compound A compound composed of a metal in combination with a polyatomic ion consisting of another element and oxygen; in all, three kinds of atoms are present.

True Formula A formula that represents the total number of atoms of each element present in one molecule or one formula unit of a compound.

THINGS TO DO

1. Study a direct combination reaction by igniting a 4-inch long strip of Mg ribbon held with forceps in a low flame from a burner. Do not look directly at the bright light produced. Collect the product on a watch glass and compare with the Mg ribbon. Write the chemical equation for the reaction.

2. Carry out a decomposition reaction. Clamp a large test tube on a ring stand at a 45° angle. Place a small amount of HgO in the test tube and apply intense heat. When it is very hot, use a glowing wood splint to determine whether O_2 is given off: The splint bursts into flame. Note also the inner sides of the test tube. Write the equation for the reaction.

3. Observe a simple replacement reaction by placing a zinc rod in a solution of $CuSO_4$. Write the equation for the reaction.

4. Watch a double replacement reaction by mixing a solution of Na_2CrO_4 with one of $PbCl_2$. Write the equation for the reaction.

5. Weigh a sample of steel wool in a balance. Then hold it with tongs in the flame of a burner. Weigh again when cool. Explain any weight change.

EXERCISES

Writing Chemical Formulas

15-1. What is the difference between an atom and an ion?

15-2. Give the symbol and oxidation number for each of the following ions.
(a) aluminum ion (b) silver ion (c) oxide ion (d) copper(II) ion (e) ferric ion

15-3. What is a polyatomic ion?

15-4. Give the formula and charge for each of the following polyatomic ions.
(a) sulfate ion (b) chlorite ion
(c) phosphate ion (d) nitrite ion
(e) hydroxide ion

15-5. What does a chemical formula represent?

15-6. How does the principle of electroneutrality apply to a chemical formula?

15-7. Write the formulas of the following binary ionic compounds.
(a) calcium oxide (b) potassium iodide
(c) zinc chloride (d) aluminum sulfide
(e) sodium phosphide

15-8. Write the formulas of the following binary ionic compounds.
(a) copper(I) oxide (b) lead(II) chloride
(c) tin(II) fluoride (d) iron(III) sulfide
(e) nickel(II) bromide

15-9. Write the formulas of the following compounds.
(a) stannic fluoride (b) cuprous oxide
(c) ferrous sulfide (d) plumbic iodide
(e) cobaltous phosphide

15-10. Write the formulas of the following ternary ionic compounds.
(a) barium sulfate (b) lead(IV) chromate
(c) stannic carbonate (d) ferrous nitrate
(e) tin(II) acetate

Naming Compounds

15-11. Give the traditional name (using -ous or -ic to indicate the charge of the cation) of each of the following binary ionic compounds.
(a) Fe_2S_3 (b) $HgCl_2$ (c) SnI_4
(d) $CuBr_2$ (e) PbO

15-12. Give the Stock name (using a Roman numeral to specify the charge of the cation) for each of the following binary ionic compounds.
(a) PbI_4 (b) SnO (c) CuS
(d) Fe_2S_3 (e) $AuCl_3$

15-13. Name each of the following ternary ionic compounds, giving both the traditional and the Stock names.
(a) $FePO_4$ (b) $CuSO_4$ (c) $Sn(NO_2)_2$
(d) $Pb(CrO_4)_2$ (e) $Au(ClO)_3$

Percent Composition of Compounds

15-14. Determine the molar mass of $C_7H_5N_3O_6$ (TNT).

15-15. What is the percent composition of cholesterol, $C_{22}H_{45}OH$?

15-16. An insecticide, parathion, has the formula $C_2H_{14}PNSO_5$. What is its percent composition?

15-17. Determine the percentage of iodine in thyroxine, $C_{14}H_{11}O_4NI_4$, a compound produced in the thyroid gland.

15-18. Tetraethyl lead, $Pb(C_2H_5)_4$, was widely used as an additive in gasoline to increase the octane rating. Determine the percentage of lead present.

15-19. Which of the following compounds contains the highest percentage of nitrogen?
(a) KNO_3 (b) NCl_3
(c) $CO(NH_2)_2$ (d) $Fe(CN)_2$

The Mole and Avogadro's Number

15-20. What does the mole have in common with the pair or the dozen?

15-21. (a) Why isn't the dozen used for counting atoms? (b) Why isn't the mole used for counting eggs?

15-22. What is the mass of one atom of platinum?

15-23. How many atoms are present in 1 kilogram of silver?

15-24. What is the mass of one molecule of sucrose, $C_{12}H_{22}O_{11}$?

15-25. How many grams of vitamin C, $C_6H_8O_6$, are there in 0.731 mol of vitamin C?

15-26. What is the mass in grams of a molecule of hemoglobin if the molar mass is about 68,000?

Empirical and Molecular Formulas

15–27. What information about a new compound must be known before its formula can be determined?

15–28. What does the empirical formula of a compound represent?

15–29. Give the empirical formula of each of the following compounds.
(a) C_4H_{10} (b) N_2O_4 (c) B_2H_6
(d) $Na_2S_2O_4$ (e) $K_2Cr_2O_7$

15–30. The empirical formula of dextrose is CH_2O, and the molar mass is 180. What is the molecular formula of dextrose?

15–31. The empirical formula of butane is C_2H_5. If the molar mass of butane is 58, what is its molecular formula?

15–32. Ethyl alcohol shows the following analysis: 52.2% C, 13.0% H, 34.8% O. What is its molecular formula if it has a molar mass of 46?

15–33. Urea shows the analysis 20.0% C, 6.7% H, 46.7% N, 26.6% O. What is its empirical formula?

15–34. What is the molecular formula of aspirin if its molar mass is 180.2 and analysis shows 60.0% C, 4.48% H, and 35.5% O?

Writing and Balancing Chemical Equations

15–35. What is the difference between a chemical reaction and a chemical equation?

15–36. Sodium forms an ionic compound with sulfur.
(a) Give the chemical formula of the compound.
(b) Name the compound.
(c) Write a balanced equation for the reaction.

15–37. Aluminum cans react with oxygen to form aluminum oxide. Write the balanced equation for this reaction.

15–38. The following reaction represents the combustion of a hydrocarbon (octane):

$$C_8H_{18} + O_2 \longrightarrow CO_2 + H_2O$$

Balance the equation.

15–39. Iron rust forms when iron reacts with oxygen to form iron(III) oxide. Write the balanced equation for this reaction.

15–40. Excess carbon dioxide is removed from the atmosphere of a spacecraft by lithium hydroxide. The products are lithium carbonate and water. Write a balanced equation for this reaction.

15–41. Aluminum sulfate reacts with calcium hydroxide to form a gelatinous compound, aluminum hydroxide, and calcium sulfate. Suspended solids in water-supply systems are removed by the aluminum hydroxide and settle to the bottom. Write a balanced equation for the reaction.

15–42. Balance the following equations.
(a) $P + O_2 \longrightarrow P_4O_{10}$
(b) $Al + O_2 \longrightarrow Al_2O_3$
(c) $SO_2 + O_2 \longrightarrow SO_3$
(d) $NH_3 \longrightarrow N_2 + H_2$
(e) $H_2O_2 \longrightarrow H_2O + O_2$
(f) $NH_4NO_3 \longrightarrow N_2O + H_2O$
(g) $Zn + H_3PO_4 \longrightarrow Zn_3(PO_4)_2 + H_2$
(h) $Al + H_2SO_4 \longrightarrow Al_2(SO_4)_3 + H_2$
(i) $NaNO_3 + H_2SO_4 \longrightarrow$
(j) $FeCl_3 + H_3PO_4 \longrightarrow$

15–43. *Multiple Choice*
A. When silicon reacts with chlorine, the compound is
(a) SiCl. (b) $SiCl_2$. (c) $SiCl_3$. (d) $SiCl_4$.
B. The formula for lead(II) sulfide is
(a) PbS. (b) PbS_2. (c) Pb_2S. (d) Pb_4S.
C. The name of the compound $CuSO_4$ is
(a) copper sulfate.
(b) cuprous sulfate.
(c) copper(I) sulfate.
(d) copper(II) sulfate.
D. A mole of sodium chloride contains
(a) 6.02×10^{23} ions.
(b) 6.02×10^{23} molecules.
(c) 1.20×10^{24} ions.
(d) 1.20×10^{24} chloride ions.
E. The numbers needed to balance the equation $Al + F_2 \longrightarrow AlF_3$ are
(a) 1, 3, 2. (b) 2, 1, 2. (c) 4, 3, 2. (d) 2, 3, 2.
F. The term "mole" refers to
(a) the number of particles in an atom of an element.
(b) the number 6.02×10^{23}.
(c) the number of atoms in a molecule.
(d) the number of electrons in an atom of carbon.
G. One mole of H_2O molecules contains
(a) 3.01×10^{23} molecules.
(b) 6.02×10^{23} atoms.
(c) 6.02×10^{23} hydrogen atoms.
(d) 1.80×10^{24} atoms.
H. An empirical formula indicates
(a) the relative number of atoms in a molecule.
(b) the three-dimensional structure of a molecule.
(c) the molar mass of a molecule.
(d) how atoms are bonded to one another in a molecule.

I. The molecular formula for a gas that has an empirical formula CH and a molecular weight of 78 is
(a) C_5H_2O. (b) C_6H_6. (c) C_3H_3. (d) C_9H_9.

J. The name of the compound Mg_3N_2 is
(a) manganese nitride.
(b) magnesium nitride.
(c) magnesium nitrite.
(d) magnesium nitrate.

SUGGESTIONS FOR FURTHER READING

Blakley, G. R., "Chemical Equation Balancing." *Journal of Chemical Education*, September, 1982.
Presents the matrix method for balancing chemical equations. The method is said to be quick and simple and has unexpected applications.

Bouma, J., "Gas Cans and Gas Cubes: Visualizing Avogadro's Law." *Journal of Chemical Education*, July, 1986.
A description of an attempt to make the gas laws "visible."

Feifer, Nathan, "The Relationship Between Avogadro's Principle and the Law of Gay-Lussac." *Journal of Chemical Education*, August, 1966.
Relates Avogadro's hypothesis to concepts introduced by Lavoisier, Dalton, and Gay-Lussac.

Fernelius, W. Conrad, "Numbers in Chemical Names." *Journal of Chemical Education*, November, 1982.
Discusses oxidation numbers and the Stock system of nomenclature.

Fernelius, W. C., and W. H. Powell, "Confusion in the Periodic Table of the Elements." *Journal of Chemical Education*, June, 1982.
The Nomenclature Committee of the American Chemical Society is reviewing the problems associated with the use of A and B to designate subgroups of elements in the periodic table. It appears that an entirely new designation is needed.

Goldwhite, Harold, "Gay-Lussac After 200 Years." *Journal of Chemical Education*, June, 1978.
His scientific legacy still makes Gay-Lussac significant to chemistry today.

Hawthorne, Robert M., Jr., "Avogadro's Number: Early Values by Loschmidt and Others." *Journal of Chemical Education*, November, 1970.
Discusses various methods used to arrive at a value for Avogadro's number.

Kolb, Doris, "The Mole." *Journal of Chemical Education*, November, 1978.
An excellent discussion of the mole concept.

Myers, R. Thomas, "Moles, Pennies, and Nickels." *Journal of Chemical Education*, March, 1989.
Demonstrates that the atomic masses are relative numbers and that early chemists knew the relative masses of atoms but not the exact mass of any atom. So they invented the idea of moles.

Poskozim, Paul S., James W. Wazorick, Permsook Tlempetpalsal, and Joyce Albin Poskozim, "Analogies for Avogadro's Number." *Journal of Chemical Education*, February, 1986.
Any true realization of the actual magnitude of Avogadro's number is impossible. The writers review analogies used to try to capture the magnitude of N_A and present new ones derived from modern technology.

Rocha-Filko, Romeu, "A Proposition About the Quantity of Which Mole Is the SI Unit," *Journal of Chemical Education*, February, 1990.
Reviews the evolution of the meaning of the term "mole" and proposes that the quantity measured is numerousness, an intrinsic property of matter.

Treptow, Richard S., "Conservation of Mass: Fact or Fiction?" *Journal of Chemical Education*, February, 1986.
Shows that conservation of mass is always strictly maintained, provided that we consider both the reacting system and its surroundings.

ANSWERS TO NUMERICAL EXERCISES

15–14. 227 amu

15–15. 81.0% C, 14.1% H, 4.9% O

15–16. 12.3% C, 7.28% H, 15.9% P, 7.28% N, 16.4% S, 41.0% O

15–17. 66.4% I

15–18. 64.1% Pb

15–19. (c) $CO(NH_2)_2$, 46.7% N

15–22. 3.24×10^{-22} g

15–23. 5.58×10^{24} atoms

15–24. 5.68×10^{-22} g

15–25. 129 g

15–26. 1.10×10^{-19} g

15–30. $C_6H_{12}O_6$

15–31. C_4H_{10}

15–32. C_2H_6O

15–33. 24.7% Ca, 14.8% C, 59.2% O, 1.2% H

15–34. $C_9H_8O_4$

ESSAY

COMPOUNDS OF THE NOBLE GASES

W I L L I A M L . M A S T E R T O N
University of Connecticut

C E C I L E N . H U R L E Y
University of Connecticut

Among the molecular structures described in this chapter were those of several compounds of the noble gas xenon. Until about 25 years ago, the noble gases were referred to as "inert" gases. They were believed to be completely unreactive. The first noble gas compound was discovered at the University of British Columbia in 1962 by Neil Bartlett, a 29-year-old chemist. In the course of his research on platinum–fluorine compounds, he isolated a reddish solid that he showed to be O_2^+ (PtF_6^-). Bartlett realized that the ionization energy of Xe (1170 kJ/mol) is virtually identical to that of the O_2 molecule (1165 kJ/mol). This encouraged him to attempt to make the analogous compound $XePtF_6$. His success opened up a new era in noble gas chemistry.

The most stable binary compounds of xenon are the three fluorides, XeF_2, XeF_4, and XeF_6. Xenon difluoride can be prepared quite simply by exposing a 1:1 mol mixture of xenon and fluorine to ultraviolet light; colorless crystals of XeF_2 (mp = 129°C) form slowly.

$$Xe(g) + F_2(g) \longrightarrow XeF_2(s)$$

The higher fluorides are formed using excess fluorine. For example, XeF_4 (Fig. A) is formed when a 1:5 mol mixture of the elements is heated to 400°C:

$$Xe(g) + 2F_2(g) \longrightarrow XeF_4(s)$$

The hexafluoride can be prepared by heating a 1:20 mol mixture of xenon and fluorine at high pressures:

$$Xe(g) + 3F_2(g) \longrightarrow XeF_6(s)$$

The fluorides of xenon are stable in dry air at room temperature. However, they react with water to form compounds in which one or more of the fluorine atoms has been replaced by oxygen. Thus, xenon hexafluoride reacts rapidly with water to give the trioxide:

$$XeF_6(s) + 3H_2O(l) \longrightarrow XeO_3(s) + 6HF(g)$$

Xenon trioxide is highly unstable; it detonates if warmed above room temperature. It has a positive heat of formation of about + 400 kJ/mol, which helps to explain why XeO_3 cannot be prepared by reacting the elements with one another.

Figure A   Crystals of xenon tetrafluoride, XeF_4. (Argonne National Laboratory)

Most of the compounds of xenon have expanded octet structures (XeO_3 is an exception). The VSEPR model has been strikingly successful in explaining the geometries of such molecules as XeF_2 (linear), XeF_4 (square planar), and XeO_3 (triangular pyramid). The XeF_6 molecule is unique among hexafluorides in that it does not have a regular octa-hedral structure. This is predictable by VSEPR theory. There are seven pairs of electrons around the xenon atom in XeF_6, one of which is unshared.

The chemistry of xenon is much more extensive than that of any other noble gas. Only one binary compound of krypton, KrF_2, has been prepared. It is a colorless solid with a positive heat of formation and decomposes at room temperature. The chemistry of radon is difficult to study because all of its isotopes are radioactive. Indeed, the radiation given off is so intense that it decomposes any reagent added to radon in an attempt to bring about a reaction.

There has been a great deal of speculation about the possibility of forming compounds of argon. Like krypton and xenon, argon has d orbitals available for bonding and so should be able to form expanded-octet molecules such as ArF_2 or ArF_4. However, all attempts to prepare such molecules have failed. It is now generally agreed that the Ar—F bond is weak, with a bond energy close to zero. In contrast, calculations suggest that the Ar—O bond might be strong enough to permit formation of the compound ArO_3, analogous to XeO_3. Unfortunately, no one knows how to make a noble gas oxide without going through the fluoride.

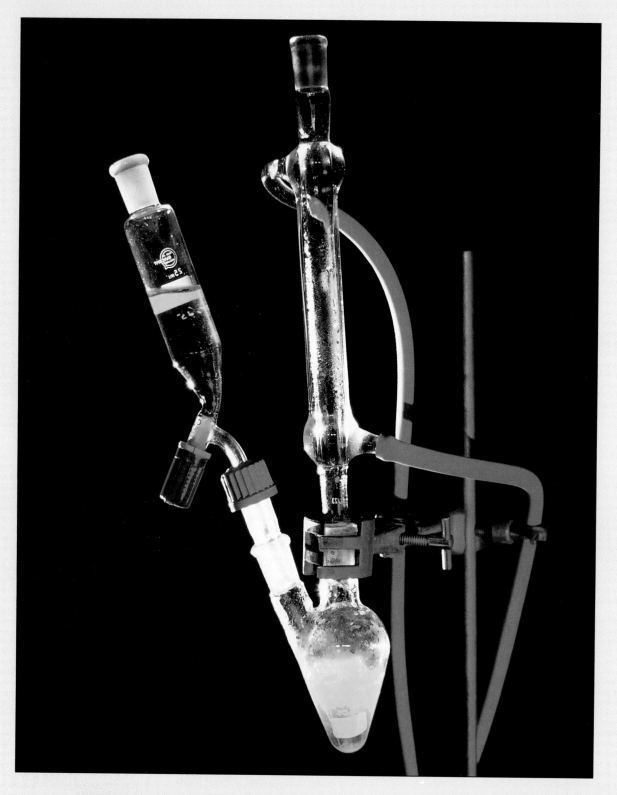

A reflux apparatus for boiling a reaction mixture without the loss of volatile compounds. The lower flask is heated and a reagent is added from the upper flask. Any vapor that boils off passes into the condenser, cools, condenses on the side of the tube, and drips back down into the flask for further reaction.
(David Taylor/Science Photo Library/Photo Researchers, Inc.)

16

CHEMICAL REACTION RATES AND EQUILIBRIUM

Although a balanced chemical equation tells us what the reactants and products are, it says nothing about *how* the reaction occurs, that is, the *mechanism* of the reaction. *Chemical kinetics* is the name given to the study of the rates and mechanisms of chemical reactions. By *rate* we mean the number of moles of a reactant converted to products per unit time. The rate of a reaction depends on a series of individual steps by which the reactants change to products. A single step may be involved or, more frequently, a sequence of steps. The single step or the sequence of steps is called the *mechanism* of the reaction. As more is learned of the mechanism of a chemical reaction, scientists acquire a greater ability to control it. Examples are the design of improved internal combustion engines and solutions to the problems of air and water pollution.

16–1 RATE OF A REACTION

Chemical reaction rates range from microseconds to centuries. Some, such as a dynamite explosion and the mixing of silver nitrate and sodium chloride, occur almost instantaneously (Fig. 16–1). Others, such as the souring of milk or the rusting of iron, proceed slowly. Even similar reactions often proceed at different rates (Fig. 16–2). Zinc releases hydrogen faster from hydrochloric acid (HCl) than from acetic acid

($HC_2H_3O_2$). Magnesium reacts faster with dilute sulfuric acid (H_2SO_4) than with concentrated sulfuric acid.

The first studies of reaction rates dealt with systems that reacted in minutes or hours. The best method is to observe the change in some physical property of the system that varies with the concentration of reactants—color changes, pressure changes, variations in electrical conductivity, and so on. The first published account of a rate measurement for a chemical reaction was by Ludwig Wilhelmy (1812–1864), a physicist at the University of Heidelberg, who studied the rate at which sucrose (cane sugar) is

Figure 16–1 A dynamite explosion in an open-pit mine occurs almost instantaneously. (© Four By Five)

Figure 16–2   Zinc metal reacts slowly with dilute sulfuric acid *(left)* but rapidly with more concentrated acid *(right)*. (Charles Steele)

Figure 16–3   White phosphorus *(above)* ignites and burns rapidly when exposed to air, so it is stored under water. Red phosphorus *(below)* reacts much more slowly with air and can be stored in contact with air. (Charles D. Winters)

converted in aqueous solution to glucose (dextrose) and fructose (fruit sugar) according to the equation

$$C_{12}H_{22}O_{11} + H_2O \longrightarrow$$

sucrose

$$C_6H_{12}O_6 + C_6H_{12}O_6 \quad (16-1)$$

glucose   fructose

To follow the course of the reaction, Wilhelmy took advantage of differences in the optical properties of the substances. Using a polarimeter, he followed the change in rotation of polarized light from *dextro* to *levo* (right to left) as the sucrose was changed to invert sugar, a mixture of glucose and fructose (Section 18–13). He observed that the rate was proportional to the concentration of sucrose, which decreases with time. The nature of the reacting substances, temperature, catalysts, and pressure are other factors that affect the rates of chemical reactions.

16–2 NATURE OF THE REACTANTS

Reaction rates depend first of all on the **nature of the reacting substances**. A given chemical change, such as oxidation in a flame, takes place very slowly with copper or silver, whereas magnesium under the same conditions burns very rapidly. White phosphorus ignites spontaneously in air; red phosphorus does not (Fig. 16–3). Some wood burns quickly and easily, but other wood is difficult to ignite and burns very slowly.

The **state of subdivision** of the reactants is another factor (Fig. 16–4). Granulated sugar dissolves

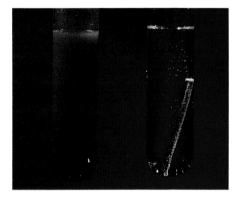

Figure 16–4   Powdered iron reacts rapidly with dilute sulfuric acid *(left)* because it has a large surface area. An iron nail *(right)* reacts much more slowly. (Charles D. Winters)

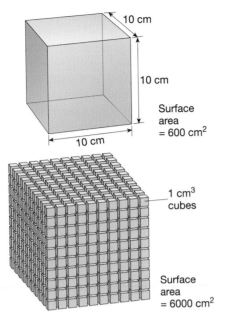

Figure 16–5 Particle size and surface area. Dividing a solid into smaller particles increases the surface area. For a cube 10 cm on each side, the surface area is 600 cm². Dividing it into smaller cubes measuring 1 cm on each side yields 1000 cubes, each having a surface area of 6 cm², for a total of 6000 cm².

in water more rapidly than does lump sugar. Powdered zinc reacts more rapidly with an acid than do lumps of zinc. For solid substances, only that portion that is exposed can react, since reactions occur only at the surface boundary between the reacting substances. Doubling the surface area doubles the rate of reaction.

When a given mass is subdivided into smaller particles, the surface area is increased and the rate of reaction increases. A finely powdered solid presents vastly more surface than a few large chunks of the same solid (Fig. 16–5). For this reason, powders react much more rapidly than larger aggregates. Thus, piles of wheat flour or coal dust burn slowly after being ignited (Fig. 16–6). If either flour or coal dust is dispersed in air, however, a mere spark can cause a disastrous explosion. In the dust cloud, the surface exposed to oxygen is much larger than in the piles, and the combustion reaction occurs much more rapidly. In this reaction, the sudden production of expanding hot gases (CO_2, H_2O) quickly increases the pressure as much as 40 atmospheres/second and produces a violent explosion. Some coal mine disasters and the explosive destruction of grain elevators are a result of such rapid reactions.

(a)

(b)

Figure 16–6 (a) A pile of lycopodium powder consisting of the spores of a common moss burns slowly in a dish. (b) The surface area is increased in a finely divided powder so that combustion is rapid when the powder is sprayed into a flame. (Charles D. Winters)

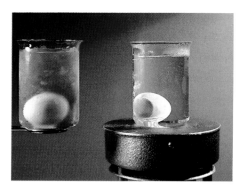

Figure 16–7   An egg in ice water *(left)* will not cook; one in boiling water *(right)* will. (Charles Steele)

16–3 EFFECT OF TEMPERATURE

Temperature has a striking effect on the rate of chemical reactions (Fig. 16–7). Reaction rates that are negligible at ordinary temperatures may become appreciable and even explosive at high temperatures (Fig. 16–8). A mixture of hydrogen and oxygen at room temperature can remain mixed for years without reacting much. At 400°C, there is some reaction; at 600°C, the reaction is fast, and at 700°C, the mixture explodes. This is typical of many reactions: At low temperatures, the chemical reaction is so slow that for all practical purposes it does not occur; in a range of intermediate temperatures, the reaction is moderately rapid; and at high temperatures, it becomes instantaneous.

A rough but useful approximation is that the rate of many chemical reactions is doubled for an increase of 10 Celsius degrees in temperature. Not all reaction rates behave in the same way with regard to temperature change, however. Most rates increase, but some actually decrease. In the latter class are enzyme reaction rates, which fall off at high temperatures as the enzyme decomposes. The reactions in which the enzyme is involved then proceed at lower rates. In some cases, a fall-off in rate occurs because of a "back reaction," the rate of which increases faster than the rate of the "forward reaction" as the temperature rises. Therefore, for the same temperature change there may be different changes in rate for different reactions.

16–4 COLLISION THEORY

The effect of temperature on reaction rates is explained by molecular motions and **collision theory**. The molecules have velocities of translation in the range of 4000 to 40,000 cm/s, and as they move they collide with other molecules. The collisions last about 10^{-11} second at room temperature.

The number of colliding molecules calculated from the kinetic theory of gases is enormous, on the order of 10^{32} molecules per liter per second at standard conditions (0°C, 1 atmosphere pressure). Each molecule experiences, on the average, 4 billion collisions each second. If collisions were the only requirement for reaction, all gaseous reactions would pro-

25°C   75°C

Figure 16–8   Antimony powder reacts slowly with bromine at room temperature (25°C) *(left)* but vigorously at 75°C *(right)*. (James Morgenthaler)

ceed at practically the same explosive rate. But experiments with gases at the same concentration at the same temperature indicate that they react at very different rates. Gaseous HI decomposes at the rate of 4.4×10^{-3} mole/liter/hour at 300°C, and gaseous N_2O_5 decomposes at the rate of 9.4×10^5 moles/liter/hour at this temperature. Therefore, collisions between molecules cannot be the only factor involved in determining the rate of a reaction.

A chemical reaction would be over within a small fraction of a second if all the collisions in which molecules are involved were effective. Instead, a reaction may require hours at this temperature, or it may not proceed discernibly at all. As the temperature increases, the kinetic energy of translation of the molecules also increases. More energy is transferred between molecules as they collide, causing them to vibrate more vigorously. Some chemical bonds that hold the atoms together in the original molecules break. At still higher temperatures, electrons are excited and molecules become ionized. An increase of 10°C at room temperature, however, produces only a small increase—about 3%—in the average kinetic energy of the molecules. Although the collision rate is increased only slightly, the reaction rate of many reactions is increased about 100%.

To account for these facts, the chemist Svante Arrhenius (1859–1927) proposed a theory that says that as molecules encounter each other, the collisions between them will result in a chemical reaction only if the energy of the molecules exceeds a certain minimum threshold to overcome the forces that tend to keep them as they are. This quantity is known as the **energy of activation**, E_a (Fig. 16–9). Every chemical reaction has a characteristic energy of activation. If the molecules collide with less energy than this critical amount, they recoil without undergoing chemical change. When the energy of activation is low (only a few kilojoules per mole), reactions proceed quickly and smoothly at room temperature. Reactions with activation energies in the range of 50 to 200 kj/mol proceed at convenient rates for experimental study at some temperature in the range of 25° to 500°C. If the energy of activation is high, the reaction might proceed at an infinitesimal rate, if at all, at room temperature.

According to Arrhenius' theory, only a small fraction of the molecules may possess enough energy of activation at a given temperature to react when they collide. Remember that although the average kinetic

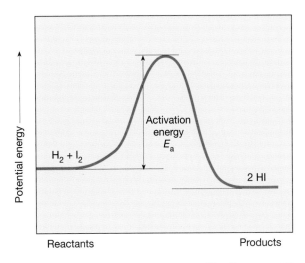

Figure 16–9 Activation energy profile. Reactants H_2 and I_2 must overcome the activation energy barrier, E_a, before they can react to form the product, HI.

energy of the molecules determines the temperature, some molecules will be moving with much greater speeds than the average and others with much lower speeds than the average. Although reaction rates are frequently doubled by increasing the temperature 10 Celsius degrees, the number of collisions does not double, and in fact, there are already many more collisions than reactions. An increase in temperature of this amount doubles the number of collisions that are effective in leading to chemical reactions, by endowing approximately double the number of molecules with the required energy of activation (Fig. 16–10).

In many reactions, however, even collisions between molecules possessing the required energy of activation do not all lead to reaction. The manner in which the molecules collide is also important (Fig. 16–11). Some molecules must be oriented in a very specific way for reaction to occur and other molecules may react when colliding in any of a number of random orientations.

This consideration led Henry Eyring (b. 1901) and others to modify classical collision theory. In the **transition state theory**, as it is known, it is assumed that colliding molecules must first achieve a specific configuration with a definite energy before they can form the products of the reaction. This "transition state" persists for only an instant (10^{-13} s) and cannot be isolated. Once it is formed, it either returns to the

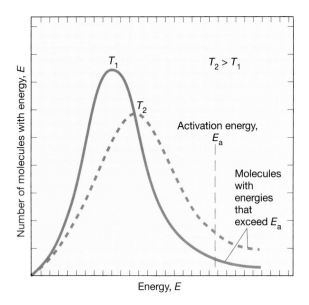

Figure 16-10 Energy distribution for gaseous molecules at two different temperatures. A small increase in temperature from T_1 to T_2 produces a large increase in the number of molecules with energies greater than the activation energy, E_a. The reaction, therefore, proceeds at a much faster rate.

original reactants or it proceeds to form products. If the two molecules approach one another with too little energy to reach the transition state, they do not react. Increasing the temperature of a reaction, then, has the effect of increasing the number of **activated complexes**, a term that describes the species present in the transition state.

16-5 EFFECT OF CONCENTRATION ON REACTION RATE

Substances that burn in air burn much more quickly in pure oxygen, which is the actual burning agent (Fig. 16-12). Since oxygen constitutes only one fifth of the air, the pure gas has a greater concentration of oxygen—five times as many oxygen molecules per cubic centimeter as air has.

The chemists Cato Maximilian Guldberg (1836-1902) and Peter Waage (1833-1900) were the first to see that the rate of a chemical reaction is usually proportional to the concentrations of the reactants expressed in moles per liter. Their generalization is

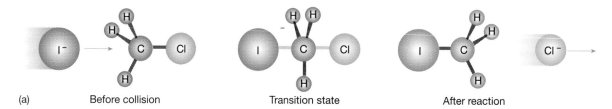

(a)   Before collision   Transition state   After reaction

Figure 16-11 (a) A collision between I^- and CH_3Cl that could lead to reaction to give CH_3I + Cl^-. The reactants possess sufficient energy and proper orientation, the I^- approaching the "back side" of the C—Cl bond. (b) Two collisions that result in no reaction because the orientation is poor.

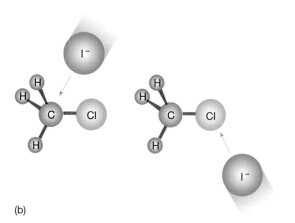

(b)

Figure 16–12 Steel wool glows when heated in air but does not burn rapidly, due to the low O_2 concentration in air (about 21%). When the glowing steel wool is put into pure oxygen, it burns vigorously. (Leon Lewandowski)

known as the **law of mass action**. For the general overall reaction

$$A + B \longrightarrow C$$

the rate of formation of C, or the rate of disappearance of A or B, is proportional to the concentration of A multiplied by the concentration of B:

$$\text{rate} \propto [A][B]$$

or, the rate is equal to a constant multiplied by these concentrations:

$$\text{rate} = k[A][B] \tag{16-2}$$

in which k is a constant determined by experiment, and the brackets [ ] denote the concentration of A or B in moles per liter. For gaseous substances, concentration is directly proportional to **partial pressures**, and partial pressures of gases may be substituted for concentrations, as in

$$\text{rate} = kp_A p_B$$

John Dalton (1766–1844), a founder of modern chemistry, was the first to observe that the total pressure of a mixture of gases is just the sum of the pres-

sures that each gas would exert if it were present alone. Each of the individual gases, if present alone under the same temperature and volume conditions as the mixture, would exert a pressure that we call the **partial pressure**. In his honor, this statement is now called **Dalton's law of partial pressures**.

$$P_{\text{total}} = P_1 + P_2 + P_3 + \cdots$$

16–6 EFFECT OF A CATALYST ON THE RATE OF A REACTION

The presence of small amounts of a foreign substance often changes the rate of a chemical reaction. The foreign substance is not changed by the reaction and may be recovered intact when the reaction is ended. Thus, hydrogen and oxygen combine rapidly to form water in the presence of finely divided platinum at low temperatures compared with those temperatures in a flame used to ignite a similar mixture, yet the platinum does not seem to change in the process. Michael Faraday studied this effect in great detail. He explained that the metal caused the gases (hydrogen and oxygen) to condense or become adsorbed, thereby bringing them into a condition such that they could readily react.

Berzelius reviewed all the examples of such effects that he could find in the literature and proposed a unifying theory. He suggested that some new force was acting in all these cases. Although he was not sure what this force was, he suggested that it be called "catalytic force," and the process "catalysis," from the Greek word for decomposition. A **catalyst** increases the rate of a chemical reaction by providing a path for the reaction that involves a lower activation energy than the one previously available (Fig. 16–13). More molecules, therefore, have the necessary energy for the new path, and the reaction speeds up (Fig. 16–14). A negative catalyst, or **inhibitor**, slows down a reaction by blocking a reaction path, thus requiring a higher activation energy. The use of catalysts has become important in industry. Many industrial processes, including the manufacture of plastics, lubricating oils, motor fuel, fibers, and detergents, would be impossible without them. Industrial catalysts are, therefore, closely guarded secrets.

Solid catalysts used with reactions of gases are called contact catalysts, or **heterogeneous catalysts**, since they exist in a different phase from the

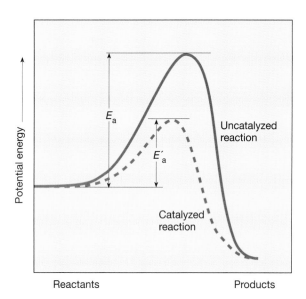

Figure 16–13   Effect of a positive catalyst on reaction rate. By lowering the required activation energy to E'_a, a positive catalyst speeds up a reaction. More molecules then have the required energy for reaction.

gaseous reactants. They provide a large surface area at which reaction occurs much more rapidly than it does in the gas phase. The **catalytic muffler**, an antipollution device installed in automobiles, is an example

(Fig. 16–15). Its function is to accelerate the reaction between oxygen and carbon monoxide, and between oxygen and unburned hydrocarbons, escaping in the exhaust (Equations 16–3 and 16–4). The solid catalyst adsorbs or binds molecules onto its surface, disrupts their bond structures to a certain extent, and brings together these activated molecules for reaction on the surface. The products of the reaction then "desorb" from the surface, and the process is repeated.

$$2CO + O_2 \xrightarrow{\text{catalyst}} 2CO_2 \qquad (16\text{–}3)$$

$$2C_8H_{18} + 25O_2 \xrightarrow{\text{catalyst}}$$
$$16CO_2 + 18H_2O \quad (16\text{–}4)$$

Finely divided platinum and nickel are among the many catalysts used in the **catalytic hydrogenation of vegetable oils** (corn oil, peanut oil, and cottonseed oil) that converts them into the solid fat called oleomargarine. In some cases, the homogeneous catalyst (a gas if the reactants are gaseous) forms an unstable intermediate compound with one of the reacting substances, which decomposes as the reaction proceeds.

Life as we know it would be impossible without the biological catalysts called **enzymes**. These high-molecular-weight protein molecules contain **active**

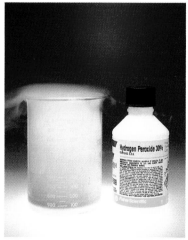

Figure 16–14   A small amount of a metal oxide catalyst speeds up the decomposition of a 30% hydrogen peroxide solution. The exothermic reaction quickly heats the solution to the boiling point of water, forming steam. The temperature increase further accelerates the decomposition. (Charles D. Winters)

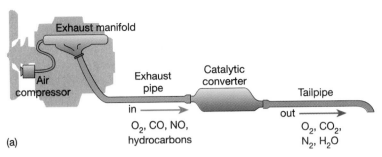

Figure 16–15   (a) The arrangement of a catalytic converter in an automobile. (b) A cutaway view of a catalytic converter, showing the pellets of the platinum/palladium/rhodium catalyst. (b, General Motors)

(b)

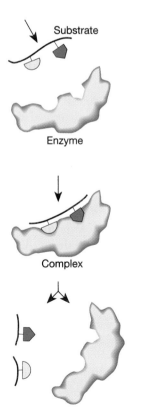

Figure 16–16   Similar enzyme and substrate shape may explain enzyme activity.

sites within their structure endowed with highly specific catalytic activity toward some chemical reaction. They are more effective as catalysts than the best catalysts produced by chemists. Thus, starch reacts with water to form sugars, a reaction that takes days. A trace of the enzyme ptyalin in saliva, however, catalyzes the conversion of starch to sugar in a fraction of this time.

The effectiveness of enzymes can be destroyed in various ways, including heating and poisoning. Heating at 80°C disrupts the protein structure and therefore the activity of enzymes. Nerve gases and insecticides can poison the active sites of enzymes and block nerve impulses, which may lead to unconsciousness and death. Many drugs used medically are believed to function by blocking or modifying the activity of enzymes in the body. Physiological poisons like mercuric chloride or rattlesnake venom react with enzymes, making them useless for essential biochemical reactions. They are negative catalysts, or inhibitors.

Enzymes are highly specific in their action. There may be as many as 30,000 different enzymes in the human body. Each enzyme is capable of catalyzing only a particular reaction or a particular substance, called the **substrate**. This implies that spatial effects between the enzymes and substrate are important. When an enzyme serves as a catalyst in the decomposition of a substrate, the enzyme shape exactly accommodates that of the substrate molecule (Fig. 16–16). A complex is formed in which chemical bonds within the substrate are weakened, and decomposition occurs more rapidly than it would without the enzyme.

16–7 REACTION MECHANISM

The process by which a reaction occurs is called the **reaction mechanism**. Knowing the mechanism of a reaction, the chemist can select conditions that will produce a good yield of the product. The study of reaction mechanisms, however, is a tremendously complex subject. Compared to the enormous number of chemical reactions that are known, only a very few reaction mechanisms have ever been identified. For this reason, the study of mechanisms is a very active area of current research.

Many reactions proceed as a **chain mechanism**. This concept appears to have originated in connection with the reaction between hydrogen and chlorine, a system that was investigated by early chemists such as John Dalton. This reaction, which appears to be simple enough to occur in one step, may actually take place in a series of steps.

$$H_2 + Cl_2 \longrightarrow 2HCl \qquad (16\text{-}5)$$

At ordinary temperatures, hydrogen and chlorine form hydrogen chloride very slowly in the dark. But if the mixture of gases is exposed to bright light, the reaction proceeds explosively. This is an example of **photochemistry**—light inducing a chemical reaction.

Einstein's introduction of the quantum theory into photochemistry led to an understanding of the hydrogen–chlorine reaction. According to Einstein's **law of photochemical equivalence**, one light quantum (or photon), hf, is absorbed for every molecule that is transformed. When Max Bodenstein (1871–1942), a physical chemist, applied this principle to experimental results, he discovered that one quantum brought about the union of 10^4 or 10^5 molecules of the reacting gases. These findings were confirmed by other experimenters and necessitated a revision of theory.

Bodenstein suggested a **chain reaction mechanism** to explain the high quantum efficiency of the reaction between hydrogen and chlorine. Walther Nernst (1864–1941) refined this idea and suggested that the light served to dissociate some of the chlorine molecules into chlorine atoms; the chlorine atoms then reacted with hydrogen molecules, forming hydrogen chloride and free hydrogen atoms. The chain finally ends with the direct union of the free chlorine and hydrogen atoms. Subsequent work has proved that this view is fundamentally correct as shown. A chlorine atom is a **free radical**—an atom or molecule containing one or more unpaired electrons. The dot on the symbol Cl· represents an unpaired electron.

$$Cl_2 + h\nu \longrightarrow Cl_2° \qquad (16\text{-}6)$$
$$Cl_2° \longrightarrow 2Cl·$$
$$Cl· + H_2 \longrightarrow HCl + H·$$
$$H· + Cl_2 \longrightarrow HCl + Cl·$$
$$H· + Cl· \longrightarrow HCl$$

(° excited molecule)

Each photon (wavelength about 4800 Å) absorbed by a chlorine molecule results in the formation of 10^5 hydrogen chloride molecules via a chain reaction. Although fast, chain reactions are not necessarily explosive. They become explosive when the heat evolved is great, thus increasing the rate of the reaction and the further liberation of heat until the reaction becomes explosive.

16-8 CHEMICAL EQUILIBRIUM

Some chemical reactions **"go to completion"** (that is, the reactants are entirely converted to products), and others do not. In the latter case, the products apparently react with one another to re-form the reactants. **Reversible reactions**, as these are called, were first carefully studied by William Williamson (1824–1904) at University College, London.

Williamson found situations in which, beginning with a mixture of A and B, the substances C and D were formed. If he began instead with a mixture of C and D, substances A and B were formed. In either case, there would be a mixture of A, B, C, and D in the end, with the proportions apparently fixed. The mixture would be at an **equilibrium**. Williamson felt that reaction did not cease at equilibrium, but that A and B were reacting to form C and D, and C and D, in turn, were reacting to form A and B at the same rate. This condition is called **dynamic equilibrium** and is represented as follows, with a double arrow to indicate reversibility:

$$A + B \rightleftharpoons C + D \qquad (16\text{-}7)$$

The idea of dynamic equilibrium did not become immediately popular with chemists. Its significance was realized by Guldberg and Waage, who studied equilibrium reactions with systems containing solids in contact with solutions. They showed that an equilibrium is reached in incomplete reactions, and they treated such reactions mathematically, expressing equilibrium conditions in terms of molecular concentration.

Let us assume that A, B, C, and D represent gases and apply the law of mass action to the reversible reaction

$$A + B \rightleftharpoons C + D$$

If the system contains only A and B molecules at the outset, then the rate of the forward reaction is expressed in terms of the concentrations of A and B as follows:

$$R_f = k_f[A][B] \qquad (16-8)$$

However, as molecules of the products C and D are formed, they begin to react, and the rate of this "back" reaction, R_b, is given by the expression

$$R_b = k_b[C][D] \qquad (16-9)$$

As the reaction between A and B proceeds, the concentrations of C and D increase, and the rate of the back reaction, R_b, increases, since it depends on the concentrations of C and D. Meanwhile, the concentrations of A and B are becoming less and less, so that the rate of the "forward" reaction, R_f, decreases. The two reaction rates, forward and back, approach each other and finally become equal (Fig. 16–17). A condition of dynamic equilibrium is established in which the opposing reactions proceed at equal rates at a certain temperature. At equilibrium, with the rate of the forward reaction equal to the rate of the back reaction,

$$R_f = R_b$$

and we may write

$$k_f[A][B] = k_b[C][D]$$

or, by rearranging terms,

$$\frac{k_f}{k_b} = \frac{[C][D]}{[A][B]}$$

Because k_f and k_b are constants, k_f/k_b is also constant, and the expression may be written as

$$K = \frac{[C][D]}{[A][B]} \qquad (16-10)$$

K is called the **equilibrium constant** for the reaction and has a specific numerical value for each reaction at a specific temperature. For the general reaction.

$$aA + bB \rightleftharpoons cC + dD \qquad (16-11)$$

the expression for the law of chemical equilibrium becomes

$$K = \frac{[C]^c[D]^d}{[A]^a[B]^b}$$

The equilibrium expressions for several typical reactions are shown below:

$$H_2 + I_2 \rightleftharpoons 2HI$$

$$K = \frac{[HI]^2}{[H_2][I_2]} = 50.2 \ (448°C)$$

$$N_2 + 3H_2 \rightleftharpoons 2NH_3$$

$$K = \frac{[NH_3]^2}{[N_2][H_2]^3} = 9.0 \ (350°C)$$

$$2SO_2 + O_2 \rightleftharpoons 2SO_3$$

$$K = \frac{[SO_3]^2}{[SO_2]^2[O_2]} = 0.99 \ (1177°C)$$

A reaction that "goes to completion," then, is a reaction that has a very large value of K. The concentration of the reactants is small, having been largely converted to products. In a reaction that has a very small value of K at equilibrium, the concentration of the products is very small. Isolated chemical systems

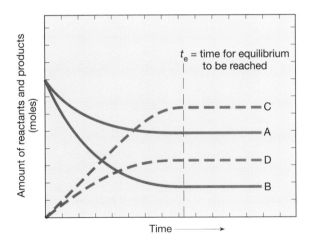

Figure 16–17   A reversible reaction: $A + B \rightleftharpoons C + D$. Neither of the initial reactants (A and B) is completely consumed. At time t_e, the forward and reverse reactions proceed with equal rates, and the moles of A, B, C, and D present stay constant.

eventually come to equilibrium, in which the forward and reverse reactions occur at the same rate and neither goes to completion.

When chemical reactions reach a state of equilibrium instead of going to completion, equilibrium may be established when a reaction is very nearly complete or at some other point. Chemical manufacturers are in business to produce high yields of products. High yields depend on reactions that go far toward completion. If a system reaches equilibrium before much of the compound has been formed, it may be possible to modify the equilibrium conditions so that the yield of the product will be increased.

16–9 LE CHÂTELIER'S PRINCIPLE

Le Châtelier's principle, formulated by Henri Le Châtelier (1850–1936), a chemist at a school of mines, is the most general principle that applies to systems at equilibrium. It states:

> **When a stress is imposed on a system at equilibrium, the equilibrium shifts in such a way as to minimize the effect of the stress.**

The equilibrium will shift in the direction that tends to oppose or counteract the change. In chemical systems, stress may be placed on the equilibrium by altering concentrations of substances, changing conditions such as temperature and pressure, and adding a catalyst.

A. Effect of Concentration on Equilibrium. In the reaction

$$H_2 + I_2 \rightleftharpoons 2HI \qquad K = \frac{[HI]^2}{[H_2][I_2]} \qquad (16\text{–}12)$$

The addition of hydrogen to the equilibrium system is considered to constitute a stress. The stress is relieved by the speeding up of the forward reaction, which consumes the added hydrogen through reaction with iodine and results in the production of a greater amount of hydrogen iodide, HI. The concentration of HI increases until the rate of the backward reaction catches up with that of the forward reaction and equilibrium is restored. When equilibrium is again achieved, there is less I_2 and more HI than initially, and some of the H_2 that had been added and constituted a "stress" will have been used up. The position of equilibrium is said to have **shifted to the right**.

Increasing the concentration of one substance in an equilibrium mixture causes the reaction to take place in the direction that consumes some of the material added (Fig. 16–18). On the other hand, decreasing the concentration of a substance causes the production of more of it by the appropriate reaction. Thus, removing a part of the HI from the system would cause equilibrium again to shift to the right, forming more HI from the reaction between H_2 and I_2. Although the concentrations of H_2, I_2, and HI are changed in both cases, the proportions are such that the law of chemical equilibrium applies as before (Table 16–1). The original value of the equilibrium constant, K, is restored for that temperature. Unless the original value of K is restored, the system is not in equilibrium, and the concentrations will undergo further change until it is. Then the rates of the forward and reverse reactions become equal again.

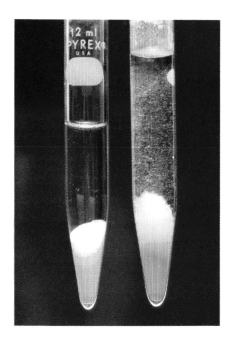

Figure 16–18 The test tube at the left contains a saturated solution of silver acetate in equilibrium with solid silver acetate. When silver ions are added, the equilibrium shifts to the right, producing more

$$Ag^+(aq) + CH_3COO^-(aq) \rightleftharpoons CH_3COOAg(s)$$

solid silver acetate, as can be seen in the test tube on the right. (Charles D. Winters)

Table 16–1 CONCENTRATIONS OF H_2, I_2, AND HI INITIALLY AND AT EQUILIBRIUM (440°C)

Experiment	Initial			Equilibrium			
	$[H_2]$	$[I_2]$	$[HI]$	$[H_2]$	$[I_2]$	$[HI]$	$\dfrac{[HI]^2}{[H_2][I_2]}$
1	1.000	1.000	0	0.218	0.218	1.564	51.5
2	0	0	2.000	0.218	0.218	1.564	51.5
3	0	0	1.000	0.109	0.109	0.782	51.5
4	1.000	0.500	0	0.532	0.032	0.936	51.5

B. Effect of Temperature on Equilibrium. Chemical changes involve the evolution or the absorption of heat (Fig. 16–19). In a system at equilibrium, an endothermic reaction and an exothermic reaction are taking place simultaneously, as in

$$H_2 + I_2 \rightleftharpoons 2HI + \text{heat} \qquad (16\text{–}13)$$

In this case, the forward reaction results in the evolution of heat and it is said to be **exothermic**: Heat flows from the system to the surroundings. The reverse reaction absorbs heat; it is **endothermic**: Heat flows into the system from its surroundings.

If we add heat to this system and increase the temperature, Le Châtelier's principle predicts that the equilibrium will shift in such a way as to absorb the added heat. An increase in the temperature speeds up both the decomposition and the formation of HI but speeds up the decomposition more than the

Figure 16–19   The middle test tube contains a purple equilibrium mixture of a pink and blue complex.

$$\underset{\text{pink}}{[Co(OH_2)_6]^{2+}} + 4Cl^- + \text{heat} \rightleftharpoons \underset{\text{blue}}{[CoCl_4]^2} + 6H_2O$$

In hot water the forward reaction is favored and the solution is blue *(right)*. At 0°C the reverse reaction is favored, so the solution is pink *(left)*. (Charles Steele)

formation. Thus, when the two rates become equal at the higher temperature, less HI is present than at the lower temperature, since more of it has decomposed. In such cases, a change in temperature results in a change, usually large, in the value of the equilibrium constant. The value of K is constant as long as the temperature does not change, but it is different for different temperatures. Thus, for the reaction

$$H_2 + I_2 \rightleftharpoons 2HI + heat$$

the value for K is 54.8 at 425°C, but it is 871 at 25°C. Therefore, the temperature at which a given value of K is valid is reported with it.

C. Effect of Pressure on Equilibrium.   In a system at equilibrium that involves a change in the total number of gaseous molecules between the reactants and the products, as in

$$\underbrace{2SO_2\ +\ \ \underbrace{O_2}_{\text{1 volume}}}_{\text{3 volumes}} \underbrace{\rightleftharpoons\ \ 2SO_3}_{\text{2 volumes}} \qquad (16-14)$$
$$\underset{\text{2 volumes}}{}$$

a change in pressure causes a shift in the position of equilibrium. In this system there are three volumes of gaseous reactants and two volumes of product. If the pressure is increased, then according to Le Châtelier's principle the system will adjust in such a way as to absorb this stress. A gaseous system may absorb the stress of increased pressure by reacting so as to decrease the number of molecules, since the pressure of a gas depends on the number of gaseous molecules present. At high pressures, the equilibrium in this system is shifted to the right, since the reduction from three to two volumes of gas lowers the pressure and thereby absorbs the stress. In a system in which there is no change in the number of molecules, as in

$$\underset{\text{2 volumes}}{2IBr} \rightleftharpoons \underbrace{I_2\ +\ Br_2}_{\text{2 volumes}} \qquad (16-15)$$
$$\underset{\text{1 volume}\quad\text{1 volume}}{}$$

the equilibrium is not affected by a change in pressure.

D. Effect of a Catalyst on Equilibrium.   The effect of a catalyst on equilibrium is to change the rate at which equilibrium is reached, without affecting the value of the equilibrium constant. A catalyst is equally effective in increasing the rates of both the forward and

reverse directions of the reaction. Thus, a catalyst that promotes the synthesis of HI from its elements is equally effective in increasing the rate of the reverse reaction, the decomposition of HI. The catalyst provides a reaction path with a lower energy barrier between the initial and final states. Its presence in a chemical system serves to accelerate the approach to equilibrium from either direction, thereby reducing the time required for the system to reach equilibrium. But the catalyst has no effect on the *composition* of the equilibrium mixture; equilibrium is merely reached more quickly.

16-10   THE HABER PROCESS: A CASE STUDY

The Haber process for the synthesis of ammonia, named after Fritz Haber (1868-1934), director of the Kaiser Wilhelm Institute for Physical Chemistry, is an outstanding example of the practical application of knowledge about the effects of temperature, pressure, concentration, and catalysis on a system in dynamic equilibrium. This process also illustrates the dual character of a major chemical discovery that may be applied to either constructive or destructive social purposes.

Nitrogen constitutes about 80% of the atmosphere, being present there in its elementary form, N_2. Although it is abundant and essential for life, nitrogen is chemically inert, reacting to form compounds only with difficulty. To most organisms, however, it is useful only in compound form as **fixed nitrogen**. Nitrogen occurs as nitrates in soil, but it is generally in short supply and must be supplied in the form of animal wastes or chemical fertilizers.

The earth's supply of nitrates is replenished partly through the activity of thunderstorms. The nitrogen and oxygen of the air in the vicinity of lightning bolts combine to form compounds that dissolve in raindrops and are brought to earth. Atmospheric nitrogen can also be "fixed" by certain soil bacteria in nitrogen compounds. The best natural source of nitrates is Chilean saltpeter, $NaNO_3$, from the Chilean desert, which is inadequate, however, to maintain the food supply of a rapidly growing world population.

At the same time, nitrogen compounds are used in very large quantities for the production of explosives: nitroglycerine for dynamite, nitrocellulose ("guncotton"), trinitrotoluene (TNT), and trinitrophenol (picric acid).

Late in the nineteenth century, Sir William Crookes (1832–1919) expressed concern that a food shortage was imminent and urged intensive cultivation to increase the yield of wheat per acre. In the absence of abundant, natural supplies of fixed nitrogen, he warned of possible depletion unless a commercial process for the fixation of atmospheric nitrogen could be developed. Otherwise, starvation appeared to be a real possibility.

Attempts were made to combine nitrogen and oxygen directly into nitric oxide, NO, as occurs in thunderstorms. But high temperatures requiring an adequate power source are required to bring about even a moderately favorable equilibrium concentration of nitric oxide. In Norway, Kristian Birkeland (1867–1917), a physicist, and Samuel Eyde (1866–1940), an engineer, did succeed in bringing such a process into commercial operation because of the availability of cheap electricity there. The compound was eventually synthesized in calcium nitrate and sold for fertilizer as "Norwegian saltpeter." Since so much electricity was required in the process, however, it did not spread to other parts of the world.

The direct combination of nitrogen and hydrogen had been studied as a possibility, then dropped. Fritz Haber took up the problem when he began his studies of the **synthesis of ammonia** in 1904. Haber's theoretical work had proved to him that the synthesis was possible, but never before had anyone observed the reaction:

$$N_2 + 3H_2 \rightleftharpoons 2NH_3 \qquad (16\text{--}16)$$

By 1908, Haber had overcome many difficulties and took out the key patent for the process that bears his name (Fig. 16–20).

The reaction of the Haber process for the synthesis of ammonia from atmospheric nitrogen is moderately exothermic:

$$N_2 + 3H_2 \rightleftharpoons 2NH_3 + 22.0 \text{ kcal} \qquad (16\text{--}17)$$

It would appear that ammonia could be produced in unlimited quantity at room temperature simply by bringing together nitrogen and hydrogen. But the reaction rate at room temperature is so slow that it is not of practical value; a mixture of nitrogen and hydrogen can be kept unchanged for years. If the reaction is carried out at high temperature, then according to Le Châtelier's principle, the endothermic part

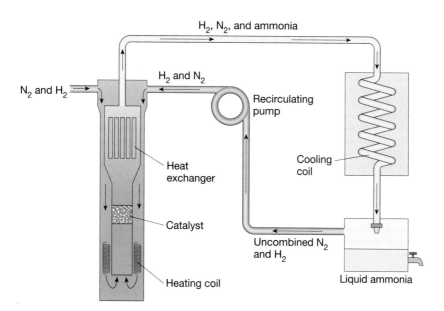

Figure 16–20 A representation of the Haber process for synthesizing ammonia.
(Source: Kenneth W. Whitten, Kenneth D. Gailey, and Raymond E. Davis, *General Chemistry,* 4th edition. Philadelphia: Saunders College Publishing, 1992, p. 677, Fig. 17–3.)

of the reaction would be favored and an equilibrium with lower values for the equilibrium constant would be established in favor of greater dissociation of NH_3. At 1000°C, the reaction proceeds almost exclusively in the opposite, endothermic direction, favoring the decomposition of ammonia into nitrogen and hydrogen, and ammonia practically ceases to exist as a chemical compound.

Haber found that at 500°C, with an iron oxide catalyst, ammonia could be produced fast enough for practical purposes, even though at equilibrium only 8% of the material was converted to ammonia. At higher temperatures, the reaction was faster, but the equilibrium too unfavorable. At lower temperatures, the reaction was too slow. Thousands of experiments were necessary to discover a suitable catalyst and appropriate conditions for a good yield of ammonia. In practice, pressures of about 200 to 600 atmospheres are necessary at a temperature of about 500°C to achieve reasonable reaction rates and yields reaching 90% (Fig. 16–21).

At equilibrium, the mixture contains ammonia, nitrogen, and hydrogen. To separate the ammonia, the gas mixture is cooled, and liquid ammonia is produced and drawn off continuously. The residual mixture of nitrogen, hydrogen, and ammonia is again compressed, heated, and passed over a catalyst for recycling. Nitrogen and hydrogen are added as they are depleted.

Much of this synthetic ammonia was oxidized to nitric acid in a process developed by Wilhelm Ostwald (1853–1932):

$$4NH_3 + 5O_2 \xrightarrow{\text{Pt}} 4NO + 6H_2O \qquad (16\text{--}18)$$
$$2NO + O_2 \longrightarrow 2NO_2$$
$$3NO_2 + H_2O \longrightarrow 2HNO_3 + NO$$

and the nitric acid went into the making of explosives. The rest was used for fertilizer.

By 1918, the production of synthetic ammonia reached 200,000 tons per year. Had it not been for the Haber process, Germany would probably not have been capable of continuing the war for more than six months. A blockade of Germany by the British fleet cut it off from its supply of Chilean saltpeter, but the Haber process—coming into commercial production just as the war was breaking out (1914)—saved Germany from premature defeat. It also saved millions of Germans from starvation during World War I. It is estimated that 100 million tons of fixed nitrogen produced by this process are consumed by the world each year, 15 million tons in the United States alone.

The worldwide development of the Haber process had an adverse effect on the economy of Chile, which for 75 years had been the world's principal source of fixed nitrogen. The specter of starvation due to a shortage of fixed nitrogen was alleviated by the development of the Haber process. Interest in developing new methods for an ammonia synthesis continues, nevertheless.

Much of the ammonia produced is shipped in liquid form in steel cylinders or tank cars and used directly as fertilizer. Some is converted to nitric acid by the Ostwald process, and part of that is reacted with ammonia to form ammonium nitrate (NH_4NO_3), which is used as a fertilizer or, when mixed with other materials, as an explosive. Explosives, of course, have peaceful civilian uses in mining and road construction, as well as military uses. The case history of the synthesis of ammonia illustrates how chemical science and technology can serve society for both military and peaceful purposes. The Haber process is an example of the neutrality of science, the fruits of which may be applied to forge weapons of war or instruments of peace. Which it shall be is a decision that society rather than science must make.

Figure 16–21   A plant for the commercial production of ammonia, NH_3, can produce up to 7000 metric tons of ammonia per day. There are nearly a hundred such plants in the world. (Dresser Industries/Kellogg Company)

SUMMARY OF EQUATIONS

Law of Mass Action

The rate of a chemical reaction is proportional to the concentrations of the reactants expressed in moles per liter.

$$\text{rate} = k[A][B]$$

$[\ \ ]$ = concentration in moles/liter

Law of Chemical Equilibrium

For the general reaction

$$aA + bB \rightleftharpoons cC + dD$$

$$K = \frac{[C]^c[D]^d}{[A]^a[B]^b}$$

KEY TERMS

Activated Complex   A temporary species formed by reactant molecules colliding before they form the product.

Active Site   A small portion of an enzyme with the specific shape and structure necessary to bind a substrate.

Catalyst   A substance that increases the rate of a chemical reaction without itself being used up or formed during the reaction.

Catalytic Hydrogenation   The addition of hydrogen, in the presence of a catalyst, to the unsaturated bonds of vegetable oils, converting them into semisolid shortenings or oleomargarine.

Catalytic Muffler   An antipollution device installed in automobiles that accelerates the reaction between oxygen and air pollutants.

Chain Mechanism   A series of reactions involving an initiation step, a series of propagation reactions, and one or more termination processes.

Chemical Equilibrium   A state in a reversible reaction in which the rates of forward and reverse reactions are equal, so that the concentrations of reactants and products no longer change.

Collision Theory   The theory that chemical reactions occur when molecules encounter each other, and

the energy of the molecules exceeds a certain minimum threshold, the energy of activation.

Dalton's Law of Partial Pressures   The total pressure of a mixture of gases equals the sum of the partial pressures of each of the gases in the mixture.

Dynamic Equilibrium   If substances A and B react to form C and D, and C and D in turn react to form A and B, a condition of dynamic equilibrium exists when the two reactions proceed at the same rate.

Endothermic Reaction   A chemical reaction in which heat is absorbed.

Energy of Activation   The minimum energy that molecules must have if collisions between them are to lead to reaction.

Enzyme   Biological catalyst. High–molecular-weight protein molecule that is endowed with highly specific catalytic activity toward some chemical reaction.

Equilibrium Constant   A numerical value that applies to a reversible chemical reaction when it is in equilibrium at a specific temperature.

Exothermic Reaction   A chemical reaction in which heat is evolved.

Fixed Nitrogen   Nitrogen in chemical combination with other elements.

Free Radical   An atom or group that has an unpaired electron.

Heterogeneous Catalyst   A catalyst that exists in a different phase (solid, liquid, or gas) from the reactants; a contact catalyst.

Inhibitor   A negative catalyst, one that slows down a chemical reaction.

Kinetics   The study of the rates of chemical reactions.

Law of Mass Action   The rate of a chemical reaction is usually proportional to some power of the concentrations of the reactants.

Law of Photochemical Equivalence   In photochemistry, light induces a chemical reaction; one photon is absorbed for every molecule that is transformed.

LeChâtelier's Principle   When a stress is imposed on a system at equilibrium, the equilibrium shifts in such a way as to minimize the effect of the stress.

Partial Pressure   In a mixture of gases, each gas exerts the same pressure that it would exert if it were present alone under the same conditions.

Photochemistry   The study of a chemical reaction induced by light.

Reaction Mechanism   A single step or sequence of steps by which reactants are changed to products in a chemical reaction.

Reversible Reaction The products of a chemical reaction react with one another to re-form the reactants.

State of Subdivision The extent to which a given mass is subdivided into smaller particles.

Substrate The reactant in an enzyme-catalyzed action.

Transition State Theory Colliding molecules may stick together to form an activated complex. The activated complex then either breaks apart to re-form the reactants or proceeds to produce the products.

THINGS TO DO

1. The iodine clock reaction is an interesting study in chemical kinetics. Two solutions, A and B, are mixed, and the time in seconds for the mixture to flash blue-black is noted. Look up specific instructions in a chemistry laboratory manual. The effects of (a) concentration and (b) temperature on the rate of a chemical reaction are nicely demonstrated.

EXERCISES

Reaction Rate

16–1. Discuss a method that might be used to measure the rate of a reaction.

16–2. Although a lump of coal burns slowly, a mine full of coal dust may explode. Explain.

16–3. Why do iron filings burn more rapidly in pure O_2 than in air?

16–4. Why does a 10°C rise in temperature have a marked effect on the rate of many chemical reactions?

16–5. Milk quickly spoils if left at room temperature but keeps for several days in a refrigerator. Explain.

16–6. Why do reaction rates usually increase with increasing temperatures?

16–7. Discuss the collision model for chemical reactions.

16–8. Describe how a catalyst may accelerate a chemical reaction.

16–9. Biochemical reactions occur in our bodies at atmospheric pressure and at about 37°C. Explain why higher pressures and temperatures are usually required to carry out similar reactions in the laboratory.

16–10. Methane burns vigorously in oxygen, yet a mixture of methane and oxygen can be kept indefinitely without reacting. Explain.

16–11. If you wished a reaction to go faster, what might you try to accomplish this?

16–12. What is the role of activation energy in the rate of a reaction?

Equilibrium

16–13. Equilibrium represents the balancing of opposing processes. Explain why a vapor pressure above a liquid in a closed container represents an equilibrium.

16–14. A chemical reaction may be reversible or irreversible. How would you classify burning a piece of paper?

16–15. A chemical equilibrium is indicated by writing two opposing arrows, $\rightleftharpoons$. Explain.

16–16. Explain why reactants are not converted completely to products in many chemical reactions.

16–17. What does it mean to say that a chemical equilibrium is a dynamic system?

16–18. Discuss the principle of dynamic equilibrium as it applies to the system.

$$3H_2 + SO_2 \rightleftharpoons H_2S + 2H_2O$$

The Equilibrium Constant

16–19. What does the equilibrium constant for a reaction represent?

16–20. What do square brackets [ ] indicate when we write an equilibrium constant?

16–21. If the reaction

$$2NO + O_2 \rightleftharpoons 2NO_2$$

is at equilibrium, are the concentrations of NO, O_2, and NO_2 necessarily equal?

16–22. Write expressions for the equilibrium constants of the following reactions.

(a) $N_2 + O_2 \rightleftharpoons 2NO$
(b) $CO_2 + H_2 \rightleftharpoons CO + H_2O$
(c) $4NH_3 + 5O_2 \rightleftharpoons 4NO + 6H_2O$

16–23. For which of the following systems does the reaction go least to completion in the forward direction: (a) $K = 10^{-10}$; (b) $K = 1$; (c) $K = 10^4$?

16–24. The rocket propellant hydrazine reacts with oxygen according to the following equation:

$$N_2H_4 + O_2 \rightleftharpoons N_2 + 2H_2O$$

Write the equilibrium expression.

16–25. Sulfur dioxide contributes to atmospheric pollution through its oxidation to sulfur trioxide:

$$2SO_2 + O_2 \rightleftharpoons 2SO_3$$

Determine the value of the equilibrium constant, K, for this reaction if the equilibrium concentrations of a mixture of the gases are

$SO_2 = 4.20 \times 10^{-7}$ mole/liter

$SO_3 = 0.500$ mole/liter

$O_2 = 2.10 \times 10^{-7}$ mole/liter

16–26. An equilibrium constant for a reaction is 3.20×10^7. Is the formation of products of reactants favored?

Le Châtelier's Principle

16–27. What does Le Châtelier's principle tell us about maximizing the yield of a reaction?

16–28. What change in conditions changes the equilibrium constant K of a reaction?

16–29. For the equilibrium system

$$2SO_2 + O_2 \rightleftharpoons 2SO_3 + heat$$

list four ways in which the concentration of SO_3 at equilibrium could be increased.

16–30. For the reaction

$$N_2 + 3H_2 \rightleftharpoons 2NH_3 + 22.0 \text{ kcal}$$

what is the effect of (a) an addition of N_2? (b) an increase in temperature? (c) an increase in pressure?

16–31. Will an increase in temperature cause the following equilibrium system to shift to the left, shift to the right, or have no effect?

$$N_2O_4 + heat \rightleftharpoons 2NO_2$$

16–32. Predict the effect of (a) an increase in the concentration of one reactant; (b) an increase in pressure; (c) an increase in temperature; and (d) the intro-

duction of a catalyst upon the following reactions at equilibrium:

1. $2CO_2 + O_2 \rightleftharpoons 2CO_2 + heat$
2. $CO_2 + H_2 \rightleftharpoons CO + H_2O + heat$

16–33. Explain how increasing the temperature would affect the equilibrium

$$H_2 + Br_2 \rightleftharpoons 2HBr + heat$$

16–34. *Multiple Choice*
   A. Which of the following would probably slow down the rate of a reaction?
      (a) increasing the temperature
      (b) increasing the concentration
      (c) using a larger beaker
      (d) lowering the temperature
   B. The principal reason for the increase in reaction rate with temperature is
      (a) molecules collide more frequently.
      (b) the pressure increases.
      (c) the activation energy increases.
      (d) the fraction of high-energy molecules increases.
   C. Which of the following changes will increase the yield of products at equilibrium?
      (a) an increase in temperature
      (b) an increase in pressure
      (c) addition of a catalyst
      (d) increasing reactant concentrations
   D. What is the effect of increasing the concentration of B in the reaction $A + B \rightleftharpoons C + D$?
      (a) the value of K decreases
      (b) the value of K increases
      (c) the equilibrium shifts to the right
      (d) the equilibrium shifts to the left
   E. A catalyst speeds up a reaction by
      (a) increasing the energy released.
      (b) decreasing the energy released.
      (c) increasing the activation energy.
      (d) decreasing the activation energy.
   F. To speed up a reaction, all of the following will work except
      (a) increasing the concentrations of the reactants.
      (b) adding a suitable catalyst.
      (c) lowering the temperature.
      (d) creating more surface area by pulverizing a solid reactant.
   G. Which one of the following changes has no effect on the position of equilibrium?
      (a) an increase in temperature
      (b) a decrease in temperature
      (c) addition of a catalyst
      (d) addition of a product

H. When a reaction has reached equilibrium,
   (a) all reaction ceases.
   (b) the rate of the forward reaction becomes zero.
   (c) the concentrations of reactants and products are equal.
   (d) the rates of the forward and reverse reactions are equal.
I. The value of K for a particular reaction changes if

(a) the temperature is changed.
(b) the concentration is changed.
(c) a catalyst is added.
(d) but K never changes.

J. Chemical kinetics is the study of
   (a) equilibrium.
   (b) reaction rates.
   (c) catalysis.
   (d) solutions.

SUGGESTIONS FOR FURTHER READING

Brice, L. K., "Rossini, William Tell and the Iodine Clock Reaction." *Journal of Chemical Education*, February, 1980.
Describes a lively variation on the stimulating iodine clock reaction.

Brill, Winston J., "Biological Nitrogen Fixation." *Scientific American*, March, 1977.
A few bacteria and simple algae are the major suppliers in nature of "fixed" nitrogen.

Goldanskii, Vitalii I., "Quantum Chemical Reactions in the Deep Cold." *Scientific American*, February, 1986.
Some reactions can take place near absolute zero even though the atoms do not have the energy required by classical chemistry. Could cold, dark clouds of galactic dust contain seeds of life?

Gruebele, Martin, and Ahmed H. Zewail, "Ultrafast Reaction Dynamics." *Physics Today*, May, 1990.
New laser techniques and molecular beam experiments make it possible to determine what happens in chemical reactions on the 10^{-13}-second time scale, the ephemeral transition state.

Haggin, Joseph, "Catalyst Development Becoming More Organized." *Chemical and Engineering News*, November 15, 1982.
Scientists are designing industrial catalysts using highly organized procedures made possible by advances in computer software and surface science.

Lewin, Roger, "RNA Can Be a Catalyst." *Science*, 26 November, 1982.
Ideas on biological catalysis and early evolution may have to be modified with the revolutionary discovery that RNA can catalyze biochemical reactions. It had been assumed that biological catalysis was the sole province of proteins.

Mickey, Charles D., "Chemical Kinetics: Reaction Rates." *Journal of Chemical Education*, September, 1980.
The practicability of a chemical reaction, particularly a commercial process, often depends on the rate and efficiency of the reaction.

Safrany, David R., "Nitrogen Fixation." *Scientific American*, October, 1974.
Life requires nitrogen that has been "fixed" through combination with other elements. Thermodynamics limits all possible methods of fixation.

Treptow, Richard S., "Le Châtelier's Principle." *Journal of Chemical Education*, June 1980.
Applying Le Châtelier's principle is not as straightforward as its simple wording implies. The author re-examines the principle and provides fresh insights.

Walling, Cheves, "The Development of Free Radical Chemistry." *Journal of Chemical Education*, February, 1986.
The field of free radical chemistry remains active with many interesting unsolved problems and applications yet to come.

Wameck, Peter, "The Formaldehyde-Sulfite Clock Reaction Revisited." *Journal of Chemical Education*, April, 1989.
Discusses the mechanism of a clock reaction, in which there is a sudden appearance of a product sometime after the initial mixing of reagents.

Zewail, Ahmed H., "The Birth of Molecules." *Scientific American*, December, 1990.
It is possible to record the motions of molecules as they form and break bonds. With lasers and molecular beams, the reaction can be seen as it proceeds from reactants through transition states and finally to products—chemistry as it happens.

ANSWERS TO NUMERICAL EXERCISES

16–23. Least: (a) $K = 10^{-10}$

16–24. $K = \dfrac{[N_2][H_2O]^2}{[N_2H_4][OH]}$

16–25. $6.75 \times 10^{18} \left(\dfrac{\text{mole}}{\text{L}}\right)^{-1}$

The solubility of oxygen gas in water supports a wide variety of aquatic life. Andrew J. Martinez/Photo Researchers, Inc.)

17

WATER, SOLUTIONS, AND POLLUTION

Before a chemical reaction can occur, particles of the reacting substances—molecules, atoms, ions—must be brought together. Substances usually do not react when dry, however. Crystalline silver nitrate and sodium chloride, for example, do not react even when they are ground together to an extremely fine state. The addition of water to the system, however, is accompanied by an instantaneous reaction and the formation of a white precipitate, silver chloride.

Many of the processes that influence our environment also involve substances in aqueous solution. Sewage, industrial wastes, and runoff from heavily fertilized land pollute our rivers. As rain falls, it picks up and dissolves gaseous impurities from the atmosphere, helping freshen the air we breathe, the air itself being a solution. Solution chemistry is important as well to an understanding of the vital processes that take place in the aqueous environment of living cells. Medicines are usually administered in solution form. In this chapter on solution chemistry we first discuss our need for water—the most common liquid and the most common solvent in our environment.

17-1 OUR WATER NEEDS

Water seems to be so abundant that many people take it for granted (Fig. 17-1). Indiscriminate disposal of wastes has polluted many of our lakes, rivers, and streams. Even the oceans are becoming so contaminated that they can no longer support some forms of marine life to the extent that they once could.

All life depends on water. Although we can survive for weeks without food, we cannot exist for more than a few days without water. Water is essential for such bodily processes as digestion, transporting nutrients and oxygen, removing carbon dioxide, and regulating body temperature. Our bodies are almost two thirds water by weight, although the water is not evenly

Figure 17–1 We are familiar with water in its three states—solid, liquid, and gas. (Tony Stone Worldwide/Tom Dietrick)

Table 17-1   WATER CONTENT OF SELECTED FOODS

Food	Water (%)
Wheat flakes	4.0
White bread	35.0
Apple pie	48.0
Hamburger	54.0
Tuna	61.0
Spaghetti	72.0
Eggs	74.0
Apples	85.0
Oranges	86.0
Carrots	88.0
Lettuce	95.0

Table 17-3   THE EARTH'S WATER

Source	Percentage of Total
Oceans	97.20
Ice caps	2.15
Subsurface	0.63
Atmosphere	0.01
Fresh-water lakes	0.009
Saline lakes	0.008
Streams	0.0001

distributed. Blood is 90% water; muscle, 75%; and bones, 20%. A loss of only 12% of the body's water content can be fatal.

Water makes up a large part of the food we eat. An average person obtains nearly half of the daily requirement of water in this manner. Bread is about one third water, and some vegetables contain as much as 95%. Table 17-1 gives the water content of selected foods.

Huge quantities of water are used in food production: one egg, 130 gallons; a quart of milk, 1000 gallons; a pound of beef, 4000 gallons. Table 17-2 indicates how important water is in maintaining our standard of living.

Seawater accounts for more than 97% of the water on earth, as seen in Table 17-3. If it could be desalted at a reasonable cost, an almost unlimited supply of usable water would be available. At present, hundreds of desalination plants are in operation around the world. With the development of better technological methods, the number should increase considerably. Table 17-4 shows some of the uses of water around the home.

17-2 SOLUTIONS

Liquid water is never found in nature as a pure substance. It dissolves so many other substances that it is known as the **universal solvent**. Natural waters are classified according to the proportion of dissolved ionic solids in them, as shown in Table 17-5. The major positive ions dissolved in seawater are sodium (Na^+), magnesium (Mg^{2+}), calcium (Ca^{2+}), potassium (K^+), and strontium (Sr^{2+}). The major negative ions are chloride (Cl^-), sulfate (SO_4^{2-}), bicarbonate (HCO_3^-), bromide (Br^-), fluoride (F^-), and iodide (I^-) (Fig. 17-2).

Table 17-2   WATER USED IN THE PRODUCTION OF VARIOUS ITEMS

Item	Water Used (gallons)
A Sunday newspaper	200
A loaf of bread	300
Gasoline (20 gal)	400
A ton of paper	43,000
An automobile	50,000
A ton of steel	88,000
A ton of aluminum	350,000
A ton of synthetic rubber	600,000

Table 17-4   DOMESTIC WATER USE

Nature of Use	Water Used (average in gallons)
Flushing a toilet	4
Washing dishes	10
A washing machine load	25
A shower (5 minutes)	25
A tub bath	35
Watering a lawn (1 hour)	300

Table 17–5   NATURAL WATERS CLASSIFIED BY
LEVEL OF DISSOLVED SOLIDS

Class	Dissolved Solids (Parts Per Million)	Examples
Fresh	0–1000	Lake Tahoe; Lake Michigan; Missouri River
Brackish	1000–10,000	Baltic Sea
Salty	10,000–100,000	Oceans
Brine	Above 100,000	Dead Sea; Great Salt Lake

How does sodium chloride dissolve in water? The surface of a sodium chloride crystal consists of positive sodium ions and negative chloride ions that attract polar water molecules (Fig. 17–3). The forces of attraction between the ions and the water dipoles are strong enough to pull the ions away from the crystal. In this process, called **solvation**, each ion becomes surrounded by a number of water molecules and the ions are said to be **solvated**. The solvated ions diffuse through the water by random molecular motion and more of the crystal dissolves. In time, if enough sodium chloride is present, the rate at which ions leave the surface equals the rate at which they return to the crystal, and the solution is **saturated** at that temperature (Fig. 17–4).

This limit, at which no additional salt goes into solution, is called the **solubility** of sodium chloride in water. It is expressed as grams of sodium chloride per 100 grams of water. A saturated solution of sodium chloride contains about 35 g NaCl per 100 g water at room temperature. A saturated solution of silver chloride, by comparison, contains 0.00009 g AgCl per 100 g water (Fig. 17–5).

A **solution** is a **homogeneous mixture**—one that is uniform throughout—of two or more nonreacting substances. In soda pop, a solution of sugar, carbon dioxide, and coloring agents, water is the solvent and the other ingredients the solutes. The component of the solution that is present in the greater amount is usually regarded as the **solvent**, and all the other components the **solutes**. When a solution involves a solid and water, the solid is the solute and the water the solvent.

Solutions that contain a relatively small amount of solute are said to be **dilute**. Solutions that contain a relatively large amount of solute are **concentrated**.

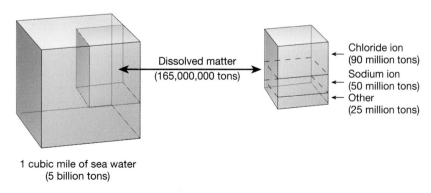

Figure 17–2   In a cubic mile of seawater, weighing nearly 5 billion tons, there are about 165 million tons of dissolved matter, mostly chloride ion (90 million tons) and sodium ion (50 million tons).

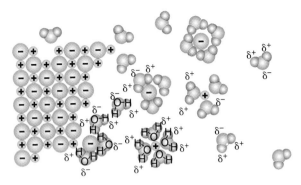

Figure 17–3 Sodium chloride being dissolved in water. The sodium and chloride ions become solvated in the solution.

Since the composition of solutions can vary between certain limits, the law of definite composition does not apply. The terms "soluble," "slightly soluble," and "insoluble" are used to describe the tendency of solutes to dissolve in a particular solvent at a given temperature.

Figure 17–5 Each beaker holds the amount of an ionic compound that will dissolve in 100 grams of water at 100°C. The compounds are *(top row, left to right)* 39 grams of NaCl, 102 grams of $K_2Cr_2O_7$, 341 grams of $NiSO_4 \cdot 6H_2O$; *(bottom row, left to right)* 79 grams of K_2CRO_4, 191 grams of $CoCl_2 \cdot 6H_2O$, and 230 grams of $CuSO_4 \cdot 5H_2O$. (Robert W. Metz)

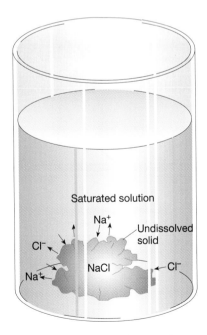

Figure 17–4 In a saturated solution, a dynamic equilibrium is established between the undissolved matter and the solution.

Figure 17–6 The nonpolar molecules in oil do not attract polar water molecules, so oil is insoluble in water. Oil is less dense than water and floats on water. (Kip Peticolas/Fundamental Photographs)

Table 17–6 RELATIVE SOLUBILITIES

Substance	Structure	Solubility in Water (Polar Solvent)	Solubility in Benzene (Nonpolar Solvent)
Sodium chloride	Ionic	Soluble	Slight soluble
Hydrogen chloride	Polar	Soluble	Slightly soluble
Sucrose (sugar)	Polar	Soluble	Slightly soluble
Chlorine	Nonpolar	Slightly soluble	Soluble
Iodine	Nonpolar	Slightly soluble	Soluble
Carbon disulfide	Nonpolar	Slightly soluble	Soluble

It is useful to remember the general rule "**like dissolves like**" for predicting solubilities. Solvents that have polar molecules, such as water, dissolve both solutes with polar molecules and solutes that are ionic, since the polar molecules and ions exert mutual attractions. Solvents having nonpolar molecules dissolve solutes with nonpolar molecules. Thus oils, which consist of nonpolar hydrocarbons, dissolve in nonpolar cleaning solvents but not in a polar solvent such as water (Fig. 17–6). Sodium chloride (ionic) is insoluble in liquid gasoline (nonpolar) but soluble in water. Table 17–6 gives the relative solubilities of several substances in a polar solvent, water, and in a nonpolar solvent, benzene.

The three states of matter—solid, liquid, and gas—can be combined in a variety of ways to form solutions. Air is a mixture of gases that are soluble in one another, so it is properly called a solution. Liquid mercury dissolves in silver and gold to form solutions called amalgams. Brass is a solid solution of zinc and copper, an **alloy**. Silver and gold can be alloyed in all proportions. Table 17–7 gives examples of possible types of solutions.

17–3 SUSPENSIONS AND COLLOIDAL SYSTEMS

When clay is mixed with water, the particles of clay slowly settle to the bottom. The particles of clay consist of groups of molecules. A mixture of this type is called a **suspension**. There is no clear line of demarcation between a suspension and a solution. A mixture is a suspension if it contains particles large enough to be visible with a high-powered optical microscope. Such particles have a diameter of 10^{-4} cm or greater. A mixture is considered to be a solution if the particles are about the size of ordinary molecules, which have a diameter of approximately 10^{-7} cm. Solutions and suspensions thus stand at the limits of a continuum with respect to particle size, as Table 17–8 indicates.

Mixtures containing groupings of molecules intermediate in size between molecules and the particles in suspensions are called colloidal suspensions, colloidal dispersions, or simply **colloids** (Fig. 17–7). Milk and blood are common examples. Although colloids appear to be homogeneous, individual particles can be seen with an optical or electron microscope and

Table 17–7 TYPES OF SOLUTIONS

	Solid	Liquid	Gas
Solid	Zinc in copper (brass; alloys)	Salt in water	—
Liquid	Mercury in gold (amalgams)	Alcohol in water	—
Gas	Hydrogen in palladium	CO_2 in water (carbonated beverages)	Oxygen in nitrogen (air)

Table 17–8 PARTICLE SIZE OF MIXTURES

	Suspensions →					← Colloids →		← Solutions	
Centimeters	10^0	10^{-1}	10^{-2}	10^{-3}	10^{-4}	10^{-5}	10^{-6}	10^{-7}	10^{-8}
Ångstrom units	10^8	10^7	10^6	10^5	10^4	10^3	10^2	10^1	10^0

Figure 17–7   Some edible colloids. (Charles Steele)

Figure 17–8   The Tyndall effect. The test tube on the left contains a true solution; the one on the right, a colloidal dispersion of starch. A laser beam from the apparatus at the far left passes through both tubes. The beam is invisible in the true solution, but the colloidal starch particles scatter the beam and make its path visible.(Beverly March)

can be detected with a light beam through the **Tyndall effect**, the scattering of light by colloidal particles, named after John Tyndall (1820—1893) (Fig. 17–8). Colloids are frequently optically opaque, whereas aqueous solutions are always transparent. The colloidal particles in a colloidal system are the **dispersed phase**, and the medium in which they occur, the **dispersing medium**. The possible types of colloidal systems are summarized in Table 17–9.

17–4 CONCENTRATION UNITS OF SOLUTIONS (MOLARITY AND MOLALITY)

Concentration refers to any measure of the amount of solute in a given amount of solvent or solution. It may be regarded as a measure of the crowdedness of the particles of solute (Fig. 17–9). The terms "concentrated" and "dilute" are nonquantitative descriptions of crowdedness. Two quantitative means of expressing concentrations are molarity and molality, both based on the mole concept. Chemists favor concentration units based on moles, since they interpret chemical phenomena in terms of numbers of atoms, molecules, and ions.

The **molarity** (**M**) of a solution is the number of moles or gram-formula masses of a solute (n) per liter of solution. Therefore,

$$\text{Molarity} = \frac{\text{Moles of solute}}{\text{Liters of solution}} \qquad (17\text{--}1)$$

$$M = \frac{n \text{ (moles of solute)}}{V \text{ (volume)}}$$

EXAMPLE 17–1

What is the concentration in molarity of 1.00 liter of an antifreeze solution prepared by dissolving 124 grams of ethylene glycol ($C_2H_6O_2$) in water?

Table 17–9   COLLOIDAL SYSTEMS

Dispersed Phase	Dispersing Medium	Example
Solid	Solid	Colored gems, pearls
Solid	Liquid	Paints, jellies
Solid	Gas	Dust, smoke, smog
Liquid	Solid	Butter, cheese
Liquid	Liquid	Milk, cream, mayonnaise
Liquid	Gas	Clouds, mist, sprays
Gas	Solid	Marshmallow, floating soap
Gas	Liquid	Whipped cream, suds, foams

SOLUTION

To express the amount of solute in moles, we determine the molar mass of $C_2H_6O_2$ and, by dimensional analysis, convert 124 g to moles of $C_2H_6O_2$. Then we substitute in the equation for molarity, and solve for M.

Volume = 1.00 liter

$$124 \text{ g } C_2H_6O_2 \times \frac{\text{mole}}{62.0 \text{ g}} = 2.00 \text{ moles } C_2H_6O_2$$

$$2C = 2 \times 12.0 = 24.0 \text{ u}$$

$$6H = 6 \times 1.0 = 6.0 \text{ u}$$

$$2O = 2 \times 16.0 = \underline{32.0} \text{ u}$$

$$\text{Molar mass}_{C_2H_6O_2} = 62.0 \text{ u}$$

Molarity = ?

$$\text{Molarity} = \frac{\text{moles of solute}}{\text{liters of solution}}$$

$$= \frac{2.00 \text{ moles } C_2H_6O_2}{1.00 \text{ liter}}$$

$$= \boxed{2.00 \ M}$$

EXTENSION

The concentration of HNO_3 in a sample of rainwater was 2.20×10^{-6} g/L. What is the molarity of the rainwater with respect to the nitric acid? (Answer: $3.50 \times 10^{-8} \ M$)

■

EXAMPLE 17 – 2

A batch of maple syrup has a concentration of 0.250 M in sucrose, its principal constituent. What

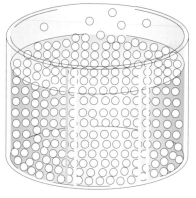

(a)   Pure solvent

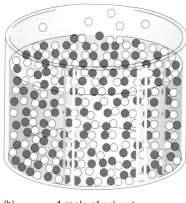

(b)   1 mole of solvent
              1 mole of solute

Figure 17–9   A pure solvent and solution compared at the molecular level.

amount of sucrose ($C_{12}H_{22}O_{11}$) is present in 1.00 liter of this maple syrup?

SOLUTION

Using the molarity relationship, we determine the number of moles of sucrose present. Then we express the number of moles of sucrose in grams.

Molarity = 0.250 M

Volume = 1.00 liter

Moles of sucrose = ?

$12C = 12 \times 12.0 = 144.0$ u

$22H = 22 \times 1.0 = 22.0$ u

$11O = 11 \times 16.0 = \underline{176.0 \text{ u}}$

Molar mass$_{C_{12}H_{22}O_{11}}$ = 342.0 u

$$\text{Molarity} = \frac{\text{moles of solute}}{\text{liters of solution}}$$

Moles of solute = molarity × liters of solution

= 0.25 M × 1.00 liter

= 0.25 mole $C_{12}H_{22}O_{11}$

$$0.250 \text{ mole } C_{12}H_{22}O_{11} \times \frac{342.0 \text{ g}}{\text{mole}}$$

$$= \boxed{85.5 \text{ g sucrose}}$$

EXTENSION

Calculate the number of grams of KOH that must be present in 100 mL of 2.50 M KOH. (Answer: 14.0 g)

■

Solutions of known molarity are prepared by weighing the solute and making up the solution in calibrated volumetric glassware (Fig. 17–10). For this reason, molarity units are convenient for labora-

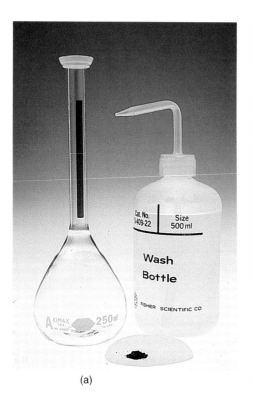

(a)   (b)   (c)

Figure 17–10 To prepare 0.250 L of a 0.0100-M solution of KMnO$_4$, 0.395 g of KMnO$_4$ is placed in a volumetric flask and dissolved in a small amount of water (b). When it is completely dissolved, sufficient water is added to fill the 0.250-L flask to the mark (c). (Charles D. Winters)

tory work. However, 1-M solutions of different solutes contain different amounts of solvent, since the amount of solvent needed to make up a given volume of solution depends on the nature of the solute.

Another way of expressing the concentration of solutions is molality, in which the amount of solvent is the same. The **molality** (m) of a solution is the number of moles of a solute in 1 kilogram of solvent.

$$\text{Molality} = \frac{\text{Moles of solute}}{\text{Kilograms of solvent}} \qquad (17\text{--}2)$$

A 1-m (read "one molal") solution of ethanol (C_2H_5OH, molar mass 46) contains 1 mole of ethanol (46 g) in 1 kilogram of water. Each kilogram of a given solvent contains the same number of moles or molecules of that solvent—for example, 1000 g/18.0 g/mole, or 55.5 moles, for water. So molality indicates the relative number of moles of both solute and solvent. Thus, a 1-m aqueous solution of ethanol consists of 1 mole of ethanol (6.02×10^{23} ethanol molecules) and 55.5 moles of water ($55.5 \times 6.02 \times 10^{23}$ water molecules). Further, since temperature has no effect on mass, molality does not change with temperature, as does molarity.

EXAMPLE 17–3

How many grams of ethanol (C_2H_5OH) must be dissolved in 500 g of water to make a 3.00-m solution?

SOLUTION
Using Equation 17–2, we calculate the number of moles of ethanol required and convert by dimensional analysis to grams.

$$\text{Molality} = 3.00 \ m$$

$$500 \text{ g } H_2O \times \frac{\text{kg}}{10^3 \text{ g}} = 0.500 \text{ kg } H_2O$$

$$\text{Moles } C_2H_5OH = \ ?$$

$$2 \text{ C} = 2 \times 12.0 = 24.0 \text{ u}$$
$$6 \text{H} = 6 \times \ 1.0 = \ 6.0 \text{ u}$$
$$1 \text{O} = 1 \times 16.0 = \underline{16.0} \text{ u}$$
$$\text{Molar mass}_{C_2H_5OH} = 46.0 \text{ u}$$

$$\text{Molality} = \frac{\text{moles of solute}}{\text{kilograms of solvent}}$$

$$\text{Moles of solute} = m \times \text{kg}$$
$$= 3.00 \ m \times 0.500 \text{ kg}$$
$$= 1.50 \text{ moles } C_2H_5OH$$

$$1.50 \text{ moles } C_2H_5OH \times \frac{46.0 \text{ g}}{\text{mole}}$$

$$= \boxed{69.0 \text{ g } C_2H_5OH}$$

EXTENSION
If 45.0 g of glucose ($C_6H_{12}O_6$) are dissolved in 500.0 g of water, what is the molality of the solution? (Answer: 0.500 m)

■

17–5 THE ARRHENIUS THEORY OF IONIZATION

The electrical conductivity of solutions can be tested with a simple apparatus (Fig. 17–11). The brightness of the light bulb is a measure of the conducting ability

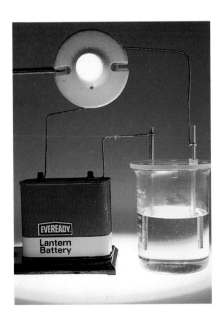

Figure 17–11   Apparatus for determining the electrical conductivity of solutions. The lamp glows brightly when the electrodes are placed in a strong electrolyte and glows dimly in a weak electrolyte. (Marna G. Clarke)

of the solution. When the electrodes are placed in distilled water, the bulb does not glow, but when they are placed in salt water, the bulb glows brightly.

A solution that conducts electricity is an **electrolyte**. This name is also given to solutes whose solutions are conducting. Acids, bases, and salts are examples. If the electrical conductivity is good, the substance is a strong electrolyte; if it conducts the current poorly, it is a weak electrolyte. Table 17–10 lists some typical electrolytes. Solutes whose solutions are nonconducting are **nonelectrolytes.** Electrolytes occur in nature mostly in seawater and to a lesser extent in fresh water. Most organic substances, such as alcohols, sugars, and starches, are nonelectrolytes.

Svante August Arrhenius (1859–1927) proposed a theory to explain the properties of aqueous solutions of electrolytes in his doctoral thesis at the University of Uppsala, Sweden. Although his **theory of ionization** met with strong opposition at first, it was to revolutionize chemistry and win a Nobel Prize for Arrhenius. Its major points are as follows:

1. Solutions of electrolytes contain ions.
2. Electrolytes dissociate or separate into ions in solvents.
3. The electrical conductivity of solutions of electrolytes is due to the presence of ions.
4. The degree of conductivity depends on the extent of dissociation of the electrolyte.

According to the Arrhenius theory, solutions of strong electrolytes are nearly completely ionized in dilute aqueous solutions. All ionic compounds are strong electrolytes, as are some covalent compounds that react completely with the solvent to produce ions. A sodium chloride solution is actually a solution of sodium and chloride ions; there are no "molecules" of sodium chloride present.

$$NaCl \longrightarrow Na^+ + Cl^- \qquad (17–3)$$

In water solution, the covalent compound HCl is also a strong electrolyte; measurements show that a 0.1-M solution of HCl is 92% dissociated at room temperature. Arrhenius assumed that most HCl molecules dissociate into hydrogen and chloride ions, which are in equilibrium with undissociated molecules:

$$\underset{\text{undissociated molecules}}{HCl} \;\rightleftharpoons\; \underset{\text{ions}}{H^+} + \underset{\text{ions}}{Cl^-} \qquad (17–4)$$

Table 17–10 SOME TYPICAL ELECTROLYTES

Strong Electrolytes

Hydrochloric acid	HCl
Hydrobromic acid	HBr
Nitric acid	HNO_3
Chloric acid	$HClO_3$
Sulfuric acid	H_2SO_4
Sodium hydroxide	NaOH
Potassium hydroxide	KOH
Calcium hydroxide	$Ca(OH)_2$
Barium hydroxide	$Ba(OH)_2$
Magnesium hydroxide	$Mg(OH)_2$

Weak Electrolytes

Sulfurous acid	H_2SO_3
Phosphoric acid	H_3PO_4
Nitrous acid	HNO_2
Acetic acid	$HC_2H_3O_2$
Carbonic acid	H_2CO_3
Boric acid	H_3BO_3
Hydrogen sulfide	H_2S
Aqueous ammonia	NH_3

Since a hydrogen ion, H^+, is a proton and cannot exist isolated in water, we now believe that the reaction of HCl in water involves reaction with water molecules in which hydronium ions, H_3O^+, or H^+ (H_2O), are formed:

$$HCl + H_2O \;\rightleftharpoons\; H_3O^+ + Cl^- \qquad (17–5)$$

17–6 COLLIGATIVE PROPERTIES

Properties of solutions that depend only on the number of particles present and not on their chemical nature are called **colligative properties**. The vapor pressure, freezing point, and boiling point are examples.

The **vapor pressure** of a liquid or solid is the pressure at which the substance in the gaseous state is in equilibrium with the liquid or solid at a given temperature (Fig. 17–12). The number of molecules in the space above the liquid increases until there are as many molecules returning to the liquid surface as are escaping from it. Each substance has a characteristic vapor pressure at that temperature. The **boiling point** of a substance is the temperature at which the

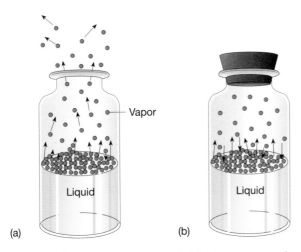

Figure 17–12 Vapor pressure. (a) In the open vessel, the molecules escape as the liquid evaporates. (b) In the closed vessel, an equilibrium is established between the escaping and returning molecules.

vapor pressure of its liquid phase equals the atmospheric pressure. For pure water, the boiling point is 100°C when the atmospheric pressure is 760 torr (1 atmosphere).

When the temperature of an aqueous solution of sugar is 100°C, the vapor pressure of the water is less than 760 torr. Because of the presence of sugar molecules, the concentration of water molecules is less in the solution than it is in pure water. In order for the solution to boil, the sugar solution must be heated to a temperature *higher* than 100°C before the vapor pressure reaches 760 torr. A surprising result is that 1 mole of *any nonelectrolyte* produces the same vapor-

pressure lowering in a given solvent and, therefore, the same boiling-point elevation. Therefore, water solutions of nonelectrolytes of the same concentration boil at the same temperature.

François Marie Raoult (1830–1901), a physicist and chemist, showed that the lowering of the vapor pressure is proportional to the molal concentration (the relative number of particles) of the solute. It follows from **Raoult's law** that the boiling point elevation of a solution, as compared with the pure solvent, is proportional to the molal concentration of the solute:

$$\Delta T_{BP} = \text{Molality} \times K_b \quad (17-6)$$

where ΔT_{BP} is the change in the boiling point and K_b is a constant for a particular solvent, the **molal boiling-point elevation constant**. For water, this constant is 0.51°C. Thus, a 1-*m* water solution of sugar boils at 100.51°C. The constants have been determined for common solvents, as shown in Table 17–11. In dilute solution, also, the lowering of the freezing point is proportional to the molal concentration of the solute:

$$\Delta T_{FP} = \text{Molality} \times K_f \quad (17-7)$$

where ΔT_{FP} is the change in the freezing point and K_f is the **molal freezing-point depression constant** for the solvent: 1.86°C for water.

Freezing-point depression is important to automobile owners in colder regions. To prevent the freezing of the water in an automobile radiator as soon as the temperature reaches 0°C, they add a solution called an **antifreeze**. The active ingredient of "permanent" antifreezes is often ethylene glycol ($C_2H_6O_2$), which also serves as a coolant in the sum-

Table 17–11 SOME MOLAL FREEZING- AND BOILING-POINT CONSTANTS (°C)

Solvent	Freezing Point (FP)(°C)	Molal Freezing-Point Depression Constant (K_f)(°C/m)	Boiling point (BP)(°C)	Molal Boiling-Point Elevation Constant (K_b)(°C/m)
Acetic acid	16.7	3.9	118.5	3.07
Benzene	5.50	5.12	80.1	2.53
Camphor	180	40	208.3	6.0
Ethyl alcohol	−115	1.99	78.4	1.19
Water	0	1.86	100.00	0.51

Table 17–12 FREEZING POINT OF WATER—
ETHYLENE GLYCOL MIXTURES

Percent (by Weight) of Ethylene Glycol	Freezing Point of Mixture (°C)	(°F)
0	0	32
10	−3.5	25.6
20	−8	17
30	−15	5
40	−24	−11
50	−36	−32

Figure 17–14 Ice cream would not harden without the addition of salt to an ice–water mixture in an ice cream freezer, which lowers the temperature of the mixture sufficiently. (Dick Frank/The Stock Market)

mertime, since it causes an elevation in the boiling point of water (Fig. 17–13). Table 17–12 gives the freezing point of water–ethylene glycol mixtures. Another practical application of this principle is the spreading of salt for the melting of ice on highways or walks. Further, salt is added to an ice–water mixture when making ice cream at home in an ice cream freezer to lower the temperature of the ice mixture sufficiently to freeze the ice cream. Otherwise, the ice cream will not harden (Fig. 17–14).

Figure 17–13 Ethylene glycol ($C_2H_6O_2$), the major component of "permanent" antifreeze, protects the water in an automobile radiator against both freezing and boil-over. The colorless glycol is made green and translucent by small amounts of corrosion inhibitors and antioxidants. (Union Carbide Corporation)

EXAMPLE 17–4

What must the concentration of an antifreeze solution be (in molality) in order to lower the freezing point of a kilogram of water to −20.0°C?

SOLUTION
From Raoult's law, we know that the change in the freezing point is proportional to the molality.

$$\Delta T_{FP} = 20.0°C \qquad \Delta T_{FP} = m \times K_f$$
$$K_f = 1.86°C$$
$$m = ? \qquad \therefore m = \frac{\Delta T_{FP}}{K_f}$$
$$= \frac{20.0°C}{1.86°C}$$
$$= \boxed{10.8\ m}$$

EXTENSION
A solution contains 513 g of sucrose ($C_{12}H_{22}O_{11}$) per kilogram of water. At what temperature would the solution freeze? (Answer: −2.79°C)

■

Figure 17–15 Citrus fruits contain sharp-tasting citric acid, $H_3C_6H_5O_7$. (Marna G. Clarke)

Figure 17–16 Among the acids found in fruit juices are quinic acid in cranberries, malic acid in apples, and tartaric acid in grapes. (Charles D. Winters)

17–7 ACIDS AND BASES

Acids and bases have been known for several hundred years and were characterized on the basis of observed properties. According to Robert Boyle and others, **acids** are compounds that, in water solution, taste sour (*acidus,* meaning "sour"), turn the color of the natural dye litmus from blue to red, neutralize bases, and react with certain metals to give free hydrogen. Although many acids, such as sulfuric and nitric, have corrosive properties, others, such as citric and tartaric acids, are common constituents of our food (Figs. 17–15 and 17–16). According to the **Arrhenius theory**, **an acid is a substance that in water solution increases the concentration of hydrogen (hydronium) ions**. Although more recent theories define acids differently, the Arrhenius theory is still applicable to common acids in water solution.

The strength of acids depends on their degree of dissociation. An acid dissociates into ions in a reversible reaction as follows:

$$HA \rightleftharpoons H^+ + A^- \qquad (17–8)$$

undissociated molecules   ions

Applying the law of chemical equilibrium to this reaction, we get

$$K_a = \frac{[H^+][A^-]}{[HA]} \qquad (17–9)$$

in which K_a is the **acid dissociation constant**. If an acid dissociates extensively, the hydronium ion concentration and the acid dissociation constant are large, and the acid is a **strong acid**. If dissociation is limited, the acid exists largely as undissociated molecules at equilibrium; K_a is small, and the acid is a **weak acid**. Table 17–13 gives some acid constants.

According to the Arrhenius theory of ionization, **a base is a substance that increases the hydroxide ion concentration of water**. When dissolved in

Table 17–13 DISSOCIATION CONSTANTS OF SOME ACIDS

Acid	Reaction	K_a (0.1 M at 25°C)	Percent Dissociated	Strength
Acetic	$HC_2H_3O_2 \rightleftharpoons H^+ + C_2H_3O_2^-$	1.8×10^{-5}	1.3	Weak
Nitrous	$HNO_2 \rightleftharpoons H^+ + NO_2^-$	4.5×10^{-4}	1.5	Weak
Sulfurous	$H_2SO_3 \rightleftharpoons H^+ + HSO_3^-$	1.3×10^{-2}	20	Moderate
Phosphoric	$H_3PO_4 \rightleftharpoons H^+ + H_2PO_4^-$	7.5×10^{-3}	27	Moderate
Sulfuric	$H_2SO_4 \rightleftharpoons H^+ + HSO_4^-$	$\sim 10^3$	61	Strong
Hydrochloric	$HCl \rightleftharpoons H^+ + Cl^-$	$\sim 10^3$	92	Strong

Figure 17–17 Most common household cleaning materials are basic. (Marna G. Clarke)

water, bases have a bitter taste and soapy feel, turn the color of litmus from red to blue, and neutralize acids. Some common bases are NaOH, KOH, NH_4OH, and $Ca(OH)_2$ (Fig. 17–17). **Neutralization** involves the combination of hydrogen ions with hydroxide ions to form water molecules. The dissociation of ammonium hydroxide in water is represented as follows:

$$NH_3(aq) + H_2O \rightleftharpoons$$
$$NH_4^+(aq) + OH^-(aq) \quad (17\text{–}10)$$

At equilibrium, the dissociation constant is:

$$K_b = \frac{[NH_4^+][OH^-]}{[NH_3]} = 1.8 \times 10^{-5} \quad (17\text{–}11)$$

where K_b is the **basic dissociation constant**. The small value of K_b shows that ammonium hydroxide is a weak base. A **neutralization reaction** is represented as follows:

$$H^+ + OH^- \rightleftharpoons H_2O \quad (17\text{–}12)$$

Water is a weak electrolyte, dissociating as follows:

$$H_2O \rightleftharpoons H^+ + OH^- \quad (17\text{–}13)$$

In pure water and aqueous solutions, the ionization constant for water, K_I, at equilibrium is

$$K_I = \frac{[H^+][OH^-]}{[H_2O]} \quad (17\text{–}14)$$

Since the concentration of undissociated water, $[H_2O]$, can be considered constant in pure water and dilute solutions, it is combined with the constant K_I to give a new constant, K_W, called the dissociation constant or **ion product of water**.

$$K_I[H_2O] = K_W = [H^+][OH^-] \quad (17\text{–}15)$$

Water is neutral, therefore, since all the hydrogen ions and hydroxide ions come from the dissociation of water molecules. For each water molecule that dissociates, one hydrogen ion and one hydroxide ion are formed. So, in pure water, the concentration of hydrogen ions equals the concentration of hydroxide ions. At 25°C, these concentrations are 1.00×10^{-7} mole/liter. Thus,

$$K_W = [H^+][OH^-] = (1.00 \times 10^{-7} \text{ mole/liter})$$
$$\times (1.00 \times 10^{-7} \text{ mole/liter})$$
$$= 1.00 \times 10^{-14} \text{ mole}^2/\text{liter}^2$$
$$(17\text{–}16)$$

Acid solutions have H^+ concentrations greater than 1.00×10^{-7} mole/liter, and basic solutions have OH^- concentrations greater than 1.00×10^{-7} mole/liter, as Table 17–14 shows.

EXAMPLE 17–5

An aqueous solution has a hydrogen-ion concentration of 0.01 mole/liter. (a) What is the concentration of hydroxide ions? (b) Is the solution acidic, basic, or neutral?

SOLUTION
The ion product for water is 1.00×10^{-14} in water solutions. Knowing $[H^+]$, we can solve for $[OH^-]$.

$$[H^+] = 0.01 \frac{\text{mole}}{\text{liter}} = 1.00 \times 10^{-2} \frac{\text{mole}}{\text{liter}}$$
$$K_W = 1.00 \times 10^{-14} \text{ mole}^2/\text{liter}^2 \quad [OH^-] = ?$$

(a) $$K_W = [H^+][OH^-]$$
$$\therefore [OH^-] = \frac{K_W}{[H^+]}$$
$$= \frac{1.00 \times 10^{-14} \text{ mole}^2/\text{liter}^2}{1.00 \times 10^{-2} \text{ mole/liter}}$$
$$= 1.00 \times 10^{-12} \text{ mole/liter}$$

(b) The solution is acidic, since $[H^+] = 1 \times 10^{-2}$ mole/liter, which is greater than 1×10^{-7} mole/liter.

EXTENSION
The concentration of hydroxide ions in a solution is 1.00×10^{-5} M. Calculate the $[H^+]$. (Answer: 1.00×10^{-9} M).

Table 17–14 RELATIONSHIP BETWEEN [H$^+$] AND
[OH$^-$] IN AQUEOUS SOLUTIONS

[H$^+$] (Moles/Liter)	[OH$^-$] (Moles/Liter)	[H$^-$][OH$^-$]	Character
$1 = 10^0$	10^{-14}	10^{-14}	Acidic
10^{-1}	10^{-13}	10^{-14}	Acidic
10^{-2}	10^{-12}	10^{-14}	Acidic
10^{-3}	10^{-11}	10^{-14}	Acidic
10^{-4}	10^{-10}	10^{-14}	Acidic
10^{-5}	10^{-9}	10^{-14}	Acidic
10^{-6}	10^{-8}	10^{-14}	Acidic
10^{-7}	10^{-7}	10^{-14}	Neutral
10^{-8}	10^{-6}	10^{-14}	Basic
10^{-9}	10^{-5}	10^{-14}	Basic
10^{-10}	10^{-4}	10^{-14}	Basic
10^{-11}	10^{-3}	10^{-14}	Basic
10^{-12}	10^{-2}	10^{-14}	Basic
10^{-13}	10^{-1}	10^{-14}	Basic
10^{-14}	$1 = 10^0$	10^{-14}	Basic

17–8 THE pH SCALE

The acidities of water solutions are expressed on a special scale. The symbol pH, introduced by the biochemist Sven P. L. Sörensen (1868–1939), was intended originally as an abbreviation for the phrase "potential of hydrogen ion." The **pH** scale has the advantage of expressing small concentrations of H$_3$O$^+$ ions more conveniently than a decimal scale and also eliminates negative powers of 10. A solution that is neutral has a pH of 7. Acidic solutions have pH values less than 7, and basic solutions pH values greater than 7 (Fig. 17–18).

The solutions of our common experience cover a considerable range of pH values (Fig. 17–19). The water content of the soil may cause it to have a value anywhere from pH 3 (very acid) to pH 10 (very alkaline), although these represent extremes; most agriculture proceeds in soil of a pH close to 6. The pH values of soil determine the solubility and thus the availability of necessary minerals to crops. Soil bacteria are also sensitive to variations in pH. Seawater has an average pH of 8. Many marine organisms die below pH 7.5; their eggs are particularly vulnerable.

The pH values of body fluids must be maintained within certain narrow limits. The normal variation of the pH of blood is from 7.30 to 7.50. If the pH goes below 7.00 or above 7.80, death follows. Enzymes are also sensitive to changes in pH. The pH values of most foods range between 2 and 8; however, an acid food may end up as an alkaline residue when metabolized in the body, and vice versa.

Rainwater is normally slightly acidic, with a pH of about 5.7, as a result of passing through atmospheric carbon dioxide. Wherever the air is polluted with

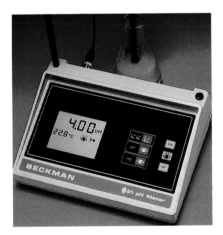

Figure 17–18 A pH meter is an electronic instrument that detects H$_3$O$^+$ concentrations and displays them as digital pH readings. The pH of this 0.10-M H$_2$S solution is 4.00 at 22.8°C. (Beckman Instruments)

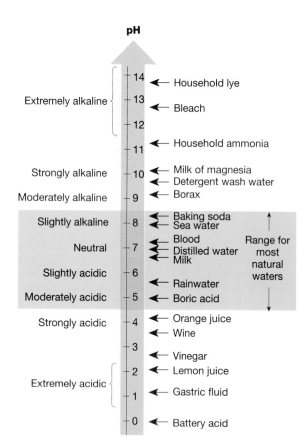

Figure 17–19 The pH scale and pH values of some common solutions.

oxides of sulfur and nitrogen, rainfall also produces sulfuric and nitric acids (Fig. 17–20). The pH of these **acid rains** may be as low as 3, acidic enough to corrode structures and threaten life by lowering the pH of forests, fields, lakes, and ponds.

17–9 ACID RAIN AND THE ENVIRONMENT

Acid rain that has a pH less than 5.6 became a major ecological issue in the 1980s. *Natural processes* such as volcanic eruptions, forest fires, and bacterial decomposition of organic matter produce damaging acidic sulfur and nitrogen compounds. The current problem, however, appears to be caused by the *burning of fossil fuels* by power plants, factories, and smelting operations, and, to a lesser extent, auto emissions.

Tall smokestacks vent their fumes, sulfur dioxide, nitrogen oxides, and traces of such toxic metals as mercury and cadmium hundreds of feet into the atmosphere (Fig. 17–21). Chemical reactions with

Figure 17–20 The brown haze above this city is caused by the pollutant NO_2. Rainfall converts it to nitric acid, a component of acid rain. (National Center for Atmospheric Research)

Figure 17–21 The burning of sulfur in high-sulfur coals forms harmful SO_2 and SO_3 gases, just as in this oxidation of sulfur carried out in a spoon lowered into an air-filled bottle. (Charles D. Winters)

Figure 17–22   The evergreens atop Camel's Hump in Vermont's Green Mountain range show the effects of acid rain. The photo on the left was taken 15 years earlier than that on the right. (U. S. Environmental Protection Agency)

water vapor in the atmosphere follow, producing dilute solutions of sulfuric acid and nitric acid. The *acidic "depositions"* can come down in almost any form: rain, snow, hail, fog, or even dry particles.

In the soil, acid rain breaks down minerals containing calcium, potassium, and aluminum, depriving plants of nutrients. Spruce and sugar maple stands in Vermont and beech and spruce forests in West Germany have shown declines, and Scandinavian scientists claim a 15% reduction in timber growth due to acid rain (Fig. 17–22). Successive rainfalls make the water increasingly acidic, causing lakes and rivers to turn clear and bluish, unable to support any but the most primitive forms of life.

In some regions, *alkaline soils and rocks* buffer the effects of damaging rain and help *neutralize the acidic water.* Much of New York, New England, eastern Canada, and Scandinavia, however, is covered with thin rocky topsoil and is vulnerable to acid rain (Fig. 17–23). In New York's Adirondack Mountains, more than 200 lakes and ponds are acidic and lifeless. In New England, acid rain has killed aquatic life in at least one tenth of the 200 largest lakes. On Cape Cod, eight of the top ten fishing ponds are too acidic for young trout to survive, and fishery biologists have stopped restocking them. In Ontario, thousands of lakes have become so acidified that they can no longer support trout and bass. A number of rivers in Nova Scotia, used as spawning grounds by Atlantic salmon,

no longer teem with fish. The problem is spreading, with damage reported in Florida, Wisconsin, Minnesota, and California. Acid rain has become, in fact, a global phenomenon. Canada may lose close to 50,000 lakes by the end of this century.

Acid rain can corrode stones statues, limestone buildings, and metal rooftops (Fig. 17–24). Athens' Parthenon and Rome's Colosseum have deteriorated severely in the past 20 years; the prime suspect is acid rain. The corrosive assault on buildings and water systems costs millions of dollars annually to correct.

Figure 17–23   Calcium carbonate (limestone) is dispersed over forests in an effort to neutralize the effects of acid rain. (Courtesy of Ohio Edison)

Figure 17–24   This bronze statue has been damaged by acid rain. (Hans Wolf/The Image Bank)

The contamination of public drinking water poses a threat to human health. Canadian industry and the winds send about half a million tons of sulfur emissions south to the United States every year. But 2 million tons annually blow north across the border from the United States, mostly from the midwest. Norway claims that its lakes are being killed by smoke emissions from the Ruhr and Saar valleys.

If nothing is done about acid rain now, the environmental costs in the future could be nightmarish. Congress considered amending the 1970 Clean Air Act requiring, in part, that industries in states east of the Mississippi reduce sulfur dioxide emissions by 8 million tons (35%) over ten years. The Environmental Protection Agency requires the installation of "scrubbers" to remove 90% of sulfur emissions at all new coal-fired power plants. Another measure that has been considered is the *"washing" of high-sulfur coal.* In this process, coal is crushed and pyrite, an iron-sulfur compound, is separated out. Ash, dirt, and rock are removed at the same time, making coal more economical to ship and less damaging to boilers.

Above all, *basic research* is necessary to fill in major gaps in our understanding of the issues, some of which are the pH of precipitation in the absence of human activity; evidence of a change in the acidity of precipitation over the past several decades; the mechanisms in the production of sulfur and nitrogen acids; and the relationship between the locations of the sources and the locations of the depositions.

17–10 WATER HARDNESS AND SOAP

Water contains dissolved impurities in the form of metal ions. It is not necessary to remove these for drinking purposes, but they may make washing difficult and cause scale to be deposited in pipes and eventually clog them. The chief problems involve calcium (Ca^{2+}) and magnesium (Mg^{2+}) ions, along with ferrous (Fe^{2+}) ions. Water containing these ions is said to be **hard**. In **temporary hardness**, water contains bicarbonate ions (HCO_3^-) as well and can be **softened**; that is, **the objectionable ions can be removed**, by heating, causing calcium carbonate to precipitate out of a solution. When chloride (Cl^-) and sulfate (SO_4^{2-}) ions are present with Ca^{2+} and Mg^{2+}, the water is said to have **permanent hardness** but may be softened by other processes: ion-exchange minerals (zeolites), in which the offending ions are exchanged for a nonoffending ion such as Na^+ in a solid held in a column, ion-exchange resins, and the addition of washing soda (Na_2CO_3). In the last case, Ca^{2+} and Mg^{2+} ions react with CO_3^{2-} ions and form insoluble precipitates of calcium and magnesium carbonates.

Ordinary **soap** is a mixture of the sodium compounds (salts) of various organic acids obtained from natural fats or oils. Perfumes, antiseptics, and other ingredients are added to produce toilet soaps and deodorizing soaps. A **floating soap** is produced if air is blown through molten soap. Potassium soaps produce a softer lather and are more soluble in water and so are used in liquid soaps and shaving creams. A **soap molecule** such as sodium stearate consists of a large nonpolar hydrocarbon portion (**hydrophobic**—repelled by water) and an ionic end (**hydrophilic**—water-soluble).

When soap is added to water, the hydrophilic ends of the molecules are attracted to the water and dissolve in it. The hydrophobic portions are repelled by the water molecules and form a thin film called a **monolayer** on the surface of the water (Fig. 17–25).

The **cleansing action of soap** involves the lowering of the **surface tension** of water, the tendency of the water surface to contract, by the monolayer of soap, and emulsification. Dirt is held to clothes by a

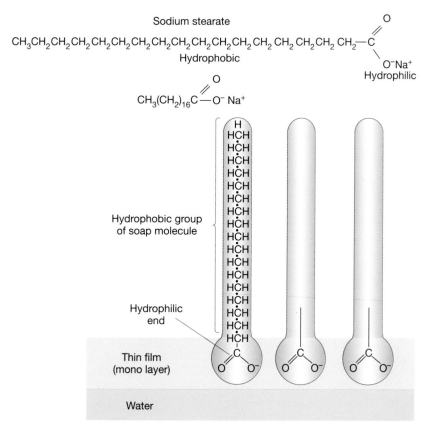

Figure 17–25 A soap molecule has a hydrophilic group and a hydrophobic group. The addition of soap to water causes a thin film, a monolayer, to form on the surface of the water.

thin film of grease or oil. The hydrophobic portions of soap molecules dissolve in the grease or oil, and the hydrophilic ends remain in the water phase. Scrubbing causes the oil or grease to **disperse into tiny droplets**, a process called **emulsification**. The oil or grease droplets become surrounded by soap molecules and are known as **micelles** (Fig. 17–26). Since the ionic ends of the soap molecules project outward, the surface of each drop is negatively charged, and because of mutual repulsion, the drops do not coalesce. The entire micelle becomes water-soluble and can be washed away by a stream of water.

Soap is an excellent cleansing agent, and it is **biodegradable**—that is, it is **capable of being broken down biologically** into harmless end products such as CO_2 and water by microorganisms in the soil or in sewage treatment plants—and relatively

nontoxic. Its main drawback is that in hard water it reacts to form a scum, or precipitate, with the Ca^{2+} and Mg^{2+} ions:

$$2CH_3(CH_2)_{16}\overset{O}{\overset{\|}{C}}-O^-Na^+ + Ca^{2+} \longrightarrow$$

sodium stearate (soluble)

$$[CH_3(CH_2)_{16}\overset{O}{\overset{\|}{C}}-O^-]_2Ca^{2+} + 2Na^+$$

calcium stearate (insoluble)

$$(17-17)$$

This reaction consumes a large amount of the soap, making it unavailable for the washing process and thereby increasing the cost. At the same time, the gummy precipitate remains behind, producing stiffness in clothes, accounting for bathtub ring, and clog-

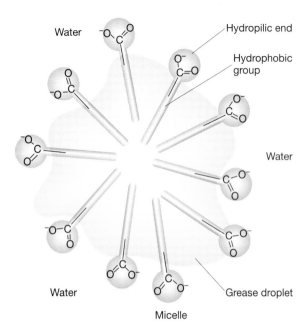

Figure 17–26   Formation of micelles when soap dissolves in grease or oil. The charged micelles are soluble in water and can be washed away.

ging washing machines. The effect of some softening agents is to precipitate the "hard" ions to form harmless granules that settle out, or to exchange the hard ions for Na^+, allowing the use of soap in hard water.

17–11 SYNTHETIC DETERGENTS

The term **detergent** denotes any **cleansing agent**; soaps fall under this broad category (Fig. 17–27). In popular usage, however, the word refers to synthetic detergents, or **syndets**, of which close to a thousand are available commercially. The new detergents avoid the problems of the curd caused by soap and hard water.

Three related developments ushered in the large-scale use of synthetic detergents: *phosphate detergents, synthetic fibers, and automatic washing machines*. The most important single factor was the introduction of sodium tripolyphosphate (STPP), $Na_5P_3O_{10}$, as a **detergent builder**—a component of detergents that softens water and prevents dirt that has already been washed out from being redeposited on clothing. The new automatic washers were redesigned to handle STPP, and synthetic fibers were formulated to be washed by the new detergent in the

Figure 17–27   Various cleaning materials used in the home. (Charles Steele)

new washing machines. As a result, easy-to-wash clothing made from synthetic fibers came into great demand. The more than 40 million washers now in use are designed almost exclusively for detergents. The use of soap, in fact, could seriously damage them.

Society paid a high environmental price when it switched from soap to synthetic detergents (Fig. 17–

Figure 17–28   Saint Martin's Bridge over the Tajo River in Toledo, Spain, was built in the fourteenth century. Non-biodegradable detergents caused the foam that pollutes the river, causing fish kills and other adverse effects. (Paolo Koch/Photo Researchers, Inc.)

$$CH_3-CH-CH_2-CH-CH_2-CH-CH_2-CH-\langle\bigcirc\rangle-SO_3^-Na^+$$
$$\quad\quad\;|\quad\quad\quad\quad|\quad\quad\quad\quad\;|\quad\quad\quad\quad|$$
$$\quad\quad CH_3\quad\quad\;\; CH_3\quad\quad\;\; CH_3\quad\quad\; CH_3$$

a sodium alkyl benzylsulfonate (ABS), not easily biodegradable

$$CH_3-CH_2-CH_2-CH_2-CH_2-CH_2-CH_2-CH_2-CH_2-CH-\langle\bigcirc\rangle-SO_3^-Na^+$$
$$\quad\quad\quad\quad\quad\quad\quad\quad\quad\quad\quad\quad\quad\quad\quad\quad\quad\quad\quad\quad\quad\quad\quad\quad\;|$$
$$\quad\quad\quad\quad\quad\quad\quad\quad\quad\quad\quad\quad\quad\quad\quad\quad\quad\quad\quad\quad\quad\quad\quad\quad CH_3$$

a sodium linear alkyl sulfonate (LAS), more easily biodegradable

28). The new products clogged sewage systems and polluted streams with foam, killing fish and wildlife and even getting into city drinking water. Some of the pollution problem was alleviated by replacing the nonbiodegradable, highly branched-chain **surfactants** (surface-active agents that provide the main source of detergent action), such as alkyl benzylsulfonates (ABS), with the more easily biodegradable but sudsless linear alkyl sulfonates (LAS). The metabolism of microorganisms is adapted to the straight-chain alkyl groups found in soaps, natural fats, and LAS; they cannot readily break down the highly branched-chain ABS syndets; see chemical structure above.

17–12 EUTROPHICATION

No other compounds that have been used as detergent builders have been as effective, safe, and cheap as the phosphates. These very phosphate builders, however, are at the center of the environmental controversy that surrounds detergents. At stake is the survival of lakes and other bodies of water, some of which are succumbing to eutrophication.

Plant nutrients—the elements essential to plant growth—increase in concentration in a lake as streams bring in dissolved compounds of nitrogen, phosphorus, potassium, and sulfur from decaying plant and animal remains. The enrichment of water by nutrients is called **eutrophication** ("well nourished") and leads to the natural aging of lakes. Ordinarily, eutrophication of a lake takes thousands of years. As nutrients build up in a lake, the plant and animal populations increase. Water weeds become abundant, and thick layers of blue-green algae crowd out other organisms. The oxygen supply is depleted, and the water is unable to support desirable species of

fish. When they die, deposits of organic matter build up on the bottom of the lake. Gradually, the lake becomes shallow and evolves into marshland, swamp, and finally dry land.

In the twentieth century, man-made pollution has been the cause of the accelerated eutrophication of Lake Zurich in Switzerland, which has gone from youth to old age in less than a century, and of many other lakes. It is estimated that during the past 50 years, Lake Erie has "aged" about 15,000 years. The catch of blue pike dropped from nearly 20 million pounds in the 1950s to less than 500 pounds in 1965, and the catch of herring and whitefish also went down sharply. During the same period, the phosphorus concentration in Lake Erie tripled, largely because of the expanded use of phosphate builders in detergents.

Most green plants, including algae, require 15 to 20 elements for their growth. Many of these elements are usually present in natural waters in sufficient concentrations for normal plant growth. However, it appears that phosphorus is a **limiting nutrient** for algae in many fresh-water lakes—the element in shortest supply with respect to the nutritional requirements of algae. Nitrogen may be a more limiting nutrient in estuaries and other coastal waters. When nitrogen and phosphorus are present in a lake in greater concentrations than normal, several species of algae grow very rapidly and produce an **algal bloom**. This unsightly, foul-smelling green scum sets off the eutrophication process.

Detergents are heavily loaded with phosphate builders (25 to 60% by weight), the source of the hundreds of millions of pounds of detergent phosphorus poured annually into rivers and lakes from municipal sewage. If eutrophication can be halted only by cutting off a critical nutrient, then phosphorus is the best candidate for reduction. Although synthetic detergents are not the only source of phosphorus in natural water, they are the largest single source. Possible substitutes for phosphates would have to be tested extensively for toxicity and ecological consequences. However, as long as phosphorus compounds continue to be used, the quantities being dumped into waterways should be reduced. Through advanced technology, it may be possible in time to remove phosphates from municipal sewage.

17–13 CLEANING UP WATER POLLUTION

Under the **Clean Water Act of 1970**, federal funding was provided in advance for sewage treatment plants in new, fast-growing industrial and residential areas. Since 1972, the Environmental Protection Agency has provided more than $33 billion in sewage treatment grants for about 22,000 projects. New legislation signed into law in 1981 concentrated on cleaning up dirty water across the United States, providing additional billions for waste water treatment plants and for protecting shorelines from sewage. It also extended the deadline for municipalities to meet the safe levels of treatment by five years, from July 1, 1983, to July 1, 1988.

After decades of abuse, the prognosis in the 1980s for the Great Lakes was that they were becoming cleaner. With 95% of the U.S. supply of fresh water in lakes and reservoirs (and 20% of the world's), they furnish the drinking water for millions of Americans (Fig. 17–29). More than a third of Canada's population also lives in the Great Lakes region. The enforcement of bans and the spending of billions of dollars to improve sewage treatment plants enabled most of the major dischargers into the Great Lakes to meet the tough new water-quality regulations.

An agreement between the United States and Canada lowered the levels of phosphates that municipalities were allowed to dump into the water. The result has been a decrease in the unsightly deposits of suds along rivers and beaches and a slowdown of eutrophication. It is even difficult to find DDT in Lake Erie, except in the sediment. One consequence of these measures is that large game fish are making a comeback. Sports fisherman are pulling in the native lake trout and also coho and chinook salmon from the Pacific, although Canada publishes a guide warning fishermen of which fish are unsafe to eat.

Figure 17–29   A marina on Lake Ontario. Pleasure boating is a source of revenue for communities and a source of pollution—from sewage, garbage, and spilled fuel. (Karen Roeder)

The environmentalists' major concern has turned to *toxic substances* in the lakes, such as mirex (an insecticide) and mercury. The problems of the Great Lakes are not all solved, although a spirit of optimism prevails that a combination of technology and policy can do the job, as it has with the *Thames River in England.*

A favorite garbage dump for centuries, the Thames had become so foul by the seventeenth century that King James had threatened to move his court to Windsor. Events such as the Industrial Revolution, the closing of London's cesspools after the introduction of the flush toilet, a swelling population, and in the 1940s detergents and other chemical effluents left the lower Thames covered with foam choking the river to death. One fish species after another disappeared.

However, pollution control measures implemented in the 1950s and 1960s halting the use of nonbiodegradable detergents and the discharge of industrial chemicals into the river and expanding the construction of municipal and private treatment plants gradually restored the river to the "sweet Thames" of the poet. More than 100 varieties of fish resumed residence there, as well as swans and other birds. The long battle with the Thames could serve as a model for river clean-ups everywhere.

SUMMARY OF EQUATIONS

$$\text{Molarity} = \frac{\text{Moles of solute}}{\text{Liters of solution}}$$

$$M = \frac{n}{V \text{ (volume)}}$$

$$\text{Molality} = \frac{\text{Moles of solute}}{\text{Kilograms of solvent}}$$

$$m = \frac{n}{\text{kg}}$$

Boiling Point Elevation

$$\Delta T_{BP} = \text{Molality} \times K_b$$

Freezing Point Depression

$$\Delta T_{FP} = \text{Molality} \times K_f$$

Dissociation Constant of an Acid

$$K_a = \frac{[H^+][A^-]}{[HA]}$$

Dissociation Constant of Water (Ion Product)

$$K_W = [H^+][OH^-]$$

KEY TERMS

Acid   A substance that, when added to water, tastes sour, turns the color of litmus from blue to red, neutralizes bases, and reacts with certain metals to give free hydrogen.

Acid Dissociation Constant   The ratio of the concentrations of the ionization products of an acid to the concentration of the un-ionized acid.

Acid Rain   Rainwater with an acidic pH, a result of air polluted with oxides of sulfur and nitrogen.

Algal Bloom   An overgrowth of tiny green plants, algae, that produces a foul-smelling scum on the surface of natural waters.

Alloy   A solid solution of two or more metals.

Antifreeze   A substance added to water in an automobile radiator to lower its freezing point.

Arrhenius Theory of Ionization   When certain substances (electrolytes) dissolve in water, they dissociate into charged particles called ions.

Base   A substance that, when added to water, has a bitter taste and soapy feel, turns the color of litmus from red to blue, and neutralizes acids.

Biodegradable   A substance that is capable of being broken down biologically into harmless end products such as carbon dioxide and

water by microorganisms in the soil or in sewage treatment plants.

Boiling Point   The temperature at which the vapor pressure of a liquid equals the pressure of the atmosphere.

Clean Water Act of 1970   Legislation that provided federal funding for sewage treatment plants in industrial and residential areas.

Colligative Properties   Properties of a solution that depend only on the number of particles present and not on their chemical nature. Examples are vapor pressure, boiling point, and freezing point.

Colloid   A mixture containing groupings of molecules intermediate in size between molecules and the particles in a suspension. Milk and blood are examples.

Concentrated   Solutions that contain a relatively large amount of solute.

Concentration   The quantity of solute dissolved in a specific quantity of solvent or solution.

Detergent   Any cleansing agent, such as soap. In popular usage, the word refers to synthetic detergents, or syndets.

Detergent Builder   A component of detergents that softens water and prevents dirt that has already been washed out from being redeposited.

Dilute   Solutions that contain a relatively small amount of solute.

Dispersed Phase   The colloidal particles in a colloid.

Dispersing Medium   The medium in which the colloidal particles in a colloid are dispersed.

Electrolytes   Solutions that conduct electricity. Also, solutes whose solutions are conducting.

Emulsification   The dispersal of oil or grease into tiny droplets.

Eutrophication   The overgrowth of vegetation caused by high concentrations of plant nutrients in bodies of water.

Floating Soap   A soap that is less dense than water owing to air blown through the molten soap.

Hard Water   Water containing a high concentration of calcium and magnesium ions. Soap does not lather in hard water and may form a scum.

Homogeneous Mixture   A mixture that is the same throughout; a solution.

Hydrophilic   Water-soluble.

Hydrophobic   Repelled by water.

Ion Product of Water   The dissociation constant of water.

Limiting Nutrient   The element in shortest supply with respect to the nutritional requirements of an organism.

Micelle   Oil or grease droplets surrounded by soap molecules.

Molal Boiling-Point Elevation Constant   The change in the boiling point of a 1-molal solution of a nonelectrolyte as compared with the pure solvent.

Molal Freezing-Point Depression Constant   The change in the freezing point of a 1-molal solution of a nonelectrolyte as compared with the pure solvent.

Molality, *m*   The number of moles of solute in 1 kilogram of solvent. A measure of the concentration of a solution.

Molarity, *M*   The number of moles of solute per liter of solution. A measure of the concentration of a solution.

Monolayer   A layer or film that is one molecule in thickness.

Neutralization   A reaction between acids and bases in which hydrogen ions from the acid are combined with hydroxide ions from the base to form water molecules.

Nonelectrolyte   Substances whose solutions do not conduct an electric current.

Permanent Hardness of Water   Water that contains chloride and sulfate ions in addition to calcium and magnesium ions.

pH   A measure of the acidity of a solution on a scale of 1 to 14, 7 being neutral, values lower than 7 acidic, and values greater than 7 basic.

Raoult's Law   The lowering of the vapor pressure of a solution is proportional to the molal concentration of the solute.

Saturated Solution   A solution that contains as much solute as can be dissolved in that solution.

Soap   A mixture of the sodium salts of various organic acids obtained from natural fats or oils that serves as a cleansing agent.

Solubility   The limit at which no additional solute goes into solution, often expressed as grams of solute per 100 grams of solvent.

Solute   A substance dissolved in a solvent.

Solution   A homogeneous mixture of two or more nonreacting substances.

Solvation   A process in which the molecules or ions of a solute become surrounded by a number of water molecules in the solution.

Solvent   The medium in which a substance is dissolved.

Strong Acid   An acid that dissociates extensively.

Surface Tension   A tension at the surface of a liquid resulting from forces of attraction acting on molecules close by, causing the surface to behave as a tense membrane.

Surfactant   A substance that has the ability to emulsify and wash away oil and grease in an aqueous suspension.

Suspension   A mixture in which particles large enough to be visible with a microscope are dispersed in a medium.

Syndet   A synthetic detergent or emulsifier that may contain a sulfate group instead of a carboxylate group.

Temporary Hardness of Water   Water that contains bicarbonate ions as well as calcium and magnesium ions and can be softened by heating.

Theory of Ionization   Electrolytes dissociate into ions in solvents; the degree of electrical conductivity of solutions depends on the extent of dissociation.

Tyndall Effect   The scattering of light by colloidal particles.

Universal Solvent   A reference to water for its ability to dissolve many substances.

Vapor Pressure   The vapor pressure of a liquid is the pressure at which the substance in the gaseous state is in equilibrium with the liquid in a confined space at a given temperature.

"Washing" of High-Sulfur Coal   The removal of sulfur from coal by crushing coal and removing an iron-sulfur compound, pyrite.

Weak Acid   An acid that dissociates to a limited extent.

THINGS TO DO

1. Measure the pH of samples of some home chemicals with strips of pH paper: vinegar, milk of magnesia, lemon juice, a soft drink, baking soda solution, and buttermilk.

2. Determine the strength of a solution of NaOH by titration. Fill one buret with a solution of NaOH and a second with a solution of HCl of known strength. Run 25.0 mL of the acid solution into a 200-mL flask, add four drops of phenolphthalein indicator, and add NaOH solution carefully until the first permanent pink coloration appears. Calculate the molarity of NaOH from

$$M_{NaOH} = \frac{M_{HCl} \cdot V_{HCl}}{V_{NaOH}}$$

3. Classify the following substances as electrolytes or nonelectrolytes by testing dilute water solutions for electrical conductivity:
   (a) water (b) table salt (c) sugar (d) alcohol (e) vinegar (f) ammonia

4. Purify water to which a little $KMnO_4$ has been added with a distillation apparatus. Note the colorless distillate as compared with the color of the solution.

5. Prepare a (a) 1-M aqueous solution of table salt; (b) 1-m aqueous solution of table salt.

6. Observe the soap-sudsing ability of (a) distilled water; (b) water containing $CaSO_4$; (c) water containing $CaSO_4$ when Na_2CO_3 has been added to it; (d) water containing $Ca(HCO_3)_2$; and (e) water containing $Ca(HCO_3)_2$ when heated.

EXERCISES

The Solution Process

17–1. Are all liquids soluble in one another? Explain.

17–2. Gasoline is a nonpolar solvent; water is a polar solvent. Is KCl more likely to be soluble in gasoline or in water?

17–3. Compare the solubility of common table salt, NaCl, with that of naphthalene (moth flakes), $C_{10}H_8$, in (a) water, (b) gasoline.

17–4. How do a mixture, a solution, and a pure compound differ in composition?

17–5. Explain what "like dissolves like" means.

17–6. Give an example of solvation involving ion–dipole interaction.

17–7. A solution is made by dissolving sugar in water. Which is the solvent and which the solute?

17–8. Suppose a solution contains either salt or sugar dissolved in water. Suggest a simple way by which you could easily tell which it is.

Solution Composition: Molarity and Molality

17–9. One mole of sugar is added to 1 liter of water. Is the concentration of the solution 1 M? Explain.

17–10. Calculate (a) the number of moles and (b) the number of grams of NaCl in 50.0 mL of 2.00 M NaCl.

17–11. Describe how you would prepare 250 mL of 0.100 M NaOH from solid NaOH and water.

17–12. A sample of rainwater has a nitric acid concentration of 3.00×10^{-8} M. How many grams of nitric acid are present in 1 liter?

17–13. Explain how you would make a 1.50-m calcium chloride solution.

17–14. How many grams of water would you have to add to 1000 g of sugar, $C_{12}H_{22}O_{11}$, in order to prepare a 1-m solution?

17–15. How many molecules of ethyl alcohol, C_2H_5OH, must be dissolved in 500 g of water to make a 1-m solution?

17–16. An analysis of rainwater revealed a concentration of nitric acid, HNO_3, of 2.2×10^{-6} g/liter. What is the molarity with respect to the nitric acid?

Colligative Properties of Solutions

17–17. What is the meaning of the word "colligative" in the context of colligative properties?

17–18. Explain why seawater has a lower freezing point than fresh water.

17–19. How can you determine experimentally whether a substance is an electrolyte or a nonelectrolyte?

17–20. Which will freeze faster, a fresh-water pond or a salt-water pond?

17–21. How does the spreading of salt on ice-covered walks and roads cause the ice to melt?

17–22. Discuss the principle whereby the addition of antifreeze to an automobile radiator exerts its action.

17–23. The solution in a car radiator froze at 0°F. Determine the molality of the solution.

17–24. A solution contains 684 g of sucrose, $C_{12}H_{22}O_{11}$, per 1000 g of water. Determine the freezing point of the solution.

17–25. A solution is made by dissolving 45.0 g of dextrose ($C_6H_{12}O_6$) in 250.0 g of water.
(a) Calculate the molality.
(b) At what temperature will the solution freeze?

17–26. Which solution would produce the brighter light in an electrical conductivity apparatus, 0.100 M NaCl or 0.100 M Na_3PO_4? Explain.

Acids and Bases

17–27. List some general properties of acids.

17–28. List some general properties of bases.

17–29. In the Arrhenius definition, (a) what characterizes an acid; (b) what characterizes a base?

17–30. What does it mean to say that an acid is strong in aqueous solution?

17–31. When aqueous solutions of sodium chloride and silver nitrate are mixed, a precipitate forms instantaneously. How does the Arrhenius theory explain this result?

17–32. Complete and balance the following equations representing neutralization reactions:
   (a) $HCl + Ca(OH)_2 \longrightarrow$
   (b) $NH_3 + H_2SO_4 \longrightarrow$
   (c) $NaOH + HNO_3 \longrightarrow$
   (d) $H_3PO_4 + LiOH \longrightarrow$

17–33. Why is the acidity of a solution expressed in terms of its pH rather than in terms of the molarity of hydrogen ion present?

17–34. How is pH defined?

17–35. Vinegar is an aqueous solution of acetic acid. Would you expect the pH of vinegar to be greater than or less than 7.00?

17–36. State whether each of the following solutions is acidic or basic. The H^+ concentrations are given.
   (a) $[H^+] = 1.00 \times 10^{-4} \, M$
   (b) $[H^+] = 1.00 \times 10^{-11} \, M$
   (c) $[H^+] = 3.50 \times 10^{-13} \, M$
   (d) $[H^+] = 0.0001 \, M$

17–37. Classify the following aqueous solutions as acidic, basic, or neutral if their pH is
   (a) 11. (b) 2. (c) 7. (d) 8.24. (e) 3.48.

17–38. Why can boric acid (H_3BO_3) be used in eyewashes, whereas hydrochloric acid (HCl) is not safe to use?

Water Pollution

17–39. How do soaps and detergents cleanse?

17–40. In regions where the water supply is rich in calcium or magnesium compounds, why are detergents better cleaning agents than soaps?

17–41. Of what importance is biodegradability to the formulators of detergents?

17–42. What are the disadvantages of hard water?

17–43. What is the environmental impact of phosphorus compounds in detergents?

17–44. Why is an algal bloom—a thick mat of blue-green algae—on the surface of a lake an undesirable sign?

17–45. How can eutrophication lead to the death of a lake?

17–46. *Multiple Choice*
   A. Which of the following would be most soluble in the polar solvent water?
      (a) I_2 (b) NaCl (c) CCl_4 (d) N_2
   B. The freezing point of H_2O is
      (a) lowered by the addition of a solute.
      (b) raised by the addition of a solute.
      (c) not affected by the addition of a solute.
      (d) does not depend on the amount of solute added.

   C. When 1 mole of NaCl is dissolved in a kilogram of water, the concentration is
      (a) 0.50 molal.
      (b) 0.50 molar.
      (c) 1.00 molal.
      (d) 1.00 molar.
   D. A solution that has a pH of 7 is
      (a) slightly acidic.
      (b) strongly acidic.
      (c) neutral.
      (d) slightly basic.
   E. K_W
      (a) equals 10^{-14}.
      (b) is the ion product of water.
      (c) equals $[H_3O^+][OH^-]$.
      (d) all of these.
   F. Which of the following is *not* a solution?
      (a) air free of smoke and dust
      (b) water free of other substances
      (c) sugar in water
      (d) HCl in water
   G. When sodium chloride dissolves in water, the following species would be found in the resulting solution:
      (a) solvated NaCl molecules
      (b) solvated Na^+
      (c) solvated Cl^-
      (d) (b) and (c)
   H. Which one of the following observations is caused by a colligative property of solutions?
      (a) Water freezes at a lower temperature under pressure.
      (b) Water boils at a higher temperature under pressure.
      (c) Salt dissolves more rapidly in hot water than in cold.
      (d) Ice melts at a lower temperature when salt is added.
   I. A small amount of a certain solute melting at 800°C is added to water. The solution will be expected to freeze
      (a) above room temperature.
      (b) at 0°C.
      (c) slightly above 0°C.
      (d) slightly below 0°C.
   J. The pH of a solution refers to
      (a) the molarity of an acidic solution.
      (b) the molality of an acidic solution.
      (c) the concentration of hydronium ions in solution.
      (d) all of these.

SUGGESTIONS FOR FURTHER READING

Carter, Henry A., "Chemistry in the Comics: Part 3. The Acidity of Paper." *Journal of Chemical Education*, November, 1989.

Perhaps one third of the 19 million books and pamphlets in the Library of Congress are too brittle for circulation, and 90 million books in American research libraries cannot be used because of brittleness due to acid attack. An understanding of the chemistry involved in the acidity and aging of paper is essential for finding the means of extending the lifetimes of comic books and library collections.

Charola, A. Elena, "Acid Rain Effects on Stone Monuments." *Journal of Chemical Education*, May, 1987.

Stone monuments made of marble, limestone, and calcareous sandstones are most susceptible to acid rain. Although pure sandstones and granites are not affected, dry deposition still occurs and causes a uniform black color to develop.

De Lange, A. M., and J. H. Potgieter, "Acid and Base Dissociation Constants of Water and Its Associated Ions." *Journal of Chemical Education*, April, 1991.

Since water can act as both an acid and a base, it has an acid dissociation constant K_a and a base dissociation constant K_b. Water is a very weak acid ($K_a = 1.80 \times 10^{-16}$) and has a K_b value of 1.80×10^{-16}.

Giguère, Paul A., "The Great Fallacy of the H⁺ Ion." *Journal of Chemical Education*, September, 1979.

The concept of the hydronium ion, H_3O^+, was not accepted at first. Definite identification of H_3O^+ ions did not come until 1951 with the discovery of nuclear magnetic resonance (nmr).

Kolb, Doris, "The pH Concept." *Journal of Chemical Education*, January, 1979.

The author attempts to clear away the confusion over the pH concept, such as "If pH is a measure of acidity, then how is it that pH goes up when the acidity goes down?"

La Rivière, J. W. Maurits, "Threats to the World's Water." *Scientific American*, September, 1989.

The world has seen rivers turn into sewers and lakes into cesspools. People die from drinking contaminated water; pollution washes ashore on beaches; fish are poisoned by heavy metals; and wildlife habitats are destroyed. A laissez-faire approach to water management will spell more of the same.

Likens, Gene E., et al., "Acid Rain." *Scientific American*, October, 1979.

The acidity of rain and snow has increased sharply in recent decades. The principal cause is the burning of fossil fuels and the release of nitrogen and sulfur oxides.

Penman, H. L., "The Water Cycle." *Scientific American*, September, 1970.

Discusses some of the properties of water and their significance in the biosphere.

Rukeyser, William Simon, "Fact and Foam in the Row over Phosphates." *Fortune*, January, 1972.

What had been a placid backwater of science—limnology, the study of fresh water—erupted into controversy in the 1970s over the role of phosphates in eutrophication.

Tarbell, D. S., and A. T. Tarbell, "The Development of the pH Meter." *Journal of Chemical Education*, February, 1980.

The "instrumental revolution" in chemical research originated with the design of the Beckman pH meter by A. O. Beckman in 1934.

Walker, Jearl, "Exotic Patterns Appear in Water When It Is Freezing or Melting." *Scientific American*, July, 1986.

An examination of how ice crystals form as water freezes contains answers to questions such as, Why is ice filled with bubbles and with tubes that resemble wormholes? When an ice cube begins to melt, why does it sputter and give off air and water?

"The Chemistry of Cleaning." *Journal of Chemical Education*, September, 1979.

Summarizes the chemical and physical principles of cleaning.

ANSWERS TO NUMERICAL EXERCISES

17–10. (a) 0.100 mole NaCl
     (b) 5.85 g NaCl

17–12. 1.89×10^{-6} g

17–14. 2924 g

17–15. 3.01×10^{23} molecules

17–16. 3.5×10^{-8} M

17–23. 9.50 molal

17–24. $-3.72°C$

17–25. (a) 1.00 molal
     (b) $-1.86°C$

The molecular structure of crambin, illuminated by its sulfur atoms. (© Irving Geis/Science Source/Photo Researchers, Inc.)

18

AN INTRODUCTION TO ORGANIC CHEMISTRY

Excluding carbon, the number of chemical compounds formed by the elements in the periodic table is probably fewer than 200,000. In contrast, carbon is the central element in more than 7,000,000 (35 times as many) compounds, which it forms with a handful of other elements, principally hydrogen, oxygen, and nitrogen. Since the ability of carbon to enter so freely into chemical combination was at first believed to be related to its source in substances of animal or vegetable origin, the term "organic chemistry" came to denote the chemistry of most carbon compounds. Today, carbon compounds are synthesized in the laboratory by the thousands and make up a large segment of our industrial production, including plastics, fuels, synthetic fibers, pesticides, detergents, and many other products.

18–1 THE ORGANIC–INORGANIC DUALISM

The belief in the difference between the organic and inorganic realms persisted for centuries. Substances that were derived from plant or animal sources — organisms — were known as organic, and all others as inorganic, and there was no apparent way to bridge the two worlds. Although the chemical elements were the same in both kinds of substances, there seemed to

be two kinds of force that governed them. "**Vital force**" was supposed to regulate the organic world, and "**chemical affinity**" the inorganic. An organic substance could be formed only through the vital force of living organisms and, therefore, could never be made in the laboratory.

Freidrich Wöhler (1800–1882), who had been a student of Berzelius, made a discovery in 1828 that led to the eventual abandonment of the organic–inorganic distinction. Wöhler **synthesized urea**, an organic compound, from ammonium chloride and silver cyanate, which were considered to be inorganic compounds. This reaction was the first evidence that a vital force was not necessary to prepare organic compounds.

$$\underset{\text{(ammonium chloride)}}{NH_4Cl} + \underset{\text{(silver cyanate)}}{AgCNO} \longrightarrow$$

$$\underset{\text{(urea)}}{NH_2-\overset{\overset{\text{O}}{\|}}{C}-NH_2} + \underset{\text{(silver chloride)}}{AgCl \downarrow} \qquad (18\text{–}1)$$

Wöhler wrote to Berzelius in triumph that he could make urea without the use of kidneys. The synthesis of urea was still inconclusive, however, since Wöhler's "inorganic" ammonia was prepared from animal hoofs and horns rather than from nitrogen and hydrogen. More decisive tests were the synthesis in 1844 of acetic acid, an organic substance, from completely inorganic materials (the elements carbon, hydrogen, and oxygen) by a student of Wöhler's, Hermann Kolbe (1818–1884), and the synthesis of scores of organic compounds—including methane, acetylene, benzene, methyl alcohol, and ethyl alcohol—by Pierre Berthelot (1827–1907) in the 1850s. It is now recognized that the same principles apply to all molecules regardless of their source, and the term **organic chemistry** is used to refer to the **chemistry of carbon compounds**.

18–2 ALKANE HYDROCARBONS: A HOMOLOGOUS SERIES

The simplest organic compounds contain only hydrogen and carbon, for which reason they are designated **hydrocarbons**. Of these, the **alkane** hydrocarbons have two kinds of bonds, carbon-to-hydrogen (C—H) and carbon-to-carbon (C—C). Since only single bonds are present, these compounds are also called **saturated** hydrocarbons, and because they are relatively stable toward many reagents, they are called **paraffins** (*parus,* meaning "too little"; *affinis,* meaning "affinity").

The alkane hydrocarbons constitute a **homologous series**—a group of related compounds in which the molecular formulas of members differ from one another by a constant increment. The simplest member of the family is one containing only one carbon. Methane, CH_4 (Fig. 18–1), is followed by ethane, C_2H_6 (Fig. 18–2), then propane,

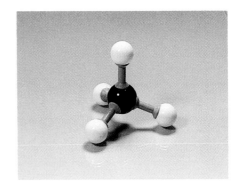

Figure 18–1   A ball-and-stick model of methane, CH_4. (Charles Steele)

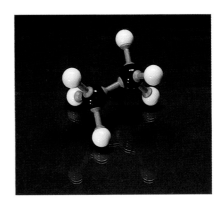

Figure 18–2   Ethane, C_2H_6 or CH_3CH_3. (Charles Steele)

Table 18–1   PHYSICAL PROPERTIES OF ALKANES

Name	Molecular Formula	Usual State	Melting Point(°C)	Boiling Point(°C)
Methane	CH_4	Gas	−182	−161
Ethane	C_2H_6	Gas	−183	−89
Propane	C_3H_8	Gas	−18	−45
n-Butane	C_4H_{10}	Gas	−138	−0.5
n-Pentane	C_5H_{12}	Liquid	−130	36
n-Hexane	C_6H_{14}	Liquid	−95	69
n-Heptane	C_7H_{16}	Liquid	−91	98
n-Octane	C_8H_{18}	Liquid	−57	125
n-Nonane	C_9H_{20}	Liquid	−51	151
n-Decane	$C_{10}H_{22}$	Liquid	−32	174
n-Undecane	$C_{11}H_{24}$	Liquid	−25.6	195
n-Dodecane	$C_{12}H_{26}$	Liquid	−9.6	215

Figure 18–3   Propane, C_3H_8 or $CH_3CH_2CH_3$. (Charles Steele)

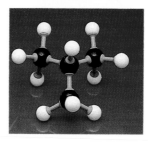

Figure 18–4   Ball-and-stick models of the two butanes, *n*-butane, $CH_3CH_2CH_2CH_3$, and isobutane, $CH_3CH(CH_3)CH_3$. (Charles D. Winters)

C_3H_8 (Fig. 18–3). Ethane differs from methane by one carbon and two hydrogens, $-CH_2$, and propane differs from ethane by the same $-CH_2$ (methylene) group.

The individual members of a homologous series, called **homologs**, are similar in structure and physical and chemical properties. Homologs conform to a **general formula**, in this case, C_nH_{2n+2}. Knowing the number of carbons present, we can readily predict the number of hydrogens. Thus, octane has 8 carbons and $(2n + 2) = 16 + 2 = 18$ hydrogen atoms per molecule. Some physical properties of the first 12 alkanes are given in Table 18–1. The symbol *n* represents "normal" and refers to a straight, unbranched chain of carbon atoms. The names and characteristics of the important classes of hydrocarbons are listed in Table 18–2.

18–3 ISOMERISM

The use of molecular formulas alone to represent organic compounds is not always sufficient. Although methane, ethane, and propane occur in just one form, there are two butanes (C_4H_{10}) (Fig. 18–4) and three pentanes (C_5H_{12}) (Fig. 18–5). There are 18 known compounds having the molecular formula C_8H_{18}. Each of these has its characteristic boiling and freezing points and other properties. The existence of more than one compound with the same molecular

Table 18–2   CLASSES OF HYDROCARBONS

Class	General Formula	Bonding of Carbons	Example
Alkane	C_nH_{2n+2}	Single bonds, no rings	CH_4 (methane)
Cycloalkane	C_nH_{2n}	Single bonds, ring	C_6H_{12} (cyclohexane)
Alkene	C_nH_{2n}	One double bond	C_2H_4 (ethylene)
Diene	C_nH_{2n-2}	Two double bonds	$CH_2=CH-CH=CH_2$ (butadiene)
Alkyne	C_nH_{2n-2}	One triple bond	C_2H_2 (acetylene)
Aromatic		At least one benzene ring	C_6H_6 (benzene)

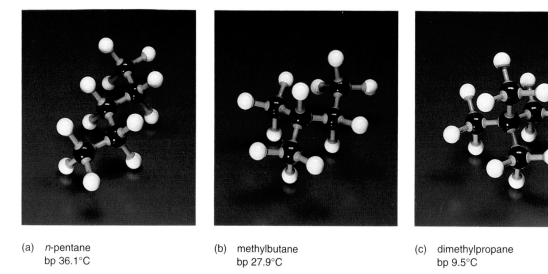

(a) *n*-pentane
     bp 36.1°C

(b) methylbutane
     bp 27.9°C

(c) dimethylpropane
     bp 9.5°C

Figure 18–5 The three isomeric pentanes, C_5H_{12}. The atoms are bonded in a different order in each of the three structural isomers. (Charles Steele)

Table 18–3 NUMBER OF POSSIBLE ISOMERS FOR CERTAIN ALKANES

Molecular Formula	Number of Possible Isomers	Molecular Formula	Number of Possible Isomers
CH_4	1	C_9H_{20}	35
C_2H_6	1	$C_{10}H_{22}$	75
C_3H_8	1	$C_{11}H_{24}$	159
C_4H_{10}	2	$C_{12}H_{26}$	355
C_5H_{12}	3	$C_{13}H_{28}$	802
C_6H_{14}	5	$C_{20}H_{42}$	366,000
C_7H_{16}	9	$C_{30}H_{62}$	4,000,000,000
C_8H_{18}	18	$C_{40}H_{82}$	63,000,000,000,000

Table 18–4 PHYSICAL PROPERTIES OF THE ISOMERS OF BUTANE AND PENTANE

Compound	Molecular Formula	Melting Point (°C)	Boiling Point (°C)	Density (g/mL, 20°C)
n-Butane	C_4H_{10}	− 135	− 0.5	0.62
2-Methylpropane (Isobutane)	C_4H_{10}	− 159	− 12	0.60
n-Pentane	C_5H_{12}	− 130	36	0.63
2-Methylbutane (Isopentane)	C_5H_{12}	− 160	28	0.62
2,2-Dimethylpropane (Neopentane)	C_5H_{12}	− 20	9.5	0.61

formula is called **isomerism**, and distinct compounds with the same molecular formula are called **isomers**. Therefore, *n*-butane and isobutane are isomers.

There are two reasons for the vast number of organic compounds: (1) the ability of carbon to bond with itself in almost endless succession; and (2) the formation of isomers. Table 18–3 gives the number of isomers possible for certain alkanes. Some properties of the two butanes and three pentanes are compared in Table 18–4.

18–4 STRUCTURAL THEORY

Since isomers have the same molecular formulas, they must differ with respect to the arrangement of the atoms within the molecules. Kekulé and Couper explained how this is possible.

In 1858 Friedrich August Kekulé (1829–1896) published a paper, and within a month, Archibald Scott Couper (1831–1892) published one of his own, independently. Each presented (1) the idea (already proposed by others) that a carbon atom forms four bonds; and (2) a new concept (original with Kekulé and Couper): the self-linking of carbon atoms to form chains. These two concepts together form the basis of

a **structural theory** that provides a simple yet useful interpretation of organic substances and their reactions. The basic assumptions of the structural theory are as follows.

1. Carbon normally forms four bonds; oxygen, two; hydrogen and the halogens, one.
2. A carbon atom may be bonded to atoms of other elements or to other carbon atoms.
3. More than one bond may connect atoms. There may be single, double, or triple bonds between two carbon atoms.

The theory was later expanded to include the concept of the **tetrahedral configuration** of carbon, proposed independently in 1874 by Jacobus Hendricus van't Hoff (1852–1911) and Joseph Achille Le Bel (1847–1930). According to this hypothesis, the four covalent bonds of the carbon atom have definite positions in space, as though the carbon atom were the center of a **tetrahedron**—a triangular-based pyramid with four faces and four corners (Fig. 18–6). Each bond is equidistant from the remaining three, and the angle between one bond and any of its neighbors is 109°28′.

Methane, CH_4, is a three-dimensional molecule with carbon at the center of a regular tetrahedron and four bonds directed toward the vertices, where hydrogen atoms are located (Fig. 18–7). Its expanded

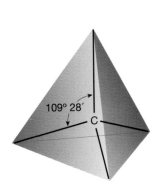

Figure 18–6   A regular tetrahedron. The valence bonds of a carbon atom are directed toward the four vertices.

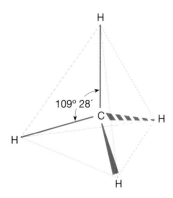

Figure 18–7   Structure of the methane molecule, CH_4. The four hydrogen atoms occupy the corners of a regular tetrahedron. The two solid lines are in the plane of the paper; the wedge-shaped line extends out toward you; and the dashed line extends back.

structural formula is usually represented as a two-dimensional Fischer projection,

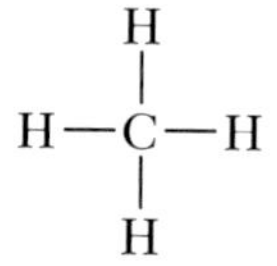

$$
\begin{array}{c}
H \\
| \\
H-C-H \\
| \\
H
\end{array}
$$

with the understanding that the H—C—H bond angles are 109°28′ (not 90°) and that each dash represents one pair of electrons, a covalent bond. Further, the groups above and below the carbon project away from the viewer and the groups to the left and right of the carbon project toward the viewer.

The expanded structural formulas of ethane, C_2H_6, and propane, C_3H_8, are

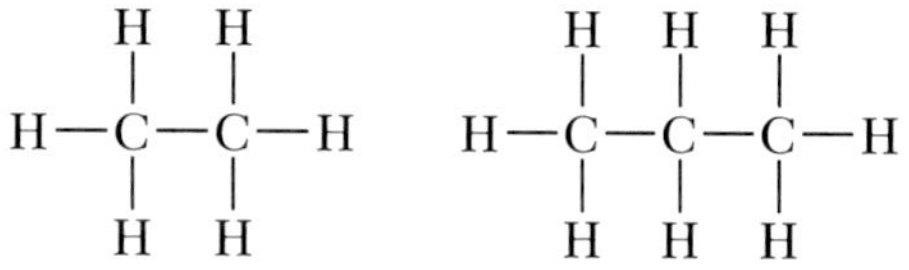

ethane   propane

Although chains of three or more carbons are represented as though they were linear, the carbon-to-carbon bond angles are really 109°28′, and the carbon skeleton is zigzag:

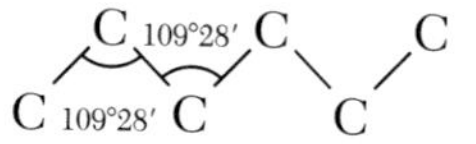

As the chain of carbon atoms lengthens, compounds are preferably represented by **condensed structural formulas**. Such formulas preserve some idea of the bonding of groups of atoms within a molecule but are more convenient to write out. They may also be written for smaller molecules, such as ethane and propane:

$$CH_3CH_3 \quad CH_3CH_2CH_3$$
ethane   propane

With butane, C_4H_{10}, it becomes possible to arrange the carbon and hydrogen atoms in more than one way, each representing a structural isomer (Table 18–5). The straight-chain isomer is designated normal, or *n*-butane. The branched structure is known as isobutane, its common or trivial name.

Table 18–5   STRUCTURAL ISOMERS OF BUTANE

Expanded Structural Formula	Condensed Structural Formula
***n*-Butane**	
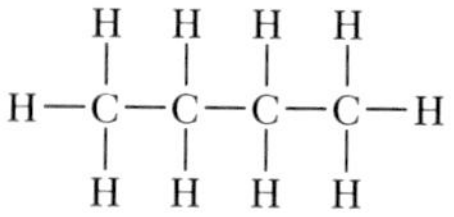	$CH_3CH_2CH_2CH_3$ or $CH_3(CH_2)_2CH_3$
Isobutane	
	CH_3CHCH_3
	CH_3
	or
	$CH_3CH(CH_3)CH_3$

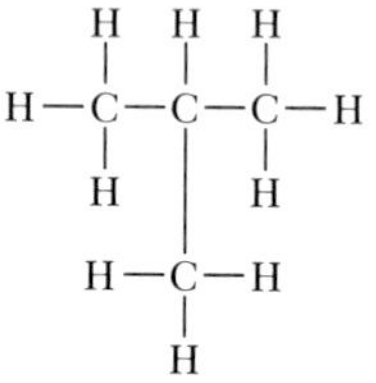

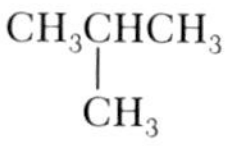

18–5 NOMENCLATURE OF ALKANES

A common suffix or other designation identifies the homologs of a series of organic compounds. The names of the **alkanes** all end with the suffix -*ane*. Beginning with **pentane**, the names indicate the **number of carbon atoms** in the molecule: *penta*, five; *hexa*, six; *hepta*, seven; *octa*, eight; and so on. To name an unbranched alkane, add the suffix -*ane* to the Greek equivalent of the number of carbons. The compound $C_{10}H_{22}$ is therefore named decane (*deka*, ten).

Compounds are referred to by two names: a common name and a systematic or IUPAC name. The International Union of Pure and Applied Chemistry (IUPAC), a group that meets periodically to consider such matters, recommends that the **systematic names** of the alkanes be derived by:

1. Determining the longest continuous chain of carbon atoms in the molecule and using the name of the alkane corresponding to this number as the basis for the name of the compound;
2. Numbering the carbon atoms of this chain from one end to the other, beginning at the end nearest the branching;

3. Naming the groups that are attached to this chain;

4. Designating each group by the number of the carbon atom to which it is attached;

5. Using a separate number for each branched group;

6. Numbering the longest continuous chain in such a way that the branched groups have the lowest possible numbers;

7. Using a prefix (*di-, tri-, tetra-*) to designate two or more like groups.

EXAMPLE 18–1

What is the systematic (IUPAC) name of the following compound?

$$
\begin{array}{c}
\overset{\displaystyle CH_3}{\underset{\displaystyle |}{}} \\
\overset{\displaystyle CH_3\ CH_2}{\underset{\displaystyle |\ \ \ |}{}} \\
\overset{1}{CH_3}-\overset{2}{CH_2}-\overset{3}{C}-\overset{4}{CH}-\overset{5}{CH_2}-\overset{6}{CH_2} \\
\underset{\displaystyle CH_3}{\underset{\displaystyle |}{}}
\end{array}
$$

SOLUTION

Since the longest continuous chain contains six carbons, the compound is a hexane. To give the substituents the lowest possible numbers, we number the chain from left to right in this case, and assign a number and a name to each branch. The two methyl groups bonded to the third carbon in the chain are designated "3,3-dimethyl." The ethyl group bonded to carbon 4 is "4-ethyl." The complete name is *4-ethyl-3, 3-dimethylhexane*. Note that a series of numbers is separated by commas, and numbers are set off from names by hyphens.

EXTENSION

Write a structural formula for 2,2,4-trimethyl hexane.

$$
\text{(Answer: } CH_3\overset{\displaystyle CH_3}{\underset{\displaystyle |}{C}}CH_2\ \overset{}{\underset{\displaystyle |}{CH}}\ CH_2CH_3)
$$
$$
\qquad\quad CH_3 \qquad CH_3
$$

∎

The words "methyl" and "ethyl" in the name of the compound in Example 18–1 are derived from the corresponding alkanes minus one hydrogen. They are examples of **alkyl groups**. Some commonly used alkyl groups are listed in Table 18–6. A **primary carbon atom** is bonded to one other carbon; **secondary** (*sec-*), to two carbons; and **tertiary** (*tert-*), to three carbons. In Example 18–1, carbon 1 is a primary carbon; carbon 2, secondary; and carbon 4, tertiary.

18–6 THE STRUCTURE OF SATURATED HYDROCARBONS

The bonding of carbon in hydrocarbons is an interesting problem. The electronic configuration of the ground state of $_6C$, $1s^2 2s^2 2p_x^1 2p_y^1$, shows two unpaired electrons in the 2p subshell (Fig. 18–8). The structure suggests a covalence of 2 for carbon. As has long been known, however, a covalence of 4 for carbon is usual. Through collisions between atoms, the 2s electrons are uncoupled, and one is promoted to the vacant $2p_z$ orbital to give an excited atom with four unpaired electrons. This structure implies that carbon should form two kinds of bonds with its valence electrons: one with the s electron and three with the p electrons. The experimental fact is that all four C—H bonds in methane, CH_4, are identical in their properties. The four outermost electrons of carbon

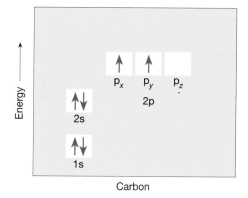

Figure 18–8   Energy-level model of the ground state of carbon.

Table 18–6 ALKYL GROUPS DERIVED FROM METHANE THROUGH BUTANE

Alkane	Alkyl Group	Alkyl Chloride
Methane	Methyl	Methyl chloride
$H-\overset{\displaystyle H}{\underset{\displaystyle H}{C}}-H$	$H-\overset{\displaystyle H}{\underset{\displaystyle H}{C}}-$	CH_3Cl
Ethane	Ethyl	Ethyl chloride
$H-\overset{\displaystyle H}{\underset{\displaystyle H}{C}}-\overset{\displaystyle H}{\underset{\displaystyle H}{C}}-H$	$H-\overset{\displaystyle H}{\underset{\displaystyle H}{C}}-\overset{\displaystyle H}{\underset{\displaystyle H}{C}}-$	CH_3CH_2Cl
Propane	n-Propyl	n-Propyl chloride
$H-\overset{\displaystyle H}{\underset{\displaystyle H}{C}}-\overset{\displaystyle H}{\underset{\displaystyle H}{C}}-\overset{\displaystyle H}{\underset{\displaystyle H}{C}}-H$	$H-\overset{\displaystyle H}{\underset{\displaystyle H}{C}}-\overset{\displaystyle H}{\underset{\displaystyle H}{C}}-\overset{\displaystyle H}{\underset{\displaystyle H}{C}}-$	$CH_3CH_2CH_2Cl$
	Isopropyl	Isopropyl chloride
	$H-\overset{\displaystyle H}{\underset{\displaystyle H}{C}}-\overset{\displaystyle H}{C}-\overset{\displaystyle H}{\underset{\displaystyle H}{C}}-H$	$CH_3\underset{\displaystyle Cl}{CH}CH_3$
n-Butane	n-Butyl	n-Butyl chloride
$H-\overset{\displaystyle H}{\underset{\displaystyle H}{C}}-\overset{\displaystyle H}{\underset{\displaystyle H}{C}}-\overset{\displaystyle H}{\underset{\displaystyle H}{C}}-\overset{\displaystyle H}{\underset{\displaystyle H}{C}}-H$	$H-\overset{\displaystyle H}{\underset{\displaystyle H}{C}}-\overset{\displaystyle H}{\underset{\displaystyle H}{C}}-\overset{\displaystyle H}{\underset{\displaystyle H}{C}}-\overset{\displaystyle H}{\underset{\displaystyle H}{C}}-$	$CH_3CH_2CH_2CH_2Cl$
	sec-Butyl	sec-Butyl chloride
	$H-\overset{\displaystyle H}{\underset{\displaystyle H}{C}}-\overset{\displaystyle H}{\underset{\displaystyle H}{C}}-\overset{\displaystyle H}{\underset{\displaystyle H}{C}}-\overset{\displaystyle H}{C}-H$	$CH_3CH_2\underset{\displaystyle Cl}{CH}CH_3$
Isobutane	Isobutyl	Isobutyl chloride
$H-\overset{\displaystyle H}{\underset{\displaystyle H}{C}}-\overset{\displaystyle H}{\underset{\displaystyle\underset{\displaystyle H-\overset{\displaystyle }{\underset{\displaystyle H}{C}}-H}{H}}{C}}-\overset{\displaystyle H}{\underset{\displaystyle H}{C}}-H$	$H-\overset{\displaystyle H}{\underset{\displaystyle H}{C}}-\overset{\displaystyle H}{\underset{\displaystyle\underset{\displaystyle H-\overset{\displaystyle }{\underset{\displaystyle H}{C}}-H}{H}}{C}}-\overset{\displaystyle H}{\underset{\displaystyle H}{C}}-$	$CH_3\underset{\displaystyle CH_3}{CH}CH_2Cl$
	tert-Butyl	tert-Butyl chloride
	$H-\overset{\displaystyle H}{\underset{\displaystyle H}{C}}-\overset{\displaystyle }{\underset{\displaystyle\underset{\displaystyle H-\overset{\displaystyle }{\underset{\displaystyle H}{C}}-H}{H}}{C}}-\overset{\displaystyle H}{\underset{\displaystyle H}{C}}-H$	$CH_3\overset{\displaystyle Cl}{\underset{\displaystyle CH_3}{C}}CH_3$

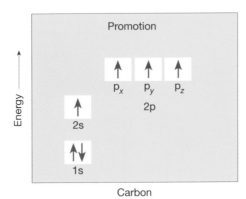

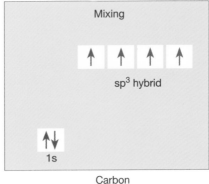

Figure 18–9   Formation of sp³ hybrid orbital. One of the 2s electrons of carbon is promoted to the vacant 2p$_z$ orbital. The remaining 2s electron and the three p electrons then mix and form four sp³ hybrid orbitals.

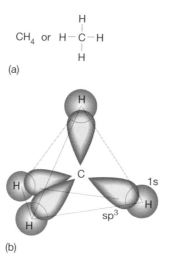

Figure 18–10.   Orbital structure for methane, CH_4. The carbon atom is shown with four sp³ hybrid orbitals in a tetrahedral arrangement and each hydrogen atom with an s orbital.

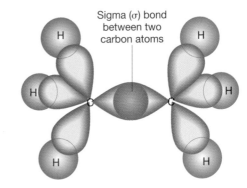

Figure 18–11   Structure of ethane, C_2H_6, a combination of two CH_3—units: CH_3CH_3.

in this case must therefore produce four equivalent orbitals, each having one electron available for a covalent bond. These four orbitals are described as **hybridized** and are called sp³ (pronounced "s-p-three") orbitals (Fig. 18–9). In the formation of CH_4, each of the four **sp³ hybrid orbitals** overlaps the 1s orbital of a hydrogen atom. With the carbon kernel in the center, each bond is directed to one of the four vertices of a regular tetrahedron, making an angle of 109°28′ with it (Fig. 18–10).

The shape of the sp³ hybrid orbitals differs from the s and p orbitals from which they are formed. The

2s orbital is spherical and therefore nondirectional. The three 2p orbitals are propeller-shaped and oriented at angles of 90° to one another. The shape of the sp³ hybrid orbitals produces a much stronger covalent bond. The best overlap is obtained if a hydrogen atom lies on the axis of an sp³ orbital. The cylindrically symmetrical orbital around the line joining carbon and hydrogen is called a **sigma (σ) bond**.

The continuous and branched chains of the higher alkanes also have sp³ hybridization. Ethane, C_2H_6, is a combination of two CH_3 — units: CH_3CH_3. The sp³ hybrid orbitals of each carbon overlap the s orbitals of three hydrogen atoms and an sp³ hybrid orbital of the other carbon, forming sigma bonds in each case (Fig. 18–11).

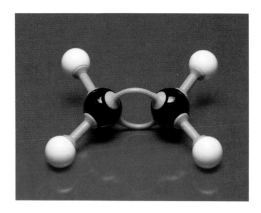

Figure 18–12. Model of ethylene, $H_2C{=}CH_2$, a planar molecule with bond angles of 120°. (Charles D. Winters)

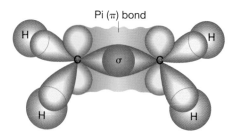

Pi (π) bond

Figure 18–14 The bonding of ethylene. Each sp^2 hybrid orbital forms three sigma (σ) bonds. The p orbitals form a pi (π) bond.

18–7 THE STRUCTURE OF UNSATURATED HYDROCARBONS

In **unsaturated hydrocarbons** there is at least one **multiple bond** between the carbon atoms. An **alkene** or **olefin** contains a **double bond**; a **polyene** (*poly*, meaning "many") contains more than one double bond; an **alkyne** contains a **triple** bond. A molecule containing alternating single and double bonds between the carbon atoms is called a **conjugated system**. Each carbon atom in an unsaturated hydrocarbon contributes two or three electrons to the formation of a double or triple bond. Since fewer electrons are available for bonding with other atoms, unsaturated hydrocarbons contain fewer hydrogens than the maximum number possible.

The simplest alkene is **ethylene**, C_2H_4. Experiments show that the **bond angles** are 120° and that ethylene is a flat, planar molecule (Fig. 18–12). We can account for this structure by assuming that the 2s electrons of carbon are uncoupled, and one is promoted to the p level in each carbon atom. Then the s and two of the p orbitals hybridize to form three **sp^2 hybrid orbitals**, which lie in one plane at angles of 120°. One p orbital is left over (Fig. 18–13). The sp^2

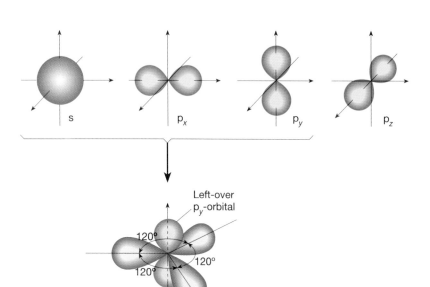

Figure 18–13 Hybridization of one s and two p orbitals yields three sp^2 hybrid orbitals. A p orbital is left over.

s p_x p_y p_z

Left-over p_y-orbital

120°   120°   120°

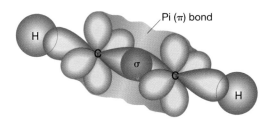

Figure 18–16   The bonding of acetylene. Each carbon forms two sigma (σ) bonds: one to the other carbon atom, the second to a hydrogen atom. The remaining p orbitals on each carbon combine to form two pi (π) bonds.

Figure 18–15.   Model of acetylene, HC≡CH. (Charles Steele)

orbitals of each carbon overlap directly and form sigma bonds with two hydrogens and with each other. The remaining propeller-shaped p orbitals of the two carbon atoms are parallel to each other and perpendicular to the plane of the sp^2 hybrid orbitals, the lobes lying above and below the plane. These p_z orbitals interact with each other laterally, producing a second, weaker bond between the carbons, called a **pi (π) bond** (Fig. 18–14). It is easier to disrupt the π portion of a double bond than a single bond.

Acetylene, C_2H_2, is the simplest hydrocarbon having a triple bond. The third bond between the carbons is formed at the expense of two bonds between carbon and hydrogen, so that acetylene is still more unsaturated than ethylene. Experiments indicate that the shape of acetylene is linear—the two carbons and two hydrogens lie in a straight line, with bond angles of 180° (Fig. 18–15).

In each carbon atom of acetylene, **sp hybridization** occurs between the s and one of the p orbitals after the s electrons have been uncoupled and one promoted to the p_z level. Two sp hybrid orbitals and the two remaining p orbitals are then available for bonding. Each of the sp orbitals is directed along the x-axis, but in opposite directions (180° apart), and each is oriented at 90° to the unhybridized p_y and p_z orbitals (Fig. 18–16). The sp orbitals of each carbon form a sigma bond with the s orbital of one of the hydrogens and with each other. The p orbitals are at right angles to each other and to the sp orbitals, and they form two π bonds between the carbon atoms. The third bond between the carbon atoms is another weak point, although the structural formula makes no distinction among the three bonds. A summary of the properties of the hybrid orbitals of carbon is given in Table 18–7, and the chemical bond characteristics of carbon, in Table 18–8.

Table 18–7   PROPERTIES OF HYBRID ORBITALS OF CARBON

Hybridization	Number of Hybrid Orbitals	Orbitals Used	Orbitals Remaining	Angle Between Hybrid Orbitals	Sigma (σ) Bonds	Pi (π) Bonds
sp^3	4	$2s, 2p_x, 2p_y, 2p_z$		109°28′	4	0
sp^2	3	$2s, 2p_x, 2p_y$	$2p_z$	120°	3	1
sp	2	$2s, 2p_x$	$2p_y 2p_z$	180°	2	2

Table 18–8   CHEMICAL BOND
CHARACTERISTICS OF CARBON

Bond	Formula	Suffix	Bond Length (Å)	Bond Energy (kcal/Mole)
Single	C—C	-ane	1.54	83.1
Double	C=C	-ene	1.34	142.0
Triple	C≡C	-yne	1.20	192.1

18–8 SOME REACTIONS OF ALKANES

Chemically, the saturated hydrocarbons are relatively inert. They react very slowly with chlorine and bromine at room temperature but quite readily at higher temperatures and in the presence of light. One or more hydrogen atoms is replaced by halogen atoms in these **substitution reactions**, and side reactions occur, so that a mixture of products is usually obtained.

$$CH_4 + Cl_2 \longrightarrow CH_3Cl + HCl \qquad (18\text{–}2)$$

Figure 18–17 The lighter contains liquid butane, C_4H_{10}. As the pressure is released, the liquid vaporizes and is ignited simultaneously as the flint wheel is struck. (Charles D. Winters)

The chlorination of methane yields not only chloromethane, CH_3Cl, but also dichloromethane, trichloromethane, nd tetrachloromethane: CH_2Cl_2, $CHCl_3$, and CCl_4.

The most important use of the saturated hydrocarbons is as **fuels** (Fig. 18–17). In complete combustion, they form carbon dioxide and water and evolve large amounts of heat, as can be seen from the combustion reactions of methane (CH_4) and octane (C_8H_{18}), a component of gasoline:

$$CH_4 + 2O_2 \longrightarrow$$
$$CO_2 + 2H_2O + 210.8 \text{ kcal/mol} \quad (18\text{–}3)$$

$$2C_8H_{18} + 25O_2 \longrightarrow$$
$$16CO_2 + 18H_2O + 1302.7 \text{ kcal/mol} \quad (18\text{–}4)$$

The composition of natural gas sold as fuel is given in Table 18–9; the compositions of other hydrocarbon fuels are given in Table 18–10. Propane and butane are gases at room temperature under normal pressure but are liquefied under pressure and sold as liquefied petroleum gas (LPG) (Fig. 18–18).

Table 18–9   COMPOSITION OF NATURAL
GAS USED AS FUEL

Component	Percent
Methane	78–80
Nitrogen	10–12
Ethane	5.9
Propane	2.9
n-Butane	0.71
Isobutane	0.26
C_5–C_7 Hydrocarbons	0.13

Table 18-10 COMPOSITION OF HYDROCARBON FUELS

Fuel	Carbon Content Of Compounds	Boiling Point Range (°C)	Uses
Acetylene	C_2H_2		Welding
Natural gas	C_1-C_4		Industrial and home fuel
Propane	C_3H_8		Bottled fuel gas
Butane	C_4H_{10}		Bottled fuel gas
Gasoline	C_5-C_{12}	35-225	Motor fuel
Kerosene	$C_{12}-C_{16}$	200-315	Jet fuel, domestic heating
Fuel oil	$C_{15}-C_{18}$	250-375	Diesel fuel, industrial fuel
Paraffin wax	$C_{20}-C_{30}$	50-60 (melting point)	Candles

If there is a lack of oxygen when a hydrocarbon is burned, carbon monoxide, CO, an invisible, deadly gas, is formed.

$$C_8H_{18} + 12O_2 \longrightarrow 7CO_2 + CO + 9H_2O \quad (18-5)$$

Millions of tons of carbon monoxide are poured into the atomosphere each year, about 80% of it from automobile exhausts (Fig. 18–19). Hydrocarbons from unburned fuel in the exhaust may also react with ozone and atomic oxygen in the air to form chemically reactive aldehydes with extremely irritating odors. **Gasoline**, a complex mixture of simple hydrocarbons plus slight amounts of additives, is very volatile and escapes into the air. With no emission controls, 60%

of the emitted hydrocarbons come from the exhaust, 20% escapes from the crankcase, 10% results from evaporation at the carburetor, and 10% evaporates from the fuel tank. The mixture of products resulting from reactions in the atmosphere—ozone, carbon monoxide, peroxyacylnitrates (PAN), aldehydes, ketones, and alkylnitrates—is called **photochemical smog**. It causes eye irritation and may interfere with respiration.

The **catalytic afterburner** uses a catalyst to allow the hydrocarbon oxidation to be completed at a lower temperature. The control of evaporation losses is brought about by a collection system that transports fuel vapors from the fuel tank and carburetor to the

Figure 18–18 This camper will use the propane tank for fuel. (Four By Five)

Figure 18–19   Automobiles are tested for carbon monoxide, nitrogen oxides, and unburned hydrocarbons. Those failing the standards established by EPA must be repaired. (Source: Joesten, et al., *World of Chemistry*. Philadelphia: Saunders College Publishing, 1991, p. 865, Fig. 18–7.)

fuel system; the vapors are then burned in the engine. Hydrocarbon emissions from crankcase "blow-by" have been eliminated by using a positive crankcase ventilation system (PCV) that recycles hydrocarbons to the engine intake.

18–9 OCTANE RATING OF GASOLINES

The efficiency of the internal combustion engine is related to the octane rating of the fuel and to the compression ratio. With some gasolines, ignition occurs more as an explosion than as a smooth burning. This not only reduces the efficiency but also produces an audible "knock" in the engine. A fuel of pure isooctane produces little knock, whereas one of pure *n*-heptane knocks badly.

The **octane rating** of a gasoline is a measure of the knock produced when the gasoline is used as an automobile fuel. Isooctane and *n*-heptane have octane ratings of 100 and 0, respectively. A mixture of 90% isooctane and 10% heptane has an octane rating of 90. A gasoline rated 90 produces the same knock as a mixture of 90% isooctane and 10% *n*-heptane. Fuels

have been developed with octane ratings greater than 100, their antiknock properties being superior to isooctane. Table 18–11 lists the octane numbers required for various engine compression ratios.

The octane rating of gasolines can be improved by various means. Heating a gasoline in the presence of catalysts such as sulfuric acid or aluminum chloride may convert a straight-chain structure into a highly branched isomer with improved burning characteristics and antiknocking properties. The addition of

Table 18–11   OCTANE NUMBER AND ENGINE COMPRESSION RATIO

Engine Compression Ratio	Octane Number Required
5:1	73
6:1	81
7:1	87
8:1	91
9:1	95
10:1	98
11:1	100
12:1	102

Table 18-12 OCTANE NUMBERS OF SOME HYDROCARBONS, WITH AND WITHOUT TETRAETHYL LEAD (TEL)

Hydrocarbon	Molecular Formula	Octane Number	
		No TEL	3.0 mL/gallon TEL
n-Butane	C_4H_{10}	93.6	101.6
n-Pentane	C_5H_{12}	61.7	88.7
2-Methylbutane	C_5H_{12}	92.6	102.0
n-Hexane	C_6H_{14}	24.8	65.3
n-Heptane	C_7H_{16}	0.0	43.5
Methylbenzene (toluene)	C_7H_8	103.2	111.8
n-Octane	C_8H_{18}	-19.0	25.0
2,2,4-Trimethylpentane (isooctane)	C_8H_{18}	100.0	115.5
Isopropylbenzene (cumene)	C_9H_{12}	113.0	116.7

small amounts of certain substances has a similar effect.

The most common additive for many years was a compound called tetraethyl lead (TEL), $(C_2H_5)_4Pb$. Its antiknock property was discovered by Thomas Midgley (1889–1944), a chemist at General Motors Research Laboratory, in 1922. When added in amounts as small as 3 milliliters per gallon, it often increased the octane number by a large amount, as Table 18–12 shows. To prevent deposits of metallic lead and lead oxides from fouling the valves and spark plugs, another compound called a scavenger, usually ethylenedibromide $(C_2H_2Br_2)$ or ethylenedichloride $(C_2H_2Cl_2)$, was added. The scavenger converted the lead to gaseous lead(II) bromide or lead(II) chloride, which escaped through the exhaust. In this way, thousands of tons of lead went out the tailpipe into our atmosphere.

$$Pb(C_2H_5)_4 + \underset{\underset{Br}{|}}{CH_2}\underset{\underset{Br}{|}}{CH_2} + 16O_2 \longrightarrow$$

$$10CO_2 + 12H_2O + PbBr_2 \quad (18\text{-}6)$$

Catalytic converters use platinum and palladium to convert harmful carbon monoxide and hydrocarbons into harmless carbon dioxide and water, as required by the Clean Air Act of 1970, which ordered a 90% reduction in these emissions. Since the catalysts can be ruined by lead in gasoline, the Environmental Protection Agency required the addition of unleaded pumps at stations. With lead being rapidly phased out, replacements had to be found.

18-10 BENZENE; AROMATIC HYDROCARBONS

Many hydrocarbons have a higher ratio of carbon to hydrogen than the alkenes or alkynes, yet have distinctly different properties from these unsaturated hydrocarbons. They are frequently fragrant themselves or can be derived from aromatic substances such as cinnamon, sassafras, and wintergreen.

Michael Faraday discovered one of these compounds in illuminating gas in 1825. Later, the same compound was obtained from coal tar, petroleum, and benzoic acid isolated from the aromatic substance gum benzoin, and it became known as **benzene**. Its molecular formula was determined to be C_6H_6. Josef Loschmidt (1821–1895) proposed that aromatic compounds, such as phenol, aniline, and toluene, could be considered derivatives of benzene, just as the alkanes were considered to be derivatives of methane, CH_4. Since then, the term **aromatic** has been applied to compounds chemically similar to benzene. **Aromatic hydrocarbons** are used extensively in the manufacture of plastics, dyes, insecticides, and many other materials.

The molecular formula of benzene, C_6H_6, suggests greater unsaturation than is present in alkenes and alkynes. Contrary to expectations, however, benzene does not show the slightest chemical similarity to these compounds. Surprisingly, benzene resembles the alkanes in this inertness toward halogens and other reagents. As with alkanes, its usual reaction is one of **substitution**:

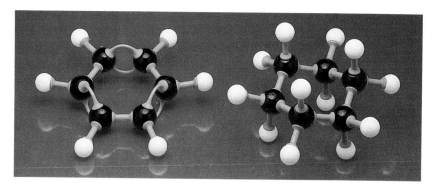

Figure 18–20 Ball-and-stick models of benzene and cyclohexane. (Charles D. Winters)

$$C_6H_6 + Br_2 \xrightarrow[\text{catalyst}]{\text{Fe}}$$

benzene   bromine

$$C_6H_5\!-\!Br + HBr \qquad (18\text{–}7)$$

bromobenzene   hydrogen bromide

$$C_6H_6 + HONO_2 \xrightarrow[\text{catalyst}]{H_2SO_4}$$

benzene   nitric acid

$$C_6H_5\!-\!NO_2 + H_2O \quad (18\text{–}8)$$

nitrobenzene   water

Benzene can be converted to cyclohexane by reacting with H_2, but under more severe conditions than the hydrogenation of alkenes and alkynes. This reaction suggests a cyclic structure of benzene (Fig. 18–20).

$$C_6H_6 + 3H_2 \xrightarrow[\text{Ni catalyst, 150°C}]{\text{high pressure}}$$

benzene

$$
\begin{array}{c}
\underset{\text{H}_2}{\text{C}} \\
H_2C \qquad CH_2 \\
| \qquad\qquad | \\
H_2C \qquad CH_2 \\
\underset{\text{H}_2}{\text{C}} \qquad (18\text{–}9)
\end{array}
$$

cyclohexane

Friedrich August Kekulé in 1865 proposed a ring structure for benzene in which three double bonds alternate with three single bonds.

$$
\begin{array}{ccc}
& \text{H} & \\
& | & \\
& \text{C} & \\
H\!-\!C & & C\!-\!H \\
\| & & \\
H\!-\!C & & C\!-\!H \\
& \text{C} & \\
& | & \\
& \text{H} &
\end{array}
\qquad
\begin{array}{ccc}
& \text{H} & \\
& | & \\
& \text{C} & \\
H\!-\!C & & C\!-\!H \\
& & \\
H\!-\!C & & C\!-\!H \\
& \text{C} & \\
& | & \\
& \text{H} &
\end{array}
$$

A second Kekulé structure can be drawn with the double bonds alternating the other way. Abbreviated Kekulé formulas are often written as follows:

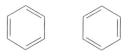

The Kekulé formula for benzene is supported in broad outline by modern studies. Electron diffraction experiments and X-ray analysis of crystalline benzene derivatives indicate that the six carbon atoms lie in a plane and that they form the corners of a regular hexagon. The hydrogen atoms lie in the same plane as the carbon atoms. Benzene is a flat, completely symmetrical molecule. However, the Kekulé formula predicts different bond lengths for the single and double bonds, but measurements show that **the six carbon-to-carbon bonds are equal in length**, 1.39Å, intermediate between the length of a single bond and a double bond.

The geometry of the benzene molecule suggests that the carbon atoms are **sp² hybridized**.

Each carbon uses its three sp² hybrid orbitals to form sigma bonds with a hydrogen atom and two carbon neighbors. The fourth valence electron of each carbon occupies a p orbital with lobes above and below the plane of the ring; there are thus six p orbitals in the molecule, all parallel to one another (Fig. 18–21). Evidently, the p orbitals do not overlap to form localized pi bonds as in alkenes and alkynes, as shown by the unvarying carbon-to-carbon bond lengths in benzene.

Instead, each p orbital must interact with each of the adjacent carbons rather than with just one. The result is two doughnut-shaped electron clouds, lying above and below the plane of the benzene ring (Fig. 18–22). This participation in the formation of more

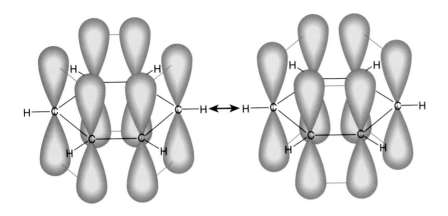

Figure 18–21 Structure of benzene. The carbon and hydrogen atoms all lie in the same plane. The p orbitals are all at right angles to the plane of the ring.

than one bond is called **delocalization**. The six p electrons of benzene are delocalized and form an **extended cyclic π bond** around all six carbons, rather than three localized π bonds between specific pairs of carbon atoms.

Delocalized bonding is stronger than ordinary covalent bonding between two atoms. Benzene is stable because the extended π bond resists breaking up; however, it is unaffected by substitution reactions. Since the benzene molecule has no definite single or double bonds, it is best represented by a **Thiele formula**, a circle in the center of a hexagon. The circle represents the π electron cloud of six electrons; the hexagon, the skeleton of six carbon atoms.

Thiele formula for benzene

Some examples of compounds derived from benzene are

Cl
para-dichlorobenzene
(moth flakes)
Cl

CH₃
NO₂ — NO₂
2,4,6-trinitrotoluene
(TNT)
NO₂

OH
Cl — CH₂ — Cl
ClCl
Cl Cl
hexachlorophene

Cl
Cl—C—Cl
C
Cl H Cl
dichlorodiphenyltrichloroethane
(DDT)

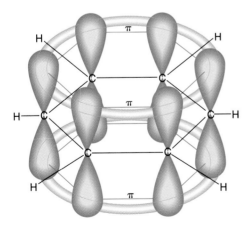

Figure 18–22 Extended bonding in the benzene molecule involves all six p orbitals, which are described as being delocalized. A cyclic charge cloud is formed above and below the plane of the carbon and hydrogen atoms.

acetylsalicylic acid
(aspirin)

saccharin

vanillin

naphthalene

18–11 FUNCTIONAL GROUPS

The replacement of a hydrogen atom of a hydrocarbon by another atom or group of atoms results in compounds called **hydrocarbon derivatives**. The atom or group of atoms is a **functional group**. The chemistry of these derivatives is mainly that of the functional group, which imparts certain common chemical and physical properties to all members of the family. Thus, the hydroxyl group, —OH, is found in all **alcohols**, represented by the general formula R—OH. The symbol R represents an alkyl group, the entity that remains following the removal of a hydrogen from an alkane, such as methyl, CH_3—, from methane, CH_4. Methyl alcohol is CH_3OH; ethyl alcohol, C_2H_5OH (Fig. 18–23).

The many thousands of organic compounds can be organized on the basis of a relatively small number of

Figure 18–23   Model of ethyl alcohol, C_2H_5OH, also called ethanol or grain alcohol. (Charles Steele)

functional groups. Some of the more common ones are listed in Table 18–13. It is not unusual for more than one such group to occur in a compound.

Each functional group is assigned a suffix in the IUPAC system; *-ol* for alcohols; *-one* for ketones; and so on. The basis of the name of a compound is the longest continuous alkane to which the functional group is attached. That alkane is considered to be the **parent hydrocarbon**, and the functional groups are derivatives of it.

EXAMPLE 18–2

Give the systematic (IUPAC) name for the following compound.

$$\underset{1}{CH_3}\overset{\overset{\displaystyle CH_3}{|}}{\underset{\underset{\displaystyle CH_3OH}{|}}{C}}\underset{3}{CH}\underset{4}{CH_2}\underset{5}{CH_2}\underset{6}{CH_3}$$

SOLUTION

We number the carbon atoms of the parent alkane starting at the end nearest the functional group, the—OH group, and designate the position of the functional group by the number of the carbon atom to which it is attached. This compound is an alcohol, specifically, a 3-hexanol. The names of any additional groups attached to the parent chain and the numbers of the carbons to which they are attached are indicated, followed by the name of the parent. Thus, the complete name is *2,2-dimethyl-3-hexanol*.

EXTENSION

Write the structure of 1-pentanol. (Answer: $CH_3CH_2CH_2CH_2CH_2OH$)

■

Alcohols are in a sense alkyl derivatives of water, in which one of the H atoms of water is replaced by an R group, as much as they are hydroxy derivatives of alkanes, in which a hydrogen atom of an alkane is replaced by an —OH group. The lower homologs are similar to water in structure:

water   an alcohol   methanol

Table 18-13   COMMON FUNCTONAL GROUPS IN ORGANIC COMPOUNDS

Type of Compound	Functional Group	General Formula	Example	Name (IUPAC)
Alkane	H	R—H	CH_4	Methane
Alkene	$\diagdown C=C \diagup$	$R_2C=CR_2$	$CH_3 \diagdown C=C \diagup CH_3$ $CH_3 \diagup \diagdown CH_3$	2,3-Dimethyl-2-butene
Alkyne	—C≡C—	RC≡CR	$CH_3—C≡C—CH_3$	2-Butyne
Halide	X : F,Cl,Br,I	R—X	CH_3Br	Bromomethane
Alcohol	—OH	R—OH	CH_3CH_2OH	Ethanol
Ether	—O—	R—O—R′	$CH_3—O—CH_3$	Methoxymethane
Aldehyde	$-\overset{O}{\overset{\|}{C}}-H$	$R-\overset{O}{\overset{\|}{C}}-H$	$CH_3\overset{O}{\overset{\|}{C}}H$	Ethanal
Ketone	$-\overset{O}{\overset{\|}{C}}-R′$	$R-\overset{O}{\overset{\|}{C}}-R′$	$CH_3\overset{O}{\overset{\|}{C}}CH_3$	Propanone
Carboxylic acid	$-\overset{O}{\overset{\|}{C}}-OH$	$R-\overset{O}{\overset{\|}{C}}-OH$	$CH_3\overset{O}{\overset{\|}{C}}OH$	Ethanoic acid
Ester	$-\overset{O}{\overset{\|}{C}}-OR′$	$R\overset{O}{\overset{\|}{C}}-OR′$	$CH_3\overset{O}{\overset{\|}{C}}-OCH_3$	Methyl ethanoate
Amine	—NH_2	R—NH_2	CH_3NH_2	Aminomethane

As in water, there are strong intermolecular attractions between alcohol molecules; these attractions arise from dipole–dipole interactions and from **hydrogen bonding**, an intermolecular force between the hydrogen of one molecule and the more electronegative oxygen in a nearby molecule (Fig. 18–24). As a result, methyl alcohol, CH_3OH, is a liquid at room temperature. Alcohols of low molecular weight are soluble in water, but as the alkyl group lengthens, the higher alcohols resemble the corresponding hydrocarbons more and more. **Polyhydric alcohols** exist in which two or more —OH groups are present. **Ethylene glycol**, commonly used as an antifreeze, is a dialcohol:

$$\underset{\text{ethylene glycol}}{\overset{CH_2CH_2}{\underset{OH\ OH}{|\ \ |}}} \qquad \underset{\text{glycerol}}{\overset{CH_2CH\ CH_2}{\underset{OH\ OH\ OH}{|\ \ |\ \ |}}}$$

and **glycerol (or glycerine)** is a tri-hydroxyalcohol (Fig. 18–25). The formulas and names of some alcohols are listed in Table 18–14.

Ethyl or grain alcohol (CH_3CH_2OH) is the most common alcohol. It is present in varying concentrations in such beverages as cider, beer, wine, champagne, and gin. Even the strongest alcoholic beverages are seldom more than 45% ethyl alcohol, or 90 **proof**. (**Proof** is twice the alcohol percentage by volume.)

Grain alcohol is often produced by fermenting sugars from various sources—corn, rice, potatoes, and sugar cane—in one of the oldest chemical proc-

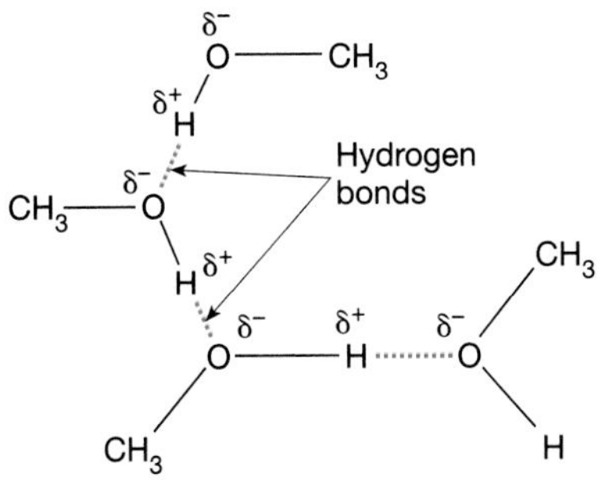

Figure 18–24 Hydrogen bonding in methyl alcohol, CH_3OH.

Figure 18–25 Polyhydric alcohols are used in permanent antifreeze and in cosmetics. (Charles Steele)

Figure 18–26 Bees making the ester beeswax, the product of the reaction of a long-chain alcohol and a long-chain carboxylic acid. (Charles D. Winters)

esses known. Fermentation takes place through certain enzymes found in yeast *(Saccharomyces cerevisiae)* that convert sugars to ethyl alcohol and carbon dioxide.

$$C_{12}H_{22}O_{11} + H_2O \xrightarrow{\text{yeast}}$$

sugar

$$4CH_3CH_2OH + 4CO_2 \quad (18-10)$$

ethyl alcohol

Most industrial alcohol is produced by reacting ethylene with water in the presence of sulfuric acid. An **addition reaction** takes place.

$$CH_2{=}CH_2 + H_2O \xrightarrow[\substack{100 \text{ atm}}]{\substack{300°C \\ H_2SO_4}}$$

$$CH_3CH_2OH \quad (18-11)$$

Since alcohol for drinking purposes is heavily taxed, industrial ethyl alcohol is **denatured** by adding poisonous compounds such as methyl alcohol (wood alcohol) to it, making it unfit to drink. Blindness can result from methanol poisoning, the optic nerve being particularly susceptible. Ethyl alcohol is used in large quantities as an industrial solvent and as a starting material for other chemical products.

Alcohols react with carboxylic acids—organic compounds containing a carboxyl group

Table 18–14 SOME COMMON ALCOHOLS

Alcohol	IUPAC Name	Common Name	Solubility in Water (g/100 g H₂O, 20°C)
CH_3OH	Methanol	Methyl alcohol	Miscible
CH_3CH_2OH	Ethanol	Ethyl alcohol	Miscible
$CH_3CH_2CH_2OH$	1-Propanol	*n*-Propyl alcohol	Miscible
$CH_3CH(OH)CH_3$	2-Propanol	Isopropyl alcohol	Miscible
$CH_3CH_2CH_2CH_2OH$	1-Butanol	*n*-Butyl alcohol	7.9
$CH_3CH_2CH(OH)CH_3$	2-Butanol	*sec*-Butyl alcohol	12.5
$CH_3CH(CH_3)CH_2OH$	2-Methyl-1-propanol	Isobutyl alcohol	10.0
$CH_3C(OH)(CH_3)CH_3$	2-Methyl-2-propanol	*t*-Butyl alcohol	Miscible

Table 18–15   SOME ESTERS: THEIR AROMAS AND FLAVORS

Ester	Formula	Aroma or Flavor
Ethyl acetate	$CH_3COOCH_2CH_3$	Apple
Ethyl butyrate	$CH_3CH_2CH_2COOCH_2CH_3$	Apricot
Ethyl formate	$HCOOCH_2CH_3$	Rum
Ethyl heptanoate	$CH_3(CH_2)_5COOCH_2CH_3$	Cognac
Isoamylacetate	$CH_3COOCH_2CH_2CH(CH_3)CH_3$	Banana
Methyl-n-valerate	$CH_3(CH_2)_3COOCH_3$	Pineapple
Octylacetate	$CH_3COO(CH_2)_7CH_3$	Orange

(—COOH)—to form the class of compounds known as **esters**. Animal and vegetable fats, oils, and waxes are largely made up of esters (Fig. 18–26). Fruit aromas have many components, of which esters are a dominant one.

$$CH_3\overset{\overset{\textstyle O}{\|}}{C}\boxed{-OH+H}\ O\ \ CH_2CH_2CH_2CH_3 \rightleftharpoons$$
$$\text{acetic acid} \qquad\qquad n\text{-butyl alcohol}$$

$$CH_3\overset{\overset{\textstyle O}{\|}}{C}-O-CH_2CH_2CH_2CH_3 + H_2O \quad (18\text{--}12)$$
$$\quad n\text{-butyl acetate} \qquad\qquad \text{water}$$

Some common esters used in perfumes and flavors are listed in Table 18–15. Natural **fats** are solid esters; **oils** are liquid esters of glycerol.

18–12 POLYMERS

Among the oldest and most familiar materials are natural **high polymers** (*poly*, meaning "many"; *meros*, meaning "part")—namely, silk, cotton, wool, and fur. Until the early decades of the twentieth century, polymers were a chemical mystery because of their complexity and fragility. Almost every aspect of our contemporary life, on the other hand, reflects in some way the advances made in the laboratory and industrial synthesis of giant molecules (Fig. 18–27). Much of our clothing is made from synthetic fibers. We walk on carpets and sit on furniture made from synthetic polymers.

Cellulose, rubber, starch, and proteins, like all organic compounds, are composed mainly of carbon, hydrogen, and oxygen. Their main distinguishing fea-

ture is the large size of their molecules. Since these **macromolecules** are made of simpler building blocks—isoprene in rubber, amino acids in proteins—they are called polymers. Molecular masses of polymers may range from 100,000 to several million. Any substance that can serve as a building unit is a **monomer**. More than 40 organic monomers derived from coal and oil can be combined in an almost limitless number of ways to form new materials.

Addition polymerization is one way by which small molecules are assembled into large ones. Many olefins polymerize in the presence of **free radicals**. Styrene is an ethylene derivative in which a benzene ring replaces one of the hydrogens of ethylene. In the first step of the process, a catalyst is dissociated by heat or light into free radicals.

$$R-R \xrightarrow[\text{or light}]{\text{heat}} 2R\cdot$$
$$\text{catalyst} \qquad\qquad \text{free radicals}$$

benzoyl peroxide

(18–13)

benzoyloxy radicals

In the **propagation** step, a free radical attacks a molecule of the monomer and reacts with its double bond by adding to it on one side and freeing an electron on the other. The new free radical is also capable

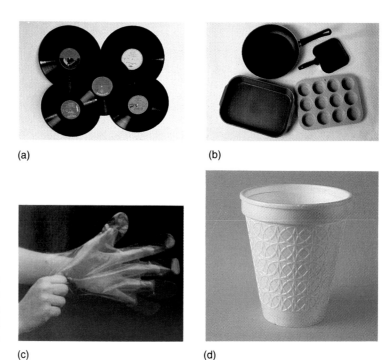

Figure 18–27 Some articles made from addition polymers. (a) Phonograph records are made from polyvinyl chloride. (b) A polyethylene glove. (c) Teflon-coated kitchen ware. (d) A polystyrene cup. (Beverly March)

of attacking a styrene molecule and maintaining the free radical character of the chain end. A large number of monomer units are added in this way to the growing chain, and a macromolecule is built up. The process **terminates** when two of the giant molecules happen to collide with each other and react to form a normal stable molecule, completing the chain reaction.

styrene a new radical

$$(18-14)$$

Polystyrene is used widely in packaging materials, food containers, electrical insulation, and molded objects. As Styrofoam it is used to make picnic coolers and other insulated containers. It has many other uses as well and can be polymerized with other materials. Polymers are so versatile in their applications that world production is millions of tons annually and rising.

The entity formed by the loss of one hydrogen from ethylene is known as the **vinyl** group. Styrene is the name given to vinyl benzene.

$$H_2C{=}CH_2 \qquad H_2C{=}CH \qquad H_2C{=}CH$$

ethylene vinyl group vinyl benzene (styrene)

Other vinyl-type addition polymers and some of their uses are listed in Table 18–16.

Another type of polymerization was discovered by Leo Hendrick Baekeland (1863–1944). An organic chemist in search of a substitute for shellac, Baekeland became interested in the gummy, tar-like liquid material that fouled up his glassware when in 1909 he carried out a reaction between phenol and formaldehyde. He found that by applying heat and pressure he could turn the liquid into a hard, transparent resin

Table 18–16 REPRESENTATIVE VINYL ADDITION POLYMERS

Monomer	Polymer	Uses
Ethylene H—C(H)=C(H)—H	Polyethylene —C—C—C—C—C—C— with H,H on each carbon	Wrapping film, tubing, molded objects
Vinyl chloride H,H–C=C–H,Cl	Polyvinyl chloride —C—C—C—C—C—C— with H above, H/Cl/H/Cl/H/Cl below	Phonograph records, raincoats, bottles
Tetrafluoroethylene F,F–C=C–F,F	Polytetrafluoroethylene —C—C—C—C—C—C— with F,F on each carbon	Teflon, chemically resistant coatings
Vinylidene chloride H,H–C=C–Cl,Cl	Polyvinylidene chloride —C—C—C—C—C—C— with H,Cl/H,Cl/H,Cl	Saran
Acrylonitrile H,H–C=C–H,C≡N	Polyacrylonitrile —C—C—C—C—C—C— with H/H, C≡N/H, C≡N/H, C≡N	Orlon, Acrilan
Propylene H,H–C=C–H,CH_3	Polypropylene —C—C—C—C—C—C— with H/CH_3 repeating	Fibers, films, carpets, heart valves, steering wheels
Isoprene H,H–C=C–CH_3, CH=CH_3	Polyisoprene —C—C=C—C— with H, CH_3, H, H	Synthetic rubber

that had interesting properties: electrical insulation and resistance to heat, moisture, chemicals, and mechanical wear. The phenol and formaldehyde monomers polymerize to form a large three-dimensional network as heat "splits out" water molecules, the hydrogen coming from the benzene ring and oxygen from formaldehyde. In effect, the phenolic rings are joined by CH_2 units.

$$\text{(18-15)}$$

Bakelite

The invention of the material, called **Bakelite,** gave birth to the synthetic plastics industry. It is used in molded objects such as telephone housings, insulators, varnishes, and many other ways.

Wallace Hume Carothers (1896–1937), a research chemist at DuPont, in 1935 prepared the first completely synthetic fiber—**nylon**—which has since replaced silk in almost all its uses. The most common type, Nylon 66, is formed by the **condensation polymerization** of a mixture of

two six-member monomers, hexamethylenediamine (NH_2—$(CH_2)_6$—NH_2) and adipic acid ($HOOC$—$(CH_2)_4$—$COOH$), in which molecules of water are split out (Fig. 18–28):

$$HOOC(CH_2)_4COOH \ + \ NH_2(CH_2)_6NH_2 \longrightarrow$$

adipic acid   hexamethylenediamine

$$\left[-\overset{\overset{\displaystyle O}{\|}}{C}-(CH_2)_4-\overset{\overset{\displaystyle O}{\|}}{C}-NH(CH_2)_6NH-\right]_n + 2nH_2O$$

nylon

$$\text{(18-16)}$$

Nylon is an example of a **polyamide,** a polymer having the repeating unit —CONH—. It comes in sheet, fiber, and bristle form and has many uses: hosiery, shirts, molded items, brushes, tires, and parachutes, to name a few.

The discovery of nylon laid the foundation for the development of all synthetic fibers. The easy-care, permanent-press **polyester** used in clothing, carpeting, and draperies is obtained from the condensation of ethylene glycol and terephthalic acid:

$$2\ \underset{\underset{\displaystyle OH}{|}}{CH_2}\ \underset{\underset{\displaystyle OH}{|}}{CH_2} + HO-\overset{\overset{\displaystyle O}{\|}}{C}-\bigcirc-\overset{\overset{\displaystyle O}{\|}}{C}-OH \longrightarrow$$

ethylene glycol   terephthalic acid

$$\left(-O-CH_2-CH_2-O-\overset{\overset{\displaystyle O}{\|}}{C}-\bigcirc-\overset{\overset{\displaystyle O}{\|}}{C}-\right)_n$$

poly(ethylene terephthalate)
(PET)

$$+\ n HOCH_2CH_2OH \quad \text{(18-17)}$$

Poly(ethylene terephthalate), or PET, is sold as Dacron, Kodel, and Fortrel (Fig. 18–29). The fabrics can be set or stabilized so that the resulting configuration is maintained during wear or use with little or no ironing required after laundering. Trouser creases and skirt pleats are there for the life of the garment. For these purposes, PET fiber is blended with cotton in a 50 : 50 or 65 : 35 ratio of PET to cotton. The consumption of synthetic fibers now exceeds that of the natural fibers cotton and wool, and polyester is one of the most versatile fibers available.

Figure 18–28 Nylon-66, a condensation polymer, is being wound onto a stirring rod after a reaction between hexamethylene diamine and a derivative of adipic acid
(Charles D. Winters)

Figure 18–29   A garment made of Dacron. (Courtesy of DuPont de Nemours and Company)

18–13 ISOMERISM REVISITED

The type of isomerism discussed in Section 18–4 is **chain isomerism**, a type of structural isomerism in which there is a difference in the sequence of carbon atoms. Examples are *n*-butane and isobutane. Consider these two halides:

$$
\begin{array}{cccc}
& H & H & \\
& | & | & \\
H-&C-&C-&Cl \\
& | & | & \\
& H & Cl &
\end{array}
\qquad and \qquad
\begin{array}{cccc}
& H & H & \\
& | & | & \\
H-&C-&C-&H \\
& | & | & \\
& Cl & Cl &
\end{array}
$$

1,1-dichloroethane   1,2-dichloroethane

Their molecular formulas are identical: $C_2H_4Cl_2$. They contain the same number and same types of bonds and the same carbon skeleton. But they differ in the point on the chain where some type of substitution occurs, and for this reason they are **positional isomers.** Another example are the alcohols 1-propanol and 2-propanol. The carbon skeleton is the same in both molecules, which differ only in the position of the —OH group.

$$CH_3CH_2CH_2OH \qquad CH_3CHCH_3$$
$$| $$
$$OH$$

1-propanol   2-propanol

A third type of structural isomerism is found in C_2H_6O. When the atom are arranged in the sequence CH_3-O-CH_3 the substance is an **ether;** in the sequence CH_3CH_2OH, the substance is an alcohol. The different sequences result in different **functional groups.** The properties of **functional group isomers** differ.

dimethyl ether	ethyl alcohol
(CH_3-O-CH_3)	(CH_2H_5OH)
1. A gas	1. A liquid
2. Practically insoluble in H_2O	2. Miscible in all proportions in H_2O

There are compounds that have the same molecular formula (the same number and kind of atoms) and the same structural formula (the same sequence of atoms) but differ in the spatial arrangement of the atoms. They are **stereoisomers**. In **geometrical** or *cis-* and *trans-* **isomers**, the compounds differ in their configuration because of lack of free rotation, most commonly about a double bond (Fig. 18–30). For example, 2-butene exists in two isomeric forms that can be represented only by their **configurational**

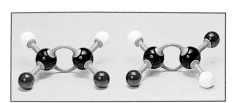

Cis and *trans* isomers.

$$
\begin{array}{ccc}
H & & H \\
\ & C=C & / \\
Cl & & Cl
\end{array}
\qquad
\begin{array}{ccc}
H & & Cl \\
\ & C=C & / \\
Cl & & H
\end{array}
$$

cis-1,2-dichloroethene   *trans*-1,2-dichloroethene
mp, – 80 °C; bp, 60 °C   mp, – 50 °C; bp, 48 °C

Figure 18–30   *Cis*-1,2-dichloroethene *(left)* and *trans*-1,2-dichloroethene *(right)* are geometrical stereoisomers. (Charles D. Winters)

Table 18–17   SOME PHYSICAL PROPERTIES OF *cis-* AND *trans*-2-BUTENE

	Boiling Point (°C)	Melting Point (°C)	Density (g/mL, 20°C)	Heat of Fusion (kcal/mol)
cis-2-Butene	+ 3.7	− 139	0.621	1.75
trans-2-Butene	+ 0.9	− 106	0.604	2.33

Table 18–18   SOME PHYSICAL PROPERTIES OF *d-* AND *l*-TARTARIC ACIDS

Acid	Melting Point (°C)	Solubility (g/100 g H$_2$O)	Density (g/mL)	Specific Rotation
d-Tartaric acid	170	139	1.76	+ 12°
l-Tartaric acid	170	139	1.76	− 12°

formulas. *Cis-* denotes that the groups that are part of the parent chain lie on the same side of an axis drawn through the double bond; *trans-* indicates that they lie on opposite sides.

$$\begin{array}{ccc} H & & H \\ \diagdown & {}^{2}\!\!\diagup\!\!{}^{3} & \diagup \\ & C \!=\! C & \\ \diagup & & {}^{4}\!\!\diagdown \\ CH_3 & & CH_3 \end{array}$$

cis-2-butene

$$\begin{array}{ccc} CH_3 & & H \\ \diagdown & {}^{2}\!\!\diagup\!\!{}^{3} & \diagup \\ & C \!=\! C & \\ \diagup & & {}^{4}\!\!\diagdown \\ H & & CH_3 \end{array}$$

trans-2-butene

Note the differences in some of the physical properties of the *cis-* and *trans*-isomers listed in Table 18–17, which reflect the differences in the shapes of the molecules and intermolecular attractions. The chemical properties of *cis-* and *trans*-isomers are similar, since the same functional group is present, but there are differences in the rates of their reactions.

A more subtle type of isomerism occurs in many compounds in which one or more **chiral** carbon atoms are present. A chiral carbon is one that is

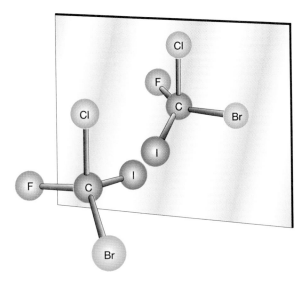

Figure 18–31 Optical isomers. Mirror images of a chiral carbon atom cannot be superimposed.

bonded to four different groups, like the carbon in CFClBrI (fluorochlorobromoiodomethane). The compound and its mirror-image isomer are **optical isomers** (Figs. 18–31 and 18–32). An estimate of the maximum possible isomers for a given structural formula is given by van't Hoff's rule: $\# = 2^n$, in which n is the number of chiral carbons. If $n = 4$ in a compound, then the maximum number of optical isomers is $2^4 = 16$. The combined effect of the various kinds of isomerism is to increase the possible number of carbon compounds enormously.

A classic illustration of optical isomerism is two forms of tartaric acid, which are mirror images of each other:

levo-(l) tartaric acid   dextro-(d) tartaric acid

$$
\begin{array}{cc}
\text{COOH} & \text{COOH} \\
| & | \\
\text{H—C—OH} & \text{HO—C—H} \\
| & | \\
\text{HO—C—H} & \text{H—C—OH} \\
| & | \\
\text{COOH} & \text{COOH}
\end{array}
$$

As Table 18–18 indicates, levo- and dextro-tartaric acids have identical physical properties, with one exception—their effect on polarized light, denoted by $[\alpha]_D^{20}$, the specific rotation. One isomer rotates the plane of polarized light 12° to the right (dextro, d); the other 12° to the left (levo, l).

Figure 18–32 Mirror images. A left hand in front of a mirror looks like a right hand. The two hands are nonsuperimposable mirror images of one another. Each hand is a chiral object. (Ray Ellis/Photo Researchers, Inc.)

KEY TERMS

Acetylene   The simplest hydrocarbon having a triple bond, $HC \equiv CH$.

Addition Polymerization   A process in which small molecules, monomers, add to one another to form a large molecule, a polymer, that contains all the atoms of the starting monomers.

Addition Reaction   A typical reaction of unsaturated hydrocarbons—alkenes and alkynes. The breaking of relatively weak pi bonds is followed by the formation of bonds with a substance that in effect adds to the hydrocarbon.

Alcohol   A hydrocarbon derivative in which one or more hydrogen atoms are replaced by —OH groups.

Alkanes   Saturated hydrocarbons. Organic compounds containing hydrogen and carbon that have two kinds of single bonds, carbon-to-hydrogen and carbon-to-carbon.

Alkenes   Unsaturated hydrocarbons containing at least one double bond between carbon atoms.

Alkyl Group   An entity derived from an alkane by the removal of one hydrogen atom.

Alkyne   A hydrocarbon that has a carbon-carbon triple bond.

Aromatic Hydrocarbons   Hydrocarbons that are chemically similar to benzene.

Bakelite   A term used for various synthetic resins and plastics.

Benzene   An aromatic hydrocarbon with the formula C_6H_6.

Bond Angle   The angle formed between three atoms in a molecule.

Catalytic Converter   A device containing a catalyst for oxidizing carbon monoxide and hydrocarbons to carbon dioxide and water.

Chain Isomerism   A type of structural isomerism in which there is a difference in the sequence of carbon atoms.

Chemical Affinity   A hypothetical force that was assumed to govern inorganic substances.

Chemistry of Carbon Compounds   Organic chemistry.

Chiral Atom An atom bonded to four different groups.

Condensation Polymerization A process by which a polymer is made; monomers react to give a polymer and some small molecule such as water.

Condensed Structural Formula A simplified structural formula in which central atoms and the atoms bonded to them are written as groups.

Configurational Formula A formula that shows the spatial arrangement of atoms.

Conjugated System A molecule containing alternating single and double bonds between the carbon atoms.

Delocalization In benzene, the participation of the p orbital in the formation of more than one bond, resulting in an extended cyclic π bond.

Denatured Alcohol Ethyl alcohol to which some noxious substance has been added to render it unfit for drinking.

Double Bond The two atoms are joined by sharing two pairs of electrons.

Ester A class of organic compounds formed when an alcohol reacts with an acid.

Ether A class of organic compounds corresponding to the general formula R—O—R.

Ethylene The simplest alkene, $H_2C=CH_2$.

Ethylene Glycol An alcohol with two —OH groups that is commonly used as an antifreeze.

Fat A solid ester.

Free Radical An atom or molecule containing an unpaired electron, for which reason it is quite reactive.

Fuel A substance that burns readily with the release of significant amounts of energy.

Functional Group An atom or group of atoms that imparts common chemical and physical properties to all members of a family containing it, such as the—OH group in alcohols.

Functional Group Isomers Isomers in which the different sequences of atoms result in different functional groups, as in dimethyl ether and ethyl alcohol.

Gasoline A liquid hydrocarbon mixture used as a fuel for internal combustion engines.

General Formula A formula that characterizes the members of a homologous series, such as CnH_2n.

Geometrical Isomers Isomers that have the same sequence of atoms, but whose orientation in space differ, such as *cis*- and *trans*-2-butene.

Glycerol (Glycerine) A trihydric alcohol.

Homologous Series A group of related organic compounds, such as the alkanes, in which the molecular formulas of the members differ from one another by a constant increment.

Homologs Members of a homologous series.

Hybrid Orbitals The result of mixing two or more orbitals to form orbitals suitable for bonding, such as sp and sp^2.

Hydrocarbon The simplest organic compounds, containing only hydrogen and carbon.

Hydrocarbon Derivative The result of replacement of a hydrogen atom of a hydrocarbon by another atom or group of atoms.

Hydrogen Bonding An intermolecular force between a hydrogen atom of one molecule and an oxygen or nitrogen atom in a nearby molecule.

Isomerism The existence of more than one compound with a given molecular formula owing to differences in structure or to the spatial orientation of the atoms.

Isomers Compounds that have the same molecular formula but different structural formulas and properties.

Macromolecule Giant molecules, or polymers.

Monomer Any substance that can serve as a building unit of a polymer.

Multiple Bond A bond with more than one pair of electrons shared.

Nylon The first completely synthetic fiber; a polyamide.

Octane Rating A measure of the antiknock properties of a gasoline used as an automotive fuel.

Oil A substance formed from glycerol and three unsaturated fatty acids.

Olefin An alkene.

Optical Isomers Two mirror-image forms of a chiral molecule.

Organic Chemistry The chemistry of the compounds of carbon.

Paraffin An alkane, or saturated hydrocarbon.

Parent Hydrocarbon In deriving the IUPAC name of a compound, the alkane corresponding to the longest continuous chain of carbon atoms in the molecule is regarded as the parent hydrocarbon.

Pentane An alkane, C_5H_{12}.

Photochemical Smog Smog created by the action of sunlight on unburned hydrocarbons and nitrogen oxides, mainly from automobiles.

Pi Bond A covalent chemical bond formed by the lateral overlap of p orbitals on adjacent carbon atoms.

Polyene An alkene that contains more than one double bond.

Polyester   A polymer consisting of repeating units of an ester used in making fibers or plastics.

Polyhydric Alcohol   An alcohol that contains two or more —OH groups.

Polymer   A large molecule formed from the combination of many small molecules.

Polystyrene   An addition polymer that has many uses such as packaging materials, electrical insulation, and insulated containers.

Positional Isomers   Isomers that have the same carbon skeleton but differ in the point on the chain where substitution occurs, such as 1-propanol and 2-propanol.

Primary Carbon   A carbon atom that is bonded to only one other carbon atom.

Proof   A concentration unit used for alcohol solutions; proof is twice the alcohol percentage by volume.

Propagation   Repeating consecutive steps in a free radical reaction.

Saturated Hydrocarbon   An alkane.

Secondary Carbon   A carbon atom that is bonded to two other carbon atoms.

Sigma Bond   A covalent bond that is cylindrically symmetrical around the line joining the nuclei.

sp Hybrid Orbitals   Formed from one s and one p orbital.

sp² Hybrid Orbitals   Formed from one s and two p orbitals.

sp³ Hybrid Orbitals   Formed from one s and three p orbitals.

Structural Theory   An interpretation of organic substances and their reactions.

Substitution Reaction   A typical reaction of alkanes, in which one or more hydrogen atoms are replaced by other atoms.

Systematic Name   A name of a compound that follows the recommendations of the IUPAC.

Termination Step   In a chain mechanism, the final step in the process.

Tertiary Carbon   A carbon atom that is bonded to three other carbon atoms.

Tetrahedral Configuration   The valance bonds of a carbon atom are directed toward the four corners of a regular tetrahedron.

Tetrahedron   A geometrical figure with four identical triangular faces.

Thiele Formula   A formula representing benzene, with a circle in the center of a hexagon.

Triple Bond   The atoms are joined by sharing three pairs of electrons.

Unsaturated Hydrocarbon   A hydrocarbon containing one or more double or triple bonds.

Vinyl Group   The entity formed by the loss of one hydrogen from ethylene.

Vital Force   A hypothetical force that was supposed to regulate substances derived from plant or animal sources.

THINGS TO DO

1. With a molecular model kit, construct models of (a) the isomeric butanes and (b) the isomeric pentanes. Name each of the isomers.

2. Construct models of the structural isomers (a) ethyl alcohol and (b) dimethyl ether. Note the different functional groups.

3. Add 1 mL of the hydrocarbon hexane to 5 mL of water in a test tube. Is hexane soluble in water?

4. Construct models of (a) ethylene; (b) acetylene; and (c) benzene.

5. Place 1 mL of hexane in a dry test tube. Add $KMnO_4(aq)$ dropwise, up to 20 drops, shaking well after each addition. Do they react?

6. Repeat (5), but with 2-hexene in place of hexane. What do you observe? This reaction is known as the Baeyer test for unsaturation.

7. Repeat (5) with toluene, an aromatic hydrocarbon. Result?

8. Examine a number of articles that consist, in part, at least of synthetic fibers. List the names of the fibers if given.

9. Study the ingredients listed on a number of food containers. Classify as many as possible into the funtional groups mentioned in the chapter.

10. If facilities are available, prepare soap. Refer to a chemistry laboratory manual for details.

EXERCISES

Structure and Bonding

18–1. What characteristics of carbon make possible the formation of the great number of organic compounds?

18–2. Why are molecular formulas alone inadequate to represent organic compounds? Illustrate with an example.

18–3. Contrast the physical properties of the ionic compound table salt, sodium chloride, with those of the hydrocarbon naphthalene ($C_{10}H_8$), moth flakes.

18–4. Which of the following are organic compounds?
(a) C_3H_8
(b) NaBr
(c) C_2Cl_6
(d) $KMnO_4$
(e) $C_6H_{12}O_6$

18–5. A compound has a melting point of 84°C, is insoluble in water, and burns. Is it likely to be an organic or an inorganic compound?

18–6. Which compound probably has the higher melting point, NaOH or CH_3CH_2OH?

18–7. What does the term "hybrid orbital" mean?

18–8. Show the overlap of orbitals in CH_3Cl.

18–9. Supply the missing hydrogens in this structural formula:

$$C-C=C-C-C$$
$$\quad\quad\quad | $$
$$\quad\quad\quad C$$

18–10. Identify the 1°, 2°, and 3° carbons in the formula given in Exercise 18–9.

Isomerism

18–11. Draw the structures of the smallest alkanes that are structural isomers.

18–12. Why are molecular formulas alone inadequate to represent organic compounds? Illustrate with an example.

18–13. (a) What is meant by a chiral carbon? (b) What unique property does it have?

18–14. Draw the structures of all the pentane isomers and name them.

18–15. Draw and label the *cis*- and *trans*-isomers of 2-butene.

Homologous Series and Functional Groups

18–16. What is meant by a homologous series of compounds?

18–17. Write structural formulas for an example of each of the following.
(a) alkene
(b) alcohol
(c) ether
(d) ester
(e) aldehyde

18–18. Identify the functional groups present in each of the following compounds.
(a) CH_2CH_2 with OH OH
(b) $C_2H_5-O-C_2H_5$
(c) $CH_3\overset{O}{\overset{\|}{C}}-O-CH_3$
(d) benzene ring with $\overset{O}{\overset{\|}{C}}-H$
(e) benzene ring with COOH and $O-\overset{O}{\overset{\|}{C}}-CH_3$

18–19. Write condensed structural formulas for the following compounds.
(a) acetylene
(b) di-*n*-propyl ether
(c) 2,2-dimethyl-3-chlorobutane
(d) bromobenzene
(e) methyl acetate

18–20. Why is methyl alcohol not used as a solvent for medicines to be taken internally?

18–21. Define and illustrate the following terms.
(a) functional group
(b) delocalization energy
(c) unsaturation
(d) aromaticity
(e) geometrical isomerism

18–22. What is denatured alcohol?

Hydrocarbons

18–23. Write the structural formula and name for the simplest (a) alkene, (b) alkyne, (c) aromatic hydrocarbon.

18-24. State what is wrong with each of the following names.
    (a) 4-methylpentane
    (b) 2-dimethylbutane
    (c) 2-ethyl-4,4-dimethylpentane
    (d) 2,3-methylhexane

18-25. Name each of the following compounds by the IUPAC system.
    (a) $CH_3CH_2CH_2CH_2CH_2CH_3$
    (b) 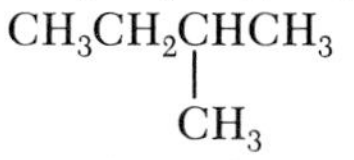 $CH_3CH_2CHCH_3$
               $|$
             CH_3
    (c)

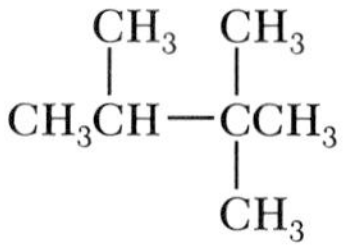

$$CH_3CH-CCH_3$$

with CH_3 and CH_3 above, and CH_3 below the second carbon.

    (d)

$$CH_3CH-CHCH_3$$

with CH_3-CH_2- above the left carbon and CH_2-CH_3 below the right carbon.

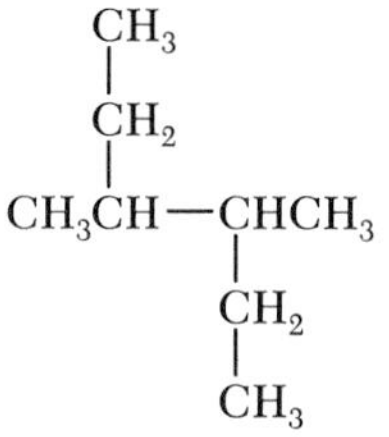

18-26. Explain the statement that the octane number of a gasoline is 94.

18-27. Why has the structure of benzene been a problem to scientists?

18-28. How do (a) benzene and (b) ethylene differ in their reactions with bromine?

18-29. Discuss the main source of the man-made atmospheric pollutants, the hydrocarbons, and the prospects for minimizing them.

18-30. Discuss the role of π bonding in benzene.

18-31. Write the structural formula for the principal reaction product when benzene reacts with HBr.

18-32. Which hybrid orbitals are involved in a $C=C$ double bond?

18-33. Draw the two resonance forms of benzene. Is the real benzene represented by either of them? Explain.

18-34. Write the structural formulas of the following groups.
    (a) methyl
    (b) ethyl
    (c) isopropyl
    (d) *t*-butyl

18-35. *Multiple Choice*
  A. Which is an isomer of 2-methylbutane?
    (a) propane
    (b) pentane
    (c) 2-methylpropane
    (d) 2-methylpentane
  B. The type of hybridization present in CH_4 is
    (a) sp.
    (b) sp^2.
    (c) sp^3.
    (d) p-p.
  C. A functional group is
    (a) one of the vertical columns in the periodic table.
    (b) a horizontal row in the periodic table.
    (c) exemplified by R in ROH.
    (d) exemplified by OH in ROH.
  D. A double bond between two carbon atoms consists of
    (a) two sigma bonds.
    (b) two pi bonds.
    (c) one sigma and one pi bond.
    (d) sp hybridized carbon atoms.
  E. Polyethylene is an example of
    (a) a monomer.
    (b) an addition polymer.
    (c) a radical.
    (d) a rubber.
  F. The normal alkane that contains seven carbon atoms is called
    (a) pentane.
    (b) septane.
    (c) hexane.
    (d) heptane.
  G. The reaction $CH_4 + Cl_2 \rightarrow CH_3Cl$ is an example of
    (a) a substitution reaction.
    (b) an addition reaction.
    (c) a polymerization reaction.
    (d) a neutralization reaction.
  H. Which of the following hydrocarbons would most likely be a component of gasoline?
    (a) C_2H_6
    (b) C_7H_{16}
    (c) $C_{12}H_{36}$
    (d) $C_{17}H_{34}$
  I. Gasoline that has the same knocking tendency as a mixture of 60% *n*-heptane and 40% isooctane has an octane number of
    (a) 20.
    (b) 40.
    (c) 60.
    (d) 100.

J. Hydrocarbons that contain a double bond are called
    (a) saturated.

(b) alkanes.
(c) alkenes.
(d) alkynes.

SUGGESTIONS FOR FURTHER READING

Brand, David J., and Jed Fisher, "Molecular Structure and Chirality." *Journal of Chemical Education*, December, 1987.

The term "chirality" is at the very heart of stereochemistry, but it has not escaped ambiguity. This essay clarifies this ambiguity.

Breslow, Ronald, "The Nature of Aromatic Molecules." *Scientific American*, August, 1972.

The surprising stability of the benzene ring is referred to as aromaticity, a property that depends on the presence of delocalized electrons. The exploration of aromaticity has both practical and theoretical significance.

Carraher, Charles E., Jr., George Hess, and L. H. Sperling, "Polymer Nomenclature—or What's in a Name?" *Journal of Chemical Education*, January, 1987.

Discusses the naming of linear organic polymers, the most common polymers used by our society. Four major types of names are emphasized: common names, source-based names, characteristic group names, and structure-based names.

Curl, Robert F., and Richard E. Smalley, "Fullerenes." *Scientific American*, October, 1991.

A new class of molecules that constitute the third form of pure carbon (the other two are diamond and graphite) was discovered in 1985. These molecules contain 60, 70, and more carbon atoms and have the structure of the geodesic dome invented by the engineer and philosopher R. Buckminster Fuller. The first such molecule, C_{60}, was named buckminsterfullerene, or buckyball for short, and the class is termed the fullerenes.

Gold, Thomas, and Steven Soter, "The Deep-Earth-Gas Hypothesis." *Scientific American*, June, 1980.

Enormous amounts of natural gas may lie deep in the earth. If they can be tapped, there would be a source of hydrocarbon fuel that could last for thousands of years.

Gray, Charles L., Jr., and Jeffrey A. Alson, "The Case for Methanol." *Scientific American*, November, 1989.

The automobile threatens the quality of life, contaminating both urban air and the global atmosphere. Vehicles operating on pure methanol would bring about dramatic decreases in urban levels of ozone and toxic substances. Methanol can be produced with current technologies from a variety of sources—gas, coal, wood, and even organic garbage.

Huffman, Donald R., "Solid C_{60}." *Physics Today*, November, 1991.

A method for producing the soccer-ball–shaped buckminsterfullerene molecule in abundance led to the discovery of a totally new form of crystalline carbon.

Kaner, Richard B., and Alan G. MacDiarmid, "Plastics That Conduct Electricity." *Scientific American*, February, 1988.

Twenty years ago plastics were rigidly categorized as insulators. The suggestion that a plastic could conduct as well as copper would have seemed even more ludicrous. Yet in the past few years this feat has been achieved.

Labianca, Dominick A., "The Chemical Basis of the Breathalyzer." *Journal of Chemical Education*, March, 1990.

The Breathalyzer is the tester of choice as a reliable, noninvasive instrument for forensic breath-alcohol analysis. Its chemistry involves the reduction of potassium dichromate by ethanol contained in the breath sample of a test subject.

Lambert, Joseph B., "The Shapes of Organic Molecules." *Scientific American*, January, 1970.

A knowledge of the shapes of molecules, which influence chemical reactivity, is valuable in the synthesis of compounds such as antibiotics of the penicillin family.

Lieber, Charles S., "The Metabolism of Alcohol." *Scientific American*, March, 1976.

Alcoholism can cause cirrhosis and death because alcohol disturbs liver metabolism and damages the liver cells.

Samulski, Edward T., "Polymeric Liquid Crystals." *Physics Today*, May, 1982.

From soft contact lenses to heat shields for spacecraft, the materials are the products of polymer science. Polymeric liquid crystals have numerous technological and biological applications.

Seymour, Raymond B., "Chemicals in Everyday Life." *Journal of Chemical Education*, January, 1987.
Few nonscientists recognize the importance of chemistry in our daily lives. We need chemicals for our food, clothing, shelter, recreation, ventilation, communication, decoration, sanitation, and education.

Wittcoff, Harold, "Nonleaded Gasoline: Its Impact on the Chemical Industry." *Journal of Chemical Education*, September, 1987.
The use of lead in gasoline to increase the octane number is rapidly being phased out. The author explores possible replacements.

C_{60} AND THE FULLERENES

MICHAEL A. DUNCAN

The University of Georgia

Some of the most unusual "organic" molecules ever imagined were discovered in the 1980s when molecular beam cluster experiments were used to vaporize solid carbon. A variety of molecules that contained only carbon atoms were produced. In the same way that molecules containing only metal atoms are called *metal clusters,* these species are called *carbon clusters.* Analysis of these carbon molecules of different sizes shows that the molecule containing 60 carbon atoms, C_{60}, is formed far more readily than other sizes and is incredibly stable.

C_{60} was first identified in molecular beam experiments by Professor Richard Smalley and his research team at Rice University. From these early experiments they could only speculate why C_{60} was so stable and so different from the other clusters. Smalley argued that the only structure that could be so stable would be a sphere of carbon formed of interconnected five- and six-membered rings (Fig. A). Smalley named the molecule "Buckminsterfullerene" after the architect Buckminster Fuller, who specialized in geodesic dome designs. The structure that he proposed has the same shape as a soccer ball, and the nickname "Buckyball" soon became associated with C_{60}. Unfortunately, because of the small amount of material produced in molecular beam experiments, Smalley could not actually measure the structure to prove that his model was right.

C_{60} might have remained a laboratory curiosity except for an exciting discovery by Donald Huff-man (University of Arizona) and Wolfgang Krätschmer (Max Planck Institute for Nuclear Physics, Heidelberg) in 1990. They found that an electrical discharge, or arc, made with graphite electrodes in a helium atmosphere generated carbon soot with unusual properties. Part of that soot dissolved in ordinary organic solvents such as benzene and toluene; normal soot does not dissolve in these solvents. After filtering away the insoluble material and evaporating the solvent, the scientists isolated a yellowish brown powder. With gram quantities available, traditional analysis using an array of specialized instrumental techniques (nuclear magnetic resonance and infrared spectroscopy, X-ray crystallography) became possible. To everyone's delight (especially Smalley's), the powder contained mostly C_{60}, and it had the structure that he had proposed almost ten years earlier! C_{60} thus became the first cluster of a pure element ever to be isolated and collected in quantities great enough for traditional chemical experiments. These same experiments also produced several larger, less symmetrical, carbon cage molecules (e.g., C_{70} and C_{84}). Taken together, C_{60} and its analogs are now referred to as the *fullerenes.*

The highly symmetrical structure of C_{60} helps to explain many of its unusual properties. It is almost a perfect sphere, with 60 atoms arranged in 20 hexagons and 12 pentagons (32 faces in all). Each of the 60 carbon atoms is sp^2 hybridized and occu-

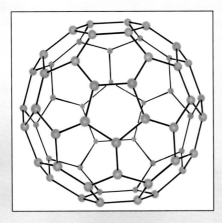

Figure A   The isosahedral structure of C_{60}.

pies an identical site in the cluster. Each has two single bonds and one double bond to its neighbors and is located at the juncture of one five-membered ring and two six-membered rings. This structure is very different from those of other forms of carbon. For example, diamond has tetrahedral bonding to four nearest neighbors. Graphite is more similar to C_{60}, with an infinite array of planar six-membered rings. However, an array of six-membered rings cannot be bent into a closed cage because of the strain on the bonds. Substitution of precisely the correct number of five-membered rings, as occurs in C_{60}, relieves this strain.

There are numerous equivalent ways to draw the chemical bonds in C_{60}. Equivalent bonding arrangements, known as *resonance structures,* are also found for molecules such as benzene and give them their "aromatic" character. C_{60} is, therefore, also an aromatic molecule. Empirical rules have been developed from chemical bonding theory to predict which other molecules similar to C_{60} might be found. In addition to C_{60}, carbon molecules with a multiple of 60 atoms, such as the cluster C_{120}, C_{240}, C_{540}, and C_{960}, are also predicted to be stable aromatic molecules with highly symmetrical open-cage structures. Attempts are underway to isolate and characterize these larger fullerenes.

The spherical cage structure makes C_{60} and the other fullerenes fascinating candidates for all kinds of applications in chemistry and for the preparation of new solid materials. For example, crystals of C_{60} have been prepared in which the molecules arrange themselves in a hexagonal close-packed structure. When alkali metal atoms such as cesium or rubidium are added into the gaps between the balls, the resulting compound is a *superconductor*. The open interior of C_{60} is a cavity about 5 Å wide, large enough to contain other atoms—in particular, metals. Numerous research groups are trying to find the conditions necessary for encapsulating atoms with C_{60} or with other fullerenes that have larger cavities. The unusual shape of the C_{60} molecule is of special interest in situations in which molecular shapes determine chemical activity, as in biological molecules, pharmaceutical drugs, and polymers. To investigate these kinds of applications, other functional groups or reactive organic systems have already been attached to C_{60}. Long chains of the form $(C_{60})-R-(C_{60})-R$ . . . have also been constructed. This peculiar new molecule, C_{60}, and other members of its family are rapidly emerging from the realm of molecular beams into the mainstream of practical chemistry.

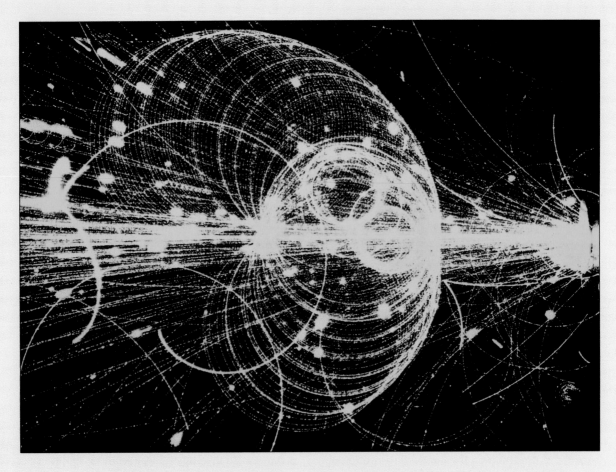

The collision of a high-energy oxygen nucleus with a nucleus in a lead target causes the spilling of 220 charged subatomic particles, as captured in this streamer chamber photograph, and perhaps 80 electrically neutral, invisible particles. (CERN/Science Photo Library/Photo Researchers, Inc.)

19

NUCLEAR REACTIONS AND STRUCTURE

The dual aspect of science as a boon and a threat to mankind is nowhere more evident than in the field of nuclear science. On the one hand, as the sources of fossil fuel—particularly oil and gas—become depleted, society must depend more and more on alternative energy forms, including, some say, nuclear energy. On the other hand, the possibility of nuclear holocaust underlies the foreign policy of the great nations of the world. In this chapter, we examine the nature of the atomic nucleus, the source of this awesome energy.

19-1 NUCLEAR STABILITY

As we explore the nature of the nucleus, there will be occasion to refer to specific nuclides. A **nuclide** is any nuclear species that has a specific atomic number and mass number. For example, uranium-235, $^{235}_{92}U$, is a nuclide; magnesium-27, $^{27}_{12}Mg$, is another nuclide. The **atomic number** (subscript) is characteristic of the element, and the **mass number** (superscript) identifies a particular nuclide of that element. The various nuclides of an element are said to be **isotopes** of that element. Thus, other isotopes of uranium exist, including the nuclides $^{234}_{92}U$ and $^{238}_{92}U$ (Fig. 19–1).

There are nuclides like $^{20}_{10}Ne$ that have equal numbers of protons and neutrons; in this case, ten of each. In most nuclides, however, the proton and neutron numbers are different, the neutron-to-proton ratio increasing to about 1.5 for heavier elements, such as $^{184}_{74}W$. A plot of the number of neutrons, N, versus the

number of protons, Z, shows that the stable (nonradioactive) nuclides fall within a narrow belt (Fig. 19–2). There are about 400 stable nuclides. No nuclides are stable beyond $Z = 83$, bismuth.

A nuclide outside the stability band tends to stabilize itself by spontaneously transforming or "decaying" to a nuclide closer to or within the stability region. If the nuclide is to the left of the stability region, it has an excess of neutrons and can achieve stability through a process that lowers the neutron-to-proton ratio. One to the right of the stability band has an excess of protons; it can achieve stability by a process that lowers the number of protons or increases the number of neutrons, thus increasing the neutron-to-proton ratio. These processes are treated in greater detail in the following section.

19-2 NUCLEAR TRANSMUTATIONS

The processes by which unstable nuclides achieve stability are called **transmutations**, since they usually involve a change in atomic number and mass number. Transmutations may be either spontaneous (as in natural radioactivity) or artificially induced in

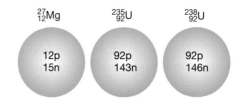

Figure 19-1 Examples of nuclides.

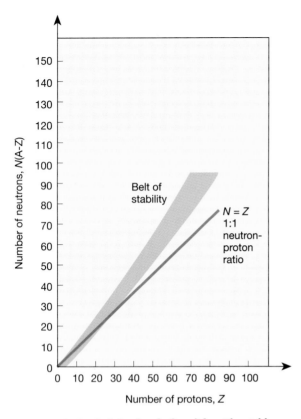

Figure 19–2   Stability band of nuclides. The stable nuclides fall approximately within the area indicted. The sloping line represents nuclides with equal numbers of protons and neutrons. Nuclides having more neutrons than protons lie above this line.

various ways to achieve stability. As a result, one element is changed, or transmuted, into another.

Like chemical reactions, nuclear reactions can be represented by equations. The same symbols are used for atoms and nuclei, although the atomic number and mass number are added to the symbols of nuclei. Just as atoms and molecules are not created or destroyed in ordinary chemical changes, so **nucleons** (protons and neutrons) are not created or destroyed in nuclear reactions, although they may interchange. Both **mass numbers** and **electric charge** are **conserved** in nuclear reactions. In a balanced nuclear equation, the sum of the atomic numbers (electric charges) of the reacting nuclei and particles will be equal to the sum of the atomic numbers of the products, and the mass numbers (total number of nucleons) of the reactants and products will balance.

Figure 19–3   β-emission. A neutron divides into a proton and a beta-particle.

EXAMPLE 19–1

When $^{82}_{35}$Br emits a beta particle, it is transmuted into another element. Identify the new element.

SOLUTION
We write a balanced equation for this transmutation, noting that there will be conservation of electric charge and conservation of mass numbers. Recall that a beta particle is an electron with a charge of -1; in nuclear equations, its mass is considered to be zero since it is 2000 times less massive than either a proton or a neutron.

$$^{82}_{35}\text{Br} \longrightarrow {}^{82}_{36}\text{X} + {}^{0}_{-1}e$$

A beta particle can be thought of as formed from a neutron that divides into a proton and an electron (beta particle). Because the proton remains in the nucleus, the positive nuclear charge is increased by one unit, from 35 to 36, but there is no change in the mass of the nucleus (Fig. 19–3). The atomic number of the new element, therefore, must be 36, and its mass number 82. The element represented by X has to be krypton, Kr.

EXTENSION
An alpha particle ($^4_2$He) is ejected by $^{239}_{94}$Pu. Determine the daughter nuclide produced in this reaction. (Answer: $^{235}_{92}$U)

EXAMPLE 19–2

In 1919, in the first artificial transmutation ever carried out, Ernest Rutherford bombarded nitrogen with alpha particles and identified hydrogen nuclei among the products. What other nuclide was produced?

SOLUTION
Recalling that alpha particles are helium nuclei, $^4_2$He, we consider them to be reactants along with $^{14}_7$N.

$$^{14}_{7}\text{N} + ^{4}_{2}\text{He} \longrightarrow ^{1}_{1}\text{H} + ^{17}_{8}\text{X}$$

The atomic number of the other product nuclide, represented by X, is 8; thus, the newly produced element must have been oxygen. In order for the mass numbers to balance, the oxygen nuclide must have had a mass number of 17.

EXTENSION

Complete the transmutation equation $^{2}_{1}\text{H} + ^{3}_{1}\text{H} \rightarrow$ $^{4}_{2}\text{He} + \text{X}$. (Answer: $^{1}_{0}n$)

■

EXAMPLE 19–3

Frédéric (1900–1958) and Irène Joliot-Curie (1897–1956) were the first to produce an artificially radioactive element. In 1933, they bombarded $^{27}_{13}\text{Al}$ with α-particles and found that neutrons and $^{30}_{15}\text{P}$ were produced. (The phosphorus, however, was radioactive and decayed into $^{30}_{14}\text{Si}$ by positron emission.) Write the equation for the first reaction.

SOLUTION

$$^{27}_{13}\text{Al} + ^{4}_{2}\text{He} \longrightarrow ^{30}_{15}\text{P} + ^{1}_{0}n$$

EXTENSION

Write the equation for the second reaction in Example 19–3. (A positron has the same mass as an electron but the opposite charge.) (Answer: $^{30}_{15}\text{P} \rightarrow$ $^{30}_{14}\text{Si} + ^{0}_{1}e$)

■

Several modes of **natural transmutation**, or *radioactivity*, are known, two of which are discussed here. **Alpha-emission**, or **α-decay**, is common for elements of atomic number greater than 82 and mass number greater than 209. Such nuclides are found to the right of the region of stability and have an excess of protons. The emission of α-particles reduces the number of protons by two and the number of neutrons by two, bringing the nuclear composition closer to the stability band. An example is

$$^{254}_{102}\text{No} \longrightarrow ^{4}_{2}\text{He} + ^{250}_{100}\text{Fm} \qquad (19\text{–}1)$$

Nuclides that have a high neutron-to-proton ratio and lie to the left of the stability band may be stabilized by **β-decay**. The net effect is a decrease in the number of neutrons by one, the result of a neutron

dividing into a proton and an electron, and an increase in the number of protons by one, thus decreasing the neutron-to-proton ratio. An example is

$$^{14}_{6}\text{C} \longrightarrow ^{0}_{-1}e + ^{14}_{7}\text{N} \qquad (19\text{–}2)$$

19–3 HALF-LIFE; CARBON-14 DATING

Consider a sample of radioactive material. Its activity, or rate of decay, is the number of transmutations or disintegrations per unit time. The SI unit of activity is the becquerel (Bq) equal to 1 decay/s. Suppose that the **decay rate** at the start is 1000 disintegrations per minute (dpm) and that we measure the decay rate over a period of time. Assume that after 10 minutes, the activity has dropped to 500 dpm; after 20 minutes, to 250 dpm; and after 30 minutes, to 125 dpm (Fig. 19–4). Every 10 minutes the activity is reduced by exactly one half. We say that the **half-life** ($t_{1/2}$) of the sample is 10 minutes.

The half-life is a useful property for identifying nuclides, since each radioactive nuclide has a characteristic half-life. If this is short, the sample loses its radioactivity rapidly; if it is long, the sample retains

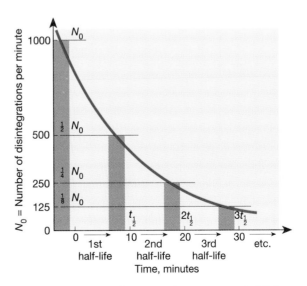

Figure 19–4 A radioactivity decay curve. A half-life is the time required for the activity to drop to one half of any value and is a constant for a given radionuclide.

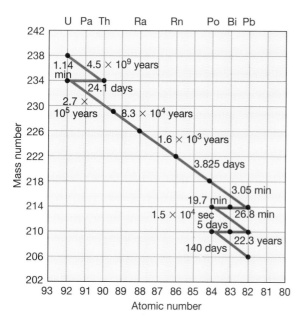

Figure 19–5   The uranium-238 radioactive decay series. The range of half-lives is from fractions of a second to billions of years.

much of its radioactivity (Fig. 19–5). The half-lives of some radionuclides are listed in Table 19–1.

■

EXAMPLE 19–4

Iodine-131, used in the treatment of diseases of the thyroid, decays by β-emission according to the following equation:

$$^{131}_{53}I \longrightarrow \, ^{131}_{54}Xe + \, ^{0}_{-1}e$$

Table 19–1   HALF-LIVES OF SELECTED RADIONUCLIDES

Element	Radionuclide	Half-life ($t_{1/2}$)	Radiation
Hydrogen	$^{3}_{1}H$	12.3 years	β
Carbon	$^{14}_{6}C$	5570 years	β
Sulfur	$^{35}_{16}S$	87.1 days	β
Cobalt	$^{60}_{27}Co$	5.3 years	α
Strontium	$^{90}_{38}Sr$	27.7 years	β
Radium	$^{226}_{88}Ra$	1620 years	α
Uranium	$^{235}_{92}U$	7.1×10^{8} years	α

The half-life for the reaction is approximately 8 days. Assuming that we start with 10.00 g of ^{131}I, how much will remain in 32 days (approximately one month)?

SOLUTION
For each half-life, the activity and therefore the amount of ^{131}I are reduced by one half.

Number of Half-Lives	Number of Days	Quantity of ^{131}I Remaining (grams)
0	0	10.00
1	8	5.00
2	16	2.50
3	24	1.25
4	32	0.612

In 32 days, 0.612 g of ^{131}I will remain.

EXTENSION
Cobalt-60, an α-emitter, has a half-life of 5.3 years. Starting with 10.00 g, how much will remain in 26.5 years? (Answer: 0.306 g)

■

In about 1950, Willard F. Libby (1908–1980) at the University of Chicago devised a method based on the half-life of carbon-14 by which archaeological discoveries can be dated. Charcoal from the Lascaux caves in France, carbon from the ashes of a prehistoric Indian campfire, the wood from a Mayan temple, or a sandal from Oregon can be dated with considerable accuracy using the **carbon-14 clock**.

Carbon-14 is produced through cosmic-ray bombardment of nitrogen in the atmosphere:

$$^{14}_{7}N + \, ^{1}_{0}n \longrightarrow \, ^{14}_{6}C + \, ^{1}_{1}H \qquad (19–3)$$

It is radioactive and decays by beta-emission:

$$^{14}_{6}C \longrightarrow \, ^{14}_{7}N + \, ^{0}_{-1}e \qquad t_{1/2} = 5570 \text{ years} \quad (19–4)$$

The concentration of carbon-14 in the atmosphere, as radioactive carbon dioxide, is approximately constant, since it is replenished at the same rate that it disintegrates. Deviations from constancy are detectable by dating tree rings. Its activity is about 14 dpm per gram of carbon.

Plants absorb radioactive carbon dioxide from the atmosphere and convert it to carbohydrates in the process of photosynthesis. The carbon-14 content of living plants becomes stabilized at approximately one carbon-14 atom for every 10^{12} carbon-12 atoms. The same ratio becomes established in animals and human beings who eat the plant materials. When a plant or animal dies, the consumption of carbon-14 ceases and the ratio of carbon-14 to carbon-12 starts to decrease. To determine its age, a sample is burned to carbon dioxide, which is then analyzed for the $^{14}C/^{12}C$ ratio. After 5570 years, the $^{14}C/^{12}C$ ratio would be only one half the normal ratio. In this way, dates can be determined quite accurately over several half-lives.

19–4 THE GEIGER-MUELLER COUNTER

Although nuclear particles are too small to be seen directly, there are ways to study their properties. The methods are based on one of the following features: (1) the ability of nuclear particles to cause ionization in matter through which they pass; (2) their ability to cause flashes of light (scintillations) when they strike certain crystals or liquids; and (3) their ability to produce tracks in specially prepared photographic emulsions. The traces or tracks that were left in various instruments led scientists to conclude that they were observing particles in action.

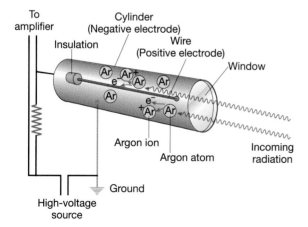

Figure 19–6 A Geiger-Mueller radiation counter detects radioactivity by ionization of argon atoms resulting from collisions with a particle of incoming radiation. The positive ions and negative electrons conduct an electric current between the oppositely charged electrodes, giving rise to a pulse in the external circuit that can be amplified.

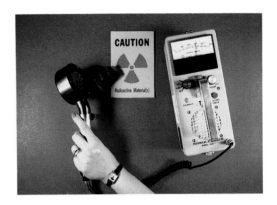

Figure 19–7 Using a Geiger counter to measure the activity in a radioactive mineral. (Photographed by Jim Lehman, James Madison University)

The **Geiger-Mueller counter** consists of a metal cylinder that contains within it a fine tungsten wire stretched along its axis and insulated from it and a small amount of gas (argon or helium) at low pressure (Fig. 19–6). A potential difference of about 1000 volts is maintained across the cylinder and wire (electrodes). When a particle from a radioactive source enters the cylinder, it collides with some of the atoms, knocking off electrons and creating positively charged gaseous ions. Under the accelerating influence of the electric field, the electrons and ions formed in the original collisions gain enough energy to produce further (secondary) ionization as they collide with neutral atoms. The process may be repeated again and again until there is an avalanche of a million or more electrons upon the wire. The discharge between cylinder and wire produces a pulse that can be amplified, can activate a clicker and be heard, or can be recorded electronically and counted (Fig. 19–7).

19–5 ENERGY IN NUCLEAR PROCESSES

Imagine assembling a carbon-12 atom from its components: six protons, six neutrons, and six electrons. The rest masses of these particles in unified atomic mass units (u) are given in Table 19–2. What mass do we anticipate for the carbon-12 atom?

Table 19–2   MASSES OF ELECTRON, PROTON, AND NEUTRON

Particle	Mass (u)
Electron	0.000549
Proton	1.007277
Neutron	1.008655

For carbon-12:

$$6 \text{ protons} = 6 \times 1.007277 = 6.043662 \text{ u}$$
$$6 \text{ neutrons} = 6 \times 1.008665 = 6.051990 \text{ ''}$$
$$6 \text{ electrons} = 6 \times 0.000549 = \underline{0.003294} \text{ ''}$$
$$\text{Mass of particles} = 12.098946 \text{ ''}$$
$$\text{Mass of carbon-12} = \underline{12.000000} \text{ ''}$$
$$\text{Mass difference} = 0.098946 \text{ u}$$

Since it is reasonable to assume that the whole is equal to the sum of its parts, it may come as somewhat of a surprise that the actual mass of a carbon-12 atom is less than the combined masses of the electrons, protons, and neutrons that compose it (Fig. 19–8). The difference, called the **mass defect**, exists to varying degrees for every nuclide except ^1_1H.

The "missing" mass is accounted for by assuming that it is converted into energy. Since mass and energy are related by the expression $E = mc^2$, we can calculate the energy equivalent of 1 atomic mass unit.

$$1 \text{ u} = 1.66 \times 10^{-27} \text{ kg}$$
$$c = \text{velocity of light}$$
$$= 3.00 \times 10^8 \frac{\text{m}}{\text{s}}$$
$$E = ?$$
$$E = mc^2$$
$$= (1.66 \times 10^{-27} \text{ kg})\left(3.00 \times 10^8 \frac{\text{m}}{\text{s}}\right)^2$$
$$= (1.66 \times 10^{-27} \text{ kg})\left(9.00 \times 10^{16} \frac{\text{m}^2}{\text{s}^2}\right)$$
$$= 14.9 \times 10^{-11} \text{ joule}$$

It is customary to express the energy of nuclear processes in units of millions of electron volts (MeV), so we convert joules to MeV. The conversion factors are

$$1 \text{ electron volt} = 1 \text{ eV} = 1.60 \times 10^{-19} \text{ joule}$$
$$1 \text{ million electron volts} = 1 \text{ MeV}$$
$$= 1.60 \times 10^{-13} \text{ joule}$$

Therefore,

$$14.9 \times 10^{-11} \text{ joule} \times \frac{1 \text{ MeV}}{1.60 \times 10^{-13} \text{ joule}}$$
$$= 931 \text{ MeV} \quad (19\text{–}5)$$

or $1 \text{ u} = 931 \text{ MeV}$

The mass defect in carbon-12 is equivalent to $0.098946 \text{ u} \times (931 \text{ MeV/u}) = 92.1 \text{ MeV}$. This energy is called the **binding energy**—the energy required to separate a nucleus into its individual nucleons, or the energy evolved when these nucleons combine to form the nucleus. The binding energy per nucleon of carbon-12 is 92.1 MeV/12 nucleons = 7.68 MeV/nucleon.

A plot of the binding energy/nucleon of the elements shows a maximum in the vicinity of iron (8.79 MeV/nucleon) and a gradual decrease toward the heavier elements such as uranium (7.59 MeV/nucleon) (Fig. 19–9). Nuclei that have the greatest binding energy/nucleon are the most stable. It appears from the curve that if heavy nuclei, such as those of uranium, could be split into lighter nuclei, energy would be released in the process. The nucleus has been split—the process is called **fission**—and its

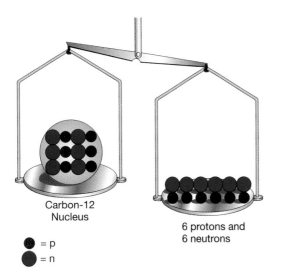

Figure 19–8   A carbon-12 nucleus compared to its components. The whole in this case is not equal to the sum of its parts.

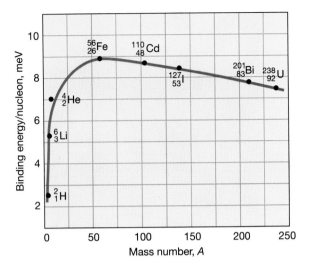

Figure 19–9 Nuclear binding energy curve. The most stable nuclei are located near iron at the maximum of the curve, where the binding energy per nucleon is greatest.

considerable energy harnessed. Further, the combination of lighter nuclei, such as hydrogen and lithium, to form heavier nuclei should also be accompanied by the release of energy. This process—**fusion**—has also been achieved.

19–6 PARTICLE ACCELERATORS: THE CYCLOTRON

Investigators of nuclear reactions soon realized that they would not get very far without the aid of suitable "hardware." To produce a nuclear reaction with a proton or an alpha particle, the Coulomb barrier of the positively charged nucleus must be penetrated. Energies in the range of millions of electron volts and higher are required.

There are a few natural sources of energetic particles. Radioactive nuclei emit alpha, beta, and gamma rays with energies of about 1 MeV. Rutherford used alpha particles emitted from polonium in his early research. High-energy nuclear particles known as cosmic rays rain in on us partly from the sun and partly from outside the solar system, especially from the center of our galaxy, with energies as high as 10^{13} MeV, but they are not available for controlled experiments.

Rutherford and others established that α-particle bombardment transmuted nearly all the lighter elements through potassium. The α-particles from naturally radioactive sources, however, did not have high enough energies to penetrate the nuclei of heavier atoms with their considerable positive charge. Since nature is limited in the energy range of particles that can be used as projectiles, some means of accelerating charged particles to any desired energy had to be found. That is the reason for developing particle **accelerators**.

Accelerators were not really new either in principle or in practice. The gas discharge tube and the X-ray tube, both widely used, accelerate electrons through potential differences of thousands of volts. The problem is to accelerate particles much heavier than electrons through voltages in the MeV range and greater. Although they differ widely in design, accelerators share the same purpose: to study the effect of bombarding atomic nuclei with particles traveling at speeds approaching that of light. These high-speed particles penetrate atomic nuclei and serve as probes in the investigation of matter.

To produce a proton with 10 MeV of energy, two approaches can be used. In the first, the proton can be accelerated through one single and large potential difference of 10 million volts to acquire the desired high energy. The **Van de Graaff generator** is based on this principle. The voltage is controllable to better than 0.1% so that all the particles in the beam have very nearly the same energy, a favorable feature for many kinds of precision work in research, industry, and medicine.

A second approach to producing high-energy particles was introduced by Ernest O. Lawrence (1901–1958) at the University of California in 1931. It is based on the concept of accelerating a particle many times in succession through a relatively low potential difference, in such a way that it picks up energy at each acceleration (Fig. 19–10). A magnetic field causes the particle to move in a circle so that it returns to the same accelerating stage of the machine. Finally, the particle arrives at the desired energy (Fig. 19–11). Lawrence's first **cyclotron** was barely 1 foot across and produced 1,200,000-electron-volt protons. It employed a circular track and a succession of small pushes activated by a magnetic field. Each time the particle went around, it was accelerated a slight amount. Accelerators have since gone to the million- and billion-volt class (Fig. 19–12). Among them are

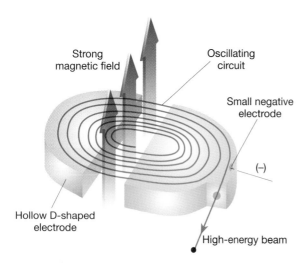

Figure 19–10 Principle of the cyclotron. Particles are accelerated many times in succession and pick up energy at each acceleration.

Figure 19–12 The main tunnel of the Tevatron, the most powerful accelerator in the world today, at Fermilab, National Accelerator Laboratory. The accelerator is four miles in circumference. (Photo Courtesy of Fermilab)

the linear accelerators, in which particles are accelerated many times along a straight-line path. The Stanford Linear Accelerator is the largest of these—over 2 miles long. It can accelerate electrons to over 20 GeV.

19–7 NEUTRON REACTIONS

A drawback in the use of charged particles such as electrons, protons, and alpha particles is that they interact strongly with the electrons of atoms and ionize atoms or molecules of the material through which they pass. In the process, they lose energy rapidly and are brought to rest and neutralized after passing through only a few centimeters of air or through thin foils. Since neutrons carry no electric charge and interact very little with electrons, they are more penetrating than charged particles of comparable energy. They are easily obtained by bombarding light elements such as beryllium, boron, and lithium with alpha particles:

$$^{9}_{4}\text{Be} + ^{4}_{2}\text{He} \longrightarrow ^{12}_{6}\text{C} + ^{1}_{0}n \qquad (19-6)$$

Enrico Fermi (1901 — 1954) and his co-workers at the University of Rome used neutrons to bombard various elements in the years 1934 to 1936 and produced many new nuclides. Since then, neutrons of low, intermediate, and high energy have been used to bombard light and heavy nuclei throughout the periodic table. When a neutron collides with a nucleus and is captured by it, the resulting compound nucleus is unstable. The nucleus may achieve stability through radioactive decay. When $^{238}_{92}\text{U}$ was bombarded with neutrons, the product nucleus was a β-emitter,

Figure 19–11 A beam of protons (bright area) from a cyclotron at the Argonne National Laboratory can cause nuclear reations to take place when it strikes the nuclei of atoms. (Argonne National Laboratory)

whereas the original uranium-238 is an α-emitter. Fermi's interpretation of the results was that neutrons had penetrated uranium nuclei, forming an unstable isotope, $^{239}_{92}U$, which then decayed by β-emission:

$$^{238}_{92}U + ^{1}_{0}n \longrightarrow (^{239}_{92}U) \longrightarrow _{-1}^{0}e + ^{239}_{93}X \qquad (19-7)$$

If this were true, it would mean the production of an element beyond the limits of the periodic table known at that time. In the same way, elements 94, 95, and other "**transuranium**" **elements**—that is, elements beyond uranium—could be synthesized. At the University of California, Edwin M. McMillan (1907–1991) succeeded in identifying element 93, called neptunium, in 1940. A year later, Glenn T. Seaborg (b. 1912), at the same university, discovered element 94, plutonium. The elements were named for the planets beyond Uranus in the solar system.

19–8 NUCLEAR FISSION

When Otto Hahn (1879–1968) and his co-worker Fritz Strassmann (b. 1902) at the Kaiser Wilhelm Institute in Berlin repeated Fermi's neutron bombardment of uranium in 1938, they expected to find uranium isotopes with greater atomic masses than that of ordinary uranium (assuming that the uranium nuclei captured neutrons) or evidence of transuranium elements. Instead, they found radioactive isotopes of much lighter elements, such as barium, krypton, strontium, and cerium. Since no particles larger than α-particles had previously been found to be ejected from nuclei, it was difficult to believe that barium or krypton nuclei were emitted from uranium following neutron bombardment. Puzzled, they sent their results to physicist Lise Meitner (1878–1969), a former co-worker, for her interpretation and published their work in a scientific journal in January, 1939.

Meitner and her nephew, Otto Frisch (1904–1979), came to the conclusion, soon confirmed by others, that in Hahn's and Strassmann's experiments, nuclei of uranium-238 were being split by captured neutrons into two nearly equal parts. Frisch coined the term **fission** for this process, comparing it to a cellular process, and Meitner calculated that about 200 MeV of energy are released from each uranium nucleus that undergoes fission. The energy appears as kinetic energy of the fission products as they fly apart and is nearly 100 million times as much energy per atom as in any common explosive. Meitner and Frisch also published their results in a scientific journal in January, 1939.

The energy released in a nuclear fission comes from the conversion of mass into energy, $E = mc^2$. In a typical event, uranium-235 captures a neutron and fissions into barium-141 and krypton-92, with the release of three neutrons (Fig. 19–13):

$$^{235}_{92}U + ^{1}_{0}n \longrightarrow ^{141}_{56}Ba + ^{92}_{36}Kr + 3^{1}_{0}n \qquad (19-8)$$

A comparison of the masses of the reactants and products shows that the combined masses of the products *after* fission is less than the sum of the

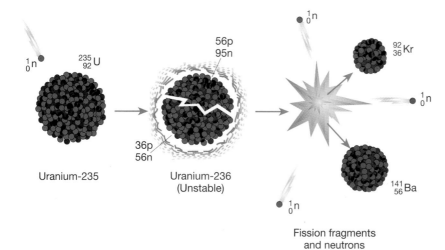

Uranium-235

Uranium-236 (Unstable)

56p / 95n

36p / 56n

$^{1}_{0}n$

$^{92}_{36}Kr$

$^{1}_{0}n$

$^{141}_{56}Ba$

$^{1}_{0}n$

Fission fragments and neutrons + ENERGY

Figure 19–13 Fission of uranium-235 following capture of a neutron. Note the formation of an intermediate unstable nuclide, ^{236}U.

masses of the reactants *before* fission. The mass difference appears as energy, approximately 200 MeV.

Before Fission

$$^{235}_{92}U = \quad 235.0439 \text{ u}$$
$$^{1}_{0}n = \quad \underline{\quad 1.0087 \text{ ''}}$$
$$236.0526 \text{ ''}$$
$$\underline{- 235.8373 \text{ ''}}$$
$$\text{Mass difference} = \quad 0.2153 \text{ u}$$

After Fission

$$^{141}_{56}Ba = 140.9139 \text{ u}$$
$$^{92}_{36}Kr = \quad 91.8973 \text{ ''}$$
$$3^{1}_{0}n = \quad \underline{\quad 3.0261 \text{ ''}}$$
$$235.8373 \text{ u}$$

$$0.2153 \text{ u} \times 931 \, \frac{\text{MeV}}{\text{u}} = 200 \text{ MeV}$$

19–9 THE CHAIN REACTION

On the average, approximately three neutrons are released during each fission of uranium-235. When uranium-235 is present in an amount exceeding a **critical size**, so that the neutrons are not lost, these neutrons may initiate fission in other nuclei, releasing more neutrons, which in turn can cause fission in still other nuclei. This process is called a **chain reaction** (Fig. 19–14). In nuclear reactors, the chain reaction is controlled and the energy is put to useful purposes. In a bomb, the chain reaction, once started, is uncontrolled and leads to an explosion.

Enrico Fermi and Leo Szilard (1898–1964) discovered that "slow" neutrons were effective in causing uranium-235 to undergo fission. They can be obtained when neutrons collide with atoms without combining, thus losing some of their kinetic energy. The first large-scale chain reaction with fissionable uranium was realized under Fermi's direction in

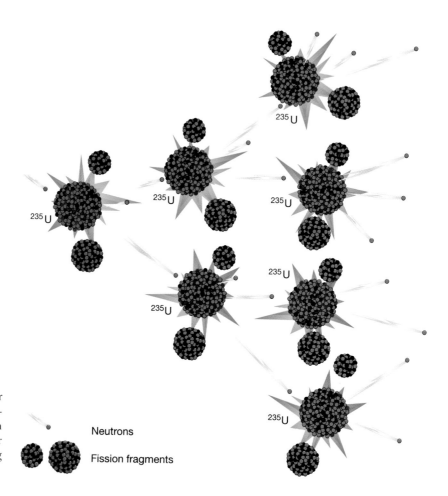

Figure 19–14 A nuclear chain reaction. The neutrons released when ^{235}U fissions in turn induce fission reactions in other ^{235}U nuclei, thereby sustaining the process.

● Neutrons

●● Fission fragments

Figure 19–15   Painting of the world's first nuclear reactor in an improvised laboratory under the west stands of the University of Chicago's Stagg Field. A self-sustained chain reaction was first achieved on December 2, 1942. There are no photographs of the reactor because of wartime secrecy. (Courtesy of Chicago Historical Society)

1941. Rods containing uranium metal and uranium oxide were embedded in a pile of graphite (carbon) blocks used as a **moderator** to slow down neutrons. The chain reaction was started by neutrons escaping from a beryllium source placed at the bottom of the pile. The reaction was not self-sustaining, however, since neutrons released by fission were lost by nonfission capture in uranium-238, or by diffusion, and the external neutron source was necessary to inject neutrons into the pile to keep the reaction going. The first **self-sustaining chain reaction** was carried out on December 2, 1942, in a pile constructed under the squash courts in the stadium of the University of Chicago (Fig. 19–15). Refinements in the nuclear pile included the introduction of cadmium rods, which could control the speed of the reaction by their ability to absorb neutrons. Today fission reactors are used for many purposes: for the generation of electricity; for medical, industrial, and agricultural research; and as an energy source to drive submarines and surface ships (Fig. 19–16).

A fission bomb is essentially a reactor, so designed that the chain reaction grows at as high a rate as possible and releases the maximum energy in the minimum time. The critical mass of uranium-235 is approximately 4 kilograms, or less than a cupful be-

Figure 19–16   The launching of the nuclear submarine *Hyman G. Rickover* into the Thames River in Connecticut on August 27, 1983. (Courtesy of E. I. DuPont de Nemours and Company)

cause of its great density (19.05 g/cm^3). The fissionable material in a bomb is separated into subcritical sections (Fig. 19–17). Triggering causes these sections to coalesce into a mass greater than the critical size, and a chain reaction is initiated by a stray neutron from cosmic radiation or from a neutron source. The

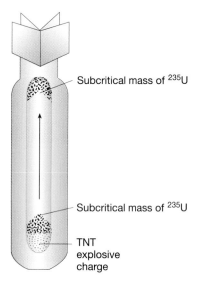

Figure 19–17   Diagram of a nuclear bomb. A high explosive (TNT) drives ^{235}U pieces together. When the mass of ^{235}U becomes supercritical, it explodes.

devastating effects of nuclear explosions are due to the intense heat, which produces fire, the strong mechanical shock waves, and the radioactivity of the fission products (Fig. 19–18).

Figure 19–18   The first fission bomb exploded in a desert in New Mexico in July, 1945, when the nuclear reaction became supercritical. (Courtesy of U.S. Army/White Sands Missile Range)

19–10 NUCLEAR POWER REACTORS

The consumption of power in the United States has doubled every 10 years for the past 30 years. If this doubling rate continues to the end of the century, the question will be not whether energy production should be increased, but how to do it with a minimum of harmful side effects.

Nuclear energy was rapidly becoming a national energy source in the 1970s. More than 70 nuclear power plants were in operation, producing about 12% of the electric power supply in the United States. Nearly 100 others were being built or designed (Fig. 19–19). They were expected to furnish about 25% of the total in the 1980s, and even more by the year 2000. The share of electricity generated by nuclear power plants was even greater in France, England, Japan, and West Germany.

Nuclear power plants are a source of thermal pollution (as are fossil-fuel plants) and low-level radiation. The advantages, according to some, are low-cost electricity; conservation of coal, oil, and gas, which are more useful as sources of organic molecules than as sources of heat; and reduction of pollution. Once the fission process generates heat, the operation of a nuclear plant is practically the same as that of a conventional plant (Fig. 19–20).

Nuclear reactors currently in use employ fuel rods containing uranium-238 and small amounts of uranium-235 and plutonium-239, the last two nuclides used as fuel. About 99.3% of natural uranium is in the form of nonfissionable uranium-238, which can be used as a fuel if it is converted to plutonium by capturing a fast neutron.

$$^{238}_{92}\text{U} + ^{1}_{0}n \longrightarrow ^{239}_{92}\text{U} \xrightarrow{^{0}_{-1}e} ^{239}_{93}\text{Np} \xrightarrow{^{0}_{-1}e} ^{239}_{94}\text{Pu}$$

$$(19\text{–}9)$$

This does not occur to a great extent in light water reactors, in which water is used as the coolant, because the fast neutrons that are necessary for the reaction are slowed down by the water. The use of liquid sodium or some other coolant could make available a high enough concentration of fast neutrons so that more nuclear fuel is produced than consumed. Such a reactor is designated a **light-metal-cooled-fast-breeder-reactor** (**LMFBR**), and nonfissionable uranium-238 that can be so transformed is described as "fertile." A relatively abundant nonfissionable material is thus transformed into fissionable fuel.

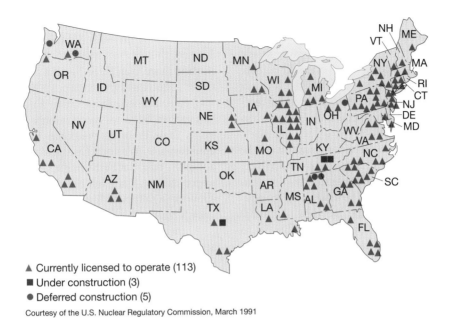

▲ Currently licensed to operate (113)
■ Under construction (3)
● Deferred construction (5)

Courtesy of the U.S. Nuclear Regulatory Commission, March 1991

Figure 19–19 Nuclear power reactors in the United States. Because of space limitations, symbols do not reflect precise locations. (Courtesy of the Office of Information Services, U.S. Nuclear Regulatory Commission)

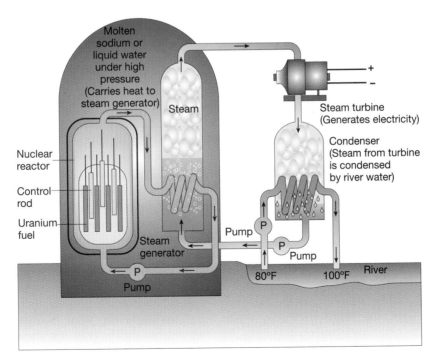

Figure 19–20 A schematic diagram of a light water nuclear power plant.

Fissionable fuels are exhaustible, and the supply of inexpensive ores containing uranium-235 could be used up in the near future. More costly fuel would then be used, but the known reserves could be consumed in a few decades by the continued deployment of inefficient nonbreeding reactors. In breeder reactors, the fuel life expectancy could be extended by hundreds of years (Fig. 19–21).

The concept of the **fast-breeder-reactor** (**FBR**) can be traced to 1944, when Enrico Fermi and Walter Zinn (b. 1906) considered the possibility of a controlled chain reaction without the moderation of neutrons that would allow effective breeding of the fissionable material. The key to good breeding is to produce an abundance of neutrons in addition to those needed to sustain the chain reaction. Fast fission produces 2.9 neutrons per fissioning nucleus, whereas slow fission produces only 2.4. This small differential produces the large surplus of neutrons that makes it possible for the FBR to produce more fuel than it consumes. By 1948, Walter Zinn, at Argonne National Laboratory, was working on the problems of the FBR, and his work led to the construction of the experimental FBR at Idaho Falls. It was there that electricity was first generated from nuclear energy in 1951.

The United States is relatively well endowed in its energy resources. Its discovered deposits of uranium alone are approximately 40% of the world reserves.

Figure 19–21   A breeder reactor at the National Reactor Laboratory in Idaho. (Courtesy of Idaho National Laboratory and the U.S. Department of Energy)

Europe and Japan are not well endowed with either fossil fuels or uranium. There is some uranium in France, though not much. In 1980, France was importing more than three fourths of its energy. All the other countries of the European Economic Community control less uranium than France. Since FBRs can extract 60 times more energy from natural uranium than their conventional light water reactor counterparts, Western Europeans were enthusiastic about the role that FBRs might play in their projected energy future. They felt that breeders would at least allow enough time to find and develop possibly better energy sources.

France developed three increasingly powerful reactors. The design of the experimental reactor Rapsodie was started in 1957; construction began in 1962; and power operation was reached in 1967. The Rapsodie provided extremely useful information about operating an FBR and for testing fuel for the next generation of FBRs. Other experimental first-generation FBRs include reactors in Idaho; the Enrico Fermi Fast Breeder Reactor, which ran from 1965 to 1972 in Michigan; the BR 5 and BOR 60, started in 1958 and 1971 and still operating in the area occupied by the former Soviet Union; and reactors in Great Britain, West Germany, Japan, Hanford (Washington), Italy, and India.

The Phenix marked the second step in the French FBR program. Design started in 1964, construction in 1968, and operation in 1973. It has performed very satisfactorily as a demonstration plant. Two other demonstration plants have been operated successfully: the British Prototype Fast Reactor, operating since 1974, and the Soviet BN 350, operating since 1972.

The third stage in the French breeder program is the noncommercial reactor, Super Phenix Mark I.

In the mid-1980s, however, Europe's enthusiasm for FBRs was dampened. Confidence in the technical potential of fast breeders as an eventual important energy source remained high, but political enthusiasm cooled. The rate of growth of electricity demand declined as oil prices rose. Public concern about safety caused licensing restrictions to be imposed. For example, after working like clockwork for eight years, Phenix sprang a leak in its cooling system in 1982, leading to a fire when the sodium came in contact with the air. The reactor was put out of operation for several weeks and stimulated a debate about fast-breeder safety. The cost of Super Phenix, jointly financed by France, West Germany, and Italy, rose alarmingly. Many economists claimed that breeders would not be competitive with light water reactors for the next 50 years. In Britain and West Germany, there was serious talk of putting a freeze on expansion plans until future needs could be perceived more clearly, if not abandoning FBR development completely. In Britain, commercial orders for FBRs were not expected until after the year 2000. As of this writing, the technology had lost its imperative, and the rapid development of FBRs was regarded as an economic and political gamble whose outcome was unpredictable.

Figure 19–22 The nuclear power plant at Chernobyl in the Soviet Union shortly after the explosion and fire in April, 1986. (© P. Angelo Simon/Phototake)

19–11 REACTOR SAFETY

A major issue involving nuclear power plants is concern over safety standards and equipment. There is some evidence that current standards for radioactive emissions are not strict enough and should be revised, since radioactivity leakage is a possibility.

The most serious safety hazard is the possibility of a **fuel meltdown**. Although a nuclear reactor cannot explode like a bomb, since there is no way to bring sufficient fissionable material together, excessive heat could melt the core and discharge large amounts of radioactive materials into the environment. This possibility exists because of the way in which pressurized water reactors are designed. Uranium atoms fission, releasing energy in the form of heat. Steam circulates around the reactor's radioactive core and absorbs this heat. The heat expands the steam, which is channeled into turbines that produce electricity. If the pressurized steam system ruptures, the heat within the reactor core cannot be dissipated immediately and could cause the reactor, which might contain 50 or 100 tons of fuel, to melt, generating more heat.

A cooling system has been designed to prevent the sequence of events leading to a fuel meltdown. Called the Emergency Core Cooling System, it would operate if a reactor's primary cooling system fails. It consists of a water supply with pumps and pipes, which would flood an overheating reactor with cooling water long enough to give the reactor operators time to shut it down. But doubts have been expressed that the system will perform as designed. Initial tests on an experimental reactor in the Idaho National Engineering Laboratory in the late 1970s did show that the emergency cooling system generally functioned as expected. But a chain of equipment malfunctions led

to a partial core meltdown at the Three Mile Island nuclear power plant near Harrisburg, Pennsylvania, in 1979. The President's Commission on the accident at **Three Mile Island** (the Kemeny Commission) concluded that the major cause that led to the accident was people-related problems and not equipment problems. History's worst nuclear accident occurred April 26, 1986, at **Chernobyl** in the Soviet Union (Fig. 19–22). A reactor at the power station exploded and burned out of control, killed 31 people, forced 135,000 to be evacuated, and spewed radiation across the world.

19–12 DISPOSAL OF RADIOACTIVE WASTES

The wastes produced by nuclear reactors are gaseous, liquid, and solid. If fission reactors supplied all the power needs of the United States, the safe disposal of radioactive ashes and gases would be a staggering problem. Each year would require the disposal of radioactive waste equivalent to that from the explosion of hundreds of thousands of nuclear bombs. Even at the present rate, the disposal problem is serious.

The waste gases are mostly radioactive isotopes of the noble gases. The most difficult problem involves krypton-85, which is now released to the atmosphere. It is long-lasting ($t_{1/2} = 10.76$ years) and unreactive chemically. Its concentration in the atmosphere is increasing, and it contributes to the general background radiation. The basic technology for trapping krypton—by cryogenic cooling, for instance—is available, but it is a very complicated process.

Tritium ($^3_1$H) enters the liquid discharge of power and fuel reprocessing plants. It is produced from lithium and boron in the coolant and combines with oxygen to form water, becoming part of the effluent. In reprocessing plants, tritium is released to the atmosphere as water vapor. In both cases, it is extremely difficult to remove. As part of a water molecule, tritium could become incorporated into biological systems. With a half-life of 12.3 years, the radiation produced by its decay would be quite damaging. No mechanism is known, however, by which tritium can concentrate in food chains, and there is no evidence to indicate that it does.

Liquid wastes are generally reduced to solid form through precipitation and added to the solid waste. In the process, the volume is reduced to one tenth of the original liquid material. Solid waste, consisting of spent fuel rods, is first stored at the reactor site while the radioactivity is reduced. It is then shipped to fuel reprocessing plants, where it is treated and stored for ten years before being shipped to a repository. Deep geological formations, salt formations in particular, are favored by some scientists. The ultimate disposal problem, however, has not been solved.

19–13 FUSION ENERGY

If a way could be found to harness nuclear fusion, another source of useful energy would be available. In this process, light nuclei are fused into heavier nuclei and energy is released (Fig. 19–23). The binding energies in the resulting nuclei are greater than in the nuclei undergoing fusion. The advantages of fusion are illustrated by the almost limitless supplies of hy-

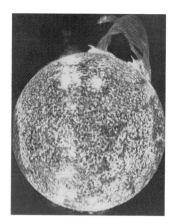

Figure 19–24   The sun derives its energy from the fusion of hydrogen into helium in its core. (Courtesy of NASA)

drogen and other light elements as compared with uranium. The radioactive waste problem would be almost nonexistent for a system relying on fusion.

The sun and the stars are believed to owe their energy to nuclear fusion (Fig. 19–24). Hans Bethe (b. 1906) at Cornell University proposed a mechanism, called the **carbon-nitrogen (CN) cycle**, which could generate energy of this magnitude. For this work he received the Nobel Prize in physics in 1967. Carbon acts as a catalyst as four protons are fused by successive capture into one α-particle (helium nucleus). Two positrons (particles with one positive charge and zero mass number) and 24.7 MeV of energy are released for each α-particle created. In fusion as in fission, mass is consumed and converted into energy. In effect, hydrogen functions as a nuclear fuel in the fiery furnace of the stars and is consumed at the rate of millions of tons each second. The mass of the sun is so great, however, that hydrogen will likely remain for billions of years. Other mechanisms, such as the **proton-proton cycle**, are also thought to operate. The carbon-nitrogen cycle is as follows:

$$^{12}_{6}\text{C} + {}^{1}_{1}\text{H} \longrightarrow {}^{13}_{7}\text{N} \tag{19–10}$$

$$^{13}_{7}\text{N} \longrightarrow {}^{13}_{6}\text{C} + {}^{0}_{1}e$$

$$^{13}_{6}\text{C} + {}^{1}_{1}\text{H} \longrightarrow {}^{14}_{7}\text{N}$$

$$^{14}_{7}\text{N} + {}^{1}_{1}\text{H} \longrightarrow {}^{15}_{8}\text{O}$$

$$^{15}_{8}\text{O} \longrightarrow {}^{15}_{7}\text{N} + {}^{0}_{1}e$$

$$^{15}_{7}\text{N} + {}^{1}_{1}\text{H} \longrightarrow {}^{12}_{6}\text{C} + {}^{4}_{2}\text{He}$$

$$\overline{4\,{}^{1}_{1}\text{H} \longrightarrow {}^{4}_{2}\text{He} + 2{}^{0}_{1}e + 24.7 \text{ MeV}}$$

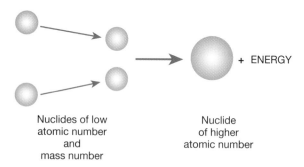

Figure 19–23   In nuclear fusion, light nuclei are fused into heavier nuclei and energy is released.

Fusion reactions occur at temperatures of millions of degrees...

Fusion reactions occur at temperatures of millions of degrees and for this reason are called **thermonuclear fusion reactions**. In order to fuse, (1) the nuclei must approach closely (about 10^{-15} m), and (2) they must possess enough kinetic energy to overcome the Coulomb repulsion of the nuclei. Since reactions between ordinary nuclei of hydrogen proceed more slowly than those between the nuclei of the heavier isotopes, deuterium or tritium, the latter are preferred (Fig. 19–25). Deuterium occurs in nature as a rare isotope of hydrogen, but since there is a lot of hydrogen, there is a lot of deuterium. Tritium is practically nonexistent and must be manufactured in nuclear reactions. A reaction between a deuteron and a triton is energetically favorable and is a key reaction in the **hydrogen bomb** and an example of an uncontrolled thermonuclear fusion reaction (Fig. 19–26).

$$\tfrac{2}{1}\text{H} + \tfrac{3}{1}\text{H} \longrightarrow \tfrac{4}{2}\text{He} + \tfrac{1}{0}n + 17.6 \text{ MeV} \qquad (19\text{–}11)$$

The ignition temperature of about 60 million degrees is provided by the detonation of a fission bomb. The first H-bomb was exploded at Eniwetok Proving Ground in the Pacific in 1952 with energy in the megaton range (10^6 tons of TNT). Since then, fission and fusion have become linked. Hydrogen, or fusion, bombs do not have a critical mass and can be made on any scale, but they depend on the explosion of a fission bomb to get started.

Thus far, temperatures of millions of degrees for periods long enough for controlled fusion reactions to occur have not been obtainable except in the case of fission bombs. Research is being carried out on methods of holding a **plasma**—a hot gas in which atoms are reduced to their nuclei and electrons—in a container by means of magnetic fields. If deuterium gas in a tube is completely ionized, it will exist as electrons and nuclei, not as atoms. A strong magnetic field could prevent the plasma from touching the sides of the container; this would be necessary because of the

<figure>

Figure 19–26   The explosion of a hydrogen (thermonuclear) bomb releases tremendous amounts of energy. (U.S. Navy/Science Photo Researchers, Inc.)

conductive loss of heat. A number of configurations have been tested, such as the **tokamak** and the **stellarator** (Fig. 19–27). Collisions between deuterons could result in fusion reactions with the release of nuclear energy in these "magnetic bottles."

Work on the problem of controlling thermonuclear reactions is continuing in many laboratories throughout the world. The Plasma Physics Laboratory at Princeton University in 1978 achieved a 60 million degree Celsius tokamak plasma temperature. This level is well above the 44 million degree Celsius temperature estimated to be needed for sustained fusion reactions, but it lasted for only 15 thousandths of a second. Further successful plasma production in the tokamak test reactor in 1982 set the stage for possible sustained fusion reactions by the 1990s. Indications are that an effective fusion reactor may become scientifically feasible before the end of the twentieth century. If it is ever developed, its environmental and other advantages will be great. For these reasons, it is being called the "downtown" reactor. The fuel supply problem would be solved, as would the problem of radioactive wastes. There is an inexhaustible supply of the basic fusion fuel, deuterium, in the oceans, and there are no significant amounts of radioactive by-products produced in fusion.

In 1989, two electrochemists, Stanley Pons of the University of Utah and Martin Fleischmann of the University of Southampton, England, made the startling announcement that they had found a way to produce fusion at room temperature. If true, it would mean that a test tube nuclear reaction could produce enough excess heat to be a viable source of cheap, virtually inexhaustible commercial power. Scientists

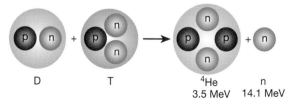

Figure 19–25   A fusion reaction with deuterium (D) and tritium (T) nuclei.

D   T ^{4}He   n
                                  3.5 MeV   14.1 MeV

p n   +   p n n $\longrightarrow$ n p p n   +   n

Figure 19–27   Photograph of the Princeton tokamak fusion test reactor, which uses magnetic field confinement. (Courtesy of Princeton Plasma Physics Laboratory, Princeton, NJ)

around the world carried out similar experiments in their laboratories in an effort to confirm these results.

The experiment was similar to one reported more than 60 years earlier. In 1927, John Tandberg, a research scientist in Sweden, placed palladium electrodes in water and ran a current across them. The cell seemed to produce more heat than could be explained by known physical processes and helium. At a time when fission had not even been discovered, Tandberg seemed to have stumbled across the fusion of hydrogen into helium and applied for a patent. It was not immediately clear whether, in their experiments, Pons and Fleischmann had achieved the breakthrough—a sustained room-temperature "fusion in a jar" reaction in a palladium electrode sitting in a jar of heavy water—or the disappointment of the decade. In the face of disbelief and often ridicule from other scientists, researchers in several countries continue their investigations into the possibility of some unknown nuclear phenomenon occurring in cold fusion.

19–14 FUSION REACTORS

One of the approaches to controlled fusion is **inertial confinement**, a process that relies on powerful laser beams to implode tiny pellets of deuterium and tritium. Instead of laser beams, intense beams of electrons (and, more recently, ions) have been employed

as well. This approach to **fuel-pellet implosion** is potentially efficient, simple, and economically competitive, but like all other approaches to fusion, it faces formidable technical problems. An intense energy source, either a laser beam or a particle beam, is

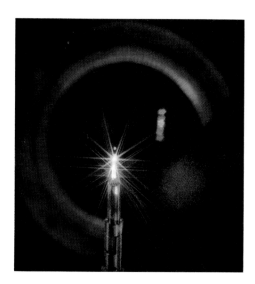

Figure 19–28   A powerful laser beam has initiated a fusion reaction in a 1-mm target capsule that contained deuterium and tritium. It is an approach to controlled fusion as a practical source of energy. (Courtesy of University of California, Lawrence Livermore Laboratory)

directed onto the outer surface of a spherical pellet, causing an implosion of the deuterium–tritium fuel mixture in the pellet (Fig. 19–28). The compression heats the fuel to the ignition temperature and increases the amount of fuel that can be burned.

Some scientists believe controlled fusion will never be achieved. Others feel that it can be done in time, perhaps a century from now, or even by the year 2000. Although fusion is worth pursuing, no one can say that it will be the dream energy that some imagine.

19–15 NUCLEAR STRUCTURE

Two models of nuclear structure, the liquid-drop model and the nuclear shell model, are discussed here.

The **liquid-drop model** portrays the nucleus as a configuration of closely packed protons and neutrons, each interacting only with its nearest neighbors. There is a striking resemblance in this model to atoms in a liquid drop, in which the forces of attraction and repulsion between particles are balanced.

Unless it is radioactive, a nucleus is stable despite the presence in it of positively charged protons that tend to disrupt it. Just as in a liquid drop, particles evaporate from the surface if they have enough energy to overcome the surface tension, so nucleons may be emitted from a nucleus by picking up enough speed through chance collisions with other nucleons and breaking through the barrier of cohesive nuclear forces. On the other hand, a particle may enter a nucleus from the outside, and by its presence and the kinetic energy that it imparts to the nucleons, it may permit the escape of a proton, a neutron, or an alpha particle. The liquid-drop model describes successfully certain nuclear phenomena—spontaneous α-particle emission, nuclear reaction, and particularly nuclear fission. But nuclei are incredibly complex, and this model yields to others for a better explanation of some effects.

As early as 1934 it was observed that nuclei having 50, 82, and 126 neutrons are unusually stable. Since then, evidence has been found that nuclei having 2, 8, 20, 28, 50, 82, or 126 nucleons of the same kind, either protons or neutrons, also have exceptional properties. These numbers are known as **magic numbers**. Helium has 2 protons and 2 neutrons and is extremely stable. So is oxygen-16, with 8 protons and 8 neutrons.

Tin, with 50 protons, has more stable isotopes than any other element. Lead-206 has 82 protons and is the end product of several radioactive series.

A **nuclear shell structure model**, analogous to that used to explain the electron structure of atoms, has been proposed to explain nuclear stability. In the Rutherford-Bohr model of the atom, as we have seen, the electrons are arranged in concentric shells, each capable of accommodating a definite number of electrons, two for the innermost shell, eight for the next, and so on. In the shell model of the nucleus, the basic assumption is that in the magic-number nuclei the shells are filled (Fig. 19–29). The excited states of nuclei are explained by nucleons being raised from one energy level to a higher energy level. The nuclear shell model successfully explains α-, β-, and γ-emission and the nature of the electric and magnetic fields that surround nuclei. The puzzling aspect that led to the designation "magic numbers" comes logically out of the theory, and the mystery surrounding these numbers is removed.

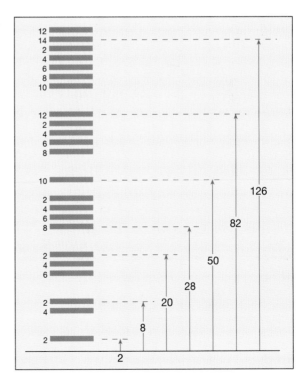

Figure 19–29 Nuclear numbers and energy levels. The magic numbers correspond to filled levels of nucleons.

19-16 BIOLOGICAL EFFECTS OF RADIATION

Radiation, whether in the form of X-rays or gamma rays, or particles such as neutrons or β-particles, is potentially damaging to cells. The radiation from nuclear reactors and fallout, therefore, is hazardous to living things. Most nuclear radiation has very high energies that can be transferred to any matter through which it passes. A single α-particle can cause the ionization of millions of atoms before its energy is dissipated. The chemistry of cells is disrupted and cells may die.

Radiation damage to the bone marrow may cause it to lose its capacity to produce enough red blood cells. When white blood cells are destroyed by radiation, the body loses its resistance to infection. The lining of the gastrointestinal tract is readily damaged, with ulcers developing in the stomach and intestines. The lymph nodes, spleen, and tonsils are most vulnerable. The skin loses its hair when exposed to heavy radiation, and the walls of the blood capillaries may be so weakened that hemorrhages occur throughout the body. Radiation-induced changes in the chromosomes of cells in the reproductive organs may result in genetic mutations. Concern over such consequences underlies the reservations that some have for the further development of nuclear power.

The most important biological measure of radiation is the *rem* (*r*adiation *e*quivalent for *m*an). A useful subdivision is the millirem (mrem): 1 mrem = 1 × 10^{-3} rem. Another unit for the biological measure of radiation is the sievert (Sv). One Sv equals 100 rem. Table 19-3 shows the effect of exposure to a single dose of radiation in rems.

Table 19-3   EFFECT OF EXPOSURE TO A SINGLE DOSE OF RADIATION

Dose (rem)	Probable Effect
0–25	No observable effect
25–50	Small decrease in white blood cell count
50–100	Lesions, marked decrease in white blood cell count
100–200	Nausea, vomiting, loss of hair
200–500	Hemorrhaging, ulcers, possibly death
500+	Fatal

Table 19-4   TYPICAL RADIATION EXPOSURES (1 mrem = 10^{-3} rem)

Sources	mrem/Yr
I. Natural	
A. External to the body	
1. From cosmic radiation	50
2. From the earth (mostly ^{222}Rn)	47
3. From building materials	3
B. Inside the body	
1. In human tissues (^{40}K, ^{226}Ra)	21
2. Inhalation of air	5
II. Man-made	
A. Medical procedures	
1. Diagnostic X-rays	50
2. Nuclear medicine	14
3. Internal diagnosis, therapy	1
B. Nuclear power industry	1
C. TV tubes, industrial wastes, consumer products	2
D. Radioactive fallout (nuclear tests)	4

Table 19-4 lists average exposures of people living in the United States to radiation from both natural and man-made sources. Notice that about two thirds of the radiation comes from natural sources. The level depends upon location. Cosmic radiation is much more intense at high elevations, in some localities twice the national average. People who work in nu-

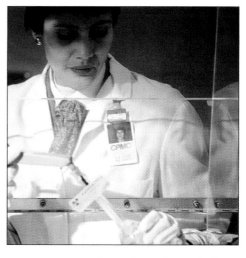

Figure 19-30   A worker with a radiation badge. (Phototake/©Yoav Levy)

clear medicine are also exposed to greater amounts than average and wear radiation badges to assure that exposures do not get too high (Fig. 19–30).

Radiation carries potential for harm, but it is impossible to escape totally. Naturally occurring radiation, or **background radiation**, is present everywhere on earth. If we stopped all forms of artificial radiation, including medical uses, we would still be exposed to background radiation.

KEY TERMS

Accelerators   Machines designed to impart high energies to electrons, protons, and other particles for the purpose of serving as probes in the investigation of properties of matter.

α-emission (α-decay).   The type of radioactive decay in which nuclei emit α-particles.

Atomic Number   The number of protons in the nucleus or the number of electrons in the neutral atom of an element.

Background Radiation   A somewhat steady level of radiation in the natural environment, as from cosmic rays or radioactivity.

Beta-decay (β-decay)   The type of radioactive decay in which nuclei emit electrons.

Binding Energy   The energy that must be supplied to separate the particles in a nucleus.

Breeder Reactor   A nuclear reactor in which the relatively abundant, nonfissionable uranium-238 is transformed into fissionable plutonium-239.

Carbon-14 Clock   A method based on the half-life of carbon-14 by which archaeological discoveries can be dated.

Carbon-Nitrogen (CN) cycle   A mechanism that has been proposed to explain the energy of the sun and stars which involves the fusion of four protons into an α-particle.

Chain Reaction   A self-sustaining reaction that, once started, provides the energy and matter necessary to continue the reaction.

Chernobyl   The site of a nuclear accident April 26, 1986.

Conservation of Electric Charge   In a balanced nuclear equation, the sum of the atomic numbers (electric charges) of the reacting nuclei and particles will be equal to the sum of the atomic numbers and charges of the products.

Conservation of Mass Number   In a balanced nuclear equation, the mass numbers (total number of nucleons) of the reactants and products will be equal.

Critical Size   The minimum quantity of fissionable material that will sustain a chain reaction.

Cyclotron   A nuclear accelerator introduced by Ernest O. Lawrence that is based on the concept of accelerating a particle many times in succession through a relatively low potential difference in such a way that it picks up energy at each acceleration.

Decay Rate   The activity of a sample of radioactive material in number of transmutations or disintegrations per unit time.

Electric Charge   Possessing a net negative or positive charge.

Fast-Breeder-Reactor (FBR)   A nuclear reactor in which the controlled chain reaction produces more fissionable material than it consumes.

Fission   The splitting of an atomic nucleus into two smaller nuclei with the release of energy and several neutrons.

Fuel Meltdown   The danger that excessive heat within the core of a nuclear reactor could cause the core to melt and discharge large amounts of radioactivity into the environment.

Fuel-Pellet Implosion   An approach to controlled fusion in which a laser beam or a particle beam is directed onto the surface of a spherical pellet and causes an implosion of the deuterium-tritium fuel mixture in the pellet.

Fusion   A process in which energy is released when light nuclei are fused into a heavier nucleus.

Geiger-Mueller Counter   A device that detects radioactivity.

Half-Life   The time required for the radioactivity of a sample to drop to one half of any value.

Hydrogen Bomb   A nuclear-fusion or thermonuclear bomb; a key reaction is the fusion of a deuteron and a triton.

Inertial Confinement   Controlled fusion; see fuel-pellet implosion.

Isotopes   Atoms having the same atomic number but different mass numbers.

Liquid-Drop Model   A model of the nucleus that portrays it as a configuration of closely packed protons and neutrons in which the forces of attraction and repulsion between particles are balanced.

Light-Metal-Cooled-Fast-Breeder-Reactor (LMFBR)   A nuclear reactor that uses liquid sodium or some coolant other than water that transforms nonfissionable ("fertile") material into fissionable fuel.

Magic Numbers   Nuclei having 2, 8, 20, 28, 50, 82, or 126 nucleons of the same kind, either protons or neutrons, have exceptional stability.

Mass Defect   The difference between the mass of a nucleus and the sum of the masses of its protons and neutrons.

Mass Number   The total number of protons and neutrons present in the nucleus of an atom.

Moderator   A material in a nuclear reactor that slows neutrons but does not absorb them.

Natural Transmutation   Radioactivity.

Nuclear Shell Structure Model   A model of the atomic nucleus that assumes that energy levels can accommodate nucleons and that the shells are filled in the magic-number nuclei.

Nucleon   Either a proton or a neutron.

Nuclide   Any nuclear species that has a specific atomic number and a specific mass number.

Plasma   A hot gas in which atoms are reduced to nuclei and free electrons.

Proton-Proton Cycle   A mechanism that attempts to explain the energy of the sun and stars through nuclear fusion.

Radioactivity   The spontaneous disintegration of atomic nuclei.

Self-Sustaining Chain Reaction   A nuclear chain reaction in which neutrons released in fission sustain the reaction and cause it to expand.

Stellarator   A device used in fusion research that holds a hot gas—plasma—in a container by means of magnetic fields.

Thermonuclear Fusion Reaction   Nuclear fusion reaction that occurs at temperatures of millions of degrees.

Three Mile Island   The site of a partial core meltdown at a nuclear power plant in 1979.

Tokamak   A device for producing controlled nuclear fusion that involves the confining and heating of a gaseous plasma by means of electric current and magnetic field.

Transmutation   The change of a nucleus of one element into a nucleus of another element either spontaneously, as in radioactivity, or by artificially induced means.

Transuranium Elements   Elements beyond uranium in the periodic table.

Van de Graaff Generator   A device that can be used to accelerate protons to high voltages with great precision; useful in research, industry, and medicine.

THINGS TO DO

1. With a Geiger counter and under supervision, determine (a) the background radiation; (b) radiation from a luminous dial watch; (c) radiation from some uranium compounds; (d) the shielding effects of various materials.

2. Tour the nuclear medicine department of a hospital.

3. If there is a nuclear research reactor in your area, arrange for an on-site visit to it.

4. Bring a radioactive material near the knob of a charged electroscope. Note the behavior of the leaves as the air becomes ionized and the charges leak off.

EXERCISES

Radioactive Decay

19–1. What does it mean to say that a nucleus is radioactive?

19–2. What does the atomic number, Z, of a nucleus represent?

19–3. What does the mass number, A, of a nucleus represent?

19–4. Define (a) nucleon; (b) nuclide.

19–5. (a) How do the isotopes of an element differ? (b) How are they similar?

19–6. Give a nuclear symbol for each of the following.
(a) an alpha particle
(b) a beta particle
(c) a proton
(d) a neutron
(e) a positron

19–7. Indicate the number of protons and the number of neutrons in each of the following nuclides.
  (a) $^{15}_{8}O$
  (b) $^{32}_{15}P$
  (c) $^{131}_{53}I$

19–8. Predict which isotope is more stable.

$^{6}_{3}Li$ or $^{9}_{3}Li$

19–9. Which isotope would you expect to be radioactive?

$^{40}_{20}Ca$ or $^{45}_{20}Ca$

19–10. How does β-emission from a nucleus affect the number of (a) neutrons? (b) protons?

19–11. Beryllium-9 has a mass of 9.01219 u. Calculate its binding energy per nucleon.

19–12. How can an electron (a β-particle) be emitted from a nucleus when it is thought that the nucleus is composed only of nucleons?

Nuclear Reactions

19–13. What are the steps in balancing nuclear equations?

19–14. What name is given to the elements with atomic numbers greater than 92?

19–15. Complete the following nuclear equations.
  (a) $^{2}_{1}H + ^{3}_{1}H \longrightarrow (\ ) + ^{1}_{0}n$
  (b) $(\ ) + ^{4}_{2}He \longrightarrow ^{42}_{20}Ca + ^{1}_{1}H$
  (c) $^{27}_{13}Al + ^{1}_{0}n \longrightarrow ^{28}_{13}Al + (\ )$
  (d) $(\ ) + ^{4}_{2}He \longrightarrow ^{35}_{17}Cl + ^{1}_{1}H$
  (e) $(\ ) \longrightarrow ^{14}_{7}N + ^{0}_{1}e$

19–16. Copper-66 decays by β-emission. Write the nuclear equation.

19–17. Nobelium-254 has been produced by the bombardment of curium-246 with carbon-12 nuclei. Four neutrons are released in the process. Write the balanced nuclear equation.

19–18. Write equations for the following induced nuclear reactions. The first symbol stands for the target nuclide. In parentheses, the symbol of the projectile particle is followed by the symbol of the ejected particle. The last symbol stands for the product nuclide.
  (a) $^{6}_{3}Li\ (n, \alpha)$ _____
  (b) _____ $(\alpha, ^{3}_{1}H)\ ^{250}_{99}Es$
  (c) $^{45}_{21}Sc\ (\alpha, p)$ _____
  (d) _____ $(^{10}_{5}B, 3^{1}_{0}n)\ ^{257}_{103}Lr$

19–19. Gold can be obtained from $^{198}_{80}Hg$ by neutron bombardment. Write a balanced equation for this reaction.

Half-Life; Radioactive Dating

19–20. The half-life of strontium-90 is 25 years. What does this mean?

19–21. Carbon from an ax handle discovered in an archaeological "dig" is only one half as radioactive as the carbon from the handle of a new ax. How old is the artifact?

19–22. Iodine-131, used in the form of sodium iodide to treat cancer of the thryoid, has a half-life of 8.05 days. If 25.0 mg of radioactive Na^{131}I is administered, how many milligrams remain after about a month (32.2 days)?

19–23. Why does an ancient cloth artifact contain less $^{14}_{6}C$ than a recent article made of similar materials?

19–24. Radioactive technetium-99 is used in bone scans and has a half-life of 6.0 hours. What fraction of a dose of ^{99}Tc remains in a patient's body after 2.0 days?

Nuclear Energy

19–25. Discuss some advantages and disadvantages of fission reactors as an energy source.

19–26. In what form is energy obtained from a nuclear reactor?

19–27. Why is it that chain reactions do not occur in natural deposits of uranium?

19–28. Why is the chance of a nuclear reactor's exploding like a bomb remote? Explain.

19–29. What becomes of the "mass-defect" in fission and fusion?

19–30. Beryllium-9 has a mass of 9.01219 u. Calculate its binding energy per nucleon.

19–31. What does a "breeder" reactor breed? Why is this significant?

19–32. In a nuclear reaction, tritium ($^{3}_{1}H$) and deuterium ($^{2}_{1}H$) fuse into helium-4 with the release of a neutron and 17.6 MeV as follows:

$$^{3}_{1}H + ^{2}_{1}H \longrightarrow ^{4}_{2}He + ^{1}_{0}n + 17.6\ MeV$$

Calculate the mass defect of this reaction in atomic mass units (u).

19–33. In a fission reactor, nuclear reactions produce heat to drive a turbine generator. How is this heat produced?

19–34. What is meant by meltdown?

19–35. Why is a fission bomb (called an A-bomb) needed to start a hydrogen bomb (H-bomb)?

19–36. How does $E = mc^2$ apply in (a) fission and (b) fusion?

19–37. Discuss the controversy surrounding "cold" fusion experiments.

19–38. *Multiple Choice*

   A. Radioactivity is caused by
      (a) too many isotopes.
      (b) short half-lives.
      (c) gamma radiation.
      (d) unstable nuclei.

   B. Which of the following detects radioactivity?
      (a) atomic pile
      (b) Geiger-Mueller counter
      (c) cyclotron
      (d) plasma

   C. The smallest amount of fissionable material that will support a self-sustaining chain reaction is called the
      (a) isotopic weight.
      (b) atomic weight.
      (c) critical mass.
      (d) mass number.

   D. The fission of uranium-235 nuclei is initiated by
      (a) electrons.
      (b) protons.
      (c) neutrons.
      (d) gamma rays.

   E. The energy obtained from nuclear reactions comes from
      (a) explosions.
      (b) helium being converted to hydrogen.
      (c) radioactive decay of uranium.
      (d) conversion of mass to energy.

   F. Naturally occurring radioactive isotopes
      (a) must have very long half-lives.
      (b) must have very short half-lives.
      (c) can have either very long or very short half-lives.
      (d) do not exist.

   G. A cyclotron is used to
      (a) accelerate neutrons.
      (b) separate the isotopes of an element.
      (c) accelerate charged particles.
      (d) increase the decay rate of a radioactive substance.

   H. Transuranium elements
      (a) are made by nuclear fusion.
      (b) result from natural radioactive decay.
      (c) have atomic numbers greater than 92.
      (d) occur in nature.

   I. Thermonuclear reactions
      (a) are fission reactions.
      (b) are fusion reactions.
      (c) are due to radioactivity.
      (d) are impossible on earth.

   J. A "breeder" reaction is one in which
      (a) neutrons are produced.
      (b) a fusion reaction occurs.
      (c) a nonfissionable nuclide is converted into a fissionable one.
      (d) electrical energy is produced.

SUGGESTIONS FOR FURTHER READING

Ahearne, John F., "Nuclear Power After Chernobyl." *Science*, Vol. 28, 8 May, 1987.
   Describes the 1986 accident at the nuclear power plant at Chernobyl in the Soviet Union and compares it with the 1979 accident at the Three Mile Island nuclear power station in Pennsylvania. Discusses implications for nuclear power in the United States.

Blomeke, John O., Jere P. Nichols, and William C. McClain, "Managing Radioactive Wastes." *Physics Today*, August, 1973.
   Although underground salt deposits are still the most favored, the arctic ice cap, deep ocean trenches, and solar orbit have been considered as storage areas.

Broyles, A. A., "Nuclear Explosion." *American Journal of Physics*, July, 1982.
   Summarizes the physics of a nuclear bomb explosion and its effects on human beings.

Donath, Fred A., "Debate on Radioactive Waste Disposal." *Physics Today*, December, 1982.
   Radioactive waste will remain active for tens of thousands of years. Do we know enough about it to dispose of it safely?

Edwards, Mike, "Chernobyl—One Year After." *National Geographic*, May, 1987.
   Focuses on the human victims of the worst nuclear power plant accident of all time.

Emmett, John L., John Nuckolls, and Lowell Wood, "Fusion Power by Laser Implosion." *Scientific American*, June, 1974.

Fusion reactions can be initiated without a magnetic field by focusing a powerful laser pulse on a frozen pellet of fuel. Laser-fusion research is comparable in scale to the magnetic-confinement approach.

Furth, Harold P., "Progress Toward a Tokamak Fusion Reactor." *Scientific American*, August, 1979.
Describes efforts to confine a superhot hydrogen plasma in a "magnetic bottle." Will this scheme be feasible for a fusion reactor?

Greiner, Walter, and Aurel Sandulescu, "New Radioactivities." *Scientific American*, March, 1990.
The two-center shell model provides a precise description of many nuclear phenomena. It has led to the prediction and discovery of both new elements and new radioactivities.

Johnson, Karen E., "Maria Goeppert Mayer: Atoms, Molecules and Nuclear Shells." *Physics Today*, September, 1986.
Mayer's early work in atomic and molecular physics gave her the ideal preparation for solving the puzzle of the nuclear "magic numbers."

Lewis, Harold W., "The Safety of Fission Reactors." *Scientific American*, March, 1980.
Considerations of reactor safety must make a considerable allowance for human behavior. Such matters as quality control, equipment maintenance, security measures, and safety precautions are as important as the technical components.

Margolis, Stanley V., "Authenticating Ancient Marble Sculpture." *Scientific American*, June, 1989.
For more than 7000 years, marble sculpture has mirrored the artistic and spiritual progress of civilization. Sophisticated new techniques based on stable and radioactive isotopes are used to authenticate ancient marble sculpture.

Rafelski, Johann, and Steven E. Jones, "Cold Nuclear Fusion." *Scientific American*, July, 1987.
The particles called muons can catalyze nuclear fusion, eliminating the need for high-temperature plasmas and powerful lasers. The process may one day become a commercial energy source.

Schoenborn, Benno P., "Neutron Scattering and Biological Structures." *Chemical and Engineering News*, 24 January, 1977.
The way molecules diffract neutrons discloses details of the structure of biological materials that neither X-ray scattering nor electron microscopy can attain.

Segrè, Emilio G., "The Discovery of Nuclear Fission." *Physics Today*, July, 1989.
Enrico Fermi's group bombarded uranium with neutrons in 1934, but it was almost five years before Hahn, Meitner, and Strassman realized what these neutrons were actually doing.

Sime, Ruth Lewin, "Lise Meitner and the Discovery of Fission." *Journal of Chemical Education*, May, 1989.
Lise Meitner's role in the discovery of fission is clarified, enabling us to gain appreciation for this remarkable scientist and new understanding of the paths to discovery.

Sparberg, Esther B., "A Study of the Discovery of Fission." *American Journal of Physics*, January, 1964.
Neither an accident nor a case of serendipity, nuclear fission represented the climax of activity in nuclear science during the 1930s.

Starke, Kurt, "The Detours Leading to the Discovery of Nuclear Fission." *Journal of Chemical Education*, December, 1979.
The process of nuclear fission was not seriously predicted and, consequently, never looked for. There was no straight path to its discovery.

Upton, Arthur C., "The Biological Effects of Low-Level Ionizing Radiation." *Scientific American*, February, 1982.
How hazardous is the low-level radiation from background and man-made sources? The evidence so far, according to the writer, indicates that the hazard is too small to be detectable.

The ultraviolet telescopes aboard the Astro mission, carried into space on a space shuttle.
(Courtesy NASA/Johnson Space Center)

III

ASTRONOMY

PART-OPENING ESSAY

PERSPECTIVE ON ASTRONOMY

Think how different our view of the universe would be if we lived in perpetual daylight. With a blue dome clamped over us, we would know of nothing beside the earth beneath our feet and the sky and clouds above. The science of astronomy might never have arisen, for what could one study other than the sun, a blinding ball of fire suspended beyond reach? Only if something eclipsed the sun would the real universe emerge.

Isaac Asimov explored this theme in his famous short story, "Nightfall." The story's world, Lagash, has six suns, at least one of which is always in the sky. Every 2000 years, the suns and an invisible moon align in such a way that the world plunges into darkness. Astronomers who warn of this event are ignored, as are the "cultists," heretics who speak of strange "stars." When the black night falls, the sky explodes with tens of thousands of stars: Lagash is in the center of a giant cluster. People lose their minds and set fires to fend off the night, burning the society to the ground.

We feel no such terror on earth, for every evening treats us to the wonders of the night sky. Those who live near cities, with their bright midnights and few sickly stars, remain within civilization's influence. But far from city lights, stars freckle the heavens and the Milky Way Galaxy spills from horizon to horizon—enough to transport one's imagination into space. It is easy at such moments to understand why the night sky has fascinated humans since their first ancient glimmers of self-awareness.

Even with this daily reminder, not many of us realize how vast and empty the universe is. Just as in physics, where the structures of atoms reveal that ordinary matter is almost entirely empty space, the facts of astronomy show that the universe essentially is a void dotted with afterthoughts of matter. From a purely physical standpoint, planets, stars, and galaxies are aberrations; nothingness is the norm.

This situation is hard to appreciate because distances in the universe are so intimidating. Most people shrug when told that the closest star other than the sun is about four light-years away. For that matter, 25 trillion miles away does not help either. Light-years are too abstract, and a trillion—the federal deficit of the United States notwithstanding—is too enormous to comprehend.

Sometimes it helps to express these distances in terms of everyday things. Shrink the sun to the size of a basketball, for instance, and the earth becomes a BB. On this scale, the distance between the earth

The outer atmosphere of the sun, the corona, as observed during a solar eclipse. (High Altitude Observatory/NCAR)

and the sun is 100 feet. The majestic planet Jupiter, a large marble, orbits 500 feet away. Tiny Pluto, a pellet of birdseed, is banished to 4000 feet away. Throw in six more planets and scatter some dust to represent asteroids and comets, and you have a fair model of our solar system: a sphere nearly two miles across with a basketball at its center, flecks of matter here and there, and lots of empty space.

Even more striking are the distances between the stars. A basketball-sized sun in New York City, for example, would have as its nearest neighbor another basketball in Honolulu, 5000 miles away. In fact, when galaxies collide—a common event in the universe—they can pass right through each other. None of the billions of stars will smash together, because the stars are so widely spaced. On the largest scale of all, some 100 billion galaxies are strewn across the universe, covering such distances that even earthly comparisons become futile.

How can one study so many objects that are so remote? Within our solar system, astronomers have hit upon a fabulously successful method: They launch robotic explorers to fly past or land on the planets and their moons. Data from the robots, beamed across space and gathered by radio antennas, allow close-up study of alien landscapes. In this way, the Pioneers, Vikings, Voyagers, and their cousins have captured the public's imagination as have few other endeavors in science.

To plan these missions and analyze their results, researchers rely heavily on the other physical sciences. The physics of gravitation certainly comes into play. *Voyager 2,* for instance, used each planet as a gravity "slingshot" to whip past Jupiter, Saturn, Uranus, and Neptune in a stunning tour of the outer solar system. To chart this complex path, planners calculated the precise gravitational pulls of each planet and their many satellites, whose orbits were painstakingly plotted.

For some missions, chemistry is crucial. The Viking missions to Mars included surface landers that searched for signs of life. Mechanical arms scooped up Martian soil and placed it into chambers; automatic devices added gases and nutrients. Despite some tantalizing initial results, these miniature labs found no signs of organic chemical reactions. Mars, at least for now, appears to be barren of life.

The earth sciences are also essential. Researchers decipher the spectacular pictures from these voyages by comparing unknown features on a planet or moon to earth's own geology. This approach has exposed ancient water courses on Mars, volcanoes on a moon of Jupiter, and plate tectonics on Venus—the same process that ruptures our planet's crust and causes earthquakes.

Beyond the solar system, astronomers must use less direct means. Telescopes collect many varieties of light, from the visible band of the human eye to radio, infrared, and ultraviolet, and energetic X-rays and gamma rays. Each reveals something different about its source. For instance, X-rays may point the way to gas spiraling into a black hole—an object so dense that even light cannot escape its intense gravity.

Many new telescopes, including the Hubble Space Telescope and the giant Keck Telescope in Hawaii, are poised to address some of astronomy's most fascinating questions. Within the decade, astronomers should nail down the age of the universe, which today they can only estimate as between 10 and 20 billion years. They also aspire to know the universe's fate: Will it expand forever, fading into a cold darkness? Or will it eventually collapse, leading to a "Big Crunch" and perhaps a completely new universe? Such questions are grounded in science, because they involve tallying all of the matter in the universe, including "dark matter" that is hidden from view. But no one argues that the answers also have profound philosophical and theological consequences.

These bold pursuits may seem irrelevant. Indeed, much of astronomy results in knowledge for its own sake, but some of that knowledge strikes at the heart of our lives. For example, of the chemical elements, only hydrogen and helium were formed in large amounts by the Big Bang. The rest came from stars, which make heavier elements like carbon and oxygen as they consume their nuclear fuel. The largest stars, once they forge iron in their cores, die in cataclysmic explosions called supernovas. These explosions create all other elements and fling them into space, providing the raw material for new suns and planets—and, perhaps, for new life.

Consider, then, the carbon in the pages of your book, the oxygen you breathe, the calcium in your bones, and the iron in your blood. As incredible as it may seem, the atoms in these molecules were born long ago in the insides of stars—the ancestors of the ones that seem so inaccessible to you when you stare into the night sky.

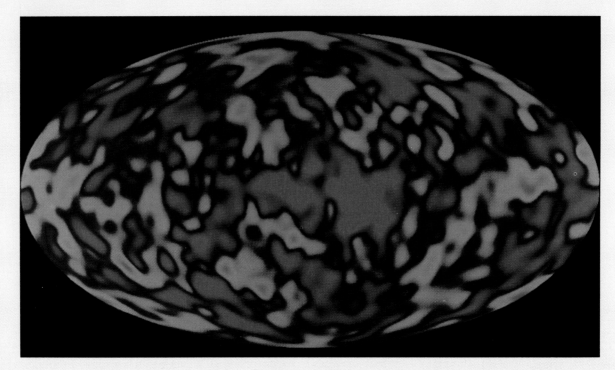

Pictured is a microwave map of the whole sky made by NASA's Cosmic Background Explorer (COBE) Differential Microwave Radiometers (DMR). The red and blue patterns indicate temperature fluctuations of one hundredth of one percent above or below the average sky temperature of 2.73 degrees above absolute zero. The map supports predictions from the "inflationary Big Bang" theory, as the amplitudes of the temperature fluctuations are consistent with explaining the birth and growth of galaxies using large amounts of invisible "dark matter." (NASA/Goddard Space Flight Center)

20

ASTRONOMY TODAY

Astronomy today is one of the most active and exciting fields of physical science. The excitement extends from astronomers and space scientists generally to the public at large, through such events and phenomena as the launching of artificial satellites, orbiting observatories, space shuttles and labs, planetary probes, the recent return of Halley's comet, the appearance of a new supernova, and the search for extraterrestrial life. Pulsars, black holes, quasars, and the discovery of scores of interstellar molecules are changing some established ideas concerning the nature of the solar system and the universe as a whole.

20–1 LIGHT AND ASTRONOMY

Almost everything we know about the planets and stars comes to us from the light and other electromagnetic radiation they send us. Until 1610, knowledge of the universe was based on observations made with the naked eye. When Galileo pointed his telescope to the moon, sun, and stars, he changed the course of astronomy dramatically. The **telescope** has since become the astronomer's single most important tool, gathering light, magnifying, and resolving. It is basically a large and powerful eye (Fig. 20–1).

Light energy originates in nuclear reactions in the interior of stars where matter is transformed into energy according to Einstein's equation $E = mc^2$. Each second, 600 million tons of hydrogen in our sun are changed into 595½ million tons of helium and energy

equivalent to 4½ million tons of matter. This energy appears first in the star's interior as gamma rays with very short wavelengths and high energy. As the gamma rays interact with matter in the star, they are gradually transformed into photons of longer wavelengths and work their way to the surface to appear as starlight.

The astronomer uses a telescope to collect light, the raw material of astronomy. Instruments attached to the telescope, like the camera and the spectroscope

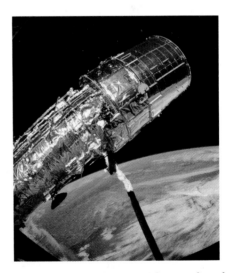

Figure 20–1 The Hubble Space Telescope, launched by NASA in 1990, is the largest telescope in space to date. It is a reflecting telescope with a main mirror 2.4 m (94 inches) in diameter. One day it may discover planets around other stars. (Hughes Danbury Optical Systems and Lockheed Missiles & Space Company, Inc.)

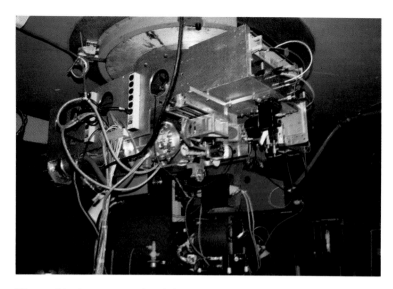

Figure 20–2 An example of the complex instrumentation in a modern telescope. This is a photo of an infrared spectrometer at the 3-m NASA Infrared Telescope Facility on Mauna Kea. The spectrometer is operated remotely from a control room and data are recorded on a computer disk. (University of Hawaii, photo by Alan Tokanaga)

or spectrograph, record and analyze the information received (Fig. 20–2). The astronomer's second most important tool, the **spectrograph**, takes starlight from a telescope, passes it through a prism, and records the resulting spectrum on a photographic plate. The amount of information that spectral analysis gives is nothing less than amazing. We have discussed spectra in Chapters 10 and 12, and in this chapter we shall study further their applications to astronomy.

The **Doppler effect** has become another powerful concept in astronomy. Again, we studied this ef-

Figure 20–3 A telescope operator at her console assists astronomers with their observations. Telescope operators are well-versed in the operation of the telescope and its systems. (Jay M. Pasachoff)

fect earlier with reference to wave motion, mainly in the realm of sound (Chapter 9). Light also behaves as a wave and therefore, also exhibits the Doppler effect. Unlike sound, the Doppler effect in light is too slight for us to notice in our daily lives. When we are dealing with the great velocities encountered in astronomy, however, changes in wavelength and frequency can be measured. If a star is moving toward us, the light waves become shorter, shifting toward the violet end of the spectrum. If the star is receding, the wavelength is lengthened, and there is a **red shift**. Such evidence is also important to any theory of the universe, as we shall see, because it is quantitatively observable.

In recent years, astronomers have not been limited to what their eyes could see or their telescopes could gather, that is, the visible region of the electromagnetic spectrum. It became possible to observe the heavens and collect radiation from nearly every region of the spectrum (Fig 20–3). **Radio astronomy** uses the waves in the radio region of the spectrum, which are about 1 million times longer than light waves, and requires different types of instruments to record and analyze data. **Infrared**, **ultraviolet**, and **X-ray astronomy** deal with wavelengths in those regions of the spectrum. Observations at these various wavelengths are being brought together and during the remaining years of the twentieth century may give the first complete picture of the universe.

20-2 THE CONSTELLATIONS

To early observers, all objects in the sky were regarded as stars. The "fixed stars" were distinguished from the "wandering stars" (planets) and with the sun and moon were believed to revolve around the earth in their respective "spheres." About 3000 stars are visible to the naked eye on any clear, moonless night, but they are not evenly distributed. The brightest are grouped into **constellations** to which the names of mythological characters have been given. There is often no apparent resemblance, however, between the shape of the mythological figure and the star grouping. There are 88 constellations, some of which are listed with their meanings in Table 20–1.

The constellations are used in astronomy mainly to indicate the region in the sky in which a celestial

Table 20–1 NAMES OF SELECTED CONSTELLATIONS

Name of Constellation	Meaning
Andromeda	Chained maiden
Aquarius	Water bearer
Aries	Ram
Cancer	Crab
Capricorn	Sea goat
Cassiopeia	Lady in a chair
Cygnus	Swan
Gemini	Twins (Castor and Pollux)
Hydra	Sea serpent
Leo	Lion
Libra	Balance
Orion	Hunter
Pegasus	Winged horse
Pisces	Fishes
Sagittarius	Archer
Scorpius	Scorpion
Taurus	Bull
Ursa Major	Big bear
Ursa Minor	Little bear
Virgo	Virgin

object may be found. The first discrete source of radio waves outside our solar system, for example, was discovered in the constellation Cygnus and designated Cygnus A. Many objects have been located in the directions of other constellations and mapped, the constellations serving as area markers.

A band of 12 constellations, called the **zodiac**, lies in the same plane as the earth's orbit around the sun (Fig. 20–4). This plane is called the **ecliptic**. These constellations are considered by some to be important in influencing human affairs, although there is no scientific basis for this belief, called astrology. Table 20–2 gives the signs of the zodiac in sequence.

Table 20–2 SIGNS OF THE ZODIAC

1. Aries	7. Libra
2. Taurus	8. Scorpius
3. Gemini	9. Sagittarius
4. Cancer	10. Capricornus
5. Leo	11. Aquarius
6. Virgo	12. Pisces

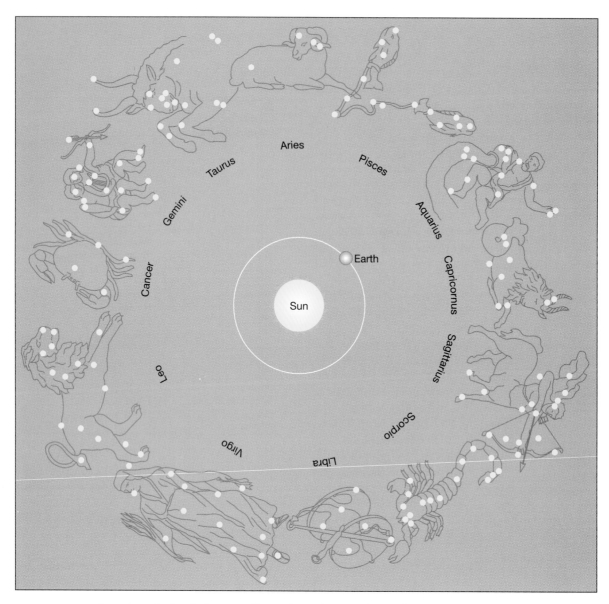

Figure 20–4 The zodiac. This band of 12 constellations lies in the same plane as the earth's orbit around the sun.

20–3 CELESTIAL DISTANCES AND POSITIONS

The star 61 Cygni is 64.8 trillion miles from the earth, and it is one of the nearest stars. For expressing distances to the stars, units more convenient than miles are needed. Even the **astronomical unit** (AU), 93 million miles—the average distance between the earth and the sun—is too short; on this scale, the distance of 61 Cygni is approxi-

mately 700,000 AU. The speed of light, however, is a useful standard for expressing stellar distances. In one year, light travels 6 trillion miles

$$\left(186,000 \ \frac{mi}{s} \times 3.20 \times 10^7 \ \frac{s}{yr} = 6.00 \times 10^{12} \ \frac{mi}{yr} \right),$$ a

distance called the **light-year (ly)**. The distance to 61 Cygni is 10.8 ly.

Although no star is close enough to have a parallax as large as 1 second of arc, the distance to such a star would be 3.26 ly. This distance is defined as 1

Table 20–3   PARALLAXES AND DISTANCES
OF SELECTED STARS

Star	Parallax (Arc-Seconds)	Distance (Parsecs)	Distance (Light-Years)
Proxima Centauri	0.76	1.3	4.2
Sirius	0.38	2.6	8.5
61 Cygni	0.30	3.3	10.8
Aldebaran	0.06	16.7	54.4

parallax-second, or 1 **parsec** (pc). If the parallax of a star is known, the distance in parsecs is its reciprocal. Alpha Lyrae, for example, has a parallax of 0.1″; its distance is therefore $1/0.1 = 10$ parsecs, or $10 \times 3.26 = 32.6$ ly. The distances of several well-known stars are given in Table 20–3.

The positions of stars and other celestial objects are specified in the same way that a position is defined on the earth's surface. The location of any spot on earth is given by its latitude and longitude. **Latitude** is measured north or south of the equator, which is assigned a value of 0°; the poles are 90° to the north and south. **Longitude** is measured east or west of the Greenwich Meridian, a line running from the North Pole to the South Pole through the former site of the Royal Greenwich Observatory in England (Fig. 20–5) and is expressed in two ways: angular measure (degrees, minutes, seconds) and time (hours, minutes, seconds). Thus, a place on earth may have a latitude of $+ 42°43'$ (a positive latitude means that the location is north of the equator) and 73°12′ western longitude (measured west from Greenwich). **Declination** (DEC) is the celestial equivalent of latitude and is specified in degrees, positive to the north and negative to the south of the extension of the earth's equator, the **celestial equator**, at 0°. **Right ascension** (RA) is similar to longitude except that astronomers measure only in the eastward direction; it is the distance measured eastward along the celestial equator from the location of the sun on the first day of spring, the vernal equinox, and is usually expressed in hours, minutes, and seconds (Fig. 20–6). An object identified as PSR 0531 + 21 (a pulsar) is located at 5^h31^mRA and 21°DEC—that is, 5 hours and 31 minutes right ascension, and $+ 21°$ declination.

Figure 20–5   The Royal Greenwich Observatory, England, is the internationally agreed upon zero point of longitude on the earth. (David Morrison)

Figure 20–6 Right Ascension (RA)-Declination (DEC) Celestial Coordinate System. Coordinates give RA in hours around the equator and DEC in degrees along the meridian.

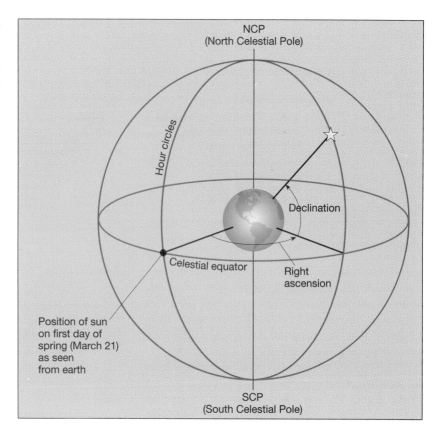

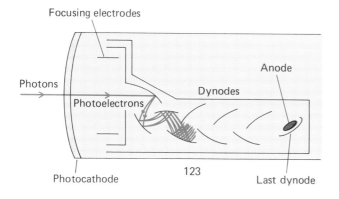

Figure 20–7 Photoelectric photometer. A photoelectric cell converts starlight into an electric current that can be amplified and recorded.

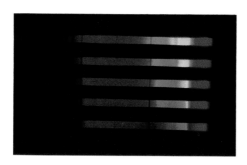

Figure 20–8 Several exposures of the spectrum of Vega, the fifth brightest star in the sky. The strongest lines in the spectrum are produced by the Balmer series of hydrogen. Stars are classified into seven spectral classes: O, B, A, F, G, K, M. (National Optical Astronomy Laboratories)

20–4 CHARACTERISTICS OF STARS

Astronomers depend on the analysis of light emitted by a star for their knowledge of its physical state. A spectrum is an indicator of temperature, luminosity (the total amount of energy a star radiates in 1 second), chemical composition, and size. Light from faint stars is measured by a **photoelectric photometer**, an instrument containing a sensitive photoelectric cell that is used to measure the intensity and color of starlight. Starlight is converted in the cell to an electric current proportional to the amount of light received; the current is then amplified and recorded electronically (Fig. 20–7).

Most spectra of stars are similar to the solar spectrum—a bright continuum crossed by dark absorption Fraunhofer lines—but there are some stars that have only bright-line spectra and no dark absorption lines (Fig. 20–8). From the hottest to the coolest stars, the spectra fall into groups designated O—B—A—F—G—K—M, in the Henry Draper (1837–1882) sequence, one that is easily remembered with the aid of the mnemonic device: *Oh, Be A Friendly Girl/Guy, Kiss Me*. Table 20–4 lists this sequence and some of its characteristics. Hundreds of thousands of stars were classified in this way by Annie Jump Cannon (1863–1941), an astronomer at the Harvard College Observatory, whose catalogue is called the Henry Draper catalogue after the benefactor who made the investigation possible. Many stars are still known by their HD (Henry Draper catalogue) numbers, such as HD 42581. Stellar spectra are compared with a particular set of stellar spectra that serve as standards.

The surface temperatures of stars vary from a relatively cool 2500°C or less to over 50,000°C. Some stars, like the sun, are rich in metals. Stars vary considerably in density and size, ranging from subdwarfs to supergiants. Their diameters range from a fraction of the diameter of the sun to a diameter 500 times as large. Stellar masses range from a hundredth the mass of the sun to about a hundred solar masses.

The Doppler effect gives information about the motion of stars (Fig. 20–9). If a star is moving away, the whole spectrum of its light is *shifted toward the red* or long-wave end. Most stars, for example, show strong absorption lines by calcium atoms at wavelengths of 3933.664Å and 3968.470Å. In the spectrum of the star delta Leporis, one of the lines of calcium is displaced 1.298Å toward the red. Assuming that the displacement is due to the Doppler effect, the speed of the star's receding motion is calculated to be 99.0 kilometers per second.

20–5 THE LIFE AND DEATH OF STARS

Despite their number and diversity, stars fall into a kind of periodic table on the basis of (1) brightness, or magnitude, and (2) temperature, or spectral class. When brightness is plotted along the vertical axis and spectral class along the horizontal axis, the resulting diagram is known as a **Hertzsprung-Russell (H-R) diagram**, after Ejnar Hertzsprung (1873–1967) and Henry N. Russell (1877–1957), who proposed it independently (Fig. 20–10). It has been called the most

Table 20–4 CLASSIFICATION OF STARS

Class	Temperature Range (°C)	Color	Spectrum	Example
O	25,000 and up	Blue-white	He^+	Lambda Cephei
B	11,000–25,000	Blue-white	H^0, He^0	Beta Centauri
A	7500–11,000	White	H^0, Mg^{2+}	Sirius A
F	6000–7500	Yellow-white	Fe^+, Ti^+	Procyon
G	5100–6000	Yellow	Ca^+, Ca^{2+}	Our sun
K	3600–5100	Yellow	Na^+	Alpha Centauri B
M	3000–3600	Reddish	AlO, TiO	Betelgeuse

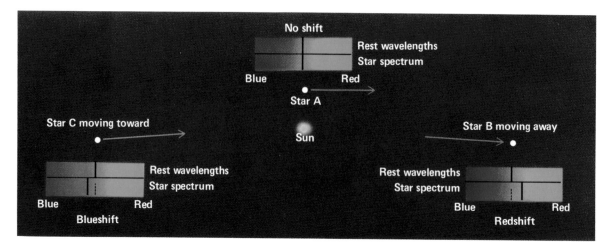

Figure 20–9 The Doppler in effect in stellar spectra. Lines from approaching stars appear blue-shifted, and lines from receding stars appear red-shifted.

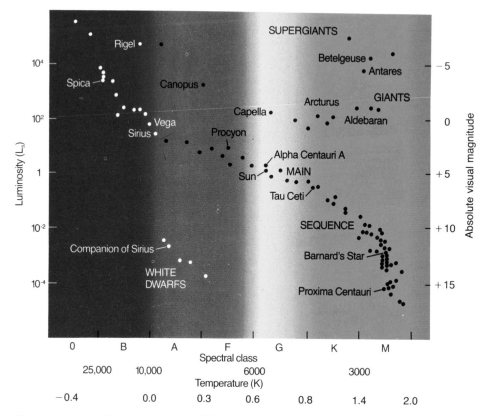

Figure 20–10 Hertzsprung-Russell (H-R) diagram for a selected sample of stars. The luminosity of stars is plotted along the vertical axis, and the spectral class along the horizontal axis. Most stars in the Milky Way occupy the "main sequence." Numerically smaller magnitudes are associated with the brighter stars. Several of the brightest stars are identified by name.

important diagram in all of astronomy, showing a pattern in the size and temperature of stars that indicates that there must be an underlying physical cause.

The most prominent region of the H-R diagram is the **main sequence**. This is a long, narrow band running diagonally across the diagram, in which over 90% of all stars near the sun fall. The more luminous stars on the main sequence are hot and bluish white in color; the dimmest stars are cool and reddish. The sun, with its predominantly yellow color, falls about three fourths of the way along the band.

A star spends most of its lifetime on the main sequence. During this time, hydrogen is converted into helium through nuclear reactions. When the supply of hydrogen is depleted, the star moves off the main sequence into a region lying above it and to the right, the **red giant** branch, the only other well-defined sequence. There, the star enters the "helium-burning" stage of development, during which nuclear reactions involving helium take place. One of these is the triple-alpha process through which three alpha particles combine to form the heavier element, carbon. The majority of stars in the solar neighborhood lie in the main sequence and red giant regions.

The lifetime of the sun as a main sequence star may be 10 billion years, half of which it has already spent. In another 5 billion years its hydrogen will have been exhausted, and it will probably evolve into a red giant. Then its volume will expand far beyond its present limits to the distance of the earth's orbit (Fig. 20–11).

Once they have passed into the helium-burning stage, stars evolve in one of three directions: **white dwarfs**, **neutron stars**, or **black holes**. By then they have virtually exhausted the nuclear fuel that enabled them to shine for billions of years. White dwarfs occupy the lower left region of the H-R diagram and represent more than 10% of all the stars in our galaxy. Stars lighter than about 1.2 solar masses can die as white dwarfs. They are relatively hot stars, but their low luminosity indicates that they must be very small. Sirius B is a white dwarf. Although it is about the size of the earth, its mass is nearly that of the sun (Fig. 20–12). The average density of a white dwarf is about a million times greater than the sun's density; a cubic inch of matter from Sirius B would weigh an amazing 1 ton on earth.

Neutron stars were first predicted theoretically in the 1930s. Their internal pressure is so great that the plasma electrons combine with protons to form neutrons. Such stars are much denser than white dwarfs; a cubic inch of a neutron star would weigh a billion tons on earth (Fig. 20–13). Only a few kilometers in diameter, a neutron star should spin very rapidly as it contracts from the size of a normal star, at rates between 30 times per second and once every few seconds, because of the conservation of angular momen-

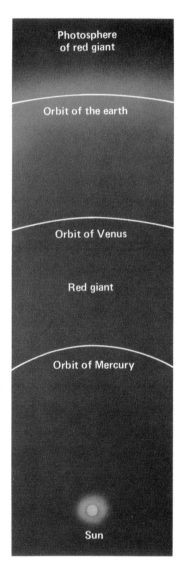

Figure 20–11 The sun may evolve into a red giant star. Its volume could then expand to the distance of the earth's orbit.

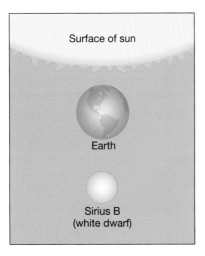

Figure 20–12   A white dwarf is about the size of the earth but contains about 300,000 times more mass.

Figure 20–14   In a black hole, the strong gravitational field would even bend light emitted at any angle away from the vertical so that it falls back onto the star.

tum (just as skaters spin more rapidly as their arms are brought closer to the body).

Astronomers now largely agree that **pulsars**, discovered in 1968 as a class of objects that emit bursts of radio noise with periods ranging from 0.033 to 3.75 seconds, are rapidly rotating neutron stars. Only neutron stars can rotate at pulsar frequencies without disruption. It is estimated that as many as 100 million of the 100 billion stars in our galaxy have burned themselves out and collapsed into neutron stars.

There has been speculation that certain stars (in which the mass considerably exceeds the mass of the sun in their final stage of evolution) may collapse to densities even greater than those found in neutron stars. The gravitational field around such a star, called

a **black hole,** would be so great that neither matter nor light could ever leave it (Fig. 20–14). The density of a black hole would be greater than 10^{16} g/cm^3, the densest matter known in the universe. Black holes are the subject of considerable research. Astronomers in 1973 detected X-ray evidence of a black hole in the constellation Cygnus. The black hole seemed to be a kind of invisible companion to a visible supergiant star called HDE-226868. Every 56 days the black hole passes behind the visible star. As it does so, it blocks X-rays that are created as gases spiral toward the black hole and are compressed and heated to 100 million degrees. A suspected black hole was also discovered in the constellation Scorpio in 1978 and named Scorpio V-861.

20–6 NOVAE AND SUPERNOVAE

Stars that at one time appeared very faint may suddenly become extremely brilliant, with a light output thousands of times greater than before. Such stars are called **novae** ("new stars") (Fig. 20–15). Novae bright enough to be visible to the naked eye occur only once every few years in our galaxy. Those that are observable by telescope occur more frequently. A nova may reach its maximum brightness in one or two days, or it may take as long as several days or weeks. Novae are exploding stars that blow off parts of their atmospheres. The optical radiation decreases gradu-

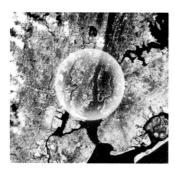

Figure 20–13   A neutron star may be the size of a city, only 20 kilometers in diameter. Even so, it may contain a solar mass or more. (NASA)

August 28, 1975
11:30 U.T.

August 29, 1975
6:45 U.T.

Figure 20–15   An amateur astronomer, Ben Mayer, photographed the eruption of Nova Cygni in 1975. Never before had a nova's brightening been so well observed. (© Ben Mayer, Los Angeles)

ally after several days or weeks, and after a few years, the star seems to return to its former state.

More rare than novae, but considerably more energy-rich, are the **supernovae**. Supernovae are major explosions, flaring up to millions of times brighter than their original state. Only two have been recorded within our galaxy by Western astronomers, the first by Tycho Brahe in 1572 in the constellation Cassiopeia, and the second by Johannes Kepler in 1604 in the constellation Serpens. At its peak, Tycho's supernova was several times brighter than Venus and could be seen even in the daytime; Kepler's was as bright as Jupiter. With powerful modern telescopes, supernovae are regularly found in other galaxies.

Chinese astronomers have recorded at least four other supernovae, including one in 1054 in the constellation Taurus, the remains of which are now recognized as the Crab Nebula (Fig. 20–16). Why Western chroniclers failed to mention a star that had been invisible and suddenly became (on July 4, 1054) one of the brightest in the sky—so bright that it could be seen in full daylight—is a mystery. The event must have been one of the most spectacular in history. Although ignored by the West, in China and Japan the "guest star" was described in detail. After glowing brilliantly for three weeks, it gradually faded and disappeared after two years. Theory predicts that some supernovae will leave behind a neutron star. In 1968, astronomers discovered a pulsar at the heart of the Crab Nebula. The Veil Nebula in Cygnus is believed to be another visual remnant of a supernova.

Ironically, the discovery of the most spectacular astronomical event since Kepler's 1604 supernova 383 years earlier—the eruption of a nearby supernova—fell to an amateur. On the night of February 23–24, 1987, Ian Shelton, then 29, was at the Carnegie Institution's Las Campanas Observatory in the Andes of Northern Chile working at the observatory's 25-cm telescope. Shelton, not a professional

Figure 20–16   The Crab Nebula, in the constellation Taurus, is the remnant of a supernova explosion which was seen in the year AD 1054. (National Optical Astronomy Observatories)

(a)   (b)

Figure 20–17   The region of the Tarantula Nebula in the Large Magellanic Cloud (a) before and (b) after February 24, 1987. Supernova 1987A shows clearly at the upper left. (© 1987 Anglo-Australian Telescope Board)

astronomer, was employed by the University of Toronto to assist researchers visiting the university's 60-cm instrument, but he had a personal interest in astrophotography.

Shelton's project that night was a 3-hour exposure of the Large Magellanic Cloud (LMC), a dwarf galaxy that lies just beyond the Milky Way, 170,000 ly from earth. Finishing that task at 2:40 AM, Shelton decided to develop the last photographic plate before retiring. Scrutinizing the plate, he saw an unfamiliar bright spot near a feature within the LMC known as the Tarantula nebula (Fig. 20–17). Shelton thought at first that it was a flaw on his photographic plate, but it was no flaw.

Shelton stepped out into the clear mountain air and looked up at the sky. In the fuzzy patch of light known as the Large Magellanic Cloud, without a telescope or binoculars, Shelton saw the spot. For more than 3 hours he tried several logical explanations and then accepted the fact that what he had just seen was a supernova, the first one visible to the naked eye since 1885 and the brightest one to appear since 1604.

Realizing the importance of Shelton's discovery, the observatory contacted the Central Bureau for Astronomical Telegrams of the International Astronomical Union at the Harvard-Smithsonian Center for Astrophysics in Cambridge, Massachusetts. The Bureau serves as the official clearinghouse for astro-nomical discoveries. Once a finding is confirmed, the Bureau alerts astronomical installations all over the world by telegram, telex, computer mail, and printed circulars. Confirmation was not a problem in this case. Almost simultaneously with the news from Chile, a call arrived from Australia reporting that an amateur astronomer in Nelson, New Zealand, Albert F. C. D. Jones, had witnessed the same event.

By the end of the day, telegrams announcing the supernova, officially designated Supernova 1987A, SN 1987A (the first supernova to be observed in 1987), were sent to 150 institutions around the world. The telegram read in part: "W. Kunkel and B. Madore, Las Campanas Observatory, report the discovery by Ian Shelton, University of Toronto, of a mag 5 object, ostensibly a supernova, in the Large Magellanic Cloud." By the next evening, nearly all major optical and radio telescopes in the Southern Hemisphere were trained on SN 1987A.

In the months following, SN 1987A became the best studied supernova in nearly 400 years. Since the supernovae of 1572 and 1604 occurred just before the invention of the telescope and modern technology, astronomers today have a sense of experiencing the find of a lifetime. The debris from SN 1987A will be observable through the telescope for decades, if not centuries. The event marks the beginning of scientific research on supernovae.

20–7 PULSATING STARS

Pulsating stars wax and wane in brightness in periods ranging from 1 to 100 days. The most familiar of these is Polaris, the pole star, which brightens and fades to a slight extent in a period of 3.97 days. One class of such variable stars is called the **Cepheid variables**, named after a pulsating star discovered in the constellation Cepheus and named delta Cephei. Because of their changing internal structure, pulsating stars are unstable. As they increase and decrease in size, their surface temperature also changes, and their light emission varies.

The Cepheid variables act as a "celestial lighthouse" and as a means of measuring distances deep into the universe where other methods of measurement are inadequate. The Cepheids were first investigated by Solon I. Bailey (1854–1931), a Harvard astronomer, who studied them in aggregations of stars called **globular clusters** within our own galaxy. Henrietta S. Leavitt (1868–1921) (Fig. 20–18) of the Harvard College Observatory later investigated the Cepheids in the Clouds of Magellan, two small companion galaxies of our own, and determined that most had periods of more than a day. At the same time, she discovered that the average apparent brightness of these Cepheids was related to their periods of pulsation. The longer the period was, the greater the luminosity. She established a **period-luminosity relation**, which became a potential yardstick for measuring great astronomical distances (Fig. 20–19).

Harlow Shapley (1885–1972), director of the Harvard College Observatory, applied the period-luminosity relation to the determination of the dis-

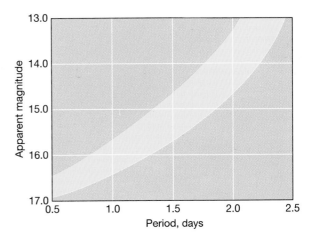

Figure 20–19 Period-luminosity relation for cepheids in the Clouds of Magellan. The longer the period, the brighter the star.

tances of the globular clusters in our galaxy and found that the Milky Way system was much larger than anyone had supposed. Edwin P. Hubble (1889–1953) at the Mt. Wilson Observatory then used it in measuring the distances to the Magellanic Clouds, to the Great Nebula in Andromeda, and to more distant galaxies up to 20 million light-years away. The relation was further used to calibrate the red shift, making possible measurement to still more distant galaxies in which the Cepheids cannot be resolved, even by the most powerful telescopes available. From such data, the size of the universe and, if the universe is expanding, its age from the origin of expansion could be estimated.

20–8 GALAXIES, NEBULAE, AND STAR CLUSTERS

When the sky is observed on a clear, moonless night, a faint milky band of light is visible across certain constellations. Ancient astronomers called it appropriately the **Milky Way**, or **galaxy** (Greek *gala*, meaning "milk") (Fig. 20–20). All the stars that can be seen with the naked eye, and most of those that can be distinguished by the most powerful telescopes, are members of the Milky Way system. Even with a pair of field glasses or a small telescope, the milky appearance can be seen to be the effect of countless stars merged to form a nebulous mist. There are more than

Figure 20–18 Henrietta S. Leavit, who established the period-luminosity relation. (Harvard College Observatory)

Figure 20–20 A drawing of the Milky Way as it appears to the eye. Seven thousand stars plus the Milky Way are shown. (Lund Observatory, Sweden)

100 billion stars in the Milky Way, the majority of them so distant that we cannot see them as individual points of light. Aside from stars, a galaxy includes such components as star clusters, clouds of gas and dust, and interstellar molecules.

Our knowledge of our place in the Milky Way and of the distribution of galaxies in the universe has been gained only in the present century. The early telescopes revealed many diffuse patches of light, which looked nearly alike and were classified as **nebulae** (Fig. 20–21). These objects were believed to lie within the Milky Way, but they did not seem to be at all like stars. Many of them are beautiful spirals and are seen in different orientations. As early as the eighteenth century, the astronomer Sir William Herschel (1738–1822) and the philosopher Immanuel Kant (1724–1804) suggested that these nebulae were actually **"island universes"** or galaxies lying far beyond the Milky Way, but this hypothesis was not generally accepted. As recently as the 1920s, the nature of nebulae was being argued.

The matter was settled in 1924 when Edwin P. Hubble, using the Cepheid variable period-luminosity relation, succeeded in measuring the distance to the Great Nebula in Andromeda and to a number of other spiral nebulae with the 100-inch telescope at Mt. Wilson, then the most powerful in the world (Fig. 20–22). Hubble established that the nearest spiral nebulae were galaxies like our own Milky Way—vast systems of stars, clusters, gas, and dust—situated a million or more light-years outside our own galaxy. The universe is populated by billions of galaxies that appear to be the building blocks of the universe. A single photograph with a telescope may easily show 1000 galaxies.

The term "nebula" now refers to a cloud of gas and dust, one of the many components of a galaxy. The Ring nebula in Lyra is an example of a planetary nebula, a small shell centered on a star that is thought to have ejected it, which shines by fluorescence stimulated by ultraviolet radiation from the central star (Fig. 20–23). The Horsehead nebula in Orion is a

Figure 20–21 The central region of the Virgo Cluster of galaxies. Virgo has hundreds of bright galaxies. (Copyright ROE/AAT Board)

Figure 20–22 The Andromeda Galaxy, also known as M31 and NGC 224, is the nearest spiral galaxy to the Milky Way and is similar to it in size and structure. This galaxy contains over 300 billion stars. (© 1959 California Institute of Technology)

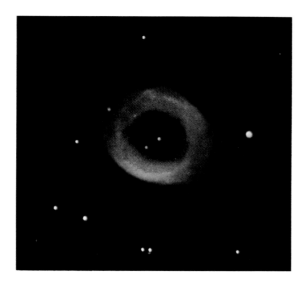

Figure 20–23   The Ring Nebula in the constellation Lyra. (© 1959 California Institute of Technology)

dark cloud of dust that hides background stars (Fig. 20–24). Some nebulae are supernova remnants produced by the explosion of a massive star; they expand at high velocities and radiate as a result of collision with interstellar gas.

If viewed from the outside, the Milky Way would probably appear like a flat disk with tightly wound spiral arms very much like the spiral Andromeda galaxy. It resembles a giant pinwheel made up of more than 100 billion stars concentrated toward a central hub and rotating slowly around a compact, brilliant nucleus (Fig. 20–25). The spiral arms consist of concentrations of stars, dust, and gas. At the speed of light it would take 100,000 years to pass from one end of the Milky Way to the other; it is 5000 to 10,000 ly in thickness. The sun is 30,000 ly from the center of the Milky Way, orbiting around it once about every 200 million years.

The space above and below the pinwheel is filled with billions of much fainter stars and some 200 fuzzy globular **star clusters**. A star cluster is a group of stars moving together through the galaxy in relative proximity. It is made up of hundreds of thousands of stars concentrated toward the center so that they appear to fuse together and suggest a continuous area of light. These globular clusters are scattered around the main nucleus of the galaxy, forming a sort of halo around the main body. None is near our sun; the closest is about 20,000 ly away. The slightly ellipsoidal shape of many of them suggests that they probably rotate.

Many of the stars in globular clusters are red giants known as **Population II** stars, a classification based primarily on their composition. They are apparently very ancient, approximately 10 billion years old, suggesting that they may have been formed before the Milky Way evolved into its present spiral structure. Hundreds of "open" star clusters, most consisting of a few hundred stars held together by their mutual gravitation, have also been found, usually in or near the spiral arms. Stars in open clusters, called **Population I stars**, are relatively young, from about 1 million to 1

Figure 20–24   The Horsehead Nebula in Orion is dark dust superimposed on glowing gas. (© 1979 Royal Observatory, Edinburgh/Anglo-Australian Telescope Board, from original U. K. Schmidt plates)

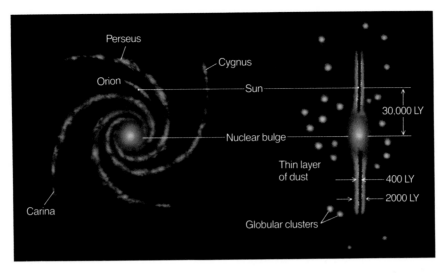

Figure 20–25 Schematic representation of the Milky Way. The face-on view shows the locations of the spiral arms. The sun is 30,000 ly from the center.

billion years old, and are often associated with interstellar matter from which new stars are believed to be forming. Little, if any, interstellar matter is found in the globular clusters.

With modern instruments, a billion or more galaxies are detectable, and several thousand are close enough for detailed inspection. They vary in size, shape, mass, and luminosity. Galaxies are usually identified by number, prefaced by initials standing for the reference in which they are catalogued. The brightest have the initial M (such as M51) for Charles Messier (1730–1817), an astronomer who listed the 103 brightest nebulae in 1784. Many of the fainter ones are designated NGC for the New General Catalogue of John L. E. Dreyer (1852–1926), based largely on discoveries of nebulae by the pioneers of galaxy studies, William, his sister, Caroline Lucretia (1750–1848), and his son John Herschel (1792–1871) (Fig. 20–26).

Figure 20–26 The Seyfert galaxy, NGC 1566, has the appearance of a normal spiral. (© 1987 Anglo-Australian Telescope Board)

Galaxies, in turn, are themselves generally found in clusters that vary in size and in the number of member galaxies. The Milky Way is a member of a cluster of 20 or so galaxies known as the **Local Group**. Andromeda, M33, and the Large and Small Magellanic Clouds are other members. Three of the members of this cluster were discovered in 1971, suggesting that there may be others as yet undiscovered. Some clusters, such as the Virgo and Coma clusters, contain thousands of members. The galaxies within a cluster are not crowded, being separated by distances several times their individual diameters. Clusters of galaxies apparently form the largest structural units in the universe.

20–9 RADIO ASTRONOMY

The recent revolution in astronomy is due in large part to a serendipitous discovery made in 1932 by Karl G. Jansky (1905–1950) of the Bell Telephone Laboratories that opened a new window on the universe. Until then, our only window into space was the visible region of the electromagnetic spectrum. To better understand the causes of interfering static, so that improved transoceanic short-wave radio-telephone systems could be designed, Jansky built a rotating radio antenna 100 feet long. He found that thunderstorms and man-made electrical interferences accounted for one kind of static, and he learned enough to solve the immediate problem at hand. At the same time, Jansky discovered a second kind of static, which sounded like a steady radio hiss on the earphones and did not appear to come from any obvious source. After months of experiment, he proved that the hiss could not possibly originate on earth and that it must come from outer space, although not from the sun, moon, or planets. The source, he established, was in the direction of the center of the Milky Way. Without knowing it, Jansky had built the world's first **radio telescope** (Fig. 20–27).

Grote Reber (b. 1911), a radio engineer and amateur astronomer, having read about Jansky's discov-

Figure 20–27 The full-scale model of the rotating antenna with which Karl Jansky discovered radio astronomy. The model is at the National Radio Astronomy Observatory's station at Green Bank, West Virginia. (Bruce Elmegreen)

Figure 20–28 Grote Reber's original radio telescope is on display at the National Radio Astronomy Laboratory. (National Radio Astronomy Laboratory)

Figure 20–29 The 1000-ft-diameter (305-m) radio dish at the National Astronomy and Ionosphere Center, Arecibo, Puerto Rico. This is the largest telescope on earth. Though fixed in the ground, some tracking is possible by moving the receiving equipment at the focus of the dish. (Cornell University Photograph)

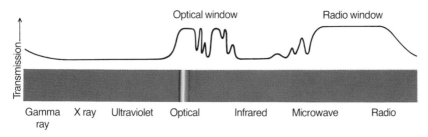

Figure 20–30 The two main "windows" in the earth's atmosphere that allow electromagnetic radiation to reach the earth's surface are the optical (visible) window and the radio window. Most other photons are blocked.

ery, built a radio telescope in his backyard (Fig. 20–28). It was movable, 31.4 feet in diameter, and resembled a giant parabolic dish, with the radio receiver held out in front on four legs. With this telescope, Reber mapped the radio sky for several years. His paper containing the first maps seemed so fantastic that it encountered resistance when it arrived at the *Astrophysical Journal.* The editor decided to publish it nevertheless, believing that it was the beginning of a new kind of astronomy. Today, radio telescopes are an indispensable tool not only for exploring the far reaches of the universe but also for providing detailed information of space closer at hand (Fig. 20–29).

If our eyes were sensitive to radio waves instead of to light, the sky would appear very different. The sun would be much less bright, and the Milky Way would shine brilliantly. Hundreds of new "stars" would cover the sky, forming unfamiliar constellations. Clouds of interstellar gas in space would become visible, as would the gaseous remnants of supernovae. One of the brightest objects would be Cygnus A.

A radio telescope, unlike an optical one, can receive from outer space waves between 1 centimeter and 20 meters in length (Fig. 20–30), which are able to penetrate the earth's atmosphere. (Wavelengths less than 1 cm are absorbed by the atmosphere; those above 20 m are reflected back into space.) Observations can be made with a radio telescope throughout a 24-hour period, since radio signals are not affected by the sunlit sky. An important advantage of a radio telescope is that it can explore behind the dust clouds of space, which are opaque to visible light. It has been estimated that this dust must hide 90% of the stars in the Milky Way from visual detection, obscuring the regions beyond about 6000 ly. Radio waves, however, penetrate this material without absorption. Further, radio has an inherent advantage over light in probing to great distances, at which the red shift is substantial. For example, the light from galaxies at a distance of 5 billion ly, moving away from us at half the speed of light, would be shifted so far to the red that only part of their spectrum in the visible range could be recorded on photographic plates; the rest of their light would be lost. With radio waves, the loss due to a wavelength shift is relatively small, and galaxies can be detected at considerably greater distances.

A basic problem of radio astronomy, on the other hand, is the much greater length of radio waves as compared with light waves, making **resolution** of the source, that is, the ability to discriminate fine detail, more difficult. Second, celestial radio signals are weak, the faintest having only a hundred-millionth of the power of a television signal. One reason for this is that the energy of a photon is inversely proportional to its wavelength, and radio waves are very long. Larger

antennas have been the solution to both problems. A large antenna has superior resolution, enabling radio sources to be narrowed down to smaller and smaller regions, and it collects more radio energy from the source, just as a larger mirror in an optical telescope collects more light.

There are many kinds of radio telescopes, each with its special use. The two principal forms of construction are (1) fixed or steerable metal-mesh or solid metal **parabolic dishes**, which collect radio signals and focus them to a radio receiver where they are amplified and then fed to recording devices, and (2) **linear arrays** of antennas arranged in various patterns.

A large radio telescope built at the Jodrell Bank station of the University of Manchester, England, is similar in shape to the one used by Reber, but it is 250 feet in diameter. Other giant radio telescopes have been built throughout the world. The largest movable, steerable paraboloid is 300 feet in diameter (Fig.

Figure 20–31   The 100-m (330-ft) radio telescope near Bonn, Germany, is the largest fully steerable radio telescope in the world. (Max Planck Institut für Radioastronomie)

20–31). The largest parabola of any kind, a dish of metal mesh 1000 feet in diameter, is built into a hole carved out of the mountains of Arecibo, Puerto Rico.

Another way to obtain better resolution of **radio sources** is through **interferometry** (Fig. 20–32). Two antennas are placed some distance apart and connected to a single radio receiver. Radio waves from space strike one slightly before the other and produce interference patterns. As the earth rotates, this radio interferometer sweeps the sky. Through interferometry, the position of a strong radio source can be determined with the same accuracy that could be obtained with a single antenna as large as the distance between the two small antennas. One of the most sensitive radio telescope systems is the National Radio Astronomy Observatory's (NRAO) three-element interferometer at Green Bank, West Virginia.

Thousands of discrete sources of radio emissions have been discovered and mapped. In many cases, no optically visible object could be identified as the source. Radio astronomers have catalogued and numbered the sources. The catalogue referred to most frequently is the Third Cambridge Catalogue. The entry designated 3C 48, for example, refers to the forty-eighth entry in this catalogue.

The first discrete source of radio waves outside our solar system was discovered in 1946. It was a spot of such intensity in the constellation Cygnus, designated Cygnus A, that it startled astronomers. Additional sources were soon found in other constellations. The first identification of a radio source with a visual object also involved Cygnus A. Rudolph Minkowski (1895–1976) and Walter Baade (1893–1960), working with photographs made with the 200-inch telescope on Palomar Mountain, then the world's largest optical telescope, showed that Cygnus A coincided with the position of a galaxy now estimated to be 700 million ly away. It was originally thought that Cygnus A represented the collision of two galaxies and that the strong radio emission came from the energy released. Astronomers no longer accept the colliding-galaxy hypothesis, and the origin of the radio energy of this and other strong sources continues to be a mystery. Many other radio sources have since been identified with visible galaxies. As resolution improved, it was shown that many individual objects were emitting radio waves. One of the brightest sources is the sun, and the planets also broadcast. Some of the nebulae within our galaxy, the remnants of supernovae, are also radio sources, as are certain classes of stars: red supergiants, red dwarfs, blue dwarfs, novae, pulsars, and X-ray stars.

Figure 20–32 Part of the Very Large Array (VLA) near Socorro, New Mexico. The VLA consists of 27 movable radio telescopes, which can be arranged in four configurations. In effect, such an array works as a large number of two-dish interferometers, all observing the same part of the sky together. (National Radio Astronomy Observatory)

20–10 HYDROGEN 21-CENTIMETER WAVES

A new dimension was added to astronomy with the discovery of radio waves emitted by hydrogen in space. It had been suggested that neutral hydrogen atoms in interstellar space might undergo a change in energy state and emit radio energy. The reasoning is as follows: A hydrogen atom in space can exist in only one of two energy states (Fig. 20–33). In the lower energy state, the electron and proton are spinning in opposite directions; in the slightly higher state, they are spinning in the same direction. Every few million years, the spin axis of the electron flips over, and a hydrogen atom in the higher state passes into the lower state. Simultaneously, a photon is emitted and carries off the energy lost in the transition.

The wavelength of the emitted photon is 21 centimeters, thousands of times longer than the wavelengths of visible light, and equivalent to a frequency of 1420 megahertz. Equipment sensitive enough to detect this radiation was not available until 1951. In that year, Edward M. Purcell (b. 1912), a physicist at Harvard University, and a graduate student named Harold Ewen detected 21-centimeter waves from hydrogen. For the first time, astronomers had a specific spectral line to work with in the radio spectrum.

The 21-centimeter radiation has become a potent astronomical tool. It can penetrate interstellar dust, because the waves are about a million times longer than the dimensions of a dust particle. Light waves tend to be scattered, their length being about the same as a dust particle. Otherwise, the two kinds of radiation behave in the same way. The discovery of this radiation made possible a complete picture of the Milky Way. The spiral shape is revealed by maps using the distribution of neutral hydrogen, which is concentrated in the spiral arms of the galaxy. The Doppler shifts in the line, which can be measured far more accurately than the Doppler shift of light, have revealed the motions of turbulent hydrogen clouds. Studies of the proportion of hydrogen in other galaxies, made possible by this radiation, provide clues to their evolution.

20–11 QUASARS

The strong radio source known as 3C 48 engaged the attention of a young astronomer, Allan Sandage (b. 1926), in 1960. With the 200-inch Palomar optical telescope, he took photographs of its position in the sky (Fig. 20–34). What he saw was one of the most puzzling objects in the universe. Previously, the sources of radio emissions had been gas clouds or galaxies, but never stars. The object Sandage photographed was not a galaxy; it resembled a faint star with a wispy line to one side. Since it was apparently not a normal star, judging from the streak of light, 3C 48 was called a quasistellar radio source, or **quasar**.

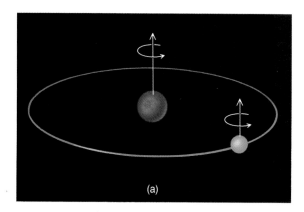
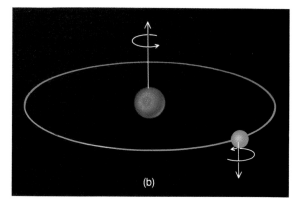

Figure 20–33 Hydrogen 21-cm radiation. When the proton and electron spins are parallel (a), the atom contains more energy than when the spins are opposed (b). When the spin of the electron flips from (a) to (b), radiation of λ = 21 cm is emitted.

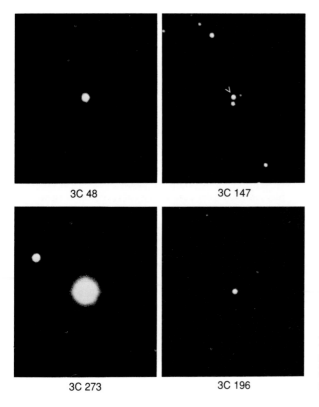

3C 48

3C 147

3C 273

3C 196

Figure 20–34 The first quasar, 3C 48, and other quasistellar radio sources photographed with the 200-inch Hale telescope at Palomar Observatory. (Palomar Observatory, California Institute of Technology)

Using a photoelectric photometer, Sandage further found that 3C 48 had an abnormal color for a star, being too bright in blue and ultraviolet. Moreover, 3C 48 had an emission spectrum, never before found in a star, rather than an absorption spectrum.

The puzzle was resolved when astronomers measured an accurate position for another radio source, 3C 273. This source coincided with another unusual star, fairly bright, with unusual colors and a bright-line spectrum. Maarten Schmidt (b. 1929) established that the lines were mostly hydrogen lines that were shifted to the red end of the spectrum. The red shift corresponded to a velocity away from the observer of approximately 50,000 km/s. When Sandage's spectrum of the first quasar, 3C 48, was re-examined, it was found that the bright lines were essentially the same lines observed in 3C 273: normal hydrogen lines, but shifted even more to the red, corresponding to a velocity of recession of 110,000 km/s.

Only a small fraction of all quasars are strong radio sources. Some quasars have little or no emission at radio wavelengths. The term "quasar" is now applied to star-like objects with large red shifts, regardless of their radio emission. Over 1500 quasars have been identified, although some astronomers have estimated that millions of quasars have existed in the universe and lived out their lifetimes and perhaps about 35,000 quasars now exist.

According to one hypothesis, quasars are very remote objects that provide a direct glimpse of the universe as it existed 8 or 9 billion years ago. According to another hypothesis, quasars are relatively nearby objects. The tremendous energy associated with quasars has been attributed to various sources: supernova explosions; gravitational collapse of a massive superstar; matter-antimatter annihilation; galaxies at an early stage of formation; condensed central cores of galaxies; and, the present consensus, giant black holes.

20–12 THE INTERSTELLAR MEDIUM

Discoveries in spectroscopy and radio astronomy have shed considerable light on the composition of the space between the stars, once believed to be largely empty. The most abundant constituent is hydrogen in various forms. Neutral atomic hydrogen (H) emits and absorbs radiation at the ratio wavelength of 21 cm; ionized hydrogen (H^+) and molecular hydrogen (H_2) have been detected in a variety of ways, including Balmer lines and radio astronomy. The elements helium, carbon, nitrogen, oxygen, sulfur, iron, chlorine, and many others are also present, as indicated by their spectral lines in nebulae.

Signals from hydroxyl radicals ($\cdot OH$) were received following a search by Sander Weinreb (b. 1936) of the Massachusetts Institute of Technology and others. Unlike the 21-cm signal of hydrogen, these signals consist of four different frequencies. When the dish of an 84-foot radio telescope was pointed toward Cassiopeia A, absorption was obtained at the 1665 and 1667 megahertz lines of the $\cdot OH$ radical. Later, two additional lines were detected in the direction of Sagittarius, at 1612.2 and 1720.6 megahertz.

Many molecules, as well as atoms and ions, have been detected in interstellar space since 1968, when molecular astronomy was born with the discovery of molecules consisting of more than two atoms. Until then, most astronomers believed that the density of the interstellar medium was too low to get more than two atoms to combine. Such ideas needed drastic revision when Charles H. Townes (b. 1915) of Stanford University and others found ammonia (NH_3) in several interstellar clouds near the center of the Milky Way. Then Townes and co-workers detected emission signals from water vapor (H_2O) in several regions of the galaxy. Many other molecules have been detected since. They have probably been formed by collisions between atoms of the interstellar gas or by combining on the surface of particles of interstellar dust.

The discovery of some very complex molecules in interstellar space, such as formaldehyde (H_2CO) and methyl alcohol (CH_3OH), shown in Table 20–5, has led to interesting speculation. According to some theories, life may occur when certain substances in a planetary atmosphere—ammonia, water, formaldehyde, and hydrogen—are irradiated by ultraviolet radiation or by lightning to produce amino acids. It appears from recent discoveries in molecular astronomy that the conditions for organized life may exist elsewhere in the universe. The discovery of hydrogen cyanide (HCN) in 1970, important in the laboratory synthesis of amino acids, and the occurrence of so

Table 20–5 SOME INTERSTELLAR MOLECULES

Year Detected	Molecule	Formula
1937	Methylidyne (ionized)	CH^+
1937	Methylidyne	CH
1939	Cyanogen radical	CN
1963	Hydroxyl radical	OH
1968	Ammonia	NH_3
1968	Water	H_2O
1969	Formaldehyde	H_2CO
1970	Carbon monoxide	CO
1970	Hydrogen cyanide	HCN
1970	X-ogen	?
1970	Cyanoacetylene	HC_3N
1970	Hydrogen (molecular)	H_2
1970	Methyl alcohol	CH_3OH
1970	Formic acid	HCOOH
1971	Carbon monosulfide	CS
1971	Formamide	NH_2CHO
1971	Carbonyl sulfide	OCS
1971	Silicon monoxide	SiO
1971	Methyl cyanide	CH_3CN
1971	Isocyanic acid	HNCO
1971	Hydrogen isocyanide	HNC
1971	Methylacetylene	CH_3CCH
1971	Acetaldehyde	CH_3CHO
1972	Thioformaldehyde	H_2CS
1972	Hydrogen sulfide	H_2S
1972	Methanimine	CH_2NH
1973	Sulfur monoxide	SO
1974	Ethyl alcohol	C_2H_5OH
1974	Silicon monosulfide	SiS
1975	Sulfur dioxide	SO_2
1975	Acrylonitrile	H_2CCHCN
1975	Methyl formate	$HCOOCH_3$
1975	Nitrogen sulfide radical	NS
1975	Cyanamide	NH_2CN
1976	Cyanodiacetylene	HC_5N
1976	Formyl radical	HCO
1976	Acetylene	C_2H_2

many biochemical compounds lend support to this possibility.

20–13 THE EXPANDING UNIVERSE

Cosmology is concerned with such general questions of the nature of the universe as: How did it originate? What is it like now? What is it likely to be like in the future? Most contemporary cosmologies stem from **Hubble's law** (derived from red-shift studies) that

Table 20–6   RECESSION SPEEDS OF
SOME GALAXIES

Position	Distance (Light-Years)	Speed (Kilometers/Second)
Virgo	6,000,000	1120
Pegasus	23,000,000	5600
Ursa Major	85,000,000	15,700
Leo	105,000,000	20,000
Gemini	135,000,000	24,500
Hydra	360,000,000	62,000

other galaxies are moving away from ours and are doing so at speeds that are proportional to the distance of the galaxy. Table 20–6 gives the recession speeds of a few galaxies from among the hundreds that have been tested. Note that the greater the distance of the galaxy, the greater its speed away from us.

Hubble's law and Einstein's general theory of relativity laid the foundation for the theory of the **expanding universe**. According to this theory, the average density of matter in the past was greater than it is now. As we go back in time, the density was enormously high, possibly 10^{14} times as dense as water. The universe started from a highly compressed and extremely hot state. About 20 billion years ago, according to the most recent estimates, the matter erupted in a vast explosion—the "**big bang**"—and the expansion of the universe began. As the matter expanded in all directions and thinned out, it cooled down and condensed to form stars and galaxies. The red shifts in the spectra of distant galaxies indicate that the universe continues to expand at the present time.

The big bang theory began with Vesto Melvin Slipher's (1875–1969) discovery in 1913 at the Lowell Observatory that about a dozen galaxies are moving away from the earth at speeds ranging up to about 1 million miles an hour. This was the first hint that the universe is expanding. When Slipher reported his finding at a meeting of the American Astronomical Society in 1914, his slides clearly revealed the red shift that indicated an enormously rapid motion of these galaxies. The audience stood up and cheered, something that almost never happens at a scientific meeting. One of the people in Slipher's audience was Edwin Hubble.

Meanwhile, Einstein published his equations of general relativity in 1917. The astronomer Willem de Sitter (1872–1934), director of the Leiden Observatory, found a solution to them that predicted an exploding universe in which the galaxies moved rapidly away from one another. This was just what Slipher had observed. After World War I, de Sitter's theoretical prediction of an expanding universe made a great impression on astronomers. Hubble said that it was mainly de Sitter's result that influenced him to take up the study of the moving galaxies where Slipher had left off.

In the 1920s, Hubble and Milton Humason (1891–1972) followed up on Slipher's work, first with the 60-inch telescope at Mt. Wilson, then with the 100-inch, the world's largest at that time. They measured the speeds and distances of many galaxies too faint to be seen by Slipher and confirmed Slipher's discovery. Some of the galaxies were moving away from us at 100 million miles an hour. In 1919, Hubble discovered that the farther away a galaxy is, the faster it moves, the relationship known as Hubble's law and one predicted by Einstein's theory of relativity.

The agreement of both theory and observation pointing to an expanding universe and a beginning in time made a tremendous impression on astronomers. Even Einstein, who resisted the new developments and held on to the idea of a static, unchanging universe, was convinced following a visit to Hubble in 1930 that the structure of the universe is not static. At present, the big bang theory has no serious competitors.

The latest evidence makes it almost certain that the big bang really did occur about 15 to 20 billion years ago. When the material of the universe was enormously hot—billions of degrees—and compressed, there occurred a "**primeval fireball**" of radiation, as described by John Wheeler (b. 1911) of Princeton University. Starting out as enormously energetic gamma rays, the radiation was "cooled" by expansion. Its wavelength increased, and it now appears mostly in the radio and microwave bands. Robert H. Dicke (b. 1916) of the same university predicted that radiation surviving from the primeval fireball ought to be detectable by radiotelescopes. Coincidentally, Arno A. Penzias (b. 1933) and Robert W. Wilson (b. 1936), both of the Bell Telephone Laboratories in New Jersey, discovered such radiation in 1965 in a manner that parallels Jansky's discovery of radio emission from space (Fig. 20–35). Adherents to

Figure 20–35 Arno A. Penzias *(left)* and Robert W. Wilson *(right)* with the horn-reflector antenna with which they discovered 3K background microwave radiation. (AT&T Bell Laboratories)

the big bang theory claim that the background microwave radiation, corresponding to the spectrum of radiation of an object at 3K, represents the remnant signals from near the origin of the big bang.

A major puzzle for cosmologists following the discovery of the 3-K cosmic background radiation was its unexpected evenness. In order to give rise to galaxies and clusters of galaxies, the big bang theory called for an almost imperceptible pattern of warm and cool patches in the radiation arising from areas of higher and lower density in the newborn universe.

A near thirty-year search for these blotches came to fruition in 1992. At the spring meeting of the American Physical Society in Washington, a research team headed by George Smoot, an astrophysicist at the University of California, Berkeley, announced that NASA's Cosmic Background Explorer satellite —COBE—had detected the elusive evidence in a swath of cosmic clouds 15 billion light-years from earth. The evidence at the Goddard Space Flight Center in Maryland was a map on the computer screens of the microwave sky with blotches indicating regions where the microwaves are an almost imperceptible 30 millionths of a degree warmer or cooler than average. These results, called the most important discovery in cosmology in 20 years, removed the biggest remaining objection to the big bang.

KEY TERMS

Astronomical Unit (AU) The average distance between the earth and the sun; approximately 93,000,000 miles.

Big Bang Theory The theory that the universe started from a highly compressed and extremely hot state that exploded and that it has been expanding ever since.

Black Hole Thought to be the final stage in the evolution of certain stars. Because it is composed of the densest matter known in the universe, the gravitational field would be so great that neither radiation nor matter could escape.

Celestial Equator The intersection of the celestial sphere with the plane that passes through the earth's equator.

Cepheid Variable A class of pulsating stars that acts as a "celestial lighthouse" and as a means of measuring distances deep into the universe.

Constellation Grouping of the brightest stars into configurations to which the names of mythological figures are attached and which designate regions in the sky.

Cosmology The study of the structure and development of the universe.

Declination (DEC) The celestial equivalent of latitude measured in degrees north or south of the celestial equator.

Doppler Effect An apparent change in wavelength of the radiation from a source when the source and the observer are moving relative to each other.

Ecliptic Plane The plane of the earth's orbit around the sun.

Expanding Universe A theory based on Hubble's law and Einstein's general theory of relativity that the expansion of the universe began with the big bang and is continuing.

Galaxy A basic structural unit of the universe that includes stars, star clusters, clouds of gas and dust, and interstellar molecules.

Globular Cluster A spherical group of thousands of stars located above and below the plane of the galaxy and in relative motion with it.

Hertzsprung-Russell Diagram A graph of brightness versus temperature or spectral class, for a group of stars.

Hubble's Law   Other galaxies are moving away from ours at speeds that are proportional to the distances of those galaxies.

Infrared Astronomy   Deals with the spectral band of radiation beyond the red, about 7000 Å to 1 mm.

Interferometry   Observations with an interferometer, a device that uses interference to measure such properties as the position or structure of an object.

Island Universes   Galaxies lying far beyond the Milky Way.

Latitude   Number of degrees north or south of the equator.

Linear Array   An arrangement of radio telescope antennas in various patterns used for interferometry to obtain high resolution.

Light-Year   The distance that light travels in a year, 6×10^{12} miles. A useful standard for expressing stellar distances.

Local Group   A cluster of 20 or so galaxies that includes the Milky Way.

Longitude   Distance measured in degrees east or west of the Greenwich prime meridian, a line running from the north pole to the south pole.

Main Sequence   A long, narrow, diagonal band on the Hertzsprung-Russell diagram on which fall over 90% of all stars near the sun.

Milky Way   The band of light across the sky from the stars and gas in the plane of the Milky Way Galaxy.

Nebula   A cloud of interstellar gas and dust.

Neutron Star   An extremely dense star composed of neutrons, believed to be closely related to pulsars.

Nova   A star that increases in brightness explosively.

Parabolic Dish   A fixed or steerable metal radio antenna which collects radio signals and focuses them to a radio receiver.

Parsec   Parallax-second. The distance of an object having a parallax of 1 second of arc. One parsec = 3.26 light-years.

Period-Luminosity Relation   The longer the period of a Cepheid variable, the greater the luminosity.

Photoelectric Photometer   An instrument used to measure the intensity and color of starlight.

Population I Stars   A classification of stars, based primarily on their composition. Population I stars have a high abundance of metals.

Population II Stars   Relatively old stars that have a very low abundance of metals.

Primeval Fireball   Radiation in the microwave band that is coming from all directions in space; interpreted to be the remnant signals of the big bang.

Pulsar   A celestial object that emits bursts of radio pulses with periods ranging from 0.033 to 3.75 seconds; believed to be a rapidly rotating neutron star.

Pulsating Star   A variable star that waxes and wanes in size and luminosity.

Quasar   A quasistellar radio source. Star-like objects that have large red shifts and extremely high energy output.

Radio Astronomy   The gathering and observation of radiation in the radio-wave region of the spectrum.

Radio Source   A location in space from which radio waves are received, including the sun, planets, nebulae, stars, and galaxies.

Radio Telescope   An antenna and reflecting disk for the detection and study of radio sources.

Red Giant   A relatively bright and relatively cool star.

Red Shift   The shift of spectral lines toward the longer, red-wavelength region of the spectrum, interpreted as a Doppler effect in receding objects.

Resolution   The ability of an optical system to distinguish detail.

Right Ascension   Celestial longitude, measured in hours of time eastward along the celestial equator.

Spectrograph   An instrument for dispersing radiation into a spectrum and photographing or mapping the spectrum.

Star Cluster   A physical grouping of many stars.

Supernova   A major star explosion, flaring up to millions of times brighter than its original state.

Telescope   An optical instrument for viewing distant objects by means of refraction of light rays through a lens or by the reflection of light waves by a concave mirror.

Ultraviolet Astronomy   Observations in the ultraviolet region of the spectrum, 100 Å to 4000 Å, wavelengths shorter than visible light and longer than X-rays.

White Dwarf   The final stage in the evolution of certain stars in which the density of matter is approximately a million times greater than that of the sun.

X-ray Astronomy   Observations in the X-ray region of the spectrum, between 1Å and 100Å.

Zodiac   The band of 12 constellations through which the sun, moon, and planets move in the course of the year.

THINGS TO DO

1. (a) Observe the sky at night. Draw a sketch of the moon, if visible, the brighter stars, and any constellations in one region, and include reference points such as a tree, chimney top, or hill. (b) Four hours later make a second sketch of the moon and the same stars and constellations from the same location and note any changes in position. (c) On the following night, at one of the times similar to the first night, repeat your observations, make a third sketch, and compare all three.

2. Partially inflate a round balloon. Glue small pieces of cotton on its surface, and inflate it further. Note how the cotton pieces move away from one another in this model of the expanding universe.

3. On a clear and moonless night, and away from city lights, observe the luminous band of the Milky Way across the sky with (a) the naked eye; (b) binoculars.

EXERCISES

Light and Astronomy

20–1. How do we know the chemical composition of stars?

20–2. (a) What information does a red shift in its spectrum yield about a star? (b) a blue shift?

20–3. Why are optical astronomers for the most part limited to nighttime observation, whereas radio astronomers can observe during the day?

20–4. Astronomically, in what way have our senses been expanded in recent years?

20–5. Which distance is greater, a light-year or an astronomical unit? Explain.

20–6. Why is bigger better for astronomical telescopes?

20–7. If the Andromeda spiral galaxy is approaching us at a velocity of 290 km/s, how long will it take to reach us if the distance is 225×10^6 ly?

20–8. A star is 45 ly away. Traveling at a speed of 100,000 mph, how long would it take to reach this star?

20–9. Indicate the close relationship of other physical sciences to astronomy.

Celestial Distances and Positions

20–10. How are the positions of celestial objects specified?

20–11. The parallax of a star is 0.001″. How far away is it in ly?

20–12. Constellations and galaxies are both groups of stars. In what ways do they differ?

20–13. Why are the constellations of the winter sky different from those of summer?

20–14. When we look up at the stars, we are looking back in history. Explain.

Characteristics of Stars

20–15. Why is the Hertzsprung-Russell diagram considered by some astronomers to be the most important diagram in astronomy?

20–16. Distinguish between pulsars and pulsating stars.

20–17. What does the sequence O, B, A, F, G, K, M represent?

20–18. What is stellar evolution?

20–19. How did the "guest star" of AD 1054 become associated with the discovery of pulsars?

20–20. What are some differences between a nova and a supernova?

20–21. What is a black hole?

20–22. Describe the possible final stages in the life cycle of a star.

Galaxies, Nebulae, and Star Clusters

20–23. What role has the radio telescope played in the recent revolution in astronomy?

20–24. What is the distinction between a nebula and a galaxy?

20–25. Galaxies have been referred to as the structural units of the universe. Explain.

20–26. How have the 21-cm radio waves emitted by neutral hydrogen extended our knowledge of the universe?

20–27. Why is the name "Milky Way" appropriate for our galaxy?

20–28. What is the difference between an open and a globular star cluster?

20–29. What is Hubble's law and what does it mean?

20–30. What is meant by "the expansion of the universe"?

20–31. What is a quasar?

20–32. Explain why astronomers of today can actually observe galaxies as they were billions of years ago.

20–33. *Multiple Choice*

A. The average distance from the earth to the sun is
 (a) a light-year.
 (b) an astronomical unit.
 (c) an angstrom unit.
 (d) a parsec.

B. A nova is
 (a) a star that suddenly increases in brightness.
 (b) a pulsar.
 (c) a quasar.
 (d) a receding galaxy.

C. The remnant of the big bang is also known as
 (a) a neutron star.
 (b) a black hole.
 (c) 3K background radiation.
 (d) the interstellar medium.

D. The energy of a main sequence star comes from
 (a) gravitation.
 (b) nuclear fission.
 (c) nuclear fusion.
 (d) helium burning.

E. A Doppler red shift for a galaxy indicates that it is
 (a) approaching.
 (b) receding.
 (c) slowing down.
 (d) speeding up.

SUGGESTIONS FOR FURTHER READING

Bethe, Hans A., and Gerald Brown, "How a Supernova Explodes." *Scientific American*, May, 1985.

A star may evolve peacefully for millions of years, but when it runs out of nuclear fuel, the core collapses in milliseconds. What follows is a supernova, an explosion more powerful than any since the big bang with which the universe began.

Bok, Bart J., "The Milky Way Galaxy." *Scientific American*, March, 1981.

Through the mid-1970s, the broad outlines of the galaxy seemed reasonably well established. Since then, a number of upheavals have occurred in the effort to understand the Milky Way.

Courvoisier, Thierry J. L., and E. Ian Robson, "The Quasar 3C 273." *Scientific American*, June, 1991.

We know more about 3C 273 than any other quasar. Yet the structure and dynamics of quasars remain a mystery in many respects years after their discovery by Maarten Schmidt in 1963.

DeVorkin, David H., "Henry Norris Russell." *Scientific American*, May, 1989.

Called the "Dean of American Astronomers," Henry Norris Russell was the Russell of the Hertzsprung-Russell diagram, of the Russell method of eclipsing binaries, and of the Russell-Saunders coupling for two-electron spectra. He helped create modern astrophysics.

Dressler, Alan, "The Large-Scale Streaming of Galaxies." *Scientific American*, September, 1987.

The Milky Way is traveling through the universe with a swarm of other galaxies. There may be a remote concentration of mass on a scale that challenges current theory.

Gott, J. Richard III, James E. Gunn, David N. Schramm, and Beatrice M. Tinsley, "Will the Universe Expand Forever?" *Scientific American*, March, 1976.

Their answer: Yes.

Halliwell, Jonathan J., "Quantum Cosmology and the Creation of the Universe." *Scientific American*, December, 1991.

When you stare out into space on a clear night and wonder, "Where did all this come from?" you are asking a question that until recently lay far outside the reach of scientific investigation. By applying quantum mechanics to the universe as a whole, though, scientists are now able to formulate and address the question in a meaningful way.

Helfand, David, "Bang: The Supernova of 1987." *Physics Today*, August, 1987.

On February 23, 1987, astronomers ended a 383-year vigil as a naked-eye supernova—Supernova 1987A—blazed forth in the Large Magellanic Cloud in the southern sky and was discovered by Ian Shelton of the University of Toronto. It has had an enormous impact on astrophysics and particle physics.

Hodge, Paul W., "The Andromeda Galaxy." *Scientific American*, January, 1981.

The Andromeda galaxy is the spiral galaxy nearest our own, and the only giant spiral close enough for us to view it in detail. It is a "laboratory" for the study of the evolution of stars, the rotation of galaxies, and the scale of distances in the universe.

Hoskin, Michael, "William Herschel and the Making of Modern Astronomy." *Scientific American*, February, 1986.

Through telescopes that he built himself, Herschel discovered thousands of stars and nebulae. The expansion in astronomical knowledge in the nineteenth century was largely brought about by his achievements.

Kellermann, Kenneth, and David Heeschen, "Radioastronomy in the 1990's." *Physics Today*, April, 1991.
In recent decades, radio galaxies, quasars, pulsars, gravitational lenses, and the microwave background radiation were discovered serendipitously because of their radio emissions. A new generation of radio telescopes to be built in the 1990s and major improvements in existing instruments may lead to equally remarkable and unexpected discoveries.

Margon, Bruce, "The Origin of the Cosmic X-ray Background." *Scientific American*, January, 1983.
It has been known since 1962 that every part of the sky emits a uniform glow of X-rays. The origin of this diffuse X-ray background is still a subject of controversy after two decades of intensive study, but one source may be distant quasars.

Mathewson, Don, "The Clouds of Magellan." *Scientific American*, April, 1985.
The nearest galactic neighbors of the Milky Way are the Large Magellanic Cloud (LMC), which lies at a distance of 160,000 ly, and the Small Magellanic Cloud (SMC), which is 200,000 ly away. The nearest neighbors of our galaxy have had a turbulent history and may have an even more violent future.

Mewaldt, Richard A., Edward C. Stone, and Mark E. Wiedenbeck, "Samples of the Milky Way," *Scientific American*, December, 1982.
The composition of samples of matter, cosmic rays, from elsewhere in the Milky Way can be determined as a result of advances in space technology and instrumentation.

O'Dell, C. R., "The Hubble Space Telescope Observatory." *Physics Today*, April, 1990.
The Hubble Space Telescope is based on the concept that an observatory placed above the distorting and absorbing terrestrial atmosphere would allow astronomers to make observations that are impossible to realize using even the largest ground-based observatories. The goal: a ten-times sharper view of the stars and galaxies.

Osmer, Patrick S., "Quasars as Probes of the Distant and Early Universe." *Scientific American*, February, 1982.
Quasars are still a great enigma in astronomy. They are a clue to how the universe looked 15 billion years ago.

Ostriker, Jeremiah P., "The Nature of Pulsars." *Scientific American*, January, 1971.
Pulsars are believed to be neutron stars in rapid rotation. Their discovery has opened new fields and illuminated old ones.

Rees, Martin J., "Black Holes in Galactic Centers." *Scientific American*, November, 1990.
The discovery of black holes in galactic centers could affect current ideas about the evolution of the universe. So far the evidence of a black hole at the center of the Milky Way is quite persuasive.

Silk, Joseph, "The Formation of Galaxies." *Physics Today*, April, 1987.
Discusses current ideas on how galaxies originated and acquired their observed forms and how this knowledge may lead to insights into the large-scale structure of the universe.

Smith, Robert W., "Edwin P. Hubble and the Transformation of Cosmology." *Physics Today*, April, 1990.
The new cosmology of the early 1930s included two key features: first, the existence of galaxies outside our own stellar system that are visible in earth-based telescopes and second, that these galaxies evince the expansion of the universe. Edwin P. Hubble is the astronomer most closely associated with both of these profound new views of the physical universe.

Solomon, Philip M., "Interstellar Molecules." *Physics Today*, March, 1973.
Molecules have been discovered in interstellar clouds. Their abundance and chemical complexity were totally unexpected.

Thompson, A. Richard, Thomas E. Gergely, and Paul A. Vanden Bout, "Interference and Radioastronomy." *Physics Today*, November, 1991.
Today's radio astronomer is often frustrated by man-made radio interference from satellites, radar, radio and television transmitters, wireless telephones, microwave ovens, computers, and even garage-door openers. Maintaining the viability of the precious radio window on the cosmos is a serious concern of physicists and astronomers.

Trefil, James, "Galaxies." *Smithsonian*, January, 1989.
For most of recorded history, the Milky Way has been the universe. We now know that the Milky Way is but one of billions of galaxies that populate it. The science of understanding galaxies, only about 70 years old, is today a field of feverish activity.

Woosley, Stan, and Tom Weaver, "The Great Supernova of 1987." *Scientific American*, August, 1989.
In much of science, a result is accepted only if it is reproducible. Yet in the case of supernova 1987A, scientists deal with an event that may not be repeated nearby for centuries. For theorists and observers collaborating to document and explain one of the heavens' grandest events, it has been the event of a lifetime.

ANSWERS TO NUMERICAL EXERCISES

20–7. 2.38×10^9 years

20–8. 3.08×10^5 years

20–11. 3260 light-years

OBSERVING THE STARS AND PLANETS

DAVID MORRISON

NASA Ames Research Center

SIDNEY C. WOLFF

National Optical Astronomy Observatories

THE STARS

One of the most beautiful and awe-inspiring sights in all of nature is the dark night sky filled with stars. The stars that are visible above the horizon change with the seasons and with the time of night. The patterns of the stars, however, are fixed, at least over the time scale of a human life. With star maps it is easy to learn your way around the sky.

All stars are assigned to 1 of 88 different groupings, or *constellations*, of stars. Most of these constellations date back to ancient times, but a few—especially in the Southern Hemisphere—were defined during the past few hundred years. Constellations usually have a story associated with them. There is Orion, the Hunter, permanently separated in the sky from the Scorpion, who caused his death (Fig. A). Andromeda is chained to a rock, offered in sacrifice to atone for the boasts of her mother Cassiopeia. There are dogs, a dragon, a ram, a dolphin, scales, crowns, and crosses. To learn to identify a constellation and to know its story is to make a friend for life.

As you learn the constellations, note that the very brightest stars are not all the same color. An especially good example is found in the winter constellation Orion. In one knee of Orion, we find the bright blue star Rigel. The eastern shoulder, diagonally across from Rigel, is marked by Betelgeuse, the brightest red star in the sky. Binoculars will permit you to see the colors of many fainter stars. Color is an indicator of the temperatures of stars, with red stars having the lowest temperatures.

Stars rise about 4 min earlier every night. Look toward the eastern horizon and note where stars are located at a specific time of night. Wait a week or two to look again, and note the difference in altitude above the horizon. Ancient peoples used the times of rising of specific stars to mark the seasons and the start of a new year.

If you look at very bright stars when they are close to the horizon, you may see them appear to change in brightness and possibly in color as well. This phenomenon is called *twinkling* and is caused by turbulence in the earth's atmosphere. As small regions in the atmosphere move about, they bend the light from a star, just as a prism does, and cause it to travel in ever-changing directions. The brightness of the star will vary, depending on what fraction of its light is directed precisely toward your eye. The color of the star may also appear to change rapidly if, for example, red light is directed toward your eye and blue light is not, and then, a

Figure A The winter constellation of Orion, the hunter, as illustrated in the seventeenth-century atlas by Hevelius. (J. M. Pasachoff and the Chapin Library)

fraction of a second later, the reverse situation happens.

Bright city lights mask all but the most brilliant stars. The next time you are out in the countryside at night, take time to look at the sky. If it is summertime, you will very likely see the Milky Way arching across the sky, displaying its greatest brilliance toward the south in the direction to the center of the Galaxy. Use binoculars to see dozens of stars in clusters like the Pleiades and the Hyades, both of which are visible in fall. In winter, use binoculars to look at the nebulosity in the sword of Orion, where star formation is vigorously in progress. The Big Dipper rises highest in the northern sky in spring; note that the star in the bend of its handle is a double star.

THE PLANETS

At almost any time of the night, and at any season, you can spot one or more bright planets visible in the sky. All five of the planets known to the ancients —Mercury, Venus, Mars, Jupiter, and Saturn—are more prominent than any but the brightest of the fixed stars, and they can be seen even from urban locations if you know where and when to look.

Venus, which appears either as an evening "star" in the west after sunset or as a morning "star" in the east before sunrise, is the brightest object in the sky after the sun and moon. It far outshines any

real star, and under the most favorable circumstances it can even cast a visible shadow. Mars, with its distinctive red color, can be nearly as bright as Venus when it is close to the earth, but normally it remains much less conspicuous. Jupiter is most often the second brightest planet, approximately equaling in brilliance the brightest of stars. Saturn is dimmer, and it varies considerably in brightness, depending on whether its rings are seen nearly edge-on (faint) or more widely opened (bright). Finally, Mercury is quite bright, but few people ever notice it because it never moves very far from the sun and is always seen against bright twilight skies. There is a story (probably apocryphal) that even the great Copernicus never saw the planet Mercury.

True to their name, the planets "wander" against the background of the fixed stars. Although their apparent motions are complex, they reflect an underlying order, which was the basis for the development of the heliocentric model of the solar system.

The two inner planets, Mercury and Venus, never appear far from the sun, since their orbits lie inside the orbit of the earth. Venus has the larger orbit, and it can achieve an angular separation from the sun of about 45°. Mercury, with its smaller orbit, never ventures more than 28° from the sun. As seen from the earth, both planets appear to move back and forth with respect to the sun, appearing alternately in the evening and morning sky.

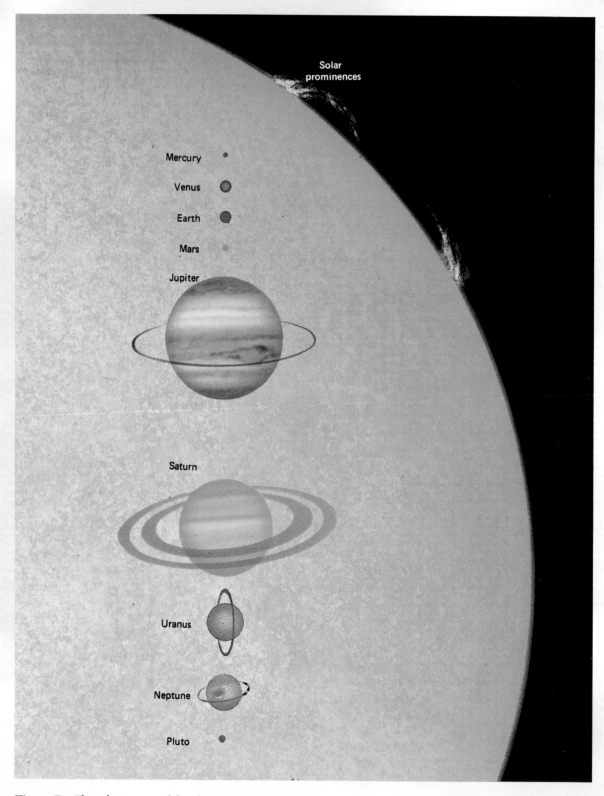

Figure B The relative sizes of the planets.

The apparent motion of planets with orbits larger than that of the earth is more complex. Since all the planets revolve about the sun in the same direction (east to west), these exterior planets normally appear to move from east to west against the stellar background. To an observer on the sun, they would always move from east to west. But from the moving earth, we get a somewhat different perspective. The earth is also moving east to west, and at a faster pace, since it is nearer the sun. Therefore, we pass each planet from time to time, and when we do, the planet appears to move in a backward direction, from west to east. The effect is similar to that of overtaking a slower car on the freeway: as you pass the other car, it seems to move backward with respect to the distant landscape.

Today we understand how the complex apparent motion of each planet is a reflection of the relative motion of the earth and the planet, each moving in its orbit about the sun. To ancient peoples, however, these phenomena remained mysterious, and they were not really understood in detail until the time of Kepler, early in the seventeenth century.

Saturn — the most beautiful object in our solar system and possibly the most beautiful object we can see in the sky. (NASA)

21

THE SOLAR SYSTEM

Explorations of earth's close neighbors in space—the moon, Mercury, Venus, Mars, and Jupiter—have given us considerable information about our small corner of the universe. And what a fascinating corner it is, containing a central star, planets and their satellites, meteors, asteroids, and comets. Although countless systems like the solar system are presumed to exist, ours is the only one known at present. Within it there are clues to the formation and evolution of the universe as a whole. The findings thus far support the uniqueness of the earth in many respects, above all, the presence of intelligent life.

21–1 THE SOLAR SYSTEM

The solar system is composed of the sun, nine known planets, at least 44 moons, thousands of "minor planets" (asteroids), meteoroids, and perhaps a billion comets. The sun, an average star, radiates its own light and differs in this respect from a planet or a moon, which shines by reflected light only. A planet and a moon have many similarities; they differ chiefly in that a **planet** revolves around the sun, whereas a **moon** revolves around a planet.

Viewed from a point far above the northern hemisphere of the earth, all of the planets revolve around the sun in elliptical orbits in a counterclockwise direction (Fig. 21–1). The orbits of the planets have been defined with such precision that the position of the earth with respect to any planet may be predicted

within hundredths of a second of arc for many years ahead. The orbits of Mars, Jupiter, Uranus, and Neptune are all tilted less than 2° out of the plane of the **ecliptic**, the plane of the earth's orbit (Fig. 21–2). The orbit of Saturn is tilted slightly more than 2°, and of Venus slightly more than 3°. The tilt of Mercury's orbit is 7°, and Pluto's, 17°. All of the planets rotate on their axes, although Venus and Uranus rotate in a direction opposite to that of the other planets. Most of the planets have some form of atmosphere, except Mercury and possibly Pluto. Only one moon—Saturn's Titan—is known to have an atmosphere.

The fact that all the planets lie nearly in the plane of the ecliptic is a puzzle, although it can be accounted for by a modern version of the **nebular theory** proposed by Immanuel Kant (1724–1804) and advanced independently by Pierre Simon de Laplace (1749–1827). Like our galaxy, the solar system is thought to have evolved slowly out of a spinning cloud of gas called a nebula, which in time would form aggregations and then condense into solid planets. According to the modern view, now also called the **dust-cloud theory**, the planets and the sun began by the condensation of a cold cloud of dust and gas (Fig. 21–3). Inertial effects set up by the rotation of the cloud opposed gravitational collapse, with the result that the cloud was flattened into a rotating disk. Further contraction of the cloud caused it to increase in density, and local instabilities broke it up into individual units. At the center was a protostar, and toward the periphery a series of protoplanets. When the density of the protoplanets exceeded a certain limit, they condensed into planets. It now appears that the cold planetary bodies condensed at about the same time, some 4.5 billion years ago.

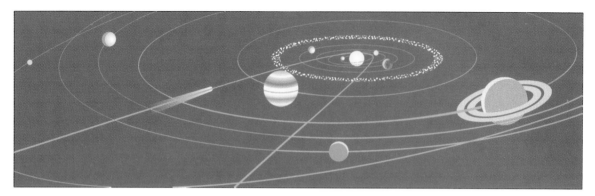

Figure 21–1   The solar system (not drawn to scale).

Radiation from the protosun, glowing from the release of gravitational energy, caused the evaporation of most (perhaps 99%) of the original mass of the protoplanets. Dissipation of material from the atmosphere into space is still going on, though at a reduced rate. Hydrogen and helium, which must have constituted the major part of the nebula and the protoplanets, have almost entirely escaped from the earth and have left behind the heavier elements. The stronger gravitational attraction of the more massive planets,

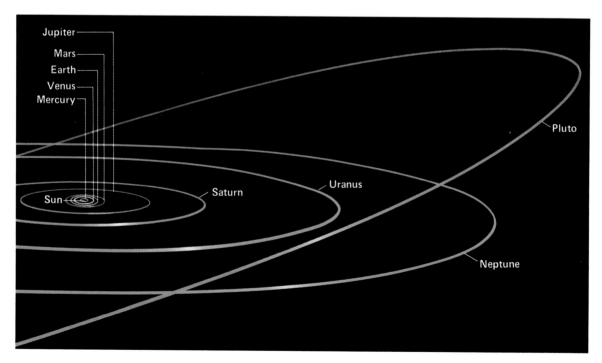

Figure 21–2   With the exception of Pluto, the orbits of the planets are only slightly inclined to the plane of the earth's orbit.

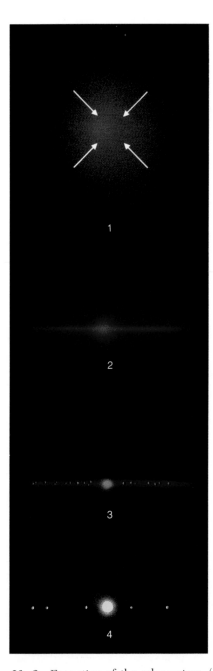

Figure 21–3 Formation of the solar system. (1) The solar nebula contracts. (2) As it contracts, its rotation causes it to flatten. (3) The nebula becomes a disk of matter with a concentration, which becomes the protosun, near the center. (4) As the nebula cools, solid particles condense into protoplanets or planetesimals, from which the planets are formed.

Jupiter and Saturn, has enabled them to retain more of the light elements, accounting for their lower density.

Although each of the planets is unique, they fall into two groups—with the exception of Pluto, about which little is known. The **terrestrial planets** are quite small, are located relatively close to the sun, and have a composition quite similar to that of the earth; they include Mercury, Venus, Earth, and Mars (Fig. 21–4). They are composed mainly of metals and are rather dense. The **Jovian** or **major planets** located in the outer regions of the solar system, are Jupiter, Saturn, Uranus, and Neptune (Fig. 21–5). They are much less dense than the earth and are enveloped in atmospheres several thousand miles deep. The nearest Jovian planet, Jupiter, has been most thoroughly studied. All the satellites (moons) are very small in comparison to their planets, with the exception of that of the earth. However, two of the moons—Jupiter's Ganymede and Saturn's Titan—are larger than the planet Mercury (Fig. 21–6, p. 540).

Estimates of the age of the solar system were little better than guesses before the discovery of radioactivity, although geologists were beginning to speak in terms of millions or hundreds of millions of years, based on studies of the earth's rocks. When it had been proved that the radioactive elements uranium and thorium decay ultimately into lead and helium, Bertram Boltwood (1870–1927), a radiochemist working with Rutherford, proposed that this fact could be used to calculate the ages of uranium- and thorium-containing minerals.

It is possible to tell how long ago the mineral had been formed by measuring the amount of lead or helium and comparing it with the content of uranium and thorium. Since the rate of decay is known, time can be measured by comparing the relative abundance of the parent elements and their noble-gas daughters present in a sample of matter. Various methods—including the decay of potassium to argon and the decay of rubidium to strontium—indicate that the cold planetary bodies of the solar system all condensed at about the same time, 4.5 or 4.6 billion years ago (Fig. 21–7, p. 541). Table 21–1 gives some of the radioactive elements used to determine the ages of minerals. These radionuclides were apparently made at the time the elements originated in stars and are not being formed on earth today.

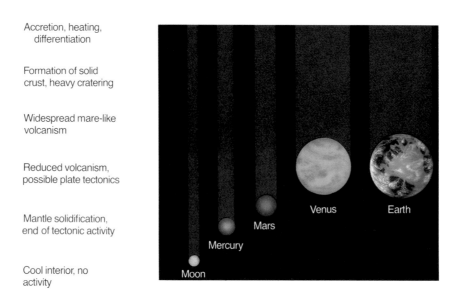

Accretion, heating, differentiation

Formation of solid crust, heavy cratering

Widespread mare-like volcanism

Reduced volcanism, possible plate tectonics

Mantle solidification, end of tectonic activity

Cool interior, no activity

Moon

Mercury

Mars

Venus

Earth

Figure 21–4 The terrestrial planets have a composition similar to that of the earth and include Mercury, Venus, earth, and Mars. The moon is shown for comparison.

EARTH

Figure 21–5 The Jovian, or major, planets include Jupiter, Saturn, Uranus, and Neptune. The earth is shown for comparison. (NASA/JPL)

Table 21–1 RADIOACTIVE "CLOCKS" USED
TO DETERMINE THE AGES
OF MINERALS

Isotope	Half-Life (Millions of Years)	Decay Products
Potassium-40	1310	Calcium-40, argon-40
Rubidium-87	50,000	Strontium-87
Thorium-232	13,900	Lead-208, helium
Uranium-235	710	Lead-207, helium
Uranium-238	4510	Lead-206, helium

Some characteristics of the solar system are summarized in Table 21–2.

21–2 THE SUN

The dominant member of the solar system is the sun, which accounts for nearly the entire mass (more than 99%). It has a diameter of 865,000 miles, 109 times the earth's diameter, and a volume over 1 million times as great as the earth's. Its average density is a little greater than that of water, 1 g/cm^3. The density increases rapidly toward the interior, where it is approximately 150 g/cm^3. The average distance between the earth and the sun is 93,000,000 miles and is called 1 astronomical unit (AU).

The sun is a typical star on the main sequence of the Hertzsprung-Russell diagram. Like other G-class stars, it is composed mainly of hydrogen and helium, and its energy is generated by nuclear reactions deep in its interior. The outer, visible layers of the sun are collectively known as its atmosphere. It consists of three layers—the **photosphere**, **chromosphere**, and **corona**—which blend into each other and have no sharp boundaries (Fig. 21–8). The chromosphere and corona can normally be seen briefly only during a total eclipse of the sun, but they can be observed at other times with instruments such as the coronagraph.

The Photosphere. The most conspicuous features of the photosphere are irregular dark spots. With the telescope, Galileo became convinced that these areas, known as **sunspots**, were actually on the surface of the sun itself, and since they apparently move across its face, he used them to demonstrate the sun's rota-

Table 21–2 CHARACTERISTICS OF THE SUN AND PLANETS°

	Sun	Mercury	Venus	Earth	Mars	Jupiter	Saturn	Uranus	Neptune
Distance from sun (millions of miles)		36	67	93	140	480	890	1800	2800
Mass (earth = 1)	333,000	0.06	0.82	1.0	0.11	318	95.2	14.5	17.2
Diameter (thousands of miles)	867	3.0	7.6	7.9	4.2	89	76	30	28
Density (water = 1)	1.4	5.4	5.1	5.5	3.9	1.3	0.7	1.3	1.6
Rotation period (earth days)	27	59	243	1	1	0.4	0.4	0.4	0.6
Length of year†		88^d	225^d	1^{yr}	687^d	12^{yr}	30^{yr}	84^{yr}	165^{yr}
Velocity of escape (miles/second)		2.2	6.2	7.0	3.1	37	22	13	14
Number of moons		0	0	1	2	14	17	15	2
Average surface temperature (K)	5800	600	750	285	240	128	105	70	55
Surface gravity (earth = 1)	28	0.37	0.89	1.00	0.38	2.74	1.14	0.96	1.15

°Excluding Pluto

†d = earth days; yr = earth years

Figure 21–6 A composite of the largest moons or planetary satellites, intermediate in size between the planets Mercury and Pluto. (Stephen Meszaros/NASA; Karen Denomy; John R. Spencer/U. Hawaii)

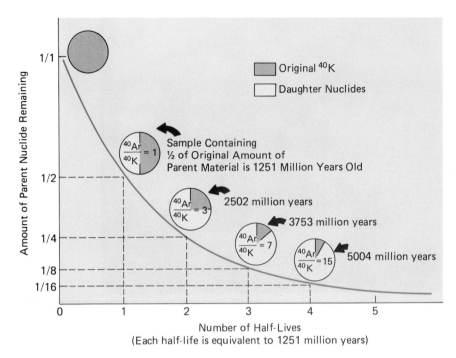

Figure 21–7 The decay curve for potassium-40 can be used for dating rocks. Its half-life is 1251 million years. If the ratio of potassium-40 to daughter nuclides such as argon-40 is 1 to 1, then the age of the sample is 1251 million years. If the ratio is 1 to 3, then another half-life has elapsed, and the rock would have an age of two half-lives, or 2502 million years.

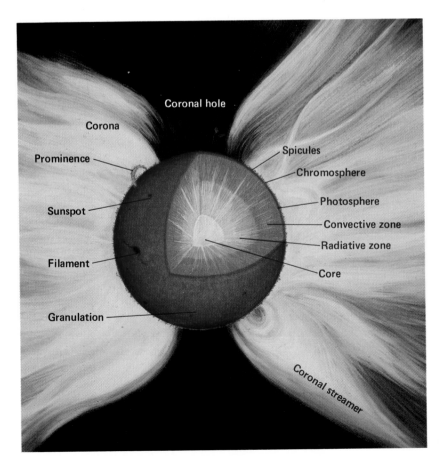

Figure 21–8 The basic structure of the atmosphere and interior of the sun.

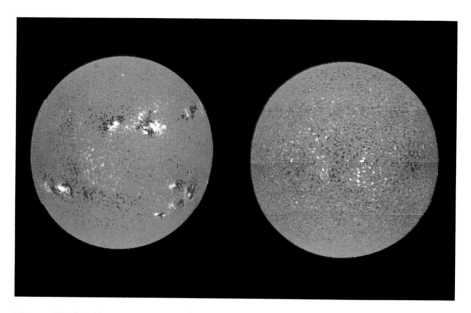

Figure 21–9   Computer-generated images compare the number of sunspots and magnetic activity on the active (*left*) and quiet sun. At solar minimum (*right*), there are no large sunspots and the magnetic fields are weak. (National Solar Observatory/National Optical Astronomy Observatories)

tion about its axis (Fig. 21–9). At the equator the period of rotation is about 25 days, whereas near the poles it is 34 days. The sunspots are regions where the gases, though hotter than the surfaces of many stars, are somewhat cooler than those of the surrounding areas and appear dark only in contrast. The largest sunspots may be over 125,000 miles across; the smallest, 1000 miles. They continually change in appearance and size. Although their nature is not fully understood, they may be due to magnetic phenomena originating in the sun's interior. George Ellery Hale (1868–1938) showed that they are regions of very high magnetic field strength.

Sunspots follow an 11-year cycle, during which they increase in number toward a maximum and then fall off to a minimum. Most of the spots appear, then disappear, within a day or so. A **sunspot cycle** begins with the appearance of small spots at latitudes of 40° to 45° in both solar hemispheres. As the cycle progresses, new and larger groups of spots form closer to the solar equator, the closest appearing 5° to 10° from the equator at the close of the cycle. As these die out, small spots reappear at the higher latitudes. Correlations have been noted between events on earth and the sunspot cycle. One link is a long-period weather

cycle; total annual rainfall varies approximately in an 11-year period, the years of maximum rainfall corresponding to the times of sunspot maxima.

The Chromosphere.   Above the photosphere is the chromosphere, a relatively thin layer 2000 to 6000 miles deep, with a temperature of 50,000°C. During an eclipse it is visible for only a few seconds, just after the moon has covered the photosphere. The brightest eruptive activity on the sun—**solar flares**—is associated with this layer. These sudden short-lived increases in the light intensity of the chromosphere release great amounts of energy, especially in the ultraviolet and X-ray regions, producing radio fade-outs and brilliant auroral displays in the earth's higher latitudes. The **aurora borealis** is the "northern lights," and the **aurora australis** is the "southern lights." Solar flares are related to the sunspot cycle, occurring mostly in the regions of sunspot groups.

The Corona.   The chromosphere blends gradually into the outermost layer, the corona. In parts of the corona, the temperature may rise as high as 3 million degrees Celsius. This layer is seen most spectacularly during a total solar eclipse (Fig. 21–10). It extends

Figure 21–10 The solar corona as seen during a total solar eclipse. The moon blocks the light from the brilliant surface (photosphere) of the sun. A filter has been used that allows the coronal streamers to be seen. (Serge Koutchmy, Institut d'Astrophysique, Paris)

more than a million miles above the photosphere. Solar prominences appear as red flame-like protuberances rising from the edge of the sun during a total eclipse. The prominences are cool, denser accumulations of hydrogen gas that form in the inner corona and may extend tens of thousands or even a million miles above the sun, then gracefully fall back. They are often geyser-like eruptions; others form huge arches (Fig. 21–11). Special telescopes called coronagraphs can study the corona from certain locations on mountain peaks on the surface of the earth, where the sky is especially clear and dust-free.

Figure 21–11 A solar prominence. Some prominences evolve rapidly, perhaps during an hour or so. Others can hover above the sun for weeks or months. (R. J. Poole/Day Star Filter Corp.)

21–3 THE MOON

Our closest permanent neighbor in space, the moon, lies at an average distance of 239,000 miles from the earth. It never comes closer than 222,000 miles or gets farther away than 253,000 miles. The moon completes one **revolution** around the earth every 27 days 7 hours 43 minutes. The force of gravity on the moon's surface is one sixth as strong as that on the earth's surface—so feeble that the moon was unable to hold an atmosphere. The moon's gravitational pull is much weaker than earth's, with an **escape velocity**—the velocity that an object must have to escape its gravitational pull—of only 2.38 km/s, compared with 11.2 km/s on earth; thus, any atmosphere that may have been there at one time would have leaked away into space long ago. The lack of an atmosphere to retain the sun's heat results in sudden and extreme temperature changes, from more than 90°C during the day to − 180°C or less when the sun sets. Any water that may have been present on the moon would have boiled away and the vapor escaped into space along with other gases. There are, therefore, no seas, lakes, rivers, or glaciers on the moon; nor are there clouds, fog, rainstorms, or snowfalls.

The moon has been known by various names, among them Diana, "she who hunts the clouds"; Selene, the sister of Helios (the sun); and Luna, the name that is often used to distinguish it from the moons of other planets. From Selene is derived the term **selenography**, the study of the same aspects of the moon that geography is of the earth, and the supposed effect of the full moon upon human behavior is reflected in the word "lunacy." The dream of reaching the moon was made a goal of national policy for the United States by President John F. Kennedy (1917–1963) during the 1960s and came to fruition in 1969.

Except for the earth and Pluto, the planets that have moons are considerably larger than their satellites. The diameter of Jupiter is 28 times as great as that of its largest moon, and Saturn's is 21 times as

Figure 21–12   The moon. The dark areas are maria, and the brighter ones, highlands. The positions of six American Apollo (A) and three Soviet Luna (L) missions are marked. (Kevin Reardon/Williams College - Hopkins Observatory)

great. The earth's diameter is only 3.6 times as great as the moon's, the two forming a sort of double planet system.

The question of the moon's origin has not yet been satisfactorily answered. There are few supporters today of the once popular planetary-capture theory, which said that the moon had once been a planet in its own right orbiting the sun and was somehow "captured" by the gravitational pull of the earth. Nor are there many adherents to the theory that the moon had once been part of the earth and broke off by centrifugal force, orbiting around the earth and leaving behind a huge basin now filled by the Pacific Ocean. The evidence today favors the theory that both the earth and the moon were formed near each other at about the same time, by the process of material accretion. Investigations by potassium-argon dating of rocks brought back from the moon show them to be equal to the earth's estimated age, 4.6 billion years.

Light and dark areas are easily distinguished, even when the moon is observed without a telescope. Through a telescope, these features are resolved in much greater detail. The largest features on the moon are dark areas, or **maria**, so called because they were originally thought to be seas (*mare*, meaning "sea") (Fig. 21–12). The maria are not seas at all, but large, roughly circular, smooth, flat, dry basins up to 700 miles in diameter, which appear dark because of their low reflectivity. When the moon is full, the dark maria suggest the appearance of the "man in the moon." Many seventeenth-century names for these features are still in use: Sea of Serenity, Sea of Tranquility, Sea of Fertility, and others. The biggest of the maria is named the Ocean of Storms. Smaller areas of darkness have such names as the Bay of Rainbows and the Lake of Dreams.

The brighter areas are the highlands—crater-covered mountains and hills. The names of the mountain chains are drawn from those of the earth: Lunar Alps, Lunar Caucasus, and so on. Some of the mountain peaks reach heights of 25,000 feet or more, as do those on earth.

In a map of the moon published in 1651, Giovanni Riccioli (1598–1671) began the custom of naming the lunar craters after famous men, mostly scientists and philosophers. He named the most conspicuous lunar crater after Tycho Brahe and honored Copernicus with a crater nearby. Craters are also named for Plato, Aristotle, Ptolemy, Archimedes, Kepler, Newton, and many others. This practice continues to be followed. An international conference was called in 1970 to consider names for craters previously identified only by number. The number of craters great and small is in the millions.

The lunar surface has been mapped in detail by several methods. Television scanning techniques were employed with the *Ranger* and *Surveyor* spacecraft, and the pictures transmitted to earth. Photographs of higher quality were obtained when spacecraft carrying astronauts orbited the moon. Finally, extremely detailed photographs were taken by the Apollo astronauts, who opened the moon as a scientific laboratory (Fig. 21–13). The smaller features of the moon are similar to the larger ones. Tiny craters have essentially the same appearance as large ones, and miniature ridges resemble the larger ranges.

The problem of the origins of the lunar craters and other features is still far from being resolved. Some selenogists have favored a volcanic origin for the craters and maria, whereas others favor an impact origin; there is evidence that both processes have contributed. Craters range in size from tiny pits to several miles, and some appear to be of more recent origin than others. The majority of the moon's craters, particularly the larger ones, may have been formed by the impact of meteorites; they closely resemble the craters left by bombs, in that their walls are much

Figure 21–13 An astronaut riding on the Lunar Rover during the *Apollo 17* mission, the last mission that carried people to the moon. (NASA/JSC)

steeper on the inside than on the outside. Many of the smaller ones may have been caused by volcanic activity; in some cases, the volcanoes may have been triggered by meteoritic impact. Lunar rocks collected by the Apollo astronauts show evidence of both forms of activity. The astronauts did resolve one question, the nature of the moon's surface. They found it to be covered with a fine, dust-like sand, which varied in depth from a fraction of an inch to several inches and clung to their space suits.

One of the surprises of the Lunar Orbiter photographic survey on the moon was the discovery of regions of dense material below the surface, called **mascons**, from the two words "mass concentration." They were detected from the strong gravitational pull they exerted on satellites, which distorted the orbits of the satellites over the maria. It has been deter-

mined that the mascons are thin, circular disks of high-density (lava-like) materials extending below the surfaces of the maria.

The orbital movement of the moon changes the visible portion of its surface as viewed from the earth. In a one-month period, the moon completes a cycle of **phases** (Fig. 21–14), the shapes of the sunlighted areas. New moon occurs when the moon is exactly in line between the earth and the sun, so that none of its illuminated surface is visible. First quarter occurs when half the moon's visible surface is illuminated, full moon when the whole disk is illuminated, and last quarter when, again, only half the moon's surface is illuminated.

By a remarkable coincidence, the apparent size of the full moon equals that of the sun, even though the moon is a much smaller body. A **solar eclipse** (*ek-*

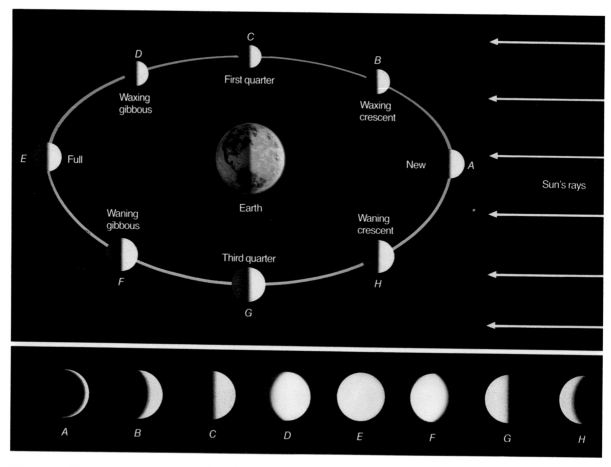

Figure 21–14   The moon's monthly cycle of phases. Note that the sun always lights half the moon. How each phase appears to us is shown below.

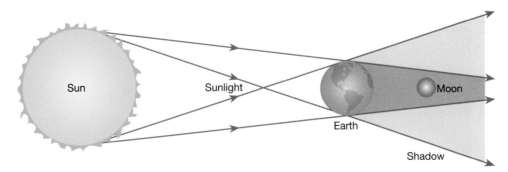

Figure 21–15   A total lunar eclipse occurs when the moon passes completely within the darkest part of the earth's shadow (the umbra). Since the moon usually passes above or below the earth's shadow, we do not have lunar eclipses most months.

leipsis, meaning "disappearance") occurs when the moon covers the sun's disk. The most spectacular kind is a total solar eclipse experienced on only a very narrow path on the earth, 150 to 170 miles wide. Although the period of totality lasts only a few minutes, a total solar eclipse has been an awesome spectacle throughout history. In modern times, total solar eclipses have provided so much information about the sun that teams of astronomers journey to remote places on the earth to observe them.

A **lunar eclipse** is much less spectacular than a solar eclipse and of less scientific value. Lunar eclipses occur when the full moon enters the shadow of the earth (Fig. 21–15). Since the earth's shadow is larger in diameter than the moon, totality can last as long as 1 hour 40 minutes. Although there are more solar than lunar eclipses, the latter can be observed simultaneously by many more observers, being visible from the earth's hemisphere facing the moon.

21–4 MERCURY

The smallest planet, Mercury, is also the closest to the sun and makes its circuit around the sun in only 88 days. The Greeks named it for Hermes, the fast-moving messenger of the gods, who was represented with winged sandals and a winged hat, and the Romans named it for their messenger, Mercury. Mercury's highly eccentric orbit takes it to within 28.3 million miles of the sun at perihelion and 43.2 million miles at aphelion; its mean distance is about 36 million miles. Its period of rotation was determined

through radar measurements in 1965. By bouncing radar waves off the surface of Mercury and catching the echoes with the radio telescope at Arecibo, Puerto Rico, Gordon Pettengill and Rolf Dyce found that Mercury rotates on its axis once in 58.65 days, or essentially two thirds the period of revolution, 87.97 days.

Mercury's orbit lies within that of the earth, making the planet invisible at night, when astronomical work is usually done. The side of the earth that is in darkness faces away from the sun, toward the planets whose orbits are outside the earth's. Only the side of the earth that is in daylight, facing toward the center of the solar system, can face Mercury (Fig. 21–16). The sun's strong glare prevents observation of Mercury with the naked eye, except for short periods of about an hour before sunrise and an hour after sunset at certain times of the year. When it is visible at all, Mercury is a very bright object low on the horizon, and through a telescope it exhibits the same phases as the moon.

From infrared studies of its surface, the temperature of Mercury is known to range from 350° to − 160°C. If an atmosphere exists (none is detectable), it must be extremely thin since the velocity of escape on Mercury is only 4.2 km/s, and since it is so close to the sun, the heat is intense. Mercury's density, 5.4 g/cm^3, suggests that it contains mainly rock and metals and that its interior may contain compounds of iron or other heavy elements. Its surface appears to be made of the same kind of rock as that found on the moon.

The spacecraft *Mariner 10* completed a five-month journey to Mercury in March, 1974. It swept

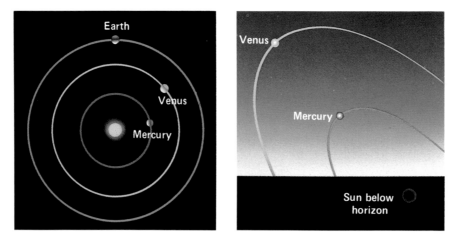

Figure 21–16 The orbits of Mercury and Venus lie within the earth's orbit. A view from the earth shows Mercury and Venus at their greatest distances from the sun. Astronomers have never had a clear view of Mercury from the earth.

NORTH

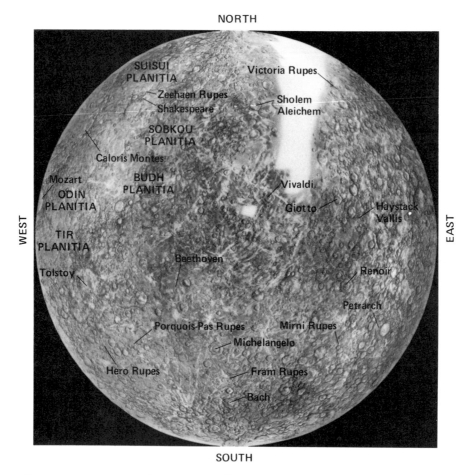

WEST

EAST

SOUTH

Figure 21–17 A map of Mercury based on the *Mariner 10* observations. The cratered surface is strikingly similar to the moon's. (U.S. Geological Survey)

over the planet at a distance of 465 miles and transmitted more than 2000 television pictures across 80 million miles of space. The photographs revealed the surface of the planet to be extraordinarily similar to that of the moon, on a regional as well as on a global basis (Fig. 21–17). Craters are the predominant surface feature, their structure being similar to lunar craters of the same size. The plains strongly resemble lunar maria. As on the moon, a heavily cratered terrain has been preserved on Mercury, without modification by volcanism or atmospheric erosion. Some of the features are probably 4 to 4.5 billion years old.

21–5 VENUS

After the sun and moon, Venus is the brightest object in the sky, reflecting more than 75% of the sunlight that strikes the dense cloud cover. Venus can often be seen in full daylight with the naked eye. It is the body most frequently referred to as the "evening star" or "morning star" when it is visible shortly after sunset or just before sunrise. It was not realized at first that both "stars" were the same object; the Greeks called it Hesperus ("west") in the evening and Phosphorus or Lucifer ("bearer of light") in the morning. The planet was almost always identified by ancient peoples as a goddess of beauty, Venus to the Romans.

Venus is 67 million miles from the sun. Its orbit is one of the least eccentric of the planets, so that its perihelion and aphelion distances are similar. At its closest approach, it is 24 million miles from the earth. Galileo was the first to observe through a telescope that Venus, like the moon, exhibits a full set of phases. Radar and the radio telescope at Arecibo (optical telescopes can probe no deeper than the outer cloud layers) established that Venus has a rotation period of 243 days, the slowest for a planet in the entire solar system, and that it rotates in a direction opposite to the typical motion of the other planets. Its period of revolution is 225 days. It has little or no magnetic field.

A number of space probes aimed at Venus in the 1960s and 1970s obtained more data about Venus than astronomers had been able to gather in the previous 400 years. Early Russian probes were apparently crushed by the pressure of the Venusian atmosphere, although *Venera 3* evidently made a crash landing in 1965—the first time a space vehicle had

reached another planet. In 1972, high-resolution radar probed through the thick clouds of Venus and for the first time resolved features on the planet's surface. The first map of a part of Venus along its equator showing discrete features—a landscape of huge, shallow craters from 20 to 100 miles wide and less than a quarter of a mile deep—was made in 1973 using a computer technique. *Mariner 10,* equipped with television cameras and radar, flew by Venus in 1974 and sent back 3400 television pictures of the cloud cover, which appeared to be moving along the equator at 350 km/h. Venus's clouds resemble terrestrial smog, but they are believed to be composed mainly of sulfuric acid droplets (Fig. 21–18).

The planet's atmosphere was the object of two *Pioneer* flights in December, 1978, the first devoted primarily to a study of the atmosphere of another planet on a global scale. *Pioneer Venus 1* was to send daily reports on the atmosphere as well as pictures and radar maps of the surface topography. It was intended to operate for up to a year in orbit around Venus. *Pioneer Venus 2,* called the transporterbus, scattered five separate probes into the atmosphere over different areas of the planet. They gathered a wide range of data, including temperatures, composi-

Figure 21–18 The circulation of the clouds of Venus, photographed from *Mariner 10.* They are different from terrestrial clouds, in spite of their similar appearance. (NASA/Ames Research Center)

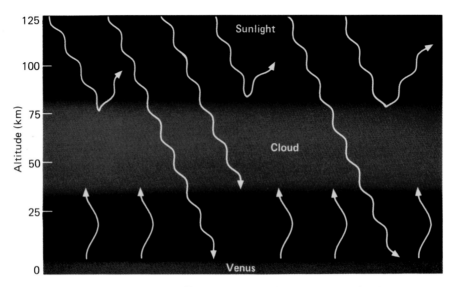

Figure 21–19 The greenhouse effect on Venus. Sunlight penetrates the clouds so the surface is illuminated with radiation in the visible part of the spectrum. Venus radiates mostly in the infrared. The infrared radiation is trapped by the carbon dioxide and other substituents in the atmosphere, and Venus heats up.

tion, density, and distribution of the atmosphere. The data were to be translated into sketches of clouds, winds, and other features, a process that would take months or years to complete (Fig. 21–19).

The atmosphere of Venus apparently consists largely of carbon dioxide (75 to 90%) and smaller amounts of nitrogen and the noble gases (5%); there is no oxygen. The cloud layer, 13 miles thick, is penetrated by some sunlight. The solar radiation heats the lower regions of the atmosphere and is then trapped by the greenhouse effect owing to the carbon dioxide. Although the temperature at the top of the cloud layer is − 50°C, at the base of the clouds 45 miles above the surface it is 100°C, and on the surface it is a furnace-like 500°C. The once popular idea of Venus as a planet having lush vegetation and being an abode for life as we know it now seems impossible. Not only does its surface seem far too inhospitable, but also no great amount of water vapor has been detected.

Three years after the *Pioneer* spacecraft began orbiting Venus, analysis of the billions of bits of data gave a portrait of the surface of Venus marked by giant mountains, plateaus, craters, valleys, earthquakes, and lava pouring from violent volcanoes as lightning bolts shower overhead (Fig. 21–20). The

planet's crust, as determined from variations in gravity, is about 30 to 36 miles deep, nearly double that of the earth's crust. Venus, however, apparently generates its crust in a way that is different from the earth's method.

The recent measurements indicate that Venus is even more different from the earth than previously imagined. The most striking differences are the slow rotation rate of Venus, the absence of a satellite, the high surface temperature, the extreme weakness or absence of a magnetic field, and the lack of water in its atmosphere.

Fundamentally, the surface of Venus is very flat. It is not without features, however, and some are spectacular. Because the planet itself is named for a goddess, the surface features are being given women's names. The only exceptions are the names of three sites that have been in use for several years.

One of these is Maxwell Montes, a great mountain range named after the physicist James Clerk Maxwell. The Maxwell range of rugged highlands is 35,000 to 40,000 feet, compared with the earth's tallest mountain, Mt. Everest, at 30,000 feet. Maxwell rises above the eastern portion of a large plateau, the Lakshmi plateau, named for a Hindu goddess. Unlike Everest,

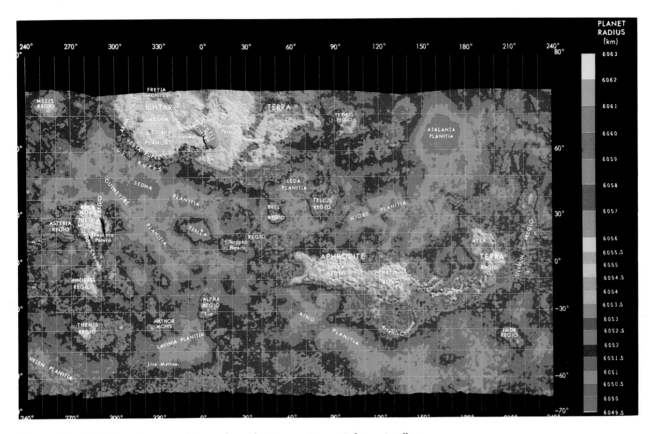

Figure 21–20   The radar map of Venus from the *Pioneer Venus Orbiter*. A rolling plain covers 60% of the surface. The highlands sit atop the plain, like continents.

NASA/Ames Spacecraft)

Maxwell towers alone above the surrounding plains and plateaus. The Lakshmi plateau is twice as large as the Tibetan plateau, the largest on earth. It is also higher, 15,000 feet compared with 14,000 feet. Maxwell and the Lakshmi plateau are on the "continent" Terra Ishtar, after the Babylonian goddess of love.

A second "continent," in the southern hemisphere, called Terra Aphrodite after the Greek goddess of love, is not as high as Ishtar but is about twice as large. Like Ishtar, it has a plateau, volcano, and complex terrain. The other elevated regions are much smaller than Ishtar and Aphrodite.

Venus has a rift valley 150 miles wide and at least 900 miles long. There are other valleys and canyons, some quite spectacular. A region called Alpha is reminiscent of the faulted ranges found in Nevada and elsewhere in the southwestern United States. To the

southwest of Alpha lies a prominent ring-shaped feature with a bright central spot that resembles many impact craters on the moon and Mercury. The feature has been named Eve, and the bright spot has been proposed as the origin of the official Venus longitude system. Another feature, Beta, is a region of apparently volcanic prominences that stretches for well over 1000 miles. One of these is comparable in size to the huge volcanoes on Mars.

Although *Pioneer Venus 1* and earth-based radar parted the clouds of Venus and gave us a preliminary view of earth's sister planet, a clearer look from a spacecraft came in the 1990s.

NASA's spacecraft *Magellan*, launched on May 4, 1989, aboard the space shuttle *Atlantis*, arrived at Venus on August 10, 1990, after traveling more than 806 million miles (Fig. 21–21). Using a technique

Figure 21–21   The *Magellan* spacecraft, a radar mapper, as it was launched from the space shuttle *Atlantis.* (NASA/ JSC)

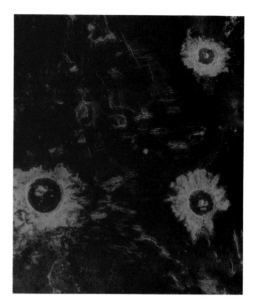

Figure 21–22   Radar images of three large impact craters on the surface of Venus, with diameters from 37 km to 50 km, taken by the *Magellan* spacecraft. (NASA)

called Synthetic Aperture Radar (SAR), *Magellan* gathered extremely detailed pictures of the planet's terrain and sent back more data on Venus than had been acquired from all previous space missions to all the planets.

Unlike typical forms of radar systems, which send out one signal at a time and process each echo before sending out another pulse, *Magellan* sent out several thousand radar pulses each second. As the spacecraft flew over the surface, its SAR system used fast computers to accumulate multiple echoes, stored the data, and returned it to earth at the end of each three-hour, nine-minute orbit. Scientists processed the data into very precise, almost three-dimensional images, about ten times sharper than any previous radar images. In 243 days, one complete rotation of Venus, *Magellan* mapped 84% of the surface and then continued to map Venus, filling in the gaps from the first probe.

Magellan photographed many new features and terrain on Venus, some never before seen in the solar system. Impact craters, formed when large meteors fell through the planet's atmosphere and struck the ground, look different from those on any other planet (Fig. 21–22). The surface is literally covered with volcanoes, hundreds of thousands of domes dotting the plains. *Magellan* revealed hundreds of *coronae,* broad circular features never before seen on any

other planet or satellite. The spacecraft also showed that the geology of Venus seems dominated by hot currents and hot spots beneath the crust, which push and stretch the surface. Venus, unlike the earth, does not seem to have plates. Further studies could lead to a better understanding of the formation of the earth and Venus.

21–6 MARS

Long known as the "red planet," Mars was associated with the Roman god of war because its reddish hue seemed to suggest images of war, blood, and flames. It is a conspicuous object in the sky, rather bright, and can easily be seen with the naked eye (Fig. 21–23). Except for the finer details, its surface can be seen with a telescope; it is the only planet whose surface is observable from the earth. Many color changes, both seasonal and nonseasonal, have been observed.

Mars has excited more interest over the years than any other planet, because at one time it was seriously considered to harbor intelligent life. The astronomer Giovanni Schiaparelli (1835–1910) in 1877 observed what appeared to be very thin lines on the surface, which he called "canali." When the term was trans-

Figure 21–23   Mars from the Hubble Space Telescope, which can be used to monitor the Martian atmosphere and surface. (NASA)

lated into "canals," it created the impression that intelligent life existed on Mars. Percival Lowell (1855–1916), who built an astronomical observatory near Flagstaff, Arizona, conceived the idea that canals had been built to irrigate the reddish desert areas of Mars by drawing water from the polar caps. But there was little resemblance between his drawings of the canals and what could be observed through the telescope or photographed.

Only in the past decade have the conditions that prevail on the planet become known. *Mariner 4* made the first close-up pictures of Mars in 1965, which revealed a surface covered with large craters reminiscent of those on the moon. *Mariners 6* and *7* in 1969 sent back pictures that showed that the Martian surface also had large areas of terrain unlike anything ever seen on the moon or the earth.

Mariner 9, however, was one of the high points in the exploration of Mars and decisively changed our view of the planet that generations had thought most closely resembled the earth. Photographs (7000 of them) sent back by *Mariner 9* in 1972 revealed surprising evidence of four enormous volcanic mountains, larger than any on earth. Perhaps Mars is just

beginning to "boil" inside, as illustrated by the well-advanced process in the Nix Olympica–Tharsis Ridge, but vulcanism has not yet spread to the planet as a whole.

The photographs also show a vast system of canyons, gullies, and channels. The huge canyons that stretch east and west along the equator of Mars are 50 to 75 miles wide and 3 to 5 miles deep, much larger than any found on earth. The largest corresponds to a feature known as Coprates, which sometimes changes in appearance with the seasons. These variable markings may have nothing to do with the surface of Coprates. The canyon is so deep that dust persists in the atmosphere between the canyon walls and makes the canyon look brighter than the surrounding landscape after the atmosphere above is comparatively free of dust. Once the canyon atmosphere is clear, there is little contrast between the interior of the canyon and the surrounding area.

Since the *Mariner* flights, it has been realized that the conditions for life on Mars are extreme by terrestrial standards. The diameter of the planet is only one half that of the earth, and its mass only one ninth. Its surface gravity, less than half that of the earth, accounts for the thin Martian atmosphere, resembling the thin air found 100,000 feet above the earth. The Martian atmosphere differs considerably from the earth's in composition, being rich in carbon dioxide and containing much less nitrogen. The thin atmosphere affords little protection against ultraviolet solar radiation, which reaches the Martian surface at an intensity that would be fatal to most forms of life. No liquid water has ever been detected on the surface, which appears to be drier than the earth's most arid deserts, but there is water vapor in the atmosphere. The temperature at the equator ranges from a daytime high of 28°C to a nighttime low of −100°C. At the poles the temperature is −215°C, compared with −89°C in Antarctica in the winter.

Two *Viking* landers touched down on the surface of Mars in 1976, while two *Viking* orbiters circled and photographed the planet with unprecedented resolution and clarity. They have provided evidence that the cratered terrain had been modified by volcanic activity, that water was a significant agent in shaping its features early in its history, and that since then, surface material has been redistributed by high-velocity wind, though causing little erosion. The surface of Mars is more like rocky volcanic deserts on the earth than the highly cratered surface of the moon (Fig.

Figure 21–24 A *Viking 2* view of the surface of Mars. This view resembles deserts in the southwest United States. (NASA)

21–24). Yet Mars seems to have little sand. The residual polar caps, it now appears, consist of water ice, not carbon dioxide ice. This discovery has led to greater acceptance of hypotheses involving water as an active agent in Mars' past. The channels on Mars are even more abundant than the *Mariner 9* photographs showed.

With a diameter of 4200 miles, Mars is midway in size between the earth and the moon. In two respects it is remarkably similar to the earth: The Martian **rotation** period is 24 hours 37 minutes 23 seconds, and its axis is tilted 23°59' to the plane of the orbit, which produces seasons similar to those on earth. It receives less radiation from the sun than does the earth. When it is closest to the earth, Mars is about 34,500,000 miles away. Mars has two small moons, Phobos and Deimos. The *Mariner* flights show Phobos to be an elongated object with a maximum diameter of about 14 miles. Deimos is probably not larger than about 4 miles. It is likely that both are captured asteroids that approached Mars too closely and were perturbed into new orbits.

21–7 JUPITER

Jupiter is by far the largest of all the planets and the most massive, more than twice as massive as the rest of the planets combined, and 318 times as massive as the earth. Aside from the sun, it is the only other significant mass in the solar system. In spite of the planet's great mass, Jupiter's density is only about 1.3 g/cm³, less than one fourth the earth's and about

the same as the sun's. Its equatorial diameter is 88,700 miles, and it exhibits marked flattening at the poles, with a diameter measured through the poles of 83,000 miles. Jupiter is the most rapidly rotating planet; its day is 9 hours 50 minutes at its equator and 9 hours 55 minutes near the poles. Its average distance from the sun is 484 million miles, and it takes 11.86 terrestrial years to complete a circuit of the sun.

The structure of Jupiter is a matter of considerable debate. Although the center of the planet probably contains a dense solid core, the outer layer appears to be composed of various gases. Methane and ammonia have been detected in its atmosphere. Water is probably present, but hydrogen and helium are very likely the predominant constituents of the planet, as they are of the universe, since Jupiter is massive enough to hold the lightest gases. The presence of helium in Jupiter's atmosphere was confirmed in 1973.

Jupiter generally appears as a very bright object, nearly equaling Venus as a reflector of sunlight. As seen through a telescope, Jupiter's most striking feature is a complex system of bands of colors. These spectacular bands, representing the outer region of the planet's atmosphere, come and go, fade and darken, widen and become narrow, and move up and down in latitude. The bright bands are called zones, and the dark bands belts. The colors that show up in the bands—red, brown, yellow, green, and blue—may be caused by various free radicals, or molecular fragments. A conspicuous atmospheric feature called the **Great Red Spot** (although it is often anything but red) appears as a huge oval shape in the southeastern quadrant. Measurements made by *Pioneer 10*, a

Figure 21–25 A montage of Jupiter and its four Galilean satellites made from the *Voyager* spacecraft. The orbits are drawn in. (NASA/JPL)

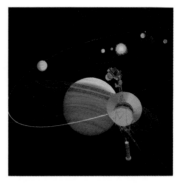

Figure 21–26 The trajectory of a *Voyager* planetary spacecraft. *Voyager* has had encounters with Jupiter, Saturn, Uranus, and Neptune. (NASA/JPL)

spacecraft that passed within 81,000 miles of Jupiter in 1973, indicate that the Great Red Spot is a huge hurricane. The red color may be due to a mix of organic compounds that form ice crystals at the tops of the clouds.

The four Jovian satellites that Galileo discovered are comparatively large bodies: Ganymede and Callisto are both larger than the moon, and Europa and Io are about the same size as the moon (Fig. 21–25). The other satellites are much smaller in diameter; the twelfth was discovered in 1951 and is only visible in the largest telescopes. The thirteenth and fourteenth were discovered in 1974 and 1975. Jupiter's outer satellites, it is believed, represent captured asteroids scooped up by the massive planet during the course of its history.

Jupiter is a source of powerful radio emissions in the decameter range. Although their origin has not yet been determined, they are probably related to Jupiter's magnetic field, which is 10 times stronger than the earth's, and to its radiation belts, 10,000 to 1

million times as intense as those surrounding the earth. In some ways Jupiter is like a small star, but it is not quite massive enough to shine on its own. Its mass is just below the mass that would have made it a star. It is the only such object that we know.

Two *Voyager* spacecraft, *Voyager 1* and *Voyager 2*, were launched in 1977 to explore the Jovian planets of the solar system (Fig. 21–26). They made their closest approach to Jupiter in 1979. Their mission: to send back pictures of Jupiter, Saturn, and their satellites and to measure the atmosphere, magnetic field, cosmic rays, and density of the planets. By studying the outer planets, it may be possible to find out more about how the earth was formed. The cold outer planets are more like the way they were when the solar system was formed. The *Voyagers* provided a quantum leap in image quality.

The analysis of data and photographs from the first *Voyager* in 1979 showed that Io is covered with volcanoes, except for the polar caps, making it the only body in the solar system besides earth that is known to have active volcanoes. Europa has a surface of ice and water and miles of long, linear structures crisscrossing it. Ganymede, the largest of Jupiter's moons, resembles our moon and may be composed of water-containing rocks. Callisto has huge dents, probably caused by meteor impacts, and is marked with bright-colored ridges (Fig. 21–27). The spacecraft also discovered a ring around Jupiter, making it the third planet in the solar system (after Saturn and Uranus) known to have such a feature. There are a very narrow outer ring and an inner ring of thinner matter that goes all the way to the surface. Jupiter's atmosphere is

Figure 21–27   Jupiter's four Galilean satellites: Io (*top left*), Europa (*top right*), Ganymede (*lower left*), Callisto (*bottom right*). Each is very different from the other moons. (NASA/JPL)

swirling and turbulent, rather than stable, and has well-ordered circulation patterns, as had been thought. The Great Red Spot's counterclockwise vortical motion is one of the most striking features of the planet.

The trajectories of the *Voyagers* were chosen so that the gravity and orbital motion of Jupiter could act as a slingshot to send them to Saturn, one arriving there in 1980, the other in 1981. Measurements were made of Saturn, its satellite Titan, and its spectacular ring system. One of the *Voyagers* was then sent to Uranus, arriving there in 1986 and going on to Neptune in 1989. Eventually, the *Voyagers* would pass beyond the solar system.

Meanwhile, the spacecraft *Galileo* was launched October 12, 1989, from the space shuttle *Atlantis* for a six-year voyage to Jupiter. The mission had three scientific objectives: probing the chemical composition and physical state of Jupiter's atmosphere; studying the same characteristics of satellites; and surveying the structure and dynamics of the planet's magnetosphere. Although *Galileo* is expected to reach Jupiter in late 1995, NASA planned scientific studies along the way, such as observations of Venus on February 10, 1990—its plasma environment, cloud patterns, and the possible existence of lightning—and observations of one or two asteroids, bodies never observed up close before.

21–8 SATURN

Saturn ranges between 839 and 937 million miles from the sun, twice as far as Jupiter. It moves so slowly along its gigantic orbit, taking 29.5 years to make one complete circuit of the sun, that the Romans named it for their god of time, Saturn. Although it is nearly the same size as Jupiter, it has less than one third as much mass; its density is only 0.7 g/cm³. Given a large enough body of water, it would float.

Saturn rotates once every 10 hours 14 minutes at the equator and once every 10 hours 38 minutes at the poles, indicating that, like Jupiter, it is not a solid surface but a dense atmosphere, the different parts of which rotate at different speeds. Any solid matter within must be very near the center, as with Jupiter. It appears to be composed largely of hydrogen and helium, followed by methane and, to a lesser extent, ammonia. Its atmosphere is deep and violent.

The most striking feature of Saturn as seen through the telescope is its ring system. Galileo was the first to view the rings, although not clearly. They appeared to him more like appendages of the planet than rings. Christian Huygens (1629–1695) saw that the rings went completely around the planet. The ring system, 171,000 miles in diameter, consists of four separate major rings, all in the plane of Saturn's equator. The two brightest are separated by the Cassini Division, named after one of its first observers, G. D. Cassini (1625–1712). In 1850, a third, dark ring lying closer to the planet was discovered. Called the Crepe Ring, because its dark appearance suggests to some a crinkly fabric, it casts its shadow across the face of the planet. In 1961 a fourth ring was discovered, which reaches almost to the planet itself.

It is generally agreed that Saturn's rings are made up of individual particles or satellites, each in its own gravitationally stable orbit around Saturn. A solid ring or "halo" of material is not physically possible. The solid appearance of the rings is an effect created by the close proximity of the tiny particles when viewed from the great distance of the earth. The particles were once believed to consist of frozen ammonia, but they are now thought to be chunks of water or ice or bits of rock coated with ice. The belt of rings is so thin, approximately 12 miles, that it disappears from sight when viewed exactly edge-on. One theory is that the rim represents material that was left over and did not accrete to form Saturn during its planet-building state.

Figure 21–28 Saturn and some of its moons, photographed from *Voyager*. Enceladus is in the foreground; Mimas is above it; Dione is at lower left; Rhea is at middle left; Titan is at top left; Iapetus is at top center; and Tethys is at upper right. (NASA/JPL)

Well beyond the ring system, Saturn has 17 known moons and perhaps many more; it is difficult to say (Fig. 21–28). Titan is the largest, being larger than Mercury and nearly as large as Mars. Titan has another distinction—it is the only moon in the solar system known to have an atmosphere. Nitrogen makes up the bulk of Titan's atmosphere, with lesser amounts of methane and hydrogen. Clouds are present, and a greenhouse effect in the atmosphere may result in Titan's absorbing more of the sun's warmth than it radiates back into space. Titan could be a miniature of what the primitive earth was like at the dawn of life.

In September, 1979, *Pioneer 11* became the first spacecraft to reach Saturn. Although the encounter was a triumph, the planet appeared to be a bland, fuzzy, yellow, washed-out sphere except for its thin, elegant rings. Its largest moon, Titan, resembled a featureless orange berry.

Voyager 1, launched from Cape Canaveral, Florida, in 1977, two weeks after an identical twin (*Voyager 2*), began its continuous coverage of Saturn in August, 1980. Its narrow-angle camera had 50 times the resolution of *Pioneer 11's* imaging photopolarimeter, and there were ten other scientific instruments aboard. Mission control at Jet Propulsion Laboratory in Pasadena, California, soon began to see ribbons and bands that gave the ring system the appearance of a celestial gold phonograph record (Fig. 21–29). Baffling "spokes"—dark, finger-like phenomena—swept randomly across the rings.

As *Voyager 1* approached more closely, Saturn's atmosphere showed much of the same turbulent complexity seen on Jupiter. The bland-looking Saturn seen through a telescope was shown to be a sphere of multicolored bands, jet streams, and violent storms. The rings were much more detailed in close-up, consisting of hundreds, perhaps 1000, ringlets. The smaller satellites, up to then seen through earthly telescopes as mere points of light bearing names out of mythology, grew into icy worlds more complex than anyone had imagined. More was learned about Saturn from *Voyager 1* than in the entire span of recorded history.

Figure 21–29 Saturn's rings from *Voyager 1.* The colors, though exaggerated, reveal different compositions from ring to ring. (NASA/JPL)

To many the world over, the achievement was the most spectacular piece of space exploration since men set foot on the moon. Later, the encounter of *Voyager 2* with Saturn in August, 1981, after a four-year, 1.4 billion-mile flight, gave scientists data that would keep many occupied for years. On August 25, 1981, *Voyager 2* rounded Saturn and set course for Uranus, four years and a billion miles away. It was scheduled to fly by Neptune in 1989.

Voyager 2 flew closer to Saturn's rings than *Voyager 1* had, with cameras that were between 50% and 100% more efficient than those on *Voyager 1.* *Voyager 1* showed that the classic A-, B-, and C-rings consisted of hundreds of smaller rings, or ringlets. *Voyager 2's* photographs revised the count to thousands. The Cassini Division between the A- and B-rings, once thought to be empty, contained a number of rings. The F-ring, discovered by *Pioneer 11,* was found by *Voyager 1* to consist of three strands of material that appeared braided or clumped, although *Voyager 2* saw no braids.

Voyager 2 took a series of time-lapse movies to study the formation and lifespans of the mysterious spokes in the B-ring. The spokes form completely in less than 20 minutes and dissipate in less than 10 hours, though some remnants persist. According to one theory, spokes are particles of fine dust electrostatically charged that are lifted above the plane of the B-ring by Saturn's magnetic field lines. Clues to the structure of the rings may come from the analysis of the experiment *Voyager 2* executed with its photopolarimeter. The device focused on the star, Delta Scorpii, and measured the star's light as it filtered through the rings to gather information about the rings' composition and structure.

As a result of the discoveries of *Voyagers 1* and *2,* the number of known moons around Saturn was increased to 17. Titan is by far the largest. Seven are intermediate-sized satellites. Eight are small moonlets. Phoebe, the outermost satellite, may be a captured asteroid.

Mimas, the moon closest to Saturn, has a crater so huge that had the object that caused it been much larger, its impact might have shattered Mimas. On its other side it is heavily pockmarked with small craters, indicating that it is relatively old. Its companion, Enceladus, shows a varied topography. There are plains but also craters, ridges, and valleys. A heat source, possibly volcanic in origin or created by gravitational stress, has softened and smoothed out its icy surface.

Tethys has the largest crater in the Saturnian system, one so large that Mimas could fit within it. It also has a strange trench that circles nearly three fourths of its circumference. Dione resembles the earth's moon; it is marked by craters, "seas," and highlands. Iapetus, with one hemisphere six times as bright as the other, shows the greatest contrast of any object in the solar system. Rhea also resembles the earth's moon but has "shoulder-to-shoulder" densely packed craters. Fascinating worlds, indeed!

When the Hubble Space Telescope turned its eye toward Saturn on August 26, 1990, the computer-corrected images of the planet were unsurprising. Saturn's calm, banded face seemed almost identical to the appearance it presented to the *Voyager* spacecraft in 1980 and 1981. But when the Hubble Space Telescope gave us a second look on November 9 and 11, 1990, it captured Saturn in a very different mood. The planet had developed a "White Spot"—a gigantic storm, at least triple the size of earth, spreading across Saturn's equatorial regions (Fig. 21–30).

The White Spot seems to be Saturn's answer to Jupiter's Great Red Spot, a hurricane that has been raging on Jupiter at least since the invention of the telescope in the seventeenth century. The difference is that such outbursts are rare and short-lived on Saturn. The previous one appeared in 1933 and survived only a few years; the last one before that was in 1870. The spot cannot be a volcanic eruption, because Saturn has no magma in its interior or a solid surface from which magma could erupt. The explanation for the White Spot may nevertheless lie deep within Saturn. With further images from the Hubble being processed, some answers should be forthcoming.

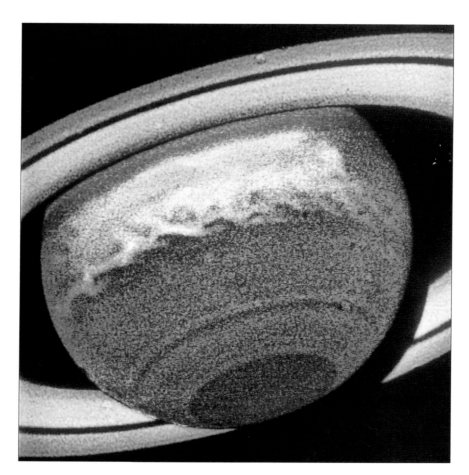

Figure 21–30 The image of a White Spot on Saturn made by the Hubble Space Telescope in 1990. Such outbursts are rare on Saturn. (NASA/STScI)

21–9 URANUS

This cold, remote planet was unknown to early astronomers, since it is rarely visible to the naked eye. Its discovery by William Herschel in 1781 added a seventh planet to the solar system and doubled the size of the solar system, since Uranus is farther from Saturn than Saturn is from the sun. Named for the father of Saturn, Uranus varies in distance from 1608 to 1956 million miles from the sun, a distance so great that the sun is probably only a faint gleam in the sky when viewed from Uranus. It takes Uranus 84.02 years to complete a circuit around the sun, and each of its seasons is 21 years long. It is believed to be covered by clouds, and its surface temperature is below − 200°C. Its rotation period is 10 hours 49 minutes. Uranus appears greenish.

A most unusual feature of Uranus is its tilt. All of the other planets spin on their axes like tops as they travel their orbits. Mercury and Jupiter stand virtually straight; Venus, earth, Mars, Neptune, and Pluto (perhaps) lean over at angles ranging from 23° to 29°. Uranus, however, is tilted at an angle of 98°. Instead of spinning like a top, it lies on its side and rolls along. The planet's direction of rotation is retrograde, the reverse of that of all other planets except Venus. Uranus has five known moons; the largest is Titania, with a diameter of 625 miles. The others are named Ariel, Miranda, Oberon, and Umbriel.

The first major structural discovery in the solar system in nearly 50 years involved Uranus. The discovery of Pluto in 1930 held claim to that distinction until 1977, when rings were discovered around Uranus (Fig. 21–31). Until then, Saturn was the only one of the nine known planets encircled by rings. While working in an airborne observatory 41,000 feet aloft, a team of astronomers spotted five rings around Uranus. Other observatories confirmed their existence, and they have been named alpha, beta, gamma, delta, and epsilon after the first five letters of the Greek alphabet. The rings are vertical to the earth's equatorial plane and were not seen earlier because light reflected from Uranus obscures the rings' lesser reflections. The rings lie in a 700-km-wide (440-mile) belt. Four of the flattened rings are about 10 km (6 miles) across, and the outermost one is 100 km (60 miles) wide. Six additional rings were discovered later.

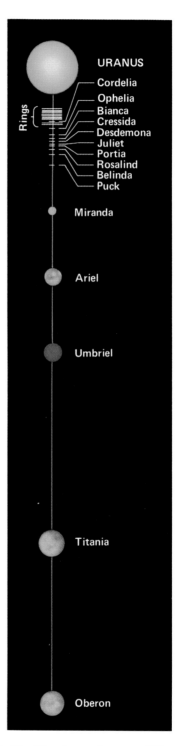

Figure 21–31 The rings and moons of Uranus.

21–10 NEPTUNE

Neptune is a twin of Uranus, which it physically resembles. It, too, is greenish, because of its atmospheric methane (Fig. 21–32). But it is a billion miles farther from the sun than Uranus (30 AU compared to about 19 AU) and is therefore much colder, −215°C or less. Light from the sun takes 4 hours to reach it. It is never visible to the naked eye but can be seen with a good pair of binoculars. Its orbital period is 164.8 years, so great that since its discovery in 1846 it has not yet made a complete revolution around the sun; that will not be until the year 2011. Although its mass is 17.2 times that of the earth, its density is only about one third as great. It has a fast rotation period of about 15 hours 30 minutes, though recent values range from 17 to 22 hours.

Neptune, you will recall (Chapter 5), was the first planet whose position was predicted before its discovery. John Couch Adams and Urbain Leverrier independently calculated its position from perturbations in Uranus's orbit, and Johann Galle made the actual discovery of the planet. It was then found that many observers, even Galileo, had previously seen Neptune just as they had seen Uranus but had mistaken it for a fixed star. Neptune has two satellites.

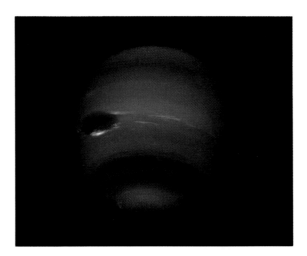

Figure 21–32   Neptune, from *Voyager 2*, with its Great Dark Spot, a giant storm in Neptune's atmosphere. (NASA/JPL)

William Lassell (1799–1880) found Triton, the brighter one, only a few weeks after the discovery of the planet; it is larger than our moon and revolves clockwise (retrograde). More than 100 years later, in 1949, Gerard Kuiper (1905–1973) discovered Nereid, which has the most eccentric of all satellite orbits in the solar system, ranging from 830,000 miles to more than 6,000,000 miles from the planet. *Voyager 2* reached the Neptune system August 24, 1989, passing close to Neptune, Triton, and Nereid.

Voyager 2 provided evidence of a number of fundamental aspects of the Neptune system. Further discoveries probably remain buried in the data that scientists will be analyzing for many years to come. The planet has a magnetic field, one with a very complicated structure, much like that of Uranus. As *Voyager* approached Neptune, an earth-sized active weather system that was soon called the Great Dark Spot became apparent. Neptune's period of rotation is 16 hours and three minutes. Neptune possesses eight moons, six of them discovered by *Voyager*. One of the six new, dark icy moons, 1989 N1, takes the place of Nereid as the second largest moon in the system. All six seem to be heavily cratered. Rings around Neptune are complete, contrary to what had been suggested by earth-based observations, but the structure of the rings has yet to be explained. Neptune has complex weather patterns, although it receives only 5% as much solar energy as does Jupiter. The mechanisms behind them have not been explained.

Neptune's major moon, Triton, in a sense stole the show at Neptune. With a temperature of 37 K, Triton is the coldest body ever observed in the solar system. It is extraordinarily bright; much of what appears dark on Triton would be bright on many other moons. Triton has a thin but extensive atmosphere composed mostly of nitrogen. It has a radius of 1360 km and a density slightly greater than 2 g per cc, suggesting that it is composed mainly of rock and water, nitrogen, and methane ices. Triton has undergone periods of major geological activity. It was once warmer than it is today. Some of its features appear to have been formed by liquid water that flowed into craters or calderas and froze.

The members of the outer solar system, such as Neptune, likely contain key chemical and physical records of the early evolution and formation of the planets. With its Grand Tour of four planets complete, *Voyager 2*, like its sister craft *Voyager 1*, followed a trajectory beyond the solar system. It may drift on, approaching Barnard's star in 6,500 years and passing Sirius, the brightest star visible from earth, in 296,036.

21–11 PLUTO

The story of the decades-long search for a "Planet X" that was known to perturb the orbits of Uranus and Neptune has been told in Chapter 5. The search ended on March 13, 1930—50 years after astronomers began to think seriously about it; the seventy-fifth anniversary of the birth of Percival Lowell; and the 149th anniversary of Herschel's discovery of Uranus—with the announcement at the Lowell Observatory of the discovery by Clyde Tombaugh of an object beyond Neptune that was apparently a planet. Other observatories soon confirmed the discovery of the planet, which was located not far from the spot originally predicted by Lowell. The name Pluto, taken from the brother of Jupiter and Neptune, was suggested.

Pluto is over 3.5 billion miles from the sun, revolving in an orbit that takes 248 years to complete. It is so faint that it can be seen only with the most powerful telescopes. Its orbit is the most eccentric of all the planets; at one point it even comes inside the orbit of Neptune, and therefore comes closer to the sun than Neptune does. It is now on that part of its orbit and will remain there until 1999. For this reason it is believed by some to be an escaped satellite of Neptune. There is no chance of collision between the two planets, because Pluto's orbit is at a relatively steep inclination of 17°.

Pluto appears to be a small body, 4000 to 6000 miles in diameter. Its rotation period as derived from variations in brightness is 6 days 9 hours 17 minutes. Pluto's surface temperature is less than −240°C. Since even methane would freeze on Pluto, it probably has no gaseous atmosphere. It was discovered in 1976 that the surface of Pluto is frozen methane. Pluto may be the only planet that has survived in a pristine state under conditions at which it was formed 4.6 billion years ago. Pluto could also have a more complex structure: a core of water ice, followed by layers of methane and ammonia, and finally pure

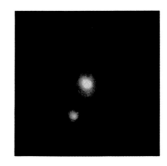

Figure 21–33 The Hubble Space Telescope's image showing Pluto and Charon as distinct objects. (NASA/STScI)

methane ice. In 1978, a moon in a 12,000-mile-high orbit was discovered around Pluto (Fig. 21–33). The great distance from the earth had prevented astronomers from resolving the planet and its moon into two separate specks of light. Tentatively named Charon, the moon circles Pluto once every 6 days 9 hours 17 minutes, an interval identical to Pluto's own period of rotation. An observer on one side of Pluto would always see the moon in the same position in the sky. To an observer on the other side, it would never be visible.

21–12 ASTEROIDS

Johannes Titius (1729–1796), a mathematician, made an interesting discovery in 1766 related to the distances from the sun of the six planets known in his time. He set down the names of the planets in order, and under each name he wrote a number: 0 for mercury, 3 for Venus, and a number twice that of the planet before it for the planets beyond Venus. Titius then added 4 to each number and divided each sum of two numbers by 10. The resulting series gave the approximate distances of the planets from the sun in astronomical units, as shown in Table 21–3.

Titius showed his calculations to the astronomer Johann Bode (1747–1826), who published them. The formula became known as Bode's law, but it is also called the **Titius-Bode law**. It works quite well for the planets then known, the distances in astronomical units coming out very close to those observed, *if* a planet is slipped in between Mars and Jupiter, as shown in Table 21–4. Further, it could be predicted from the Titius-Bode law that an imaginary planet beyond Saturn would have a distance of 19.6 AU.

Table 21–3 PLANETARY DISTANCES IN AU USING THE TITIUS-BODE LAW

	Mercury	Venus	Earth	Mars	—	Jupiter	Saturn	Uranus	Neptune	Pluto
	0	3	6	12	24	48	96	192	384	768
Add:	4	4	4	4	4	4	4	4	4	4
Sum:	4	7	10	16	28	52	100	196	388	772
Divide by 10:	0.4	0.7	1.0	1.6	2.8	5.2	10.0	19.6	38.8	77.2

When Uranus was discovered, its distance turned out to be 19.19 AU, in fairly close agreement.

The "missing" planet between Mars and Jupiter seemed to be discovered on New Year's Day, 1801, when Giuseppe Piazzi (1746–1826) found a small body in orbit around the sun between Mars and Jupiter. He sent data on the object to a number of scientists, including Bode and Karl Friedrich Gauss (1777–1855), who later calculated its orbit, showing that it was nearly circular, like that of a planet. Bode pointed out that the body's average distance from the sun was 2.77 AU, remarkably close to the predicted distance of the "missing" planet of 2.8 AU.

The "planet" was very small, only 480 miles in diameter, not much more than an orbiting chunk of rock, but a planet by definition nevertheless. Piazzi named the new planet Ceres, after the Roman goddess of agriculture. A year later, Wilhelm Olbers (1758–1840) found another "planet" in the same distance range; it was even smaller, and he named it

Pallas. By 1807, two others had been discovered, Juno and Vesta. The four objects were at first called "planetoids," meaning planet-like bodies, and were later referred to as minor planets. For some reason, however, the name **"asteroids,"** meaning star-like bodies, became attached to them, and the name stuck.

For about 30 years it was assumed that Ceres, Pallas, Juno, and Vesta were the only asteroids. Then each year saw the discovery of five or six more. With a new technique involving the use of a camera as well as a telescope, astronomers found hundreds of asteroids. Scientists at the Lowell Observatory in Flagstaff, Arizona, had discovered a number of asteroids, seven of which they have since named after the seven astronauts who died in the explosion of the space shuttle *Challenger* (Fig. 21–34). Several thousand are now catalogued, and the total number is estimated to be about 100,000. Ceres is still the largest, while the great majority are much less than 1 mile in diameter. Between them, Ceres and Pallas probably account for half the total mass of all asteroids. The asteroid belt between Mars and Jupiter is at an average distance from the sun of 250 million miles. It is not known whether asteroids are the debris of a former planet or planets that disintegrated for some reason or whether they are part of the original matter of the solar system that never became a planet.

On October 29, 1991, as the spacecraft *Galileo* was moving out beyond Mars, it became the first spacecraft to have a close encounter with an asteroid. The target was the asteroid Gaspra. Seen as a speck of reflected light in telescopes, Gaspra has a diameter of no more than eight miles and appears to have an elongated shape like a lumpy potato.

The *Galileo* spacecraft flew within 1000 miles of Gaspra, providing the first chance to observe an asteroid from nearby. It was instructed to snap 150 color

Table 21–4 PLANETARY DISTANCES FROM THE SUN

Planet	Distance Calculated from Titius-Bode Law (AU)	Actual Distance (AU)
Mercury	0.4	0.39
Venus	0.7	0.72
Earth	1.0	1.0
Mars	1.6	1.52
—	2.8	—
Jupiter	5.2	5.20
Saturn	10.0	9.55
Uranus	19.6	19.2
Neptune	38.8	30.1
Pluto	77.2	39.5

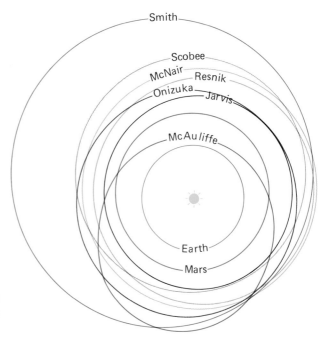

Figure 21–34 Asteroids have been named for the seven astronauts who died in the explosion of the space shuttle *Challenger*. (Edward Bowell/Lowell Observatory)

and black-and-white photographs of Gaspra and gather remote-sensing data on the asteroid's chemical composition.

However, instead of getting the results almost immediately, scientists had to wait. *Galileo's* main antenna for transmitting images and data was jammed in the wrong position. Instead, the results were recorded on board and were to be played back as soon as the antenna could be deployed, or in November, 1992 at the latest, when *Galileo* was to return to the vicinity of earth for one final gravitational boost on its journey to Jupiter. *Galileo* would then be close enough for its smaller antennas to transmit the missing pictures. Researchers were eager to see what *Galileo* could turn up, hoping to find clues in Gaspra's composition as to how planets are born and how they have changed over time.

21–13 METEORITES

On February 12 1947, thousands of people in the environs of the Siberian village of Novoprovka were eyewitnesses to a "rain of iron," one of the rare occasions when the earth has collided with another member of the solar system. Against the blue of the sky, they saw a ball of light as brilliant as the sun and about the size of the full moon. On the slopes of the nearby mountains, explorers later found more than 100 craters, some of them 30 to 40 feet deep and as wide as 75 feet at the top. The ground over an area of several square miles was strewn with pieces of iron, ranging from several hundred pounds in weight to tiny specks. Before it entered the earth's atmosphere, the **meteorite** is estimated to have weighed 1000 tons and to have had a diameter of 30 feet. After it broke up into millions of pieces, each piece flew through the air at supersonic speed. The shock waves felled trees and completely shattered the rocks of the mountain slopes.

In prehistoric times, a vastly more destructive collision caused the Barringer "meteor crater" near Flagstaff, Arizona, a huge hole about a mile across and several hundred feet deep (Fig. 21–35). Countless pieces of iron, the largest of which weigh hundreds of pounds, are embedded in the ground for a distance of several miles from the crater. On September 28, 1969, an object exploded over the town of Murchison in Australia, showering fragments over an area 7 miles long and 2 miles wide. The noise lasted about a minute and sounded like thunder or a series of sonic booms. At least 82 kilograms of meteorite fragments from the Murchison event are now in public collections (Fig. 21–36).

Figure 21–35   The Barringer "meteor crater" (actually a meteorite crater) near Flagstaff, Arizona. Dozens of other terrestrial craters are now known, many from space photographs. (Allan E. Morton)

The idea that meteorites were of extraterrestrial origin enjoyed roughly the same reputation in the year 1800 that UFOs (unidentified flying objects) do today. Primitive peoples had made objects of stones believed to have fallen from heaven, but in more sophisticated times this source was put down as an old wives' tale. In 1803, however, after the village of L'Aigle in France was pelted by a dense shower of falling stones, the Academy of Sciences appointed a commission headed by Jean Baptiste Biot (1774–

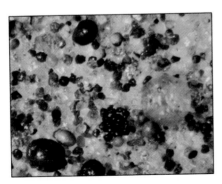

Figure 21–36   Close-up of a part of the Murchison meteorite.

1862) to investigate the event. The commission's report eliminated the possibility that the stone shower was a terrestrial phenomenon and produced evidence that satisfied the Academy that the stones came from outside our planet.

Subsequently, museums began to collect meteorites. The largest in "captivity," one found by Eskimos in Greenland and named Ahnigito, was brought to the United States by the explorer Robert E. Peary (1856–1920) in 1897. Weighing 33 tons, it is housed in the Hayden Planetarium in New York City.

There is little doubt that meteorites are small asteroids. When the Pribram meteorite fell in Czechoslovakia in 1959, and the Lost City meteorite in the United States in 1970, their orbital trajectories were photographed in precise detail. There was enough information to show that they were typical asteroids with paths overlapping that of the earth. Further studies indicated that they were very old, at least comparable to the age of the earth.

There are two principal types of meteorites, **stony** and **metallic**, the latter consisting mainly of iron and some nickel. Because the iron meteorites survive the trip through the atmosphere better than the stony ones and are more easily distinguished from terrestrial rocks, more of them are found. The chemist Edward C. Howard seems to have been the first to examine the internal structure of stony meteorites (around 1800), and he found small masses ranging in size from a pinhead to a small pea. Later, the mineralogist Gustav Rose (1798–1873) established a classification of meteorites, naming this class **chondrites**, after their internal structure (*chondros,* meaning "grain of seed"). The small rounded bodies in chondrites came to be called **chondrules**. There is evidence that chondrites are pieces of planetary matter in a primitive state and that the sun and chondrites were made from the same parent material, the metal content being the same in both. This has opened the possibility of analyzing in detail the stuff of which the planets are made, since even on our own planet we can examine only the crust and know very little about the interior.

As early as 1834, Berzelius extracted complex organic substances from a meteorite that had fallen in a village in France. By 1962, it had been clearly shown that meteorites contain hydrocarbons, and by 1970 compounds such as alcohols, carboxylic acids, sugars, and amino acids were reported in meteorites. There has been some reluctance to accept many such reports, since some of the material in carbonaceous chondrites—those with an appreciable amount of carbonaceous material other than free carbon— could be biological contaminants that entered the meteorites from the air or ground. A number of fragments of the Murchison meteorite, however, were collected soon after impact and are believed to have been exposed to very little contamination. The Murchison meteorite contains a number of organic compounds never before encountered in meteorites; the compounds seem to have been synthesized by non-biological processes and are of extraterrestrial origin. It has been suggested that because of the environment of their formation, the process of chemical evolution stopped rather than proceeded on to life.

21–14 COMETS

Unlike the planets, which move across the sky with great regularity, new comets appear and disappear unpredictably. They are surrounded by an air of mystery partly for this reason and also because the visible ones often develop a long, spectacular luminous tail (*cometes,* hairy, from resemblance to female tresses) and were regarded as an omen of war and pestilence. Although comets have not yet been fully explained, it is believed that they are members of the solar system rather than interlopers from outer space. Astronomers know the orbit of Halley's comet very well (Fig.

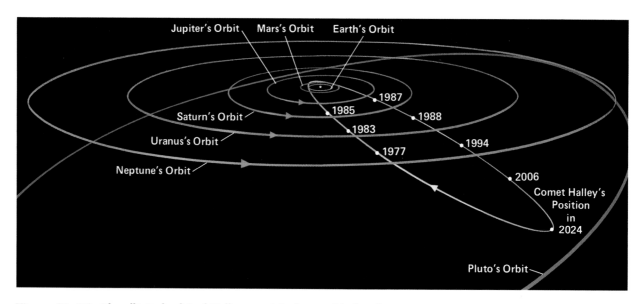

Figure 21–37   The elliptical orbit of Halley comet is shown with the planetary orbits. The direction of the comet's motion is opposite to that of the revolution of the planets.

21–37). When it reappeared in 1986, however, it was the faintest as seen from earth in any appearance in the last 2000 years (Fig. 21–38). It is due back in 2061.

Very little conclusive evidence is available concerning the nature or origin of comets. Normally, a **comet** is an aggregation of solids that has a nucleus 1 mile to 50 miles in diameter. According to a model nicknamed the "**dirty snowball**," the nucleus consists mainly of solid water (H_2O), methane (CH_4), and ammonia (NH_3), with molecules and particles of heavier elements interspersed among them. There may be an enormous number of comets, possibly 100 billion. A comet moves around the sun in a huge orbit, one circuit of which may take millions of years. Gravitational attractions of the planets may distort the orbit so that the comet makes a complete turn around the sun in as little as a few years.

When the comet is distant from the sun, it remains frozen, but when it approaches the sun it begins to vaporize. The escaping gases surround the nucleus, forming an envelope called the head, or **coma**, ranging from 10,000 to more than 100,000 miles in diameter. Ultimately, depending on how close the comet comes to the sun and the chemical composition of the comet, these gases (and some dust) may form the comet's **tail**, which points away from the sun and in some cases extends 100 million miles into space. It is believed that **meteors**, or "shooting stars," come from the tails of comets. The periods of maximum intensity of certain meteor showers coincide with the times when certain comets pass closest to the earth and the earth is passing through their dust.

The vast majority of comets observed are not visible to the naked eye. A few are large and brilliant and can often be seen in broad daylight. When a comet is discovered, it is identified by the name of its discoverer and by an alphabetical letter designating the sequence of comets discovered in any particular year. Comet Ikeya-Seki 1965 f designates the comet discovered by these amateur astronomers, and it was the sixth discovered in the year 1965.

21–15 THE SEARCH FOR EXTRATERRESTRIAL LIFE

A driving force in the space effort has been the search for life elsewhere in the solar system and in the universe. **Exobiology** is the name given to this new discipline, the study of extraterrestrial life. The reasoning is that, given the right conditions, life may arise spontaneously and evolve.

These conditions for life as we know it include a surface temperature low enough to provide water in the liquid state and one that is suitable for the chemical reactions of life processes; a store of the chemical elements required for these processes, such as carbon and oxygen; and a source of radiant energy. Even if life is not found on other planets, various stages of **chemical evolution**, the transformation of simple gases to complex compounds that are presumed to sustain life, may be found.

It was learned from the Apollo program that life did not evolve on the moon. There is no life there today or any indication of life having been there in the past. Mercury and Venus seem most inhospitable to life as we know it. Some of the outer planets, such as Jupiter, and some large satellites are candidates for life. Mars seems to present the only possibility of life among the terrestrial planets besides the earth.

Figure 21–38 Halley's comet in the sky seen from Georgia in March, 1986. (James R. Westlake, Jr.)

If there is life on Mars, it must exist in a world drier than the driest deserts of earth. It must withstand ultraviolet radiation stronger than on any mountaintop of earth. And it must contend with temperatures that dip to hundreds of degrees below zero each night. Harsh as these conditions may be, laboratory experiments with earth organisms under simulated Martian conditions have shown that such adaptation is possible.

The *Viking* search for life on Mars involved two spacecraft. *Viking 1* set down safely on Chryse Planitia in the northern hemisphere of Mars on July 20, 1976. *Viking 2* landed at Utopia Planitia on September 3, 1976 (Fig. 21–39). The *Viking* landers scooped up soil from the Martian surface and performed experiments with the Automatic Biological Laboratory until June 1, 1977, when the laboratories were switched off from earth. They had run out of the nutrient liquids and helium gas needed to conduct the experiments. Although the two spacecraft did not detect life in the samples of Martian soil they tested, the findings may not negate the possibility of life on Mars. The *Vikings* only searched a few square meters of the planet. There may be life on Mars that has simply escaped detection so far.

Within the universe there may be stars similar to our sun with a planet that has an environment like earth's. The Harvard astronomer Harlow Shapley suggested that one star in a million has the necessary conditions for the existence of life. In the Milky Way Galaxy alone there may be 100,000 planets similar to the earth among the billions of planets believed to exist there. But even our largest optical telescopes are not capable of detecting a planet in orbit around another star. Evidence supporting the hypothesis comes from a close neighbor of our sun, Barnard's star, 6 light-years away. From a wiggle in is path, astronomers have concluded that two planets about the size of Jupiter and Saturn are orbiting the star and exerting a gravitational pull that affects its course. Smaller planets may also be orbiting Barnard's star.

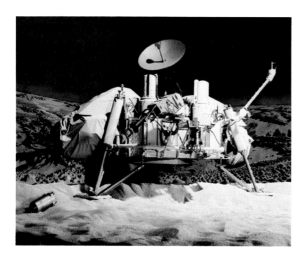

Figure 21–39   Engineering model of a *Viking* lander in a Mars simulation laboratory. (NASA/JPL)

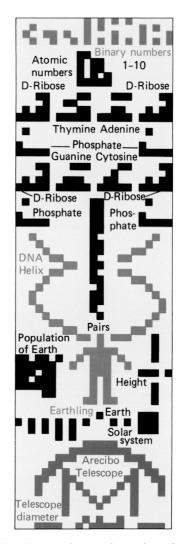

Figure 21–40   A radio signal sent from the radio telescope at Arecibo, Puerto Rico, in 1974 bore a message from the people on earth. It was directed at the globular cluster M13. An answer could not be expected before 48,000 years. (Courtesy of Cornell University)

Exobiologists are confident that the creation of life anywhere in the universe is more the rule than the exception. In any event, there is growing interest among scientists not only in the search for extraterrestrial life but also in the search for extraterrestrial intelligence (SETI). Space travel as we know it would be out of the question for contacting a civilization in another solar system because of the vast distances involved. Traveling at rocket speeds of 8 km/s, a spaceship would take 160,000 years or more to reach Proxima Centauri, the nearest star to the sun, only 4.3 light-years away.

Our capability for detecting extraterrestrial intelligence (ETI) lies rather in radio astronomy and involves radio telescopes; advanced computers for processing data; and human imagination, skill, and commitment. The first such effort was organized in 1959 by Frank Drake (b. 1930) at the National Radio Astronomy Observatory in Greenbank, West Virginia. Drake directed an antenna toward two nearby stars, Epsilon Eridani and Tau Ceti, and for two weeks listened for possible radio signals but heard none.

There is no hard scientific evidence that life of any kind exists on far-off worlds. The dozens of searches that radio astronomers have conducted have not turned up the slightest hint of a signal that might have been given off by a planetary civilization orbiting a remote star (Fig. 21–40). But until 1983 the search efforts concentrated on only small parts of the sky and tuned in on only a few of the radio frequencies that might be used for transmissions.

In that year, the most sophisticated search yet for evidence of extraterrestrial intelligent life was begun at the Harvard University 84-ft-wide radio telescope northwest of Boston. In a program designed by Paul Horowitz of Harvard in collaboration with scientists at Stanford University, the University of California at Berkeley, and the NASA Ames Research Center, the heavens were to be scanned for four years in the hope of picking up some orderly signal that would indicate ETI.

A compact multichannel receiver was developed that could be hooked to a large antenna and listen to 131,072 channels simultaneously. The telescope's signals could be fed into a computer and analyzed for the blips that might mean a message. An added feature was the capability of the telescope to focus its listening on certain frequencies, like those of the 21-cm hydrogen signal, that would presumably be preferred by an advanced civilization for broadcast because there is not so much interference in that part of the radio spectrum. NASA was expected to begin its own intensive searches using even larger radio "ears" such as the 1000-ft Arecibo antenna in Puerto Rico. If indeed there are extraterrestrials (ETs) out there, some earthlings hoped that they would call.

KEY TERMS

Asteroids   "Minor planets," most of whose orbits around the sun lie in a belt between the orbits of Mars and Jupiter.

Aurora Australis   The southern aurora, visible over the southern hemisphere. It displays glowing lights in the sky resulting from processes in the earth's upper atmosphere.

Aurora Borealis   The northern aurora, visible over the northern hemisphere.

Chemical Evolution   The transformation of simple gases in the primitive atmosphere and ocean to complex compounds that are presumed to sustain life.

Chondrite   A stony meteorite.

Chondrule   A small crystalline spherical mass in the core of a stony meteorite.

Chromosphere   A relatively thin atmospheric layer of the sun lying between the photosphere and the corona.

Coma   The region surrounding the head of a comet.

Comets   Members of the solar system that orbit the sun in highly elongated orbits and when near the sun show a coma and a tail.

Corona   The outermost layer of the sun's atmosphere seen most spectacularly during a total solar eclipse.

Dirty Snowball Model   A theory explaining comets as amalgams of ices, dust, and rocks.

Dust-Cloud Theory   The theory that the solar system formed out of a cloud of gas and dust.

Ecliptic   The plane of the ecliptic is the plane of the earth's orbit around the sun.

Escape Velocity   The velocity that an object must have to overcome the gravitational pull of a body and escape into space.

Exobiology   The search for extraterrestrial life.

Great Red Spot   A conspicuous feature of the planet Jupiter.

Jovian or Major Planets   The giant planets: Jupiter, Saturn, Uranus, and Neptune.

Lunar Eclipse   An eclipse of the moon; this occurs when the moon passes into the earth's shadow.

Maria   The name originally given to the dark areas of the moon because they were thought to be seas (*mare,* meaning "sea").

Mascons   Regions of dense material under the surface of the moon; mass concentrations.

Metallic Meteorite   An interplanetary chunk of rock composed mainly of iron and nickel that impacts on a planet or moon.

Meteor   A track of light in the sky from rock or dust burning up as it falls through the earth's atmosphere.

Meteorites   Small asteroids that have orbits overlapping that of the earth and impact the earth.

Moon   A planetary satellite.

Nebular Theory   The theory that the sun and planets formed out of a cloud of gas and dust.

Phase   The varying shape of the lighted part of a planet or moon.

Photosphere   The visible surface of the sun.

Planet   A relatively large body revolving around the sun.

Revolution   The orbiting of one celestial body around another.

Rotation   The turning of a body around its own axis.

Satellite   An object that orbits around an astronomical body.

Selenography   The study of the same aspects of the moon that geography studies of the earth.

Solar Eclipse   An eclipse of the sun; this occurs when the moon partially or totally covers the sun's disk.

Solar Flare   Sudden increases in the light intensity of the chromosphere of the sun that release great amounts of energy.

Stony Meteorite   A meteorite composed mostly of stony material; a chondrite.

Sunspot   A region of the solar photosphere that is relatively cool and appears dark by contrast against the surrounding hotter photosphere.

Sunspot Cycle   An 11-year cycle during which sunspots increase in number toward a maximum and then fall off to a minimum.

Terrestrial Planets   The inner planets Mercury, Venus, earth, Mars, and sometimes Pluto. The moon is also included among the terrestrial bodies.

Titius-Bode Law   A formula that gives the radii of the orbits of the seven innermost planets and the radius of the asteroid belt.

THINGS TO DO

1. Note the phases of the moon and its position in the sky over a period of one month. Record the time of each observation.

2. Observe the surface of the moon with binoculars.

3. Visit an observatory and observe the moon and planets through a telescope.

4. If Venus is visible at this time of year, observe its motion.

5. Simulate an eclipse with a lamp, a globe of the earth, and a tennis ball.

6. Map the features of the full moon when it is rising.

7. Observe a meteor shower.

8. Observe the four largest moons of Jupiter with binoculars or a telescope.

EXERCISES

The Solar System

21–1. What is the basis for the statement that our solar system is not unique in the universe?

21–2. How is a solar system believed to have been formed?

21–3. From the information now available, discuss the possibility of life as we know it on other planets in our solar system.

21–4. What are some essential differences between planets and stars?

21–5. Which of the planets would appear to be most suitable for human exploration? In your opinion, should such exploration be encouraged?

21–6. On which planet do we see the surface rather than the atmosphere?

21–7. Which planets are likely to have their original atmospheres? Explain.

The Sun

21–8. Discuss the nature of sunspots and the chronology of a sunspot cycle. What influence, if any, has a sunspot cycle on human affairs?

21–9. What makes the sun shine?

21–10. On a sketch of the sun, show the locations of the chromosphere, the corona, and the photosphere.

21–11. Why didn't the planets become stars, assuming that they condensed out of the same primeval nebula as the sun?

Our Moon

21–12. What information has been obtained about the moon, first by means of improved instrumentation such as the telescope, camera, and lunar orbiter, and finally through astronaut landings, that was not available previously?

21–13. In your opinion, has the price tag in the billion-dollar range for lunar exploration been worth the return?

21–14. Why are the "seas," "oceans," and "bays" on the moon misnomers?

21–15. Discuss the structure and possible origin of Saturn's rings and their relation to the structure and origin of the solar system.

21–16. What can studying rocks from the moon tell us about the early history of the earth?

21–17. Why can we never see one hemisphere of the moon from the earth?

The Terrestrial Planets

21–18. List several differences between the terrestrial planets and the Jovian planets.

21–19. Why do radar observations of Venus provide more data about the surface structure than a flyby with close-up camera?

21–20. Why do we say that Venus is our "sister planet"?

21–21. Why does Mars vary greatly in brightness as seen from the earth?

21–22. What evidence is there that there is, or has been, water on Mars?

21–23. Compare the temperature ranges on Mercury, Venus, earth, and Mars.

21–24. How might the atmosphere of Venus have evolved through a runaway greenhouse effect?

21–25. Do radar observations of Venus study the surface or the clouds? Explain.

The Major Planets

21–26. Discuss the structure and possible origin of Saturn's rings and their relation to the structure and origin of the solar system.

21–27. Of all the natural moons, what is unique about Saturn's Titan that makes it of particular interest?

21–28. Is Jupiter more like a star than a planet? Explain.

21–29. What distinction does Titan hold among the moons in the solar system?

21–30. What is strange about the direction in which Uranus rotates?

21–31. What fraction of its orbit has Neptune traversed since it was discovered?

21–32. What evidence suggests that Pluto is not a "normal" planet?

21–33. Why are the moons of the giant planets more appealing for direct exploration than the planets themselves?

Asteroids, Meteorites, and Comets

21–34. What significance may meteorites have concerning the composition of the earth's interior?

21–35. What causes light to be emitted by a meteorite?

21–36. Discuss the structure of a comet.

21–37. What is the connection between meteorites and asteroids?

21–38. Discuss the changes that occur as a comet approaches the sun.

21–39. Was the 1986 passage of Halley's comet as spectacular as anticipated? Explain.

The Search for Extraterrestrial Life

21–40. What are the prospects for the existence of life elsewhere in the solar system?

21–41. Discuss the value of supporting exobiology.

21–42. How probable is life elsewhere in the Milky Way Galaxy?

21–43. Do you think there is extraterrestrial intelligence?

21–44. What would it matter to humanity if the existence of other technological civilizations was definitely proved or disproved?

21–45. Is space travel to other solar systems a realistic goal?

21–46. How might communication between worlds occur?

21–47. Do you think it likely that human beings will ever colonize other planets?

21–48. *Multiple Choice*
   A. Sunspots appear dark because they are
      (a) hotter than the surrounding surface.
      (b) cooler than the surrounding surface.
      (c) flares.
      (d) coronal streamers.
   B. Galileo discovered all but
      (a) sunspots.
      (b) the phases of Venus.
      (c) the moons of Mars.
      (d) the moons of Jupiter.
   C. The lunar maria are presumed younger than the highlands because
      (a) they are lighter.
      (b) they are darker.
      (c) they have water.
      (d) they have fewer craters.
   D. The greenhouse effect heats a planet because
      (a) infrared radiation is trapped.
      (b) more sunlight gets in.
      (c) the winds don't blow too fast.
      (d) it removes CO_2 from the atmosphere.
   E. The Great Red Spot of Jupiter is probably
      (a) a continent.
      (b) a storm.
      (c) a hole in the clouds.
      (d) a thick cloud layer.

SUGGESTIONS FOR FURTHER READING

Arvidson, Raymond E., Alan B. Binder, and Kenneth L. Jones, "The Surface of Mars." *Scientific American*, March, 1978.
The *Viking* spacecraft have viewed Mars from the ground and from orbit, adding evidence of how it has been shaped by volcano, meteorite impact, water, and wind.

Balsiger, Hans, Hugo Fechtig, and Johannes Geiss, "A Close Look at Halley's Comet." *Scientific American*, September, 1988.
A spectacular rendezvous occurred in March, 1986. The Japanese probes *Sakigake* and *Suisei*, the Soviet probes *Vega-1* and *Vega-2*, and the European probe *Giotto* encountered Halley's comet to analyze for the first time the gases and dust in the immediate vicinity of a comet and to photograph its nucleus.

Binzel, Richard P., "Pluto." *Scientific American*, June, 1990.
Pluto is so small and distant that it appears as a feature-less blob even through the largest earth-based telescopes. Although it is the only planet that has not yet been visited by a scientific spacecraft, a new picture of Pluto has begun to emerge during the past decade.

Binzel, Richard P., M. Antonietta Barucci, and Marcello Fulchignoni, "The Origins of the Asteroids." *Scientific American*, October, 1991.
The spacecraft *Galileo,* bound for Jupiter, was scheduled to investigate the asteroid 951 Gaspra in October, 1991. The current view is that asteroids are remnants of a planet that failed to form and, as such, provide clues to conditions in the early solar system.

Black, David C., "Worlds Around Other Stars." *Scientific American*, January, 1991.
A number of astronomers have possibly detected planets orbiting some nearby stars, although all the sightings are highly tentative. A new generation of detectors and telescopes should improve the situation.

Bork, Alfred, "Newton and Comets." *American Journal of Physics*, December, 1987.
Newton's correspondence was much concerned with comets in the period just before 1687, when two bright comets were seen. These comets seem to have been a major stimulation to Newton's work on mechanics.

Brandt, John C., and Malcolm B. Niedner, Jr., "The Structure of Comet Tails." *Scientific American*, January, 1986.
The years 1985–86 may one day be regarded as a golden age for cometary astronomy. Observations of comets Giacobini-Zinner and Halley may help to clarify the origin, composition, and dynamics of comets and their tails.

Cuzzi, Jeffrey N., and Larry W. Esposito, "The Rings of Uranus." *Scientific American*, July, 1987.
The *Voyager 2* encounter with Uranus in January, 1986, determined precisely the location of the known rings, searched for new rings, observed the size and distribution of particles in the rings, and searched for small ring moons orbiting between the rings.

El-Baz, Farouk, "Naming Moon's Features Created 'Ocean of Storms.'" *Smithsonian*, January, 1979.
A glimpse of a Space Age twist in a centuries-old problem: classifying and naming the features of the moon.

Foukel, Peter V., "The Variable Sun." *Scientific American*, February, 1990.

The sun's output of radiation and particles varies. Investigators are working to determine more accurately the connections between conditions on the sun and those on earth—auroras, power lines, the ozone layer, and climate, for example.

Grossman, Lawrence, "The Most Primitive Objects in the Solar System." *Scientific American*, February, 1975.

There is evidence that the meteorites known as carbonaceous chondrites are mixtures of minerals that have survived unaltered since they condensed out of the nebula that gave rise to the sun and planets.

Haberle, Robert M., "The Climate of Mars." *Scientific American*, May, 1986.

It started out much like the earth's early climate, but it evolved differently. Once warm enough to support flowing water, Mars is now so cold that carbon dioxide freezes at the poles every winter.

Ingersoll, Andrew P., "Uranus." *Scientific American*, January, 1987.

The giant blue-green planet was visited by *Voyager 2* in January, 1986. It has one pole pointing toward the sun, an atmosphere that is dense and icy, and winds that resemble the earth's.

Kuiper, Thomas B. H., "Resource Letter ETC-1: Extraterrestrial Civilization." *American Journal of Physics*, February, 1989.

A guide to the literature of the possibility of intelligent life elsewhere in the universe. In addition to popularity, the subject has gained respectability in recent years.

Miner, Ellis D., "*Voyager 2's* Encounter with the Gas Giants." *Physics Today*, July, 1990.

Voyager's epic journey to Jupiter, Saturn, Uranus, and Neptune is one of humanity's most productive exploratory missions. Barring unforeseen catastrophic failures, *Voyager 1* and *Voyager 2* should continue to collect field, particle, and wave data until at least the year 2015.

Murray, Bruce C., "Mars from Mariner 9." *Scientific American*, January, 1973.

Mariner 9 provided about 100 times the amount of information accumulated by all previous flights to Mars, decisively changing our view of the planet.

Owen, Tobias, "Titan." *Scientific American*, February, 1982.

Titan is Saturn's largest moon and the only moon in the solar system with a substantial atmosphere. It may resemble the earth's atmosphere before life arose.

Pasachoff, Jay M., "The Solar Corona." *Scientific American*, October, 1973.

Until this century, the nature of the corona was a mystery. Recently, some of the mystery has begun to disappear, thanks to telescopes, computers, and satellites.

Pollack, James B., and Jeffrey N. Cuzzi, "Rings in the Solar System." *Scientific American*, November, 1981.

The rings of Saturn have a wealth of detail—bands, spokes, and braids. It emerges that Jupiter also has a ring, and Uranus has at least nine discrete rings.

Runcorn, S. K., "The Moon's Ancient Magnetism." *Scientific American*, December, 1987.

The moon is now a dead body, but it seems once to have generated its own magnetic field and even to have had a system of satellites early in its history.

Russell, Dale A., "The Mass Extinctions of the Late Mesozoic." *Scientific American*, January, 1982.

The dinosaurs and many other species of plants and animals suddenly died out about 63 million years ago. The cause may have been the fall of an asteroid.

Saunders, R. Stephen, "The Surface of Venus," *Scientific American*, December, 1990.

The silvery clouds that make Venus such a lovely sight in our skies also completely obscure the planet itself. NASA's *Magellan* spacecraft, with its sophisticated synthetic aperture radar, mapped virtually the entire surface of Venus much as *Mariner 9* mapped Mars 20 years earlier.

Schultz, Peter H., "Polar Wandering on Mars." *Scientific American*, December, 1985.

Regions at the planet's equator seem once to have been near a pole. Many puzzling features and processes may be explained by this theory.

Soderblom, Laurence A., and Torrence V. Johnson, "The Moons of Saturn." *Scientific American*, January, 1982.

Before November, 1980, the moons of Saturn that were known were no more than dots of light in a telescope. Now, thanks to the spacecraft *Voyager 1* and *Voyager 2*, the moons of Saturn have been transformed into an array amounting to 17 new worlds.

Stephenson, F. Richard, "Historical Eclipses." *Scientific American*, October, 1982.

Astronomers, historians, and poets have been recording eclipses of the sun and moon for more than 2500 years. Their data may make significant contributions to a solution of problems in geophysics.

Whipple, Fred L., "The Spin of Comets." *Scientific American*, March, 1980.

Like an atom, a comet has a nucleus. Just as atomic nuclei spin and gyrate when subjected to a magnetic field, so do comet nuclei have rotation periods and a spin axis that gyrates.

Mt. Hood, Oregon, in the spring as seen from Bold Mt. (Craig Tuttle/The Stock Market)

IV

EARTH SCIENCE: GEOLOGY, OCEANOGRAPHY, AND METEOROLOGY

PART-OPENING ESSAY

PERSPECTIVE ON EARTH SCIENCE

At first glance, the earth sciences seem the most tangible of all the physical sciences. After all, you can reach down and grab a handful of earth to examine as you please. You can also collect fossils, study rocks under microscopes, measure earthquake faults, and stomp along glaciers or the slopes of a volcano. These explorations involve things you can see and touch, unlike the forces and particles of physics, the arrangements of invisible atoms in chemistry, and astronomy's waves of light stretched thin across the universe.

But earth scientists face as many intellectual challenges as their collagues in kindred disciplines. Will a slow buildup of gases from human industry drastically alter the planet's climate? Can clouds and winds tell us what the weather will be like a month from now? Why do some creatures thrive in the intense pressures of the ocean's black depths? And how can one possibly learn about the core of the earth, a hellish environment shielded from us by thousands of miles of rock?

These questions make it clear that the modern earth sciences embrace far more than classical geology. They concern all of the planet's workings, from atmosphere to ocean, from land to earth's inner fires. There is a good reason for this: Scientists now believe that one cannot understand any part of the earth in isolation. Its components are inextricably linked in a dynamic earth "system"—an intricate web of cause and effect. Some examples of current research will illustrate this claim.

The "greenhouse effect," a topic much in the news, concerns the growing levels of carbon dioxide (CO_2) and other gases in the atmosphere. Molecules of CO_2 and water vapor act like the windows of a greenhouse, letting sunlight warm the planet's surface but preventing some heat from radiating back into space. Earth's natural amounts of CO_2 and water vapor create a comfortable greenhouse, without which the planet would be far colder and unsuitable for life. However, humans are causing problems. When burned, fossil fuels such as oil and coal—the driving forces of civilization—spew CO_2 into the air. Other products of industry, such as the chlorofluorocarbons that erode the ozone layer, also intensify the greenhouse effect.

Predicting the consequences of this requires input from a broad range of scientists. Oceanographers know that the world's oceans absorb CO_2 via chemical reactions and biological use by marine plants. But it is not clear whether the oceans can absorb enough extra CO_2 to make a difference. Meteorologists make computer models of how the greenhouse effect might alter the planet's clouds. If more clouds form, they may reflect sunlight into space and prevent a dramatic warming. Geologists investigate what the earth's atmosphere and temperatures were like in the past by extracting cores of ice from the thick polar icepacks; marine scientists can read the same information in the shells of small animals buried in ocean sediments. They have found evidence of natural long-term cycles in temperature and CO_2 levels. Ominously, however, the concentration of CO_2 is building more rapidly today than ever before.

It is not surprising that scientists disagree on how to sort all this out. Some feel that the earth's average temperature will climb a few degrees in the next several decades. That could melt enough polar ice to raise sea level a few feet—a potential disaster for coastal cities and low-lying countries. Others think it is premature to sound the alarm, because a definite increase in temperature is not yet apparent.

A similar mixture of disciplines emerges when scientists consider the earth's interior. The deepest drill holes penetrate just a few miles, which does not help if you want to know what lurks hundreds or thousands of miles beneath your feet. So scientists must use seismic waves from earthquakes as "X-rays" of the planet's hidden structures. Just as a light beam bends when it passes from air into water, seismic waves speed up, slow down, and distort in earth's various layers. Large earthquakes may cause the most damage, but because they vibrate the entire planet they are also the most valuable tools for researchers. By compiling thousands of such records, geophysicists have created detailed three-dimensional maps of the earth. (One of the few benefits of underground nuclear explosions is that they also are useful as seismic probes.)

These maps are informative, but they are static "snapshots." We now know that the planet's innards are anything but static. Giant cells of pliable rock churn through the thick mantle, like boiling water rising and falling in a spaghetti pot. Plumes of heat from the earth's core may power these motions. In turn, the mantle drives the wanderings of continents across the surface, causing earthquakes when fragments of the earth's crust grind past one another. An exciting new approach is giving earth scientists a way to explore these internal seethings: mineral physics.

Experiments in mineral physics occur entirely in the laboratory. The main tool is the "diamond-anvil cell," a high-tech vise containing two small diamonds. Researchers insert minerals between the flat faces of the diamonds and squeeze them together, creating extremely high pressures. Then they scorch the minerals with a laser beam, mimicking the brutal combination of temperature and pressure that makes materials act strangely in the earth's mantle or core. The results offer glimpses on a tiny scale of the earth's innermost secrets.

These pursuits may show you why our home planet inspires a particular scientific passion. We want to understand our planet so we feel more secure about our place in the universe, and we want to know enough about it so we can help it survive the slings and arrows of misfortune cast by our species. But this passion can easily turn to dogma. Throughout history, scientists and society have harshly greeted new theories about the earth and have accepted them only with reluctance: It is spherical, not flat; it revolves around the sun, not vice versa; its continents drift and crash rather than stand still. The Gaia hypothesis, a fascinating proposal of today, is thus far suffering the same skepticism and condemnation.

Named for the Greek goddess of the earth, the Gaia hypothesis maintains that living things reworked the earth's hostile atmosphere over billions of years and now actively keep the planet fit for life. If the planet gets too hot, marine organisms may release a gas that increases cloud cover. If the planet gets too cold, plants may grow darker on average to retain more heat. These changes would happen over long periods of time, says James Lovelock, creator of this view. However, they would happen quickly enough to regulate changes in the planet's climate. Lovelock, in fact, envisions Gaia as the largest living organism in the solar system. This is a radical alternative to conventional wisdom, which holds that life has adapted simply as the planet's physical evolution has taken its own path.

We may never know whether Lovelock is our era's Galileo or a misguided romanticist. But a healthy scientific debate of his thoughts, rather than callous dismissal, might expose entirely new aspects of life's relationship with its delicate habitat. Perhaps we would realize, as we carefully considered the causes of one ecological tragedy after another, that Gaia is trying to tell us something.

Lightning over London. (Zefa/The Stock Market)

The Grand Canyon, Hopi Point, at sunrise as seen from Powell Point, South Rim, Grand Canyon National Park, Arizona. (© Jeff Gnass/The Stock Market)

22

THE EARTH'S CRUST AND INTERIOR— INVITATION TO EARTH SCIENCES

Recent discoveries have changed the way in which most geologists view the evolution of the earth's surface. The central idea of a major new theory— that the earth's surface consists of a small number of rigid plates in motion relative to each other—has gained acceptance. According to *plate tectonics*, as the new theory is called, plate motions are responsible for the present positions of the continents, for the formation of many mountain ranges, and for essentially all major earthquakes. Data gathered over the past 30 years in marine geology—the sounding of the deep ocean, the samples and photographs of the ocean floor, the measurements of heat flow and magnetism—are being reinterpreted according to the new concepts.

Plate tectonic theory has unified widely separated fields in the earth sciences, from volcanology and seismology to sedimentary geology. The forces that cause the plates to move, however, are not yet understood. The main difficulty is a lack of knowledge about the earth's mantle, the shell of rock separating the core from the thin surface crust, where the driving mechanisms are believed to originate. Nevertheless, many long-standing problems within the field of geology are being reconsidered in the light of this revolution in fundamental concepts.

22–1 STRUCTURE OF THE SOLID EARTH

It seems paradoxical that at a time when the universe is yielding so many of its secrets to boundless human curiosity, knowledge of our own earth remains inaccessible. For example, the world's deepest mine (in South Africa) goes down only about 3.4 km. The deepest hole ever drilled in the earth (in a Texas oil field) was 7.7 km, about one tenth of 1% of the earth's radius (6400 km). Although the mantle comprises about 67% of the total mass of the earth, no one has yet obtained an uncontaminated sample of it.

Nonetheless, the basic divisions of the earth have been known for some time (Table 22–1). From **earthquake (seismic) waves**, a series of vibrations or shock waves set in motion by the sudden movement of crusted blocks along a fault, geophysicists have inferred that the solid earth consists of three principal concentric zones (Fig. 22–1): (1) a thin **crust** varying from 5 km in ocean basins to 50 km in thickness beneath mountain ranges; (2) a **mantle** about 2900 km thick; and (3) a **core** about 3400 km thick that accounts for more than half of the earth's radius and nearly a third of its mass.

The mantle, in turn, is subdivided into an upper mantle, a transition zone, and a lower mantle, and the core is divided into a liquid outer core and a solid inner core (Fig. 22–2). The crust and part of the upper mantle, a zone 50 to 200 km thick, together form the **lithosphere**, which is thicker beneath continents than beneath oceans. The lithosphere, in turn,

Table 22–1   THE EARTH'S STRUCTURE

	Distance from Surface of the Earth (km)	Thickness (km)	Density (g/cm³)
Crust	0–60		
Continental		32	2.8
Oceanic		5	3.0
Upper mantle	400		
Transition zone	1000		
Lower mantle	2900	2900	3.3–5.6
Outer core	5000	2100	9.5–12
Inner core	6371	1400	13 +

overlies the **asthenosphere**, a plastic layer 100 km thick. A partially molten zone separates the lithosphere from the asthenosphere. The percentage of molten material is very small, probably less than 2%. This is enough, however, to reduce the strength of rocks in the zone and allow the lithosphere to slide over the asthenosphere. Directly beneath the asthenosphere lies the **mesosphere**, a zone where pressures are sufficiently great to impart greater strength and rigidity to the rock.

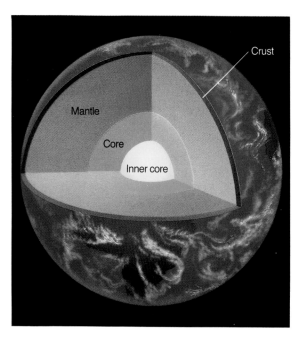

Figure 22–1   The structure of the earth's interior. The major zones are the crust, the mantle, and the core.

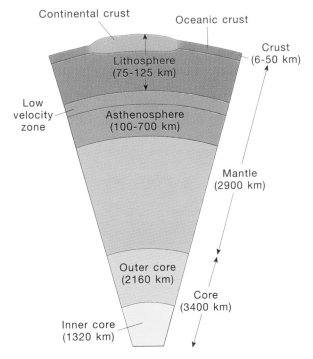

Figure 22–2   Details of the structure of the earth's interior.

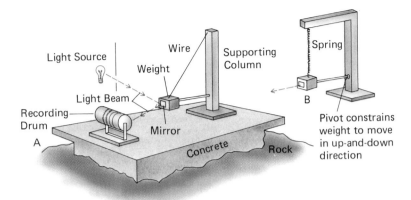

Figure 22–3 A pendulum seismograph. Earth motions are greatly amplified by the use of accessories. Horizontal (a) and vertical (b) earth motion can be recorded.

22–2 SEISMIC WAVES

More has been learned about the earth's interior by studying seismic waves than by any other means. The velocities of earthquake waves and of waves produced by controlled underground explosions depend on the rigidity and density of materials, which in turn depend on their chemical composition and structure. Depending on the medium through which they pass, the seismic waves are bent, speeded up, slowed down, and sometimes stopped altogether. By studying the transmission of waves through the earth, seismologists learn much about the physical properties of the interior.

The faint echoes emerging at the earth's surface actuate the sensitive pendulum recorders known as **seismographs** (Fig. 22–3). When the earth shakes, the recording drum and support of the inertial unit move relative to the weight. The relative motion is shown on the drum. Modern seismographs are far more complex than the simple instrument shown. They employ electronic devices to pick up, amplify, and record the motions of the earth, and digital processing greatly speeds analysis and interpretation (Fig. 22–4).

The major recent improvement in seismology has been the deployment of seismographic arrays in various localities, such as the Large Aperture Seismic Array in Montana, which covers 200 km and contains 525 seismographs. **Seismic arrays** have become remarkably effective probes for studying the earth's interior. Just as radio telescopes have revealed celestial objects that were once invisible, the new generation

Figure 22–4 A modern seismograph. (© Mark E. Gibson/The Stock Market)

of seismic instruments has detected fine details of the earth's structure that were once unobservable. Considerable refinement has also been possible with the advent of the worldwide network of standardized seismic stations recording the effects of large underground explosions. The records of seismographs are known as **seismograms**.

Four basic types of seismic waves are analyzed—two kinds of **body waves (P,S)** and two corresponding **surface waves (L, R)**. The first signal on a conventional seismograph is usually due to a fast-moving **P (for primary) wave**. The P waves are compressional or longitudinal waves (like sound waves; see Chapter 9) in which the displacement of the particles in the ground is along the waves' direction of travel. They are the fastest of the seismic waves, traveling at about 6.0 to 6.7 km/s in the crust and 8.0 to 8.5 km/s in the upper mantle.

The second type of body waves is called **S (for secondary)** waves. They are transverse waves (like light waves; see Chapter 10) in which the direction of

ground motion is perpendicular to the direction of travel. The velocity of S waves is lower than that of P waves: 3.0 to 4.0 km/s in the crust and 4.4 to 4.6 km/s in the upper mantle. Unlike P waves, S waves cannot be transmitted through liquids. Since S waves do not travel through the outer core, this region is believed to be molten (Fig. 22–5).

The two types of surface waves are **L (Love)** and **R (Rayleigh)** waves (Fig. 22–6). They travel near the earth's surface at speeds of about 3.5 to 5.2 km/s but are not confined to the crust. L waves cause sideways motion along the surface; R waves cause the particles to execute elliptical motion. (It has been found that the ratio of body-wave magnitude to surface-wave magnitude can be used to distinguish an earthquake from an explosion, such as a nuclear detonation, eliminating in principle the need for on-site inspection to monitor underground testing of nuclear weapons.)

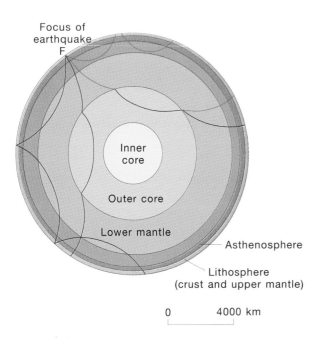

Figure 22–5   Effect of the earth's core on seismic waves. The core bends and concentrates the P waves (black) but does not transmit S waves (red). Since a liquid medium would have such an effect, the outer core is believed to be molten. (From U.S. Geological Survey publication, *The Interior of the Earth*)

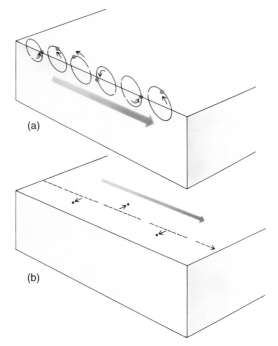

Figure 22–6   Surface seismic waves travel through the outer crust of the Earth. R (Rayleigh) surface waves (a) have an elliptical motion that is opposite to that in propagation. L (Love) surface waves (b) vibrate horizontally and perpendicular to the direction of wave propagation.

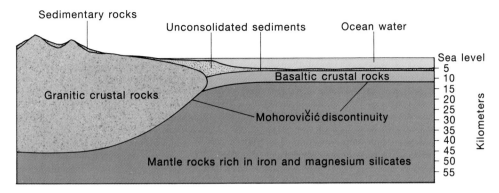

Figure 22–7   A cross-sectional view of the earth's crust. The oceanic and continental portions vary in thickness and composition.

22–3 THE CRUST: ROCKS AND MINERALS

The part of the solid earth with which we are most familiar, the crust, constitutes less than 1% of the earth's total mass. It consists of material with an average density of only about 2.8 g/cm³, compared with 4.5 g/cm³ for the mantle and 11.0 g/cm³ for the core (see Table 22–1). Its overall average thickness is 17 km, but under the oceans it averages 5 km, under the continents about 35 km, and under some mountains, 60 km (Fig. 22–7). Both oceanic and continental crust "float" in mantle rocks in much the same way that icebergs float in water. Continents stand higher because they are thick and are composed of relatively low density rocks. The thinner, heavier oceanic crust does not stand as high. Table 22–2 gives the ten most abundant elements of the earth's crust, comprising more than 99% by weight.

The crust contains a great variety of **rocks**, heterogeneous aggregates of minerals in varying proportions. A **mineral** is a naturally occurring inorganic homogeneous substance with a definite chemical composition and crystal structure and characteristic physical properties, such as a particular color, luster, hardness, and density (Figs. 22–8 through 22–11). Some common rock-forming minerals and their

Table 22–2   THE TEN MOST ABUNDANT ELEMENTS OF THE EARTH'S CRUST

Element	Weight (%)
Oxygen	46.60
Silicon	27.72
Aluminum	8.13
Iron	5.00
Calcium	3.63
Sodium	2.83
Potassium	2.59
Magnesium	2.09
Titanium	0.44
Hydrogen	0.14

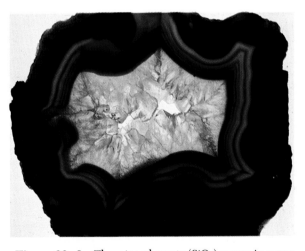

Figure 22–8   The mineral quartz (SiO_2) occurs in many families of rocks. Agate, a variety that exhibits distinct banding, is shown here with crystalline quartz in the center.

chemical composition are listed in Table 22–3. A solid rock layer is called **bedrock**; a layer of decomposed rock, called **regolith**, is usually above the bedrock; and above this may be a soil. Rocks are subdivided into three categories, depending on their origin: *igneous*, *sedimentary*, and *metamorphic* (Fig. 22–12).

Figure 22–9 Calcite ($CaCO_3$) is the main constituent of limestone and marble. (Courtesy of Wards Natural Science Establishment, Inc., Rochester, N.Y.)

Figure 22–10 A potassium feldspar [$K(AlSi_3O_8)$]. The feldspars are the most abundant constituents of rocks and compose about 60% of the total weight of the earth's crust.

Figure 22–11 Biotite mica [$K(Mg,Fe)_3(AlSi_3O_{10})(OH)_2$]. The micas are common constituents of igneous and metamorphic rocks.

Table 22–3 COMMON ROCK-FORMING MINERALS

Chemical Group	Mineral	Chemical Composition
Carbonates	Calcite	$CaCO_3$
	Dolomite	$CaMg(CO_3)_2$
Halides	Fluorite	CaF_2
	Halite	$NaCl$
Oxides	Hematite	Fe_2O_3
	Magnetite	Fe_3O_4
	Quartz	SiO_2
Sulfates	Gypsum	$Ca(SO_4) \cdot 2H_2O$
Sulfides	Galena	PbS
	Pyrite	FeS_2
	Chalcopyrite	$CuFeS_2$
Silicates	Quartz	SiO_2
	Olivine	$(Mg,Fe)_2SiO_4$
Amphiboles	Hornblende	$Ca_4Na_2(MgFe)_8$ $(AlFe)_2(Al_4Si_{12}O_{44})$ $(OHF)_4$
Feldspars	Albite	$Na(AlSi_3O_8)$
	Anorthite	$Ca(Al_2Si_2O_8)$
	Microcline	$K(AlSi_3O_8)$
	Orthoclase	$K(AlSi_3O_8)$
	Plagioclase	Mixture of albite and anorthite
Micas	Biotite	$K(Mg,Fe)_3(AlSi_3O_{10})(OH)_2$
	Chlorite	$(MgFeAl)_6(AlSi_4O_{10})(OH)_8$
	Kaolinite	$Al_2(Si_2O_5)(OH)_4$
	Muscovite	$KAl_2(AlSi_3O_{10})(OH)_2$
	Talc	$Mg_3(Si_4O_{10})(OH)_2$
Pyroxenes	Augite	$Ca(MgFeAl)(AlSi)_2O_6$

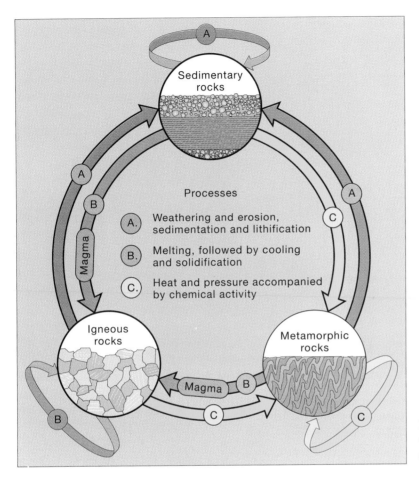

Figure 22–12 The rock cycle. Geological processes change one type of rock into another.

Igneous rocks are by far the most commonly found rocks in the earth's crust. Among them are granite, basalt, gabbro, and pumice. They have been formed from molten matter (*ignis*, meaning "fire"), possibly when magma (fluid rock beneath the earth's surface) poured forth from a volcano as lava and then solidified. Such rocks are called **extrusive igneous rocks** (Fig. 22–13). Or they may have solidified within the earth's interior through the process of crystallization, without breaking through to the surface, with widely differing chemical compositions. Such rock is known as **intrusive igneous rock** (Fig. 22–14). Among the most common minerals in continental igneous rocks are quartz, feldspars, and micas. Pyroxenes and amphiboles are also quite common.

Sandstone, siltstone, and shale are examples of **sedimentary rocks**, products formed from the deposition of weathering and erosion of older rocks at the earth's surface. These sediments may be carried away

Figure 22–13 Basalt, an extrusive igneous rock, forms the huge hexagonally jointed columns of Devil's Tower National Monument, Wyoming. (R. Gabler)

Figure 22–14   Granite, a coarse-grained intrusive igneous rock, forms the terrain of the Sierra Nevada in California, including these cliffs in the Yosemite Valley. El Capitan is the cliff on the left. On the right one can see Bridal Veil Falls.

by water, wind, or ice and deposited elsewhere on land, a lake bottom, or the ocean floor (Fig. 22–15). Or the material may be dissolved in water and later precipitated elsewhere. Some organic sedimentary rocks, such as limestones, have been formed from organisms growing in lakes and seas. The loose material is eventually compressed into sedimentary rocks by the weight of the overlying mass; it may also be cemented by material dissolved and reprecipitated between the grains. Some of the common sedimentary materials are gypsum, calcite, halite, clays, quartz, and dolomite.

Metamorphic rocks are igneous, sedimentary, or metamorphic rocks that have been changed or recrystallized by enormous heat and pressure deep in the earth's crust, without melting. In this way, limestone is converted to marble, shale to slate, and sandstone to quartzite (Fig. 22–16). The resulting rock is usually harder and more compact and is more resistant to weathering than before.

Figure 22–15   The upper strata of the Grand Canyon, Arizona, consist of different types of sedimentary rocks laid down at various time periods. (R. Sager)

Figure 22–16 This pink marble from Georgia was derived from limestone. Marble may be snowy white in its purest form. Impurities in the limestone produce a beautiful variety of colors. (The 1.0-cm square is for scale.)

Figure 22–17 Olivine, an important mineral of the upper mantle, can be cut and polished into attractive gemstones called *peridots*.

22–4 THE MANTLE AND CORE

A sharp boundary between the **crust** and the mantle, discovered by Andrija Mohorovičić (1857–1936), a seismologist at the University of Zagreb, Yugoslavia, provides a means of defining the depth of the earth's crust. This boundary is known as the **Mohorovičić discontinuity**, or the Moho for short, and constitutes the transition between the crust and the mantle. The speed of P waves here increases abruptly from approximately 6.7 km/s, typical of the bottom of the lower crust, to about 8.0 km/s, characteristic of the top of the upper mantle. The sudden change in seismic velocity indicates that the material above the Moho is different from the material below.

It is generally accepted that the mantle is made up predominantly of denser silicate materials that are rich in magnesium and iron. By comparison, stony meteorites of the chondrite variety, which are believed to have been formed at about the same time as the earth, contain 80% to 90% silicates, mainly magnesium-iron silicates. Compounds of four elements —magnesium, iron, silicon, and oxygen—are believed to comprise more than 90% of the mantle. In the upper mantle, these elements are mainly in the form of silicates. The important minerals of the upper mantle are olivine, pyroxene, and garnet (Fig. 22–17).

The **mantle** constitutes about 85% of the earth's total volume. It consists of a relatively thin (50-km) high-velocity layer; a second layer, 100 km thick, the **asthenosphere**; a **transition zone**, 400 to 1000 km thick; and the **lower mantle**, 1000 to 2900 km thick. The transition region is characterized by several abrupt increases in velocity, which laboratory studies indicate can be attributed to changes in crystal structure induced by high pressures. The main component of the mantle, olivine, is transformed at pressures corresponding to a depth of about 400 km to a structure in which the atoms rearrange themselves into a tighter packing arrangement, thereby increasing the density of the material.

Like the crust, the mantle is believed to be solid, with the exception of isolated regions containing small pockets of magma. Large regions of the mantle may be very close to their melting point at the prevailing high temperatures and pressures, however, and may have a fluid-like, plastic character. The lower mantle is believed to be relatively homogeneous, consisting of mixtures of such high-density silicates and oxides as enstatite ($MgSiO_3$), stishovite (SiO_2), and periclase (MgO).

The **outer core** is believed to be liquid, since S waves cannot penetrate it. It has the properties of molten iron (80% to 85%) mixed with nickel, cobalt, sulfur, or silicon. The origin of the earth's magnetic field is believed to be in the outer core. Fluid motions in the electrical conducting core are thought to cause it to act like a dynamo. Although the **inner core** is believed to be solid, there is evidence from wave velocities suggesting that it is near the melting point or is partly molten. At the center of the earth, the density is estimated to be 12.86 g/cm^3, the pressure about 3.5 million atmospheres, and the temperature between 3500° and 6000°C.

22–5 VOLCANOES

The dynamic nature of the apparently solid earth is evident in the violent eruptions of volcanoes. More than 500 volcanoes have erupted in the past four centuries alone. The eruptions at the surface are only small manifestations of great events going on in the earth's interior. So far, there is only speculation about these events, for no single theory has yet embraced all of the details of volcanic behavior.

A volcanic eruption can be an awesome spectacle. Krakatoa, a small volcanic island in Indonesia, had been dormant for 200 years when it suddenly exploded in 1883. Detonations were heard in Australia, nearly 5000 km away, as the volcano ejected 17 cubic kilometers of rock and pumice into the air and de-

stroyed two thirds of the island. A catastrophic tidal wave that followed drowned 36,000 people on the adjacent coasts of Sumatra and Java. For years, the volcanic ash circled the earth, creating spectacular blood-red sunsets. One clue to the origin of volcanoes is their location. Active volcanoes are concentrated in parts of the world where earthquakes are most common (Fig. 22–18). Furthermore, volcanoes are commonly found in young mountain belts (although there are young mountains, such as the Himalayas, which do not have them).

Volcanic activity has been most widespread during the periods of adjustment that follow the formation of mountain ranges. Extremely hot material located tens of kilometers below the surface becomes liquefied if the pressure is reduced or the temperature rises. The pressure may be reduced by the bending or cracking of the rocks above, or the temperature may be increased by radioactive heating. The liquefied material, called magma, is lighter than the rocks lying above it and tends to rise wherever it finds an opening (Fig. 22–19). If it rises directly to the surface of the earth through fractures in the rock, it may come out quietly as fluid lava or as cinders, glowing ash, and

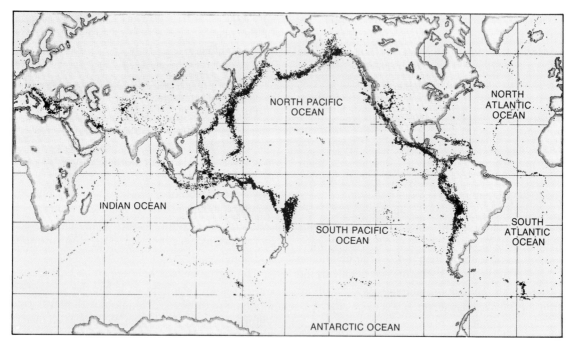

Figure 22–18 Distribution of earthquakes around the world. The major volcanic regions correlate strongly with earthquake belts.

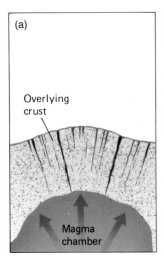 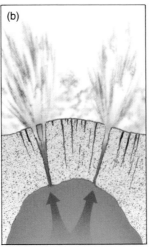 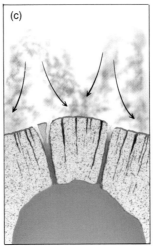 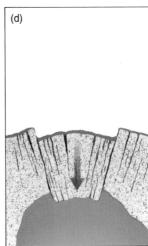

(a)

Overlying
crust

Magma
chamber

(b)

(c)

(d)

Figure 22–19   (a) Extremely hot material in the earth's interior may become liquefied as magma and rise toward the surface. (b) A column of magma, rock fragments,
and gas explodes upward through fractures in the rock. (c) When the gas is used up,
the vertical column expands outward as an ash. (d) The roof of the magma chamber
collapses downward to form a caldera.

partially solid plugs. In the event that it reaches a roof
of solid rock some distance below the surface, it may
spread sideways and eventually erupt into one or
more volcanoes where there are cracks in the roof
(Fig. 22–20).

Lava is the molten material with its associated
gases that flows out of volcanic vents and fissures; the
term is also applied to the rock that solidifies from this
material. **Magma** is the name for this material when it
occurs below the surface, where the gas content tends

Figure 22–20   The eruption of Galung-
gung volcano, West Java, Indonesia, December 3, 1982. (Photo by Ruska Hadian, volcanology survey of Indonesia; courtesy of U.S. Geological
Survey)

Figure 22–21   A lava flow from a basaltic eruption in Hawaii. The liquid red-orange lava cools and freezes into black igneous basaltic rock. (David Morrison)

Table 22–4   COMMON IGNEOUS ROCKS°

Coarsely Crystalline Intrusive Igneous Rocks	Approximate Extrusive Equivalents	Common Silicate Mineral Components	Average Specific Gravity

Granite	Rhyolite	Quartz
		Potash feldspar
		Sodium plagioclase
		(minor biotite,
		    amphibole,
		    magnetite)
			2.7

Diorite	Andesite	Sodium-calcium
		    plagioclase
		Amphibole, biotite
		    (minor pyroxene)
			2.8

Gabbro	Basalt	Calcium plagioclase
		Pyroxene
		Olivine
		    (minor amphibole,
		    ilmenite)
			3.0

° From Levin, H. L., *The Earth Through Time*, 3rd edition. Philadelphia: Saunders College Publishing, 1988, p. 36.

Figure 22–22 Mount Lassen, California, is an example of a volcano that extrudes extremely stiff lava, forming steep slopes of solid lava rock. It was last active between 1914 and 1921. (Philip Wallick/The Stock Market)

to be greater. One type of magma is represented by basalt—a dark, heavy, relatively fluid material rich in iron and magnesium minerals and low in silica (SiO_2) (Fig. 22–21). Basalt is typical of oceanic volcanoes and some continental volcanoes (Table 22–4). Rhyolite, a lighter material at the other end of the spectrum, is low in iron and magnesium minerals and high in silica; it is extrusive granite and relatively viscous. The more silica there is in magma, the more resistant it will be to flow. An increase in pressure also tends to increase viscosity. Heating has the reverse effect. A rise in temperature results in a decrease in the viscosity of the magma. Lavas of intermediate composition are typical of island arcs and young mountain ranges. Lavas range in temperature from 600° to 1200°C. Magma temperatures below the surface must be higher than those for surface lavas, because of the greater pressures. The basaltic magma probably forms because localized regions of the upper mantle

are somehow heated and melt. Most granitic magma may involve the melting of the lower part of the earth's continental crust.

Volcanoes are found in a great variety of shapes and sizes. Not all are explosive; some, like many of the volcanoes of Hawaii, are sluggish and gentle. Basaltic lavas are fluid and tend to flow; viscous "granitic" lavas tend to erupt explosively. When the flow of lava is rapid, the structure of the volcano tends to be flat, like an inverted saucer, and explosive eruptions are few and weak. If the lava is squeezed out in a very viscous condition, like toothpaste from a tube, and moves at little more than a snail's pace, a very steep-sided mountain, such as Lassen Peak in California, is produced (Fig. 22–22). Volcanoes such as Mt. Rainier in the United States and Fujiyama in Japan have a graceful profile, rising from a wide base to a tall, slender peak (Fig. 22–23). A million years may be required to build a composite volcano such as these,

Figure 22–23 Mount Rainier, seen here from Mirror Lake, Washington, is a composite volcano, as are most of the well-known volcanoes of the world. It is composed of both lava and ash. (Gerry Ellis/The Wildlife Collection)

involving in part the outpouring of lava and in part explosive discharges. The explosive, viscous varieties, on the other hand, may grow spectacularly. In just one year, the Mexican volcano of Paricutin, born in 1943 in a cornfield, grew to a height of 1200 feet.

The Cascade volcanoes of the Pacific Northwest are part of the so-called **Ring of Fire** that rims the Pacific from the tip of South America north to the Aleutians and Japan and down to New Zealand. A string of 15 major volcanoes runs from Mt. Garibaldi in British Columbia to Lassen Peak in northern California. Many Americans have assumed that most of these volcanoes are "dead." Although eight have erupted in the past 200 years, only Lassen Peak had erupted in the twentieth century (1915), and that was a relatively minor eruption.

In an effort to evaluate the hazards presented by the dormant volcanoes of the Cascade Range, Dwight Crandell and Donal Mullineaux of the U.S. Geological Survey studied each volcano, guided by the geologist's dictum that "what has happened before can happen again." Their 1978 report on Mt. St. Helens, a relatively little known volcano in southern Washington, indicated that it had been more active and more explosive during the last 4500 years than any other volcano in the contiguous 48 states (Fig. 22–24). Although it was dormant since 1857, they stated that

Mt. St. Helens would erupt again, perhaps before the end of this century.

The volcanic ranges of the Ring of Fire are thrust up at subduction zones where the moving tectonic plates of the Pacific plunge under the confining plates to the east, north, and west. The subduction of the Pacific plates provides the molten rock and pressure that powers the volcanoes. Unlike most of the volcanoes on the margins of the Pacific Basin, the Cascades are not fed magma by the Pacific or one of the other major plates. Instead, the small Juan de Fuca plate slowly bends downward, slips beneath the Washington coast, into the mantle, and melts enough to supply magma to the Cascade volcanoes.

The forecast concerning Mt. St. Helens was fulfilled on May 18, 1980, when the volcano erupted violently with a force equivalent to 10 megatons of TNT. The eruption caused the loss of nearly 100 lives, and destruction was estimated at over $2 billion. The death toll would have been much larger had warnings not been issued and had the eruption not occurred early on a Sunday morning. The explosion blew the top off Mt. St. Helens, transforming it from a symmetrical cone 9677 ft high to a flat top 1300 ft lower (Fig. 22–25). A number of further significant eruptions occurred on the volcano later. Shortly thereafter, Mt. Baker began steaming, and unusual seismic activity occurred near Mt. Hood and Mt. Shasta. Mt.

Figure 22–24   A view of Mount St. Helens in the Cascade Range of Washington, Spirit Lake, and surrounding forests prior to the eruption of 1980. (Weyerhaeuser Company)

Figure 22–25   Mount St. Helens erupted violently on May 18, 1980. The blast and landslide removed over one cubic mile (4.1 cubic kilometers) of material from the mountain's north slope, leaving a crater 1300 feet (400 meters) deep. (Bob Dieter/Weyerhaeuser Company)

Bailey became more active than it had been in the previous 100 years.

The May 18, 1980, eruption of Mt. St. Helens was one of the most closely monitored and most fully documented volcanic eruptions in history. By 1982, volcanologists were predicting moderate eruptions on the volcano quite accurately. The U.S. Geological Survey proposed to expand its monitoring of the Cascade Range volcanoes by emplacing seismometers, tiltmeters (highly sensitive levels, with associated electronic devices that record changes in slope that often precede an eruption as magma moves upward), and distance-measuring facilities on the more active peaks. Studies of earlier ashfalls were to be made, including carbon-14 dating to determine the frequency and characteristics of earlier eruptions. Although precise forecasts of eruptions are not possible, greater knowledge should lead to the saving of many lives.

Advance warnings of volcanic eruptions in Japan and the Philippines in 1991 did save many lives. On June 3, 1991, Mount Unzen, a volcano in southwestern Japan, about 49 miles east of Nagasaki on the island of Kyushu, erupted, killing 38 people. Many of those killed were journalists and volcanologists, drawn to the mountain by warnings of an impending eruption, who were caught in the red-hot avalanches hurtling down the mountain's slopes. Most residents of the area heeded the warnings and fled to safety. In 1792, the last time the volcano erupted, 15,000 people were killed by the landslides and tidal waves it created.

It has been estimated that volcanoes have killed 315 persons per year on average from 1600 to 1900. In the twentieth century the rate jumped to 845 per year. It is not that the number of eruptions is abnormal, but that the populations near the volcanoes have been growing rapidly.

Volcanologists are using various high-technology tools to make their predictions. Besides seismometers and tiltmeters, they can use chemical sensors, mounted on airplanes, that detect increases in sulfur dioxide emissions, indicating that magma has reached the surface. Laser-based devices can pick up minute bulges on a mountain slope that are about the width of a nickel and invisible to the naked eye but that are a sign of explosive potential. Video cameras monitor the shape and color of fumes at a number of volcanoes. Software packages for the acquisition and analysis of seismic data enable these cameras, with lap-top

computers, to monitor the creaking and bulging of the mountain as fresh magma wells up within it. Such monitoring has convinced some volcanologists that Mount Fuji has entered an active phase and that a giant eruption may occur on this volcano only 62 miles from Tokyo (Fig. 22–26). Modern instrumentation enables Mexico to monitor Popocatépetl near Mexico City. Of the 22 eruptions at Mount St. Helens since 1980, 19 were predicted by the U.S. Geological Survey.

A week after the eruption of Japan's Mount Unzen, another volcano erupted 1500 miles away on Luzon Island in the Philippines. After sleeping quietly for 635 years and appearing like a nice little hill covered with lush vegetation, Mount Pinatubo erupted in a series of explosions that shot plumes of steam and ash as high as 80,000 feet into the air. A giant mushroom cloud was visible 60 miles away in Manila. Nearly 500 Filipinos died, but several hundred thousand had been at grave risk before fleeing the area at the government's orders. Before the first blast, the U.S. Air Force evacuated most of the 16,000 Americans from nearby Clark Air Base to the Subic Bay Naval Station. When Subic was also pelted by stones from the volcano and shaken by earthquakes, many of the evacuees were sent home to the United States.

The response to the volcanic crisis of Mount Pinatubo became a historic case study marked by close international cooperation, high-technology gadgetry, and good relations between scientists and policy

Figure 22–26 Mount Fuji, Japan, may be entering an active phase. (Miko Yamashita/Woodfin Camp & Assoc.)

makers. The volcano itself cooperated, giving scientists time to put their technology to work. Pinatubo took more than two months to build to its climax, while some destructive volcanoes have reached theirs in a week. Once the volcanologists reached a consensus as to what the mountain was doing, they expressed their concerns to elected officials and the media. The officials translated the scientific concerns into an evacuation of more than 80,000 people, which stayed ahead of the threat of Pinatubo's eruption. In this case, the path was shown to reversing the trend toward higher volcanic death tolls.

22–6 CONTINENTAL DRIFT

Until recently, most geologists conceived the earth's crust as being a fairly stable layer enveloping the mantle and core. The only kind of motion perceived in this framework was the tendency of the crust as a rigid shell of varying rock density to float on a plastic layer of greater density. A scientific revolution involving a change in view from a static earth to a dynamic one, however, is profoundly changing our understanding of the earth. The concept of a static earth held that the continents and ocean basins were permanent features that had existed in their present form since early in the earth's history. The basis for a dynamic earth is the idea that continents are constantly moving and ocean basins are opening and closing.

The theory of **continental drift** proposes that the continents did not always exist where they are today but were part of a supercontinent, which the geophysicist Alfred L. Wegener (1880–1930) called **Pangaea** (all-earth) (Fig. 22–27). About 200 million to 250 million years ago, this single continent began splitting apart at what are now the rifts in the oceanic ridges, and the pieces (continents) drifted to their present locations. Until the 1960s, the idea that the continents had drifted apart was regarded with considerable skepticism. Since then, as a result of many new findings, the theory has gained much support. The evidence today favors the temporary formation of two large land masses: **Gondwanaland** in the Southern Hemisphere and **Laurasia** in the Northern Hemisphere.

The geological and paleontological evidence for the theory of continental drift is strong. Wegener himself said that he began taking the idea of drifting continents seriously after learning of the fossil evidence for a former land connection between Brazil

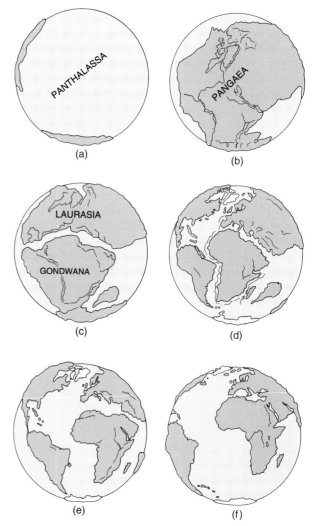

Figure 22–27   (a, b) Alfred L. Wegener's model of the earth about 250 million years ago has one ocean (Panthalassa) and one continent (Pangaea). (c) The continent split into a northern Laurasia and southern Gondwanaland. (d) Further division occurred, until the earth resembled (e). (f) Further widening of the Atlantic and northward migration of India bring the earth to its current state.

and Africa. The similarities and differences between fossils in various parts of the world not only support the continental drift concept but also provide a reasonably precise timetable for a number of the key events before and after the break-up. Following the separation of the continents, evolution proceeded on different paths, leading to the present biological diversity observed on the various continents.

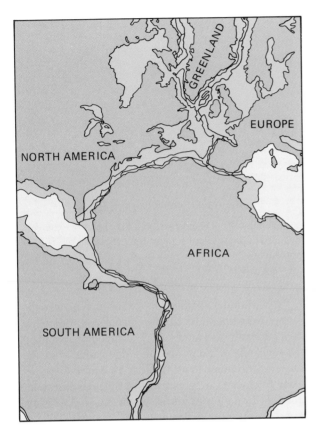

Figure 22-28 Fit of the continents. The fit between South America and Africa is excellent. (Adapted from E. C. Bullard, et al., *Philosophical Transactions of the Royal Society of London*, 258:41, 1965.)

On the geological side, Patrick M. Hurley (b. 1912) at the Massachusetts Institute of Technology studied rocks in Africa and Brazil that were 2 billion and 550 million years old. When the maps of Africa and South America were fitted together, the regions consisting of rocks of the same age appeared to be extensions of one another on both continents. An additional correlation in the southern hemisphere was found in a number of glaciations that left a distinctive record in the southern parts of South America, Africa, Australia, Antarctica, India, and Madagascar, which suggested that they were joined into a single unit—Gondwanaland—until about 200 million years ago, when it began to break up.

Many attempts have been made to guess precisely how South America, Africa, Australia, Antarctica, and India were once joined as Gondwanaland, but there is as yet no general agreement. The fit between the coast of South America and Africa is excellent, and the fit between Australia and Antarctica is good (Fig. 22–28). But the fit of all five major units is not settled, and the original position of Madagascar is unknown. Unlike the original Wegener idea that the continents alone are drifting across the ocean floor, the current view is that the continents, the ocean floor, and the upper part of the upper mantle are drifting.

22-7 SEA-FLOOR SPREADING

Opposition to the history of continental drift was centered on the belief that the earth's crust and mantle were too rigid to permit such motions. It seemed unlikely that a large continent could cross an entire ocean basin and leave no trace of its passage. On the other hand, the existence of great young mountain ranges, deep ocean trenches, and volcano and earthquake belts that in some cases are continuous over distances of several thousand kilometers indicates the large-scale motion of material forming the earth's crust. Furthermore, the relatively thin layers of sediment on the sea floor suggest that it has been regularly rejuvenated, and the oceanic lands and submerged volcanoes are all relatively young.

When oceanographers explored and mapped the oceanic ridges and rifts in meticulous detail, it became clear that the ocean floor was relatively young compared with the continents. Arthur Holmes (1890—1965) and Harry H. Hess (1906—1969) then proposed **sea-floor spreading** as a mechanism for continental drift. According to this hypothesis, the continents do not plough their way through the ocean floor but move with it, much like material on a conveyor belt. Molten rock from the mantle below rises up along a crack-like valley on the crest of the **mid-oceanic ridge**, a mountain range that runs for thousands of kilometers through the major ocean basins, then spreads outward and quickly hardens into new ocean floor (Fig. 22–29). In this way, the sea floor is continuously rejuvenated, sweeping along with it the layer of sedimentary material so that no part of it remains truly ancient. Elsewhere on the earth, the ocean floor plunges into a trench and then sinks deep into the mantle.

Support for the hypothesis of sea-floor spreading has come from the rocks of the ocean floor, which carry a permanent record of their own history. The oceanic crust is not uniformly magnetized. Magnetometer surveys of the ocean floor reveal a remark-

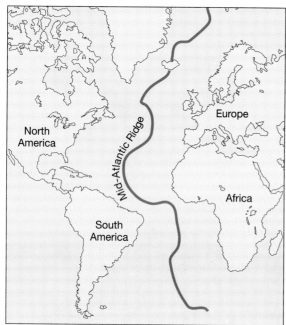

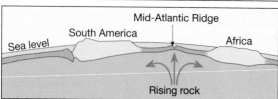

Figure 22–29   The Mid-Atlantic Ridge. This ridge is slowly expanding and pushing the continents apart.

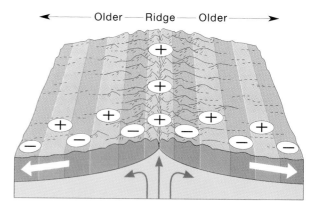

Figure 22–30   The normal $(+)$ and reversed $(-)$ magnetizations of the sea floor support the hypothesis of sea-floor spreading.

Verification of sea-floor spreading has changed the status of continental drift from speculative conjecture to a generally accepted fact. The speed of spreading on each side of a mid-oceanic ridge varies from less than a centimeter to as much as 18 centimeters per year. Such rates are geologically fast and can account for the formation of entire ocean floors. At 15 cm/ year, the floor of the Pacific Ocean (about 15,000 km wide) could be produced in 100 million years. The directions and rates of motion for both sea-floor spreading and continental drift are in complete agreement. It is now apparent that virtually all of the present area of the oceans has been created by sea-floor spreading during the past 200 million years.

22–8   PLATE TECTONICS

As a result of marine magnetic and earthquake studies, the theory of plate tectonics has gained increasing acceptance among geologists. **Plate tectonics** unites the concept of sea-floor spreading with the earlier idea of continental drift in a single unifying theory. It is believed that relative plate motions are responsible for the positions of the continents, for the formation of many mountain ranges, and for essentially all major earthquakes.

"Tectonics" is a geological term that refers to earth movements. Plate tectonics is the theory that the surface of the earth is divided into a number of large, rigid, plate-like regions that move with respect to one another. North America is on one such plate; Africa is on another (Fig. 22–31). The plates are believed to

ably regular pattern of symmetric magnetic strips on either side of the ocean ridges. Strips of material with normal polarization, with north pointing in the north direction, alternate with strips of reverse, or antinormal, polarization. The entire ocean floor is covered with long strips of alternating magnetic polarity, each strip at least 10 km wide (Fig. 22–30).

To account for the symmetrical pairs of parallel strips that have the same direction of magnetization, it has been proposed that when lava wells up in the ocean ridge, it is magnetized as it cools in the direction of the earth's magnetic field. The alternations of polarity indicate that the earth's magnetic field reverses at widely varying intervals, from 50,000 years to 20 million years. Unless this magnetism is disturbed by reheating or physical distortion, it becomes a permanent record of the direction and polarity of field at the time the rock was formed.

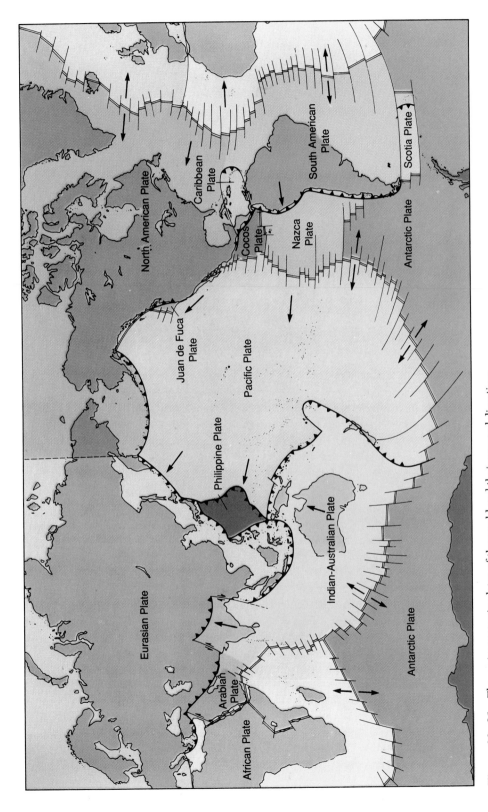

Figure 22–31 The major tectonic plates of the world and their general directions of movement. Plates often include both oceans and continents. Tectonic activity occurs along the plate boundaries where plates separate, collide, or slide past one another.

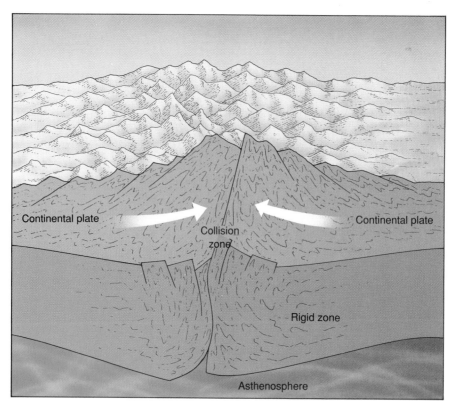

Figure 22–32   Convergent plate boundaries occur when two continental plates collide. Massive mountain building occurs as the crust is thickened. The Himalayas were formed when India collided with Eurasia.

be 50 to 100 km thick and to slide on the asthenosphere, the warmer, less rigid material of the upper mantle. The boundaries of the plates are not limited to continental boundaries; plates often include both oceans and continents.

Three types of **plate boundaries** are recognized: spreading or **divergent**, in which new crust is created as two adjacent plates move away from each other; **convergent**, when two plates converge and one plate overrides the other, pushing it into the mantle where it is heated and assimilated; and **transcurrent fault boundaries**, which occur when two adjacent plates slide past each other. Mid-oceanic ridges are the most common example of divergent boundaries. Trenches and young mountain ranges mark convergent zones (Fig. 22–32). Long linear faults, such as California's San Andreas fault, mark the contact where plates slide side by side (Fig. 22–33). In the San Andreas case, the edge of the Pacific plate is moving north

westerly at the rate of a few centimeters per year, carrying with it past the mainland a narrow strip of the California coast.

According to plate tectonics, the vast Himalayan range was created when a plate of the earth's crust carrying the land mass of India collided with the plate carrying Asia some 45 million years ago, having traveled 5000 km across what is now the Indian Ocean. The northern edge of the Indian plate crumpled up the many layers of shallow-water sediments laid down over millions of years on the continental shelf that bordered the southern edge of Asia, and formed the Himalayas.

It is believed that plate tectonics has been an active mechanism for much longer than 200 million years. During this time, virtually all the present oceans were created and others destroyed. Where the plates floated apart, the continents embedded in them also drifted away from one another. The processes we see

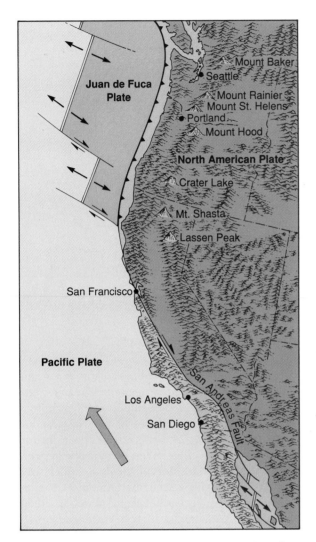

Figure 22–33 A transcurrent, or lateral, plate boundary occurs when two adjacent plates slip past each other. The San Andreas Fault system marks such a boundary. The Pacific plate to the west of this fault is moving northwestward at the edge of the North American plate, occasionally causing major earthquakes.

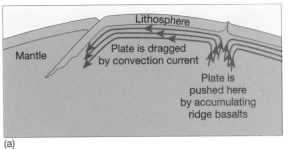

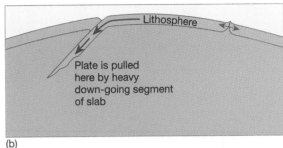

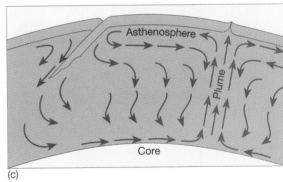

Figure 22–34 Three models of possible driving mechanisms for plate movements.

today may have been in action all through geological time; continents may have split many times and formed new oceans and may have collided and been welded together. Although the geometry and kinematics of plates are fairly well known, the dynamics —the driving mechanisms of the plate motions—is a subject of considerable research. Most models assume some form of thermal convection within the mantle, but there is as yet little quantitative evidence available on the composition and properties of the mantle (Fig. 22–34).

22–9 EARTHQUAKES

The great San Francisco earthquake of 1906 comes to mind when many Americans think of earthquakes. Yet three of the greatest earthquakes in American history occurred not on the Pacific coast but in New Madrid, Missouri, near a meander loop of the Mississippi River (Fig. 22–35). The dates were December 16, 1811, and January 23 and February 7, 1812. The effects were felt throughout the northeastern United States and Canada. Even today, the New Madrid re-

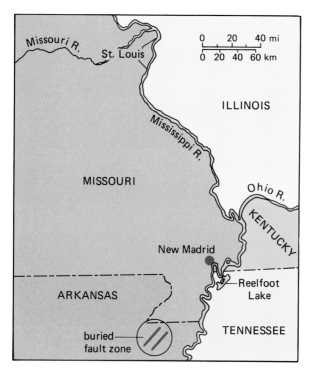

Figure 22–35 New Madrid, Missouri, is one of the most earthquake-prone regions in the United States.

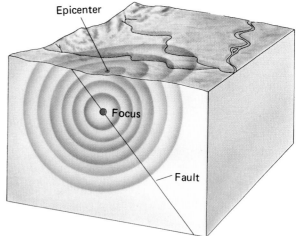

Figure 22–36 The focus of an earthquake is the point of initial movement. The epicenter is the point on the earth's surface directly above the focus and the point at which the strongest shock is normally felt.

gion experiences small earthquakes every 48 hours on the average.

Other major earthquakes in the central and eastern United States occurred in Charleston, South Carolina in 1886; off Cape Ann, Massachusetts, in 1638 and 1755; in the St. Lawrence Valley in 1663 and 1925; and in the Grand Banks off Nova Scotia in 1929. The point at which an earthquake originates is the **focus**. The strongest shock is normally felt at the earthquake **epicenter**, the point on the earth's surface that lies directly above the focus (Fig. 22–36).

Earthquakes are vibrations of the earth that occur when strain in the crust is suddenly released by displacement along a **fault**. They vary greatly in the **magnitude** of the energy released. The widely used **Richter scale**, devised in 1935 by Charles F. Richter of the California Institute of Technology, describes the amplitude of the earthquake waves and is related to the amount of energy released at the source. It is a measure of ground motion as recorded on seismographs. Magnitude on the Richter scale is proportional to the logarithm of the amplitude of a recorded seismic wave. On a scale from 1 to 10, a magnitude of

6 indicates a disturbance 10 times as large as a magnitude of 5. Earthquakes of magnitude 8 or larger thus are millions of times as energetic as one of magnitude 1. The estimated magnitudes of the New Madrid earthquakes are 8.6, 8.4, and 8.7, respectively. **Intensity** measures the effect of the earthquake on structures and people. The **Mercalli Scale** is a widely used standard for judging earthquake intensity.

The Good Friday earthquake of March 27, 1964, that struck Alaska had a Richter magnitude of 8.5. Releasing twice the energy of the great San Francisco earthquake, it was felt over an area of 500,000 square miles. Avalanches and landslides caused tremendous devastation, and **tsunamis**—seismic sea waves— were sent as far as California, Hawaii, and Japan.

The San Fernando earthquake of February 9, 1971, with a magnitude of 6.4 on the Richter scale, caused considerable damage in a densely populated area. Sixty-four persons lost their lives, and property damage was estimated at half a billion dollars. Highways, hospitals, old structures, freeways, reservoirs, bridges, utilities, and pipelines sustained severe damage. The potential for a catastrophe of major proportions when an earthquake occurs in an urban area has stimulated increased research into earthquake prediction and prevention. Two networks of modern seismic instruments that monitor the New Madrid zone 24 hours per day can detect and locate disturbances as small as magnitude 1.0. The siting and construction of nuclear power plants are determined in large part on the basis of the estimated earthquake hazard. For this reason, no nuclear plants have been built in the area of New Madrid.

The theory of plate tectonics explains earthquakes such as those along the San Andreas fault. The San Andreas is a **plate-boundary fault**, the boundary between the Pacific plate and the North American plate. The Pacific plate slides northwest along the San Andreas fault at approximately 5.5 centimeters per year. Where the movement is impeded, as it is in northern and southern California, the fault is locked and is storing up strain energy. Ultimately, this energy will be released. The seismic activity in the New Madrid region is more difficult to explain. There are no prominent surface traces; the faults are deeply buried under river sediments, and it is not a plate boundary. The earthquakes are of the type known as **midplate quakes**, disturbances that have little significance on a worldwide scale.

The most thoroughly observed moderate quake ever to occur in the United States struck near Coyote Lake on the Calaveras fault, a branch on the San Andreas, in August, 1979, with a magnitude of 5.7. A dense network of monitoring equipment was centered near Holister, 25 kilometers away. The data were disappointing to those who hoped to learn more about how to predict earthquakes. There were no obvious phenomena that could have allowed a prediction, such as clusters of small quakes, tilting of the earth's surface, or anomalies in the earth's magnetic field. Even the microearthquakes (magnitude less than 3) that formed before the magnitude 7.7 quake in Oaxaca, Mexico, in 1978 were lacking. The ability to predict earthquakes is no longer believed to be right around the corner.

Nonetheless, scientists warned in 1988 that Americans living east of the Rocky Mountains, the eastern two thirds of the nation, could experience a significant earthquake within 20 years. The probability of one with a reading of 6.5 and above occurring somewhere in the eastern United States before the year 2000 was considered better than 75% to 95%; before the year 2010, nearly 100%. Although few earthquakes have occurred in the eastern states during this century, they were relatively common in the seventeenth, eighteenth, and nineteenth centuries. Tension is likely building deep underground that could be released suddenly in a damaging tremor.

While California has adopted strict building codes to limit earthquake damage, no other area is so well prepared. A major quake in the East would affect more people and probably cause greater damage and injuries. The unreinforced highways and bridges could collapse, as would even solid houses of brick and stone. The infrastructure in the East and Midwest has never been tested by a major earthquake in recent times.

Unlike the western United States, where hot rock close to the surface provides a shock-absorbing cushion, the middle of the continental plate is cold, hard, and thick. A large quake in the Midwest or East would likely resonate far and wide, as did its precursors in the past century. The shock waves from the New Madrid tremors in the nineteenth century pulsed outward for hundreds of miles, wrecked boats in Charleston, South Carolina, cracked masonary in Cincinnati, and caused buildings to shake and church bells to peal as far away as New York City and Boston.

Two recent earthquakes focused considerable public attention on earthquake safety. The 6.9 Armenian earthquake of December, 1988, resulted in 25,000 deaths and half a million people left homeless. Economic losses approached $16 billion. A great many buildings were not seismically resistant, even though the area has been known from antiquity to be geologically unstable.

The Loma Prieta, California, earthquake of October 17, 1989—the "World Series Earthquake"— aroused an even more intense reaction. As the eyes of America were turned toward Game 3 of the World Series at Candlestick Park, San Francisco, the scenario of an actual damaging earthquake was played out visually before a wide audience.

At 5:04 PM, the largest earthquake in northern California since 1906 struck the San Francisco Bay area. Television viewers looked on in surprise and horror as the Goodyear Blimp transmitted pictures of the collapsed section of the Bay Bridge, part of the Cypress I-880 viaduct in Oakland, and the Marina District fire in San Francisco (Figs. 22–37 and 22–38). The earthquake took hundreds of lives and caused at least $6 billion in damage.

The Loma Prieta earthquake measured 7.1 on the Richter scale and was centered about 70 miles south of San Francisco in the rugged Santa Cruz Mountains where a segment of the San Andreas fault was ruptured. Loma Prieta is a peak 1157 m high on the east side of the fault. The earthquake began suddenly and

Figure 22–37   Collapse of part of Interstate Highway 880 in Oakland, California, during the October 17, 1989, earthquake. (Courtesy of Paul Scott/Sygma)

without foreshock activity at a point 17.6 km beneath the earth's surface. The rupture was complete 7 to 10 seconds later. The net movement in the earthquake advanced a small part of the Pacific plate to the northwest and lifted up the southern Santa Cruz mountains locally.

One reason that Northern California did not suffer even greater damage in the Loma Prieta quake was undoubtedly the quality of its building construction.

Figure 22–38   The Marina District fire in San Francisco following the 1989 Loma Prieta earthquake. (Courtesy of Michael Williamson/Sygma)

A second reason was that the quake was over in only 7 to 10 seconds (the Armenian earthquake shook the ground for 30 seconds). Had it gone on for much longer, damage would have been much more severe. Soils in many regions around the Bay Area would have been liquefied, probably bringing down many more buildings. Water-saturated sandy soils require several seconds of jostling before the sand particles become surrounded by water and transformed into a liquid slurry. Rather than liquefaction, the amplification of the quake by the soft soils of the Marina District led to its devastation. The resulting strong shaking acted on a structural weakness in the wood frame houses. Their first stories—garages—had few if any internal walls or supports and crumpled under the weight of the two to four stories above them. In a study of the Loma Prieta earthquake, the U.S. Geological Survey Staff concluded that earthquake hazards cannot be reduced through science alone. A well-informed public must insist upon the mitigation of hazards before the next earthquake strikes.

22–10 EARTH SCULPTURING

Ever since it was formed, the earth's crust has been subjected to processes leading to the destruction of land forms through **erosion**, the removal of parts of the crust to new locations by various agents. Most of the features of the topography are the results of sculpting by erosional forces, which have their origin in the sun, the oceans, and the atmosphere.

The preliminary, or static, phase of erosion involves **weathering**, of which two broad types are identified: *chemical* and *mechanical*. Wherever rocks are exposed to air, water, and organisms, weathering reduces them to a soft, loosened debris that can be carried away by active agents of erosion, such as running water, wind, ice, and waves.

Chemical weathering results in the decomposition of rock. Through the chemical activity of water, oxygen, or carbonic acid, new minerals, usually softer and of lower density, are produced from the original ones; for example, feldspar is reduced to clay. Since it usually involves water, chemical weathering proceeds more rapidly in moist than in arid climates (Fig. 22–39). A striking example is a monument that stood nearly unscathed for 3500 years in Egypt but deteriorated badly in a relatively few years when brought to

Figure 22–39 The Parthenon, in Athens, Greece, has become severely damaged. Air pollutants dissolved into atmospheric moisture have speeded up chemical weathering, causing the exposed surfaces of the marble to flake off. (R. Sager)

Figure 22–40 Frost wedging at Devil's Lake, Wisconsin. The pressure of the expansion of water as it freezes in rock wedges can eventually split apart the rock. (R. Gabler)

the more humid and polluted environment of New York City.

Mechanical weathering causes rocks to disintegrate into smaller pieces, as when quartzite breaks into smaller fragments and becomes sand. In regions that experience frequent alternation of freezing and thawing, ice wedgework becomes a dominant disintegration process. Water seeps into rock openings and expands as it freezes; the resultant pressure on the surrounding rock often splits it apart. **Root weather-**

ing also causes rocks to disintegrate. The roots and trunks of trees growing in rock crevices develop enormous splitting pressures on the surrounding rocks (Fig. 22–40).

When weathering has caused rocks to disintegrate or decay, gravity may force the debris to move downslope from higher to lower ground (Fig. 22–41). This process, called **mass-wasting**, may be carried out in several ways: rockfalls, landslides, mudflows, and

Figure 22–41 Piles of rock debris formed by weathering and rockfalls occur at the base of steep slopes, as in the Canadian Rockies. (B. Bradley, NOAA, National Geophysical Data Center)

creep. Important results of weathering are the formation of soil and mineral ore deposits.

Ultimately, streams transport the rock material to the sea. Incidental to their flow to the sea, streams are by far the dominant agent of erosion on the earth's surface today, more important than wind and ice. In the process, they are responsible for producing many landforms, consuming uplands by eroding their valleys. The most important erosional process of streams involves **hydraulic action**—the removal and transportation of loose material from the channel sides and bottom. The swifter the current, the larger the size of the particles that can be transported. The solvent action of water is less significant as an erosive force.

As the velocity of the stream decreases, the stream deposits the heavier particles first and then the finer materials. The sediments accumulate and may form a **delta** near the junction of the river with a lake or the ocean. Some deltas, such as those of the Mississippi and Nile rivers, extend their coastal areas for hundreds of square miles into the sea (Fig. 22–42).

Figure 22–42   A satellite view of the Nile Delta, as the Nile River enters the Mediterranean Sea. The fertile Nile Delta, about 500,000 square miles in area, is surrounded by light-colored deserts. (NASA)

The flat, fertile alluvium, the material deposited by the running water, makes excellent farm soil.

22–11 MATERIALS OF THE EARTH'S CRUST

The new ideas of plate tectonics have had an impact on economic geology on a global scale. Many ore deposits occur at the boundaries, present or past, of the crustal plates. What ores are formed and where they are placed in the crust may depend on the tectonic history of a region. This can be important in a resource-hungry world.

Many major classes of mineral deposits fit the new conceptual framework. They occur in volcanic or igneous rocks and were formed at the same time as these rocks. According to plate tectonics, volcanism occurs at diverging plate boundaries such as mid-oceanic ridges, where mantle material rises to form new crust; at converging boundaries, where plates descend into the mantle through subduction, leading to volcanism that forms chains of mountains or island arcs; and over hot spots caused by ascending plumes of mantle material. Each process may give rise to characteristic types of ore deposits.

The mineral deposits of the island of Cyprus, a rich source of copper, are a prime example of the first type. The copper sulfide ore occurs in a distinctive progression of rock types known as an ophiolitic sequence. The sequence is exactly that which should be formed at a mid-oceanic ridge: on top, sediments of a type formed on the ocean floor; beneath the sediments, pillow lavas formed when molten volcanic material erupts into seawater; farther down, sheets of basalt formed as cracks in the ocean floor are filled from below with volcanic material; and on the bottom, rocks rich in iron and magnesium that are believed to be characteristic of the mantle. The minerals include sulfides of copper, iron, and sometimes zinc, chromium ore, and asbetos deposits. Mineral deposits of the Cyprus type are found in many parts of the world, including the northeastern United States and eastern Canada.

A second type of mineral deposit, ores known as porphyry coppers, is associated with converging plate boundaries. The copper deposits in the Andes are an example. There the eastward-moving oceanic crust of the Pacific plunges under the lighter material of the westward-moving South American continent. Partial melting of the oceanic plate produces magmas that rise through the overlying continental rocks, sometimes forming volcanoes. The upper parts of these magma intrusions often contain copper and molybdenum and sometimes gold and silver as well.

Plate tectonic mechanisms explain the source of much of the world's mineral wealth—iron ores, gold deposits, copper, zinc, lead, and silver ores. Still, the new models of ore formation are far from complete and not entirely accepted. Mineral exploration can be guided by an understanding of how, and perhaps where, ores are formed and deposited, but far more exhaustive studies are required.

22–12 GEOLOGICAL EVIDENCE OF LIFE

Fossils—the remains of organisms that lived in the geological past—preserved in sedimentary rocks are the basis for dividing geological time into eons, eras, periods, and epochs (Fig. 22–43). The fossil deposits form layers that can be identified in geological formations. The ages of the deposits can be calculated from the constant rate of decay of radioactive nuclides in the earth's crust. A date can be assigned to a rock unit and to nearby fossil-bearing strata by determining how much of a nuclide has decayed since the minerals in the rock unit crystallized.

A dramatic boundary in the rock record separates the **Cambrian period** from all that came before. The 11 periods of geological time since the start of the Cambrian are referred to as the **Phanerozoic eon**—the era of manifest life. The preceding era is simply called the **Precambrian**. It has been established that the Phanerozoic eon began about 570 million years ago. Radionuclide dating places the age of the earth and the rest of the solar system at 4.6 billion years. Thus, the Precambrian eon involves approximately 90% of the earth's entire history.

Cambrian strata contain many fossils of marine plants and animals such as seaweeds, sponges, mollusks, and the early arthropods called **trilobites** (Fig. 22–44). It was thought for many years that the underlying Precambrian strata contained no fossils. Life seemed to come into existence abruptly during the Cambrian period. Yet it did not seem reasonable to assume that life began with organisms as complex as trilobites.

In the 1950s it became evident that many Precambrian rocks *are* fossil-bearing. These fossils had escaped detection earlier because they are the remains of microscopic forms of life called **microfossils**. Precambrian fossils have even been found in some of the most ancient sedimentary deposits known.

The paleobotanist (student of fossil plants) Elso S. Barghoorn (1915–1984) of Harvard, working with the sedimentologist (student of sedimentary rocks and their inclusions) Stanley A. Tyler (b. 1911) of the University of Wisconsin, in 1954 discovered fossil microscopic plants in an outcropping of Precambrian rocks called the Gunflint Iron formation near Lake Superior in Ontario. Most of the Gunflint fossils resemble modern bacteria and blue-green algae (cyanobacteria). The rocks have been dated to an age of about 2 billion years.

Since then, microfossils have been identified in many places and in even older rocks. The oldest are the Fig Tree and Onverwacht deposits of South Africa, which are 3.2 and 3.4 billion years old, respectively. Both deposits contain microfossils that resemble bacteria. The world's oldest known multicellular organisms left their record in rocks that were to be called the Torrowangee Group of Australia about 1 billion years ago.

The search for geological evidence of ancient life is a laborious process. Thousands of rock specimens have to be sawed into thin slices and polished so that they can be studied under the light microscope and the electron microscope. But the fossils reveal a surprising amount of information about the size, shape, structure, and even the internal structure of the cells of the organisms.

The geological record of the evolution of life on earth shows that life on earth appeared within about 1 billion years after its formation. We have no direct geological evidence to tell us when the transition from nonliving to living occurred. For perhaps 3 billion years all life on earth was at or below the unicellular level of bacteria or unicellular blue-green algae. The step from nonbiological organic matter to life seems to have been easier than the step from one-celled bacteria to many-celled organisms, a step that took twice as long.

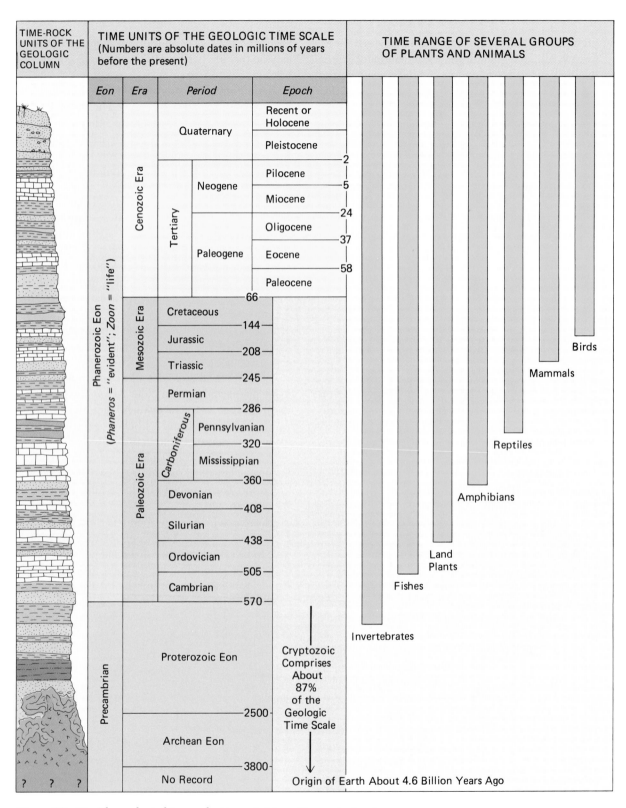

Figure 22–43 The geological time scale. (From A. R. Palmer, "The Decade of North American Geology, Geologic Time Scale." *Geology* 11:503–504, 1983.)

Figure 22–44 *Ellipsocephalus,* a variety of trilobite, from Bohemia. Its average length is 2.8 cm.

KEY TERMS

Asthenosphere The portion of the upper mantle below the lithosphere. It consists of weak, plastic rock where magma may form.

Bedrock The solid rock that underlies soil or regolith.

Body Seismic Waves Earthquake waves that travel through the interior of the earth.

Cambrian Period Designates the lowest systems of Paleozoic rocks.

Chemical Weathering The chemical decomposition of rocks and minerals by exposure to air, water, and other chemicals in the environment.

Continental Drift The theory that the continents were once part of a single supercontinent that began splitting apart millions of years ago, and that the pieces — continents — drifted to their present locations.

Convergent Plate Boundary A boundary where two tectonic plates collide head-on.

Core Central part of the earth that lies beneath the mantle.

Crust The outer part of the lithosphere.

Delta The nearly flat, fan-shaped tract of land where a river flows into a still body of water, such as a sea or lake.

Divergent Plate Boundary The boundary or zone where two tectonic plates separate from each other.

Earthquake A motion or trembling of the earth caused by the sudden movement of crustal blocks along a fault.

Epicenter The point on the earth's surface directly above the focus of an earthquake.

Erosion The removal of weathered rocks by moving water, wind, ice, or gravity.

Fault A fracture in the earth's crust along which rocks on one side have been displaced relative to rocks on the other side.

Focus The point within the earth's crust where an earthquake occurs.

Fossils The remains of organisms that lived in the geological past.

Gondwanaland In continental drift theory, a temporary large land mass in the Southern Hemisphere.

Hydraulic Action The mechanical loosening and removal of material by flowing water.

Igneous Rock Rock formed by the cooling and solidification of magma.

Intensity (of an Earthquake) A measure of the effects of an earthquake on buildings and people.

Laurasia In continental drift theory, a temporary large land mass in the northern hemisphere.

Lava Molten material that flows out of volcanic vents as hot streams or sheets and solidifies as rock.

Lithosphere The outer shell of the earth, comprising the crust and upper mantle.

Love Wave An earthquake wave that travels along the surface of the earth or along a boundary between layers within the earth.

Magma The name given to lava when it occurs below the earth's surface.

Magnitude (of an Earthquake) A quantitative measure of the strength of an earthquake determined from seismic recordings.

Mantle A layer of dense rocky material about 2900 km thick that lies between the earth's crust and core and occupies about 80% of the volume of the earth.

Mass-wasting The movement of earth material downslope primarily under the influence of gravity.

Mechanical Weathering The

disintegration of rock into smaller pieces by physical processes.

Mercalli Scale   A qualitative scale of earthquake intensity that measures its effect on buildings and people.

Mesosphere   The lower part of the earth's mantle.

Metamorphic Rock   Rock that has been changed by heat and pressure in the earth's crust so that its original characteristics have been lost.

Microfossils   Remains of microscopic forms of life.

Mid-oceanic Ridge   A continuous submarine mountain chain that delineates the belt along which tectonic plates move apart and new oceanic crust is created.

Mineral   A naturally occurring solid substance with a definite chemical composition and crystal structure.

Mohorovičić Discontinuity (Moho)   The sharp boundary between the earth's crust and underlying mantle.

Pangaea   In the continental drift theory, the original supercontinent.

Phanerozoic Eon   The periods of geological time since the start of the Cambrian period 570 million years ago.

Plate Boundary   A boundary between two lithospheric plates.

Plate Tectonics   The theory that the earth's surface consists of a number of rigid plates that move with respect to one another.

Precambrian Eon   The first 90%

of the earth's history; the first 4 billion years.

Primary (P) Wave   A seismic wave formed by the alternate compression and expansion of rock. It is the fastest of the seismic waves.

Rayleigh (R) Wave   A seismic wave propagated along the earth's surface and involved in ground roll during an earthquake.

Regolith   The loose, weathered earth materials that usually cover bedrock.

Richter Scale   A numerical scale that measures the energy released by an earthquake.

Ring of Fire   The belt of subduction zones and volcanism that borders the Pacific Ocean along the continental margins of Asia and the Americas.

Rock   Heterogeneous aggregates of minerals in varying proportions that form an appreciable part of the earth's crust.

Root Weathering   The disintegration of rocks caused by the pressures of the growth of tree roots and trunks in rock crevices.

Sea-Floor Spreading   A mechanism for continental drift, according to which molten rock from the mantle rises up along the mid-oceanic ridge, then spreads outward in both directions and hardens into new ocean floor.

Secondary (S) Wave   A seismic body wave in which the oscillation is perpendicular to the direction of wave travel.

Sedimentary Rocks   Rocks formed from the debris made by the weathering and erosion of older rocks that has been carried elsewhere by water or wind and lithified.

Seismic Array   The deployment over a large area of a large number of seismographs with which to probe the earth's interior.

Seismic Wave   Waves in the earth produced by earthquakes or underground explosions.

Seismogram   The record made by a seismograph.

Seismograph   An apparatus that registers the shocks and motions of earthquakes and underground explosions.

Surface Seismic Wave   A seismic wave that travels near the earth's surface.

Transcurrent Fault Boundary   A boundary between two lithospheric plates where the plates are sliding past one another.

Transition Zone   A layer of the earth's mantle lying between the asthenosphere and the lower mantle.

Trilobites   Paleozoic marine arthropods that were especially abundant during the early Paleozoic period.

Tsunami   A large sea wave produced by a submarine earthquake, a volcanic eruption, or landslide and characterized by long wavelength and great speed.

Weathering   The decomposition and disintegration of rocks and minerals at the earth's surface by mechanical and chemical processes.

THINGS TO DO

1. Visit a mineralogical collection in a museum of natural history, earth science department, or other place with such a display.

2. Observe rock specimens and geological formations in a rock quarry. Collect specimens, if permitted.

3. Collect interesting rocks at the seashore, lake, stream, and backyard. Label them with the locality from which they came. Try to identify them with a field book of rocks and minerals.

4. Study freshly broken rocks with a magnifying glass and try to examine crystals of various minerals.

5. Examine sand with a magnifying glass or under a microscope. Identify crystals of quartz and other minerals.

6. Test rock samples for limestone or marble with lemon juice or vinegar. Limestone will effervesce where the acid reacts with it, releasing carbon dioxide.

7. Examine samples of various types of soil with a magnifying glass and/or microscope: sandy, loam, and clay, a soil rich in humus.

8. Observe exposed layers of bedrock along road cuts.

9. Study the mineral components of granite, the rock most commonly used in road construction and buildings: quartz, mica, and feldspar.

EXERCISES

Structure of the Solid Earth

22–1. How is information concerning the earth's interior obtained?

22–2. What is the evidence for the fluidity of the earth's outer core and for the solidity of the inner core?

22–3. Discuss the structure of the earth's interior?

22–4. What is the Mohorovičić discontinuity? How was it discovered?

22–5. Identify the major zones of the earth from the surface to the center.

Earth's Crust

22–6. If a planet had no sedimentary or metamorphic rocks, what could you say about its geological history?

22–7. Distinguish between rocks and minerals.

22–8. Describe the major classifications of rock.

22–9. Why are the rocks found on the ocean floors relatively young?

22–10. How do impacts by comets and asteroids influence the earth's geology?

Mantle and Core

22–11. How do geophysicists determine the location of the top of the mantle?

22–12. What is the most obvious difference between the inner core and the outer core?

22–13. Which is denser, the mantle or the core?

22–14. What is the evidence for the structure and composition of the mantle?

Volcanoes

22–15. Describe the world distribution of volcanic activity and earthquake epicenters.

22–16. Explain the relationship between magma and lava.

22–17. Discuss the varieties of volcanic eruptions and their relations to the shapes of volcanoes.

22–18. What types of volcanoes are especially dangerous?

22–19. Why does magma tend to rise to the surface?

22–20. What is the status of the science of predicting a major eruption of a volcano?

Plate Tectonics

22–21. How might the Himalayan Mountains have originated, according to plate tectonic theory?

22–22. Discuss some evidence in support of the concept of continental drift.

22–23. How is sea-floor spreading related to plate tectonics?

22–24. Why is the ocean floor youthful in relation to the age of the earth?

22–25. What evidence exists for the idea that the earth's magnetic field has reversed at intervals over the ages?

22–26. In what way is the comparison of continents to floating icebergs valid?

22–27. Discuss the kinematics of three types of plate boundaries.

22–28. What evidence has been found to support Wegener's theory of continental drift?

Earthquakes

22–29. What causes an earthquake?

22–30. How are regions of active volcanoes related to earthquake belts?

22–31. Explain the geologist's description of an earthquake in terms of its intensity and magnitude.

22–32. Discuss the major kinds of earthquake waves.

22–33. How great are the chances of a destructive earthquake east of the Rockies?

Earth Sculpting

22–34. Why is the rate of weathering greater in a humid than in an arid climate?

22–35. Why is none of the original crust of the earth now visible?

22–36. How does physical weathering encourage chemical weathering in rock?

22–37. What types of rocks best resist all types of weathering?

22–38. Why have the forces of weathering and erosion not worn the continents down to low-lying plains?

22–39. *Multiple Choice*
A. The oldest rock on the earth's surface is
   (a) sedimentary.
   (b) igneous.
   (c) tectonic.
   (d) metamorphic.
B. An instrument that measures earthquake activity is a
   (a) polarimeter.
   (b) refractometer.
   (c) seismograph.
   (d) micrometer.
C. The theory that continents float on layers of mantle rock is called
   (a) sea-floor spreading.
   (b) plate tectonics.
   (c) mass wasting.
   (d) volcanism.
D. The Mid-oceanic Ridge is
   (a) a region where tectonic plates are colliding.
   (b) bordered by young igneous rocks.
   (c) bordered by old sedimentary rocks.
   (d) a region where earthquakes are common.
E. An agent of significant erosion is
   (a) oxidation of minerals in air.
   (b) falling snow.
   (c) wave action.
   (d) radioactive decay.

SUGGESTIONS FOR FURTHER READING

Anderson, Don L., "Where on Earth Is the Crust?" *Physics Today*, March, 1989.
It is ironical that astronomers speak confidently of the chemistry of stars, galaxies, and interstellar space, but we are grossly ignorant of the composition of our own planet.

Badash, Lawrence, "The Age-of-the-Earth Debate." *Scientific American*, August, 1989. During the past three centuries, the earth has "aged" 4.5 billion years. When the radioactive dating methods pioneered by Ernest Rutherford, Bertram B. Boltwood, and Arthur Holmes at last received the blessing of geologists, they found that they had a prospect for dating all of our geological history.

Bonatti, Enrico, "The Rifting of Continents." *Scientific American*, March, 1987.
It begins above a hot zone in the mantle. Upwelling molten rock weakens the continental crust, pierces it, and finally rifts it in two, creating an ocean.

Brimhall, George, "The Genesis of Ores." *Scientific American*, May, 1991. When 80,000 "forty-niners" rushed to California in the mid-nineteenth century to find gold, many of them knew where to look. Only recently, however, have scientists begun to understand the processes by which gold and other ores—silver, iron, copper, and tin—are formed and where metals originate.

Brush, Stephen G., "Discovery of the Earth's Core." *American Journal of Physics*, September, 1980.
Until 1896, scientists generally believed that the entire earth was a solid as rigid as steel. Emil Wiechert's model of the earth with an iron core and stony shell was destined to revolutionize the study of the earth.

Cook, Frederick A., Larry D. Brown, and Jack E. Oliver, "The Southern Appalachians and the Growth of Continents." *Scientific American*, October, 1980.
The petroleum industry's exploration technique of seismic-reflection profiling is applied intensively to the study of the deep crust and upper mantle. The technique has mapped new details of the structure of the continental basement.

Decker, Robert, and Barbara Decker, "The Eruptions of Mount St. Helens." *Scientific American*, March, 1981.
The violent eruption of Mount St. Helens on May 18, 1980, was the most recent of at least 20 eruptions of this volcano over the past 4500 years. It was one of the most fully documented volcanic eruptions in history.

Dewey, John F., "Plate Tectonics." *Scientific American*, May, 1972.
As the plates making up the earth's surface shift, new crust and mountains are formed and continents drift.

Francis, Peter, and Stephen Self, "Collapsing Volcanoes." *Scientific American*, June, 1987.

A catastrophic collapse is a "normal" event in the life cycle of many volcanoes. Otherwise, volcanoes might be the highest mountains on the earth, as they are on Mars and probably on Venus.

Frohlich, Cliff, "Deep Earthquakes." *Scientific American*, January, 1989.
In most earthquakes the earth's crust cracks like porcelain as stress builds up, a fracture forms at a depth of a few kilometers, and slip relieves the stress. But some earthquakes, one in five, occur hundreds of kilometers down in the earth's mantle, where it is thought that rock is prevented from cracking. Geophysicists are still struggling to solve the puzzle.

Grieve, Richard A., "Impact Cratering on the Earth." *Scientific American*, April, 1990.
In the past billion years, an asteroid or a comet has struck the earth thousands of times, each event lasting only a few seconds. Most of the impact craters on the earth disappeared long ago. Some impacts may have altered biological as well as geological evolution.

Hallam, A., "Alfred Wegener and the Hypothesis of Continental Drift." *Scientific American*, February, 1975.
The most important points in Wegener's hypothesis of 1912, neglected or scorned for 50 years, have been substantiated by discoveries in geophysics and oceanography.

Howell, David G., "Terranes." *Scientific American*, November, 1985.
They are blocks of crust that accrete to the ancient cores of continents. The concept of terranes emerged in the 1970s as a revision of plate-tectonic theory, which failed to account for the geology of Alaska.

Hurley, Patrick M., "The Confirmation of Continental Drift." *Scientific American*, April, 1968.
The evidence now favors the idea that the continents were once combined into two land masses: Gondwanaland and Laurasia.

Jones, David L., Allan Cox, Peter Coney, and Myrl Beck, "The Growth of Western North America." *Scientific American*, November, 1982.
New evidence shows that the growth of continents is not slow and steady but episodic. The process involves collision and accretion, grafting onto the continent by the addition of large blocks of crust carried east and north from their origin in the Pacific basin hundreds or thousands of kilometers.

Jordan, Thomas H., "The Deep Structure of the Continents." *Scientific American*, January, 1979.
There are new insights into processes that control continental evolution and tectonics.

Lay, Thorne, Thomas J. Ahrens, Peter Olson, Joseph Smyth, and David Loper, "Studies of the Earth's Deep Interior: Goals and Trends." *Physics Today*, October, 1990.
We are only in the early stages of achieving a true understanding of the deep interior of the planet on which we live. Successful theories will need to synthesize data and concepts from a variety of disciplines, ranging from computer science, geochemistry, and mathematics, to all branches of geophysics.

Maxwell, John C., "What Is the Lithosphere?" *Physics Today*, September 1985.
The stony outer portion of our highly dynamic earth participates in a massive convection of matter that continually builds, destroys, and moves plates of continental and oceanic crust.

Molnar, Peter, "The Structure of Mountain Ranges." *Scientific American*, July, 1986.
Enormous forces are required not only to build but also to support a mountain range. Some stand on plates of strong rock; others are buoyed up by crustal roots reaching deep into the mantle.

Molnar, Peter, and Paul Tapponnier, "The Collision between India and Eurasia." *Scientific American*, August, 1977.
For 40 million years, the Indian subcontinent has been pushing northward against the Eurasian land mass.

Mutter, John C., "Seismic Images of Plate Boundaries." *Scientific American*, February, 1986.
Since 70% of the earth's surface is covered by oceans, most of the plate boundaries, where plates collide and rift apart and where geology on a large scale really happens, are deeply submerged. Structural images of the crust can be made at the boundaries by bouncing sound off rock layers under the sea floor and recording the reflections.

Pollack, Henry N., and David S. Chapman, "The Flow of Heat from the Earth's Interior." *Scientific American*, April, 1977.
The heat-flow pattern of the earth is interpreted in terms of plate tectonics.

Smith, Robert B., and Robert L. Christiansen, "Yellowstone Park as a Window on the Earth's Interior." *Scientific American*, February, 1980.
Yellowstone National Park is a unique natural laboratory for studying the interior of the earth.

Wyllie, Peter J., "The Earth's Mantle." *Scientific American*, March, 1975.
Although the surface of the earth is shaped by the action of the mantle, the mantle is inaccessible. Nevertheless, much has been learned about it by indirect means.

Nonsuch Island and surrounding ocean, Bermuda. (R. Sager)

23

THE EARTH'S HYDROSPHERE—
OCEANOGRAPHY

Earth's abundant supply of water makes it unique among the planets. The large planets—Jupiter, Saturn, Uranus, and Neptune—have only a small, solid core surrounded by enormous atmospheres. Among the terrestrial planets, Mercury has practically no atmosphere, and the side facing the sun is hot enough to melt lead. Venus has a thick atmosphere and a surface that may be even hotter than the surface of Mercury. Mars and the moon have a surface marked with craters formed by meteorites and perhaps by volcanoes. Since much of the surface of the earth is covered by a liquid shell, or *hydrosphere*, earth is called the "water planet." Without the oceans, the earth would probably be as desolate as the surface of the moon. *Oceanography* is the application of techniques and principles from many sciences—chemistry, geology, physics, and biology —to the study of the sea.

23–1 THE WORLD OCEAN

The ocean waters cover four fifths of the southern hemisphere and more than three fifths of the northern hemisphere, in all about 71% of the earth's surface. They are joined in essentially a world ocean containing about 1350 million cubic kilometers of water at an average depth of 4 kilometers. There are 3 square kilometers of water for each square kilometer of land of the earth's surface area. The continents project like islands above the vast surrounding sea, at an average elevation of 875 meters.

This continuous body of water is subdivided, sometimes arbitrarily, into oceans and seas (Fig. 23–1). The relatively circular Pacific Ocean, covering about a third of the earth's surface, is by far the largest ocean. The S-shaped Atlantic Ocean is about half the size of the Pacific but has more coastline than the Pacific and Indian oceans combined. The Atlantic and Pacific oceans are often subdivided at the equator into northern and southern parts. The third main ocean is the Indian Ocean. The Arctic and Antarctic oceans are sometimes regarded as separate oceans.

Seas are salt-water bodies that are smaller than oceans and are somewhat enclosed by land. Deep seas such as the Mediterranean and the Gulf of Mexico occur between continents. Shallow seas such as the North Sea and Hudson Bay occur within depressed parts of a continent. Marginal seas such as the Bering Sea may be deep or shallow and are separated from the main oceans by island arcs. Although somewhat enclosed by land, seas always interchange water with the oceans and differ from lakes in this respect. Other seas are the Baltic, Caribbean, Coral, Black, Yellow, and Red Seas.

23–2 ORIGIN OF SEAWATER

Unlike the moon and Mercury, which were too weak gravitationally to retain their hydrospheres or atmospheres, or Mars, which held on to only a trace of each, the earth was able to keep its water and air. Yet gases such as neon and argon were almost completely

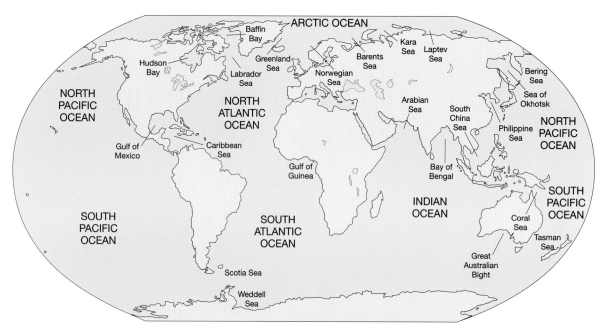

Figure 23–1   The earth's hydrosphere, a continuous body of water, is subdivided into oceans and seas.

lost. The earth's water was bound up in more stable compounds. Several mechanisms have been proposed for the origin of water and other "excess volatiles"—substances such as carbon dioxide, chlorine, nitrogen, and sulfur—that are more abundant in the ocean and atmosphere than can be accounted for by the weathering of rocks alone.

Most scientists consider the degassing of the earth's interior, as proposed by William Rubey (1898–1974), of the United States Geological Survey and the University of California, to have been the best mechanism for producing the excess volatiles, and thereby the atmosphere and the ocean. Soon after it was formed, the earth cooled and solidified. When molten magma cools, solid minerals are formed. During the crystallization process, some of the gaseous constituents are freed. Gaseous and volatile substances, including water vapor, escaped to the surface in a process known as **degassing**. The rate at which this process occurred is not known. One possibility involves rapid degassing followed by reaction between the ocean and the earth's crust. Or degassing may have been slow, with the ocean evolving over most of the earth's history.

The water vapor released from the surface of the earth was caught in the gaseous envelope surrounding it. As the earth's crust continued to cool, water condensed and fell to the earth's irregular surface as a liquid, accumulating in depressions and forming the oceans. The release of water into the oceans was gradual during the first billion years of earth's history. In addition, oceans received water from volcanoes and geysers. For the past 1.5 billion years, the oceans have essentially maintained their present chemical composition.

23–3   SALT IN THE SEA; DESALINATION

For billions of years chemical matter dissolved from the land has been deposited in the sea. The evaporation of 100 kg of seawater yields about 96.5 kg of water. The rest, 3.5 kg, consists of dissolved solids such as common salt (NaCl) and small amounts of dissolved gases such as oxygen (O_2). The amount of dissolved solids in seawater is termed the **salinity**, measured in percent or in parts per thousand. If all of the oceans could be evaporated into space, enough salt would remain to cover the earth with a layer close to 70 m in thickness.

Certain elements are more abundant than others in seawater because some substances are more soluble than others. For example, sodium chloride (NaCl) is very soluble. But calcium carbonate ($CaCO_3$), as in limestone and clam shells, is much less soluble, and silicon dioxide (SiO_2), as in quartz and glass, is almost insoluble.

From 1872 to 1876, the ship HMS *Challenger* carried scientists on a round-the-world cruise to study the biology of the sea, collect water samples and bottom samples, chart the ocean depths, measure movements of water masses, and measure water temperatures (Fig. 23–2). Over a nine-year period, William Dittmar (1833–1892), a chemist at the University of Glasgow, analyzed 77 water samples taken on this voyage. He determined the percentage composition of the dissolved solids, such as chloride, sulfate, and sodium, in each of the samples. In another study, Georg Forchhammer (1794–1865), a geologist at the University of Copenhagen, spent 20 years analyzing 200 samples of seawater collected for him by friends, ship captains, and naval officers.

These analysts and current analysts working with modern analytical techniques and instrumentation reached essentially the same conclusion (Fig. 23–3): About 95% of the dissolved solids are made up of only six major elements (Cl, Na, Mg, S, Ca, and K) (Table 23–1). In fact, 86% of all dissolved salts consist of only two elements, sodium and chlorine. At least 35 other elements have also been detected in seawater. Although actual concentrations of dissolved substances may vary from place to place, the ratio of any one of the major elements to the total dissolved solids is nearly constant. This illustrates the effectiveness of the mixing processes within the oceans on a worldwide basis.

The increasing demand for fresh water for agricultural, domestic, and industrial use is focusing attention on the sea. Nearly 4000 desalination plants are in

Figure 23–2 The HMS *Challenger,* the research tool of a team of scientists, sailed into all of the major ocean areas during a 3.5-year expedition. Its mission was to chart the ocean depths, describe its biology, and examine the ocean's chemistry and bottom deposits. (From "Report of the Scientific Results of the Exploring Voyage of HMS *Challenger* During the Years 1873–1876," Narrative, Part II, 1885.)

Figure 23–3 The *JOIDES Resolution* is a floating oceanographic research center with a seven-story laboratory stack. (Courtesy of the Ocean Drilling Program, National Science Foundation)

Table 23–1 MAJOR CONSTITUENTS OF DISSOLVED SOLIDS IN SEAWATER

Ion		Percentage by Weight
Chloride	Cl^-	55.06
Sodium	Na^+	30.58
Sulfate	$SO_4^=$	7.68
Magnesium	Mg^{++}	3.74
Calcium	Ca^{++}	1.20
Potassium	K^+	1.12
Bicarbonate	HCO_3^-	0.41
Bromide	Br^-	0.19

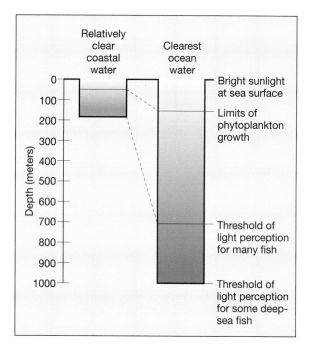

Figure 23–4 Light penetration in clear coastal water and the clearest ocean water on a bright sunny day. Blue and green penetrate the deepest and are scattered the most. The sea is a dark world.

operation throughout the world, producing about 3.4 billion gallons of pure water each day. The oldest desalination technique is distillation. Seawater is evaporated using solar energy or the combustion of hydrocarbons and fresh water is condensed. More elaborate methods than solar stills are also used, such as multi-stage flash (MSF) distillation operated in conjunction with thermal electric power plants, but they are more energy intensive. Key West, Florida, is the first U.S. city to derive its municipal water supply from the sea through desalination.

23–4 WHY IS THE OCEAN BLUE?

Although intense sunlight strikes the ocean surface, the amount of light that enters the water depends on such factors as the sun angle, the surface condition, and the clarity of the water. Light traveling through the water becomes progressively dimmer through absorption and scattering by particles in the water and the seawater itself (Fig. 23–4). Less than 50% penetrates below 1 meter, and only 2% may be left at a depth of 30 m. Thus, the sea is a dark world.

Most of the red light is filtered out in the upper few meters. Blue light penetrates the deepest, more than 500 m in the clearest ocean water, followed by green light. The color of seawater is due to the light that is reflected from the surface and the light that is scattered within the water. Since blue and green penetrate the deepest, they are scattered the most. The sea is deep blue in the clearest open ocean water. It is green or yellowish in turbid coastal water that con-

tains a considerable amount of suspended particulate matter.

23–5 HEAT IN THE SEA

On the whole, the ocean is a tremendous reservoir of cold water that is not much above freezing. More than 99% of the heat entering the sea comes from the absorption of sunlight and its conversion into heat. Since most light is absorbed near the surface, that is where the greatest warming occurs. Warm water tends to stay at the surface, being less dense than cold water, resulting in a stratification of warm surface water over cold deeper water.

Sea-surface temperatures are highest near the equator and lowest near the poles. They are measured directly using an ordinary thermometer in a sample of water scooped in a bucket or a **Nansen bottle**, named for Arctic explorer and oceanographer Fridtjof Nansen (1861–1930). Another method is used with artificial weather satellites. Photographs taken with infrared-sensitive film show differences in ocean shading that are directly related to the sea-surface temperature.

The temperature of a deep-water sample is taken with a special thermometer attached to a Nansen bottle. The thermometer in the bottle is lowered into the water upside down and then reversed at the desired depth, registering the temperature at that depth. The mercury column breaks, leaving a column of mercury that is proportional to the temperature. After the temperature is read on board ship, the thermometer is tipped back to the original position. The mercury column is rejoined and can be used again. Another instrument, the **bathythermograph** (*bathys,* meaning "deep"), using a coated glass plate, records the temperature and pressure continuously from the surface to the desired depth.

Sea-surface temperatures vary from about $-2°C$ (ice) around Antarctica and Greenland to about 30°C (86°F) at the equator. The average is about 16°C (61°F). The temperature of the uppermost layer is relatively constant, changing little between day and night and between summer and winter. It is called the **mixed layer**, from the effects of wind stirring and radiation cooling, and may be up to 250 m deep (Fig. 23–5). The ocean reacts very slowly to changes in air temperature and therefore serves as a heat regulator for the atmosphere. The **thermocline**, a zone of rapid decrease in temperature with depth, extends to a depth of 1.2 to 1.4 km. Below this level the temperature decreases only slightly to the bottom. At a depth of 1.6 km and lower, the temperature of all ocean water is less than 4°C (39°F). Thus, most seawater, even the bottom water in tropical latitudes, is only slightly above freezing.

The amount of heat entering the sea, most of it from solar radiation with minute amounts from the earth's interior, is in balance with that leaving the sea (Fig. 23–6). Evaporation accounts for 55% of the outgoing heat, radiation 40%, and conduction 5%. The **heat budget** equation for the oceans relates these aspects of heat transfer as follows:

$$\text{Solar radiation} = \text{Evaporation} + \text{Re-radiation} + \text{Conduction} \qquad (23\text{–}1)$$

23–6 TIDES

Twice each day the level of the ocean changes in a regular and predictable way. These changes are called **tides**. Tides are complex phenomena that were not adequately understood until Sir Isaac Newton presented his law of universal gravitation (Chapter 5). Tides are caused by the gravitational forces of moon, sun, and earth.

The moon revolves around the earth every 29.5 days. The gravitational attraction between the moon and the earth causes the ocean waters on the side of the earth facing the moon to bulge toward the moon (Fig. 23–7). At a point on the far side of the earth, which is farthest from the moon, gravitational attraction is weakest (Newton's inverse-square law), and the moon tends to pull the entire earth away from matter, including ocean water; the result is the second tidal bulge. These two bulges, 180° of longitude apart, represent **high tides**. They are separated by about 12 hours, or half a day, because 180° is half the earth's circumference. Midway between the high tides, 90° of longitude, there are two **low tides**. Here the waters recede as they are pulled toward the areas of high tide.

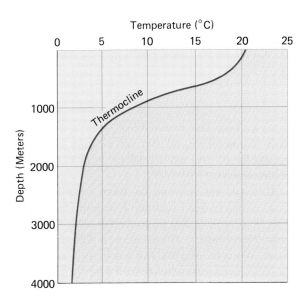

Figure 23–5 Temperature distribution in the oceans. The temperature of ocean water tends to decrease with depth. The thermocline is a zone of rapid decrease in temperature with depth. At a depth of 2000 meters, the temperature remains virtually constant at 1° to 3°C. Thus, the oceans are a reservoir of cold water.

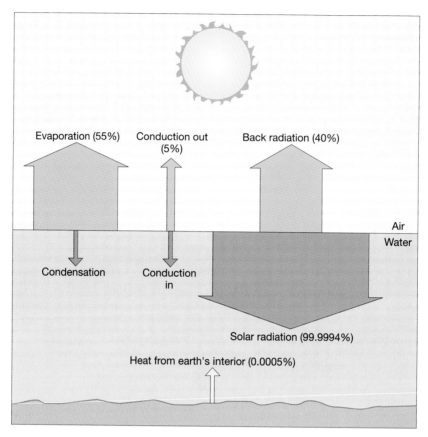

Figure 23–6   The heat budget of the ocean. The amount of heat entering the ocean, most of it from solar radiation, is in balance with that leaving the ocean, much of it through evaporation.

Theoretically, coastal locations experience two high tides and two low tides during a **tidal day**. The length of a tidal day is 24 hours, 50 minutes. As the earth completes one rotation in 24 hours, the moon moves 12° eastward along its monthly 360° orbit around the earth. The earth, therefore, must turn for an extra 50 minutes to reach its previous day's position relative to the moon. The time between two high tides is called the **tidal interval**, and it averages 12 hours, 25 minutes.

The difference in sea level at high tide and low tide is called the **tidal range**. The tidal range varies from place to place in response to factors such as the shape of the coastline, the depth of the water, and the topography of the ocean floor. The average tidal range along the Pacific Coast of the United States, an open ocean coastline, is 2 to 5 meters. In restricted seas, such as the Mediterranean or Baltic, it is usually 0.7 meter or less. The Bay of Fundy on Canada's east coast, a funnel-shaped bay off a major ocean, is famous for its enormous tidal range of as much as 15–21 meters (Fig. 23–8).

Although the sun has far more mass than the moon, it plays a less important role in producing our tides because it is 390 times farther away from the earth. Nevertheless, its tidal effect is 46% that of the moon. When the sun and the moon are aligned with the earth, as they are when there is a new or full moon, the gravitational attracton of the sun reinforces that of the moon, causing abnormally high and low tides, increasing the tidal range (Fig. 23–9). This situation occurs every two weeks and is called **spring tide** (no relation to the season).

A week after a spring tide, at quarter moon, the moon has revolved a quarter of the way around the earth; its gravitational pull on the earth is then exerted

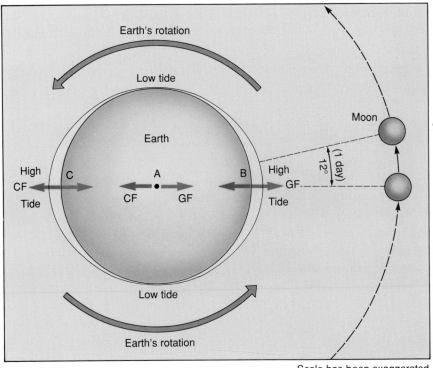

A. Gravitational force (GF) and centrifugal force (CF) are equal. Thus separation between earth and moon remains constant.

B. Gravitational force exceeds centrifugal force causing ocean water to be pulled toward moon.

C. Centrifugal force exceeds gravitational force causing ocean water to be forced outward away from moon.

Scale has been exaggerated

Figure 23–7   Ocean tides are caused by the gravitational forces of moon, sun, and earth. The waters on the side of the earth facing the moon bulge toward the moon, while a second bulge occurs on the opposite side of the earth as the moon pulls the earth away from the water. These two bulges represent high tides. Midway between the high tides there are two low tides.

Figure 23–8   The tidal range in the Bay of Fundy on Canada's east coast is noted for its extreme values. The left photo was taken at low tide, the right photo at high tide.

(Nova Scotia Tourism and Culture)

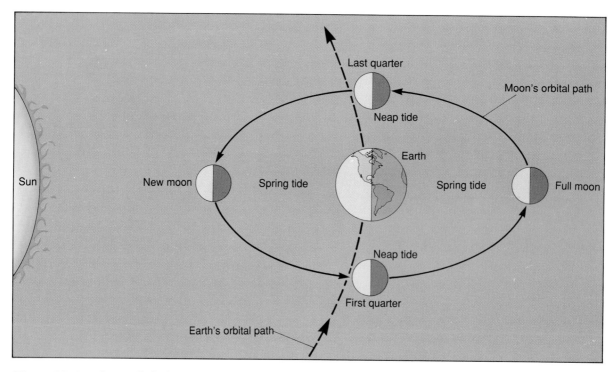

Figure 23–9 Abnormally high and low tides, spring tides and neap tides, respectively, occur at intervals when the sun and the moon are aligned with the earth (spring tides) or when they are acting at right angles (neap tides).

at an angle of 90° to that of the sun. At this time, the moon's attraction is reduced by the counteracting force of the sun's gravitational pull. The result is that the high tides are not as high, and the low tides are not as low. This situation is called **neap tide** and occurs every two weeks. Computers today can prepare tide tables of local coastal areas years in advance, based on long years of observation and many local factors, which are useful to sailors, fishermen, and others.

23–7 SOME TOOLS OF THE OCEANOGRAPHER

Recent advances in oceanography have come from the use of advanced instrumentation. Echo-sounding instruments—sonar—give a continuous trace of ocean depths. Topographic features of the sea floors, and even materials or structures several kilometers

below them, are recorded in detail (Fig. 23–10). Far from being monotonous flat plains, the ocean floors are as irregular as the surfaces of the continents (Fig. 23–11).

Probing devices lowered from oceanographic ships photograph and sample the sea floors. Long "cores" of sediment are removed from depths of several kilometers, revealing details of sea-surface temperature and salinity spanning millions of years. Radioactive decay methods applied to the dating of lavas indicate the age of different parts of the sea floor.

More recently, direct human observation of the sea has yielded valuable information about its composition and deep-sea life. Diving operations range from simple aqualung or scuba equipment (*self-*contained *u*nderwater-*b*reathing *a*pparatus) used to explore coral reefs to the deep-sea-diving saucers developed by explorer Jacques-Yves Cousteau (b. 1910) (Figs. 23–12 and 23–13). People have been lowered into some of the greatest depths in the seas. Others have

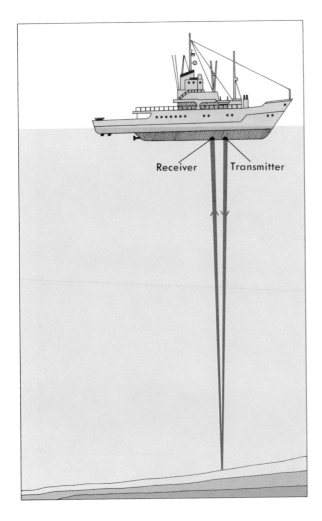

Figure 23–10 Ocean depths are measured with echo-sounding devices, based on the speed with which sound travels in water. Sound waves from a sharp signal emitted at 1/4-second intervals reach the sea floor and are reflected back to the research vessel. The results are automatically recorded and give a continuous profile of the sea floor.

drifted along while embedded in the flow of deep ocean currents.

A particularly valuable tool for studies of the sea bed has been the development of deep-diving **manned submersibles,** which virtually established the science of ocean bottom exploration. A geologist in the U.S. research submersible *Alvin,* a deep submergence vehicle (DSV) operated by the Woods Hole Oceanographic Institution, can collect rock samples and document the setting of each rock (Fig. 23–14). For the first time, a marine geologist can have maps of a site as precise as those of a geologist on land. Remotely operated cameras, high-resolution sonar scanners, and other sensing devices guide the *Alvin* to key geological sites to gather information. It has been used to investigate cycles of volcanic activity

and patterns of geomagnetic reversal along the Mid-Atlantic Ridge, and such properties as the magnetization and the electrical conductivity of the crustal rocks, the velocity of seismic waves under, and gravitational anomalies over the East Pacific Rise.

Robotics will probably be the way of future ocean exploration. Sophisticated unmanned submersibles, or **remotely operated vehicles** (ROVs), such as the ***Argo/Jason*** built by Robert D. Ballard and a team of engineers at Woods Hole, provide greater safety and improved access to the ocean floor. Ballard located the site of the ***Titanic*** with the *Argo/Jason*. This two-part system permits establishment of a submersible platform, from which exploration in the vicinity of the base is conducted with precise maneuverability by a tethered robot equipped with high-intensity lights,

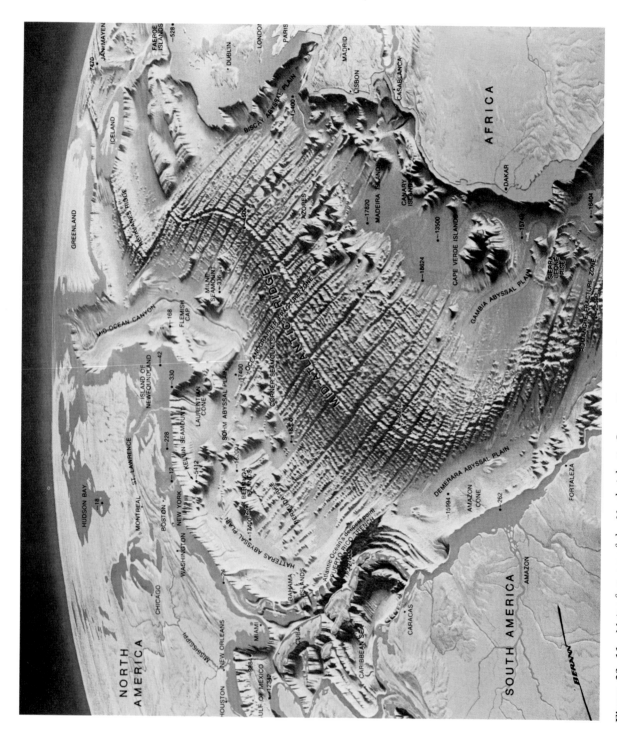

Figure 23–11 Major features of the North Atlantic Ocean. (Courtesy of Aluminum Company of America)

622

Figure 23–12 A scuba diver off Key Largo, Florida.
(Alese & Mort Pechter/The Stock Market)

Figure 23–13 Cousteau's deep-sea-diving saucer in La Jolla, California. (Shirley/Richards/Photo Researchers, Inc.)

telemetry, color video cameras, and manipulator arms. Longer, safer, and better-controlled dives are realized.

23–8 THE CONTINENTAL MARGINS

The edges of the continents are covered by seawater. There are two zones that vary in slope, depth, and width: (1) the gently sloping continental shelf and (2) the steeper continental slope descending to the deep-sea floor. Together, they constitute the **continental margins** (Fig. 23–15).

During the last glacial period, the **continental shelves** were largely dry land. In a geologic sense, then, these shallow areas are not part of the oceanic crust but resemble the continents in their structure and composition. About 70% of the area, as a result, is covered by terrestrial deposits laid down on river and coastal plains 10,000 to 25,000 years ago. Thirty percent is covered with marine deposits—muds, silts, biological sediments such as shell debris, and chemical precipitates. Sand deposits occur near the shore zone to depths of 10 to 20 m (Fig. 23–16). The continental shelf varies from 0 to 1500 km in width, with an average of 80 km, and 20 to 550 m in depth, with an average of 133 m. Over 90% of the world's seafood is captured in the waters of the shelves. Enormous oil and gas reserves lie beneath this part of the sea floor, as photographs of offshore drilling rigs remind us daily (Fig. 23–17). Vast quantities of tin, gold, diamonds, and other minerals are also stored in the shelf sediments.

The steeper **continental slope** extends from the continental shelf to the deep-sea floor, merging with

Figure 23–14 *Alvin,* a deep-diving manned research submersible, takes scientists to the ocean depths. It has made numerous scientific discoveries. (Rod Catanach/Woods Hole Oceanographic Institute)

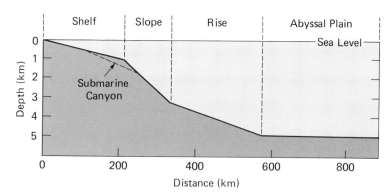

Figure 23–15   Profile of the continental margin. The continental shelf is the relatively shallow part of the sea floor that extends from the coast to the steeper continental slope, which descends to the deep-sea floor.

the floor at depths of 3000 to 4500 m. Sixty percent of the material of the slope surface is mud; 25% is sands; 10% is rock and gravel; and 5% is shell beds and organic oozes.

The most conspicuous features of the continental slope are **submarine canyons**, which are V-shaped, rock-walled, winding valleys. Most of the large canyons are near the mouths of large, ancient rivers such as the Congo or Hudson. The Monterey Canyon off California, the largest, is deeper and wider than the Grand Canyon. Some submarine canyons owe their origins to **turbidity currents**—fast-flowing bodies of water loaded with suspended mud and sand. The currents flow downslope along the bottom of the ocean. The suspended particles, under the high speed of the water, apparently can cut canyons in hard rocks.

Millions of years are needed for most canyons to form. Although turbidity currents have never been observed directly, there is indirect evidence for their existence. A turbidity current off the Grand Banks of Newfoundland in 1929 sequentially broke a series of

Figure 23–17   An oil-drilling platform off the coast of Nigeria. Oil and gas reserves lie beneath the continental shelves. (Visuals Unlimited/API)

Figure 23–16   A sandy beach, Virgin Gorda Beach, Virgin Islands. (Patrick Montagne/Photo Researchers, Inc.)

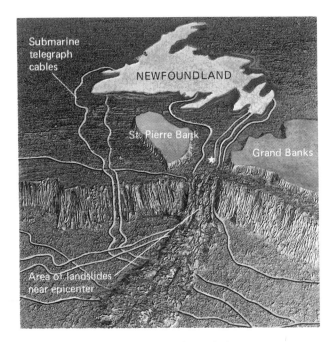

Figure 23–18 Evidence of a turbidity current. A star shows the epicenter of an earthquake that occurred on the Great Banks in the northwest Atlantic in 1929 and triggered the current. The turbidity current snapped several communication cables strung across the floor of the Atlantic. The times of the breaks were recorded, enabling scientists to determine the speed and distance of the current.

Trans-Atlantic Telegraph cables. The time each cable was broken was automatically recorded, indicating the current was traveling 100 km per hour over the continental slope (Fig. 23–18).

23–9 UNDER THE DEEP SEAS

Beyond the continental slope, the deep seas account for about 84% of ocean area. They were once thought to be entirely floored by flat, featureless surfaces, the sea-floor plains. Instead, the ocean floor is almost as complex as the land surface.

Sea-floor or **abyssal plains** are best developed between the continental slope and the mid-oceanic ridge at depths of 3000 to 6000 m. Many are flat and featureless and covered with organic oozes consisting of the remains of microscopic marine organisms and clays. Others form **deep-sea fans** from silt and sand derived from rivers with much suspended matter; the largest are beyond the Ganges and Indus deltas. Undersea volcanoes called **seamounts** rise abruptly 3 to 4 km above the sea floor. They are present in all oceans but are most common in the Pacific. Some, like the Hawaiian and the Canary Islands, rise above sea level and appear as oceanic islands. Many are found along chains, with the oldest at one end and the youngest at the other (Fig. 23–19).

As the volcanic activity subsides, the peaks are eroded by wave action. Eventually, the flattened sea-

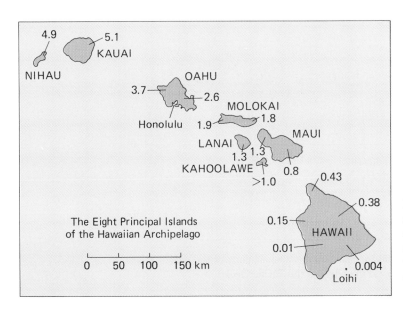

Figure 23–19 The Hawaiian Islands are a chain of undersea volcanoes called seamounts that rise above sea level and appear as oceanic islands. The ages, in millions of years, are for the oldest volcanic rocks found on each island or seamount.

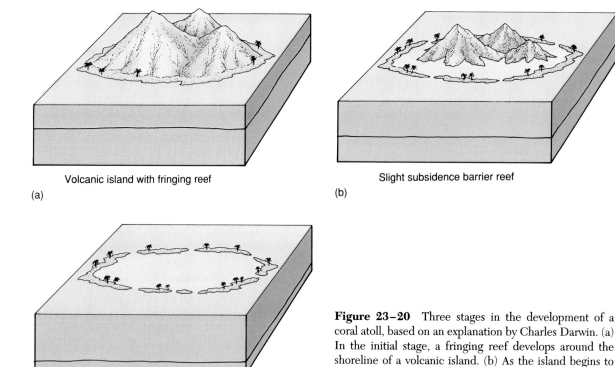

Volcanic island with fringing reef

(a)

Slight subsidence barrier reef

(b)

An atoll

(c)

Figure 23–20   Three stages in the development of a coral atoll, based on an explanation by Charles Darwin. (a) In the initial stage, a fringing reef develops around the shoreline of a volcanic island. (b) As the island begins to erode and subside, the coral builds upward and a barrier reef forms. (c) The volcanic center of the island is completely submerged below a central lagoon, forming an atoll.

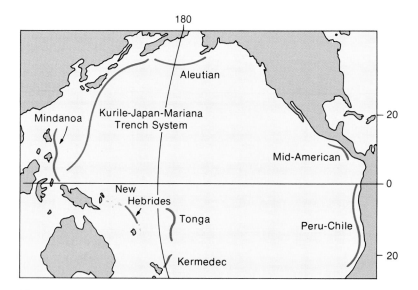

Figure 23–21   Oceanic trenches are the deepest parts of the ocean. Most trenches are located around the edges of the Pacific Ocean.

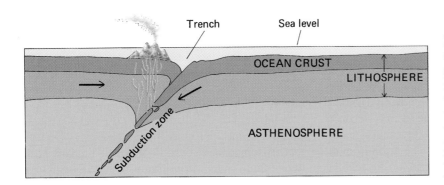

Trench   Sea level

OCEAN CRUST

LITHOSPHERE

Subduction zone

ASTHENOSPHERE

Figure 23–22 Along the leading edges of tectonic plates that have oceanic crust at their converging margins, deep-sea trenches form. Ocean floor rock is destroyed in the trenches by subduction, and volcanic island arcs are developed parallel to the trenches. An example is the Aleutian Trench west of Alaska.

mounts sink below sea level and are then called **tablemounts** or **guyots**, discovered during World War II. The great naturalist Charles Darwin (1809–1882) suggested that **coral atolls**, ring-like islands made up of coral reefs, are built over guyots (Fig. 23–20). This view is still accepted.

The deepest parts of the ocean, at depths of 7500 to 11,500 m, are oceanic **trenches**. The Challenger Deep in the Marianas Trench in the North Pacific Ocean is deeper (11,500 m) than Mt. Everest is tall (8730 m). These long, narrow, arc-shaped and steep-sided depressions frequently run parallel to **island arcs**, curving chains of volcanic islands such as the Aleutians, and to continents. Most trenches are located around the edges of the Pacific Ocean (Fig. 23–21). The deepest part of the Atlantic is the Puerto Rico Trench (8648 m), while the Java Trench (7125 m) is the deepest point in the Indian Ocean. Trenches form belts of intense earthquake and volcanic activity, like the mid-oceanic ridges, and are closely linked to continental drift. Ocean floor rock is being created at the oceanic ridges and destroyed in the trenches by subduction, so that the oceanic crust is recycled over many millions of years (Fig. 23–22). Deep-sea trenches are a major global feature that forms along the leading edges of tectonic plates.

23–10 OCEAN SEDIMENTS

The most economically important feature of the ocean floor is **sediments**, matter that settles out of the water onto the floor. Many originate on land and are brought to the oceans by streams or wind. Others are the remains of organisms that live in the oceans —protozoans, tiny mollusks, and algae.

Ancient sediments are the source of petroleum and various mineral resources. Gold and diamonds are found on continental shelves off land areas rich in these precious minerals. **Manganese nodules** are just beginning to be mined. They are formed by successive coatings of minerals rich in manganese and other metals precipitated around objects such as rock particles. The world is turning more and more to the oceans as the source of valuable natural resources.

23–11 HYDROTHERMAL VENTS

The recent discovery of **hot springs** at the bottom of the ocean on the crest of the mid-oceanic ridge has been one of the most exciting developments in oceanography. Known technically as **hydrothermal vents**, their existence had been predicted by J. W. Elder soon after the formulation of the theory of plate tectonics. Their place in that theory is quite straightforward. Like Yellowstone National Park and other areas of volcanic activity on the continents, the tectonic spreading centers at the mid-oceanic ridges are volcanically active. It was believed, therefore, that they also ought to have hot springs, the sign of hydrothermal activity. Now that prediction is being confirmed. The discovery of a system of hot springs along the 30,000-mile length of the mid-oceanic ridge is leading to new theories of ocean chemistry, biology, and ore formation.

The first attempts at finding hot springs under at least 2.5 kilometers of water were made in the early 1970s. At that time, knowledge of the sea floor was rudimentary. The sonar devices that were then in service, although adequate for flat terrain, were frequently damaged or lost at the rugged topography of the ridge axes. With the introduction of new technologies in the mid-1970s, changes began to come rapidly.

The U.S. Navy, for example, gave oceanographers access to techniques it had developed for mapping the ocean floor. High-precision, deep-sea navigation systems were then placed in routine service. Second, a large-field survey camera replaced the flimsy cameras that were difficult to control. A 1.5-ton camera vehicle, *Angus,* was developed at the Woods Hole Oceanographic Institution to house color cameras, stroboscopic lamps, power supplies, and sonar. The vehicle or "gorilla cage" is towed at elevations of about 20 meters above the sea floor. A color photograph is made every 10 seconds and carries a record of when it was made, enabling researchers to locate the feature exactly. Third, research submarines that can work at the depths of the ridge axis are available. *Alvin,* operated by Woods Hole, carries two researchers and a pilot. Its role is in visiting sites already identified in photographs. It can land on the sea floor very close to the designated target.

In the spring of 1977, an expedition using this equipment found the first hot-spring field in the Galápagos Rift Zone, 300 miles northeast of the Galápagos Islands. Warm water at a maximum temperature of 17°C streamed from every opening in the sea floor over a circular area about 100 meters in diameter, an "oasis" in an otherwise bleak, near-freezing terrain. Then, surrounded by an environment in which organisms otherwise are rare, entire ecosystems living without sun became apparent: fields of giant clams,

reefs of mussels, crabs, large pink fish, anemones, giant red tube-worms, and bioluminescent organisms. *Alvin* surfaced eventually, its sample basket loaded with clams and mussels. During the expedition, *Alvin* made a total of 15 dives to various targets.

Analysis of the hot-spring water showed that it had a high concentration of hydrogen sulfide (H_2S), a fact that explained the ocean-floor oases. Bacteria that derive their energy from the oxidation of hydrogen sulfide are at the base of this ecological system. Chemical reactions between the hydrogen sulfide and oxygen taken up by the bacteria release energy that drives the bacteria's metabolisms. The bacteria in turn nourish other species, suggesting that hot springs must be common on the ridge axes throughout the world if such a highly developed ecological system is to be sustained.

Many more hot springs were found in the Galápagos ridge axis when an expedition returned in 1979. Later, *Alvin* and *Angus* joined an expedition on the ridge axis at the East Pacific Rise off the tip of Mexico's Baja California. *Angus* photographed fields of hot springs, and *Alvin* made an amazing discovery among them: jets of black water called "**black smokers**" (from their precipitating iron sulfide) spewing from mineralized chimneys up to several meters tall (Fig. 23–23). The temperature of the vent water, determined by a new temperature probe, was 350°C in vent after vent, as predicted by the evolving model. Recently, a forest of black smokers has been discovered from *Alvin* near the Galápagos fields, and sulfide

Figure 23–23   A black smoker rising from the East Pacific Rise off the coast of Mexico. It is so called because of its ion sulfide. The hot (350°C), nutrient-rich water, heated by hot basalt, sustains thriving plant and animal communities. (Courtesy of Dudley Foster/Woods Hole Oceanographic Institute)

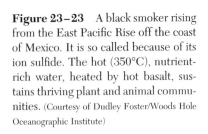

mounds have been found in the Guaymas Basin in the Gulf of California.

The metal-rich deposits of the mid-oceanic ridges and the floor of the ocean seem to be the result of the mixing of hot, acidic, reducing hydrothermal water with cold, alkaline, oxidizing ambient water as the hot solution rises to the surface. Even ore deposits that lie on the continents are proving to have been placed there by plate-tectonic forces after they had been manufactured under the sea by hydrothermal activity. Although many questions about them remain to be answered, the ocean's hot springs have been called at least the discovery of the decade.

23–12 ENVIRONMENTAL OCEANOGRAPHY

We now have the ability to affect the ocean on a broad scale, either by design or by accident. Industrial and domestic wastes are often dumped directly into the sea. Industrial wastes are often toxic to marine and human life. Domestic sewage provides nutrients for marine plants, including organisms that cause disease. Areas near metropolitan centers become more polluted, but the wastes are not restricted to the area where they are dumped. They are carried by currents often far from the site of intervention. Sewage sludge sinks and pollutes the bottom sediments.

Oil pollution is a particular problem in the coastal environment because oil floats on water and does not mix with it. Thousands of tons of crude oil are released into the sea off the coasts in large-scale oil spills from tankers. They pollute beaches and kill bird and marine life. With the increase in off-shore drilling for oil, rigorous safeguards will have to be instituted if accidents that cause pollution are to be kept at a minimum.

One of the more spectacular oil spills occurred in Prince William Sound, Alaska, on Good Friday, 1989. The *Exxon Valdez* tanker, loaded with 1.2 million barrels of crude oil, left Alaska's south coast, heading for California. Twenty-five miles out it ran into a reef. The rocks at Bligh Island tore five gashes in the hull, one of them 6 feet wide by 20 feet long, creating the worst oil spill in U.S. waters. Within a week, about 240,000 barrels of oil—10 million gallons—had spread over 900 square miles of water. The clean-up continues to this day at a cost of billions of dollars (Fig. 23–24). The toxic effects of the oil, deposited in marshes, bottom sediments, and the gravel of beaches, are expected to be felt for many years. The accident took place in good weather, in the absence of traffic, with good radio contact between Coast Guard radar operators and the ship, and with clear visibility of 10 miles. Even sophisticated technology cannot prevent an accident where there is gross human error, as there was in this case.

Figure 23–24 The *Exxon Valdez* oil spill caused this devastation in Prince William Sound, Alaska, in March, 1989. The toxic effects of the oil are expected to be felt for many years. (James Wilson/Woodfin Camp & Associates)

The Mediterranean Sea has a striking case history of marine pollution and the efforts to combat it. During the summer months, upwards of 100 million people visit the area. In a typical pattern, many come down with an intestinal ailment, or a skin or eye infection. Some contract dysentery, typhoid, poliomyelitis, or cholera.

In the 1980s the countries that rim the Mediterranean took steps to reverse that scenario. They adopted the first treaty to deal directly with pollution of the Mediterranean that originates on land, pledging themselves to take action that would prevent and control pollution discharged from rivers or emanating from any other land-based sources within their territories.

Such pollution constitutes about 85% of all pollutants entering the sea. It comprises industrial wastes, municipal sewage, fertilizers, and pesticides. Some of the chemicals originate at sources hundreds of miles inland, even from land-locked countries. Although the clean-up was expected to cost billions of dollars, the countries realized that they had to do it in order to protect their tourist industry and the health of their own people.

One part of the treaty, called the "blacklist," lists materials that are prohibited from entering the sea because of their toxicity, persistence in the environ-ment, or accumulation in the food chain. The black list includes organic chemicals containing halogens, phosphorus, or tin; mercury, cadmium, and their compounds; waste lubricating oils; nondegradable synthetic products; substances that are carcinogenic, teratogenic, or mutagenic; and radioactive isotopes.

A second list, the "gray list," includes substances generally accepted as being less noxious than the first. Their release into the Mediterranean from land-based sources is strictly limited. Measures are taken to check the discharges, which are subject to a licensing procedure. The gray list includes 21 elements, ranging from antimony and arsenic to vanadium and zinc, as well as compounds made from them; cyanides and fluorides; hydrocarbons and crude oil; thermal discharges; chemicals that can cause eutrophication; acids and alkalies in large quantities; and nontoxic products that may become harmful to the marine environment if discharged in large quantities. The signatories to the treaty hoped that by the end of the decade they would have reversed the trend of pollution and could control it, making the Mediterranean a clean and safer place. Following a four-year study of the oceans in general, the United Nations concluded that the ocean was healthier in 1982 than it was in 1972.

KEY TERMS

Abyssal Plain   A nearly level featureless part of the ocean floor between the mid-oceanic ridge and the continental slope.

Bathythermograph   A device that measures and records the change in temperature with depth of water.

Black Smoker   A jet of black water rich in iron sulfide spewing from mineralized chimneys up to several meters tall in hot-spring fields.

Continental Shelf   The gently sloping surface around the margin of a continent or island.

Continental Slope   The steeply sloping surface reaching from the continental shelf to the ocean bottom.

Coral Atoll   A circular coral reef that surrounds a lagoon and is bounded on the outside by the deep water of the open sea.

Deep-Sea Fan   Submerged fan-shaped deposit of sediment often located at the seeward margin of a submarine canyon.

Degassing.   The expulsion of gases during the crystallization of minerals from cooling magma.

Guyot   A flat-topped volcanic feature on the ocean floor.

Heat Budget   The equilibrium between the heat absorbed by the earth in one year and the amount of heat radiated back into space in one year.

High Tide   The tide when it is at its greatest elevation.

Hot Spring   A spring formed where hot groundwater flows to the surface; also found at the bottom of the ocean on the crest of the mid-oceanic ridge.

Hydrosphere   The earth's water shell, including the oceans, lakes, streams, glaciers, ground water, and similar features.

Hydrothermal Vent   Submarine hot spring.

Island Arc   A curved chain of volcanic islands along a deep oceanic trench; found near tectonic plate boundaries where subduction of one plate beneath the other occurs.

Low Tide   The farthest ebb of the tide.

Manganese Nodules   Lumps containing oxides of manganese, iron, nickel, and copper found scattered over the ocean floor.

Manned Submersible   A deep-diving oceanic research vehicle.

Nansen Bottle   A device used by oceanographers to obtain samples of ocean water.

Neap Tide   A tide of minimum range occurring at the first and the third quarters of the moon.

Oceanic Trench   A long, narrow, deep depression on the ocean floor.

Oceanography   The scientific study of the sea.

Remotely Operated Vehicle (ROV)   An unmanned submersible equipped for longer, safer, better-controlled dives.

Salinity   The amount of dissolved solids in seawater.

Sea   A body of salt water that is more or less landlocked.

Seamount   An undersea volcano.

Sediment   Finely divided particles of organic or inorganic origin that accumulate in loose form.

Sonar   An echo-sounding instrument that detects the presence and location of a submerged object by means of sonic and supersonic waves, and records topographical features of the seafloor.

Submarine Canyon   A V-shaped rocky canyon cut into the continental shelf or slope.

Tablemount   A flat-topped seamount.

Thermocline   A zone in the ocean that is characterized by a rapid decrease in temperature. It extends from a depth of 250 to 1400 m.

Tidal day   A period that encompasses two high and two low tides: 24 hours, 50 minutes.

Tidal interval   The time between two high tides, averaging 12 hours, 25 minutes.

Tidal range   The difference in sea level between high tide and low tide.

Tide   The periodic rise and fall of sea level in response to the gravitational interaction of moon, sun, and earth.

Turbidity Current   A dense, fast-flowing body of sediment-laden water.

THINGS TO DO

1. Observe the change of water pressure with depth, using a tall cylinder of water and a pressure gauge. Take measurements at various depths.

2. A graphic demonstration of the variation of water pressure with depth: Punch holes up the side of a tall can about 2 inches apart and cover the row of holes with plastic tape. Fill the can with water above the top hole, hold the can over a sink, and strip the tape from the holes beginning at the bottom. Observe the distance the jets travel outward from the can.

3. Study a globe of the earth to see what proportion of it is covered with water.

4. Distill natural salt water, if available, or a synthetic brine.

EXERCISES

The Global Ocean

23–1. Why is earth called the "water planet"?

23–2. What are the oceans used for?

23–3. How does the average depth of the oceans compare with the average elevation of the continents?

23–4. What is meant by the "heat budget" of the ocean?

23–5. Why is the sea cold?

23–6. Why is the sea dark?

Oceanography

23–7. Discuss the interdisciplinary nature of oceanography.

23–8. Explain why our knowledge of the ocean has increased by leaps and bounds in recent years.

23–9. Discuss some of the tools that are now available for observing the ocean.

23–10. How is a knowledge of the speed of sound in ocean water used to determine the topography of the ocean floor?

The Ocean Floor

23–11. How are continental shelves different from continental slopes?

23–12. How are submarine canyons formed?

23–13. What are some features of the deep-sea bottom?

23-14. How are seamounts and tablemounts related?

23-15. Where is the youngest part of the ocean floor?

23-16. Why are turbidity currents effective as agents of submarine erosion?

23-17. In what way may mid-oceanic ridges and deep-sea trenches be related to plate tectonics?

23-18. How are features of the continental shelf related to adjacent land masses?

23-19. What made the discovery of hydrothermal vents the oceanographic discovery of the decade?

Economic and Environmental Oceanography

23-20. Can anything be done to reduce marine pollution? Explain.

23-21. What are the attractions of offshore drilling and mining?

23-22. Discuss some instruments now available to observe the ocean.

23-23. What are the most valuable resources presently obtained from beneath the sea floor?

23-24. Is desalination of seawater a viable means of obtaining water that is fit to drink? Discuss.

23-25. *Multiple Choice*

    A. The oceans cover about what percentage of the earth's surface?

        (a) 25
        (b) 50
        (c) 70
        (d) 90

    B. Undersea volcanoes are known as

        (a) submarine canyons.
        (b) seamounts.
        (c) sea-floor plains.
        (d) trenches.

    C. A theory that explains the origin of the ocean is

        (a) degassing.
        (b) oceanography.
        (c) desalination.
        (d) turbidity current.

    D. A zone of the ocean in which the temperature decreases rapidly with depth is called the

        (a) continental shelf.
        (b) continental slope.
        (c) thermocline.
        (d) heat budget.

    E. Features of the ocean floors can be determined by

        (a) sonar.
        (b) Nansen bottle.
        (c) bathythermograph.
        (d) telescope.

SUGGESTIONS FOR FURTHER READING

Childress, James J., Horst Felbeck, and George N. Somero, "Symbiosis in the Deep Sea." *Scientific American*, May, 1987.
The symbiosis of invertebrate animals and sulfide-oxidizing bacteria explains the remarkable density of life at deep-sea hydrothermal vents.

Dolan, Robert, and Harry Lins, "Beaches and Barrier Islands." *Scientific American*, July 1987.
A severe northeaster along the Atlantic coast on March 7, 1962, now serves as the benchmark for workers in coastal geology and engineering. As a result of work done since then, it is now understood how beaches and barrier islands were created, how external forces change them, and why they will continue to change in spite of human efforts to halt the natural processes.

East Pacific Rise Study Group, "Crustal Processes of the Mid-Ocean Ridge." *Science*, 3 July, 1981.
The mid-oceanic ridge is the largest single geological feature on the surface of the earth. What is the nature of the magma chamber system that underlies it and is thought to be responsible for its volcanism, crustal structure, magnetic character, and other processes?

Gass, Ian G., "Ophiolites." *Scientific American*, August, 1982.
Ophiolites seem to be fragments of oceanic crust on land. They are clues to how oceanic crust forms and spreads.

Greenberg, David A., "Modeling Tidal Power," *Scientific American*, November, 1987.
The highest tides in the world are found in the upper Bay of Fundy, which lies between New Brunswick and Nova Scotia. What would be the environmental and economic cost of harnessing tidal power to generate electricity?

Heezen, Bruce C., and Ian D. MacGregor, "The Evolution of the Pacific." *Scientific American*, November, 1973.
Deep-sea drilling shows differences in the floor of the Pacific Ocean. Does the movement of the crust under the basin account for them?

MacDonald, Ken C., and Bruce P. Luyendyk, "The Crest of the East Pacific Rise." *Scientific American*, May, 1981.

Undersea exploration has revealed much about how new segments of the earth's crust emerge. At least 70% of the crust has formed at the mid-oceanic rifts.

MacDonald, Kenneth C., and Paul J. Fox, "The Mid-Ocean Ridge." *Scientific American*, June, 1990.

Even though the 75,000-kilometer-long formation known as the mid-oceanic ridge is by far the longest structure on the earth, less was known about its features in 1982 than about the craters on the dark side of the moon. Equipped with a new type of sonar called Sea Beam, the Scripps Institution of Oceanography research vessel *Thomas Washington* surveyed the East Pacific Rise, a volcanic mountain chain that lies under the Pacific Ocean and is part of the mid-oceanic ridge.

Nelson, C. Hans, and Kirk R. Johnson, "Whales and Walruses as Tillers of the Sea Floor." *Scientific American*, February, 1987.

Gray whales and Pacific walruses modify the Bering Sea floor in the course of gathering food on the continental shelf underlying the sea. Their feeding activities seem to be beneficial to the area, enhancing its productivity.

Penney, Terry R., and Desikan Bharathan, "Power from the Sea." *Scientific American*, January, 1987.

The temperature difference between warm surface water and cold bottom water can be exploited to generate electricity. The technology will be useful when oil supplies become scarce.

Revelle, Roger, "The Ocean." *Scientific American*, September, 1969.

Introductory article in an entire issue devoted to the topic.

Rona, Peter A., "Mineral Deposits from Sea-Floor Hot Springs." *Scientific American*, January, 1986.

Seawater circulating through fractured volcanic rock above sources of heat participates in chemical exchanges with the rock. A major result is significant deposits of metal, which may lead to mining in the sea on a commercial scale. The finding also elucidates the origin of some of the major mineral deposits on land.

Spindel, Robert C., and Peter F. Worcester, "Ocean Acoustic Tomography." *Scientific American*, October, 1990.

Like the computed tomography (CT) used in medicine and seismic tomography used in geology, ocean tomography employs beams of energy to create three-dimensional images of the ocean's interior—its temperatures, salinities, densities, and current speeds. Widespread applications of acoustic tomography may result in important new understanding of the ocean in the decade ahead.

Whitehead, John A., "Giant Ocean Cataracts," *Scientific American*, February, 1989.

The world's tallest waterfall is Angel Falls in Venezuela, with a height of just over a kilometer. The Guaira Falls along the Brazil–Paraguay border has the largest average flow rate, about 13,000 cubic meters per second. But below the Denmark Strait there is a giant ocean cataract that carries 5 million cubic meters of water per second through a descent of 3.5 kilometers. Undersea cataracts, through their influence on the salinity, temperature, and biology of the ocean have an effect on the climate and ecology of the entire earth.

ESSAY

THE RISING SEA

KAREN ARMS

In 1987, the National Research Council reported that sea levels have been rising worldwide for at least the last century.

The sea level is rising partly because coastal land is subsiding, but most of the rise is due to heat, which not only melts ice but also causes sea water to expand (Figure A). Scientists predict that the sea level on the East Coast of the United States will rise between 40 centimeters and 1.5 meters before 2085 and by up to 7 meters in the following cen-

Figure A   Sheets of ice and snow melting and about to tumble into the ocean. More than 99% of the fresh water on earth is locked up in permanent ice caps, snowfields, and glaciers. When these melt, they add a huge volume of water to the ocean. (Carolina Biological Supply Company)

tury. As an example of what this would mean, Figure B shows that a rise of about 5 meters in the sea level would affect 40% of Florida's population, putting Miami, St. Petersburg, and Jacksonville under water. We should be planning now to move several major cities inland, or at least to permit them to expand only away from the sea.

Another unfortunate effect of the rising sea level will be the flooding and disappearance of coastal wetlands, such as marshes and mangrove swamps. These wetlands, such as those on the southern and eastern coasts of the United States, are home to innumerable species of birds, and they are vital to the survival of seafood species such as oysters, shrimp, crabs, and many species of fish. They also serve an important role in breaking down pollutants that wash down rivers from farther inland. Without them, we shall lose much of our seafood, many species of plants and animals, and much of our ability to protect the ocean from pollution (Figure C).

One complication we encounter in discussions of the greenhouse effect is that no one is sure how much warming will produce how much rise in sea level. Another is the prediction that the sea level will rise more in some places than others. The level has risen about 30 centimeters along the Atlantic Coast of the United States. On the other hand, the sea level has fallen around Alaska. How can this be? Surely one can calculate how much heat it will take to melt a particular ice cap or glacier and how much the volume of the ocean will increase as the water is heated and expands. The difficulties arise because the volume of the ocean is not the only thing that determines sea level. We measure sea level by the level of the land, and land rises and falls for several reasons.

The outer shell of the earth is made up of huge **tectonic plates**, which float on the fluid core of the earth. These plates move slowly in various directions (sometimes causing earthquakes). On the West Coast of the United States, the Pacific Plate, under the ocean, is sliding under the American Plate on which the land lies, pushing it up. So the West Coast is rising. (The San Andreas Fault marks this collision of plates.) The East Coast, on the other hand, is sinking on the eastern side of the American Plate. So any increase in the volume of the ocean will raise the sea level on the East Coast of the United States more than on the west.

Movement of tectonic plates is not the only reason land rises and falls. We have noted that Alaska is rising, possibly because some of the ice that

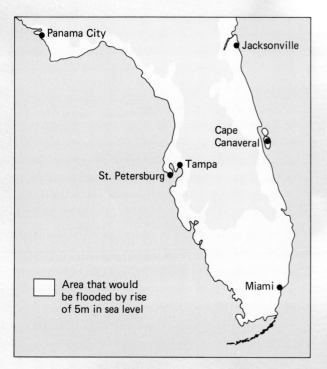

Figure B A map of Florida showing the approximate area that would be flooded if the sea level rose about 5 metres.

Figure C   Wetlands in the southeastern United States threatened by the rising sea level. As a Georgia bumper sticker says, "Without wetlands there would be no seafood." (Karen Roeder)

weighs it down has melted and let the land rise. More common are human activities that cause the land to sink. For instance, as we pump or dig oil, water, or coal out of the earth, the land above may sink (Figure D).

Deltas are peculiarly vulnerable to human disruption. A **delta** is the land that forms at the mouth of a river as the river deposits sediment that it has picked up farther inland. This sediment is often fertile soil, so many deltas are important agricultural regions. Bangladesh lies almost entirely on the delta of the Ganges River. The dreadful floods in that country in recent years result from deforestation in other countries higher up the river. Because the vegetation that used to absorb some of the heavy monsoon rains is gone, the water now runs down unchecked, carrying its load of soil far out to sea (together with much of Bangladesh's soil) and flooding Bangladesh. The combination of floods and rising sea level is washing Bangladesh away.

Another example is the Nile delta, which has produced food and prosperity for Egypt for thousands of years. The delta used to be fertilized by annual floods that overflowed the banks of the river and dropped sediment on the surrounding fields. But now, in the name of flood control, the upper Nile has been dammed. The sediment that once fertilized the delta accumulates behind the dam. Soil erosion removes soil from the delta and it is not replaced. To make matters worse, irrigation projects pump water to the surface and the delta subsides into the space left beneath it. A slight rise in sea level threatens to flood Egypt's main agricultural area forever, this time with sea water.

Louisiana, which contains most of the delta of the Mississippi River, is losing land faster than either Bangladesh or Egypt. It loses about 70 square kilometers of land every year to the rising sea, more than any other state or country in the world.

Figure D   The sea rises and the land sinks. This oil pumping station and refinery at Whitfield, California, lies behind massive seawalls because it is below sea level. The land it stands on subsides as oil and water are pumped out of the earth beneath. (Carolina Biological Supply Company)

We cannot predict the rise in sea level with any accuracy, partly because we cannot predict human behavior, which can change very rapidly. Instead of continuing to rise, our consumption of fossil fuel might very well fall during the next century, as it did during the energy crisis of the 1970s. Equally, the pattern of deforestation may change. At the present rate, there will be essentially no tropical forest left by 2085. But many countries have programs to reforest their land, and this might happen faster than we now anticipate. In China, for instance, more of the land is now covered by forest than was the case 20 years ago. The only thing we can say for sure is that we are heating up the atmosphere and that this is causing changes in temperature and rainfall throughout the world.

Earth and sky, Masai Mara National Reserve, Kenya. (© Don Mason/The Stock Market)

24

THE EARTH'S ATMOSPHERE AND BEYOND—METEOROLOGY

In the most general terms, the earth's atmosphere—the gaseous envelope surrounding the earth—is divided into lower and upper regions. The lower atmosphere usually is considered to extend to the top of the stratosphere, an altitude of about 50 km. Everything above that is the upper atmosphere. The lower and upper atmospheres, in turn, are commonly subdivided on the basis of temperature distribution into the troposphere, tropopause, stratosphere, stratopause, mesosphere, thermosphere, and exosphere (Fig. 24–1).

Weather refers to the state of the atmosphere for a short period of time, a day or so, and for a specific area, such as New York City. It is described in terms of temperature, humidity, clouds, precipitation, pressure, and wind. *Meteorology* is the science of the atmosphere and of weather and depends on an understanding of basic natural laws. Rockets, balloons, satellites, and ground stations are all used to gather weather information, and sophisticated computers perform in minutes calculations that might otherwise take months. Despite modern instruments and methods, weather forecasting is at best an inexact although improving science.

24–1 EVOLUTION OF THE EARTH'S ATMOSPHERE

The most common atoms in the universe are the two simplest: hydrogen (92%) and helium (7%). Most of the atoms of the other common elements consist of carbon, nitrogen, oxygen, sulfur, phosphorus, neon, argon, silicon, and iron. When a planet forms out of dust and gas, it ought to be mostly hydrogen and helium, and these gases would make up most of the original atmosphere. The composition of the large Jovian planets—Jupiter, Saturn, Uranus, and Neptune—is close to that of the universe at large. The small terrestrial planets—Mercury, Venus, earth, and Mars—have more of the heavier elements and are poorer in such gases as helium that could escape from their weaker gravitational pull.

Helium atoms do not combine with other atoms, but hydrogen atoms do. They are so numerous that any atom that can combine with hydrogen will do so. Each carbon atom combines with four hydrogen atoms to form methane (CH_4); nitrogen forms ammonia (NH_3); oxygen forms water (H_2O); and sulfur forms hydrogen sulfide (H_2S). These compounds would all be found in the primitive atmosphere of the earth, **Atmosphere I**, and the ocean would have contained much ammonia in solution (Fig. 24–2).

Water vapor released from the interior of the earth by degassing processes such as volcanism would have

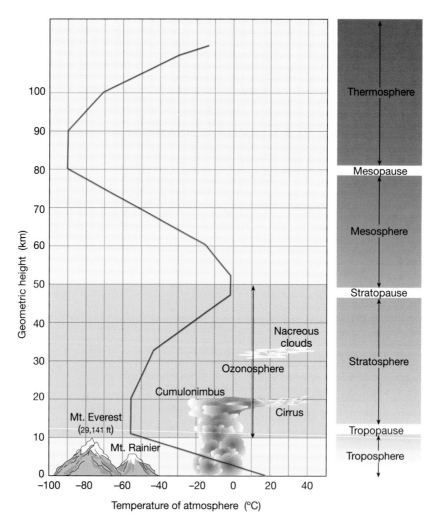

Figure 24-1   The structure of the earth's atmosphere. The curving line shows the temperature (scale at bottom of figure). In the troposphere, the zone where our weather occurs, the temperature decreases as you go up in altitude, since a major source of heat is infrared radiation emitted from the ground. (U.S. Navy Research Facility)

been a component of a secondary atmosphere as well, **Atmosphere II**. The decomposition of metal carbides, binary compounds of carbon and other elements, could give rise to methane (CH_4), carbon dioxide (CO_2), and carbon monoxide (CO); nitrides, binary ammonia (NH_3), compounds of nitrogen and nitrogen (N_2); and sulfides, hydrogen sulfide (H_2S). Atmosphere II would be composed chiefly of nitrogen (N_2), carbon dioxide (CO_2), and water (H_2O)

vapor, and the ocean would contain much carbon dioxide (CO_2) in solution.

The present atmosphere of the earth, **Atmosphere III**, consists mainly of nitrogen (N_2), oxygen (O_2), and water vapor (H_2O), with only small quantities of gas dissolved in the ocean. The oxygen was put there only after life had developed by early organisms making glucose through **photosynthesis**, a process by which carbohydrates are synthesized from carbon

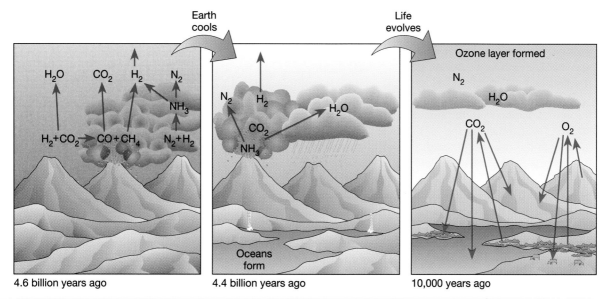

Figure 24–2 The evolution of the earth's atmosphere. (*Left*) The primitive atmosphere, nearly 5 billion years ago, Atmosphere I, was rich in hydrogen compounds: CH_4, NH_3, H_2O, and H_2S. (*Middle*) Atmosphere II was composed chiefly of N_2, CO_2, and H_2O as the earth cooled. (*Right*) Atmosphere III, the present atmosphere, consists mainly of N_2, O_2, and H_2O, the oxygen produced by plants and the source of the ozone layer.

dioxide and water, and releasing oxygen as a by-product. The oxygen released by photosynthesis would screen out the ultraviolet radiation with an ozone layer it would form in the atmosphere, and gradually the atmosphere would turn from reducing to oxidizing. Life, then, must have originated in either Atmosphere I or Atmosphere II.

24–2 THE TROPOSPHERE

The lowest region of the atmosphere, the **troposphere**, extends to a height of about 18 km over the equator and 8 km over the poles, with an average height of 11 km. It is within the troposphere that people live and work and all the earth's weather takes place. It is a turbulent region of swirling air masses, cloud formations, warm and cold fronts, and storms. It is the source of the air we breathe. Two gases, nitrogen and oxygen, account for 99% of dry, clean air, remaining essentially unchanged with increasing altitude (Table 24–1).

The water-vapor content of the atmosphere is concentrated in the troposphere and varies from about 20 grams per kilogram of air in the tropics to less than 0.50 g/kg at the poles. It is virtually nonexistent in the layers of the atmosphere above the troposphere.

Table 24–1 COMPOSITION OF UNCONTAMINATED DRY AIR NEAR SEA LEVEL

Component	Mole (%)
Nitrogen	78.084
Oxygen	20.947
Argon	0.934
Carbon dioxide	0.0314
Neon	0.00182
Helium	0.000524
Krypton	0.000114
Hydrogen	0.00005
Xenon	0.0000087

Dust, smoke, salt, pollen grains, volcanic ash, organic materials, and a variety of other solids are also present. They are involved in cloud and fog formation and rainfall, acting as nuclei around which water droplets and ice crystals develop.

Water vapor and carbon dioxide strongly absorb infrared radiation from the sun and earth and play a key role in maintaining the earth's heat balance by holding in the heat of the earth's lower atmosphere through the **greenhouse effect** (Fig. 24–3). As the chemist Svante August Arrhenius recognized, the atmosphere is transparent to incoming visible radiation but is opaque to the longer-wavelength infrared radiation (heat) that is reradiated by the surface of the earth; such radiation cannot escape back into space because it is trapped by carbon dioxide. The effect of the carbon dioxide is the same as that of glass over a greenhouse, and it leads to a warming of the climate; the more carbon dioxide, the more warming.

The temperature in the troposphere decreases with increased altitude at a nearly uniform rate of about 6.5 Celsius degrees per kilometer, falling to an average temperature of $-56°C$ at the top. You can find snow for skiing in Southern California if you go to an altitude of 2500 m. The vertical temperature change, or gradient, is called the normal or environmental **lapse rate of temperature** (Fig. 24–4). The main reason for the decrease in temperature is the

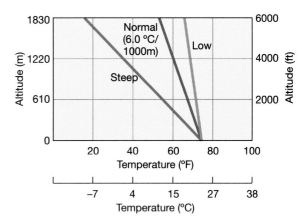

Figure 24–4   The decrease in temperature in the troposphere with increased altitude is called the lapse rate of temperature. Steep, normal, and low lapse rates are illustrated.

decrease in air density with increasing elevation. The higher one is above sea level, as on a mountain top, the less dense the air. Since the source of heat is the earth's surface, the dense air at lower levels receives heat by conduction and convection more efficiently than less dense air. Further, water vapor and carbon dioxide, the chief infrared absorbers, are more concentrated at lower levels. The **tropopause** is the altitude at which the temperature ceases to fall with increased altitude and separates the troposphere from the stratosphere.

24–3 THE STRATOSPHERE; THE OZONE LAYER

The **stratosphere**, the second layer of the atmosphere, extends from an altitude of 11 to 50 km. It was so named in the belief that it consisted of an arrangement of layers ("strata") that did not mix. The temperature of the atmosphere, contrary to popular belief at the turn of the century, stops falling at about 11 km at about $-56°C$ and remains fairly constant with increasing altitude to 20 km. In the range from 20 to 50 km, the temperature increases progressively, reaching a maximum of about $-20°C$ at 50 km.

Except for **ozone**, the three-atom molecule of oxygen, O_3, the chemical composition of the stratosphere is essentially constant. The concentration of ozone, however, increases from about 0.04 parts per

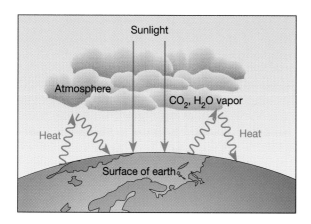

Figure 24–3   The operation of the greenhouse effect. Visible light passes through atmospheric H_2O and CO_2 and reaches the earth's surface. But heat radiated from the surface is in the infrared and is absorbed by these gases, thus trapping heat in the atmosphere and resulting in an elevated surface temperature.

million in the troposphere to a maximum of about 10 parts per million at an altitude of 25 to 30 km in the stratosphere. Although it is only a trace constituent, ozone exerts a profound influence on life on earth by absorbing the sun's ultraviolet radiation from 2100 to 2900 Å. Were this radiation not absorbed, many forms of life would cease to exist and others would be imperiled. In the process of absorbing ultraviolet radiation and decomposing into diatomic oxygen molecules and oxygen atoms, ozone releases heat to the stratosphere. Because of this ozone layer, temperatures increase in the upper parts of the stratosphere. Temperatures at the **stratopause**, about 50 km above the earth, are about the same as temperatures found on the earth's surface. The temperature then decreases progressively from $-20°C$ at 50 km to approximately $-92°C$ at 85 km. In this region, the **mesosphere**, the ozone concentration becomes lower and lower.

$$O_2 \xrightarrow{UV} O + O \qquad \text{(Dissociation of oxygen molecules)} \qquad (24-1)$$

$$O + O_2 \longrightarrow O_3 \qquad \text{(Formation of ozone)} \qquad (24-2)$$

$$O_3 \xrightarrow{UV} O_2 + O \qquad \text{(Dissociation of ozone)} \qquad (24-3)$$

Chlorofluorocarbons (CFCs), such as $CFCl_3$ and CF_2Cl_2, also known as Freons or halocarbons, are gases that were used widely as propellants in aerosol cans and as refrigerants (Fig. 24–5). A theory was advanced in 1974 by Mario J. Molina and F. Sherwood Rowland of the University of California at Irvine that these compounds represent a definite hazard to the earth's ozone layer. They are ideal as propellants to spray cosmetic products, cleaning agents, and some medications because they do not react chemically with whatever is being sprayed. But this same chemical inertness causes them to persist in the atmosphere, making their way into the stratosphere, where they are broken down by ultraviolet light, releasing free chlorine atoms that react with and deplete the ozone (Fig. 24–6).

$$CFCl_3 \xrightarrow{UV} \qquad (24-4)$$
$$CFCl_2 + Cl \qquad \text{(Dissociation of Freon-11)}$$

$$Cl + O_3 \longrightarrow \qquad (24-5)$$
$$ClO + O_2 \qquad \text{(Reaction of chlorine atom with ozone)}$$

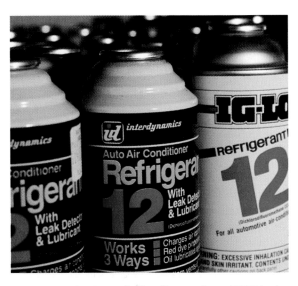

Figure 24–5 Cans of chlorofluorocarbons (CFCs), also known as Freons or halocarbons, for use in automobile air conditioners. (Karen Roeder)

A panel of scientists concluded that the continued release of halocarbons would eventually produce a reduction between 2% and 20% in stratospheric ozone. A reduction of this magnitude would lead to an increase in the amount of damaging ultraviolet radiation reaching the surface of the earth. This would lead, in turn, to a larger increase in all forms of skin cancer and would also have harmful effects on plants and animals. As a result, the use of halocarbons as propellants has been banned, except in cases of special need, as in medicine.

Atmospheric scientists of the British Antarctic Survey in 1985 made the completely unanticipated discovery of an **ozone hole** in the polar atmosphere over Halley Bay, Antarctica. The springtime amounts of ozone had decreased by more than 40% between 1977 and 1984. Other groups soon confirmed the report and showed that ozone has disappeared rapidly from the lower stratosphere (12 to 24 km) every spring in a column of air covering the entire continent and beyond and should reach 50% depletion in the next few years. Such loss levels are completely unprecedented in the history of ozone observations.

Recent data are regarded as strong circumstantial evidence that CFCs and the ozone hole are connected; that is, that chlorine causes the hole. Because

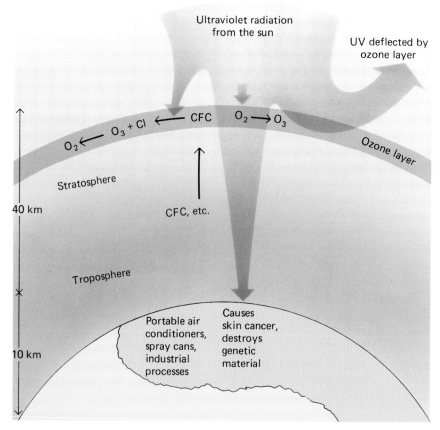

Figure 24–6   The ozone layer is formed when ultraviolet radiation decomposes diatomic oxygen molecules, which subsequently reorganize into triatomic molecules of ozone. It is being depleted by the action of chlorofluorocarbons (CFCs).

these effects are so serious, many investigators are studying them. Until we know why the hole forms, we will not know whether it has global implications or will remain confined to the Antarctic stratosphere, where meteorological conditions are unique.

The ozone hole is not actually a hole but a region that contains an unusually low concentration of ozone (Fig. 24–7). Total ozone loss over Antarctica averages 50% or higher. Recent work indicates that **polar stratospheric clouds** (PSCs) trigger ozone depletion; ice particles in the clouds catalyze chemical reactions that free chlorine from CFCs. The ozone hole occurs near Antarctica during spring because the formation of the hole requires the presence of PSCs, which form only during the coldest times of the year.

An international mission in 1989, the Airborne Arctic Stratospheric Expedition (AASE), sought to determine whether the chlorine chemistry that destroys ozone in Antarctica also is operating in the Arctic. The answer turns out to be yes. Large regions in the Arctic stratosphere above 18 kilometers experienced 10% to 20% ozone depletion. Relatively milder winter conditions in the north polar region, however, have so far saved the Arctic from the massive seasonal ozone losses of Antarctica. The regions at altitudes below 18 kilometers were too warm for PSCs to form. A change in climate in the Arctic or an unusually cold Arctic winter could lead to a sudden large depletion of ozone not very far away from populated northern latitudes.

Is there a quick technological means to stop ozone depletion, which is known to be widespread, even over the equator? There seems to be no quick fix for this problem. Large quantities of CFCs remain in

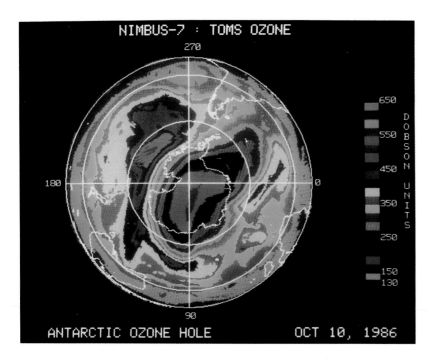

Figure 24–7 A computer map of ozone concentrations over the southern hemisphere, based on measurements made aboard the satellite Nimbus-7. This map shows the ozone hole (purple) in the ozone layer over Antarctica. (NASA/Goddard Space Flight Center)

refrigerators, air conditioners, and foams, much of which will eventually be released into the atmosphere. Fifty-six nations signed the 1987 Montreal Protocol on Substances That Deplete the Ozone Layer, which requires them to reduce their production and use of CFCs by 50% by the year 2000. As the chemistry of ozone depletion and the role that chlorine plays became clear, and new scientific data emerged showing that the loss of ozone was greater than estimated, the same nations in 1990 agreed to phase out CFC production completely by the turn of the century and to reduce the production of other agents. It appears that chlorine may not return to levels that existed before the discovery of the ozone hole until the middle of the next century, or even later.

24–4 THE IONOSPHERE; RADIO TRANSMISSION

The **ionosphere** extends from about 50 km to thousands of kilometers beyond the earth's surface. It is a region that has a relatively large concentration of free electrons and positively charged ions, formed by the absorption of the sun's ultraviolet radiation and to a lesser extent by X-rays, and by collisions with cosmic

rays, which strip electrons from molecules and atoms of nitrogen and oxygen. The concentrations of charged particles are still so low, however, that they do not recombine to a significant extent.

The existence of the ionosphere was independently proposed by Arthur E. Kennelly (1861–1939), an electrical engineer, and Oliver Heaviside (1850–1925), a physicist, to explain engineer and inventor Guglielmo Marconi's (1874–1937) successful transmission of radio waves over long distances around the earth. Since radio waves travel in straight lines, it was thought that they would tend to be lost in space rather than follow the curvature of the earth. Kennelly and Heaviside explained that long-distance radio transmission occurs because an electrically conducting region in the upper atmosphere reflects radio waves back to earth, and their idea was confirmed experimentally. When a radio wave enters the ionosphere, it induces a current among the charged particles there. The upper part of the wave front, being in a more highly charged area, is speeded up. The wave pivots and thus is bent downward. The repeated bouncing of the radio wave between the ionosphere and the ground accounts for the reception of radio signals around the curvature of the earth (Fig. 24–8).

The ionosphere is divided into D, E, F_1, and F_2 regions according to increasing altitude and electron

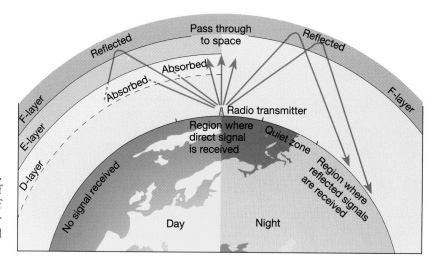

Figure 24–8 The ionosphere, or Kennelly-Heaviside layer of the atmosphere. Reflections of radio waves from the F-layer allow radio transmission around the curvature of the earth.

concentration, the electron concentration depending on a number of factors, such as time of day, season of the year, latitude, and phase of the solar cycle. The D region between 50 and 85 km has a maximum concentration at noon of about 1.5×10^4 electrons/cm³, which almost vanishes at night; for this reason radio reception from distant stations is usually better at night than during the day. At night the base of the ionosphere rises, the reflection occurs at higher levels, and the radiowaves are transmitted over long distances. The electron concentration reaches a maximum of about 1×10^6 electrons/cm³ at an altitude of 250 km, then gradually falls off.

24–5 VAN ALLEN BELTS AND AURORAS

A region called the **thermosphere** overlaps part of the ionosphere, beginning at an altitude of about 85 km and extending to an average altitude of about 500 km. In the lower thermosphere, the temperature increases rapidly with increasing altitude as molecular oxygen absorbs solar radiation up to 2000 Å and dissociates with the release of heat. Although the temperature in the upper thermosphere averages about 1300°C, the low concentration of matter in this region does not cause the gas to be hot in the conventional sense of being hot to the touch. With increasing altitude, the number of oxygen atoms exceeds the number of oxygen molecules as a result of dissociation until, at 400 km, less than 1% of the oxygen is in the molecular form. Nitrogen, on the other hand, is not easily dissociated by solar ultraviolet radiation and remains predominantly diatomic throughout the upper atmosphere.

Figure 24–9 The composition of the atmosphere. Information is based on the *Explorer* satellite program of the National Aeronautics and Space Administration. (Courtesy of NASA)

The outermost region of the atmosphere, above an average altitude of 500 km, is known as **exosphere**. Atomic oxygen is the most abundant atom in the lower exosphere, and helium and atomic hydrogen at higher altitudes, although the concentrations of all are very low (Fig. 24–9). Atomic hydrogen, the main constituent of the upper exosphere, is produced mainly by the dissociation of water vapor and methane by ultraviolet light at altitudes below 100 km. The light hydrogen atoms diffuse rapidly upward to the top of the atmosphere, where many are freed from the earth's gravitational pull and escape. The ions of hydrogen, helium, and oxygen are also present in the exosphere but cannot escape, being confined by the earth's magnetic field. (The problem of how some helium manages to acquire sufficient energy to escape has not yet been solved.)

With the advent of artificial satellites, a region of the atmosphere called the **magnetosphere** has been recognized. Extending outward for thousands of miles, this region consists of a thin, electrified gas of electrons and protons that gyrates along the lines of force of the earth's magnetic field and forms a vast region of radiation around the earth (Fig. 24–10). The **Van Allen belts**, named for James A. Van Allen (1914–1992), a physicist at the University of Iowa, were discovered and mapped with instrumentation placed aboard the first American satellites. A belt of

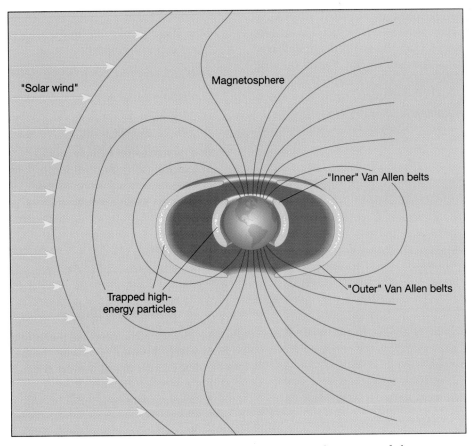

Figure 24–10 The Van Allen belts in the earth's magnetosphere surround the earth like a small and a large doughnut. Though often called radiation belts, they are actually regions of charged particles trapped by the earth's magnetic field. The outer belt is mostly particles from the sun, arriving in the solar wind. The inner belt contains particles set free in the earth's atmosphere by cosmic rays.

high-energy protons is trapped in the region of the magnetic field about 3200 km above the earth at the magnetic equator. Another belt, containing high-energy electrons, surrounds the magnetic equator about 16,000 km out from earth. The earth's magnetic field acts as a trap for charged particles, deflecting them from their course and causing them to pursue a spiral path around the lines of force. According to one theory, cosmic-ray protons collide with atmospheric nuclei, producing neutrons; the neutrons stream back into the magnetosphere where they decay into protons, electrons, and neutrinos. The charged protons and electrons spiral around the magnetic lines of force and move around the earth. The outer regions of the magnetosphere interact with hot, ionized gas streaming from the sun, generating auroras and magnetic storms that disrupt communications.

The spectacular phenomena called **auroras** are produced directly or indirectly by fast-moving electrons and protons from the sun that are deflected, because of their electrical charge, toward the magnetic poles by the earth's magnetic field (Fig. 24–11). They are observed most frequently near the magnetic poles, generally at magnetic latitudes from 60° to 70°. Auroras take a great variety of forms and colors and usually occur at altitudes from 75 to 400 km. The colors are produced when solar electrons excite the atoms and molecules of the earth's atmosphere, causing them to emit light. Excited atomic oxygen emitting at 5577 Å accounts for the green; atomic oxygen emitting at 6300 Å and 6364 Å accounts for the red; and excited ionized molecular nitrogen is the source of the blue and violet.

24–6 WEATHER LORE

Long years of human observation of weather patterns have produced a body of weather superstitions that are part of our folklore. Although some have a grain of truth, they are for the most part often broad generalizations that have limited accuracy:

> *Red sky in the morning, sailor's warning;*
> *Red sky at night, sailor's delight.*
>
> *Rain long foretold, long last;*
> *Short notice, soon past.*
>
> *Rain before seven, shine before eleven.*
>
> *A rainbow in the morning is the shepherd's warning,*
> *A rainbow at night is the shepherd's delight.*
>
> *But I know ladies by the score*
> *Whose hair, like seaweed, scents the storm;*
> *Long, long before it starts to pour*
> *Their locks assume a baneful form.*

24–7 OBSERVING THE WEATHER

A wide variety of instruments has been developed for the study of the atmosphere. **Windvanes** are the simplest instruments for determining wind direction (Fig. 24–12). They usually consist of a thin metal vane that is moved by the wind and points in the direction from which the wind is blowing. Some produce an electrical current that is converted to a reading on a dial in a weather station.

An **anemometer** is a series of cups attached to a vertical shaft that measures wind velocity on an indicator. The wind strikes the cups, causing them to spin and vary an electrical current that gives wind velocities. The wind vane and anemometer are combined in the **aerovane**.

Pilot balloons are the most common method of gathering upper-air data, to 20 km. These helium-

Figure 24–11 The aurora borealis was visible over much of the United States on March 13–14, 1989. It is seen here over Williams College in Massachusetts. Excited atomic oxygen accounts for the yellow-green color and also the red color that was sometimes seen. Nitrogen glows blue and violet as well as red. (Kevin Reardon/Williams College-Hopkins Observatory)

Figure 24–13   A helium-filled weather balloon. (National Center for Atmospheric Research/National Science Foundation)

Figure 24–12   A wind vane points in the direction from which the wind is blowing. Some vanes produce a reading on a dial in a weather station. (David Pollack/The Stock Market)

filled bags are carried aloft and moved by winds and air currents. Their motion can be tracked by radar (Fig. 24–13). A **radiosonde** is an instrument package carried by a balloon that measures temperature, pressure, and humidity with miniature electronic devices. These data are radioed continuously to a receiver at a ground station. At a predetermined level, the balloon bursts and the radiosonde is parachuted to earth. A **dropsonde** is a variation of a radiosonde released by aircraft to record data as it falls through the air.

Rocket-borne instrument packages are used at heights between 30 and 90 km above the earth's surface. The electronic equipment measures such variables as pressure, temperature, wind, and density. **Radar** is used for tracking storms and weather patterns over large areas. **Computers** process vast amounts of data in a fraction of the time required by people and draw synoptic weather maps almost instantaneously.

Air temperature is the most fundamental of the weather elements. Variations in atmospheric heat energy are the ultimate driving force of all weather changes. The **thermograph** makes a continuous record of temperature. Changes in temperature cause changes in the shape of a bimetallic unit or liquid-filled tube that are magnified by a lever arrangement to which a pen arm is connected. Minute temperature changes of 0.01°C or less can be measured with instruments containing a **thermistor**, a metallic unit whose electrical resistance varies with temperature. They are also involved in instrument packages designed to telemeter to earth the temperatures of the surface of the planets and the moon.

The **sling psychrometer** measures relative humidity, the ratio of the actual amount of water vapor present in the air to the capacity of the air to hold water vapor. It consists of two thermometers, one of which is covered with muslin that is moistened when used. Evaporation from the wet bulb lowers its temperature below that of the dry bulb, the extent depending on the relative humidity, which is read from standard meteorological tables. A **hygrograph** (*hygros*, meaning "moist") keeps a permanent record of relative humidity by magnifying the changes in

Figure 24–14 An aneroid barometer. Some of the air has been removed from the airtight box, which is made of thin, flexible metal. When the pressure of the atmosphere changes, the remaining air in the box expands or contracts (Boyle's law), moving the flexible box surface and an attached pointer along a scale. (Taylor Scientific Instruments)

length of blond human hair, which is particularly sensitive to water vapor in the air.

Pressure distribution is a primary characteristic of weather systems migrating over the earth's surface. The **mercury barometer** is the standard instrument for measuring air pressure. The mercury rises to a height such that its weight balances the weight of a column of air on an area of the well surface equal to the area of the tube. As the air pressure changes with changes in weather, the height of the mercury column changes. Air pressure is stated in length, or pressure units. At sea level the average height of the mercury column is 76 cm (29.92 in.). The **bar** is the common pressure unit (10^6 dynes per square centimeter). Average sea-level pressure is 1.0132 bars or 1013.2 millibars.

The **aneroid barometer** is portable and easier to read than a mercury barometer (Fig. 24–14). It contains an evacuated chamber supported by ribbed walls. Variations in air pressure cause the chamber to expand or contract and are reflected by a dial indicator through gears and levers. The face of the dial is calibrated in centimeters, inches, or millibars. A **barograph** is a recording aneroid barometer that produces a continuous record of pressure.

One type of **rain gauge** consists of a cylinder 10 or 20 cm in diameter, leading into a funnel that collects the water in a long, narrow tube. A calibrated stick indicates the depth of water that would be caught in a

Figure 24–15 A satellite image of a storm system centered over Kansas and Oklahoma. Air pressure, wind speed, and wind direction are superimposed on the image. (NOAA)

straight-sided, flat-bottomed pan. Snow is measured by sampling the snow depth, then converting the average into the equivalent of water, which is about one tenth the depth of the snow.

A **ceilometer** measures ceiling, the height of the lowest cloud cover of the sky. The device projects a beam of light to the cloud base. The base of the cloud reflects the beam, which is then picked up by a detector that gives a reading of the ceiling or cloud height.

TIROS **weather satellites** (*T*elevision and *I*nfra *R*ed *O*bservation *S*atellites) were first placed in orbit in 1960 to analyze infrared energy and cloud covers. Over the years they produced vast quantities of data that were analyzed by computers. **Nimbus satellites**, first launched in 1964, photograph still larger areas of the earth's surface. They are larger and more complex weather satellites than TIROS and can make night-

time photographs of cloud cover. Data recorded on magnetic tape for readout upon command are used in weather analysis and forecasting (Fig. 24–15). **Synchronous orbit satellites** that keep pace in orbit with the earth's own rotation continuously monitor the atmosphere within their field of view.

24–8 THE HYDROLOGIC CYCLE

The average water-vapor content of the earth's atmosphere is rather constant. Evaporation from oceans, lakes, rivers, and moist soil replaces the moisture lost by rain, dew, snow, sleet, and hail. The process of maintaining this steady state is known as the **hydrologic cycle** (Fig. 24–16).

Air is saturated with water vapor at a particular temperature when the maximum possible amount,

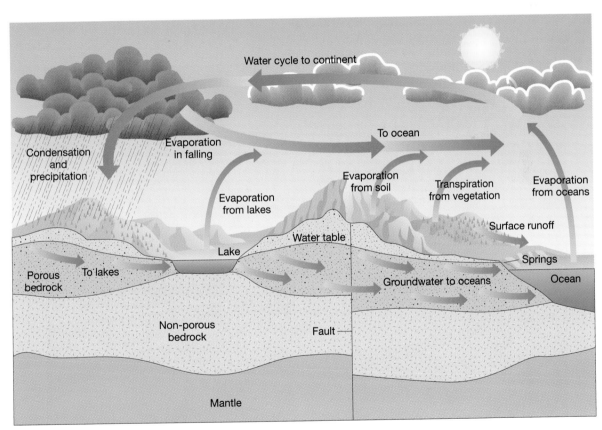

Figure 24–16   The hydrologic cycle. Water is cycled between the atmosphere, the soil, lakes and streams, plants and animals, subsurface storage, glacial ice, and, principally, the oceans. For the world as a whole, the total amount of evaporation equals the total amount of precipitation.

Figure 24–17 An advection fog, a thick fog blanket, caused by warm, moist air passing over colder water. (R. Sager)

Figure 24–18 City dwellers have learned to enjoy, as well as to cope with, the beauty of snow, as in Central Park, New York City. (J. Sapinsky/The Stock Market)

called the **capacity**, is present. **Specific humidity** refers to the weight of water vapor (grams) per weight of air (kilograms). **Relative humidity** is the ratio of the actual amount of water vapor present in the air to the capacity at that temperature.

$$\text{Relative humidity} = \frac{\text{Specific humidity}}{\text{Capacity}} \times 100\% \quad (24\text{--}6)$$

Saturation is achieved primarily by a lowering of the temperature of the air until its capacity is reached. The temperature at which saturation occurs is called the **dew point**. At or below the dew point, water vapor is condensed to liquid water, provided that condensation particles or nuclei are present.

Dew is deposited as fine beads of moisture on grass and other matter close to the earth's surface. It forms mainly on clear, calm nights when the ground cools rapidly by radiation, and particularly in summer when there may be considerable water vapor in the air. When the dew point is below freezing, **frost** forms instead through sublimation. **Fog** develops when there is condensation throughout a layer of air near the land or water surface cooled below the dew point. A thick fog blanket called an **advection fog** often results from the movement of warm, moist air over a cooler surface (Fig. 24–17). The famous "pea soup" fogs of Britain occur during the cold season when relatively warm, moist air from the Atlantic is carried over the colder British Isles.

Precipitation occurs when large masses of moist air are cooled rapidly below the dew point. Condensation continues until water drops or ice particles are formed that are too large to remain suspended in air and fall earthward. **Drizzle** is precipitation composed of tiny droplets, each less than 0.5 mm in diameter. **Rain** consists of water droplets from 0.5 to 5 mm in diameter. Above this size the drop breaks apart as it falls. **Snow** is a form of ice in hexagonal (six-sided) crystals. It is produced when water vapor sublimes at temperatures well below the freezing point. It is never "too cold to snow" (Fig. 24–18). Raindrops falling through cold air may freeze, forming small pellets of ice called **sleet**. **Hail** consists of rounded pieces of ice, often in concentric layers like the layers of an onion. The ice layers may result from the repeated lifting of hailstones upward through warm, moist air layers. An **ice storm** or **glaze** is a coating of clear ice that forms on branches, wires, and other surfaces. It occurs when rain falls through a layer of cold air close to the ground, causing the droplets to freeze as they touch exposed surfaces (Fig. 24–19). The weight of the ice may cause branches and wires to snap.

24–9 CLOUD TYPES

Clouds are composed of tiny water droplets or ice crystals. Those above 6 km high usually are composed of ice crystals because air temperatures there are below freezing. Below this region, most clouds are composed of water droplets. The brilliant snowy appearance of clouds in bright sunlight is due to their high **albedo**, that is, their capacity to reflect sunlight over the entire visible spectrum. Dense clouds appear gray or black because sunlight is absorbed rather than reflected.

Stratiform clouds are clouds formed into layers. **Cumuliform** clouds have a flat base and a massive, globular shape. Clouds are grouped into classes according to height and form, whether stratiform or

Figure 24–19 The aftermath of an ice storm in Chicago. (David W. Hamilton/The IMAGE Bank)

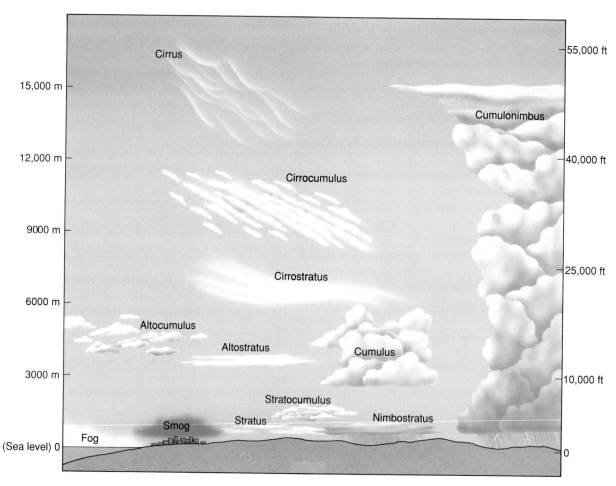

Figure 24–20   Clouds are grouped into classes and subclasses according to height and form.

Figure 24–21   Cirrus clouds are high wisps of thin clouds composed of ice crystals. They usually form streaks called mares' tails. (D. Wise)

Figure 24–22   Altocumulus clouds are middle-level cloud masses that lie in a distinct layer. Blue sky is visible in breaks between the masses. (Courtesy Kenneth R. Martin Collection)

cumuliform (Fig. 24–20). An international system of classification lists four classes: high clouds, middle clouds, low clouds, and clouds of vertical development.

High clouds above 6 km are cirrus, cirrostratus, and cirrocumulus, all composed of ice crystals (Table 24–2). **Cirrus** clouds are thin and feather-like and usually form streaks called mares' tails (Fig. 24–21). **Cirrostratus** clouds are also thin and often cause a halo around the sun or moon. This is evidence that the particles consist of ice crystals rather than water droplets, since the clouds lie above the freezing level.

Cirrocumulus clouds are composed of small cumulus masses called a "mackerel sky."

Middle clouds extend from 2 to 6 km in height. **Altostratus** clouds form a thick, gray blanket with a smooth underside that shuts out sunlight but causes the sun to be seen as a bright spot through the cloud. **Altocumulus** clouds are grayish cumuliform cloud masses that lie in a distinct layer and are associated with fair weather (Fig. 24–22). Blue sky is visible in breaks between the masses.

Low clouds occur from ground level to a height of 2 km above the earth's surface. **Stratus** clouds form a

Table 24–2   CLOUD TYPES AND WEATHER

Altitude over Middle Latitude (in Feet)	Name, Abbreviation, and Symbol	Description	Composition	Possible Weather Changes
High Clouds 18,000–45,000	Cirrus (Ci)	Mares' tails Wispy and feathery	Ice crystals	May indicate a storm, showery weather close by
	Cirrostratus (Cs)	High veil Halo cloud	Ice crystals	Storm may be approaching
	Cirrocumulus (Cc)	Mackerel sky	Ice crystals	Mixed significance, indication of turbulence aloft, possible storm
Middle Clouds 6500–18,000	Altocumulus (Ac)	Widespread, cotton ball	Ice and water	Steady rain or snow
	Altostratus (As)	Thick to thin; overcast; high, no halos	Water and ice	Impending rain or snow
Low-Family Clouds Sea Level to 6500	Stratocumulus (Sc)	Heavy rolls, low, widespread Wavy base of even height	Water	Rain may be possible
	Stratus (St)	Hazy cloud layer, like high fog Somewhat uniform base	Water	May produce drizzle
	Nimbostratus (Ns)	Low, dark gray	Water, or ice crystals	Continuous rain or snow
Vertical Clouds Few hundred to 65,000	Cumulus (Cu)	Fluffy, billowy clouds Flat base, cotton-ball top	Water	Fair weather
	Cumulonimbus (Cb)	Thunderhead Flat bottom and loft top Anvil at top	Ice (upper levels) Water (lower levels)	Violent winds, rain, hail are all possible Thunderstorms

° From Navarra, J. G., *Atmosphere, Weather and Climate.* Philadelphia: W.B. Saunders Company, 1979, p. 189.

Figure 24–23 Nimbostratus clouds are dense, dark gray, low-level clouds that produce rain, snow, and overcast skies. (D. Wise)

Figure 24–24 Cumulonimbus clouds, or thunderheads, are clouds of vertical development that may tower thousands of meters. They are associated with violent thunderstorms and occasionally hail. (Courtesy Kenneth R. Martin Collection)

low, uniform cloud sheet that completely covers the sky. **Nimbostratus** clouds are dense, dark gray stratus clouds that produce rain or snow (Fig. 24–23). **Stratocumulus** clouds are low cylindrical cloud masses, dark gray on the shaded side but white on the illuminated side. Blue sky is visible between the narrow breaks. They are best developed during a clearing period following a storm.

Clouds of vertical development are all of the cumuliform type. The smallest are the **cumulus of fair weather** clouds, found when the weather is fair and there is much sunshine. They are snow white and cotton-like, with flat bases and round tops. Congested cumulus clouds are larger and denser than small cumulus and have tops resembling heads of cauliflower. They may grow into gigantic **cumulonimbus** clouds, or thunderheads, that extend to heights of 20 km and are the source of violent thunderstorms with rain, hail, wind, thunder, and lightning (Fig. 24–24).

24–10 AIR MASSES

We can often detect the presence of air masses by our senses. After suffering through a hot, oppressively sticky summer heat wave, we appreciate the arrival of cool, dry air. A large, hot, humid air mass was simply replaced by a cool, dry air mass. A knowledge of air masses helps in understanding our day-to-day weather changes.

An **air mass** is a large body of air that may extend over a large part of a continent or ocean. It is fairly uniform in temperature and humidity at a given altitude level. Vertically, it includes the lower part of the troposphere from ground level to heights of 3 to 6 km. It usually has different properties from those of adjacent air masses and has distinct boundaries.

An air mass acquires its properties of temperature and humidity from its **source region,** the land or sea surface over which it originates (Fig. 24–25). Air-mass source regions usually coincide with large bodies of land or water. For example, air over a warm, tropical ocean will be warmed by radiation and will absorb water vapor by evaporation from the sea surface. The entire mass will acquire a high moisture content. When an air mass breaks loose from its source region, it moves in a generally eastward direction, steered by strong winds in the upper troposphere. Most of the United States does not lie in any air-mass source region but is affected by the passage

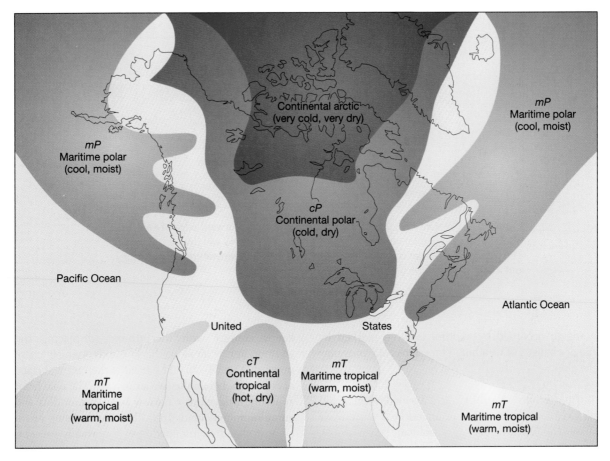

Figure 24–25   Source regions of North American air masses. Air-mass movements carry the temperature and humidity features of these source regions into distant areas. (After Trewartha)

of air masses from source regions lying to both the north and south.

Cold northern air masses are called **polar air masses**, designated P on weather maps, or **arctic air masses**, A. Depending on whether they form over land or water, the air masses will be dry or humid. This leads to two subdivisions, **continental** and **maritime**. The polar and arctic continental air masses (cP and cA) are cold and dry (0.5 g of H_2O/kg air); of the two, the arctic air mass is the colder. They originate in a vast source region from Hudson Bay to Alaska. Their influence is greatest in winter, when they bring cold waves to much of the country. Polar maritime air (mP) is of oceanic origin, the North Atlantic, and is cold and moist, bringing drizzle to the northeastern United States in the warmer seasons and often heavy snowfalls from storms called northeasters in the winter.

Warm air masses originating in low latitudes are called **tropical air masses**, T. Maritime tropical air masses (mT) originate in the Caribbean Sea, Gulf of Mexico, and neighboring parts of the Atlantic Ocean. They are warm and moist (15 to 17 g H_2O/kg air). Their encounter with cP, cA, and mP air masses from the north produces frequent weather disturbances in the central and eastern United States.

24–11 WEATHER FRONTS

Air masses of different properties do not mix easily and tend to be separated by distinct boundaries called **weather fronts**. The movement of one air mass into

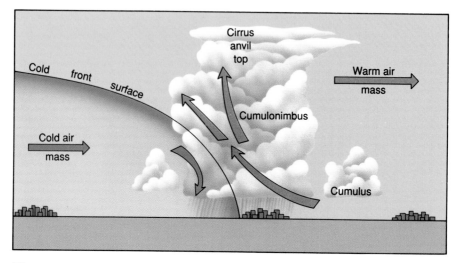

Figure 24–26 Cross section of a cold front. The dense forward edge of cold air displaces warmer, lighter air upward, often producing showers and thundershowers. The passage of the front is generally followed by a drop in temperature and humidity and a shift in wind direction.

a region occupied by another is often marked by major changes of weather. Three types of fronts may develop.

1. Cold Front. A **cold front** is formed when a cold air mass invades a region occupied by a warmer air mass. The forward edge of the cold air, being denser, keeps contact with the ground, displacing the warmer, lighter air along this line and forcing it to rise upward (Fig. 24–26). The front is the entire surface of contact between the two air masses. The cold front zone often develops a narrow band of cumulus and cumulonimbus clouds from the steeply rising warm air just ahead of it. Cloudiness is normally present,

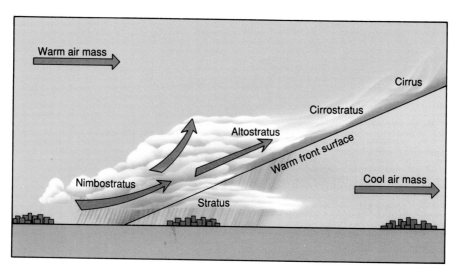

Figure 24–27 Cross section of a warm front. Warm fronts advance more slowly than cold fronts and replace (rather than displace) cold air by gently rising over it. A period of cloudiness and precipitation in the form of drizzle, rain, or sleet may follow. The temperature may then rise gradually, with the wind shifting to the south or west.

and precipitation, if it occurs, takes the form of showers and thundershowers. The passage of the front often brings a marked drop in temperature and humidity and a shift in wind direction.

2. Warm Front. A relatively warm air mass moving into a region occupied by colder air forms a **warm front**. The warm air rises over the denser, cold air that is close to the ground over a broad, gently sloping front and forms stratus-type clouds (Fig. 24–27). A long period of cloudiness and precipitation may follow. If the cold air beneath a warm front is below freezing, rain falling through it may form sleet or freeze upon contact with the ground, producing glaze or an ice storm. As the warm front passes, the temperature rises gradually and the wind shifts to the south or west.

3. Occluded Front. A cold front may overtake a warm front and completely lift the warm air off the ground. The warm front is said to be **occluded** and is then located at higher altitudes, where it produces precipitation. A cold front travels fairly rapidly because cold air can more easily displace warm air than vice versa.

4. Stationary Front. A **stationary front** has no forward motion along a line of transition between two different air masses.

24–12 WIND

In the meteorologic sense, **wind** is air in a predominantly horizontal motion. The terms "updraft" and "downdraft" that describe vertical motions of air do not qualify as wind in this sense.

Winds are named after their sources. Thus, a wind that blows from the northwest is called a northwest wind. Wind direction is determined with a weathervane. The term **windward** refers to the direction from which the wind comes. The side of a mountain, for example, that faces the direction from which the wind is coming is called the windward slope (Fig. 24–28). **Leeward**, on the other hand, means the direction toward which the wind is blowing. Where the winds are coming out of the west, the leeward slope of a mountain would be the east slope.

Wind velocity is measured by devices called anemometers. Velocity is usually expressed in miles per hour, kilometers per hour, or nautical miles per hour (knots). (An international nautical mile is equal to 6076.115 feet or 1852 meters and has been used officially in the U.S. since 1959.)

The sun is the ultimate source of energy for wind. When the sun warms one part of the earth more than another, the warm air expands and is displaced upward by cooler, denser air, which moves in as wind along the earth's surface from adjacent regions (Fig. 24–29). The air molecules are closer together in cool air, so that a column of cool air weighs more than a

Figure 24–28 The term "windward" means facing into the wind, and "leeward" facing away from the wind.

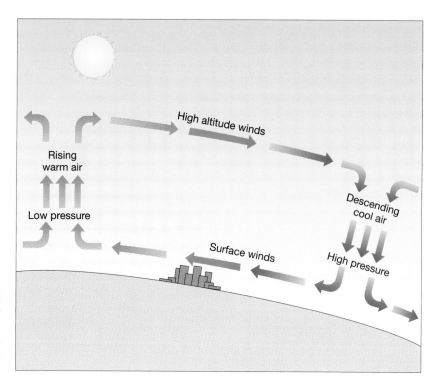

Figure 24–29 Wind caused by the unequal heating of the earth's surface. Warm air expands and is displaced by cooler, denser air, which moves in as wind.

column of warm air of the same volume. A mass of cool air above the ground, therefore, is called a **high-pressure area**. A warm air mass, on the other hand, constitutes a **low-pressure area**.

The pressures developed in highs and lows are affected by moisture and altitude as well as by temperature. Water vapor has a lower density than dry air and, when present, makes air lighter because the water vapor replaces air. Masses of moist air tend to rise, as a result, until they reach cooler regions where **condensation** may occur. Atmospheric pressure decreases as one ascends to higher altitudes because the column of air above land is then shorter and has less mass than near sea level. The pressure difference between two areas is called a **pressure gradient**.

Land and sea breezes are caused by the differential heating of land and water, which have unequal heat capacities (Fig. 24–30). Salt water has a heat capacity of 0.9 calorie per gram, compared to less than 0.2 calorie per gram for most common minerals. For this reason, seawater warms and cools much more slowly than solid ground.

During the day, when the land heats quickly under the action of the sun's rays, the air above the land is also heated, expands, and rises. This local area of low

pressure is replaced by cooler, denser air flowing in from the sea. Thus, there is a **sea breeze** of cool, moist air blowing in over land during the day. The cooling winds help alleviate the heat of summer days and help explain why seashores are so popular in summer. The reduction in temperature along the coast can be as much as 9°F to 16°F.

At night, these conditions are reversed. The land and the air above it cool more quickly and to a lower temperature than the nearby water body and the air above it. As a result, the pressure builds up over the land and air flows from the cool land surface toward the lower pressure over the water, which has retained its warmth, creating a **land breeze**.

In mountainous regions, there is a diurnal mountain breeze–valley breeze cycle that resembles the land breeze–sea breeze cycle along coastal areas. During the day, the mountain slopes and valleys exposed to the sun's rays are heated, and the air is warmed, expands, and rises up the sides of the mountains as a warm **valley breeze** (Fig. 24–31). The clouds that often hide mountain peaks are the visible evidence of condensation in warm air rising from the valleys. At night, the mountain walls cool rapidly because of the relatively low heat capacity of rock and

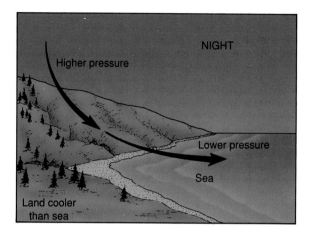
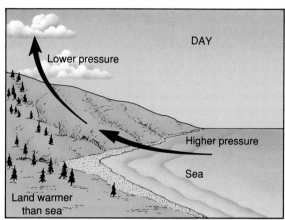

Figure 24–30   Land and sea breezes are a consequence of the different rates of heating and cooling of land and water areas. The land becomes warmer than the sea during the day and cooler than the sea at night. Air flows from the cooler to the warmer area.

soil. Air in contact with the cooled surfaces is chilled, becomes denser, and flows down the mountains into the lowlands as a cool **mountain breeze**.

We all know that a hot but windy day is not nearly so unpleasant as a hot day without wind. The reason is that winds increase the rate of evaporation. In order for evaporation to take place, heat must be removed from our bodies. For the same reason, the wind on a cold day increases our discomfort.

Global wind patterns are quite complex. The earth is heated unevenly for a number of reasons, including differential heating of land and water, different albedos of surfaces, and unequal distribution of insolation—energy received from the sun. In an area of warm air, such as the equator, the air expands in volume and decreases in density during the day, and a zone of low pressure occurs. High pressure, on the other hand, occurs at the poles, where the air is colder

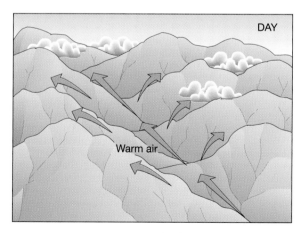
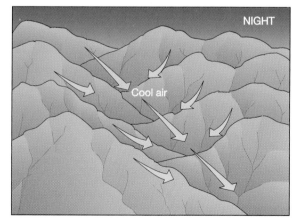

Figure 24–31   Mountain and valley breezes are caused by the heating of mountain slopes during the day and their cooling at night. Warm air rises up slopes during the day as a valley breeze, and cold air drains slopes at night as a mountain breeze.

and denser. From this we might expect a gradual thermally induced increase in pressure from the equator to the poles.

However, it is known that pressure does not increase latitudinally in a regular fashion from the equator to the poles. Instead, there are belts of high pressure in the subtropics and belts of low pressure in the subpolar regions. If the earth did not rotate on its axis, the pressure differences between the equator and the poles would generate two large air mass movements. In the northern hemisphere, the surface winds would blow southward, while in the upper atmosphere warm air would flow toward the North Pole. The flows would be reversed in the southern hemisphere.

The atmospheric circulation is more complex than this because the earth does rotate. The warm air masses that rise at the equator divide and move at high altitudes toward the poles, but they do not move directly north or south. The poleward-flowing air masses are deflected in response to the earth's rotation, to the right in the northern hemisphere and to the left in the southern hemisphere, an effect called the **Coriolis effect**.

By the time the air masses reach about 30° north and south latitude, the air is flowing from west to east (Fig. 24–32). This bending of poleward-flowing air causes the air to pile up over the subtropics, where it is now somewhat cooled and begins to descend, producing the subtropical high-pressure areas known as the **horse latitudes**. The winds in these regions are light and variable, rainfall is slight, and the skies are generally clear.

The air masses that descend at the horse latitudes return to the equator. Again, because of the earth's rotation, they are deflected and blow from the northeast in the northern hemisphere and from the southeast in the southern hemisphere. These winds are called the **trade winds** because they are reliable and once provided the trade routes for merchant vessels.

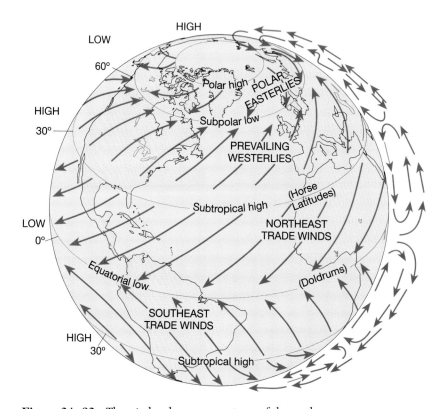

Figure 24–32 The wind and pressure systems of the earth.

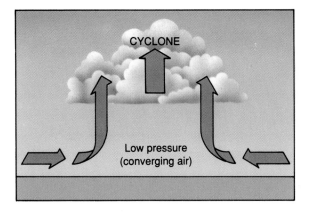
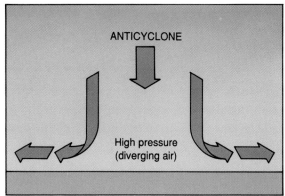

Figure 24–33 Winds converge in a cyclone (low-pressure center) and diverge from an anticyclone (high-pressure center).

The cycle is complete when the air reaches the equator and the convection cycle begins anew.

Poleward of the subtropical highs in both the northern and southern hemispheres are the wind belts known as the **westerlies**. Subpolar lows, large belts of low pressure that extend through the upper middle latitudes, are developed until about 65° latitude. In these regions, the prevailing wind directions are from the southwest in the northern hemisphere and from the northwest in southern regions.

In the polar regions, the air is cooled and compressed to form high-pressure systems, the **polar highs**. Cold air in these regions moves generally southward in the northern hemisphere and northward in the southern hemisphere. The deflection of the winds causes them to be designated the **polar easterlies**. They blow from the polar regions to the subpolar low-pressure zones.

A center of low atmospheric pressure is called a **cyclone** (Fig. 24–33). It is usually formed in the zone separating masses of cold and warm air. As winds blow inward from all sides to the center of the cyclone, they are deflected. The warm air in the center is displaced upward to higher and cooler levels where clouds and precipitation may occur. Hurricanes are several tropical cyclones.

An **anticyclone** is a high-pressure system in which atmospheric pressure decreases toward the outer limits of the system. Circulation of air in an anticyclone is opposite to that in a cyclone. In the northern hemisphere, air moves downward through the center of the anticyclone and spirals outward with a clock-

wise deflection. The descending air tends to be cool and dry, so that anticyclones are generally associated with pleasant weather.

24–13 WEATHER MAPS

A weather map gives a broad view of weather conditions over a large region. Specialized maps show specific data such as temperature, pressure, and precipitation of a region (Fig. 24–34).

Frontal systems are represented by lines indicating the character of the air mass. A cold front is drawn with a series of triangles on the side of the line toward which the air is moving (Fig. 24–35). A warm front is drawn with a series of half-circles. An occluded front is shown with alternating triangles and half-circles pointing in the same direction. These symbols and others are designed and set by international agreement.

Equal temperature regions are connected by lines called **isotherms**. Regions of equal pressure are connected by lines called **isobars**. When isobars are close together, the horizontal change in pressure, or pressure gradient, is great (Fig. 24–36). When isobars are relatively far apart, the pressure gradient is weak. Roughly circular isobar patterns enclosing highs and lows are particularly important features of a weather map. Estimates of the direction of movement of weather systems are possible. **Station models** containing data about cloud cover, temperatures, precipitation, and other variables for different stations throughout the country may also be shown.

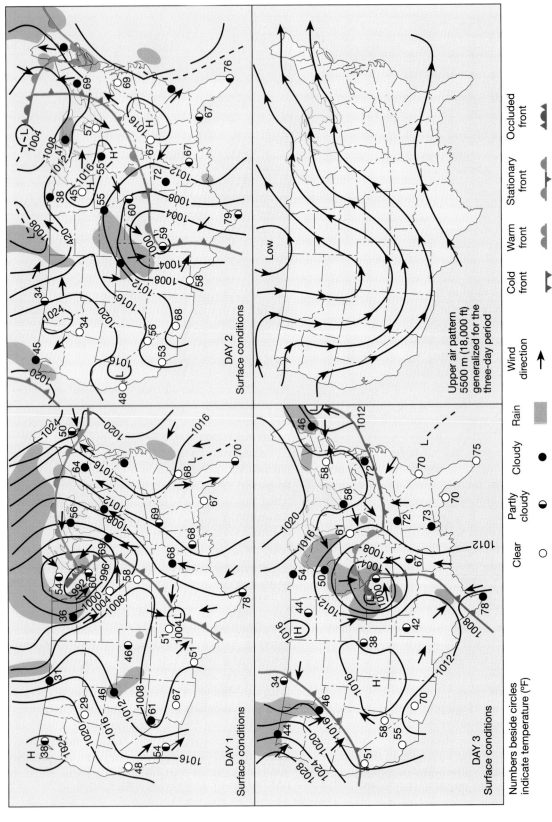

Figure 24–34 A three-day sequence of daily weather maps illustrating the movement of weather systems across the United States. The upper air pattern for the same period is also shown.

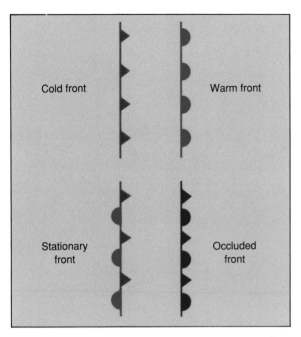

Figure 24–35 The four major frontal symbols used on weather maps.

24–14 VIOLENT WEATHER— THUNDERSTORMS, TORNADOES, AND HURRICANES

Thunderstorms are found along both cold and warm fronts (Fig. 24–37). Those along a cold front are often particularly violent. When dissimilar air masses meet, large masses of warm, moist air may be forced to rise several kilometers in altitude (Fig. 24–38). As it rises, the air expands, since there is less air pressing on it from above. As the air expands, it cools with no decrease in heat energy; the same quantity of heat is spread over a greater volume, and the temperature therefore drops. The rising convection column of warm, less dense air resembles the updraft in a chimney in a fireplace.

When the air arrives at a level at which the dew point is reached, a cumulus cloud begins to form, then builds rapidly into the cumulonimbus cloud from which heavy showers will fall. Condensation occurs at the upper levels with the release of latent heat and produces downdrafts. Violent updrafts and down-

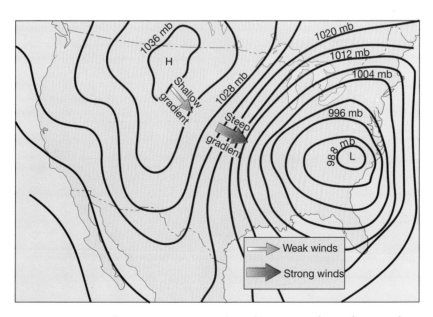

Figure 24–36 Isobars connect regions of equal pressure. When isobars are close together, the horizontal change in pressure, or pressure gradient, is steep. When far apart, the pressure gradient is weak. The steeper the pressure gradient, the faster and stronger the resulting wind from a region of high pressure to one of low pressure.

Figure 24–37 A thunderstorm over Salt Lake City, Utah. (Tim Thompson/The Stock Market)

drafts sweep through the storm cells, rain falls on the ground level, and there are lightning and thunder. In the final stage of a thunderstorm, downdrafts occur over the entire cell, updrafts decrease, rain slows because the storm is cut off from its source of warm, moist air, then stops, and the cloud evaporates.

Tornadoes, the most violent form of weather, are related to very severe thunderstorms. They are produced when the greatest contrast exists in temperature and moisture between two air masses. Although they are most common in the central United States from early spring to late summer, they have been known to occur in nearly every state (Fig. 24–39). According to the National Severe Storms Forecast Center in Kansas City, Missouri, in 1990 there were 1132 tornadoes nationwide, which took 53 lives.

Dark tapering funnel clouds extend from a cumulonimbus cloud to the ground (Fig. 24–40). The winds composing the funnel have been estimated to have velocities as high as 500 mph. Devastation follows along a swath 300 to 400 m wide and several kilometers long when a funnel cloud touches the

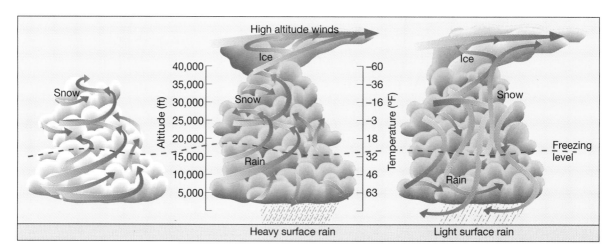

Figure 24–38 The stages of a thunderstorm. (Adapted from NOAA/PA 75009)

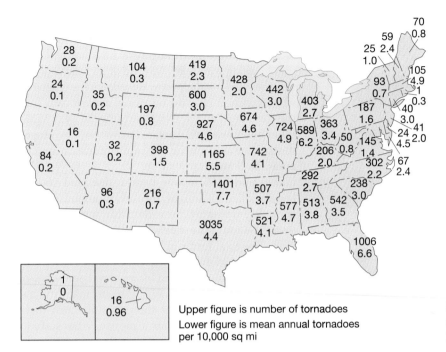

Upper figure is number of tornadoes
Lower figure is mean annual tornadoes
per 10,000 sq mi

Figure 24–39 Tornado frequency by state from 1953 to 1982.

ground, both from the great wind velocity and from the sudden lowering of air pressure as the funnel passes over. Buildings may "explode" from the expansion of the air within, especially if the doors and windows are closed. **Waterspouts** are essentially tornadoes that develop over the ocean. When the funnel reaches the surface of the sea, air, water, and sea spray are sucked in and it develops into a waterspout. "**Dust devils**" are swirling masses of air that are less extensive and violent than tornadoes.

Hurricanes, or **typhoons**, as they are called in East Asia, are severe tropical disturbances that usually develop on the western side of oceans and spend their entire lives there. If their paths take them over islands and coastal lands, the results can be devastating destruction of property and sometimes loss of life from winds, rains, and high seas.

For example, on September 21, 1938, a great hurricane hit Long Island with little warning and raged through New England and into Canada with winds that blew at an average of 121 mph and reached 163 mph at the Mount Washington Observatory in New Hampshire. The nameless hurricane left 682 people dead, nearly 20,000 building destroyed, thousands of boats smashed, church steeples toppled, and some 250 million trees uprooted or snapped off, many of which had taken over a century to attain their stature. A hurricane that struck Galveston, Texas, in 1900 took 6000 lives.

Since the turn of the century, 137 hurricanes have made direct hits on the U.S. coastline. Hurricane Ca-

Figure 24–40 A tornado setting down in Enid, Oklahoma. (The IMAGE Bank)

Figure 24–41   Devastation left by Hurricane Hugo in Charleston, South Carolina. (Jan Staller/The Stock Market)

mille, whose 200 mph winds ravaged the Mississippi Gulf Coast and killed 256 people in 1969, was one of the most powerful.

By September 21–22, 1989, Hurricane Hugo was judged to be the tenth most intense hurricane, having devastated the Caribbean islands of Guadeloupe, Montserrat, St. Croix, St. Thomas, and Puerto Rico

earlier in the week, and then smashing Charleston, South Carolina (Fig. 24–41). Improvements in predictions brought about largely by satellites and computers, along with television and other means of getting the word out, helped to keep down the death toll, which reached 71 as the hurricane made its way from the Caribbean up the East Coast (Fig. 24–42). While other hurricanes had damaged more territory, Hugo's destruction was among the most intense in specific areas. In South Carolina, the storm destroyed 3785 homes and 5185 mobile homes, damaged 100,000 homes, killed millions of trees and thousands of small animals and tens of thousands of fish, and steamrolled over 3 million of the state's 12 million acres of timberland. The total losses in South Carolina were put at more than $3.7 billion.

Considerable time and effort have been spent on studies of the development, growth, maturity, and paths of hurricanes, but there is still much to be learned. Even though a hurricane can be tracked with radar and studied through the use of planes and weather satellites, it is not possible to predict its path. Further, meteorologists can list factors that are favorable for the development of a hurricane, among them a warm ocean surface of about 25°C (77°F) along with warm, moist, overlying air, but they cannot say that a hurricane will definitely develop and travel along a particular path.

A hurricane is a circular low-pressure system with wind speeds of at least 75 mph. It has a diameter from 100 to 400 mi and extends upward to heights of 12 to

Figure 24–42   The tracking of Hurricane Hugo by satellite and computer. (Frank P. Rossotto/The Stock Market)

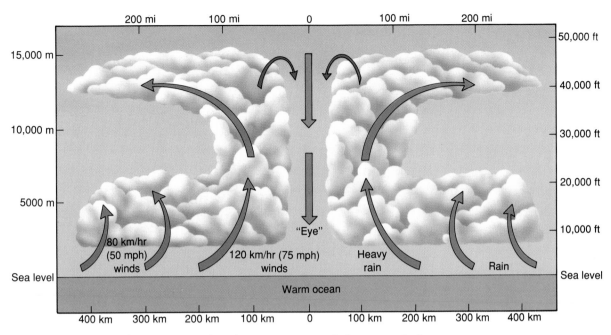

Figure 24–43   Circulation pattern within a hurricane. Moisture-laden air is sucked in at the center, spirals inward, rises to the top and spirals outward.

14 km (40,000 to 45,000 ft) (Fig. 24–43). Air laden with moisture is sucked in at the center and spirals inward, then rises rapidly to the top and spirals outward. This movement produces the enormous amounts of rain during a hurricane. At the center of the hurricane is an area of calm, clear, usually warm and humid air, the **eye of the hurricane**. Strong pressure gradients are the reason behind the powerful winds of the hurricane. Generally, a hurricane does not last long over land, since its source of moisture, and therefore energy, is cut off, and friction with the land surface produces a drag on the whole system. Even when it has lost some of its power, however, the storm can still do great damage.

24–15 WEATHER FORECASTING

The U.S. National Weather Service operates the National Meteorological Center in Camp Springs, Maryland, where weather forecasting based on computers is carried on. In Europe, 17 member countries receive their forecasts from the European Center for Medium Range Weather Forecasts, with headquarters in Reading, England. Medium-range weather forecasting involves predicting on Monday or Tuesday what the weekend weather will be like.

At each weather center, millions of bits of weather data are received each day and are transformed into a picture of that day's weather over the entire globe. Computers perform billions of mathematical operations to predict the behavior of the atmosphere several days later. The more precise the picture of the present weather and the more realistic the computer model of the atmosphere, the better the forecast will be.

Weather observations are received from thousands of surface weather stations, hundreds of sites from which radiosondes are launched, ships at sea, commercial planes in flight, weather buoys, and balloons. Some of the gaps in the observation network over vast ocean areas are filled in by weather satellites, which provide temperature soundings and wind speeds, although with lower accuracies than radiosonde observations.

Computers analyze the observations to create an accurate picture of the atmosphere (Fig. 24–44). With this picture as a starting point, they simulate

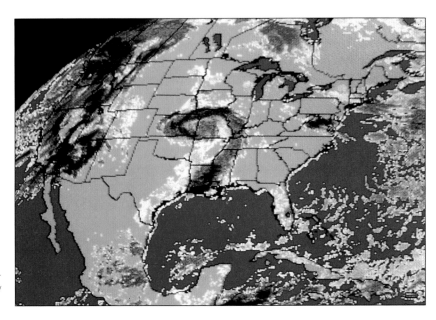

Figure 24–44 A computerized weather chart. (Walter Bibikow/The IMAGE Bank)

future atmospheric behavior with equations stating basic physical principles such as the gas law and Newton's second law of motion. At the present time, forecasts of atmospheric pressure—the highs and lows that generally correspond to fair weather and foul— are useful to five or six days. Temperature forecasts are not quite as good, and precipitation forecasts are the most difficult to make. Up to a limit, medium-range forecasting may be extended.

The National Weather Service has embarked upon an ambitious modernization program. It plans to spend $2.25 billion in the next five to ten years to replace its aging network of radar, ground instruments, computers, and satellites. There is a drastic need for renovation if forecasters, for example, are to pin down more precisely the movement of severe storms so that better and more timely warnings can be made.

Forecasters still launch old-fashioned balloons twice a day to take readings in the atmosphere. They use refrigerator-size computers that have less power than the average personal computer. And they depend on radar equipment that runs on World War II-type vacuum tubes. This antiquated system is prone to breakdowns.

A network of 160 radar stations is proposed that will eventually blanket the United States with high-power radar, greatly improving the accuracy of predictions. The new radar detection system—**Nexrad** (for Next Generation Radar)—is powerful enough to track a swarm of insects moving across a wheat field 50 km away (Fig. 24–45). The technology at the heart of the system is **Doppler radar**.

Like conventional weather radar, Doppler radar can reveal where rain, snow, and hail are falling and provide some measure of how heavy the precipitation is (Fig. 24–46). Doppler radar does more, though. By bouncing microwaves off the tiny droplets in the center of a cloud and picking up the echoes, Doppler systems can map out wind speeds and directions from

Figure 24–45 A Next Generation Radar (Nexrad) computer. (Brownie Harris/The Stock Market)

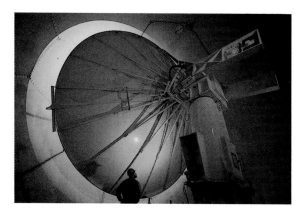

Figure 24–46 Next Generation Doppler weather radome. (Brownie Harris/The Stock Market)

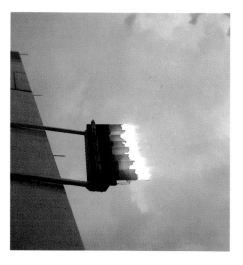

Figure 24–47 Finely dispersed silver iodide for cloud seeding experiments is generated in burners in the wings of an aircraft. (Atmospherics, Inc.)

the frequency shifts imparted by precipitation or dust. High-velocity winds and a high level of organization can signal the development of severe weather, including tornadoes.

Other facets of modernization include a network of 1000 automated temperature and rainfall sensors that will take readings every five minutes instead of once an hour, as is now done with weather balloons. Five new weather satellites are planned to replace three that are either out of commission or nearing the end of their life cycle. A supercomputer for weather models is in place, and the software for its radar and satellite stations has been upgraded. A new communications system, the Advanced Weather Interactive Processing System (AWIPS) will speed the dissemination of data and forecasts between the central office in Camp Springs and weather stations around the country. Most of us have an interest in accurate weather predictions—travelers, picnickers, airlines, utilities, thousands of businesses, farmers, fishermen, snowmakers, moviemakers, and baseball umpires. It appears that, given time, we will get them.

24–16 WEATHER MODIFICATION

In their research into artificially produced rainfall in 1946, Vincent Schaefer (b. 1906) and Irving Langmuir (1881–1957) of the General Electric Research Laboratory induced precipitation by seeding a super-cooled cloud with dry ice. The temperature in the vicinity of the dry ice fell to below $-40°C$, ice crystal nuclei were formed from the water vapor in the cloud, and rain fell.

It has since been found that silver iodide will produce the same reaction at a temperature of $-6°C$ (Fig. 24–47). It causes a rapid growth of ice crystals in the cloud that leads to precipitation. But weather modification is not yet an important process on a large scale. Much more research is needed. Besides the scientific questions, complex legal and moral questions must be addressed.

KEY TERMS

Advection Fog Fog produced by the movement of warm moist air across a cold sea or land surface.

Air Mass A large body of air that is relatively homogeneous in temperature and moisture content.

Albedo The ratio of the amount of the sun's radiation reflected by a body to the amount striking it.

Altocumulus Clouds Grayish cloud masses that lie in a distinct layer with blue sky visible in breaks between the masses.

Altostratus Clouds Clouds forming a thick, gray blanket that very often totally covers the sky; the sun may be seen as a bright spot through the cloud.

Anemometer An instrument for measuring wind speed.

Aneroid Barometer A portable barometer. Variations in air pressure cause the chamber to expand or contract and are reflected by a dial indicator.

Anticyclone An area of high atmospheric pressure; also known as a high.

Arctic Air Mass A cold northern air mass.

Atmosphere I The primitive atmosphere of the earth, containing hydrogen compounds: CH_4, NH_3, H_2O, and H_2S.

Atmosphere II The secondary atmosphere of the earth, composed chiefly of nitrogen, carbon dioxide, and water vapor.

Atmosphere III The present atmosphere of the earth, consisting mainly of nitrogen, oxygen, and water vapor.

Aurora Display of bands of light in the sky caused by charged particles from the sun interacting with the earth's magnetic field and upper atmosphere.

Bar A unit of pressure equal to 1000 millibars or 29.53 inches of mercury.

Barograph A recording barometer, usually of the aneroid type.

Barometer An instrument for measuring atmospheric pressure.

Ceilometer An instrument for measuring cloud height.

Cirrocumulus Clouds A thin, white patch of cloud composed of small cumulus masses called a "mackerel sky."

Cirrostratus Clouds Thin clouds that may totally cover the sky and often cause a halo around the sun or moon.

Cirrus Clouds High, thin, white, feathery clouds that usually form streaks called mares' tails.

Cloud A mass of suspended water droplets or ice crystals in air above ground level.

Cold Front The leading edge of a relatively cold air mass that replaces warmer air.

Condensation The change in state from a vapor to a liquid.

Continental Air Mass An air mass that forms over land.

Coriolis Effect Effect of the earth's rotation on horizontally moving bodies, such as wind and ocean currents, causing them to be deflected to the right in the northern hemisphere and to the left in the southern hemisphere.

Cumuliform Clouds Cloud that have a flat base and a massive, globular shape.

Cumulonimbus Clouds Appear as huge towers or mountains that are the source of violent thunderstorms.

Cumulus A cauliflower-shaped cloud type.

Cyclone A center of low atmospheric pressure, also known as a low.

Dew Tiny droplets of water on ground surfaces formed by condensation when air at the surface reaches the dew point.

Dew Point The temperature at which condensation will occur in a parcel of air.

Doppler Radar A system that can reveal where precipitation is falling and how heavy it is, and can determine wind speeds and directions.

Drizzle Droplets of precipitation less than 0.5 mm in diameter.

Dropsonde A radiosonde that is dropped by parachute from an aircraft to take measurements of the atmosphere below.

Dust Devil A small but vigorous whirlwind, usually of short duration, rendered visible by dust, sand, and debris picked up from the ground.

Exosphere The outermost region of the atmosphere, above an average altitude of 500 km.

Eye of the Hurricane An area of calm, clear air at the center of a hurricane.

Fog Visible, minute water droplets suspended in the atmosphere near the earth's surface, reducing visibility below 1 kilometer (0.62 mile).

Frost Frozen condensation that occurs when air at ground level is cooled to a dew point of 0°C (32°F) or below.

Glaze A coating of ice that develops when rain strikes a freezing surface.

Greenhouse Effect Warming of the atmosphere that occurs because short-wave solar radiation heats the planet's surface, but the loss of long-wave infrared radiation is hindered by gases such as carbon dioxide.

Hail Precipitation in the form of balls or lumps of ice.

High Clouds Cirrus, cirrostratus, and cirrocumulus clouds, all composed of ice crystals.

High Pressure Area A mass of cool air above the ground.

Horse Latitudes Regions near 30° north and south latitude characterized by high pressure, light winds, and clear skies.

Hurricane A severe tropical cyclone of great size that produces torrential rains and high-velocity winds.

Hydrologic Cycle Circulation of water within the earth system from evaporation to condensation, precipitation, runoff, storage, and re-evaporation.

Hygrograph An instrument for automatically recording variations in atmospheric humidity.

Ice Storm Glaze.

Ionosphere A layer of the atmosphere, extending from about 50 km beyond the earth's surface, that has a large concentration of free electrons and positively charged ions. The ionosphere reflects radio waves to the earth's surface.

Isobar   A line on a chart connecting points of equal pressure.

Isotherm   A line connecting points of equal or constant temperature.

Land Breeze   Air flow at night from the land toward the sea, caused by a movement of air from a zone of higher pressure associated with cooler nighttime temperatures over land.

Lapse Rate   A temperature decrease with height.

Leeward   Located on the side facing away from the wind.

Low Pressure Area   A warm air mass.

Magnetosphere   A region of the atmosphere consisting of electrons and protons that gyrate along the lines of force of the earth's magnetic field and form the Van Allen belts.

Maritime Air Mass   An air mass that develops over an extensive water surface and has a high moisture content.

Mercury Barometer   An instrument for measuring atmospheric pressure by balancing it against a column of mercury.

Mesosphere   The layer of the atmosphere above the stratosphere characterized by temperatures that decrease regularly with altitude.

Meteorology   The study of the atmosphere and its processes.

Mountain Breeze   Air flow downslope from mountains toward valleys during the night.

Nexrad   Next Generation Radar, a new radar weather detection system based on Doppler radar.

Nimbostratus Clouds   Dense, dark gray stratus clouds that produce rain or snow.

Nimbus Satellite   A weather satellite that can photograph large areas of the earth's surface and make nighttime photographs of cloud cover.

Occluded Front   The boundary between a rapidly advancing cold air mass and an uplifted warm air mass cut off from the earth's surface.

Ozone   A form of oxygen, O_3.

Ozone Hole   A region in the ozone layer of the atmosphere that contains an unusually low concentration of ozone.

Photosynthesis   A process by which plants synthesize carbohydrates from carbon dioxide and water.

Pilot Balloon   A helium-filled bag carried aloft and moved by winds and air currents for gathering upper-air data to 20 km.

Polar Air Mass   A type of air, the characteristics of which are developed over high latitudes.

Polar Easterlies   Easterly surface winds that move out from the polar highs toward the subpolar lows.

Polar Highs   High-pressure systems located near the poles where air is settling and diverging.

Polar Stratospheric Clouds   Stratospheric clouds that form over the poles and are believed to trigger ozone depletion.

Precipitation   Moisture in any form falling from the atmosphere, such as drizzle, rain, snow, and hail.

Pressure Gradient   The rate of change of atmospheric pressure horizontally with distance.

Radiosonde   A balloon-borne instrument package that measures and transmits meteorological data.

Rain   Falling droplets of liquid water.

Rain Gauge   An instrument that measures rainfall.

Relative Humidity   The ratio of water vapor in the air compared with the amount held if the air were saturated at the same temperature.

Sea Breeze   Air flow by day from the sea toward the land.

Sleet   Precipitation in the form of ice pellets, or a mixture of rain and snow.

Sling Psychrometer   An instrument that measures humidity.

Snow   Precipitation in the form of ice crystals.

Source Region   A nearly homogeneous surface of land or ocean over which an air mass acquires its temperature and humidity characteristics.

Specific Humidity   The mass of water vapor in the air compared with the total mass.

Station Model   A pattern for entering, on a weather map, the meteorological symbols that represent the state of the weather at a particular observing station.

Stationary Front   A frontal system between air masses of nearly equal strength that produces stagnation over one location for an extended period of time.

Stratiform Clouds   Clouds formed into layers.

Stratocumulus Clouds   Low cylindrical cloud masses, dark gray where shaded but white on the illuminated side, best developed during a clearing period following a storm.

Stratopause   The upper limit of stratosphere, separating it from the mesosphere.

Stratosphere   The layer of the earth's atmosphere from an altitude of 11 to 50 km.

Stratus Clouds   A uniform layer of low sheetlike clouds, frequently grayish in appearance.

Synchronous Orbit Satellite   Earth satellites that keep pace in orbit with the earth's own rotation.

Thermistor   An electrical resistor making use of a semiconductor whose resistance varies sharply in a known manner with the temperature.

Thermography   The detection and recording of variations in body temperature on thermograms.

Thermosphere   A region in the atmosphere beginning at an altitude of about 85 km and extending to an average altitude of about 500 km.

Thunderstorm   A local storm produced by a cumulonimbus cloud accompanied by heavy wind, rain, lightning, and thunder.

Tornado   A violent, destructive storm nearly always observed as a "funnel cloud," with wind speeds estimated at 100 to more than 300 mph.

Trade Winds   Surface winds blowing in low latitudes from the subtropical highs toward the equator.

Tropical Air Mass   A type of air, the characteristics of which are developed over low latitudes.

Tropopause   The boundary between the troposphere and stratosphere.

Troposphere   The lowest region of the earth's atmosphere, with an average height of 11 km. It is a region of cloud formations, warm and cold fronts, and storms.

Typhoon   A severe tropical cyclone in the western Pacific.

Valley Breeze   Air flow upslope from valleys to mountains during the day.

Van Allen Belts   Regions of high-energy protons and high-energy electrons that surround the earth in the outer atmosphere.

Warm Front   A front that moves in such a way that warmer air replaces colder air.

Waterspout   A tornado that develops over the ocean.

Weather   Short-term variations of the atmosphere, usually thought of in terms of temperature, humidity,

precipitation, wind, cloudiness, and visibility.

Weather Front   A boundary separating two different air masses.

Weather Satellite   Satellites such as TIROS and Nimbus that are designed to analyze infrared energy and cloud covers for use in weather analysis and forecasting.

Westerlies   Surface winds flowing from the polar portions of the subtropical highs, carrying weather conditions from west to east through the middle latitudes.

Wind   Air in generally horizontal movement from areas of higher pressure to areas of lower pressure.

Wind Vane   An instrument for determining wind direction.

Windward   Location on the side that faces toward the wind; the direction from which a wind is blowing.

THINGS TO DO

1. Experience the cooling effect of evaporating (a) water, (b) alcohol on the skin.

2. Follow atmospheric pressure changes with a barometer for 30 days. Correlate the pressure with weather changes.

3. Chart the mean daily temperature over a period of 30 days.

4. Record the precipitation in your area over a 30-day period with a rain gauge.

5. Observe the expansion of air when heated by placing a balloon over the neck of a flask and placing the flask in a container of warm water.

6. Collect some snowflakes, if possible, on a piece of dark cloth and study their hexagonal structure with a hand lens or microscope.

7. If there is an opportunity, collect some hailstones. Take a cross section and observe the layered arrangement of the ice.

8. Follow national weather systems with the daily weather map in a newspaper over a 30-day period.

9. Observe the varieties of cloud patterns over a 30-day period.

EXERCISES

The Atmosphere

24–1. What are the major gases comprising the atmosphere and the percentage of the total that each supplies?

24–2. Why is it useful to think of the atmosphere as an ocean of air?

24–3. Why are hydrogen and helium scarce in the earth's atmosphere?

24–4. List four regions of the atmosphere that are based on thermal structure.

24–5. How is the temperature profile of the stratosphere related to ozone?

24–6. Discuss the threat to the survival of the ozone layer.

24–7. (a) What is the greenhouse effect?
(b) Why is there so much discussion of it in the media?

24–8. What important role does ozone play in the atmosphere?

24–9. How does the troposphere differ from the stratosphere?

24–10. How is it possible for radio waves, which travel in straight lines, to follow the curvature of the earth?

24–11. What is the difference between a meteorologist and a climatologist?

24–12. Explain how auroras form.

24–13. Why were the radiation belts encompassing the earth discovered only recently?

Atmospheric Pressure and Winds

24–14. Give the common units in which air pressure is expressed.

24–15. Why does the atmosphere become progressively more dense from its outer limits to sea level?

24–16. How can you apply your knowledge of pressure and winds in everyday life?

24–17. Why is it difficult to cook foods in boiling water at high elevations?

Moisture, Condensation, and Precipitation

24–18. What is meant by the "hydrologic cycle"?

24–19. Discuss five meteorological instruments.

24–20. Discuss the major classes of clouds.

24–21. What is relative humidity?

24–22. Explain the meaning of dew point.

24–23. Why does the relative humidity rise at night?

24–24. Distinguish between condensation and precipitation.

24–25. How do fronts cause clouds and precipitation?

24–26. (a) How many inches of precipitation have fallen in your area this year?
(b) Is that average or unusually high or low?

24–27. List the various forms of precipitation and the manner in which each is produced.

24–28. What is the role of nuclei of condensation in cloud formation?

Air Masses and Atmospheric Disturbances

24–29. Discuss the characteristics of an air mass.

24–30. How are weather fronts related to air masses?

24–31. Compare the amounts of snowfall and rainfall.

24–32. How does a thunderstorm develop?

24–33. What makes a tornado so life-threatening?

24–34. What is the current status of weather modification?

24–35. *Multiple Choice*
A. Atmospheric pressure is usually recorded in
(a) grams.
(b) meters.
(c) millibars.
(d) isobars.
B. The relative humidity of the air is measured with a
(a) barometer.
(b) sling psychrometer.
(c) thermograph.
(d) ceilometer.
C. The lowest region of the atmosphere is called the
(a) troposphere.
(b) stratosphere.
(c) ionosphere.
(d) exosphere.
D. Rain or snow is most likely to be produced by which of the following clouds?
(a) cirrostratus
(b) nimbostratus
(c) stratocumulus
(d) altocumulus
E. A boundary that separates air masses of different properties is known as a (an)
(a) weather front.
(b) cold front.
(c) warm front.
(d) occluded front.

SUGGESTIONS FOR FURTHER READING

Akasofu, Syun-Ichi, "The Dynamic Aurora." *Scientific American*, May, 1989.
The aurora borealis and aurora australis are lights emitted when atoms and molecules in the ionosphere are struck by electrons blowing in from the sun. Where is the power supply of the "cathode tube" of the auroras, and why does it seem to fluctuate from time to time, causing the auroras to ebb and flow across the sky?

Baum, Rudy M., "Stratospheric Science Undergoing Change." *Chemical and Engineering News*, 13 September, 1982.
Stratospheric chemistry came into its own only in the late 1960s and early 1970s as a result of perceived threats to the ozone layer. Airplanes, gondolas of balloons, rockets, computers, and ground-based measurement techniques have been placed in service.

Chameides, William L., and Douglas D. Davis, "Chemistry in the Troposhere." *Chemical and Engineering News*, 4 October, 1982.
Since the 1950s, advances in chemical analytical techniques and high-speed computers have led to an explosion of knowledge concerning the composition of the atmosphere. Field sampling platforms range from ships to satellites. A multitude of exciting research opportunities await scientists trained in a wide variety of fields.

Elliott, Scott, and F. S. Rowland, "Chlorofluorocarbons and Stratospheric Ozone." *Journal of Chemical Education*, May, 1987.
Circumstantial evidence suggests that the ozone hole discovered over the continent of Antarctica is caused by products of the photodecomposition of synthetic chlorofluorocarbons.

Graedel, Thomas E., and Paul J. Crutzen, "The Changing Atmosphere." *Scientific American*, September, 1989.
In the past 200 years, the atmosphere's composition has changed faster than at any time in human history. The activities of human beings account for most of the rapid changes: the combustion of fuels, other industrial and agricultural practices, biomass burning, and deforestation. Some effects, such as acid rain and smog, are already evident; unwelcome surprises, such as the Antarctic ozone hole, may be lurking.

Grove, Noel, "Air: An Atmosphere of Uncertainty." *National Geographic*, April, 1987.
Where our air is concerned, we live in an age of uncertainty. Air pollution is modern man's wolf at the door.

Jones, Philip D., and Tom M. L. Wigley, "Global Warming Trends." *Scientific American*, August, 1990.
The world's climate has become generally warmer during the past century, and this trend shows no signs of abating. The causes of global warming are less certain than the trend itself.

Kasting, James F., Owen B. Toon, and James B. Pollack, "How Climate Evolved on the Terrestrial Planets." *Scientific American*, February, 1988.
Why is Mars too cold for life, Venus too hot, and the earth just right? Largely because they differed in their ability to cycle carbon dioxide between the crust and the atmosphere.

Kerr, Richard A., "Cloud Seeding: One Success in 35 Years." *Science*, August 6, 1982.
Proof of the ability to increase rain or snow eluded researchers for 30 years. Only a single set of experiments appears to have confirmed an increase in precipitation after cloud seeding.

LaBastille, Anne, "Acid Rain: How Great a Menace." *National Geographic*, November, 1981.
Acid rain has eliminated fish in thousands of lakes in Scandinavia and hundreds in the United States and Canada. Some scientists see the year 2000 as the earliest that emissions from the burning of fossil fuels can be stabilized and then slowly reduced.

Lansford, Henry, "The Frightening Mystery of the Electrical Storm." *Smithsonian*, August, 1979.
Thunder and lightning bring disasters with them: hail, downbursts, flash floods, and tornadoes. There is still much to do to understand thunderstorms.

Lynch, David K., "Atmospheric Halos." *Scientific American*, April, 1978.
Rings around the sun and moon are caused by ice crystals. Physicists still want to know precisely how.

Matthews, Samuel W., "Is Our World Warming? Under the Sun." *National Geographic*, October, 1990.
Humanity has become a more important agent of environmental change than nature. What we are doing to the earth's atmosphere is ominous.

Nahin, Paul J., "Oliver Heaviside." *Scientific American*, June, 1990.
Born in a London slum and lacking a university education, Oliver Heaviside became one of the leading Victorian physicists. He extended the electromagnetic theory of James Clerk Maxwell, discovered the circuit principle that made long-distance telephony possible, and foresaw television and over-the-horizon radio.

Ramage, Colin S., "El Niño." *Scientific American*, June, 1986.
El Niño, an anomalous warming of surface water in the equatorial Pacific, occurs at irregular intervals in conjunction with the Southern Oscillation, a massive seesawing of atmospheric pressure. Both phenomena are linked to global changes in weather patterns.

Revelle, Roger, "Carbon Dioxide and World Climate." *Scientific American*, August, 1982.
It seems well established that the carbon dioxide content of the atmosphere has increased in the past century from the burning of fossil fuels and the clearing of forests. Slow, pervasive environmental shifts are likely.

Roble, Raymond G., "Chemistry in the Thermosphere and Ionosphere." *Chemical and Engineering News*, June 16, 1986.

The thermosphere and ionosphere have been observed and studied for decades, first by means of radio-wave probing and radar soundings of the ionosphere and since the 1950s by rockets and satellites. Because of this probing, they are better known than the middle atmospheric regions below — the stratosphere and the mesosphere — which primarily are accessible only by balloons, rockets, or remote-sounding techniques.

Ruddiman, William F., and John E. Kutzbach, "Plateau Uplift and Climatic Change." *Scientific American,* March, 1991.
Prior to 40 million years ago most of the world was warmer and wetter than it is now. Then this warm, wet climate largely disappeared, persisting today only in limited regions, such as Southeast Asia, the Gulf Coast of the United States, and the tropics. Research indicates that the formation of giant plateaus in Tibet and the American West helped shape modern-day climatic trends.

Schneider, Stephen H., "Climate Modeling." *Scientific American,* May, 1987.
Will the greenhouse effect bring on another Dust Bowl? Would nuclear war mean "nuclear winter"? At present we are altering our environment faster than we can understand the resulting climatic changes. Climate models do not yield definite forecasts of what the future will bring.

Shaw, Robert W., "Air Pollution by Particles." *Scientific American,* August, 1987.
Acid rain, rain or snow carrying dissolved acids, is not the only way pollutants reach the earth from the atmosphere. Acidic gases and particles find their way to the ground even under dry conditions. The main source of these particles is the combustion of fossil fuels.

Shell, Ellen Ruppel, "Solo Flights into the Ozone Hole Reveal Its Causes." *Smithsonian,* February, 1988.
A NASA expedition sent a plane 10 miles into the stratosphere to find what is causing the ozone hole. Conclusion: a clear connection between high chlorine levels and ozone depletion.

Stolarski, Richard S., "The Antarctic Ozone Hole." *Scientific American,* January, 1988.
Is the ozone depletion an anomaly, or is it a sign that the ozone layer is in jeopardy globally?

Toon, Owen B., and Richard P. Turco, "Polar Stratospheric Clouds and Ozone Depletion." *Scientific American,* June, 1991.
Research strongly indicates that three kinds of clouds that form over the poles, called polar stratospheric clouds, trigger ozone depletion in the Arctic stratosphere. In the Antarctic stratosphere they help to create the ozone hole.

White, Robert M., "The Great Climate Debate." *Scientific American,* July, 1990.
Some claim that global climate warming threatens the very habitability of the planet. Others hold that the predictions of environmental collapse are not well founded. It is likely that humanity will have to adapt to some climate changes.

ESSAY

OUR DELICATE ATMOSPHERE

ROBERT E. GABLER
Western Illinois University

ROBERT J. SAGER
Pierce College, Washington

DANIEL L. WISE
Western Illinois University

THE GREENHOUSE EFFECT

The term *atmospheric effect* refers to the fact that certain gases—primarily carbon dioxide (CO_2), water vapor, and other trace gases: methane (CH_4), nitrous oxide (NO_2), and chlorofluorocarbons (CFCs)—absorb very little of the solar energy that moves toward the earth but do absorb reradiated earth energy. This process helps to limit wild fluctuations in our atmospheric temperature and is therefore a vital and necessary atmospheric process. However, as we noted earlier, human activity is augmenting the atmospheric effect. To differentiate this human-induced component, we refer to it as the *greenhouse effect*.

The greenhouse effect is increasing for two reasons. First, industrial activity, largely the burning of fossil fuels, is adding additional carbon dioxide and other trace gases (methane and chlorofluorcarbons) to the atmosphere. The second reason for an increase in the greenhouse effect is that we are eliminating one of two major sinks, or reservoirs, for CO_2. One major sink for CO_2 is the world ocean. Oceans, primarily through storage in phytoplankton, are large CO_2 reservoirs and play a significant role in regulating atmospheric CO_2 content. It is estimated that the oceans hold 50 times the CO_2 of the atmosphere.

A second major sink for atmospheric CO_2 is forested land. Human activity—deforestation—is having a significant impact upon this sink. Trees take in CO_2 and give off oxygen. Thus, when you cut down and burn trees, two negative results occur. First, you add CO_2 to the atmosphere by the burning, and second, you take away a vehicle for removing CO_2 from the atmosphere.

There is fairly conclusive evidence that the CO_2 content of the atmosphere is increasing. In 1988 the average annual CO_2 content of the atmosphere was approximately 350 parts per million. This represents a 25% increase over the past 100 years, and even more alarming is the projection that atmospheric CO_2 content will possibly double in the next 25 to 50 years! What might the consequences of such an increase in CO_2 be?

We use computers to run very complex mathematical models of the atmosphere to stimulate what might happen if CO_2 content does double. Because of the complexity of this task, the models must use fairly simplified representations of the energy processes that go on in the atmosphere and at the earth's surface. This results in some variations in predictions from the models, but for the most part they seem to agree that a doubling of CO_2 in our atmosphere would result in: (1) a rise in global mean temperature of 1.5 to 5.5°C (2.7 to 10.0°F); (2) warming more pronounced at higher latitudes than at lower latitudes; (3) warming greater in winter than summer; and (4) midcontinental areas drier and coastal areas wetter.

What could be the impact of such climatic changes?

Agriculture. To compensate for the increased temperatures, agriculture would shift north. Irrigation would become much more widespread, especially in the Corn Belt, to hedge against the more frequent dry periods. It is estimated that the United States would continue to produce enough food for its own use but would drastically cut its exports.

Coastal Areas. The greenhouse warming would result in a rise in sea level of 1.0 meters (3.281 ft) by the year 2050. Such a rise would result not only in loss of beaches, but estuaries and wetlands would be destroyed, greatly affecting species that use these habitats as spawning areas, feeding grounds, or residences.

Natural Hazards. Greenhouse warming will alter natural hazards in terms of their frequency, magnitude, and spatial and temporal distribution. For example, midwest rainfall, while reduced in total amount, would occur in more intense convective cells with fewer gentle large-scale showers. This could increase flooding. Hurricanes would occur over a longer season, would be more intense, and would make landfall more often.

Forests. Forests would shift north in response to the increased warming. A major concern is that the poleward shift of temperature zones could occur more rapidly than most plants and trees are thought capable of naturally migrating. This could result in widespread depletion of our middle latitude forests.

Health. The hotter temperatures would increase the chance of heart attack and encourage the spread of parasitic diseases from Central America into the southern United States. In addition, the drier and warmer weather would cause more crop failure. This in turn would increase famine and death, especially in less developed countries where populations already exist on the borderline of starvation.

Would anything good result from the increased warming? Perhaps. It could result in higher latitude locations, such as South Saskatchewan, having their growing season increased 4 to 9 weeks and the time it takes for wheat to mature being reduced 4 to 14 days. Some of the industrial cities around the Great Lakes that have seen a decline in population and the quality of living as industry moved south could see a rejuvenation as adequate water supplies become a major consideration for industrial location. However, the negative results from greenhouse warming could be catastrophic and would far outweigh any pluses.

Will the greenhouse warming take place? It is hard to tell because of the numerous variables involved in assessing climate change. However, reducing CO_2 input from fossil fuel burning and stopping wholesale deforestation should be attempted anyway—for environmental considerations as well as to reduce the impact of the greenhouse effect on society.

DEPLETION OF THE OZONE LAYER

Ozone, the triatomic form of oxygen (O_3), is primarily concentrated in a layer high above the earth within the stratosphere. While the actual amount of ozone in the atmosphere is small (0.00005% by volume), its ability to absorb harmful ultraviolet rays from the sun makes it vital to humans.

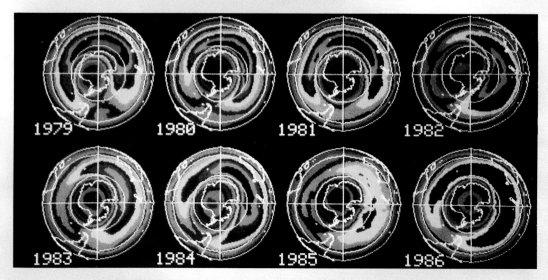

Figure A   Progressive thinning of the Antarctic ozone layer documented by space observations: October monthly averages, 1979–1986. Purple-violet: low levels; yellow, brown, green: high levels. (NASA)

In 1974 scientists first documented that man-made gases can interact with ozone and convert it to diatomic oxygen (O_2), which offers no protection from ultraviolet radiation. While several gases have been linked to this ozone depletion process, the primary culprit seems to be chlorofluorocarbons (CFCs), which are gases involved in the green-house effect as well. CFCs are used in refrigeration, air conditioning, insulation, packing materials (fast-food containers), and as a propellant for aerosol cans.

These man-made chemicals appear to have already upset the state of equilibrium that has existed in the ozone concentration for millions of years. Recent studies by NASA detected a thinning of the ozone layer above the midlatitudes. More alarming has been an increase in the ozone hole over Antarctica. Every year (around September) a hole appears in the ozone layer over Antarctica, grows in size, and then disappears. This hole, which is related to the unusual weather conditions of this region, grows larger each year (Fig. A).

Scientists from around the world are studying the situation over Antarctica since this region most likely can serve as a gauge of depletion of the ozone layer in other regions of the world. An increase in ultraviolet radiation resulting from a depletion of the protective ozone layer would have a significant impact upon our society.

The Environmental Protection Agency estimates that every 1% decrease in stratospheric ozone could result in 24,000 to 57,000 more cases of cataracts and 43,000 new cases of skin cancer each year in the United States. Increased ultraviolet radiation would also increase the effect of many diseases that affect the skin or start on the skin: leprosy, small-pox, herpes, bubonic plague, and some types of diphtheria. Furthermore, excessive ultraviolet radiation causes cell and tissue damage in many plants. This could result in a significant reduction in crop yields throughout the world.

What can be done? In 1977 the United States banned all nonessential aerosol products using CFCs. While this was a step in the right direction, more needed to be done. Finally, in 1987, 38 countries met in Montreal to sign an agreement—the so-called Montreal Protocol—to cut CFC use in half by the end of the century. This historic first effort toward international control of an air pollutant is hopefully only the beginning.

A P P E N D I X

SOME USEFUL CONVERSION FACTORS

Length

1 in. = 2.54 cm
1 m = 39.37 in. = 3.281 ft = 1.1 yd
1 ft = 0.3048 m = 12 in
3 ft = 1 yd = 0.9144m
1 km = 0.621 mi
1 mi = 5280 ft = 1.61 km
1 Å = 10^{-10} m = 10^{-8} cm
1 μm = 1 μ = 10^{-6} m = 10^4 Å
1 lightyear = 9.46×10^{12} km

Area

1 m^2 = 10^4 cm^2 = 10.76 ft^2
1 ft^2 = 0.0929 m^2 = 144 in.^{2}
1 in.^{2} = 6.452 cm^2

Volume

1 m^3 = 10^6 cm^3 = 6.102×10^4 in.^{3}
1 ft^3 = 1728 in.^{3} = 2.83×10^{-2} m^3
1 liter (L) = 1000 cm^3 = 1.0576 qt = 0.0353 ft^3
1 ft^3 = 7.481 gal = 28.32 liters
1 milliliter (mL) = 1 cubic centimeter (cm^3)
1 gal = 3.786 L = 4 qt = 231 in.^{3}

Mass

1000 kg = 1 t (metric ton)
1000 g = 1 kg = 6.85×10^{-2} slug
1 u = 1.66×10^{-27} kg
1 slug = 14.59 kg

Time

1 year = 365 days = 3.16×10^7 s
1 day = 24 h = 1.44×10^3 min = 8.64×10^4 s

Speed

1 mi/h = 1.47 ft/s = 0.447 m/s = 1.61 km/h
1 m/s = 100 cm/s = 3.281 ft/s = 2.237 mi/h
1 mi/min = 60 mi/h = 88 ft/s

Acceleration

1 m/s^2 = 3.28 ft/s^2 = 100 cm/s^2
1 ft/s^2 = 0.3048 m/s^2 = 30.48 cm/s^2

Force

1 N = 10^5 dyn = 0.225 lb
1 lb = 4.45 N
1 ton = 2000 lb

Pressure

1 Pa = 1 N/m^2 = 1.45×10^{-4} lb/in.^{2}
1 bar = 10^5 N/m^2 = 14.50 lb/in.^{2}
1 atm = 760 mmHg = 76.0 cmHg = 760 torr
1 atm = 14.7 lb/in.^{2} = 1.013×10^5 N/m^2 (or Pa)

Energy and Work

1 J = 0.738 ft·lb = 10^7 erg
1 cal = 4.186 J
1 Btu = 252 cal = 778 ft·lb = 1.054×10^3 J
1 eV = 1.60×10^{-19} J
931.5 MeV is equivalent to 1 u

Power

1 hp = 550 ft·lb/s = 0.746 kW = 746 W
1 W = 1 J/s = 0.738 ft·lb/s = 10^7 erg·sec
    = 0.00134 horsepower
1 BTU/h = 0.293 W

Temperature

°F = 9/5 C + 32
°C = 5/9 (F − 32)
K = C + 273.15
0 K = − 273.15°C

SI DERIVED UNITS

Physical Quantity	Name of Unit	Symbol	Base Units	Expression in Terms of Other SI Units
Frequency	hertz	Hz	s^{-1}	
Force	newton	N	$kg \cdot m/s^2$	
Pressure	pascal	Pa	$kg/m \cdot s^2$	N/m^2
Energy; work	joule	J	$kg \cdot m^2/s^2$	$N \cdot m$
Power	watt	W	$kg \cdot m^2/s^3$	J/s
Electric charge	coulomb	C	$A \cdot s$	
Electric potential	volt	V	$kg \cdot m^2/A \cdot s^3$	W/A
Electric resistance	ohm	Ω	$kg \cdot m^2/A^2 \cdot s^3$	V/A
Area	square meter	m^2		
Volume	cubic meter	m^3		
Density	kilogram per cubic meter	kg/m^3		

SI BASE UNITS

Physical Quantity	Name of Unit	Symbol
Length	meter	m
Mass	kilogram	kg
Time	second	s
Temperature	kelvin	K
Electric current	ampere	A
Luminous intensity	candela	cd
Amount of substance	mole	mol

SI PREFIXES

Factor	Prefix	Symbol	Factor	Prefix	Symbol
10^{18}	exa	E	10^{-1}	deci	d
10^{15}	peta	P	10^{-2}	centi	c
10^{12}	tera	T	10^{-3}	milli	m
10^{9}	giga	G	10^{-6}	micro	μ
10^{6}	mega	M	10^{-9}	nano	n
10^{3}	kilo	k	10^{-12}	pico	p
10^{2}	hecto	h	10^{-15}	femto	f
10^{1}	deka	da	10^{-18}	atto	a

PERIODIC TABLE OF THE ELEMENTS

State: S Solid L Liquid G Gas X Not found in nature

Legend:
- Metals
- Transition Metals
- Nonmetals
- Noble gases
- Lanthanide series
- Actinide series

Atomic number
Symbol
Atomic mass

Example:
92 [S]
U
Uranium
238.03

Group	IA	IIA	IIIB	IVB	VB	VIB	VIIB	VIIIB			IB	IIB	IIIA	IVA	VA	VIA	VIIA	VIIIA
1	1 [G] H Hydrogen 1.01																	2 [G] He Helium 4.00
2	3 [S] Li Lithium 6.94	4 [S] Be Beryllium 9.01											5 [S] B Boron 10.81	6 [S] C Carbon 12.01	7 [G] N Nitrogen 14.01	8 [G] O Oxygen 16.00	9 [G] F Fluorine 19.00	10 [G] Ne Neon 20.18
3	11 [S] Na Sodium 22.99	12 [S] Mg Magnesium 24.31											13 [S] Al Aluminum 26.98	14 [S] Si Silicon 28.09	15 [S] P Phosphorus 30.97	16 [S] S Sulfur 32.06	17 [G] Cl Chlorine 35.45	18 [G] Ar Argon 39.95
4	19 [S] K Potassium 39.10	20 [S] Ca Calcium 40.08	21 [S] Sc Scandium 44.96	22 [S] Ti Titanium 47.90	23 [S] V Vanadium 50.94	24 [S] Cr Chromium 52.00	25 [S] Mn Manganese 54.94	26 [S] Fe Iron 55.85	27 [S] Co Cobalt 58.93	28 [S] Ni Nickel 58.71	29 [S] Cu Copper 63.55	30 [S] Zn Zinc 65.38	31 [S] Ga Gallium 69.72	32 [S] Ge Germanium 72.59	33 [S] As Arsenic 74.92	34 [S] Se Selenium 78.96	35 [L] Br Bromine 79.90	36 [G] Kr Krypton 83.80
5	37 [S] Rb Rubidium 85.47	38 [S] Sr Strontium 87.62	39 [S] Y Yttrium 88.91	40 [S] Zr Zirconium 91.22	41 [S] Nb Niobium 92.91	42 [S] Mo Molybdenum 95.94	43 [X] Tc Technetium 97	44 [S] Ru Ruthenium 101.07	45 [S] Rh Rhodium 102.91	46 [S] Pd Palladium 106.4	47 [S] Ag Silver 107.87	48 [S] Cd Cadmium 112.40	49 [S] In Indium 114.82	50 [S] Sn Tin 118.69	51 [S] Sb Antimony 121.75	52 [S] Te Tellurium 127.60	53 [S] I Iodine 126.90	54 [G] Xe Xenon 131.30
6	55 [S] Cs Cesium 132.91	56 [S] Ba Barium 137.34	71 [S] Lu Lutetium 174.97	72 [S] Hf Hafnium 178.49	73 [S] Ta Tantalum 180.95	74 [S] W Tungsten 183.85	75 [S] Re Rhenium 186.21	76 [S] Os Osmium 190.2	77 [S] Ir Iridium 192.22	78 [S] Pt Platinum 195.09	79 [S] Au Gold 196.97	80 [L] Hg Mercury 200.59	81 [S] Tl Thallium 204.37	82 [S] Pb Lead 207.2	83 [S] Bi Bismuth 208.98	84 [S] Po Polonium 209	85 [S] At Astatine 210	86 [G] Rn Radon 222
7	87 [S] Fr Francium 223	88 [S] Ra Radium 226.03	103 [X] Lr Lawrencium 260	104 [X] Unq 261	105 [X] Unp 262	106 [X] Unh 263	107 [X] Uns 264	108 [X] Uno 265	109 [X] Une 266									

Lanthanide series:

57 [S] La Lanthanum 138.91	58 [S] Ce Cerium 140.12	59 [S] Pr Praseodymium 140.91	60 [S] Nd Neodymium 144.24	61 [X] Pm Promethium 145	62 [S] Sm Samarium 150.4	63 [S] Eu Europium 151.96	64 [S] Gd Gadolinium 157.25	65 [S] Tb Terbium 158.93	66 [S] Dy Dysprosium 162.50	67 [S] Ho Holmium 164.93	68 [S] Er Erbium 167.26	69 [S] Tm Thulium 168.93	70 [S] Yb Ytterbium 173.04

Actinide series:

89 [S] Ac Actinium 227	90 [S] Th Thorium 232.04	91 [S] Pa Protactinium 231.04	92 [S] U Uranium 238.03	93 [S] Np Neptunium 237.05	94 [S] Pu Plutonium 244	95 [X] Am Americium 243	96 [X] Cm Curium 247	97 [X] Bk Berkelium 247	98 [X] Cf Californium 251	99 [X] Es Einsteinium 254	100 [X] Fm Fermium 257	101 [X] Md Mendelevium 258	102 [X] No Nobelium 259

ELEMENTS, THEIR SYMBOLS, AND THEIR ATOMIC MASSES*

Element	Symbol	Atomic Number	Atomic Mass	Element	Symbol	Atomic Number	Atomic Mass
Actinium	Ac	89	(227)	Neon	Ne	10	20.18
Aluminum	Al	13	26.98	Neptunium	Np	93	(237)
Americium	Am	95	(243)	Nickel	Ni	28	58.69
Antimony	Sb	51	121.8	Niobium	Nb	41	92.91
Argon	Ar	18	39.95	Nitrogen	N	7	14.01
Arsenic	As	33	74.92	Nobelium	No	102	(259)
Astatine	At	85	(210)	Osmium	Os	76	190.2
Barium	Ba	56	137.3	Oxygen	O	8	16.00
Berkelium	Bk	97	(247)	Palladium	Pd	46	106.4
Beryllium	Be	4	9.012	Phosphorus	P	15	30.97
Bismuth	Bi	83	209.0	Platinum	Pt	78	195.1
Boron	B	5	10.81	Plutonium	Pu	94	(244)
Bromine	Br	35	79.90	Polonium	Po	84	(210)
Cadmium	Cd	48	112.4	Potassium	K	19	39.10
Calcium	Ca	20	40.08	Praseodymium	Pr	59	140.9
Californium	Cf	98	(251)	Promethium	Pm	61	(145)
Carbon	C	6	12.01	Protactinium	Pa	91	(231)
Cerium	Ce	58	140.1	Radium	Ra	88	(226)
Cesium	Cs	55	132.9	Radon	Rn	86	(222)
Chlorine	Cl	17	35.45	Rhenium	Re	75	186.2
Chromium	Cr	24	52.00	Rhodium	Rh	45	102.9
Cobalt	Co	27	58.93	Rubidium	Rb	37	85.47
Copper	Cu	29	63.55	Ruthenium	Ru	44	101.1
Curium	Cm	96	(247)	Samarium	Sm	62	150.4
Dysprosium	Dy	66	162.5	Scandium	Sc	21	44.96
Einsteinium	Es	99	(254)	Selenium	Se	34	78.96
Erbium	Er	68	167.3	Silicon	Si	14	28.09
Europium	Eu	63	152.0	Silver	Ag	47	107.9
Fermium	Fm	100	(257)	Sodium	Na	11	22.99
Fluorine	F	9	19.00	Strontium	Sr	38	87.62
Francium	Fr	87	(223)	Sulfur	S	16	32.07
Gadolinium	Gd	64	157.3	Tantalum	Ta	73	180.9
Gallium	Ga	31	69.72	Technetium	Tc	43	(99)
Germanium	Ge	32	72.59	Tellurium	Te	52	127.6
Gold	Au	79	197.0	Terbium	Tb	65	158.9
Hafnium	Hf	72	178.5	Thallium	Tl	81	204.4
Helium	He	2	4.003	Thorium	Th	90	232.0
Holmium	Ho	67	164.9	Thulium	Tm	69	168.9
Hydrogen	H	1	1.008	Tin	Sn	50	118.7
Indium	In	49	114.8	Titanium	Ti	22	47.88
Iodine	I	53	126.9	Tungsten	W	74	183.9
Iridium	Ir	77	192.2	Unnilennium	Une	109	(266)
Iron	Fe	26	55.85	Unnilhexium	Unh	106	(263)
Krypton	Kr	36	83.80	Unniloctium	Uno	108	(265)
Lanthanum	La	57	138.9	Unnilpentium	Unp	105	(262)
Lawrencium	Lr	103	(260)	Unnilquadium	Unq	104	(261)
Lead	Pb	82	207.2	Unnilseptium	Uns	107	(262)
Lithium	Li	3	6.941	Uranium	U	92	238.0
Lutetium	Lu	71	175.0	Vanadium	V	23	50.94
Magnesium	Mg	12	24.31	Xenon	Xe	54	131.3
Manganese	Mn	25	54.94	Ytterbium	Yb	70	173.0
Mendelevium	Md	101	(258)	Yttrium	Y	39	88.91
Mercury	Hg	80	200.6	Zinc	Zn	30	65.39
Molybdenum	Mo	42	95.94	Zirconium	Zr	40	91.22
Neodymium	Nd	60	144.2				

* Approximate values of atomic masses for radioactive elements are given in parentheses.

PHYSICAL CONSTANTS

Quantity	Symbol	Value and SI Units
Acceleration due to gravity at the surface of the earth	g	9.80 m/s^2
Atomic mass unit	u	$1.66 \times 10^{-27} \text{ kg}$
Avogadro's number	N_A	$6.02 \times 10^{23} \text{ particles/mol}$
Charge of the electron	e	$1.60 \times 10^{-19} \text{ C}$
Gravitational constant	G	$6.67 \times 10^{-11} \text{ N} \cdot \text{m}^2/\text{kg}^2 = \text{m}^3/\text{kg} \cdot \text{s}^2$
Rest mass of the electron	m_e	$9.1 \times 10^{-31} \text{ kg}$
Rest mass of the proton	m_p	$1.67 \times 10^{-27} \text{ kg}$
Rest mass of the neutron	m_n	$1.67 \times 10^{-27} \text{ kg}$
Planck's constant	h	$6.63 \times 10^{-34} \text{ J} \cdot \text{sec}$
Speed of light in vacuum	c	$3.00 \times 10^8 \text{ m/s}$

PLANETARY PHYSICAL DATA

Planet	Diameter (km)	Diameter (Earth = 1)	Mass (Earth = 1)	Mean Density (g/cm³)	Rotation Period (days)	Inclination of Equator to Orbit (°)	Surface Gravity (Earth = 1)	Velocity of Escape (km/s)
Mercury	4878	0.38	0.055	5.43	58.6	0.0	0.38	4.3
Venus	12,104	0.95	0.82	5.24	− 243.0	177.4	0.91	10.4
Earth	12,756	1.00	1.00	5.52	0.997	23.4	1.00	11.2
Mars	6794	0.53	0.107	3.9	1.026	25.2	0.38	5.0
Jupiter	142,800	11.2	317.8	1.3	0.41	3.1	2.53	60
Saturn	120,540	9.41	94.3	0.7	0.43	26.7	1.07	36
Uranus	51,200	4.01	14.6	1.2	− 0.72	97.9	0.92	21
Neptune	49,500	3.88	17.2	1.6	0.67	29	1.18	24
Pluto	2200	0.17	0.0025	2.0	− 6.387	118	0.09	1

INDEX